DEVELOPMENTAL BIOLOGY

Developmental Biology

SCOTT F. GILBERT
SWARTHMORE COLLEGE

Sinauer Associates, Inc. • Publishers
Sunderland, Massachusetts

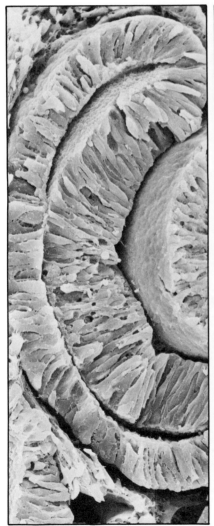
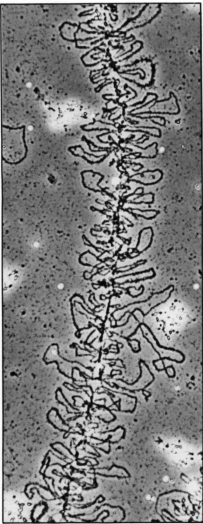
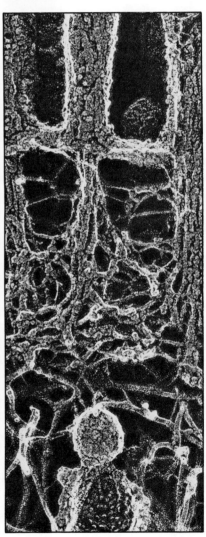

To Daniel and Sarah

THE COVER
Pluteus larva (front cover) and 32-cell stage (back cover) of the sea urchin *Lytechinus pictus*. Courtesy of George Watchmaker, Lawrence Livermore National Laboratory, Livermore, California.

DEVELOPMENTAL BIOLOGY

Copyright © 1985 by Sinauer Associates, Inc. All rights reserved. This book may not be reproduced in whole or in part for any purpose whatever, without permission from the publisher. For information address Sinauer Associates Inc., Sunderland, Mass. 01375.
Printed in U.S.A.

Library of Congress Cataloging in Publication Data

Gilbert, Scott F.,   1949–
  Developmental biology.

  Includes bibliographies and index.
   1. Embryology.   I. Title.
QL955.G48   1984   574.3   84-10658
ISBN 0-87893-246-1

8 7 6 5 4 3

CONTENTS

77772

Chapter 3:
Cleavage: creating multicellularity   73

Chapter 4:
Gastrulation: reorganizing the embryonic cells   110

Chapter 5:
Early vertebrate development: neurulation and ectoderm   149

Chapter 6:
Early vertebrate development: mesoderm and endoderm   190

PART II:
MECHANISMS OF CELLULAR DIFFERENTIATION

Chapter 7:
Progressive determination 237

Chapter 8:
Determination by cytoplasmic specification 269

Chapter 9:
Genomic equivalence and differential gene expression: embryological investigations 297

Chapter 14:
Translational and posttranslational regulation of developmental processes   433

PART III:
CELL INTERACTIONS IN DEVELOPMENT

Chapter 15:
Spatial development: the role of the cell surface   471

Chapter 19:
Sex determination

638

Chapter 20:
The saga of the germ line

664

PREFACE

If there is anyone more foolish than the student who thinks he or she can learn all developmental biology in one semester, it is the professor who believes he or she can teach it. How much more foolish, then, is the person who writes a book for them? The motivation for writing this volume was the challenge to put recent advances in developmental biology into a coherent historical and conceptual order. This is not the only order possible. One of the strengths of contemporary developmental biology is that it attracts people from diverse backgrounds and training. Every developmental biologist has a preferred way of integrating the hypotheses and observations concerning animal development, and there is no one "best" way to introduce students to the field. Similarly, there is no one best way to write a textbook, and I feel that I should outline my basic principles at the outset.

First, this is a textbook written with juniors and seniors in mind. It does not assume a level of biology more sophisticated than that given by most good introductory biology textbooks, but a familiarity with cell biology and genetics will certainly make the going easier.

Second, the book is divided into three sections. The first section analyzes the basic phenomena and patterns of animal development from fertilization through the formation of the major embryonic organs. This section is placed first to provide the context for the more molecularly-oriented chapters that follow. This allows the analyses of differential gene activity and cell interactions to be placed within the framework of microscopically observable development. The second section focuses on the mechanisms of cell determination and differentiation, discussing the ways by which cells of an embryo become different from one another and from their progenitors. The third section concentrates on the other major problem of development—morphogenesis, the processes by which cells interact to form highly complex tissues and organs.

I have also tried to order the chapters in a manner that allows the student to have a conceptual framework in which to understand each successive chapter. Thus, the order of certain chapters differs from what might be expected. This is most obvious in that this book begins with fertilization and ends with gametogenesis. Here, the fertilization chapter includes a large section on the structure of the gametes, while the details of gametogenesis are saved until the end of the book, where the intricate gene activities and cell-cell interactions can be appreciated. (Thus, one can discuss the expression of vitellogenin genes and the interactions of the oocyte and follicle cells after first having a general understanding of the translational control of gene expression and the importance of gap junctions.) Putting gametogenesis last not only allows it to be used as a culminating example of development but also provides a circular return to the beginning.

Third, this book is organized by principles, not by "systems." This allows one to see the major themes and variations of animal development. In the chapter on cleavage, the earliest developmental stages of various different organisms are contrasted and compared, while in the chapter on translational control of gene expression, the regulation of oocyte messages, globin messages, and casein messages are discussed together. So, while there is no one chapter on insect development, such information can be found in nearly every chapter of the book. Similarly, there are no chapters entitled "History of Embryology" or "Development of the Immune System." Having worked in these fields, I have too much respect for them to confine them to isolated (and often unassigned) chapters. I have introduced these topics, throughout the book, within their appropriate contexts.

Fourth, the text attempts to present developmental biology as a dynamic human endeavor based on repeated observations and controlled experiments. Since the data herein do not exist independently of the people who obtained them, I have included numerous citations so that interested individuals can read the original publications.

Fifth, I have tried to respect both the organismal and the molecular approaches to development, integrating them in various degrees in each section. I believe that the integration of molecular, cellular, and organismal approaches, if successful, should provide a richer understanding of development than any of these approaches taken individually.

Sixth, this book attempts to blur some of the lines separating developmental biology, genetics, and evolution. I have emphasized the use of genetic mutants to analyze normal developmental processes and have tried to show that some of the most intricate events on the molecular level are responsible for enabling the developing organism to survive in a particular environment. Moreover, when we say that one species has evolved into another, what we mean is that its development has changed. Genetic mutations affect evolution by working through development, and the elucidation of the mechanisms by which this is accomplished promises to be one of the most exciting chapters of modern biology.

Seventh, I have tried to increase the flexibility of this book by including "Sidelights and Speculations" sections. Their roles are twofold. Some take a concept introduced in the text and relate it to another field of biology. Others explore concepts which are too new or controversial to be included in the main body of the book, but which may be very exciting areas in the next few years. These latter sections are my attempts to keep the book from being rapidly outdated in such a quickly moving field.

Acknowledgments

The creation of a textbook is not unlike the creation of a complex organ or organism. During its four-year gestation, this book has developed in ways barely hinted at in the first drafts. As in organ formation, the interactions enabling these developments can be roughly divided into permissive ones and instructive ones. Those individuals having instructive interactions with the text (that is, where the information herein would be different, had they not intervened) include R. Auerbach, G. Florant, R. Freter, R. Herlands, J. Jenkins, J. Lilien, A. Mahowald, R. Marchase, M. Oster-Granite, S. Roth, R. Savage, B. Shur, and D. Sonneborn. D. Kirk, J. Gerhart, L. Iten, R. Raff, and the late R. Briggs have been especially helpful in forming the current state of this book. The readability of this text has been greatly improved by comments of students in my developmental biology and developmental genetics courses, and a large debt of gratitude (by all those who are assigned its chapters) should go to them and especially to my teaching assistants, L. Klein, R. Nelson, W. Kirby, D. Sivitz, and T. Kushner. Since my own views of development and how to teach it have come in good measure from certain excellent teachers and advisers, I wish to acknowledge their continued presence in this book.

But instructive interactions alone neither an organ nor a book make. The entire project would have come to a halt were it not for the numerous scientists who sent me their photographs and the extremely supportive biology department of Swarthmore College. H. Ewing and G. Flickinger must be singled out for working beyond the call of duty when I needed immediate typing or photographic development. The continuous updating of these chapters has been made possible by the computer wizardry of two students, B. Datloff and R. Mahajan, who felt that the best way to do well in my course was to type the text used in it. From the conception of this book through its delivery, Andy Sinauer and his staff have gently and effectively assured its proper development. My special thanks to C. Wigg and J. Simpson for their editing and to L. Meszoly, J. Vesely, and J. Woolsey for their art and production work on the book. Moreover, this book would never have been completed were it not for the encouragement of my wife, Anne Raunio, who, as an obstetrician, knows the joys of developmental biology and who felt that I should go through the experience of labor, if only with a book. My thanks to them all.

SCOTT F. GILBERT

PART I

Patterns of development

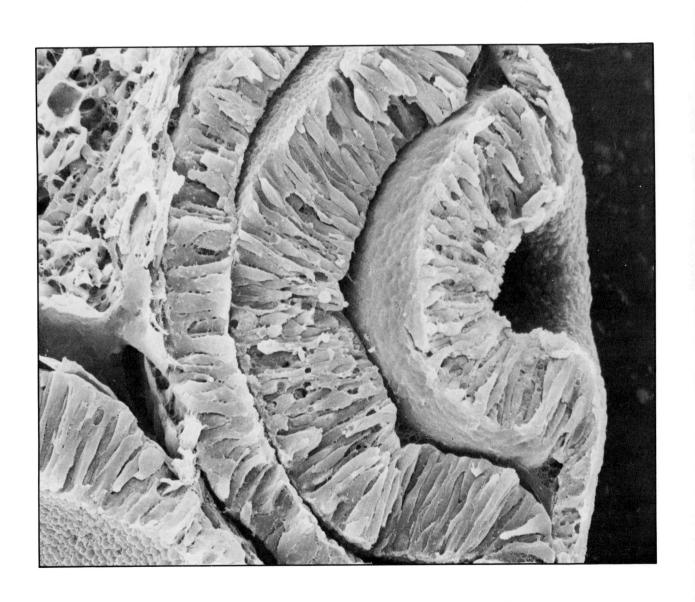

CHAPTER 1

An introduction to animal development

It now remains to speak of animals and their nature. So far as we can, we will not exclude any one of them, no matter how mean; for though there are animals which have no attraction for the senses, yet for the eye of science, for the student who is naturally of a philosophic spirit and who can discern the causes of things, nature which fashioned them provides joys that cannot be measured.
—ARISTOTLE (CA. 330 B.C.)

Joy in looking and understanding is nature's most beautiful gift.
—ALBERT EINSTEIN (1953)

Introduction

According to Aristotle, the first embryologist known to history, science begins with wonder. "It is owing to wonder that people began to philosophize, and wonder remains the beginning of knowledge." The development of animals has been a source of wonder throughout human history, and it has constantly stimulated individuals to seek the causes for such remarkable, yet commonplace, phenomena. The simple procedure of cracking open a chick egg on each day of its 3-week development period provides a remarkable experience—a thin band of cells is seen to give rise to an entire bird. Aristotle performed this experiment and noted the formation of the major organs. Most any person can wonder at this phenomenon, but it is the scientist who seeks to discover how development actually occurs. Rather than dissipating wonder, our new understanding increases it.

Multicellular organisms on earth do not spring forth fully formed. Rather, they arise by a relatively slow process of progressive change that we call DEVELOPMENT. In nearly all cases, the development of a multicellular organism begins with a single cell—the fertilized egg, or ZYGOTE—which divides mitotically to produce all the cells of the body. The study of animal development has traditionally been called EMBRYOLOGY, referring to the fact that between the stage of the fertilized egg and birth, the developing organism is known as an EMBRYO. However, development does not stop at birth—or even at adulthood. Most organisms never cease developing. Each day, we replace over a gram of skin cells (the older cells being sloughed off as we walk), and our bone marrow sustains the development of millions of new erythrocytes every minute of our lives. Therefore, in recent years, it has become customary to speak of DEVELOPMENTAL BIOLOGY as the discipline that involves studies of embryonic and other developmental processes.

Developmental biology is one of the most exciting and fast-growing fields of biology. Part of its excitement comes from its subject matter, for we are just beginning to understand the molecular mechanisms of animal development. Another part of the excitement comes from the unifying role that developmental biology is beginning to assume in the biological sciences. Developmental biology is creating a framework that integrates molecular biology, physiology, cell biology, anatomy, cancer research, immunology, and even evolutionary and ecological studies. The study of development has become essential for understanding any other area of biology.

Principal features of development

Development accomplishes two major functions. It generates the cellular diversity and order within each generation, and it assures the continuity of life from one generation to the next. The first function involves the production and organization of all the diverse types of cells in the body. A single cell, the fertilized egg, gives rise to muscle cells, skin cells, neurons, lymphocytes, blood cells, and all the other cell types. This generation of cellular diversity is called DIFFERENTIATION; the processes that organize the differentiated cells into tissues and organs are called MORPHOGENESIS (creation of form and structure) and GROWTH (increase in size). The second major function of development is REPRODUCTION: the continued generation of new individuals of the species.

The major features of animal development are illustrated in Figure 1. The life of a new individual is initiated by the fusion of genetic material from the two GAMETES—the sperm and the egg. This fusion, called FERTILIZATION, stimulates the egg to begin development. The subsequent sequence of stages is collectively called EMBRYOGENESIS. Throughout the animal kingdom an incredible variety of embryonic types exists, but most patterns of embryogenesis are variations on four themes.

1. Immediately following fertilization, CLEAVAGE occurs. Cleavage is a series of extremely rapid mitotic divisions wherein the enormous volume of cytoplasm is divided into numerous smaller cells. These cells are called BLASTOMERES, and by the end of cleavage they generally form a sphere known as a BLASTULA.

2. After the rate of mitotic division has slowed down, the blastomeres undergo dramatic movements wherein they change their positions relative to one another. This series of extensive cell rearrangements is called GASTRULATION. As a result of gastrulation, the typical embryo contains three cell regions called GERM LAYERS. The outer layer—the ECTODERM—produces the cells of the epidermis and nervous system; the inner layer—the ENDODERM—produces the lining of the digestive tube and its associated organs (pancreas, liver, and so on); and the middle layer—the MESODERM—gives rise to several organs (heart, kidney, gonads), connective tissues (bone, muscles, tendons), and the blood cells.

3. Once the three germ layers are established, the cells interact with each other and rearrange themselves to produce the bodily organs. This process is called ORGANOGENESIS. (In vertebrates, organogenesis is initiated when certain mesodermal cells interact with overlying ectodermal cells to cause the latter to form neural tube. This tube will become the brain and spinal cord.) Many organs contain cells from more than one germ layer, and it is not unusual for the outside of an organ to be derived from one layer and the inside from another. Also during organogenesis certain cells undergo long migrations from their place of origin to their final location. These migrating cells include the precursors of blood cells, lymph cells, pigment cells, and gametes.

4. As seen in Figure 1, a portion of egg cytoplasm gives rise to the precursors of the gametes. These cells are called GERM CELLS, and they

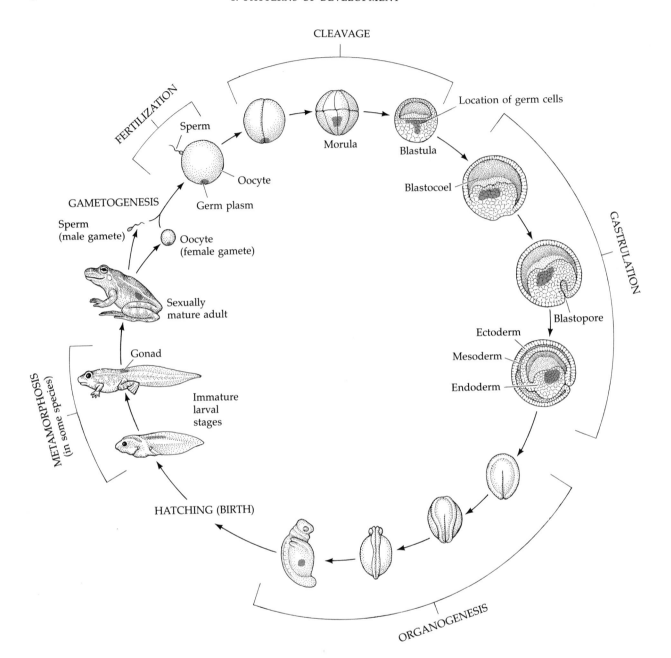

FIGURE 1
Developmental history of a representative organism, the frog. The stages from fertilization through hatching (or birth) are known collectively as embryogenesis. The region of the embryo set aside for producing germ cells is shown in gray. Gametogenesis, which is complete in the sexually mature adult, begins at different times during development, depending upon the species.

are set aside for their reproductive function. All the other cells of the body are called SOMATIC CELLS. This separation between somatic cells (which give rise to the individual body) and the germ cells (which contribute to the formation of a new generation) is typically one of the first differentiations to occur during animal development. The germ cells eventually migrate to the gonads, where they differentiate into gametes—sex cells capable of participating in fertilization to create a new individual. The development of gametes, called GAMETOGENESIS, is usually not completed until the organism has become physically mature. At this time, the gametes may be released and undergo fertilization to begin a new life. Meanwhile, the adult organism eventually undergoes senescence and dies.

Our eukaryotic heritage

Organisms are divided into two major groups, their classification depending on whether or not they have a nuclear envelope. On one side of this classification line are the PROKARYOTES (karyon = nucleus), which include the bacteria and the blue-green algae. These cells lack a true nucleus. On the other side are the EUKARYOTES, which include protozoans, animals, plants, and fungi. Cells of eukaryotes have a well-formed nuclear envelope surrounding their chromosomes. This fundamental difference between eukaryotes and prokaryotes influences how they arrange and utilize their genetic information. In both groups, the information needed for development and metabolism is encoded in the deoxyribonucleic acid (DNA) sequence of chromosomes. The prokaryotic chromosome is a small, circular, double helix of DNA, having approximately a million base pairs. The eukaryotic cell usually has several chromosomes, and the simplest eukaryotic protozoans have over ten times the amount of DNA found in the most complex prokaryotes. Moreover, the structure of eukaryotic genes is more complex than those of prokaryotes. The amino acid sequence of a prokaryotic protein is a direct reflection of the DNA sequence in the chromosome. The protein-coding DNA of a eukaryotic gene, however, is frequently divided up such that the amino acid sequence of a protein is derived from discontinuous segments of DNA (Figure 2). It is believed that without intervening DNA sequences the messenger RNA (mRNA) that codes for the protein would not get through the nuclear envelope (as will be detailed in Chapter 13).

Eukaryotic chromosomes are very different from prokaryotic chromosomes; eukaryotic DNA is wrapped around specific proteins called HISTONES, which can organize the DNA into compact structures. In bacteria, there are no histones. Moreover, eukaryotic cells undergo mitosis wherein the nuclear envelope breaks down and the replicated chromosomes are equally divided between the daughter cells (Figure

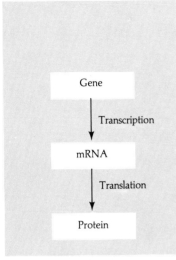

(A) PROKARYOTIC CELL

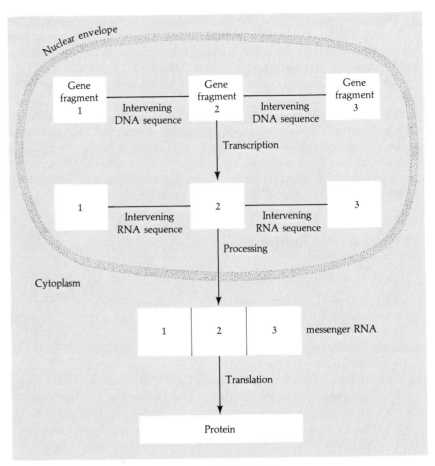

(B) EUKARYOTIC CELL

FIGURE 2
Summary of steps by which proteins are synthesized from DNA. (A) Prokaryotic (bacterial) gene expression: the coding regions of DNA are colinear with the protein product. (B) Eukaryotic gene expression: the genes are discontinuous, and a nuclear envelope separates the DNA from the cytoplasm.

FIGURE 3
Diagrammatic representation of mitosis in animal cells. During interphase, the DNA is doubled in preparation for cell division. During prophase, the nuclear envelope breaks down and a spindle forms between the two centrioles. At metaphase, the chromosomes align at the equator of the cell and, as anaphase begins, the duplicated chromosomes (called chromatids) are separated. At telophase, the chromosomes reach the mitotic poles and the cell begins to pinch in. At each pole are the same number and type of chromosomes as were present in the cell before it divided.

3). In prokaryotes, cell division is not mitotic; no mitotic spindle develops, and there is no nuclear envelope to break down. Rather, the daughter chromosomes remain attached at adjacent points on the cell membrane and are separated and pass into different cells by the growth of the cell membrane between these attachment points.

Prokaryotes and eukaryotes also have different mechanisms of gene regulation. In both prokaryotes and eukaryotes, DNA is transcribed by enzymes called RNA polymerases to make ribonucleic acid (RNA). When messenger RNA is produced in prokaryotes, it is immediately translated into a protein. In fact, one end of the mRNA is busily synthesizing protein while the other end of it is still being transcribed from

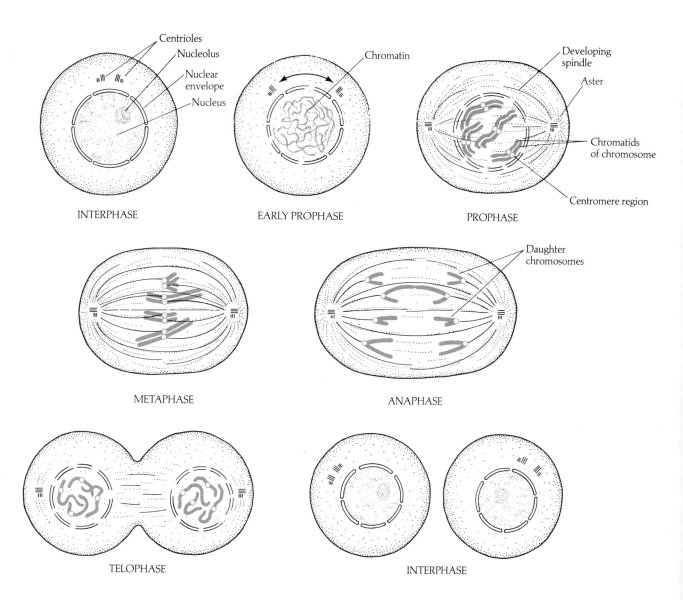

INTERPHASE

EARLY PROPHASE

PROPHASE

METAPHASE

ANAPHASE

TELOPHASE

INTERPHASE

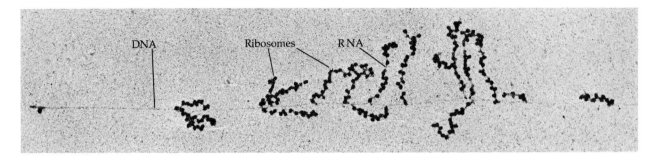

FIGURE 4

Concurrent transcription and translation in prokaryotes. A portion of *E. coli* DNA runs horizontally across this electron micrograph. On either side, transcripts of mRNA can be seen. Ribosomes have attached to the mRNA and are synthesizing proteins (which cannot be seen). The mRNA can be seen increasing in length from left to right, indicating the direction of transcription. (Courtesy of O. L. Miller, Jr.)

the DNA (Figure 4). Thus, transcription and translation are coordinated events in prokaryotes. The existence of the nuclear envelope in eukaryotes provides the opportunity for an entirely new type of cell regulation. The ribosomes, which are responsible for translation, are on one side of the nuclear envelope, whereas the DNA and RNA polymerases needed for transcription are on the other side. Therefore, transcription (from DNA to RNA) must be finished before translation (from RNA to protein) can commence. In between transcription and translation, the transcribed RNA must be processed so that it can pass through the nuclear envelope. By regulating which mRNAs can get into the cytoplasm, the cell is able to select which of the newly synthesized messages will be translated. Thus, a new level of complexity has been added, one that we shall see is extremely important for the developing organism.

Development among the unicellular eukaryotes

All multicellular eukaryotic organisms have evolved from unicellular protists. It is in these protists that the basic features of development first appeared. Simple eukaryotes give us our first examples of the nucleus directing morphogenesis, the use of the cell surface to mediate cooperation between individual cells, and the discovery of sexual reproduction.

Control of developmental morphogenesis in *Acetabularia*

At the turn of the century, it had not yet been proved that the nucleus contained hereditary or developmental information. Some of the best evidence for this theory came from studies in which unicellular organisms were fragmented into nucleate and anucleate pieces (reviewed in Wilson, 1896). When various protozoans were cut into numerous fragments, nearly all the pieces died. However, the fragments containing nuclei were able to live and to regenerate entire, complex, cellular structures (Figure 5).

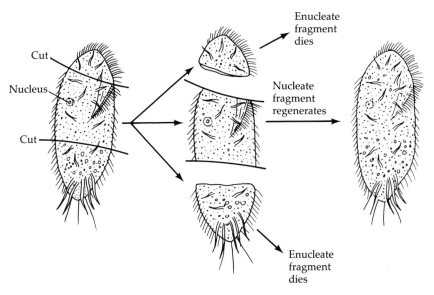

FIGURE 5
Regeneration of the nucleate fragment of the unicellular protist *Stylonychia*. The anucleate fragments survive for a time but finally die.

(A)

——————
1 cm

Nuclear control of cell morphogenesis and the interaction of nucleus and cytoplasm are beautifully demonstrated by studies of *Acetabularia*. This enormous single cell (1–2 inches) consists of three parts: a cap, a stalk, and a rhizoid (Figure 6). The rhizoid is located at the base of the cell and holds it to the substrate. The single nucleus of the cell resides within the rhizoid. The size of *Acetabularia* and the location of its nucleus allow investigators to remove the nucleus from one cell and replace it with a nucleus from another cell. In the 1930s, J. Hämmerling took advantage of these unique features and transferred nuclei from one species—*A. mediterranea*—into the enucleated rhizoid of another species—*A. crenulata*. As Figure 6 shows, these two species have very different cap structures. Hämmerling found that when the nucleus from one species was transplanted into the stalk of another species, the cap eventually assumed the form associated with the *donor* nucleus (Figure 7). Thus, the nucleus is seen to control *Acetabularia* development.

The formation of a cap is a complex morphogenic event involving the synthesis of numerous proteins, the products of which must be placed in a certain portion of the cell. The transplanted nucleus does indeed make its species-specific cap, but it takes several weeks to do

FIGURE 6
Two species of the giant unicellular protist *Acetabularia*. (A) *Acetabularia mediterranea* cell. (B) *Acetabularia crenulata* cell. The rhizoid contains the nucleus. (Courtesy of H. Harris.)

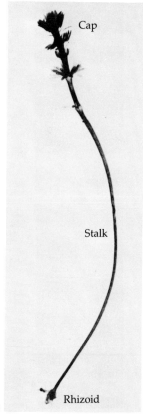

(B)

FIGURE 7

Effect of exchanging nuclei between two species of *Acetabularia;* nuclei were transplanted into enucleated rhizoid fragments. *A. crenulata* structures are shaded; *A. mediterranea* structures are shown in white.

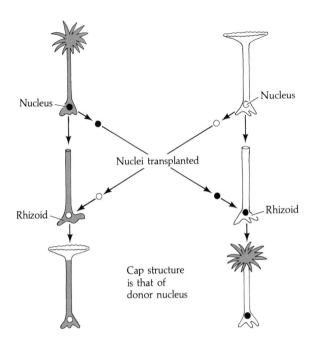

so. Moreover, if the nucleus is removed at an early time in *Acetabularia* development, before the cap is formed, a normal cap is formed weeks later, even though the organism will eventually die. This suggests that (1) the nucleus contains information specifying the style of cap produced (i.e., it contains the genetic information that specifies the proteins responsible for the production of a certain type of cap), and (2) the information enters the cytoplasm long before cap production occurs. The information in the cytoplasm is not used for several weeks.

One hypothesis to explain this observation was that the nucleus synthesizes a stable mRNA, which lies dormant in the cytoplasm until the time of cap formation. More recent evidence based on an observation that Hämmerling published in 1934 supports this hypothesis. Hämmerling fractionated young *Acetabularia* into several parts (Figure 8). The portion with the nucleus formed a cap as expected; so did the apical tip of the stalk. The remaining portion of the stalk did not form a cap. Thus, Hämmerling postulated (nearly 30 years before the existence of mRNA was known) that the instructions for cap formation were somehow stored at the tip. More recent studies (Kloppstech and Schweiger, 1975) have shown that nucleus-derived mRNA does indeed accumulate at this region. Ribonuclease, an enzyme that cleaves RNA, completely inhibits cap formation when added to the seawater in which *Acetabularia* is growing. In enucleated cells, this effect is permanent; once the RNA is destroyed, no cap formation can occur. In nucleate cells, however, the cap can form after the ribonuclease is washed away—presumably because new mRNA is being made by the nucleus.

It is obvious from the preceding discussion that nuclear transcription plays an important role in the formation of the *Acetabularia* cap. But note that the cytoplasm also plays an essential role in cap formation. The mRNAs are not translated for weeks, even though they are in the cytoplasm. Something in the cytoplasm controls when the message is utilized. Hence, the expression of the cap is controlled, not only by nuclear transcription, but also by cytoplasmic factors influencing translation. In this unicellular organism, "development" is controlled at both the transcriptional and the translational levels.

Differentiation in the amoeboflagellate *Naegleria*

One of the most remarkable cases of protist "differentiation" is that of *Naegleria gruberi.* This organism occupies a special place in protist taxonomy because it can change its form from that of an amoeba to that of a flagellate (Figure 9). During most of its life cycle, *N. gruberi* is a typical amoeba, feeding on soil bacteria and dividing by fission. However, when the bacteria are diluted (either by rainwater or by water added by an experimenter), each *N. gruberi* rapidly develops a streamlined body shape and two long anterior flagella. Thus, instead of having several differentiated cell types in one organism, this one cell has different cell structures and biochemistry at different times of its life.

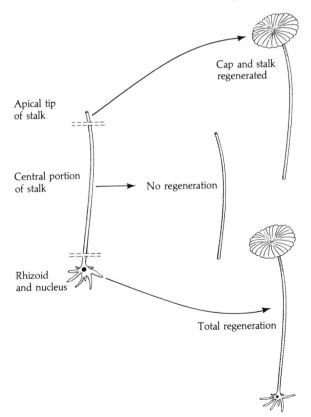

Cap and stalk
regenerated

Apical tip
of stalk

Central portion
of stalk → No regeneration

Rhizoid
and nucleus

Total regeneration

FIGURE 8
Regeneration ability of different fragments of *Acetabularia mediterranea.*

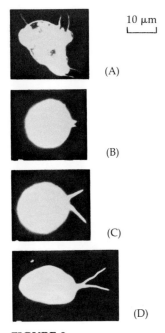

10 μm

(A)

(B)

(C)

(D)

FIGURE 9
Transformation from ameboid to flagellated phase in *Naegleria*. (A) Amoeba. (B) Spherical cell with two short flagella. (C) Spherical cell with longer flagella. (D) Flagellated cell. (From Fulton and Dingle, 1967; photographs courtesy of C. Fulton.)

Differentiation into the flagellate stage occurs within an hour (Figure 10). During this time the amoeba has to create a kinetic apparatus to serve as the basal plates of the flagella as well as create the flagella themselves. Flagella are made from many proteins, the most abundant of which is TUBULIN. The tubulin proteins are organized into microtubules, and the microtubules are further organized into an arrangement that will permit flagellar movement (Figure 11). Fulton and Walsh (1980) showed that the tubulin for the *Naegleria* flagella does not exist in the amoeba stage. It is made de novo ("from scratch"), starting with new transcription from the nucleus. To show this, the investigators manipulated transcription at various stages with ACTINOMYCIN D, an antibiotic drug that selectively inhibits RNA synthesis. When added before the dilution of the food supply, this antibiotic prevented tubulin synthesis. However, if the actinomycin D is added 20 minutes after dilution, tubulin is made. Therefore, it appears that the mRNA for tubulin is made during the first 20 minutes after dilution and is used shortly thereafter. This interpretation was confirmed when it was shown that mRNA extracted from amoebae does not contain any detectable messages for flagellar tubulin, whereas mRNA extracted from differentiating cells contains a great many such messages.

Here, then, is an excellent example of the TRANSCRIPTIONAL CONTROL of a developmental process: The *Naegleria* nucleus responds to environmental changes by synthesizing the mRNA for flagellar tubulin. We also see another process that remains extremely important in the

FIGURE 10
Differentiation of the flagellate phenotype in *Naegleria*. Amoeba that had been growing in a bacteria-enriched medium are washed free of bacteria at time zero. By 80 minutes, nearly the entire population has developed flagella.

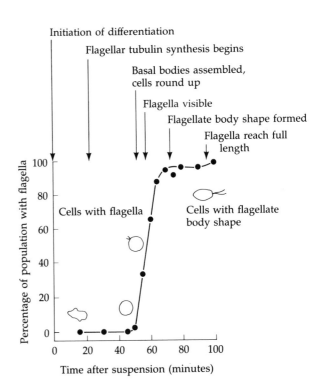

development of all other animals and plants, namely, the assembly of tubulin molecules to produce a flagellum. This arrangement, whereby tubulin is polymerized into microtubules and the microtubules assembled into an ordered array is seen throughout nature. In mammals it is evident in the sperm flagellum and in cilia of the spinal cord and respiratory tract. This POSTTRANSLATIONAL CONTROL, whereby proteins interact with one another to generate organized structures within the cell, will be discussed more fully later. We see, then, that unicellular eukaryotes can control their development at the transcriptional, translational, and posttranslational levels.

Origins of sexual reproduction

Sexual reproduction is one other invention of the protists that has had a profound effect on more complex organisms. It should be noted that sex and reproduction are two distinct and separable processes. Reproduction involves the creation of new individuals. Sex involves the combining of genes from two different individuals into new arrangements. *Reproduction in the absence of sex* is characteristic of those organisms that reproduce by fission; there is no sorting of genes when an amoeba divides or when a hydra buds off cells to form a new colony. *Sex without reproduction* is also common among unicellular organisms. Bacteria are able to transmit genes from one individual to another by means of sex pili (Figure 12). This transmission is separate from reproduction. Protists are also able to reassort genes without reproducing. Paramecia, for instance, reproduce by fission, but sex is accomplished by CONJUGATION (Figure 13). When two paramecia join together, they link their oral apparatuses and form a cytoplasmic connection. Each macronucleus (which controls the metabolism of the organism) degenerates while each micronucleus undergoes meiosis followed by mitosis to produce eight haploid micronuclei, of which all but one degenerate. The remaining micronucleus divides once more to form a stationary micronucleus and a migratory micronucleus. Each migratory micronucleus crosses the cytoplasmic bridge and fuses with ("fertilizes") the stationary micronucleus, thereby creating a new diploid nucleus in each cell. This diploid nucleus then divides to give rise to a new micronucleus and a new macronucleus as the two partners disengage. No reproduction has occurred, only sex.

The union of these two distinct processes, sex and reproduction,

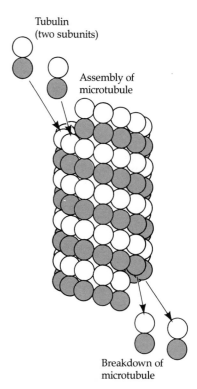

Tubulin (two subunits)

Assembly of microtubule

Breakdown of microtubule

FIGURE 11
Assembly and disassembly of microtubules by the polymerization or depolymerization of tubulin dimers.

FIGURE 12
Sex in bacteria. The bacterial cell covered with numerous appendages (PILI) is capable of transmitting genes to a recipient cell (lacking pili) through a SEX PILUS. In this figure, the sex pilus is highlighted by viral particles that bind specifically to that structure. (Photograph courtesy of C. C. Brinton, Jr. and J. Carnahan.)

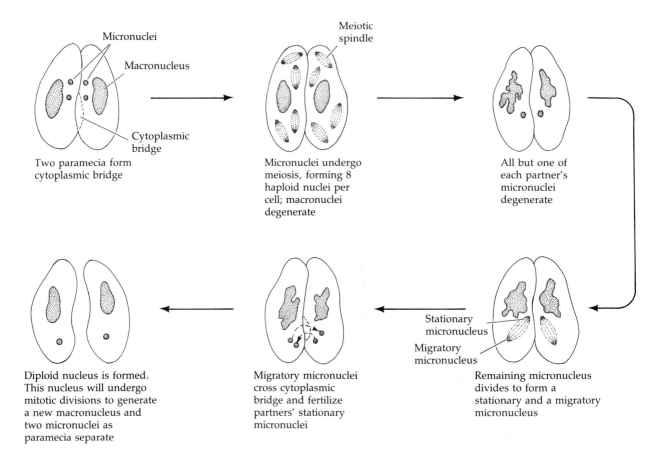

Micronuclei

Macronucleus

Cytoplasmic bridge

Two paramecia form cytoplasmic bridge

Meiotic spindle

Micronuclei undergo meiosis, forming 8 haploid nuclei per cell; macronuclei degenerate

All but one of each partner's micronuclei degenerate

Stationary micronucleus

Migratory micronucleus

Remaining micronucleus divides to form a stationary and a migratory micronucleus

Migratory micronuclei cross cytoplasmic bridge and fertilize partners' stationary micronuclei

Diploid nucleus is formed. This nucleus will undergo mitotic divisions to generate a new macronucleus and two micronuclei as paramecia separate

FIGURE 13
Conjugation in paramecia. Two paramecia can exchange genetic material, leaving them with genes that differ from those with which they started.

into SEXUAL REPRODUCTION is first seen in other unicellular eukaryotes. Figure 14 shows the life cycle of *Chlamydomonas*. This organism is usually haploid, having just one copy of each chromosome (like a mammalian gamete). The individuals of each species, however, are divided into two mating groups: *plus* and *minus*. When these meet, they join their cytoplasms together and their nuclei fuse to form a diploid zygote. This zygote is the only diploid cell in the life cycle, and it eventually undergoes meiosis to form four new *Chlamydomonas*. Here is sexual reproduction, for chromosomes are reassorted during the meiotic divisions and more individuals are formed. Note that in this early trial of sexual reproduction, the gametes are morphologically identical—the distinction between sperm and egg has not been made yet.

In evolving this sexual reproduction, two important advances had to be achieved. The first is the mechanism of MEIOSIS (Figure 15), whereby the diploid complement of chromosomes is reduced to the

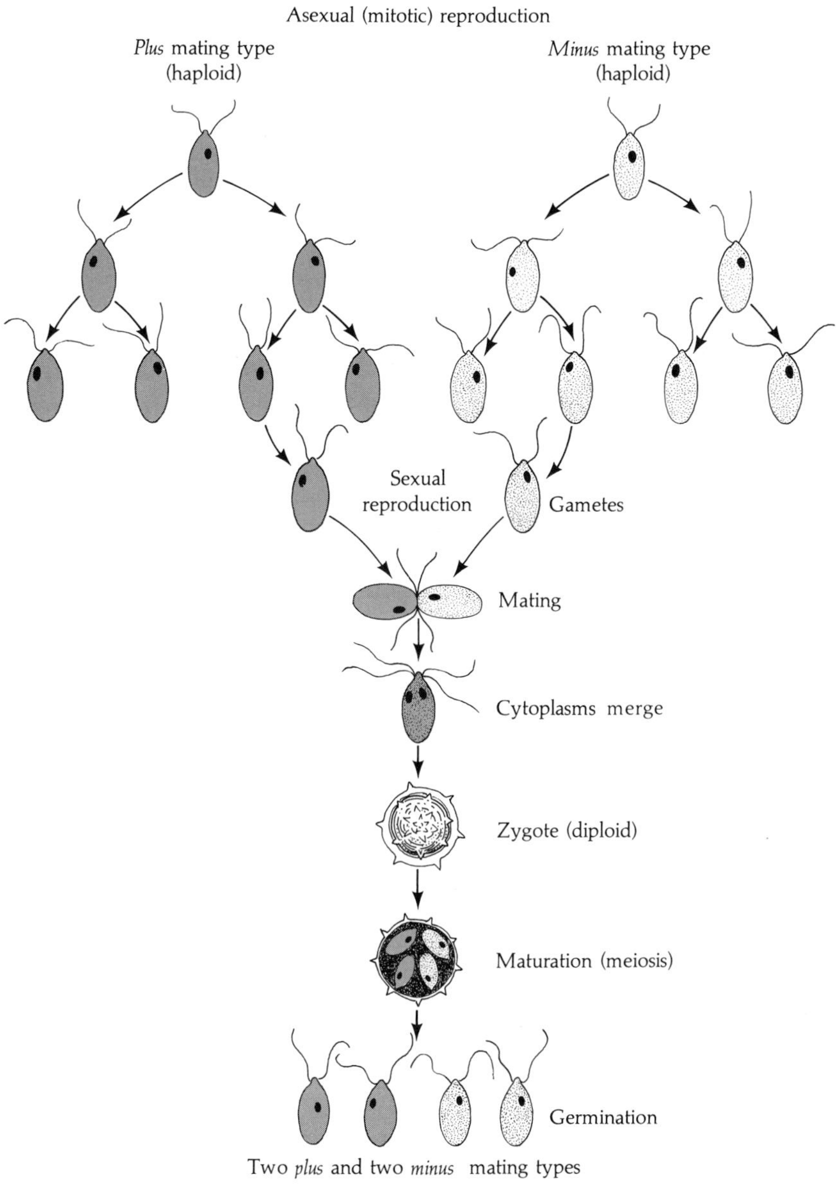

Asexual (mitotic) reproduction

Plus mating type
(haploid)

Minus mating type
(haploid)

Sexual
reproduction   Gametes

Mating

Cytoplasms merge

Zygote (diploid)

Maturation (meiosis)

Germination

Two *plus* and two *minus* mating types

FIGURE 14
Sexual reproduction in *Chlamydomonas*. Two strains, both haploid, can reproduce asexually when separate. Under certain conditions, the two strains can unite to produce a diploid cell that can undergo meiosis to form four new haploid organisms. (After Strickberger, 1968).

haploid state. (This will be discussed in detail in Chapter 20.) The other advance is the mechanism whereby the two different mating types recognize each other. Recognition occurs first on the flagellar membranes (Figure 16; Goodenough and Weiss, 1975; Bergman et al., 1975). The agglutination of flagella allows specific regions of the cell membrane to come together. These specialized sectors contain mating type-specific components that allow the cytoplasms to fuse. Following agglutination, the *plus* individuals initiate the fusion by extending a "fertilization tube"

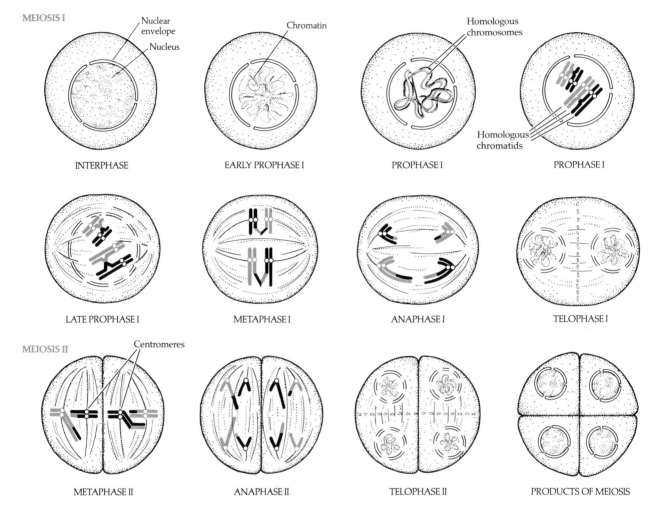

MEIOSIS I

Nuclear envelope
Nucleus
INTERPHASE

Chromatin
EARLY PROPHASE I

Homologous chromosomes
PROPHASE I

Homologous chromatids
PROPHASE I

LATE PROPHASE I

METAPHASE I

ANAPHASE I

TELOPHASE I

MEIOSIS II

Centromeres
METAPHASE II

ANAPHASE II

TELOPHASE II

PRODUCTS OF MEIOSIS

FIGURE 15
Summary of meiosis. The DNA has already replicated during interphase. During prophase, the nuclear envelope breaks down and homologous chromosomes (each chromosome being double and each chromatid joined at the centrosome) pair together. Chromosomal rearrangements can occur between the four homologous chromatids at this time. After the first metaphase, the two original chromosomes are separated into two cells. During the second division, the centromere splits such that each new cell has one copy of each chromosome.

not unlike that which we will see in sperm. This tube contacts and fuses with a specific site on the minus individual. Interestingly, the mechanism used to extend this tube—the polymerization of the protein actin—is the same that will extend the acrosomal process of sea urchin sperm; and in the next chapter, we shall see that the recognition and fusion of sperm and egg occur in an amazingly similar manner.

We can readily see that unicellular eukaryotes have the basic elements of the developmental processes that characterize the more complex organisms of other phyla: Cellular synthesis is controlled by transcriptional, translational, and posttranslational regulation; a mechanism for processing RNA through the nuclear membrane exists; the structures of individual genes and chromosomes are as they will be throughout eukaryotic evolution; mitosis and meiosis are perfected; and sexual reproduction exists, showing cooperation between individual cells. Such intercellular cooperation becomes even more important with the evolution of multicellular organisms.

Colonial eukaryotes: The evolution of differentiation

One of evolution's most important experiments was the creation of multicellular organisms. There appear to be several paths by which single cells evolved multicellular arrangements; we shall only discuss two of these ways. The first path involves the orderly division of the reproductive cell and the subsequent differentiation of the new cells. This path to multicellularity can be seen in a remarkable series of multicellular organisms collectively referred to as the Volvocales.

The Volvocales

The simpler organisms among the Volvocales are collections of numerous *Chlamydomonas*-like cells; but the more advanced members of this group have developed a second, very different, cell type. The simplest Volvocales genus—*Oltmannsiella*—contains four *Chlamydomonas*-like cells in a row, embedded in a gelatinous matrix. In the genus *Gonium* (Figure 17), a single cell divides to produce a flat plate of 4 to 16 cells, each with its own flagella; all the flagella are on the same side of the gel matrix. In a related genus—*Pandorina*—the 16 cells form a sphere; and in *Eudorina*, the sphere contains 32 or 64 cells arranged in a regular pattern. Already, then, a very important developmental principle has been worked out: the ordered division of one cell to generate a number of cells that are organized in a predictable fashion. This is the precursor to embryonic cleavage.

(A)

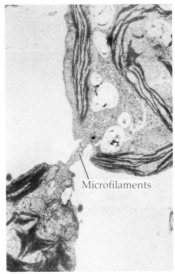
Microfilaments

(B)

FIGURE 16
Two-step recognition in mating *Chlamydomonas*. (A) Scanning electron micrograph (×7000) of mating pair. The interacting flagella twist about each other, adhering at the tips (arrows). (B) Transmission electron micrograph (×20,000) of a cytoplasmic bridge connecting the two organisms. The microfilaments extend from the donor cell (lower) to the recipient (upper) cell. (From Goodenough and Weiss, 1975; and Bergman et al., 1975; reproduced by permission of U. Goodenough.)

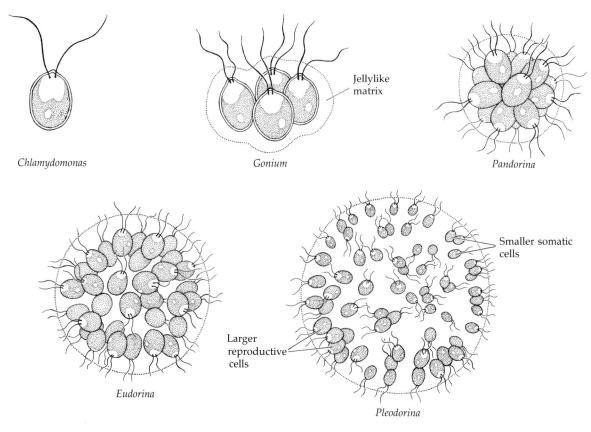

Chlamydomonas

Gonium   Jellylike matrix

Pandorina

Eudorina

Larger reproductive cells   Smaller somatic cells

Pleodorina

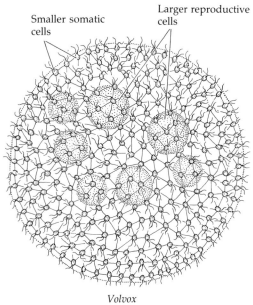

Smaller somatic cells   Larger reproductive cells

Volvox

FIGURE 17
Representatives of the order Volvocales. Complexity increases from the single-celled *Chlamydomonas* to the multicellular *Volvox.*

The next two genera of the Volvocales series exhibit another important principle of development. Here we have the differentiation of cell types within an individual organism. The reproductive cells become differentiated from the somatic cells. In *Eudorina*, for instance, every cell can, and normally does, produce a new organism by mitosis (Figure 18). In the genera *Pleodorina* and *Volvox*, however, relatively few cells can reproduce. In *Pleodorina californica*, the cells in the anterior region are restricted to a somatic function. Only those cells on the posterior side can reproduce. In this species, a colony usually has 128 or 64 (rarely 32) cells. The ratio of the number of somatic cells to the number of reproductive cells is usually 3:5; thus, a 128-cell colony has 48 somatic cells and a 64-cell colony has 24.

In *Volvox*, almost all the cells are somatic, and very few of the cells are able to produce new individuals. In the simpler members of the genus *Volvox*, reproductive cells, as in *Pleodorina*, are derived from cells that originally look and function like somatic cells before they enlarge and divide to form new progeny. However, in more complex members of the genus, such as *V. carteri*, there is a complete division of labor: The reproductive cells that will create the next generation are set aside during the division of the reproductive cells that are forming a new individual. The reproductive cells never develop functional flagella and never contribute to motility or other somatic functions of the individual; they are entirely specialized for reproduction. Thus, although the simpler Volvocales may be thought of as colonial organisms (since each cell is capable of independent existence and of perpetuating the species), in *Volvox carteri* we have a true multicellular organism with two distinct and interdependent cell types (somatic and reproductive), both of which are required for perpetuation of the species. In a later chapter we shall see that the pattern exhibited by *V. carteri*—setting aside a separate line

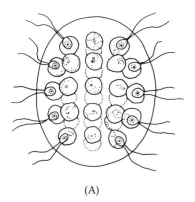
(A)
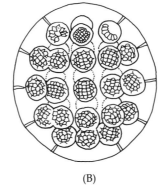
(B)

FIGURE 18
Asexual reproduction in *Eudorina elegans*. (A) Mature colony. (B) Entire colony in division. (After Hartmann, 1921.)

of reproductive cells very early in development—is also exhibited by most animal embryos.

Although the multicellular Volvocales reproduce predominantly by asexual division of cells (as described earlier), they are also capable of sexual reproduction involving production and fusion of haploid gametes (Figure 19). We see here that there are two different types of gametes: large eggs and small motile sperm. This type of sexual reproduction is called HETEROGAMY, as opposed to the ISOGAMY of *Chlamydomonas*. Here,

FIGURE 19
Sexual reproduction in *Volvox*. (A) Subsurface view of bisexual *Volvox* colony, showing cytoplasmic connections uniting each cell. The sperm-producing cell and the egg-producing cell are so labeled. (B) Formation of haploid sex cells, which unite to form diploid zygotes. As with *Chlamydomonas,* meiosis occurs after fertilization. (A from Berrill, 1961.)

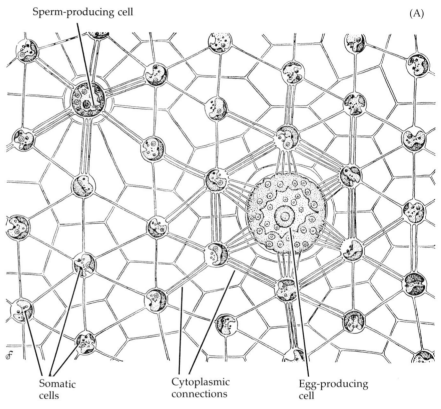

(A)

Sperm-producing cell

Somatic cells

Cytoplasmic connections

Egg-producing cell

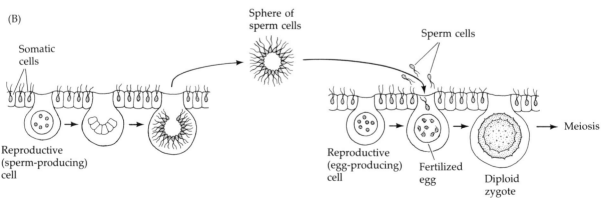

(B)

Somatic cells

Reproductive (sperm-producing) cell

Sphere of sperm cells

Sperm cells

Reproductive (egg-producing) cell

Fertilized egg

Diploid zygote

Meiosis

too, we see gametogenesis for the first time. There are distinct developmental pathways for the formation of sperm and eggs. The fertilization reaction seems to be similar to that of *Chlamydomonas*, and it results in the formation of a diploid zygote, which undergoes meiosis to form a new haploid individual.[1]

Dictyostelium

Another type of multicellular organization derived from unicellular organisms is found in the slime mold *Dictyostelium discoideum*. The life cycle of this fascinating and unusual organism is illustrated in Figure 20. In its vegetative cycle, solitary haploid amoebae (called myxamoebae

[1]It is important to note that the separation of the germ line from the somatic cells occurs early in evolution. This differentiation is among the first that take place in normal development. This separation seems to be a long-standing principle: Even in ciliated protists, there are often two nuclei, a somatic (macro-) nucleus to direct metabolism and a germ (micro-) nucleus for reproduction.

FIGURE 20
Life history of the slime mold *Dictyostelium discoideum.* Haploid spores give rise to myxamoebae, which can reproduce asexually to form more haploid myxamoebae. As the food supply diminishes, aggregation occurs at central points and a migrating pseudo-plasmodium is formed. Eventually it stops moving and forms a fruiting body that releases more spores. (Photographs courtesy of K. Raper.)

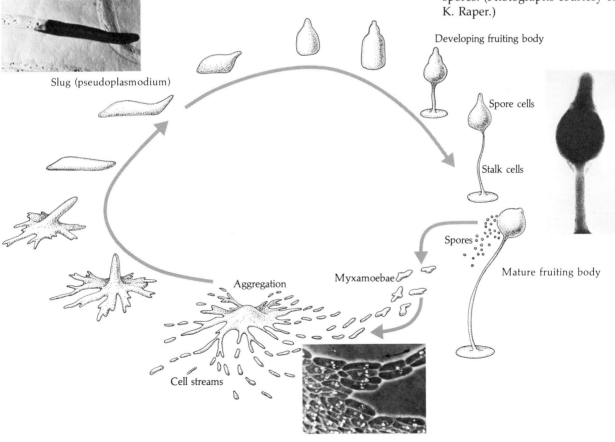

Slug (pseudoplasmodium)

Developing fruiting body

Spore cells

Stalk cells

Spores

Myxamoebae

Mature fruiting body

Aggregation

Cell streams

to distinguish them from the true amoeba species) live on decaying logs, eating bacteria and reproducing by binary fission. When they have exhausted their food supply, tens of thousands of these amoebae join together to form moving streams of cells that converge at a central point. Here they form a conical mound, which eventually absorbs all the streaming cells and bends over to produce the migrating slug. The slug (often given the more dignified titles of pseudoplasmodium or grex) is usually 2–4 mm long and is encased in a slimy sheath. The grex begins migration (if the environment is dark and moist) with its anterior tip slightly raised. When migration ceases, these anterior cells, representing 15–20 percent of the entire cellular population, move downward to form a stalk while the posterior cells are brought upward. The stalk cells die, but the posterior cells, elevated above the stalk, become spore cells. These spore cells disperse, each one becoming a new myxamoeba. In addition to this asexual cycle, there is a possibility for sex in *Dictyostelium*. Two cells can fuse to create a giant cell, which digests all the other cells of the aggregate. When it has eaten all its neighbors, it encysts itself in a thick wall and undergoes meiosis and mitotic division; eventually new myxamoebae are liberated.

Dictyostelium has been a wonderful experimental organism for developmental biologists, for it is an organism wherein initially identical cells are differentiated into one of two alternative cell types, spore or stalk. It also is an organism wherein individual cells come together to form a cohesive structure composed of differentiated cell types, akin to tissue formation in more complex organisms. The aggregation of thousands of amoebae into a single organism is an incredible feat of organization.

The first question is what causes these cells to aggregate. Time-lapse microcinematography has shown that no directed movement occurs during the first 4–5 hours following nutrient starvation. During the next 5 hours, however, the cells are seen moving at about 20 μm/min for 100 seconds. This movement ceases for about 4 minutes, then resumes. Although the movement is directed toward a central point, it is not a simple radial movement. Rather, cells join with each other to form streams; the streams converge into larger streams; and eventually all streams merge at the center. Bonner (1947) and Shaffer (1953) showed that this movement was due to CHEMOTAXIS; that is, the cells were guided to aggregation centers by a soluble substance. This substance was later identified as cyclic adenosine 3′,5′-monophosphate (cAMP) (Konijn et al., 1967; Bonner et al., 1969), the chemical structure of which is shown in Figure 21A.

Aggregation is initiated by the periodic secretion of cAMP by a few randomly distributed cells (Cohen and Robertson, 1971). Neighboring cells respond to the cAMP in two ways: They initiate a movement toward the source of the cAMP pulse, and they release cAMP of their own (Robertson et al., 1972; Shaffer, 1975). After this, the cell is unresponsive to further cAMP pulses for several minutes. The result is a wave of cAMP that gets propagated throughout a population of cells

(A)

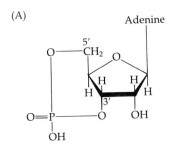

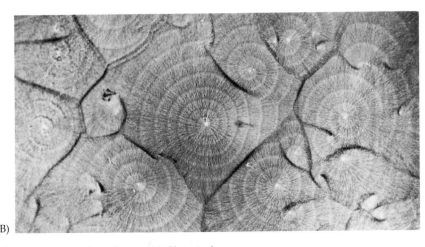

(B)

FIGURE 21

Chemotaxis due to waves of cyclic adenosine monophosphate. (A) Chemical structure of cAMP. (B) Visualization of the cAMP waves through the medium. The centers secrete cAMP at regular intervals and each secretion is relayed so that the wave propagates outward. Waves were charted by saturating filter paper with radioactive cAMP and placing it on an aggregating colony. The cAMP from the colony would dilute the radioactive cAMP. When the radioactivity on the paper was recorded (by placing it in X-ray film), the regions of high cAMP concentration in the culture showed up lighter than those of low cAMP concentration. (B from Tomchick and Devreotes, 1981; courtesy of P. Devreotes.)

(Figure 21B). As each wave passes, the cells take another step toward the center.

Slime molds have also evolved a mechanism whereby the cAMP-stimulated cells can stick to each other. Normally, individual amoebae will not adhere to each other, but they will stick together after 5 hours of starvation. A set of cell surface proteins (called DISCOIDINS) develops at this time to mediate this intercellular cohesiveness (Rosen et al., 1973; Ray et al., 1979).

The differentiation of individual amoebae into either stalk or spore cells is a complex matter. Bonner (1957) demonstrated that in all cases the anterior cells become stalk, whereas the remaining cells are destined to form spores. However, surgically removing the anterior part of the slug will not change this situation. The new anterior cells, which had been destined to produce spores, will now become the stalk (Raper, 1940). Somehow a decision has to be made such that whichever cells are anterior become the stalk cells and whichever cells are posterior become spores. This ability of a cell to change its developmental fate is called REGULATION, and we shall see this phenomenon in many embryos, including those of mammals. Again, one finds, as in Volvocales, a differentiation between somatic (stalk) and reproductive (spore) cells.

Slime molds, however, are an evolutionary dead end. More complex multicellular organisms do not form from the aggregation of formerly independent cells. However, the ability of individual cells to sense a gradient (as in the amoeba's response to cAMP) is probably very important for certain events during animal development, and the role of cell surface proteins for cell cohesiveness is also found throughout the animal kingdom. So here we see the differentiation and morphogenesis of an organism comprising two major cell types. The principles established in such "simple" systems of development are still those used by the more complex animal species.

Developmental patterns among metazoans

Since the remainder of this book concerns the development of META-ZOANS—multicellular organisms that pass through embryonic stages of development—we will present an overview of their developmental patterns. Figure 22 illustrates the major evolutionary trends of metazoan development. The most striking observation is that life has not evolved in a straight line, but rather there are several evolutionary paths. We can furthermore see that most of the species of metazoans belong to one of two major branches of animals: protostomes and deuterostomes.

The colonial protozoans are thought to have given rise to two groups of metazoans, both of which pass through embryonic stages of development. One of these groups is the Porifera (sponges). These animals develop in a manner so different from that of any other animal group that some taxonomists do not consider them metazoans at all. The sponge has three different types of somatic cells, but one of these, the ARCHEOCYTE, can differentiate into all the other cell types in the body. Sponges whose cells are passed through a sieve can regenerate new sponges from individual cells. Sponges contain no mesoderm, so there are no true organ systems in the Porifera; nor do they have a digestive tube or circulatory system, nerves or muscles. Thus, even though they pass through an embryonic and larval stage, they are very unlike most metazoans.

Protostomes and deuterostomes

The other group of metazoans is characterized by having three germ layers during development. Some members of this group constitute the RADIATA, because they have a radial symmetry, like that of a tube or wheel. The Radiata include the coelenterates[2] (jellyfish, corals, and hydra) and the ctenophores (comb jellies). In these animals, the mesoderm is rudimentary, consisting of sparsely scattered cells in a gelatinous matrix. Most metazoans, however, are bilateral species (called

[2]Some taxonomists prefer the name Cnidaria for the phylum formerly called Coelenterata. The cnidarians include hydroids and anemones but not the comb jellies (which develop in a somewhat different manner).

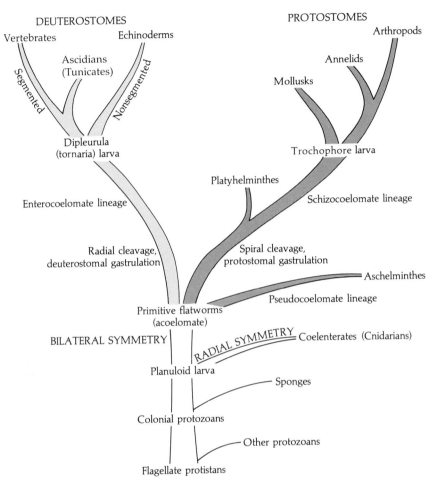

DEUTEROSTOMES

PROTOSTOMES

Vertebrates

Echinoderms

Arthropods

Annelids

Ascidians
(Tunicates)

Mollusks

Segmented

Nonsegmented

Dipleurula
(tornaria) larva

Trochophore larva

Platyhelminthes

Enterocoelomate lineage

Schizocoelomate lineage

Radial cleavage,
deuterostomal gastrulation

Spiral cleavage,
protostomal gastrulation

Aschelminthes

Pseudocoelomate lineage

Primitive flatworms
(acoelomate)

BILATERAL SYMMETRY

RADIAL SYMMETRY Coelenterates (Cnidarians)

Planuloid larva

Sponges

Colonial protozoans

Other protozoans

Flagellate protistans

FIGURE 22
A diagrammatic representation of major evolutionary divergences. (Other models are possible, but the general schemes are all similar to the one shown here.)

Bilateria) and are classified as either protostomes or deuterostomes. All Bilateria are thought to have descended from a primitive type of flatworm. These flatworms were the first to have a true mesoderm (although it was not hollowed out to form a body cavity), and they are thought to have resembled the larvae of certain contemporary coelenterates. The differences in the two major divisions of the Bilateria are illustrated in Figure 23. PROTOSTOMES ("mouth first"), which include the mollusc, arthropod, and worm phyla, are so called because during gastrulation the mouth regions are formed first. The body cavity of these animals forms from the hollowing out of a previously solid cord of mesodermal cells.

The other great division of the animal kingdom is the DEUTEROSTOME lineage. Phyla in this division include chordates and echinoderms. Although it may seem strange to classify humans and wolverines in the same group as starfish and sea urchins, certain embryological features stress this kinship. First, in deuterostomes ("mouth second"), the

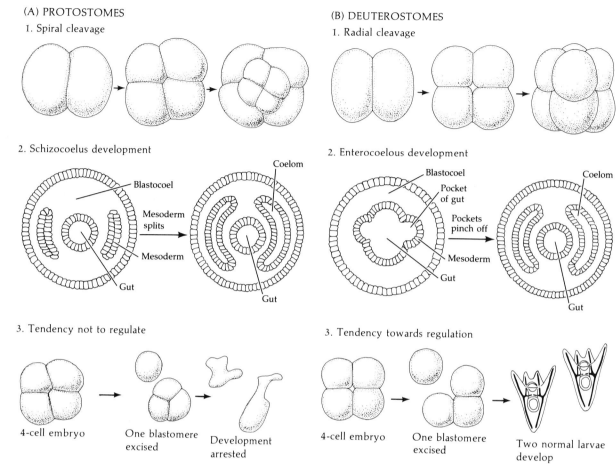

(A) PROTOSTOMES

1. Spiral cleavage

2. Schizocoelus development

3. Tendency not to regulate

4-cell embryo   One blastomere
                       excised   Development
                                        arrested

(B) DEUTEROSTOMES

1. Radial cleavage

2. Enterocoelous development

3. Tendency towards regulation

4-cell embryo   One blastomere
                       excised   Two normal larvae
                                        develop

FIGURE 23
Principal tendencies of the deuterostomes and protostomes.

mouth region is formed after the anal region. Also, whereas protostomes form their body cavity by hollowing out a solid mesodermal
block ("schizocoelous" formation of the body cavity), most deuterostomes form their body cavity from mesodermal pouches extending
from the gut ("enterocoelous" formation of the body cavity).

In addition, protostomes and deuterostomes differ in the way they
undergo cleavage. Most deuterostomes undergo cleavage such that the
blastomeres are perpendicular or parallel to each other. This is called
RADIAL CLEAVAGE. Protostomes, on the other hand, form blastulae composed of cells that are at acute angles to the polar axis of the embryo.
Thus, they are said to have SPIRAL CLEAVAGE. Furthermore, the cleavage-stage blastomeres of most deuterostomes have a greater ability to
regulate development than those of protostomes. If a single blastomere
is removed from a 4-cell sea urchin or human embryo, that blastomere
will develop into an entire organism and the remaining ¾ embryo will
also develop normally. However, if the same operation were performed

on a snail or worm embryo, both the single blastomere and the remaining ones would develop into partial embryos, each one lacking what was formed in the other.

The evolution of organisms depends upon changes in their development. One of the greatest evolutionary advances—the amniote egg—occurred among the deuterostomes. This type of egg, exemplified by that of a chick (Figure 24), first appeared in reptiles about 30 million years ago.

The amniote egg liberated vertebrates from any dependence upon water during their development; they were free to roam on land, far from existing ponds. Whereas amphibians must return to water to breed and to enable their eggs to develop, the amniote egg carries its own water and food supply. The egg is fertilized internally and contains YOLK to nourish the developing embryo. Moreover, it contains two sacs: the AMNION, which contains the fluid bathing the embryo, and the ALLANTOIS, in which waste materials from embryonic metabolism collect. The entire structure is encased in a shell, which allows the diffusion of oxygen but which is hard enough to protect the embryo from environmental assaults. A similar development of egg casings enabled arthropods to be the first terrestrial invertebrates. Thus, the final crossing of the boundary from water to land occurred with the modification of the earliest stage in development, the egg.

Developmental biology provides an endless assortment of fascinating animals and problems. It is this author's task to select a small sample of them to illustrate the major principles of animal development. Yet this is but an incredibly small collection. Right now, we are merely observing the small tidepool within our reach while the whole ocean of developmental principles lies before us. Our study of animal development will begin, then, with the early stages of animal embryogenesis: fertilization, cleavage, gastrulation, and the establishment of the body plan. Although an attempt has been made to survey the important variations throughout the animal kingdom, a certain deuterostome chauvinism may be apparent.

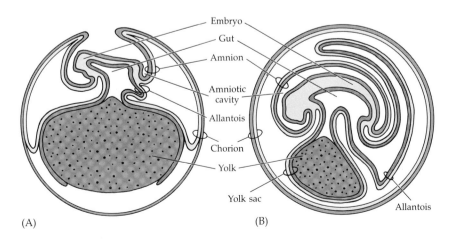

Embryo
Gut
Amnion
Amniotic cavity
Allantois
Chorion
Yolk
Yolk sac
Allantois

(A) (B)

FIGURE 24
Diagram of the amniote egg of the chick showing the development of membranes enfolding the embryo. (A) 3-Day incubation. (B) 7-Day incubation.

Literature cited

Bergman, K., Goodenough, U. W., Goodenough, D. A., Jawitz, J. and Martin, H. (1975). Gametic differentiation in *Chlamydomonas reinhardtii*. II. Flagellar membranes and the agglutination reaction. *J. Cell Biol.* 67: 606–622.

Berrill, J. (1961). *Growth, Pattern and Form*. Freeman, San Francisco.

Bonner, J. T. (1947). Evidence for the formation of cell aggregates by chemotaxis in the development of the slime mold *Dictyostelium discoideum*. *J. Exp. Zool.* 106: 1–26.

Bonner, J. T. (1957). A theory of the control of differentiation in the cellular slime molds. *Q. Rev. Biol.* 32: 232–246.

Bonner, J. T., Barkley, D. S., Hall, E. M., Konijn, T. M., Mason, J. W., O'Keefe, G. and Wolfe, P. B. (1969). Acrasin, acrasinase and the sensitivity to acrasin in *Dictyostelium discoideum*. *Dev. Biol.* 20: 72–87.

Cohen, M. H. and Robertson, A. (1971). Wave propagation in the early stages of aggregation of cellular slime molds. *J. Theor. Biol.* 31: 101–118.

Fulton, C. (1977). Cell differentiation in *Naegleria gruberi*. *Annu. Rev. Microbiol.* 31: 597–629.

Fulton C. and Dingle, A. D. (1967). Appearance of flagellate phenotype in populations of *Naegleria* amebae. *Dev. Biol.* 15: 165–191.

Fulton, C. and Walsh, C. (1980). Cell differentiation and flagellar elongation in *Naegleria gruberi*: Dependence on transcription and translation. *J. Cell Biol.* 85: 346–360.

Goodenough, U. W. and Weiss, R. L. (1975). Gametic differentiation in *Chlamydomonas reinhardtii*. III. Cell wall lysis and microfilament associated mating structure activation in wild-type and mutant strains. *J. Cell Biol.* 67: 623–637.

Hämmerling, J. (1934). Über formbildende Substanzen bei *Acetabularia mediterranea*, ihre räumliche und zeitliche Verteilung und ihre Herkunft. *Wilhelm Roux Arch. Entwicklungsmech. Org.* 131: 1–81.

Hämmerling, J. (1953). Nucleo-cytoplasmic relationships in the development of *Acetabularia*. *Int. Rev. Cytol.* 2: 475–498.

Hartmann, M. (1921). Die dauernd agame Zucht von *Eudorina elegans*, experimentelle Beiträge zum Befruchtungs- und Todproblem. *Arch. Protistk.* 43: 223–286.

Kloppstech, K. and Schweiger, H. G. (1975). Polyadenylated RNA from *Acetabularia*. *Differentiation* 4: 115–123.

Konijn, T. M., van der Meene, J. G. C., Bonner, J. T. and Barkley, D. S. (1967). The acrasin activity of adenosine-3′, 5′-cyclic phosphate. *Proc. Natl. Acad. Sci. USA* 58: 1152–1154.

Raper, K. B. (1940). Pseudoplasmodium formation and organization in *Dictyostelium discoideum*. *J. Elisha Mitchell Sci. Soc.* 56: 241–282.

Ray, J., Shinnick, T. and Lerner, R. A. (1979). A mutation altering the function of a carbohydrate binding protein blocks cell–cell cohesion in developing *Dictyostelium discoideum*. *Nature* 279: 215–221.

Robertson, A., Drage, D. J. and Cohen, M. H. (1972). Control of aggregation in *Dictyostelium discoideum* by an external periodic pulse of cyclic adenosine monophosphate. *Science* 175: 333–335.

Rosen, S. D., Kafka, J. A., Simpson, D. L. and Barondes, S. H. (1973). Developmentally regulated carbohydrate binding protein in *Dictyostelium discoideum*. *Proc. Natl. Acad. Sci. USA* 70: 2554–2558.

Shaffer, B. M. (1953). Aggregation in cellular slime moulds: *In vitro* isolation of acrasin. *Nature* 171: 975.

Shaffer, B. M. (1975). Secretion of cyclic AMP induced by cyclic AMP in the cellular slime mould *Dictyostelium discoideum*. *Nature* 255: 549–552.

Tomchick, K. J. and Devreotes, P. N. (1981). Adenosine 3′,5′-monophosphate waves in *Dictyostelium discoideum*: A demonstration by isotope dilution-fluorography. *Science* 212: 443–446.

Wilson, E. B. (1896). *The Cell in Development and Inheritance*. Macmillan, New York.

CHAPTER 2

Fertilization: beginning a new organism

Urge and urge and urge,
Always the procreant urge of the world
Out of the dimness opposite equals advance,
Always substance and increase, always sex,
Always a knit of identity, always distinction,
Always a breed of life.
—WALT WHITMAN (1855)

Introduction

Fertilization is the process whereby two sex cells (GAMETES) fuse together to create a new individual with genetic potentials derived from both parents. Fertilization, then, accomplishes two separate activities: sex (the combining of genes derived from the two parents) and reproduction (the creation of new organisms). Thus, the first function of fertilization involves the transmission of genes from parent to offspring; and the second function of fertilization is to initiate in the egg cytoplasm those reactions that permit development to proceed.

Although the actual details of fertilization vary enormously from species to species, the events of conception generally consist of four major activities:

1. *Contact and recognition between sperm and egg.* This is the quality control step. The sperm and egg must be of the same species.
2. *Regulation of sperm entry into the egg.* This is the quantity control step. Only one sperm can ultimately fertilize the egg. All other sperm must be eliminated.
3. *Fusion of the genetic material of sperm and egg.*
4. *Activation of egg metabolism to start development.*

Structure of the gametes

Before proceeding to investigate each of the fertilization activities, we must first discuss the structures of the sperm and the egg, the two cell types specialized for fertilization.

Sperm

It is only within the past century that the sperm's role in fertilization has been known. Anton van Leeuwenhoek, the Dutch microscopist who codiscovered sperm in 1678, believed them to be parasitic animals living within the semen (hence the term *spermatozoa*—"sperm animals"). He believed that they had nothing at all to do with reproducing the organism in which they were found. The other codiscoverer of sperm, Nicolas Hartsoeker, put forth an alternative hypothesis, namely, that the entire embryonic individual lay preformed within the head of the sperm (Figure 1). This belief that the sperm contained the entire embryonic organism never gained much popularity as it implied an enormous waste of potential life. Most scientists favored Leeuwenhoek's explanation.

In a series of experiments performed in the late 1700s, Lazzaro Spallanzani demonstrated that filtered toad semen devoid of sperm would not fertilize eggs. He concluded, however, that the viscous fluid retained by the filter paper, and not the sperm, was the agent of fertilization. He, too, felt that the spermatic animals were clearly parasitic. Yet, this was the first evidence suggesting their importance.

The combination of better microscopic lenses and the cell theory led to a new appreciation of spermatic function. In 1821, J.B. Dumas and J.L. Prévost claimed that sperm were not parasites but rather were the active agents of fertilization. They noted the universal existence of sperm in sexually mature males and their absence in immature and aged individuals. These observations, coupled with the known absence of spermatozoa in the sterile mule, convinced them that "there exists an intimate relation between their presence in the organs and the fecundating capacity of the animal." They claimed that the sperm actually penetrated the egg and contributed materially to the next generation.

These claims were largely disregarded until the 1840s when A. von Kölliker described the formation of sperm from testicular cells. He concluded that sperm were greatly modified cells that came from the testes of adult males. He ridiculed the idea that the semen could be normal while serving to support an enormous number of parasites. Even so, von Kölliker denied that there was any physical contact between sperm and egg. He believed that the sperm excited the egg to develop much like a magnet communicated its presence to iron. It was only in 1876 that Oscar Hertwig demonstrated the penetration of the egg by the sperm and the union of their nuclei. Hertwig had sought an organism suitable for detailed microscopic observations; and he found that the Mediterranean sea urchin, *Toxopneustes lividus*, was perfect. Not only was it common throughout the region and sexually mature throughout most of the year, but its eggs were available in large numbers and were transparent even at high magnifications. After mixing together sperm and egg suspensions, Hertwig observed the sperm entering the egg and saw the two nuclei unite. He also noted that only one sperm was seen to enter each egg and that all the nuclei of the embryo were derived from the fused nucleus created at fertilization. Hertwig's observations were soon extended by Herman Fol, who detailed the mechanism of sperm entry. Fertilization was at last recognized as the union of sperm and egg, and the sea urchin remains the best-studied animal regarding fertilization.

Each sperm is known to consist of a haploid nucleus, a propulsion system to move the nucleus, and a sac of enzymes that enable the nucleus to enter the egg. Most of the cytoplasm of the sperm has been eliminated during its maturation, leaving only certain organelles that have been modified for their spermatic function (Figure 2). During the course of sperm maturation, the haploid nucleus has become severely streamlined and its DNA has become tightly compressed. Above this compressed haploid nucleus lies the ACROSOMAL VESICLE. This vesicle is derived from the Golgi apparatus and contains enzymes that digest proteins and complex sugars. Thus, it can be considered to be a modified lysosome. These stored enzymes will be used to penetrate the outer

FIGURE 1
The human infant preformed in the sperm: Nicolas Hartsoeker's homunculus. (From N. Hartsoeker, 1694. *Essai de Dioprique.*)

FIGURE 2
The modification of a germ cell to form a sperm. The centriole produces a long flagellum at what will be the posterior end of the sperm and the Golgi apparatus forms the acrosomal vesicle at the future anterior end. The mitochondria (dots) collect about the flagellum near the base of the haploid nucleus and become incorporated into the midpiece of the sperm. The remaining cytoplasm is jettisoned, and the nucleus condenses. The size of the mature sperm has been enlarged relative to the other figures. (Adapted from Clermont and Leblond, 1955.)

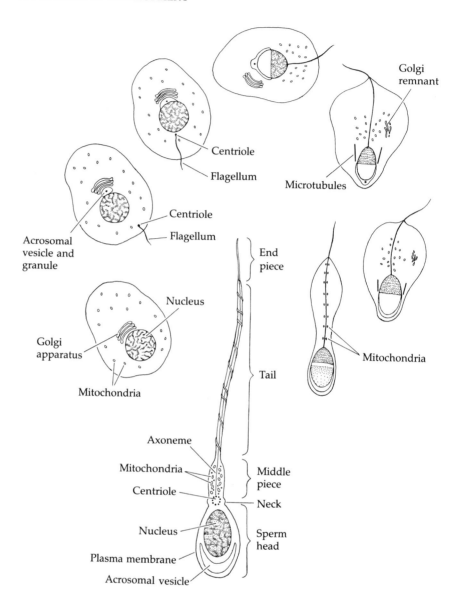

coverings of the egg. In many species, such as sea urchins, a region of globular actin molecules lies between the nucleus and the acrosomal vesicle. These proteins will be used to extend a fingerlike process during the early stages of fertilization. In these species, recognition between sperm and egg involves molecules on this ACROSOMAL PROCESS. Together, the acrosome and nucleus constitute the head of the sperm.

The means by which the sperm are propelled varies according to how the species had adapted to environmental conditions. In some cases (such as the parasitic roundworm *Ascaris*), the sperm travel by the amoeboid motion of PSEUDOPODS—local extensions of the cell membrane. In most species, however, each sperm is able to travel long distances by whipping its FLAGELLUM.

Flagella are complex structures. The major motor portion of the flagellum is called the AXONEME. An axoneme is formed by the MICROTUBULES emanating from the centriole at the base of the sperm nucleus (Figure 3B). The core of the axoneme consists of two central microtubules surrounded by a row of nine doublet microtubules. Actually, only one microtubule of each doublet is complete, having 13 protofilaments; the other is C-shaped and has only 11 protofilaments (Figure 3A–C). A three-dimensional model of a complete microtubule is shown in Figure 3D. Here one can see the 13 interconnected protofilaments, which are made exclusively of the dimeric protein TUBULIN.

Another protein, DYNEIN, is attached to the microtubules. This protein can hydrolyze molecules of adenosime triphosphate (ATP) and can

FIGURE 3
The motile apparatus of the sperm. (A) Cross section of the flagellum of a mammalian spermatozoon showing the central axoneme and the external fibers. (B) Interpretive diagram of the axoneme, showing the "9 + 2" arrangement of the microtubules and other flagellar components. (C) Schematic diagram of the association of tubulin protofilaments into a microtubule doublet. The first (i) portion of the doublet is a normal microtubule comprising 13 protofilaments. The second (ii) portion of the doublet contains only 11 (occasionally 10) protofilaments. (D) A three-dimensional model of microtubule (i). The α- and β-tubulin subunits are similar but not identical; and the microtubule can change size by polymerizing or depolymerizing tubulin subunits at either end. (A courtesy of D. M. Phillips; B from De Robertis et al., 1975; C from Tilney et al., 1973; D from Amos and Klug, 1974; photograph courtesy of the authors.)

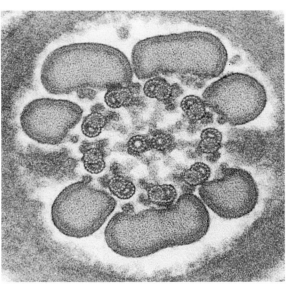

(A)

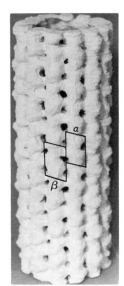

(D)

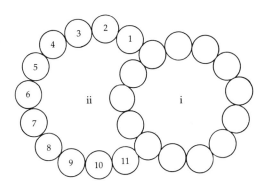

(B) AXONEME

Radial spoke
Spoke head
Central sheath
Dynein arms
Central microtubule
Microtubule doublet
Interdoublet link

(C) MICROTUBULE DOUBLET

convert the released chemical energy into the mechanical energy that propels the sperm (Ogawa et al., 1977). The importance of dynein protein can be seen in those individuals with the genetic syndrome called Kartagener's triad. These individuals lack dynein on all their ciliated and flagellated cells; therefore, these structures are rendered immotile. Males with this disease are sterile (immotile sperm), are susceptible to bronchial infections (immotile respiratory cilia), and have a 50 percent chance of having their heart on the right side of their body (Afzelius, 1976).

The "9 + 2" microtubule arrangement with the dynein arms has been conserved throughout the protist, plant, and animal kingdoms, suggesting that this arrangement is extremely well suited for transmitting the energy for movement. The energy to whip the flagellum and thereby propel the sperm comes from rings of mitochondria located in the neck region of the sperm (Figure 2).

In many species (notably mammals), a layer of dense fibers has interposed itself between the mitochondrial sheath and the axoneme. This fiber layer stiffens the sperm tail. Because the thickness of this layer decreases toward the tip, it probably functions to prevent the sperm head from being whipped around too suddenly. Thus, the sperm has undergone extensive modification for the transmission of its nucleus to the egg.

Egg

All the material necessary for the beginning of growth and development must be stored in the egg. Therefore, whereas the sperm has eliminated most of its cytoplasm, the developing egg not only conserves its material but is actively involved in accumulating more. It either synthesizes or absorbs proteins, such as yolk, that act as food reservoirs for the developing embryo. Thus, birds' eggs are enormous single cells that have become swollen with their accumulated yolk. Even eggs with relatively sparse yolk are comparatively large. The volume of a sea urchin egg is 1.9×10^5 cubic micrometers, more than 10,000 times the volume of the sperm! In the diagram of the sea urchin egg during fertilization (Figure 4), one can see how large the egg is compared to the sperm. The figure also shows the various components of the mature oocyte. So, while sperm and egg have equal haploid nuclear components, the egg has an enormous cytoplasmic storehouse that it has accumulated during its maturation. This cytoplasmic trove includes proteins, ribosomes and transfer RNA (tRNA), messenger RNA, and morphogenic factors.

- *Proteins*. It will be a long while before the embryo is able to feed itself or obtain food from its mother. The early embryonic cells need some storable supply of energy and amino acids. In many species, this is accomplished by accumulating yolk proteins in the egg. Many of the yolk proteins are made in other organs (liver, fat body) and travel to the egg.

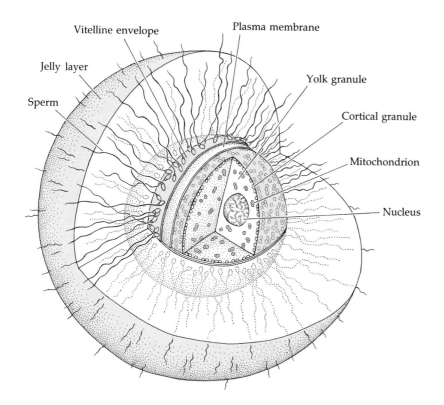

Jelly layer
Vitelline envelope
Plasma membrane
Sperm
Yolk granule
Cortical granule
Mitochondrion
Nucleus

FIGURE 4
Structure of the sea urchin egg during fertilization. (After Epel, 1977.)

- *Ribosomes and tRNA.* As we will soon see, there is a burst of protein synthesis soon after fertilization. This protein synthesis is accomplished by ribosomes and tRNA, which preexist in the egg. The developing oocyte has special mechanisms to synthesize ribosomes, and certain amphibian oocytes produce as many as 10^{12} ribosomes during their meiotic prophase.

- *Messenger RNA.* In most organisms, the instructions for early development are already packaged in the oocyte. It is estimated that the eggs of sea urchins contain 25–50 thousand different types of mRNAs. This mRNA, however, remains dormant until after fertilization.

- *Morphogenic factors.* These are molecules that direct the differentiation of cells into certain cell types. They appear to be localized throughout the egg and become segregated into different cells during cleavage (Chapters 7 and 8).

Within this enormous volume of cytoplasm resides a large nucleus. In some species (sea urchins, for example) the nucleus is already haploid at the time of fertilization. In other species (including many worms and some mammals) the egg nucleus is still diploid and the sperm enters before the meiotic divisions are completed. The stage of the egg nucleus at the time of sperm penetration is illustrated in Figure 5.

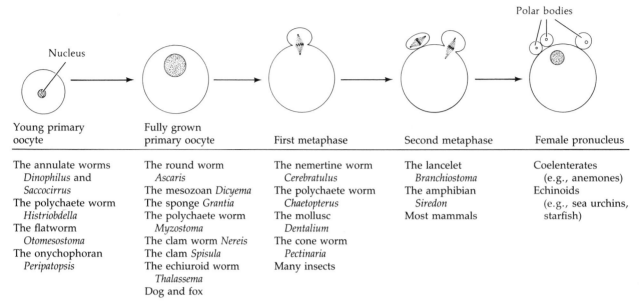

Young primary oocyte	Fully grown primary oocyte	First metaphase	Second metaphase	Female pronucleus
The annulate worms *Dinophilus* and *Saccocirrus*	The round worm *Ascaris*	The nemertine worm *Cerebratulus*	The lancelet *Branchiostoma*	Coelenterates (e.g., anemones)
The polychaete worm *Histriobdella*	The mesozoan *Dicyema*	The polychaete worm *Chaetopterus*	The amphibian *Siredon*	Echinoids (e.g., sea urchins,
The flatworm *Otomesostoma*	The sponge *Grantia*	The mollusc *Dentalium*	Most mammals	starfish)
The onychophoran *Peripatopsis*	The polychaete worm *Myzostoma*	The cone worm *Pectinaria*		
	The clam worm *Nereis*	Many insects		
	The clam *Spisula*			
	The echiuroid worm *Thalassema*			
	Dog and fox			

FIGURE 5

Stages of egg maturation at the time of sperm penetration in different animals. (Modified from Austin, 1965.)

Enclosing the cytoplasm is the oocyte PLASMA MEMBRANE. This membrane must regulate the flow of certain ions (especially Na^+) during fertilization and must be capable of fusing with the sperm plasma membrane. Immediately adjacent to and above the plasma membrane is the VITELLINE ENVELOPE (Figure 6). This glycoprotein membrane is essential for the species-specific binding of sperm. In mammals, the vitelline envelope is very thick and is called the ZONA PELLUCIDA. The mammalian egg is also surrounded by a layer of cells, the CORONA RADIATA (Figure 7). These represent ovarian follicular cells, which were nurturing the egg at the time of its release from the ovary. Sperm have to get past these cells, too, in order to fertilize the egg.

Lying immediately beneath the plasma membrane of the sea urchin egg is the cortex. The cytoplasm in this region is more highly ordered than the internal cytoplasm and has high concentrations of globular actin molecules. During fertilization, these actin molecules polymerize to form long cables of actin known as MICROFILAMENTS. The microfilaments are used to extend the egg surface into the microvilli, which surround the sperm (Figures 6 and 18). The cortical cytoplasm is much less fluid than the internal regions and is not displaced by centrifugation. Within this cortex are the CORTICAL GRANULES (Figures 4 and 6). These membrane structures are homologous to the spermatic acrosomal vesicle, being Golgi-derived organelles containing proteolytic enzymes. However, whereas each sperm contains one acrosomal vesicle, each sea urchin egg contains approximately 15,000 cortical granules. Moreover, in addition to containing the digestive enzymes, the cortical granules also contain MUCOPOLYSACCHARIDES and HYALINE PROTEIN. The enzymes and mucopolysaccharides are active in preventing other sperm from entering the egg after the first sperm has entered, and the hyaline protein surrounds the early embryo and provides support for the cleavage-stage blastomeres.

FIGURE 6
The sea urchin egg cell surface. (A) Scanning electron micrograph of an egg before fertilization. The plasma membrane is exposed where the vitelline layer has been torn. (B) Transmission electron micrograph of an unfertilized egg showing microvilli and plasma membrane, which is closely covered by the vitelline envelope. A cortical granule lies directly beneath the plasma membrane of the egg. (From Schroeder, 1979; photographs courtesy of T. E. Schroeder.)

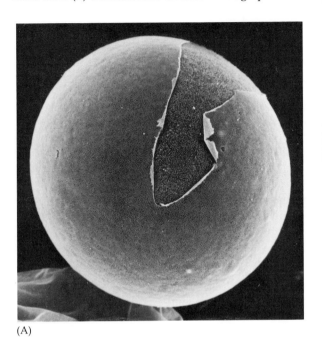

(A)

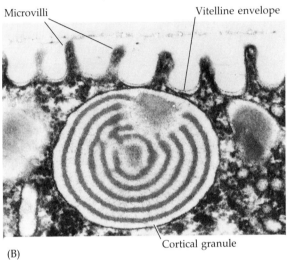

Microvilli

Vitelline envelope

Cortical granule

(B)

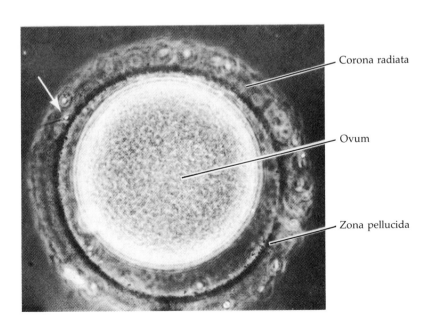

Corona radiata

Ovum

Zona pellucida

FIGURE 7
A human egg immediately preceding fertilization. The egg, or ovum, is encased in the zona pellucida. This, in turn, is surrounded by the cells of the corona radiata. The sperm (arrow) has penetrated the corona radiata but must still get through the zona pellucida in order to reach the ovum and achieve fertilization. (Courtesy of R. Yanagimachi.)

Many types of eggs also secrete an egg jelly outside their vitelline envelope. This glycoprotein meshwork can have numerous functions. Most often, though, it is used to either attract or activate sperm. The egg, then, is a cell specialized for receiving sperm and initiating development.

Recognition of sperm and egg: Action at a distance

Many marine organisms release their gametes into the local environment. This environment may be as small as a tidepool or as large as the ocean. Moreover, this environment is shared with other species that may shed their sex cells at the same time. These organisms are faced with two problems: (1) How can sperm find eggs in such a dilute concentration? (2) What mechanism prevents starfish sperm from trying to fertilize sea urchin eggs? Two major mechanisms have evolved to solve these difficulties: species-specific attraction of sperm and species-specific sperm activation.

Sperm attraction

Species-specific sperm attraction (CHEMOTAXIS) has been best documented in coelenterates by R. L. Miller (1966). He showed that in two species of *Campanularia*, the gonangium (where the eggs are stored) secretes a substance that causes sperm to swim toward it rather than randomly in all directions. Sperm of the same species are thus directed to swim into the funnel opening to reach the eggs within (Figure 8).

In 1978, Miller showed that the eggs of another coelenterate, *Orthopyxis caliculata*, also synthesized a chemotactic substance. Developing eggs (oocytes) at various stages in their maturation were fixed onto microscope slides, and sperm were released at a certain distance from the eggs. Miller found that when sperm were added to eggs that had not yet completed their second meiotic division, there was no attraction of sperm to eggs. However, after the second meiotic division was finished and the eggs were ready to be fertilized, the sperm migrated toward them. Thus, the oocyte not only controls the type of sperm it will attract but also the time at which it will attract them. The mechanisms for chemotaxis are different in other species (reviewed in Metz, 1978).

Although sperm attraction is not thought to be widespread in nature, other animals' eggs, such as those of herring, chitins, and tunicates, also have chemotactic properties. In herring (as in fish eggs generally), sperm must find the opening in the thick CHORION that surrounds the oocyte. This opening is called the MICROPYLE. The chorion releases a substance that increases sperm motility in the immediate vicinity of the micropyle, improving the chances that the sperm will

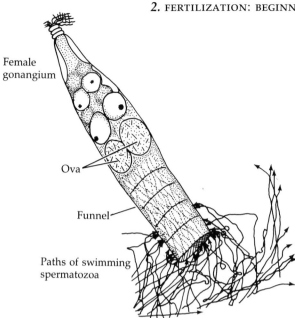

Female
gonangium

Ova

Funnel

Paths of swimming
spermatozoa

FIGURE 8
Sperm attraction at a distance shown by the paths of spermatozoa swimming in the region of an egg-bearing gonangium of *Campanularia flexuosa*. Except for a few spermatozoa on the right, almost all move in a manner directed toward the gonangium and increase in speed as they approach it. (Adapted from Miller, 1966.)

enter the micropyle (Yanagimachi, 1957). Furthermore, the same reaction will occur if the egg is removed: The sperm cluster around the micropyle and enter the empty chamber (Figure 9).

The acrosome reaction

The second interaction involving sperm and egg involves the activation of sperm by egg jelly. In most marine invertebrates, this acrosome reaction has two components: the rupture of the acrosomal vesicle and

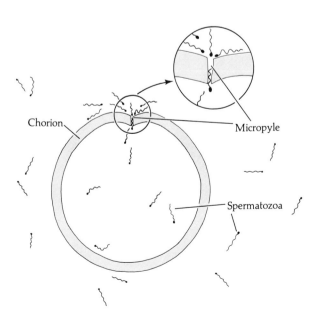

Chorion

Micropyle

Spermatozoa

FIGURE 9
Sperm attraction by chorion micropyle. Behavior of spermatozoa near the chorion of the herring (*Clupea*) egg after the egg itself has been removed. Spermatozoa merely drift until they are brought near the surface of the micropyle. There they show vigorous activity and swim into and through the micropyle (inset). (Adapted from Yanagimachi, 1957.)

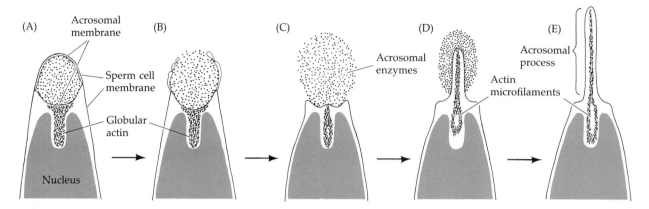

FIGURE 10
Acrosome reaction in sea urchin sperm. (A–C) The portion of the acrosomal membrane lying directly beneath the sperm cell membrane fuses with the cell membrane to release the contents of the acrosomal vesicle. (D, E) As the actin molecules assemble to produce microfilaments, the acrosomal process is extended outward. (Adapted from Saunders, 1970.)

the extension of the acrosomal process (Figure 10). The acrosome reaction can be initiated by soluble egg jelly, the egg jelly surrounding the egg, or even by contact with the egg itself in certain species. It can also be activated artificially by increasing the calcium concentration of seawater.

In sea urchins, contact with egg jelly causes breakdown of the acrosomal vesicle and release of protein-digesting enzymes that are responsible for digesting a path through the jelly coat to the egg surface (Dan, 1967; Franklin, 1970; Levine et al., 1978). The sequence of events is outlined in Figure 11. The acrosome reaction is initiated by a sulfated

FIGURE 11
Model of the possible interrelationships in the process of the sea urchin acrosome reaction. (Modified from Schackmann and Shapiro, 1981.)

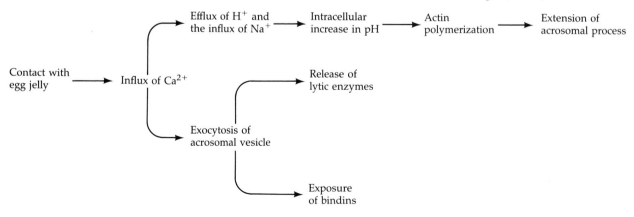

polysaccharide in the egg jelly that allows calcium and sodium ions to enter the sperm head and replace potassium and hydrogen ions (SeGall and Lennarz, 1979; Schackmann and Shapiro, 1981). The rupture of the acrosomal vesicle is caused by the calcium-mediated fusion of the acrosomal membrane with the adjacent sperm plasma membrane (Figures 10 and 12). This is essentially an exocytosis reaction wherein a vesicle is brought to the cell surface and fuses with the cell membrane to release its contents. Such exocytotic reactions are seen in the release of insulin precursors from pancreatic cells, in the release of neurotransmitters from synaptosomes, and in the release of histamine from mast cells. In all cases, there is a calcium-mediated fusion between the secretion vesicle and the cell membrane.

The second part of the acrosome reaction involves the extension of the acrosomal process (Figure 13). This protrusion arises from the polymerization of globular actin molecules into actin filaments. The polymerization of these actin molecules appears to be dependent upon the release of hydrogen ions from the sperm head (Tilney et al., 1978; Schackmann et al., 1978). It may be that a regulatory protein is responsible for blocking actin polymerization and that the rise in intracellular pH interferes with this function.

The acrosomal reactions of some marine invertebrates are often highly specific. The sperm of sea urchins *Arabacia punctulata* and *Strongylocentrotus drobachiensis* will react only with jelly of their own eggs. However, *S. purpuratus* sperm can also be activated by *Lytechinus variegatus* (but not *A. punctulata*) egg jelly (Summers and Hylander, 1975). Therefore, egg jelly may provide species-specific recognition in some species but not in others.

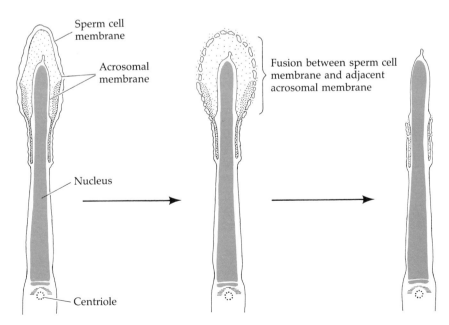

Sperm cell membrane

Acrosomal membrane

Fusion between sperm cell membrane and adjacent acrosomal membrane

Nucleus

Centriole

FIGURE 12
The acrosome reaction in hamster spermatozoa. Interpretive diagram of electron micrographs showing the fusion of the acrosomal and cell membranes in the sperm head. (From Yanagimachi and Noda, 1970.)

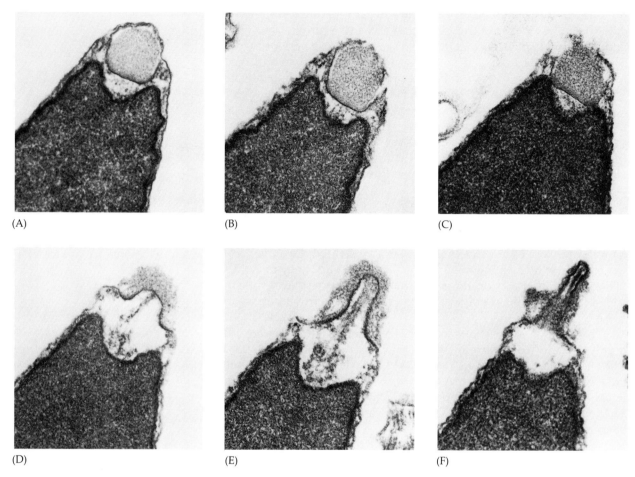

FIGURE 13
Acrosome reaction in sea urchin sperm. (A–C) Acrosomal vesicle ruptures and releases enzymes. (D–F) Actin polymerizes to extend the acrosomal process. (Courtesy of G. L. Decker and W. J. Lennarz.)

Recognition of sperm and egg: Contact of gametes

Species-specific recognition in sea urchins

Once the sea urchin sperm has penetrated the egg jelly, the acrosomal process of the sperm contacts the outer layer of the oocyte vitelline envelope (Figure 14). The major species-specific recognition step occurs at this point. The acrosomal protein mediating this recognition is called BINDIN. In 1977 Victor Vacquier and his co-workers isolated this non-

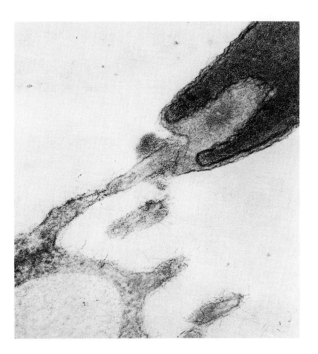

FIGURE 14
Contact of a sea urchin sperm acrosomal process with an egg microvillus. (From Epel, 1977; photograph courtesy of F. D. Collins and D. Epel.)

soluble 30,500-d protein from the acrosome of *Strongylocentrotus purpuratus*. This protein is capable of binding to dejellied eggs on the isolated vitelline envelopes from *S. purpuratus* (Vacquier and Moy, 1977; Figure 15). Further, its interaction with eggs is species-specific (Glabe and Vacquier, 1977; Glabe and Lennarz, 1979); bindin isolated from the acrosomes of *S. purpuratus* agglutinates its own dejellied eggs, but not those of the closely related species *S. franciscanus*.

In 1979, Moy and Vacquier demonstrated that bindin is located specifically on the acrosomal process, exactly where it should be for sperm–egg recognition. The immunochemical staining procedure they used is illustrated in Figure 16A. Antibodies to purified sea urchin bindin were made by injecting the bindin into rabbits. This rabbit anti-bindin was then bound to sea urchin sperm that had undergone the acrosome reaction. After any unbound antibody was washed off, the sperm were treated with swine antibodies that could bind to rabbit antibodies. These swine antibodies were covalently linked to peroxidase enzymes. In such a fashion, peroxidase molecules were placed wherever bindin was present. These enzymes can catalyze the formation of a dark precipitate from diaminobenzidine and hydrogen peroxide. When these two substrates were added to the treated sperm, the precipitate was found to cover the acrosomal process (Figure 16B). Moreover, bindin could not be detected on the sperm surface until after the acrosomal reaction, and it was found at the sperm–egg junction (Figure 16C).

Biochemical studies have found that the bindins of closely related sea urchin species are indeed different. This finding implies the exis-

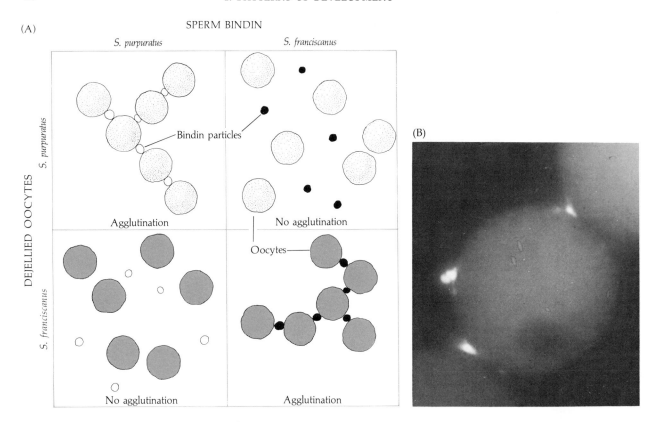

FIGURE 15

Species-specific agglutination of dejellied ova by bindin. (A) Agglutination was promoted by adding 212 μg bindin to a plastic well containing 0.25 mL of a 2 percent (volume:volume) suspension of eggs. After 2–5 minutes of light shaking, the wells were photographed. (The interpretive diagram is based on photographs of Glabe and Vacquier, 1977.) (B) Fluorescence micrograph of *S. purpuratus* eggs bound together by fluorescence-labeled *S. purpuratus* bindin particles. Bindin particles were invariably present at the places where two eggs came together. (From Glabe and Lennarz, 1979; photograph courtesy of the authors.)

tence of a species-specific bindin receptor on the vitelline envelope. Such a receptor was also suggested by the experiments of Vacquier and Payne (1973), who saturated sea urchin eggs with sperm. As seen in Figure 17, sperm binding does not occur over the entire egg surface. Even at saturating numbers of sperm (approximately 1500) there appears to be room on the oocyte for more sperm heads. This implies that there is a limiting number of sperm binding sites. A large glycoprotein complex from the vitelline envelopes of sea urchin eggs has been isolated and shown to bind radioactive bindin in a species-specific manner (Glabe and Vacquier, 1978; Glabe and Lennarz, 1979). This glycoprotein is also able to compete with eggs for the sperm of the same species. Thus, if *S. purpuratus* sperm is mixed with the bindin receptor from *S. purpuratus* eggs, the sperm bind to it and will not fertilize the eggs. The

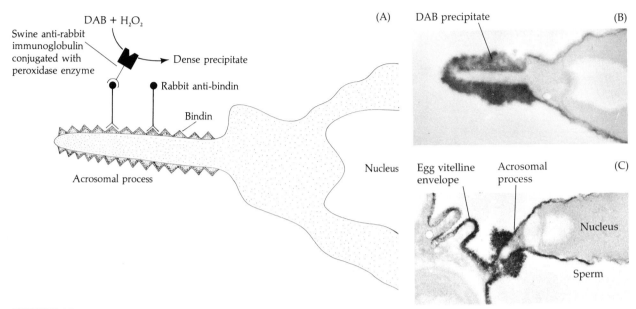

FIGURE 16

Localization of bindin on the acrosomal process. (A) Immunochemical localization technique places a rabbit antibody wherever bindin is exposed. A swine antibody that reacts against rabbit antibody can then bind to this first antibody. The swine antibody has been covalently linked with a molecule of peroxidase, an enzyme that can catalyze the reaction of DAB and peroxide to form an electron-dense precipitate. Thus, this precipitate will form only where bindin is present. (B) Localization of bindin to the acrosomal process after the acrosome reaction (×133,200). (C) Localization of bindin to the acrosomal process at the junction of the sperm and the egg. (B and C from Moy and Vacquier, 1979; photographs courtesy of V. D. Vacquier.)

isolated bindin receptor from *S. purpuratus*, however, does not interfere with the fertilization of other related sea urchins. Thus, species-specific recognition of sea urchin gametes occurs at the levels of acrosome activation and sperm adhesion to the vitelline envelope.

Gamete binding in mammals

So far we have been focusing our discussion on those organisms for which fertilization takes place externally. In many species, fertilization is internal, and the fertilization process has been adapted to this environment. One important conclusion has been that the reproductive tract of female mammals plays a very active role in the fertilization process. Newly ejaculated mammalian sperm are unable to undergo the acrosomal reaction without residing for some amount of time in the female reproductive tract. This requirement for CAPACITATION varies from spe-

FIGURE 17
Scanning electron micrograph of sea urchin sperm bound to the vitelline envelope of an egg. (Courtesy of C. Glabe, L. Perez, and W. J. Lennarz.)

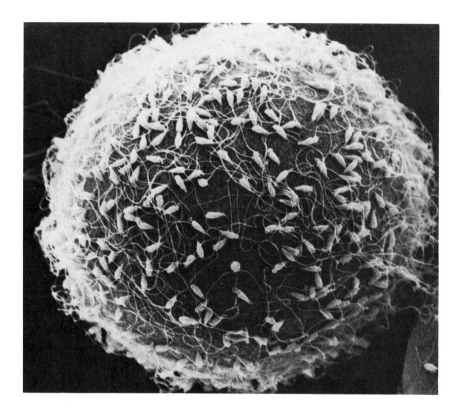

cies to species (Gwatkin, 1976) and can be mimicked in vitro by incubating sperm in fluid from oviducts or uterus. One hypothesis postulates that capacitation represents changes in the lipid structure of the sperm cell membrane. Support for this view comes from recent studies (Davis et al., 1980) showing that the membrane cholesterol:phospholipid ratio in the sperm membrane decreases with capacitation, and that albumin molecules found in the female reproductive tract are capable of transferring cholesterol from the sperm. The species-specific differences in timing the capacitation event correlate closely with the time it takes to change the cholesterol:phospholipid ratio (Davis, 1981). Removal of cholesterol is thought to destabilize the sperm cell membrane of the acrosomal vesicle. Without such lipid changes, the fusions of the acrosome reaction could not occur.

The acrosome reaction in mammals is still a controversial subject, and sperm location at the time of the acrosome reaction differs among species. This difference may reflect differences in the ability of sperm to penetrate the cells of the corona that surround the egg. In some species, the acrosome reaction occurs while the sperm is at a distance from the egg. In rabbits, for instance, the acrosome reaction appears to be induced by soluble egg-derived substances (Bedford, 1968), whereas the guinea pig acrosome reaction seems to be a timed event that takes place whether or not any egg is present (Yanagimachi and Usui, 1974).

Human sperm are also seen to undergo the acrosome reaction at some distance from the egg. Here, the mammalian acrosome reaction releases enzymes involved in lysing the adhesions holding together the cells of the corona radiata (Zaneveld and Williams, 1970; Williams, 1972). These enzymes, called LYSINS, include HYALURONIDASE, which hydrolyzes the hyaluronic acid that constitutes a major part of the extracellular matrix, and CORONA-DISPERSING ENZYME, which specifically disrupts the attachments of the corona cells. These enzymes enable the sperm to pass through the corona to reach the zona pellucida of the egg.

In the best-studied mammal—the mouse—the acrosome reaction occurs *after* the sperm has bound to the zona pellucida (Saling et al., 1979; Florman and Storey, 1982). In mammals, the zona pellucida plays a role analogous to that of the vitelline envelope. The binding of sperm to the zona is relatively, but not absolutely, species-specific (species-specificity should not be a major problem when fertilization occurs internally), and the binding of mouse sperm to the mouse zona can be inhibited by first incubating the mouse sperm with zona glycoproteins. Bleil and Wassarman (1980) have isolated from the zona pellucida a 83,000-d glycoprotein that is the active competitor in this inhibition assay. Thus, there is a specific glycoprotein in the zona pellucida to which the mouse sperm bind. Moreover, this same protein has been seen to initiate the acrosome reaction after sperm have bound to it. The mouse sperm can thereby concentrate its proteolytic enzymes directly at the point of attachment at the zona pellucida.

All mammalian sperm contain a lysin that digests away the zona pellucida at the point of attachment (Stambaugh and Buckley, 1969). This lysin, called ACROSIN, has an amino acid sequence very similar to that of trypsin; and like trypsin, it must be activated in order to work. It has been shown (Wincek et al., 1979; Stambaugh and Mastroianni, 1980) that the acrosomal form of this enzyme is inactive but that it becomes activated by a glycoprotein from the female reproductive tract. Here, too, the female reproductive tract is seen to be not merely a passive conduit for sperm but an active agent in the process of fertilization.

The molecular mechanism by which the zona pellucida recognizes the mammalian sperm is presently being studied. Shur and Hall (1982a,b) have shown that the capacitation of mouse sperm involves the unmasking of a specific enzyme on the sperm cell surface. This enzyme, a GLYCOSYLTRANSFERASE, recognizes an N-acetylhexosamine carbohydrate (probably N-acetylglucosamine) on the mouse zona pellucida. Fertilization can be inhibited by enzymatically digesting the N-acetylhexosamine structures from the glycoproteins on the zona surface. It appears, then, that sperm–egg binding is accomplished by numerous enzyme–substrate interactions on the zona pellucida. The zona pellucida provides the substrate—N-acetylhexosamine molecules—and the sperm provides the glycosyltransferase enzymes that bind to them. The lock-and-key binding of zona N-acetylhexosamine by sperm glycosyltransferase links the sperm and egg together. Once this binding is complete, the acrosome reaction is initiated.

Gamete fusion and prevention of polyspermy

Fusion between sperm and oocyte cell membranes

Recognition of sperm by the vitelline envelope or zona is followed by the enzymatic lysis of that portion of the envelope in the region of the sperm head (Epel, 1980). This is followed by the fusion of the sperm cell membrane with the cell membrane of the egg.

The entry of a sperm into the sea urchin egg is illustrated in Figure 18. The egg surface is covered with small extensions called MICROVILLI; gametic recognition and fusion usually occur at this point. The fertilizing sperm are seen to attach to the microvilli. Schatten and Schatten (1980) have seen that about 15 neighboring microvilli elongate to engulf the sperm head and cell to form the FERTILIZATION CONE. This cytoplasmic protruberance averages 6.7μm in length and is 2μm wide. Homology between the egg and the sperm is again demonstrated as the transitory fertilization cone appears to be extended by the polymerization of actin molecules in the egg cortex (Figure 18C). The sperm is then pulled into the egg by the contraction of the fertilization cone (Figure 18D). The fusion of egg and sperm membrane can be seen as the sperm is taken into the egg.

FIGURE 18
Scanning electron micrographs of the entry of sperm into sea urchin eggs. (A) Contact of sperm head with egg microvillus through the acrosomal process. (B) Formation of fertilization cone as microvilli rise to fuse with incoming sperm. (C) Sperm becomes internalized within the egg. (D) Transmission electron micrograph of sperm internalization through the fertilization cone. (A–C from Schatten and Mazia, 1976; photographs courtesy of Gerald Schatten; D courtesy of F. J. Longo.)

(A)

(B)

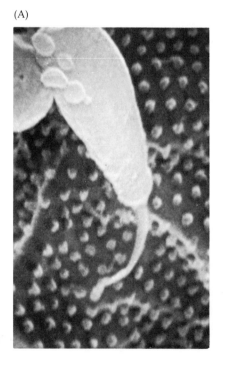

In the sea urchin, all parts of the egg are capable of fusing with sperm; in certain amphibians and in *Orthopyxis* (p. 40), there are specialized regions of the membrane for sperm recognition and fusion (Vacquier, 1979). Although not much is known about the mechanism for gamete fusion, Hirao and Yanagimachi (1978) have demonstrated that membrane phospholipids are extremely important in this reaction. This is in accord with present models of membrane fusion, which emphasize the role of membrane lipid molecules (Figure 19).

The portion of the sperm cell membrane that fuses with the egg has been visualized by fluorescent dyes. Gabel et al., (1979) fertilized sea urchin eggs with sperm containing fluorescent molecules within their membranes. After fertilization, these membranes were visible as discrete patches of fluorescent material (Figure 20). Only one patch was seen per oocyte, an observation indicating that one sperm fertilized each egg. Surprisingly, this patch maintained its integrity throughout early development, including the larval stage. Whereas in most cells membrane components diffuse quickly throughout the membrane and have a relatively fast rate of turnover (Johnson and Edidin, 1978; Wolf et al., 1981; Schimke, 1975), the sperm membrane remains a discrete entity for several days. This rigidity of the spermatic membrane portion has been postulated to play a role in directing the plane of the first cell cleavage and initiating cytoplasmic movements (Manes and Barbieri, 1976, 1977).

(C)

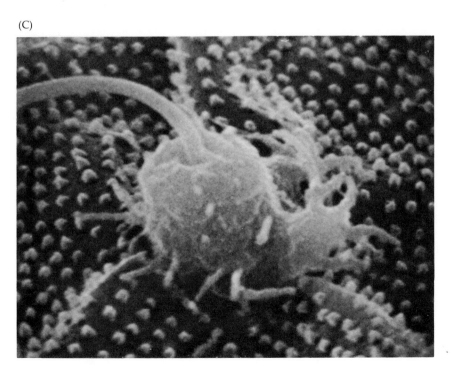

(D)

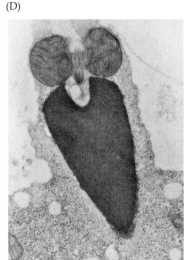

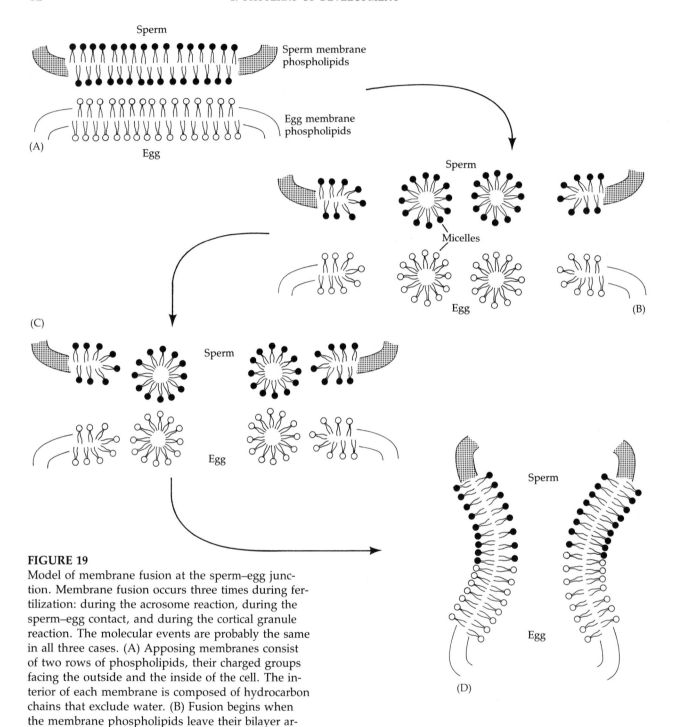

FIGURE 19

Model of membrane fusion at the sperm–egg junction. Membrane fusion occurs three times during fertilization: during the acrosome reaction, during the sperm–egg contact, and during the cortical granule reaction. The molecular events are probably the same in all three cases. (A) Apposing membranes consist of two rows of phospholipids, their charged groups facing the outside and the inside of the cell. The interior of each membrane is composed of hydrocarbon chains that exclude water. (B) Fusion begins when the membrane phospholipids leave their bilayer arrangement and form "micelles" (spherical aggregates having their charged groups on the outside and their hydrocarbons within). By rearranging these micelles (C), the membranes can reform (D), thereby connecting the two previously separated cells.

FIGURE 20
Fluorescence photomicrographs of the newly fertilized egg (A) and late gastrula stage embryo (B) of the sea urchin *S. purpuratus.* The arrows point to the regions of the cell and the embryo contributed by the fluorescent sperm membrane. (Photographs courtesy of B. Shapiro.)

(A)

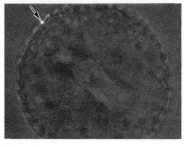

(B)

Prevention of polyspermy

As soon as one sperm has entered the egg, the fusibility of the egg membrane, which was so necessary to get the sperm inside the egg, becomes a dangerous liability. In sea urchins, as in most animals studied, any sperm that enters the egg can provide a haploid nucleus and a centriole to the egg. In normal MONOSPERMY, where one sperm enters the egg, a haploid sperm nucleus and a haploid egg nucleus combine to form the diploid nucleus of the fertilized egg (ZYGOTE), thus restoring the chromosome number appropriate for the species. The centriole, coming from the sperm, will divide to form the two poles of the mitotic spindle during cleavage.

The entrance of multiple sperm—POLYSPERMY—leads to disasterous consequences in most organisms. In the sea urchin, fertilization by two sperm would result in a triploid nucleus wherein each chromosome would be represented three times rather than twice. Worse, it means that instead of a bipolar mitotic spindle that would separate the chromosomes into two cells, the triploid chromosomes would be divided into four cells. Since there is no mechanism to ensure that each of the four cells receives the proper number and type of chromosomes, the chromosomes would be apportioned unequally. Some cells would receive extra copies of certain chromosomes and other cells would lack them (Figure 21). Theodor Boveri demonstrated in 1902 that such cells either die or develop abnormally.

Species have evolved ways to prevent the union of more than two haploid nuclei. The most common way is to prevent the entry of more than one sperm into the egg. The sea urchin egg has two mechanisms to avoid polyspermy: a fast reaction accomplished by an electrical change in the egg plasma membrane and a slower reaction caused by the bursting of the cortical granules.

The fast block to polyspermy. The egg cell membrane has to be able to fuse with sperm, but it must lose that ability immediately after the entry of the first sperm. The fast block to polyspermy involves changes in the egg cell membrane such that it loses its fusibility (Just, 1919). This change in membrane fusibility involves changing the electrical potential of the egg membrane.

The cell membrane provides a selective barrier between the egg cytoplasm and the outside environment, and the ionic concentration of the egg is greatly different from that of its surroundings. This is especially true in the case of sodium ions (Na^+). Seawater has a particularly high sodium ion concentration whereas the egg cytoplasm has hardly

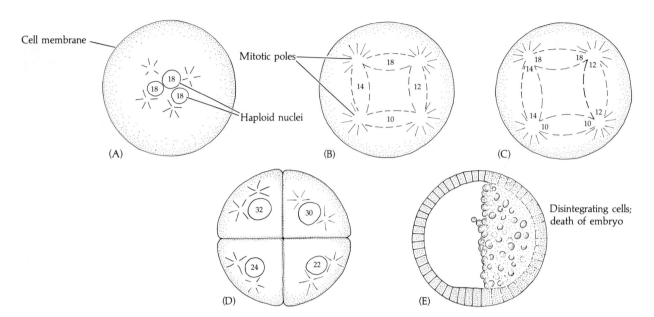

FIGURE 21
Aberrant development in a dispermic sea urchin egg. (A) Fusion of three haploid nuclei, each containing 18 chromosomes, and the division of the two sperm centrioles to form four mitotic poles. (B) The 54 chromosomes randomly assort on the four spindles. (C) At anaphase of the first division, the duplicated chromosomes are pulled to the four poles. (D) Four cells containing different numbers and types of chromosomes are formed, thereby causing the early death of the embryo (E). (After Boveri, 1907.)

any free sodium. This condition is maintained by the cell membrane, which steadfastly inhibits the entry of sodium into the oocyte. If an electrode is inserted into the egg and a second electrode is placed on the oocyte surface, one can measure the constant potential difference across the egg plasma membrane. This RESTING MEMBRANE POTENTIAL is usually about 70 millivolts, usually expressed as −70 mV because the inside of the cell is negatively charged with respect to the exterior.

Within a tenth of a second after the entry of the first sperm, this resting potential disappears. A small influx of sodium ions into the egg has been permitted, thereby bringing the potential difference to 0 or even +20 mV (Figure 22). Although sperm can fuse with membranes having a resting potential of −70 mV, they cannot readily fuse with membranes having a resting potential more positive than −10 mV.

Jaffe (1976) and her co-workers have found that polyspermy can be induced in oocytes if the oocytes are artificially supplied with electric current, which keeps the resting potential at −70 mV. Conversely, fertilization can be prevented entirely by artificially keeping the resting potential of oocytes positive (Cross and Elinson, 1980). The fast block to polyspermy can also be circumvented by lowering the concentration

FIGURE 22
Membrane potential of sea urchin eggs before and after fertilization. Before the addition of sperm, the egg membrane separates a potential difference of some −70 mV (the inside of the cell is more negatively charged than the environment). As a sperm enters the egg, the potential rises within 0.1 seconds, thereby demonstrating that the potential difference has been eliminated. (Adapted from Jaffe, 1980.)

of sodium ions in the water (Figure 23). If the sodium ions are not sufficient to neutralize the resting potential, polyspermy occurs (Gould-Somero et al., 1979; Jaffe, 1980).

The slow block to polyspermy. Sea urchin eggs (and many others) have a second mechanism to ensure that sperm do not push their way through the cell membrane. Carroll and Epel (1975) demonstrated that polyspermy can still occur if the sperm bound to the vitelline envelope are not somehow removed. This removal is accomplished by the CORTICAL GRANULE REACTION.

Directly beneath the sea urchin oocyte membrane are 15,000 granules about 1 μm in diameter (Figure 6B). Upon sperm entry, these cortical granules burst, releasing their contents into the area between the egg plasma membrane and the vitelline envelope. The proteins linking the vitelline envelope to the egg are dissolved by the proteolytic enzymes, and the mucopolysaccharides cause an osmotic rush of water

FIGURE 23
Polyspermy in eggs of the sea urchin *L. pictus* induced by fertilization in low-sodium seawater. (A) Control eggs developing in 490 mM Na$^+$. (B) Polyspermy in eggs fertilized in 120 mM Na$^+$ (choline was substituted for sodium). The eggs were photographed during the first cleavage. (C) Table showing rise of polyspermy with decrease in sodium ions. (From Jaffe, 1980; photographs courtesy of L.A. Jaffe.)

(A)

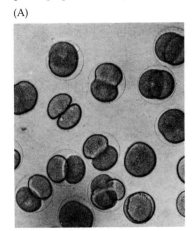

(B)

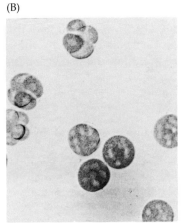

(C)

[Na](mM)	Percentage of polyspermic eggs
50	100
120	97
360	26
490	22

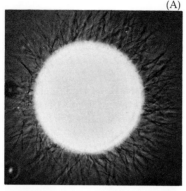

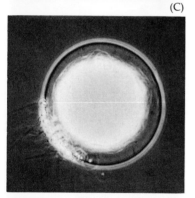

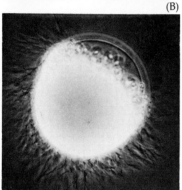

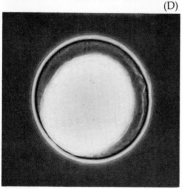

into the space between the cell membrane and the vitelline envelope. The vitelline envelope is thus elevated, whereupon it is called the FERTILIZATION MEMBRANE. The cortical granule discharge changes this membrane, too. First, the proteases modify or strip off the bindin receptor and any sperm attached to it (Glabe and Vacquier, 1978). Second, a peroxidase hardens the fertilization membrane by crosslinking the tyrosine residues on adjacent proteins (Foerder and Shapiro, 1977). As shown in Figure 24, the fertilization membrane starts to form at the site of sperm entry and continues its ascent around the oocyte. As this membrane forms, sperm are released. This process starts about 20 seconds after sperm entry and is complete by the first minute of fertilization. In mammals, the cortical granule reaction does not create a fertilization membrane, but the effect is the same. The released enzymes modify the zona pellucida sperm receptor such that it can no longer bind sperm (Bleil and Wassarman, 1980). This is called the ZONA REACTION.

The mechanism for the cortical granule reaction is similar to that of the acrosome reaction. In the presence of calcium ions, the cortical granule membranes fuse with the oocyte plasma membrane, causing the exocytosis of their contents (Figure 25). Following the breakdown of the cortical granules about the point of sperm entry, a wave of cortical granule exocytoses propagate through the cortex to the opposite side of the egg.

More recent experiments have demonstrated that calcium ions are directly responsible for propagating the cortical reaction and that the calcium ions are stored in the cortical granules themselves. The first evidence came in 1974 when unfertilized eggs were treated with the ionophore A23187. This drug transports calcium ions across membranes, allowing them to traverse the otherwise impermeable barriers. Placing sea urchin eggs into seawater containing A23187 caused the cortical granule reaction and the elevation of the fertilization membrane. Moreover, this reaction occurred without any calcium ions present in the seawater. Therefore, A23187 caused the release of calcium ions already bound within the egg (Chambers et al., 1974; Steinhardt and Epel, 1974).

Vacquier (1975) demonstrated that this chain reaction of cortical granule eruption was caused by the release of free calcium ions from the cortical granules themselves. First, he "glued" the eggs to petri

FIGURE 24
Formation of the fertilization membrane and removal of excess sperm. Sperm was added to sea urchin eggs and the suspension was fixed in formaldehyde to prevent further reactions. (A) Ten seconds after sperm addition. Sperm are seen surrounding the egg. (B, C) Twenty-five and 35 seconds after insemination, a fertilization membrane forms around the egg, starting at the point of sperm entry. (D) Fertilization membrane is complete and excess sperm are removed. (From Vacquier and Payne, 1973; photographs courtesy of V.D. Vacquier.)

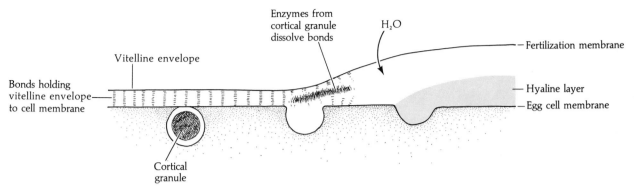

FIGURE 25
Schematic diagram showing the events leading to the formation of the fertilization membrane and the hyaline layer. As cortical granules burst, they release peptidases, which cleave the bonds joining the vitelline envelope to the cell membrane. Mucopolysaccharides released by the cortical granules form an osmotic gradient, thereby causing water to enter and swell the space between the vitelline envelope and the cell membrane. Other enzymes released from the cortical granules act to harden the fertilization membrane and to release sperm bound to it. (Modified from Austin, 1965.)

dishes by treating the dishes with protamine sulfate. He then lysed the eggs and washed away most of the contents with calcium-free seawater containing ethylenediaminetetraacetic acid (EDTA). EDTA inactivates ionic calcium by forming a tightly bound complex with it. Any calcium ions from the lysed cells can be eliminated by this treatment. This procedure allows one to obtain petri dishes full of membrane fragments and cortical granules (Figure 26A). Adding calcium to such a dish produces a cortical granule reaction wherein all the cortical granule membranes fuse (Figure 26B). Moreover, when a small number of granules in the center of a dish are broken with a glass rod, an expanding circle of cortical granule lysis ensues. Both calcium ions and proteases are released. Thus, calcium ions not only cause the rupture of cortical granules but are stored in these granules as well. When one granule ruptures, it releases calcium to initiate the rupture of other granules.

Further studies (Fulton and Whittingham, 1978; Hollinger and Scheutz, 1976) have shown that calcium ions will initiate cortical granule reactions when injected into sea urchin, mouse, and frog eggs, and that procaine, an anesthetic that prevents the release of bound calcium, can inhibit the sea urchin oocyte cortical granule reaction.

Although the oocytes of most species use both fast and slow blocks to polyspermy, variations exist throughout nature. In mammals, polyspermy is minimized by the small number of sperm that reach the site of fertilization (Braden and Austin, 1954). In the hamster, however, the egg plasma membrane can fuse with sperm even through the 4-cell stage (Usui and Yanagimachi, 1976). The block to polyspermy in hamsters appears to be controlled by the release of sperm binding sites on

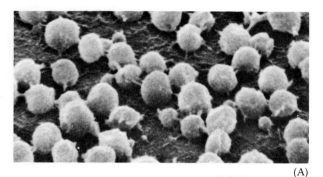
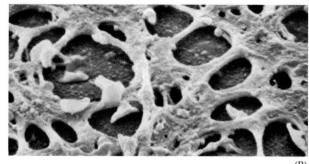

(A)   (B)

FIGURE 26
Calcium-mediated fusion of cortical granules. (A) Scanning electron micrograph of cortical granules bound to the inner surface of the plasma membrane. (B) Scanning electron micrograph of fused cortical granule material
after calcium-mediated discharge. (Photographs courtesy of V.D. Vacquier.)

the zona pellucida. Rabbits rely completely on the fast block to polyspermy and nobody will argue with their success. Lastly, certain animals
have defenses to polyspermy about which we know very little. In the
yolky eggs of certain birds, reptiles, and amphibians, several sperm
actually do enter the egg cytoplasm. In some unknown way, all but one
of these sperm are disintegrated in the cytoplasm after the fusion of the
egg pronucleus with one of the sperm pronuclei (Epel, 1980). Whatever
the mechanism, one haploid nucleus from the sperm is allowed to fuse
with the haploid nucleus of the egg.

Fusion of the genetic material

In sea urchins, the sperm nucleus enters the egg through the "fertilization cone" of elongated microvilli (Figure 18). Entry of the sperm nucleus is perpendicular to the oocyte surface. After fusion of the sperm
and egg membranes, the sperm nucleus and its centriole separate from
the mitochondria and the flagellum. The mitochondria and the flagellum
disintegrate, and sperm-derived mitochondria are not found in developing or adult organisms (Dawid and Blackler, 1972; Giles et al., 1980).
Thus, mitochondria are transmitted solely through the maternal parent.
   Once within the egg cytoplasm, the sperm nucleus decondenses to
form the MALE PRONUCLEUS. The egg nucleus, once haploid, is called
the FEMALE PRONUCLEUS. Sperm nuclear membrane vesiculates into
small packets, thereby enabling the chromatin to decondense. Remnants of this membrane are carried by the chromatin. Soon, new membranous vesicles aggregate along the periphery of the chromatin mass
and connect with the fragments of the old membrane to produce the
membrane of the male pronucleus.
   The male pronucleus rotates 180 degrees so that the sperm centriole

now faces the egg pronucleus. The microtubules of the centriole of the male pronucleus connect to the female pronucleus and the two nuclei migrate toward each other (Bestor and Schatten, 1981). Their fusion together forms the diploid ZYGOTE NUCLEUS. The initiation of DNA synthesis can occur either in the pronuclear stage (during migration) or after the formation of the zygote nucleus (Figure 27).

In mammals, the process of nuclear fusion takes considerably

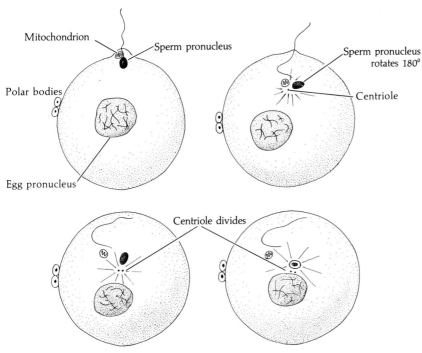

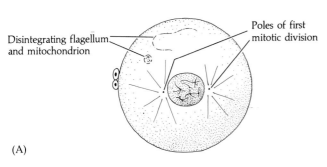

(A)

FIGURE 27
Nuclear events in fertilization in the sea urchin *Arbacia*. (A) Diagram of fertilization. The entire sperm enters the egg cytoplasm; the nucleus and centriole detach from the mitochondria and flagellum. After the sperm pronucleus turns 180 degrees so that the centriole faces the egg pronucleus, nuclear fusion takes place. The mitochondria and flagellum disintegrate. The sperm centriole divides to become the poles of the first mitotic division. (B) Fusion of pronuclei in the sea urchin egg. (A after C. R. Austin, 1965; B courtesy of F. J. Longo.)

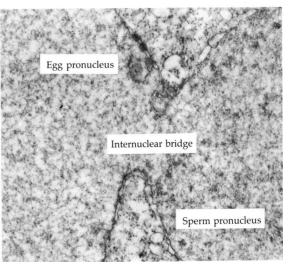

(B)

longer—about 12 hours as compared to 1 hour in the sea urchin (Figure 28). The mammalian sperm nucleus "burrows" in almost parallel to the plane of the egg rather than approaching perpendicularly. As with sea urchins, the sperm is drawn into the egg by numerous microvilli. The

FIGURE 28

Diagram of fertilization in the hamster. (A) The sperm attaches to the zona. (B) The sperm digests its way into the perivitelline space between the zona and the egg. Upon attaching to the egg (C), it places its head parallel to the egg cell membrane (D). The firm attachment of the sperm to the egg and the whipping motion of the tail causes the egg to rotate (E) while the entire sperm enters the perivitelline space (F). The sperm and egg membranes fuse (G) and the sperm head decondenses (H). After both pronuclei swell and meet in the center of the egg (I, J), the pronuclear membranes disintegrate and the first mitosis begins (K, L). (Modified from R. Yanagimachi, 1981.)

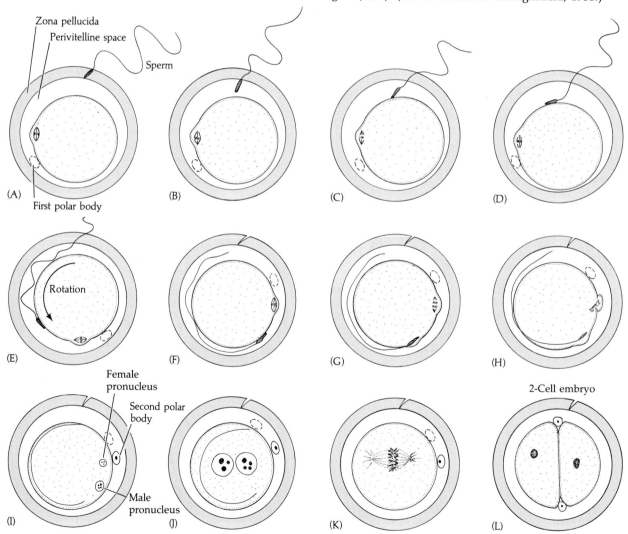

mammalian sperm nucleus also breaks down as its chromatin decondenses and is then reconstructed by the coalescing of vesicles. Because the decondensation of chromatin can be induced artificially by treating the rabbit sperm nuclei with reagents that disrupt disulfide bonds, it is likely that the oocyte contains a substance that also disrupts these linkages (Calvin and Bedford, 1971).

SIDELIGHTS & SPECULATIONS

The nonequivalence of mammalian pronuclei

Although male and female pronuclei of mammals can be considered genetically equivalent, studies have suggested that they may be functionally different. In humans there exists a placental abnormality called the hydatidiform mole. The fetus is absent and the placental tissue is abnormally enlarged. A majority of such moles have been shown to arise from a haploid sperm fertilizing an egg in which the female pronucleus is absent. After entering the egg, the sperm chromosomes duplicate themselves, thereby restoring the diploid chromosome number. Thus, the entire genome is derived from the sperm (Jacobs et al., 1980; Ohama et al., 1981). Here we see a situation in which the cell can survive and have a normal chromosome number, but its development is abnormal.

The hypothesis that male and female pronuclei are different gains support from pronuclear transplantation experiments (Table 1; McGrath and Solter, 1984). Pronuclei of recently fertilized mouse eggs can be removed and added to other recently fertilized oocytes. The two pronuclei can be distinguished (the female pronucleus being the one beneath the polar bodies); and oocytes with two male or two female pronuclei can be constructed. Although embryonic cleavage occurs, neither of these types of eggs develops to birth, whereas some control eggs (containing one male pronucleus and one female pronucleus from different zygotes) undergoing such transplantation develop normally. Thus, though the two pronuclei are seen to be equivalent in most animals, in mammals there may be important functional differences between them.

TABLE 1
Pronuclear transplantation experiments

Class of reconstructed zygotes	Operation	Number of successful transplants	Number of progeny having the specified phenotype
Bimaternal		339	0
Bipaternal		328	0
Control		348	18

Source: McGrath and Solter, 1984.

The mammalian male pronucleus enlarges while the oocyte nucleus completes its second meiotic division. Afterward, each pronucleus migrates toward the other, replicating its DNA as it travels. Upon meeting, the two nuclear membranes meet and break down. However, instead of producing a common zygote nucleus, the chromatin condenses into chromosomes that orient themselves upon a common mitotic spindle. Thus, a true diploid nucleus in mammals is first seen, not in the zygote, but at the 2-cell stage.

Activation of egg metabolism

So far, then, we have discussed the mechanisms by which sperm and egg recognize each other, fuse together, and merge their respective haploid nuclei. But for fertilization to lead to development, changes must occur in the egg cytoplasm.[1]

The mature sea urchin oocyte is a metabolically sluggish cell that is activated by the entering sperm. This activation is merely a stimulus, however, which sets into action a preprogrammed set of metabolic events. The responses of the egg to the sperm can be divided into those "early" responses that occur within seconds of the cortical reaction and those "late" responses that take place several minutes after fertilization begins (Figure 29).

[1]In certain species of salamanders, this developmental function of fertilization has been totally divorced from the genetic function. The silver salamander (*Ambystoma platineum*) is a species consisting solely of females. Each female produces an egg with an unreduced chromosome number. This egg, however, is metabolically inert. So the silver salamander mates with the male Jefferson salamander (*A. jeffersonianum*). The sperm from the male Jefferson salamander only stimulates the egg's development (Uzzell, 1964). It does not contribute genetic material.

FIGURE 29
Time course of events during the fertilization of the sea urchin embryo. Time axis is presented logarithmically. (After Epel, 1980.)

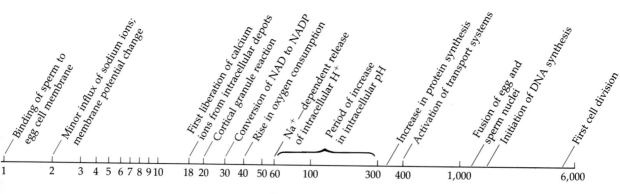

Early responses

As we have seen, the entry of the sperm into the egg activates the two major blocks to polyspermy: the fast block, initiated by sodium influx into the cell; and the slow block, initiated by intracellular release of calcium ions. Several experiments have shown that this release of calcium ions is essential for activation of egg metabolism.

The rise of intracellular calcium and its subsequent binding by oocyte components can be monitored visually by the dye aequorin. Aequorin is a protein that can be isolated from a luminescent jellyfish and will emit light when it binds free calcium ions. Eggs are injected with this dye and then fertilized. Figure 30 shows the striking wave of

FIGURE 30
Wave of free calcium ions propagated in the egg of the fish *Medaka*. The egg has been injected with aequorin and oriented with its micropyle (the point of sperm entry) to the left. Photographs taken 10 seconds apart reveal the release of calcium ions in successive parts of the egg. The last frame traces the leading edges of the 11 wave fronts. (From Gilkey et al., 1978; photographs courtesy of L. F. Jaffe.)

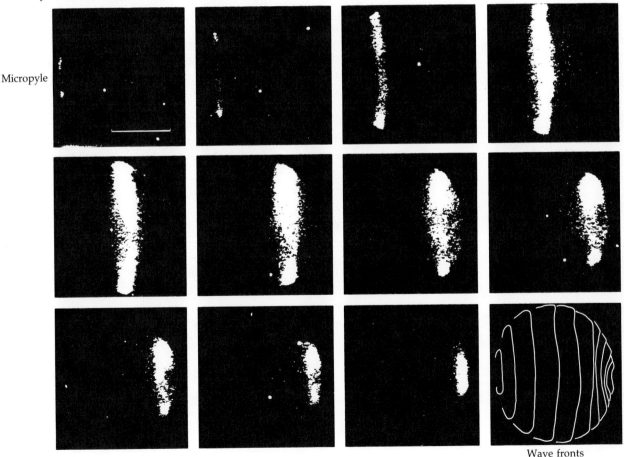

Micropyle

Wave fronts

calcium release that propagates across the egg. Starting at the point of sperm entry, a band of light traverses the cell (Ridgeway et al., 1975; Steinhardt et al., 1977). The entire release and binding of calcium ions is complete in 1 to 2 minutes. Note that free calcium ions are bound almost as soon as they are released. The calcium wave has been observed in several species, including sea urchins and amphibians, and appears to reflect the release of bound calcium from the endoplasmic reticulum as well as from the cortical granules (Jaffe, 1983).

This brief calcium flux is essential for development. If EDTA is injected into the eggs, there is no flux of free calcium ions across the egg and no activation of egg metabolism. More importantly, eggs can be activated artificially in the absence of sperm by procedures that release free calcium into the oocyte. Richard Steinhardt and David Epel (1974) found that micromolar amounts of calcium ionophore A23187 elicit most of the oocyte responses characteristic of a normally fertilized egg. Membrane elevation, a burst of oxygen utilization, and increases in protein and DNA synthesis are each generated in their proper order. Moreover, this activation takes place in the total absence of calcium ions in the sea water. In most cases, development ceases at the first mitosis, because the eggs are still haploid and lack the sperm centriole. The stimulation of oocyte development in the absence of sperm is called ARTIFICIAL PARTHENOGENESIS.

This essential calcium flux is responsible for activating a series of metabolic reactions (Figure 31). One of these is the increase in oxygen utilization by the egg. Strangely, this burst of oxygen use does not take place in the mitochondria and probably has little or nothing to do with the generation of ATP. Rather, this oxygen is used to oxidize various oocyte components. One of these components is the fertilization membrane. Foerder et al. (1978) have shown that this rather fluid envelope is hardened by the oxidation of protein-bound tyrosine residues. Several

FIGURE 31

Model of the possible interrelationships in the process of sea urchin fertilization. (After Epel, 1980.)

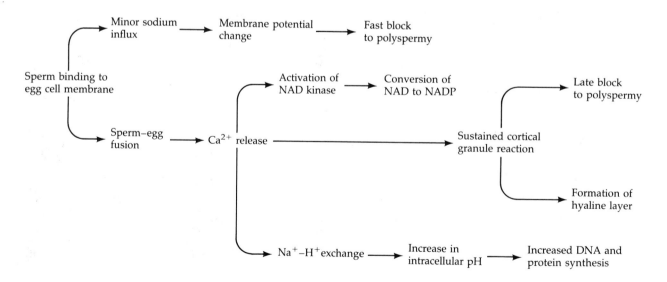

lipids in the oocyte cytoplasm are also oxidized at this time (Perry and Epel, 1977).

Another calcium-mediated effect is the activation of the enzyme NAD$^+$ kinase, which converts NAD$^+$ to NADP$^+$. This change may have important consequences for the metabolism of the cell. NADP$^+$ (but not NAD$^+$) can be used as a coenzyme for lipid biosynthesis. Thus, the switch from NAD$^+$ to NADP$^+$ may be important in the construction of the many new cell membrane components required during cleavage.

Late responses

Another of the early events of egg activation is a second influx of sodium ions, causing a 1:1 exchange between sodium ions from the seawater and hydrogen ions from the egg. In the sea urchin egg, nearly 75 percent of the hydrogen ions are lost. This loss of about 2 mM hydrogen ions causes the pH to rise from 6.8 to 7.2, and it brings about enormous changes in egg physiology (Shen and Steinhardt, 1978). Although this change is believed to be caused by a calcium-mediated reaction, it has been found that any means of raising the intracellular pH can initiate the late fertilization responses.

The late responses of fertilization include activation of DNA synthesis and protein synthesis. The burst of protein synthesis usually occurs within several minutes after sperm entry and is not dependent on new messenger RNA synthesis (Figure 32). Rather, new protein synthesis utilizes mRNAs already present in the oocyte cytoplasm. (Much more will be said about this in Chapter 14.) Such a burst of

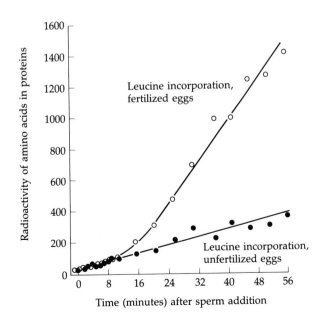

Time (minutes) after sperm addition

FIGURE 32
Protein synthesis in the newly fertilized sea urchin egg. [^{14}C]Leucine incorporation into the proteins of unfertilized eggs (filled circles) or fertilized eggs (open circles). (After Epel, 1967.)

protein synthesis can be induced artificially by placing unfertilized eggs into a solution containing ammonium ions (Johnson et al., 1976). These ions lose protons outside the cell, diffuse through the plasma membrane as NH_3, and once inside the cell, pick up protons to become NH_4^+. The resulting loss of free hydrogen ions is reflected in the increase of intracellular pH (Figure 33). Thus, bypassing the sodium–hydrogen exchange, ammonium ions are able to raise the pH of the eggs. In so doing, they activate protein and DNA synthesis.

Conversely, agents that block the rise in pH also block these late fertilization events. Thus, the activation of protein synthesis can be stopped by placing newly fertilized eggs into seawater containing low concentrations of sodium ions or into seawater containing amiloride, a drug that inhibits sodium ion transport.

The stimulus to initiate DNA synthesis is still unknown. In mature sea urchin oocytes, all the enzymes and substrates needed for DNA synthesis are present (De Petrocellis and Monroy, 1974), but no DNA replication occurs before fertilization. One possibility is that the events of fertilization unite the chromatin with the DNA synthetic enzymes. Infante et al. (1973) demonstrated that although the DNA polymerase of the sea urchin oocyte was present on the nuclear envelope, the DNA was not. Immediately after fertilization, however, replicating DNA also appeared to reside at the nuclear envelope.

Rearrangement of egg cytoplasm

Fertilization also initiates the processes by which cell differentiation takes place. This sometimes demands a radical displacement of cytoplasmic materials. As we shall see in Chapters 7 and 8, the cytoplasm of the oocyte frequently contains MORPHOGENIC DETERMINANTS that become segregated into specific cells during cleavage. These determinants ultimately lead to the activation or repression of specific genes and

FIGURE 33
Alkalation of egg by ammonia. (A) Continuous recordings of intracellular pH as an egg is placed into various concentrations of NH_4Cl. (B) Activation of development by ammonia-treated eggs. Within a minute after fertilization, eggs were washed in sodium-free seawater so that the Na^+–H^+ exchange could not occur. Washed eggs would resume cleavage if placed into normal seawater or into seawater containing 50 mM Na^+ or 5mM NH_4Cl. Eggs transferred to sodium-free seawater without ammonium ions do not cleave. (A after Shen and Steinhardt, 1978; B after Johnson et al., 1976.)

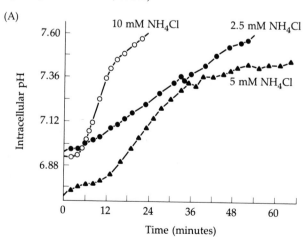

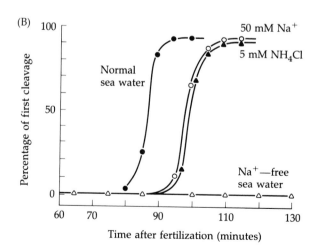

thereby confer certain properties to the cells that incorporate them. The correct spatial arrangement of these determinants is crucial for proper development.

In some species, the rearrangement of these determinants into their needed orientation can be visualized because of the presence of cytoplasmic pigment granules. One such example is the egg of the tunicate *Styela partita* (Conklin, 1905). The unfertilized egg of this animal is seen in Figure 34A. Here a central gray cytoplasm is enveloped by a cortical layer containing yellow lipid inclusions. During meiosis, the breakdown of the nucleus releases a clear substance that accumulates in the animal (upper) hemisphere of the egg. Within 5 minutes of sperm entry, the inner clear and cortical yellow cytoplasms migrate into the vegetal (lower) hemisphere of the egg. As the male pronucleus migrates from the vegetal pole to the equator of the cell along the future posterior side of the embryo, the lipid inclusions migrate with it. This forms a YELLOW CRESCENT extending from the vegetal pole to the equator. This migration brings the yellow plasm into the area where muscle cells will later form in the tunicate larva.

Cytoplasmic movement is also seen in amphibian oocytes. The cortex of the unfertilized amphibian egg contains dark melanin granules in the animal hemisphere only. A single sperm can enter any place on the animal hemisphere; and when it does, it changes the cytoplasmic pattern of the egg. The most obvious change can be seen in the cortical cytoplasm adjacent to the cell membrane. Originally, a covering of dark pigment envelops the animal hemisphere of the egg. After sperm entry, however, some of this cortical cytoplasm shifts relative to the inner cytoplasm. As a result, one region of the egg that had formerly been covered by dark cortical cytoplasm is now exposed (Figure 35). This underlying cytoplasm, located near the equator on the side exactly opposite the point of sperm entry, contains diffuse pigment granules and therefore appears gray (Roux, 1887; Ancel and Vintenberger, 1948). Thus, this region has been referred to as the GRAY CRESCENT. As we

FIGURE 34
Cytoplasmic rearrangement in the egg of the tunicate *Styela partita*. (A) Before fertilization, yellow cortical cytoplasm surrounds gray yolky cytoplasm. (B) At the moment of sperm entry, the yellow cortical cytoplasm and the clear cytoplasm derived from the breakdown of the occyte nucleus stream vegetally toward the sperm. (C) As the sperm pronucleus migrates animally toward the egg pronucleus, the yellow and clear cytoplasms move with it. (D) The final positions of the clear and yellow cytoplasms. They mark the positions where the cells give rise to the mesenchyme and muscles, respectively. (After Conklin, 1905.)

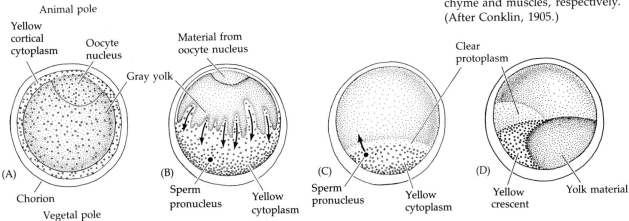

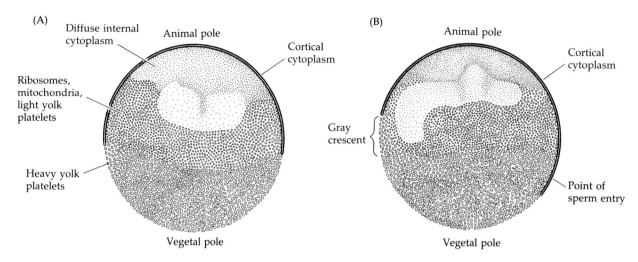

FIGURE 35

Cytoplasmic rearrangements accompanying fertilization in the toad *Discoglossus pictus*. Eggs were fertilized and sectioned at various times. (A) Diagram of an egg sectioned 15 minutes after fertilization. (B) An egg 135 minutes after fertilization. (The first cleavage in this species begins at 150 minutes.) The cortical cytoplasm has moved relative to the internal cytoplasm and the positions of the internal components have also changed. (After Klag and Ubbels, 1975.)

shall see in subsequent chapters, the gray crescent is the region where gastrulation is initiated in amphibian embryos.

The cytoplasmic rearrangements affect the internal cytoplasm as well. The unfertilized frog egg has a stratified arrangement of cytoplasmic components. The animal hemisphere contains, in addition to its pigmented cortex, (1) a layer of clear and lightly pigmented cytoplasm containing much of the material that had been in the huge oocyte nucleus prior to meiosis; (2) a layer of common cytoplasmic elements that includes the ribosomes and mitochondria; and (3) small nutrient reserves including glycogen granules, fat inclusions, and small yolk platelets. The vegetal hemisphere contains large yolk platelets with relatively few other cytoplasmic constituents. After fertilization, this layered arrangement is severely perturbed. On the side giving rise to the gray crescent, the most animal layer of cytoplasm is brought into contact with the most vegetal region of the cytoplasm, whereas the other side retains the layered arrangement characteristic of the unfertilized egg (Klag and Ubbels, 1975; Gerhart et al., 1983).

Cleavage, the event that separates fertilization from embryogenesis, has a special relationship to these cytoplasmic regions. In tunicate embryos the first cleavage bisects the egg into mirror-image duplicates. From that stage on, every division on one side of that cleavage furrow has a "mirror-image" division on the opposite side. Similarly, the gray crescent is bisected by the first cleavage furrow in amphibian eggs.

Thus, the position of the first cleavage is not random but tends to be specified by the point of sperm entry and the subsequent migration of egg cytoplasm. The coordination of cleavage plane and cytoplasmic rearrangements is probably mediated through the microtubules of the sperm aster. When amphibian eggs are incubated in anti-microtubule drugs such as colchicine, the asters do not form and cytoplasmic displacements fail to occur. Thus, the spermatic centrioles not only organize the mitotic apparatus but also rearrange the zygote cytoplasm (Gerhart et al., 1981). The stage is now set for the development of a multicellular organism.

Literature cited

Afzelius, B. A. (1976). A human syndrome caused by immotile cilia. *Science* 193: 317–319.

Ancel, P. and Vintenberger, P. (1948). Recherches sur le determinisme de la symmetrie bilatérale dans l'oeuf des amphibiens. *Bull. Biol. Fr. Belg.* [Suppl.] 31: 1–182.

Amos, L. A. and Klug, A. (1974). Arrangement of subunits in flagellar microtubules. *J. Cell Sci.* 14: 523–549.

Austin, C. R. (1965). *Fertilization.* Prentice-Hall, Englewood Cliffs, NJ.

Bedford, J. M. (1968). Ultrastructural changes in the sperm head during fertilization in the rabbit. *Am. J. Anat.* 123: 329–357.

Bestor, T. M. and Schatten, G. (1981). Anti-tubulin immunofluorescence microscopy of microtubules present during the pronuclear movements of sea urchin fertilization. *Dev. Biol.* 88: 80–91.

Bleil, J. D. and Wassarman, P. M. (1980). Mammalian sperm in egg interaction: Identification of a glycoprotein in mouse-egg zonae pellucidae possessing receptor activity for sperm. *Cell* 20: 873–882.

Boveri, T. (1902). On multipolar mitosis as a means of analysis of the cell nucleus. (Translated by S. Gluecksohn-Waelsch.) In B. H. Willier and J. M. Oppenheimer (eds.), *Foundations of Experimental Embryology.* Hafner, New York, 1974.

Boveri, T. (1907). Zellenstudien VI. Die Entwicklung dispermer Seeigeleier. Ein Beitrag zur Befruchtungslehre und zur Theorie des Kernes. *Jena. Zeit. Naturwiss.* 43: 1–292.

Braden, A. W. H. and Austin, C. R. (1954). The number of sperms about the eggs in mammals and its significance for normal fertilization. *Aust. J. Biol. Sci.* 7: 543–551.

Calvin, H. I. and Bedford, J. M. (1971). Formation of disulfide bonds in the nucleus and accessory structures of mammalian spermatozoa during maturation in the epididymus. *J. Reprod. Fertil.* [Suppl.] 13: 65–75.

Carroll, E. J. and Epel, D. (1975). Isolation and biological activity of the proteases released by sea urchin eggs following fertilization. *Dev. Biol.* 44: 22–32.

Chambers. E. L., Pressman, B. C. and Rose, B. (1974). The activity of sea urchin eggs by the divalent ionophores A23187 and X-537A. *Biochem. Biophys. Res. Commun.* 60: 126–132.

Clermont, Y. and Leblond, C. P. (1955). Spermiogenesis of man, monkey, and other animals as shown by the "periodic acid-Schiff" technique. *Am. J. Anat.* 96: 229–253.

Conklin, E. G. (1905). The orientation and cell-lineage of the ascidian egg. *J. Acad. Nat. Sci. Phila.* 13: 5–119.

Cross, N. L. and Elinson, R. P. C. (1980). A fast block to polyspermy in frogs mediated by changes in the membrane potential. *Dev. Biol.* 75: 187–198.

Dan, J. C. (1967). Acrosome reaction and lysins. In C. B. Metz and A. Monroy (eds.), *Fertilization,* Vol. I. Academic, New York, pp. 237–367.

Davis, B. K. (1981). Timing of fertilization in mammals: Sperm cholesterol/phospholipid ratio as determinant of capacitation interval. *Proc. Natl. Acad. Sci. USA* 78: 7560–7564.

Davis, B. K., Bryne, R. and Bedigan, K. (1980). Studies on the mechanism of capacitation: Albumin-mediated changes in plasma membrane lipids during *in vitro* incubation of rat sperm cells. *Proc. Natl. Acad. Sci. USA* 77: 1546–1550.

Dawid, I. B. and Blackler, A. W. (1972). Maternal and cytoplasmic inheritance of mitochondria in *Xenopus. Dev. Biol.* 29: 152–161.

DePetrocellis, B. and Monroy, A. (1974). Regulatory processes of DNA synthesis in the embryo. *Endeavour* 33: 92–98.

De Robertis, E. D. P., Saez, F. A. and De Robertis, E. M. F. (1975). *Cell Biology*, Sixth Edition, Saunders, Philadelphia.

Deuchar, E. (1975). *Cellular Interactions in Animal Development*. Halstead, New York.

Epel, D. (1967). Protein synthesis in sea urchin eggs: A "late" response to fertilization. *Proc. Natl. Acad. Sci. USA* 57: 899–906.

Epel, D. (1977). The program of fertilization. *Sci. Am.* 237(5): 128–138.

Epel, D. (1980). Fertilization. *Endeavour N.S.* 4: 26–31.

Florman, H. M. and Storey, B. T. (1982). Mouse gamete interactions: The zona pellucida is the site of the acrosome reaction leading to fertilization *in vitro*. *Dev. Biol.* 91: 121–131.

Foerder, C. A. and Shapiro, B. M. (1977). Release of ovoperoxidase from sea urchin eggs hardens the fertilization membrane with tyrosine crosslinks. *Proc. Natl. Acad. Sci. USA* 74: 4214–4218.

Foerder, C. A., Kubanoff, S. J. and Shapiro, B. M. (1978). Hydrogen peroxide production, chemiluminescence, and the respiratory burst at fertilization: Interrelated events in early sea urchin development. *Proc. Natl. Acad. Sci. USA* 75: 3183–3187.

Frankhauser, G. (1955). The role of the nucleus and cytoplasm. In B. J. Willier, P. A. Weiss and V. Hamburger (eds.), *Analysis of Development*. Saunders, Philadelphia, p. 139.

Franklin, L. E. (1970). Fertilization and the role of the acrosomal reaction in non-mammals. *Biol. Reprod.* [Suppl.] 2: 159–176.

Fulton, B. P. and Whittingham, D. G. (1978). Activation of mammalian oocytes by intracellular injection of calcium. *Nature* 273: 149–151.

Gabel, C. A., Eddy, E. M. and Shapiro, B. M. (1979). After fertilization surface components remain as a patch in sea urchin and mouse embryos. *Cell* 18: 207–215.

Gerhart, J., Ubbels, G., Black, S., Hara, K. and Kirschner, M. (1981). A reinvestigation of the role of the grey crescent in axis formation in *Xenopus laevis*. *Nature* 292: 511–516.

Gerhart, J., Black, S., Gimlich, R. and Scharf, S. (1983). Control of polarity in the amphibian egg. In W. R. Jeffery and R. A. Raff (eds.) *Time, Space and Pattern in Embryonic Development*. Alan R. Liss, New York, pp. 261–286.

Giles, R. E., Blanc, H., Cann, H. M. and Wallace, D. C. (1980). Maternal inheritance of human mitochondrial DNA. *Proc. Natl. Acad. Sci. USA* 77: 6715–6719.

Gilkey, J. C., Jaffe, L. F., Ridgeway, E. B. and Reynolds, G. T. (1978). A free calcium wave traverses the activating egg of the medaka, *Oryzias latipes*. *J. Cell Biol.* 76: 448–466.

Glabe, C. G. and Lennarz, W. J. (1979). Species-specific sperm adhesion in sea urchins: A quantitative investigation of bindin-mediated egg agglutination. *J. Cell Biol.* 83: 595–604.

Glabe, C. G. and Vacquier, V. D. (1977). Species-specific agglutination of eggs by bindin isolated from sea urchin sperm. *Nature* 267: 836–838.

Glabe, C. G. and Vacquier, V. D. (1978). Egg surface glycoprotein receptor for sea urchin sperm bindin. *Proc. Natl. Acad. Sci. USA* 75: 881–885.

Gould-Somero, M., Jaffe, L. A. and Holland, L. Z. (1979). Electrically mediated fast polyspermy block in eggs of the marine worm, *Urechis caupo*. *J. Cell Biol.* 82: 426–440.

Gwatkin, R. B. L. (1976). Fertilization. In G. Poste and G. L. Nicolson (eds.), *The Cell Surface in Animal Embryogenesis and Development*. Elsevier North-Holland, New York, pp. 1–53.

Hirao, Y. and Yanagimachi, R. (1978). Effects of various enzymes on the ability of hamster egg plasma membrane to fuse with spermatozoa. *Gamete Res.* 1: 3–12.

Hollinger, T. G. and Schuetz, A. W. (1976). "Cleavage" and cortical granule breakdown in *Rana pipiens* oocytes induced by direct microinjection of calcium. *J. Cell Biol.* 71: 395–401.

Infante, A. A., Nauta, R., Gilbert, S., Hobart, P. and Firshein, W. (1973). DNA synthesis in developing sea urchins: Role of a DNA–nuclear membrane complex. *Nature New Biol.* 242: 5–8.

Jacobs, P. A., Wilson, C. M., Sprenkle, J. A., Rosenshein, N. B. and Migeon, B. R. (1980). Mechanism of origin of complete hydatidiform moles. *Nature* 286: 714–716.

Jaffe, L. A. (1976). Fast block to polyspermy in sea urchins is electrically mediated. *Nature* 261: 68–71.

Jaffe, L. A. (1980). Electrical polyspermy block in sea urchins: Nicotine and low sodium experiments. *Dev. Growth Differ.* 22: 503–507.

Jaffe, L. F. (1983). Sources of calcium in egg activation: A review and hypothesis. *Dev. Biol.* 99: 265–276.

Johnson, J. D., Epel, D. and Paul, M. (1976). Intracellular pH and activation of sea urchin eggs after fertilization. *Nature* 262: 661–664.

Johnson, M. and Edidin, M. (1978). Lateral diffusion in plasma membrane of mouse egg is restricted after fertilization. *Nature* 272: 448–450.

Just, E. E. (1919). The fertilization reaction in *Echina-*

rachnius parma. Biol. Bull. 36: 1–10.

Klag, J. J. and Ubbels, G. A. (1975). Regional morphological and cytochemical differentiation of the fertilized egg of *Discoglossus pictus* (Anura). *Differentiation* 3: 15–20.

Levine, A. E., Walsh, K. A. and Fodor, E. J. B. (1978). Evidence of an acrosin-like enzyme in sea urchin sperm. *Dev. Biol.* 63: 299–306.

Manes, M. E. and Barbieri, F. D. (1976). Symmetrization in the amphibian egg by disrupted sperm cells. *Dev. Biol.* 53: 138–141.

Manes, M. E. and Barbieri, F. D. (1977). On the possibility of sperm aster involvement in dorso-ventral polarization and pronuclear migration in the amphibian egg. *J. Embryol. Exp. Morphol.* 40: 187–197.

McGrath, J. and Solter, D. (1984). Completion of mouse embryogenesis requires both the maternal and paternal genome. *Cell* 37: 179–183.

Metz, C. B. (1978). Sperm and egg receptors involved in fertilization. *Curr. Top. Dev. Biol.* 12: 107–148.

Miller, R. L. (1966). Chemotaxis during fertilization in the hydroid *Campanularia. J. Exp. Zool.* 162: 23–44.

Miller, R. L. (1978). Site-specific agglutination and the timed release of a sperm chemo-attractant by the egg of the leptomedusan, *Orthopyxis caliculata. J. Exp. Zool.* 205: 385–392.

Moy, G. W. and Vacquier, V. D. (1979). Immunoperoxidase localization of bindin during the adhesion of sperm to sea urchin eggs. *Curr. Top. Dev. Biol.* 13: 31–44.

Ogawa, K., Mohri, T. and Mohri, H. (1977). Identification of dynein as the outer arms of sea urchin sperm axonomes. *Proc. Natl. Acad. Sci. USA* 74: 5006–5010.

Ohama, K., Kajii, T., Okamoto, E., Fukada, Y., Imaizumi, K., Tsukahara, M., Kobayashi, K. and Hagiwara, K. (1981). Dispermic origin of XY hydatidiform moles. *Nature* 292: 551–552.

Perry, G. and Epel, D. (1977). Calcium stimulation of a lipoxygenase activity accounts for the respiratory burst at fertilization of the sea urchin egg. *J. Cell Biol.* 75: 40a.

Ridgeway, E. B., Gilkey, J. C. and Jaffe, L. F. (1975). Free calcium increases explosively in activating medaka eggs. *Proc. Natl. Acad. Sci. USA* 74: 623–627.

Roux, W. (1887). Beiträge zur Entwicklungsmechanik des Embryo. *Arch. Mikrosk. Anat.* 29: 157–212.

Saling, P. M., Sowinski, J. and Storey, B. T. (1979). An ultrastructural study of epididymal mouse spermatozoa binding to zonae pellucida *in vitro*: Sequential relationship to acrosome reaction. *J. Exp. Zool.* 209: 229–238.

Saunders, J. W., Jr. (1970). *Principles and Patterns of Animal Development.* Macmillan, New York.

Schackmann, R. W., Eddy, E. M. and Shapiro, B. M. (1978). The acrosome reaction of *Strongylocentrotus purpuratus* sperm: Ion requirements and movements. *Dev. Biol.* 65: 483–495.

Schackmann, R. W. and Shapiro, B. M. (1981). A partial sequence of ionic changes associated with the acrosome reaction of *Strongylocentrotus purpuratus. Dev. Biol.* 81: 145–154.

Schatten, G. and Mazia, D. (1976). The penetration of the spermatazoan through the sea urchin egg surface at fertilization: Observations from the outside on whole eggs and from the inside on isolated surfaces. *Exp. Cell Res.* 98: 325–337.

Schatten, H. and Schatten, G. (1980). Surface activity at the egg plasma membrane during sperm incorporation and its cytochalasin B sensitivity: Scanning electron microscopy and time-lapse video microscopy during the fertilization of the sea urchin *Lytechinus variegatus. Dev. Biol.* 78: 435–449.

Schimke, R. T. (1975). Turnover of membrane proteins in animal cells. *Methods Membr. Biol.* 3: 201–236.

Schroeder, T. E. (1979). Surface area change at fertilization: Resorption of the mosaic membrane. *Dev. Biol.* 70: 306–326.

SeGall, G. K. and Lennarz, W. J. (1979). Chemical characterization of the component of the jelly coat from sea urchin eggs responsible for induction of the acrosome reaction. *Dev. Biol.* 71: 33–48.

Shen, S. S. and Steinhardt, R. A. (1978). Direct measurement of intracellular pH during metabolic depression of the sea urchin egg. *Nature* 272: 253–254.

Shur, B. D. and Hall, N. G. (1982a). Sperm surface galactosyltransferase activities during *in vitro* capacitation. *J. Cell Biol.* 95: 567–573.

Shur, B. D. and Hall, N. G. (1982b). A role for mouse sperm surface galactosyltransferase in sperm binding for the egg zona pellucida. *J. Cell Biol.* 95: 574–579.

Stambaugh, R. and Buckley, J. (1960). Identification and subcellular localization of the enzymes effecting penetration of the zona pellucidae by rabbit spermatozoa. *J. Reprod. Fertil.* 19: 423–432.

Stambaugh, R. and Mastroianni, L. (1980). Stimulation of rhesus monkey (*Macaca mulatta*) proacrosin activation by oviduct fluid. *J. Reprod. Fertil.* 59: 479–484.

Steinhardt, R. A. and Epel, D. (1974). Activation of sea urchin eggs by a calcium ionophore. *Proc. Natl. Acad. Sci. USA* 71: 1915–1919.

Steinhardt, R., Zucker, R. and Schatten, G. (1977). Intracellular calcium release at fertilization in the sea urchin egg. *Dev. Biol.* 58: 185–196.

Summers, R. G. and Hylander, B. L. (1975). Species-specificity of acrosome reaction and primary gamete binding in echinoids. *Exp. Cell Res.* 96: 63–68.

Tilney, L. G., Bryan, J., Bush, D. J., Fujiwara, K., Mooseker, M. S., Murphy, D. B. and Snyder, D. H. (1973). Microtubules: Evidence for thirteen protofilaments, *J. Cell Biol.* 59: 267–275.

Tilney, L. G., Kiehart, D. P., Sardat, C. and Tilney, M. (1978). Polymerization of actin. IV. Role of Ca^{++} and H^+ in the assembly of actin and in membrane fusion in the acrosomal reaction of echinoderm sperm. *J. Cell Biol.* 77: 536–560.

Usui, N. and Yanagimachi, R. (1976). Behaviour of hamster sperm nuclei incorporated into eggs at various stages of maturation, fertilization, and early development. *J. Ultrastruct. Res.* 57: 276–288.

Uzzell, T. M. (1964). Relations of the diploid and triploid species of the *Ambystoma jeffersonianum* complex. *Copeia* 1964: 257–300.

Vacquier, V. D. (1975). The isolation of intact cortical granules from sea urchin eggs: Calcium ions trigger granule discharge. *Dev. Biol.* 43: 62–74.

Vacquier, V. D. (1979). The interaction of sea urchin gametes during fertilization. *Am. Zool.* 19: 849–849.

Vacquier, V. D. and Moy, G. W. (1977). Isolation of bindin: The protein responsible for adhesion of sperm to sea urchin eggs. *Proc. Natl. Acad. Sci. USA* 74: 2456–2460.

Vacquier, V. D. and Payne, J. E. (1973). Methods for quantitating sea urchin sperm in egg binding. *Exp. Cell Res.* 82: 227–235.

Vacquier, V. D., Tegner, M. J. and Epel, D. (1973). Protease release from sea urchin eggs at fertilization alters the vitelline layer and aids in preventing polyspermy. *Exp. Cell Res.* 80: 111–119.

Williams, W. L. (1972). Biochemistry of capacitation of spermatozoa. In K. S. Moghissi and E. E. S. Hafez (eds.), *Biology of Mammalian Fertilization and Implantation.* Charles Thomas, Springfield, IL.

Wincek, T. J., Parrish, R. F. and Polakoski, K. L. (1979). Fertilization: A uterine glycosaminoglycan stimulates the conversion of sperm proacrosin to acrosin. *Science* 203: 553–554.

Wolf, D. E., Kinsey, W., Lennarz, W. and Edidin, M. (1981). Changes in the organization of the sea urchin egg plasma membrane upon fertilization: Indications from the lateral diffusion rates of lipid soluble fluorescent dyes. *Dev. Biol.* 81: 133–138.

Yanagimachi, R. (1981). Mechanisms of fertilization in mammals. In L. Mastroianni and J. D. Biggers (eds.), *Fertilization and Embryonic Development in Vitro.* Plenum, New York, pp. 81–182.

Yanagimachi, R. and Noda, Y. D. (1970). Electron microscope studies of sperm incorporation into the golden hamster egg. *Am. J. Anat.* 128: 429–462.

Yanagimachi, R. and Usui, N. (1974). Calcium dependence of the acrosome reaction and activation of guinea pig spermatozoa. *Exp. Cell Res.* 89: 161–174.

Zaneveld, L. J. D. and Williams, W. L. (1970). A sperm enzyme that disperses the corona radiata and its inhibition by decapitation factor. *Biol. Reprod.* 2: 363–368.

CHAPTER 3

Cleavage: creating multicellularity

*One must show the greatest respect towards
any thing that increases exponentially,
no matter how small.*
—GARRETT HARDIN (1968)

*The history of a man for the nine months preceding his birth
would probably be far more interesting, and contain events of far
greater moment, than all the three-score and ten years that
follow it.*
—SAMUEL TAYLOR COLERIDGE (1885)

Introduction

Remarkable though it is, fertilization is but the initiatory step in development. The zygote, with its new genetic potential and its new arrangement of cytoplasm, now begins the production of a multicellular organism. In all animal species known, this begins by a process called CLEAVAGE, a series of mitotic divisions whereby the enormous volume of egg cytoplasm is cleaved into numerous smaller nucleated cells. These cleavage-stage cells are called BLASTOMERES.

In most species, there is no net increase in embryonic volume during cleavage. This is surprising, for in most other cases of cell proliferation, there is a period of cell growth between mitoses; a cell will expand to nearly twice its volume, then divide. This growth produces a net increase in the total volume of cells while maintaining a relatively constant nucleus:cytoplasm ratio. During embryonic cleavage, however, cytoplasm volume does not increase. Rather, the enormous volume of zygote cytoplasm is divided into increasingly smaller cells. First the egg is divided in half, then quarters, then eighths, and so forth. This division of egg cytoplasm without growth is accomplished by abolishing the interphase growth period between divisions, while the cleavage of nuclei occurs at a rate never seen again (even in tumor cells). A frog egg, for example, can divide into 37,000 cells in just 43 hours; and mitosis in cleavage-stage *Drosophila* occurs every 10 minutes for over 2 hours.

This increase in cell number can be appreciated by comparing cleavage with other stages of development. Figure 1 shows the logarithm of cell number in a frog embryo plotted against the time of development (Sze, 1953). It illustrates a sharp discontinuity between cleavage and gastrulation. One consequence of this rapid division is that the ratio of cytoplasmic volume to nuclear volume gets increasingly smaller as cleavage progresses. In amphibian embryos, this 1000-fold decrease in the cytoplasmic volume to nuclear volume ratio is crucial in timing the activation of certain genes.

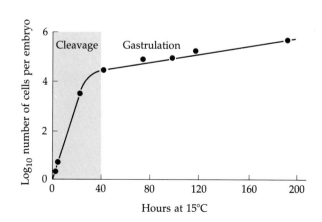

FIGURE 1
Formation of new cells during the early development of the frog *Rana pipiens*. (After Sze, 1953.)

In the frog *Xenopus laevis*, transcription of new messages is not activated until after 12 divisions. At that time, the rate of cleavage decreases, the blastomeres become motile, and nuclear genes begin to be transcribed. It is thought that some factor in the egg is being titrated by the newly made chromatin, as the time of this transition can be changed by altering the amount of chromatin in each nucleus: The more chromatin present in the embryo, the earlier the transition. If the amount of chromatin present is double the normal amount of chromatin, the transition occurs one division earlier (Newport and Kirschner, 1982). Thus, cleavage begins soon after fertilization and ends when the embryo achieves a new balance between nucleus and cytoplasm.

PATTERNS OF EMBRYONIC CLEAVAGE

Cleavage is a very well coordinated process under genetic regulation. The pattern of embryonic cleavage particular to a species is determined by two major parameters: (1) those factors in the egg cytoplasm influencing the angle of the mitotic spindle and the timing of its formation, and (2) the amount and distribution of yolk protein within the cytoplasm.

The amount and distribution of yolk determines where cleavage can occur and the relative size of the blastomeres. When one pole of the egg is relatively yolk-free, the cellular divisions occur there at a faster rate than at the opposite pole. The yolk-rich pole is referred to as the VEGETAL POLE; the ANIMAL POLE is relatively low in yolk concentration. The zygote nucleus is frequently displaced toward the animal pole. Table 1 provides a classification of cleavage types and shows the influence of yolk on the early developmental stages. In general, yolk inhibits cleavage. In zygotes with relatively little yolk (isolecithal and moderately telolecithal eggs), cleavage is HOLOBLASTIC, meaning that the cleavage furrow extends through the entire egg. Zygotes containing large accumulations of yolk protein undergo MEROBLASTIC cleavage, wherein only a portion of the cytoplasm is cleaved. The cleavage furrow does not penetrate into the yolky portion of the cytoplasm. Meroblastic cleavage can be DISCOIDAL, as in birds' eggs, or SUPERFICIAL, as in insect zygotes, depending on whether the yolk deposit is located to one side (telolecithal) or in the center of the cytoplasm (centrolecithal).

Yolk is an evolutionary adaptation that enables an embryo to develop in the absence of an external food source. Animals developing without large yolk concentrations, such as sea urchins, usually develop a larval stage fairly rapidly. This larval stage can feed itself, and development then continues from this free-swimming larva. Mammalian eggs, which also lack large quantities of yolk, adopt another strategy, namely, to make a placenta. As we shall see, the first differentiation of the mammalian embryo sets aside the cells that will form the placenta.

TABLE 1
Classification of cleavage types

Cleavage pattern	Position of yolk	Cleavage symmetry	Representative animals
Holoblastic (complete cleavage)	Isolecithal (oligolecithal) (sparse, evenly distributed yolk)	Radial Spiral Bilateral Rotational	Echinoderms, *Amphioxus* Most molluscs, annelids, flatworms, and roundworms Ascidians Mammals
	Mesolecithal (moderately telolecithal)	Radial Bilateral	Amphibians Cephalopod molluscs
Meroblastic (incomplete cleavage)	Telolecithal (dense yolk concentrated at one end of egg)	Discoidal (bilateral)	Reptiles, fishes, birds
	Centrolecithal (yolk concentrated in center of egg)	Superficial	Most arthropods

This organ will supply food and oxygen for the embryo during its long gestation.

At the other extreme are the eggs of fishes, reptiles, and birds. Most of their cell volume is yolk. The yolk must be sufficient to nourish these animals, as they develop without a larval stage or placental attachment. The correlation between heavy yolk concentration and lack of larval

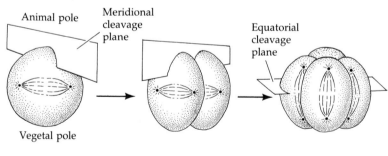

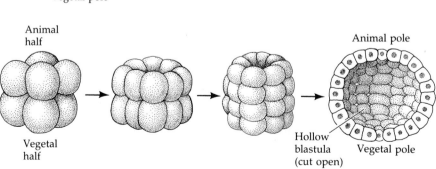

FIGURE 2
Holoblastic cleavage in the echinoderm *Synapta digita* leading to the formation of hollow blastula, as shown in the cutaway view in the last panel. (After Saunders, 1982.)

forms is seen in certain species of frogs. Certain tropical frogs such as *Liopelma* and *Arthroleptella* lack a tadpole stage. Rather, they provision their eggs with an enormous yolk concentration (Lutz, 1947). This enables these frogs to extend their ranges into regions lacking ponds. The eggs do not need to be laid in water because the tadpole stage has been eliminated. The accumulation of large quantitites of yolk has allowed these frogs to evolve in a new fashion.

However, the yolk is just one factor influencing a species' pattern of cleavage. There are also inherited patterns of cell division that are superimposed upon the restraints of the yolk. This can readily be seen in isolecithal eggs, in which very little yolk is present. In the absence of a large concentration of yolk, four major cleavage types can be observed: radial holoblastic cleavage, spiral holoblastic cleavage, bilateral holoblastic cleavage, and rotational holoblastic cleavage.

Radial holoblastic cleavage

The sea cucumber, *Synapta*

Radial holoblastic cleavage is the simplest form of cleavage to understand. It is characteristic of echinoderms and the protochordate *Amphioxus*. The furrows in this type of cleavage are oriented parallel to and perpendicular to the animal–vegetal axis of the egg. Figure 2 shows the cleavage patterns of the sea cucumber, *Synapta digita*. After the union of the pronuclei, the axis of the first mitotic spindle is formed perpendicular to the animal–vegetal axis of the egg. Therefore, the first cleavage furrow passes directly through the animal and vegetal poles, creating two equal-sized daughter cells. This cleavage is said to be meridional because it passes through the two poles like a meridian on a globe. The mitotic spindles of the second cleavage are at right angles to the first but are still perpendicular to the animal–vegetal axis of the egg. The cleavage furrows appear simultaneously in both blastomeres and also pass through the two poles. Thus, the first two divisions are both meridional and are perpendicular to each other. The third division is equatorial: The mitotic spindles of each blastomere are now positioned parallel to the animal–vegetal axis and the resulting cleavage furrow separates the two poles from each other, dividing the embryo into eight equal blastomeres. Each blastomere in the animal half of the embryo is now directly above a blastomere of the vegetal half.

The fourth division is again meridional, producing two tiers of 8 cells each. Successive divisions produce 32-, 64-, 128-, and 256-cell embryos, with meridional divisions alternating with equatorial divisions. The resulting embryo consists of blastomeres arranged in horizontal rows along a central cavity. At both poles of the embryo, blastomeres move toward each other to create a hollow sphere composed of a single cell layer. This hollow sphere is called the BLASTULA, and the central cavity is referred to as the BLASTOCOEL.

At any time during the cleavage of *Synapta*, an embryo bisected through any meridian will produce two mirror-image halves. This type of symmetry is characteristic of a sphere or cylinder and is called radial symmetry. Thus, *Synapta* is said to have radial holoblastic cleavage.

Sea urchins

Sea urchins also have a radial holoblastic cleavage, but with some important modifications. The first and second cleavages are very similar to those of *Synapta*; both are meridional and are oriented perpendicular to each other. Similarly, the third cleavage is equatorial, separating the two poles from one another (Figure 3). In the fourth cleavage, however, events are very different. The four cells of the animal tier split meridionally into eight blastomeres, each having the same volume. These cells are called MESOMERES. The vegetal tier, however, undergoes an unequal equatorial cleavage to produce four large cells, the MACRO-MERES, and four smaller MICROMERES at the vegetal pole. As the 16-cell embryo cleaves, the eight mesomeres divide equatorially to produce two "animal" tiers, an_1 and an_2, one above the other. The macromeres divide meridionally, forming a tier of eight cells below an_2. The micro-meres also divide, producing a small cluster beneath the larger tier. All the cleavage furrows of the sixth division are equatorial; and the seventh cleavage is meridional, producing a 128-cell blastula.

In 1939, Sven Hörstadius performed a simple experiment that dem-onstrated that the timing and placement of each sea urchin cleavage is independent of preexisting cleavages. He showed that if he inhibited the first one, two, or three cleavages by shaking the eggs or placing them into hypotonic seawater, the unequal (fourth) cleavage division that forms the micromeres still occurs at the appropriate time (Figure 4). Thus, Hörstadius concluded that there are three factors that deter-mine cleavage in the 8-cell embryo: (1) There are "progressive changes in the cytoplasm which cause spindles formed after a certain time after fertilization to lie in a certain direction"; (2) there must be micromere-forming material in the vegetal cytoplasm; (3) there must be some mechanism by which the micromere-forming material is activated at the correct time (Hörstadius, 1973).

Although the nature of the "micromere clock" is still unknown, Ikeda (1965) has presented evidence that the cell may keep track of its cleavage by cyclic changes in its cytoplasm. He found that sulfhydryl (—SH) groups on proteins became exposed in a cyclic manner. This alternation between oxidized (—S—S—) and reduced (—SH HS—) conformations probably reflects basic changes in cell metabolism. Ikeda has shown that maximum sulfhydryl group exposure occurs during cleavage and that the minimum number of sulfhydryl groups are pres-

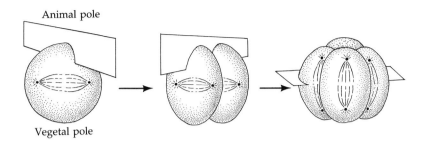

Animal pole

Vegetal pole

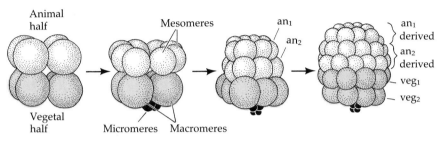

Animal half

Mesomeres

an_1

an_2

an_1 derived

an_2 derived

veg_1

veg_2

Vegetal half

Micromeres   Macromeres

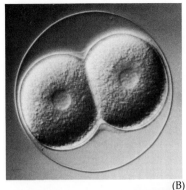

(B)

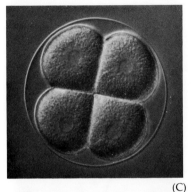

(C)

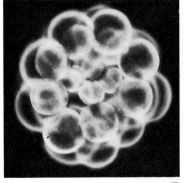

(D)

(A)

FIGURE 3
(A) Cleavage in sea urchin development. The figure shows the formation of particular tiers of cells. (B–D) Photomicrographs of live embryos of the sea urchin *Lytechinus pictus,* looking down upon the animal pole. (B) Two-cell stage; (C) 4-cell stage of sea urchin development. (D) The 32-cell stage is shown without the fertilization membrane to allow visualization of the animal pole mesomeres, the central macromeres, and the vegetal micromeres, which angle into the center. (Photographs courtesy of George Watchmaker.)

ent when the cells are not cleaving (Figure 5A). This cycle of reduction–oxidation is not affected by those treatments that prevent cleavage. The micromere cleavage always occurs at the fourth —SH peak after fertilization, no matter how many cleavages have been prevented previously (Figure 5B). Conversely, if the sulfhydryl peak is prevented (by adding ether to the seawater), the micromere division is delayed. It is possible that the cyclical reduction and oxidation may play a role in intracellular timekeeping. Sakai (1968) has extracted a protein from sea urchin eggs that contracts upon oxidation of its free sulfhydryl groups. Such proteins could possibly respond to these cycles.

The blastula stage of sea urchin development begins at the 128-cell stage. Here the cells form a hollow sphere surrounding a central blastocoel (Figure 6). Every cell is in contact with the proteinaceous fluid of the blastocoel and with the hyaline layer within the fertilization membrane. During this time, contacts between the cells are tightened. Dan-Sohkawa and Fujisawa (1980) have analyzed this process in starfish embryos and have shown that the closing of the hollow sphere is contemporaneous with the formation of tight junctions between the

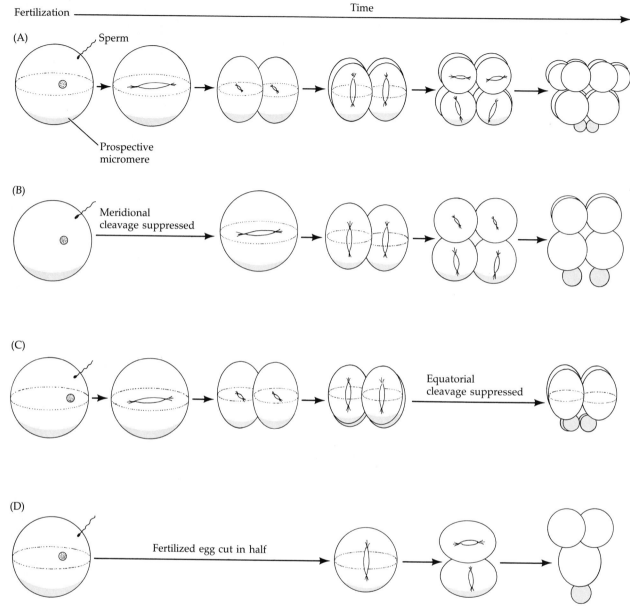

FIGURE 4
Micromere formation is dependent on time of development, not on the number of divisions. (A) Normal sequence of sea urchin cleavage, the micromeres forming at the fourth cleavage. (B, C) The first or second cleavage is suppressed by hypotonic seawater. (D) A fertilized egg was cut in half meridionally, so that half the normal cleavage pattern occurs. Micromeres (shaded) appear at approximately the same time regardless of the number of cleavages. (After Hörstadius, 1939.)

(A)

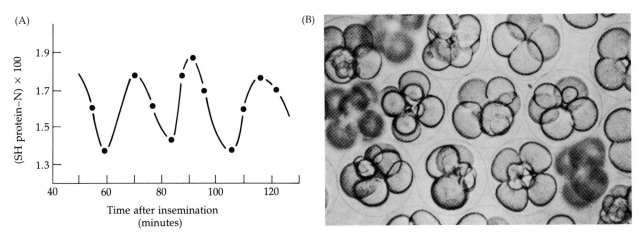

(B)

FIGURE 5
(A) Cyclical variation of exposed protein sulfhydryl groups. Maximum peaks coincide with the midmitosis period of the cleavages and occur regardless of whether cleavage is blocked or not. (B) Micromere formation in 8-cell sea urchin embryos after a division is suppressed by ultraviolet irradiation. (From Ikeda, 1965.)

blastomeres. These junctions unite the loosely connected cells into a tissue and seal off the blastocoel from the outside environment (Figure 7). As the cells continue to divide, the cell layer expands and thins out. This creates a situation where the blastula remains one cell-layer thick.

Two theories have been offered to explain the concomitant expansion and blastocoel formation. Dan (1960) hypothesized that the motive force of this expansion is the blastocoel itself. As the blastomeres secrete proteins into the blastocoel, the blastocoel fluid becomes syrupy. This blastocoel sap absorbs large quantities of water by osmosis, thereby swelling and putting pressure on the blastomeres to expand outward. This pressure would also align the long axis of each cell so that division would never be inward toward the blastocoel. This would create further expansion by having the population oriented in one plane only. Wolpert and Gustafson (1961) and Wolpert and Mercer (1963) think that pressure

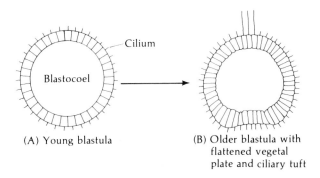

FIGURE 6
Sea urchin blastulae. (A) Early sea urchin blastula shows a single layer of rounded cells surrounding a large blastocoel. (B) As division continues, the cells of the late blastula show differences in shape as the vegetal plate cells elongate. (From Guidice, 1973.)

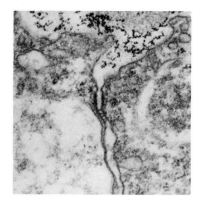

FIGURE 7

Area of contact at the junction of two cells of a 1024-cell starfish blastula. Membrane regions cross over at these junctions to form tight seals. (From Dan-Sohkawa and Fujisawa, 1980; photograph courtesy of the authors.)

from the blastocoel is not needed to get this effect. They emphasize the role of differential adhesiveness of the cells to each other and to the hyaline layer. They find that as long as the cells remain strongly attached to the hyaline layer, they have no alternative but to expand. This expansion creates the blastula rather than the other way around.

The blastula cells develop cilia on their outer surfaces (Figure 8), thereby causing the blastula to rotate within the fertilization envelope. Soon afterward, the cells secrete a HATCHING ENZYME that enables them to digest the fertilization membrane. The embryo is now a free-swimming HATCHED BLASTULA.

FIGURE 8

Ciliated blastula cells. Each cell develops a single cilium. (Photographs courtesy of W. J. Humphreys.)

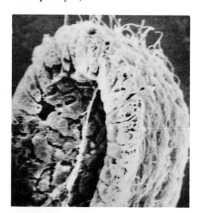

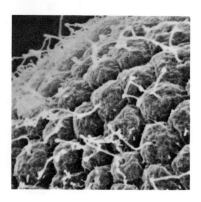

Amphibians

Cleavage in frog and salamander embryos is radially symmetrical and holoblastic, just like echinoderm cleavage. The amphibian egg, however, has much more yolk. This yolk, which is concentrated in the vegetal hemisphere, acts as an impediment to cleavage. Thus, the first division begins at the animal pole and slowly extends down into the vegetal region (Figure 9). In the axolotl salamander, the cleavage furrow extends through the animal hemisphere at a rate close to 1 mm/min. The cleavage furrow bisects the gray crescent and then slows down to a mere 0.02–0.03 mm/min pace as it approaches the vegetal pole (Hara, 1977).

Figure 10 shows a scanning electron micrograph of the first cleavage in a frog egg. One can see the folds in the cleavage furrow and the difference between the furrows in the animal and vegetal hemispheres. While the first cleavage furrow is still trying to cleave the yolky cytoplasm of the vegetal hemisphere, the second cleavage has already started near the animal pole. This cleavage is at right angles to the first one and is also meridional. The third cleavage, as expected, is equatorial. However, because of the vegetally placed yolk, this cleavage furrow in amphibian eggs is much closer to the animal pole. It divides the frog embryo into four small animal blastomeres (micromeres) and four large blastomeres (macromeres) in the vegetal region (Figure 11). This UNEQUAL HOLOBLASTIC CLEAVAGE establishes two major embryonic regions: a rapidly dividing region of micromeres near the animal pole and a more slowly dividing macromere area. As cleavage progresses, the animal region becomes packed with numerous small cells while the vegetal region contains a relatively small number of large, yolk-laden macromeres (Figure 9).

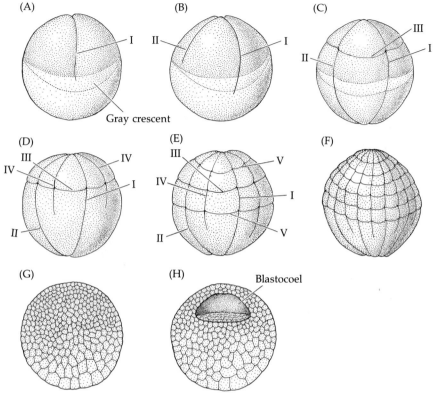

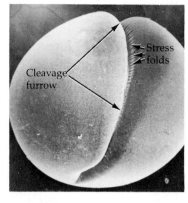

FIGURE 10
Scanning electron micrograph of the first cleavage of a frog egg. (From Beams and Kessel, 1976; photograph courtesy of the authors.)

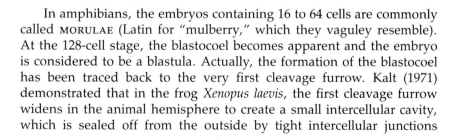

FIGURE 9
Cleavage of a frog egg. Cleavage furrows are designated by Roman numerals in order of appearance. The vegetal yolk impedes the cleavage such that the second division begins in the animal region of the egg before the first division has divided the vegetal cytoplasm. The third division (C) is displaced toward the animal pole. The vegetal hemisphere ultimately contains longer and fewer blastomeres than the animal half. (After Carlson, 1981.)

In amphibians, the embryos containing 16 to 64 cells are commonly called MORULAE (Latin for "mulberry," which they vaguley resemble). At the 128-cell stage, the blastocoel becomes apparent and the embryo is considered to be a blastula. Actually, the formation of the blastocoel has been traced back to the very first cleavage furrow. Kalt (1971) demonstrated that in the frog *Xenopus laevis*, the first cleavage furrow widens in the animal hemisphere to create a small intercellular cavity, which is sealed off from the outside by tight intercellular junctions

FIGURE 11
Scanning electron micrographs of 4-cell and 16-cell frog embryos, showing the size discrepancy between the animal and vegetal blastomeres after the third division. (Photograph courtesy of L. Beidler.)

FIGURE 12
Formation of the blastocoel in a
frog egg. (A) First cleavage
plane, showing a small cleft
where blastocoel will enlarge.
(B) Eight-cell embryo showing a
small blastocoel (arrow) at the
junction of the three cleavage
planes. (From Kalt, 1971; photographs courtesy of M.R. Kalt.)

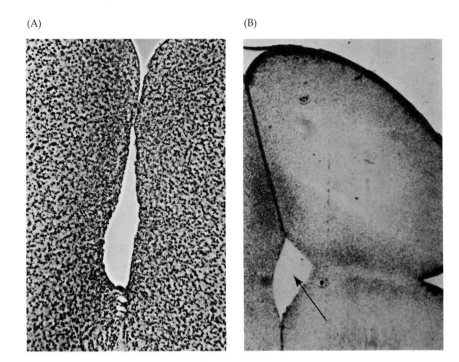

(A)   (B)

(Figure 12). This cavity expands during subsequent cleavages to become
the blastocoel.

The blastocoel probably serves two major functions. First, it is a
cavity that permits cell migration during gastrulation. It has also been
suggested that it prevents the cells beneath it from interacting prematurely with the cells above it. Nieuwkoop (1973) took embryonic newt
cells from the roof of the blastocoel and placed them next to the vegetal
yolky mass. Such cells produced mesoderm tissue instead of ectoderm.
Because mesodermal tissue is normally formed from those animal cells
that are adjacent to the endoderm precursors, it seems plausible that
the vegetal cells influence adjacent cells to differentiate into mesodermal
tissues. Thus, the blastocoel would preserve the integrity of those cells
fated to give rise to the skin and nerves.

Spiral holoblastic cleavage

Spiral cleavage is characteristic of annelids, turbellarian flatworms, nem-
ertean worms, and all molluscs except cephalopods. It differs from
radial cleavage in numerous ways. First, the eggs do not divide in
parallel or perpendicular orientations to the animal–vegetal axis of the
egg. Rather, cleavage is at oblique angles, forming the "spiral" arrangement of daughter blastomeres. Second, the cells touch each other at
more places than do those of radially cleaving embryos. In fact, they

FIGURE 13
Diagram showing the arrangement of four and eight soap bubbles in a slightly concave dish. The thermodynamic arrangement maximizes contact and is very reminiscent of spirally cleaving embryos.

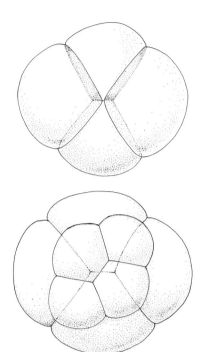

take the most thermodynamically stable packing orientation, much like that of adjacent soap bubbles (Figure 13). Third, spirally cleaving embryos usually undergo fewer divisions before they begin gastrulation. This makes it possible to know the fate of each individual cell of the blastula. When the fates of the individual cells from annelid, flatworm, and mollusc embryos were compared, the same cells were seen in the same places, and their general fates were identical (Wilson, 1898). The blastulae so produced have no blastocoel and are called STEREOBLAS-TULAE.

Figure 14 depicts the cleavage of the mollusc *Trochus*. The first two cleavages are nearly meridional, producing four large macromeres (labeled A, B, C, and D). In many species, the blastomeres are different

FIGURE 14
Spiral cleavage viewed from the animal pole (top sequence) and from one side (lower sequence). In the latter case, the cells derived from the A blastomere are shaded. In the earliest stages shown here, the mitotic spindles are represented and are seen to divide the cells unequally and at an angle to the vertical and horizontal axes.

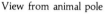

View from animal pole

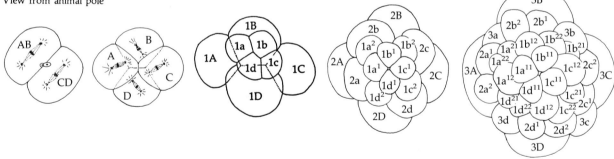

Side view

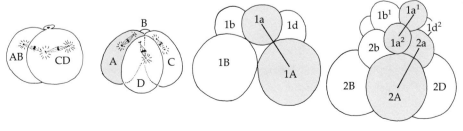

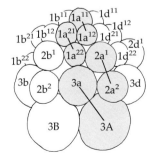

sizes (D being the largest), a characteristic that allows them to be individually identified. In each successive cleavage, each macromere buds off a small micromere at its animal pole. Each successive quartet of micromeres is displaced to the right or to the left of its sister macromere, creating the spiral relationship characteristic of the cleavage. Looking down on the embryo from the animal pole, the upper ends of the mitotic spindle appear to alternate clockwise and counterclockwise. This causes alternate micromeres to form obliquely to the left and to the right of their macromere. At the third cleavage, the A macromere will give rise to two daughter cells, macromere 1A and micromere 1a. The B, C, and D cells behave similarly, producing the first quartet of micromeres. In most species, the micromeres are to the right of their macromeres, an arrangement indicating a dextral (as opposed to sinestral) spiral. At the fourth cleavage, the 1A macromere divides to form macromere 2A and micromere 2a; and micromere 1a divides to form two more micromeres, $1a^1$ and $1a^2$. Further cleavage yields blastomeres 3A and 3a from the 2A macromere; and micromeres such as $1a_2$ divide to produce cells such as $1a^{21}$ and $1a^{22}$.

The orientation of the cleavage plane to the left or to the right is controlled by cytoplasmic factors within the oocyte. This was discovered by analyzing mutations of snail coiling. Some snails have their coils opening to the right of their shells, whereas other snails have their coils opening to the left. Usually, the rotation of coiling is the same for all members of a given species. Occasionally, though, mutants are found. For instance, in the species where the coils open on the right, occasional individuals will be found where the coil opens on the left. Crampton (1894) analyzed the embryos of such aberrant snails and found that their early cleavage differed from the norm. The orientation of the cells after second cleavage was different (Figure 15). This was the result of a different orientation of the mitotic apparatus in the sinestrally coiling snails. All subsequent divisions in left-coiling embryos are mirror images of those of dextrally coiling embryos. in Figure 15B, one can see that the position of the 4d blastomere (which is extremely important, as its progeny will form the mesodermal organs) is different in the two types of spiraling embryos. Eventually, the two snails are formed with their bodies on different sides of the coil opening.

The direction of snail shell coiling is controlled by a single pair of genes (Sturtevant, 1923; Boycott et al., 1930). In the snail *Limnaea peregra*, most individuals are dextrally coiled. Rare mutants exhibiting left-handed coiling were found and mated with wild-type snails. These matings show that there is a "right-handed" allele (*D*), which is dominant to the "left-handed" allele (*d*). However, the direction of cleavage is *not* determined by the genotype of the developing snail but by the genotype of the snail's mother. A *dd* female snail can only produce sinestrally coiling offspring, even when the offspring's genotype is *Dd*. Thus, a *Dd* individual will coil either left or right depending on the genome of its mother.

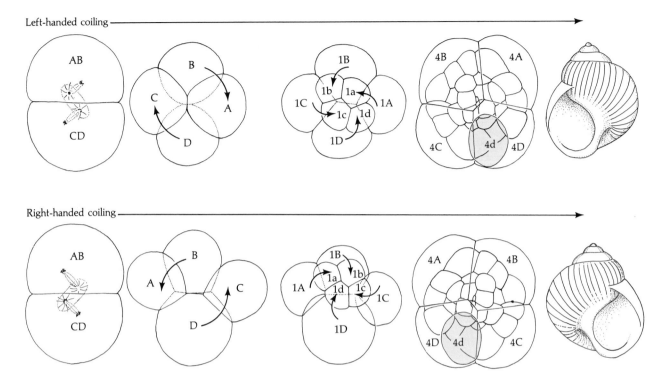

FIGURE 15
Looking down upon the animal pole of right-handed and left-handed snails.
The origin of right-handed and left-handed snail coiling can be traced to the
orientation of the mitotic spindle at the second cleavage. The left-handed
(top) and right-handed (bottom) snails develop as mirror-images of each
other. (After Morgan, 1927.)

The matings produce a chart like this:

DD (female)	×	dd (male)	→	Dd (all right-coiling)
DD (male)	×	dd (female)	→	Dd (all left-coiling)
Dd	×	Dd	→	1 DD:2 Dd:1 dd (all right-coiling)

Thus, the genetic factors involved in snail coiling are brought to the
embryo in the oocyte cytoplasm. It is the genotype of the ovary in
which the oocyte develops that determines which orientation the cleavage will take.

Another exciting discovery concerning molluscan cleavage is that
certain blastomeres communicate with each other. In those molluscs
with equal-sized blastomeres at the 4-cell stage, the determination of
which cell will give rise to the mesodermal precursor cell is accomplished between the fifth and sixth cleavages. At this time, the 3D
macromere extends inwardly and contacts the micromeres of the animal
pole. Without this contact, the 4d cell given off by the 3D macromere

does not produce mesoderm (Biggelaar and Guerrier, 1979).[1] Laat and co-workers (1980) demonstrated that at the time of contact (and not before) small molecules are capable of diffusing between the 3D macromere and the central micromeres. They injected the dye Lucifer Yellow (molecular weight, 457) into one of the macromeres. When added before the 32-cell stage, the dye resides only in that single macromere and its progeny. However, when the dye is injected into the 3D blastomere at the beginning of the 32-cell stage, the first tier of micromeres picks up the yellow color (Figure 16). Low-molecular-weight material is transferred from one cell to the other at just that time when the micromeres alter the developmental potential of the 3D blastomere. Transmission electron microscopy shows that at this time, GAP JUNCTIONS appear on the surfaces of these cells. These modifications of the cell membrane facilitate the intercellular transfer of small molecules and ions between neighboring cells, and they are found in many tissues in both embryonic and adult animals. (The microanatomy of these junctions will be discussed in Chapter 15.)

Bilateral holoblastic cleavage

Bilateral holoblastic cleavage is found primarily in ascidians (tunicates). Figure 17 shows the 'cleavage pattern of a tunicate, *Styela partita*. The most striking phenomenon in this type of cleavage is that the first cleavage plane establishes the only plane of symmetry in the embryo. Each successive division orients itself to this plane of symmetry, and the half-embryo formed on one side of the first cleavage is the mirror image of the half-embryo on the other side. The second cleavage is meridional, like the first division, but unlike the first division, it does not pass through the center of the egg. Rather, it creates two large anterior cells (A and D) and two smaller posterior cells (B and C). Each side now has a large and a small blastomere. At the next three divisions, differences in cell size and shape highlight the bilateral symmetry of

[1]Do not worry about those mollusc embryos with unequal-size blastomeres at the 4-cell stage. We will see much of them in Chapter 8.

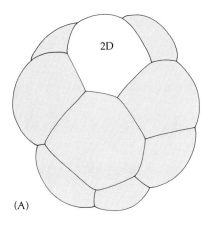

(A)

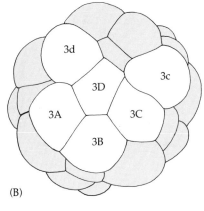

(B)

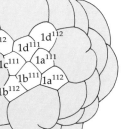

(C)

FIGURE 16
Intercellular communication during mollusc cleavage. Diagrammatic representations show the extent to which the fluorescent dye Lucifer Yellow (white area) has spread after injection into specific macromeres of *Patella vulgata*. (A) Labeling pattern 20 minutes after injection of the 2D macromere of the 16-cell embryo shows no transfer of dye to adjacent cells. (B) Vegetal view of a 32-cell embryo after the injection of dye into the 3D macromere shows that dye spreads to adjacent cells (except $2d^{22}$). (C) Animal view of the same embryo about 45 minutes after injection shows that the 3D macromere has transferred dye to the central group of animal micromeres. (After de Laat et al., 1980.)

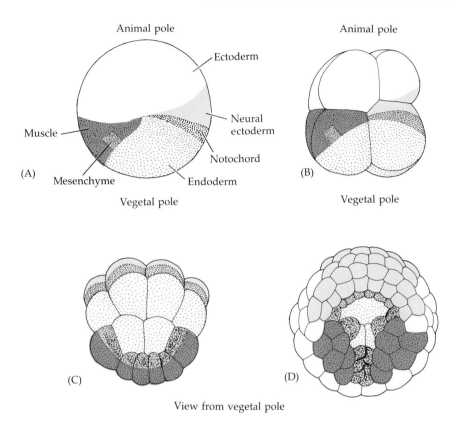

View from vegetal pole

FIGURE 17
Bilateral symmetry in a tunicate egg. (A) Uncleaved egg. (B) Eight-cell embryo showing the blastomeres and the fates of various cells. It can be viewed as two 4-cell halves; from here on each division on the right side of the embryo has a mirror-image division on the left. (C,D) Views of later embryos from the vegetal pole. [The regions of cytoplasm destined to form particular organs are labeled in panel (A). and are coded by shading throughout the diagram.] (After Balinsky, 1981.)

these embryos. At the 32-cell stage, a small blastocoel is formed and gastrulation begins.

As mentioned in the last chapter, certain tunicates, such as *S. partita*, contain colored cytoplasmic regions. These colored plasms are localized bilaterally around the plane of symmetry. During cleavage, these plasms become partitioned into different cells. Moreover, the type of cytoplasm the cells receive prefigures their eventual fate. Cells receiving clear cytoplasm become ectoderm; those containing yellow cytoplasm give rise to mesodermal cells; the cells that incorporate the slate-gray inclusions become endoderm; and the light gray cells become the neural tube and notochord.

Rotational holoblastic cleavage

It is not surprising that mammalian cleavage has been the most difficult to study. Mammalian eggs are among the smallest in the animal kingdom, making them hard to manipulate experimentally. The human zygote, for instance, is only 100 μm in diameter. Also, mammalian zygotes are not produced in numbers comparable to sea urchin or frog embryos. Usually fewer than ten eggs are ovulated by a female at a given time, so it is difficult to obtain enough material for biochemical studies. As a final hurdle, the development of mammalian embryos is accomplished within another organism rather than outside it. Only recently has it been possible to duplicate some of these internal conditions and observe development in vitro.

With all these difficulties, knowledge of mammalian cleavage was worth waiting for. What we finally learned about mammalian cleavage was strikingly different from most other patterns of embryonic cell division. The mammalian oocyte is released from the ovary into the oviduct (Figure 18). Fertilization occurs in the ampulla of the oviduct, a region close to the ovary. Meiosis is completed at this time, and first cleavage begins about a day later. Cleavages in mammalian eggs are among the slowest cleavages occurring in the animal kingdom—about 12 to 24 hours apart. Meanwhile, the cilia in the oviduct are pushing the embryo toward the uterus; the first cleavages occur along this journey.

There are several features of mammalian cleavages that distinguish it from other cleavage types. The first feature mentioned was the relative slowness of the divisions. The second fundamental difference is the unique orientation of mammalian blastomeres in relation to each other. The first cleavage is a normal meridional division; however, in the second cleavage one of the two blastomeres divides meridionally and the other divides equatorially (Figure 19). This type of cleavage is called ROTATIONAL CLEAVAGE (Gulyas, 1975).

The next major difference between mammalian cleavage and that of most other embryos is the marked asynchrony of early division. Mammalian blastomeres do not all divide at the same time. Thus, as shown in Figure 20, mammalian embryos do not increase evenly from 2- to 4- to 8-cell stages but frequently contain odd numbers of cells.

FIGURE 18
Development of a human embryo from fertilization to implantation. The egg "hatches" from the zona upon reaching the uterus, and it is probable that the zona prevents the cleaving cells from sticking to the oviduct rather than traveling to the uterus. (After Tuchmann-Duplessis et al., 1972.)

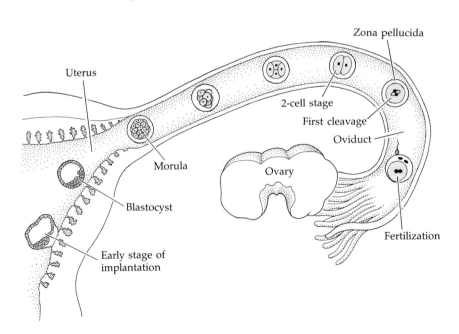

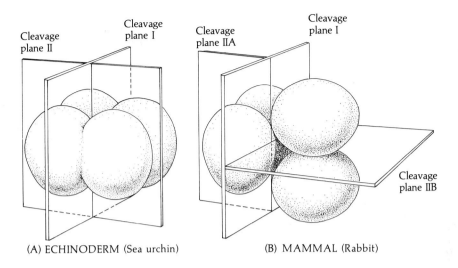

(A) ECHINODERM (Sea urchin) (B) MAMMAL (Rabbit)

FIGURE 19
Comparison of early cleavage of (A) echinoderms and (B) mammals. (After Gulyas, 1975.)

Perhaps the most crucial difference between mammalian cleavage and all other types is the phenomenon of COMPACTION. As seen in Figure 20, mammalian blastomeres through the 8-cell stage form a loose arrangement with plenty of space between them. However, following the third cleavage, the blastomeres undergo a spectacular change in

FIGURE 20
The cleavage of a single mouse embryo in vitro. (A) 2-cell stage; (B) 4-cell stage; (C) early 8-cell stage; (D) compacted 8-cell stage; (E) morula; (F) blastocyst. (From Mulnard, 1967; photograph courtesy of J. G. Mulnard.)

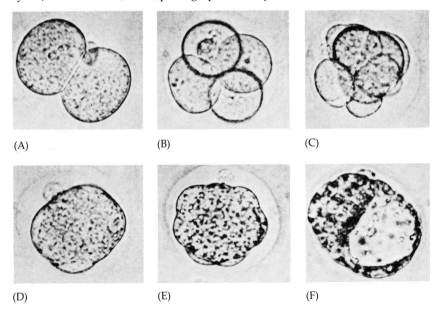

(A) (B) (C)

(D) (E) (F)

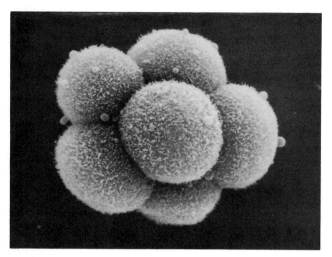 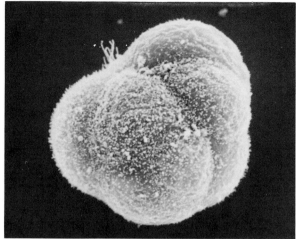

FIGURE 21

Scanning electron micrograph of uncompacted (left) and compacted (right) 8-cell mouse embryo. (Photographs courtesy of C. Ziomek.)

their behavior. They suddenly huddle together, maximizing their contact with the other blastomeres and forming a compact ball of cells (Figure 21). This tightly packed arrangement is stabilized by tight junctions that form between the outside cells of the ball, sealing off the inside of the sphere (Figure 22). The cells within the sphere form gap junctions, enabling small molecules and ions to pass between the cells.

Figure 22 illustrates compaction diagrammatically. It shows that compaction causes the formation of two cell regions: the outer cell layer and a few internally placed cells. As cleavage continues, the compacted cells are called a MORULA. The early morula does not have an internal cavity, although most of the morula cells are in the outer cell layer, with only a few (3–5 at the 32-cell stage) being internal. Fluid secreted into the morula creates a blastocoel, and the internal cells are positioned on one side of the ring of external cells. This structure is called a BLASTOCYST

FIGURE 22

Schematic diagram of cell shape changes and compaction in early mouse development. (After Dulcibella, 1977.)

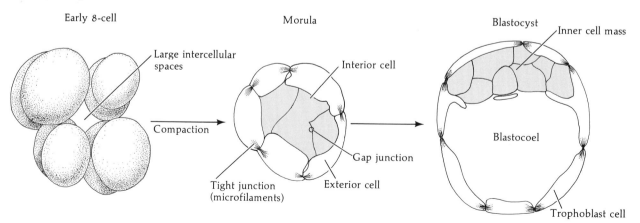

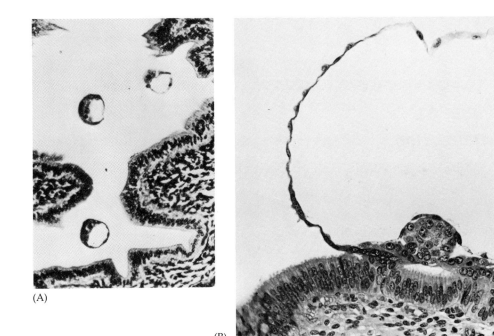

(A)

(B)

FIGURE 23
(A) Mouse blastocysts entering uterus. (From Rugh, 1967.) (B) Initial implantation of the blastocyst onto the uterus in a rhesus monkey. (Courtesy of the Carnegie Institution of Washington; Chester Reather, photographer.)

(Figures 22 and 23), and it is another hallmark of mammalian cleavage. The inner and outer cells of the blastocyst differ from one another morphologically, functionally, and biochemically. The predominant cell type constitutes the outer cell layer. These are called TROPHOBLAST (or TROPHECTODERM) cells. This group of cells produces no embryonic structures. Rather, they form the tissue of the CHORION, the outer portion of the placenta. These are the tissues that enable the fetus to get oxygen and nourishment from the mother, that secrete hormones so that the mother's uterus will keep the fetus, and that produce regulators of the immune response so that the mother will not reject the embryo as she would an organ graft. They do not, however, produce any cell of the developing embryo. Once this trophoblast is formed, the embryo is capable of IMPLANTING in the uterine wall—another function of the trophoblast cells, because only they can attach to the lining of the uterus and anchor the embryo to that site (Figure 23). It is the INNER CELL MASS (ICM) that will generate the embryo. These cells not only look different from the trophoblast cells, they are synthesizing different products even at this early stage of development. Thus, the distinction between trophoblast and inner cell mass blastomeres represents the first differentiation event in mammalian development.

SIDELIGHTS & SPECULATIONS

The cell surface and the mechanism of compaction

Compaction creates the circumstances whereby the first differentiation in mammalian development—the separation of trophoblast from inner cell mass—is effected. How is this done? It appears that compaction is mediated by events occurring at the cell surfaces of adjacent blastomeres. First, specific cell surface proteins are seen to play a role. There exists a type of tumor, called a teratocarcinoma, whose cells resemble those of the inner cell mass. Compaction is inhibited when 8-cell mouse embryos are incubated in antiserum made against teratocarcinoma stem cells (Figure 24) (Kemler et al., 1977; Johnson et al., 1979). Thus, there is a molecule or set of molecules that is present on the teratocarcinoma cell and on 8-cell blastomeres and that is necessary for the occurrence of compaction. One such molecule is UVOMORULIN, a 120,000-d glycoprotein found on the cell surfaces of these cell types. Antibodies to this molecule cause the decompaction of the morula and inhibit the cell–cell attachments of teratocarcinoma cells (Hyafil et al., 1981; Peyriéras et al., 1983). The carbohydrate portion of this glycoprotein may be essential to its function, as TUNICAMYCIN (a drug that inhibits the glycosylation of proteins) also prevents compaction.

Second, the cell membrane may be modified during compaction by cytoskeletal reorganization. Microvilli, extended by actin microfilaments, appear on adjacent cell surfaces and attach one cell to the other. These microvilli may be the sites where uvomorulin is functioning to mediate intercellular adhesion. The flattening of the blastomeres against one another may

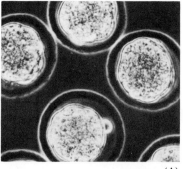

(A)

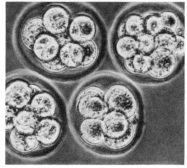

FIGURE 24   (B)
Prevention of compaction by antiserum directed against teratocarcinoma stem cells. (A) Normal compaction occurring in the absence of antiserum. (B) Proliferation without compaction occurring in the presence of antibodies to teratocarcinoma stem cells. (Photograph courtesy of C. Ziomek.)

therefore be brought about by the shortening of the microvilli through actin depolymerization (Pratt et al., 1982; Sutherland and Calarco-Gillam, 1983). Thus, there is growing evidence that compaction is caused by changes in the architecture of the blastomere cell surface.

Formation of the inner cell mass

The creation of an inner cell mass distinct from the trophoblast is *the* crucial process of early mammalian development. How is a cell directed into one or the other of these paths? How is a cell informed that it is either to give rise to a portion of the adult mammal or that it is to give rise to a rather remarkable supporting tissue that will be discarded at birth? Observations of living embryos suggest that this momentous decision is merely a matter of a cell's being in the right place at the right time. Up through the 8-cell stage, there are no obvious differences in the biochemistry, morphology, or potency of any of the blastomeres. However, compaction forms inner and outer cells with vastly different properties. By labeling the various blastomeres, numerous investigators have found that the cells that happen to be on the outside will form the trophoblast whereas the cells that happen to be inside will generate the embryo (Tarkowski and Wróblewska, 1967).[2] Hillman and co-workers (1972) have shown that when each blastomere of a 4-cell mouse embryo is placed on the outside surface of a mass of aggregated blastomeres, the external, transplanted cells will only give rise to trophoblast tissue. Therefore, it seems that whether a cell becomes trophoblast or embryo depends on whether it was an external or an internal cell after compaction.

If most of the cells of the blastocyst give rise to the trophoblast, exactly how many cells actually form the embryo? Our best answers to this question come from observations of allophenic mice. Allophenic mice are the product of two early-cleavage (usually 4- or 8-cell) embryos that have been aggregated together to form a composite embryo. As shown in Figure 25, the zona pellucidae of two different genetically marked embryos are removed and the embryos brought together to form a common blastocyst. These prepared blastocysts are implanted into the uterus of the foster mother. When they are born, the allophenic offspring have some cells from each embryo. This is readily seen when the aggregated blastomeres come from mouse strains that differ in their coat colors. When blastomeres from white and black strains are aggregated, the result is commonly a mouse with black and white bands (Figure 25B). If there were only one cell in the blastocyst that gave rise to the embryo, this result would not be possible; the offspring would be either all white or all black. If two cells of the blastocyst were responsible for producing the embryo, we would expect the allophenic pattern to be expressed only half the time (1WW:2WB:1BB). Should there be three embryo-producing cells, the chance that the embryo would have an allophenic pattern would then increase to 75 percent (1WWW:3WWB:3WBB:1BBB), and the four cells would give an 88 percent chance of providing two-color mice. The experimental data of Mintz (1970) is that 75 percent of the double embryos yield allophenic mice, thus suggesting that three blastomeres of the blastocyst produce the entire embryo. Markert and Petters (1978) have shown that three 4-cell embryos can unite to form a common compacted morula (Figure 26) and that the resulting mouse can have the coat colors of the three different strains. Therefore, while it is not certain that three cells is the absolute number of blastomeres that forms the embryo, we can be fairly certain that the number is not much greater and that most of the cells of the blastocyst never contribute to the adult organism.

[2]The inner cells have been found to come most frequently from the first cell to divide at the 2-cell stage. This cell usually produces the first pair of blastomeres to reach the 8-cell stage, and these cells usually divide so that they are inside the loosely aggregated cluster of blastomeres (Graham and Kelly, 1977).

(A)

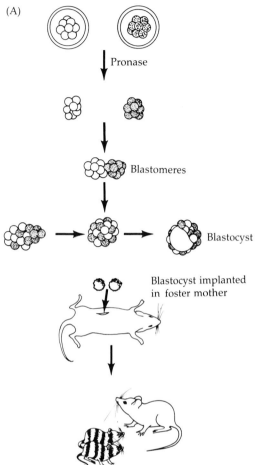

Pronase

Blastomeres

Blastocyst

Blastocyst implanted
in foster mother

(B)

FIGURE 25

Production of allophenic mice. (A) The experimental
procedures used to produce allophenic mice. Early
8-cell embryos of genetically distinct mice (here,
those with coat-color differences) are isolated from
the mouse oviducts and brought together after their
zonas are removed by proteolytic enzymes (pronase).
The cells form a composite blastocyst, which is implanted into the uterus of a foster mother. (B) An
adult allophenic mouse showing contributions from
the pigmented (black) and unpigmented (white) embryos. (Photograph courtesy of B. Mintz.)

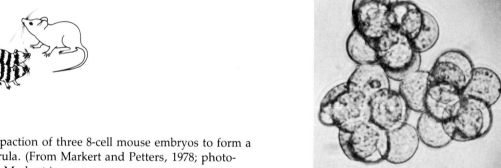

i

FIGURE 26

Aggregation and compaction of three 8-cell mouse embryos to form a
single compacted morula. (From Markert and Petters, 1978; photographs courtesy of C. Markert.)

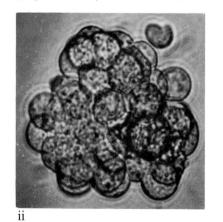

ii

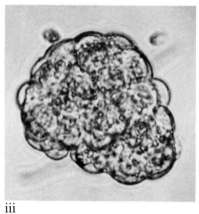

iii

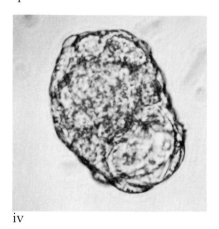

iv

Twins

Human twins are classified into two major groups: monozygotic (one-egg; identical) twins and dizygotic (two-egg; fraternal) twins. It has long been thought that fraternal twins were the result of two separate fertilization events whereas identical twins were formed from a common embryo whose cells somehow dissociated from one another. This would mean that an isolated mammalian blastomere could give rise to an entire embryo. In 1952, Seidel supported this notion by destroying one cell of a 2-cell rabbit embryo. The resulting blastomere was able to develop into a complete adult. Even one remaining blastomere of an 8-cell embryo was reported to develop successfully. Kelly (1977) has shown that each blastomere of the 4- and 8-cell embryos can give rise to all of the cell types of the embryo (Figure 27). She isolated each of the four blastomeres of a mouse embryo and allowed them to divide once in culture before surrounding them with blastomeres from a different mouse strain. These allophenic composites produced blastocysts, which were implanted into foster mothers. The inner blastomeres contributed to every tissue of the resulting mouse. Similarly, Gardiner and Rossant (1976) have shown that if ICM cells (but not trophoblast cells) are injected into blastocysts they contribute to the new embryo. Thus, it is possible that identical twins can be produced by the separation of early blastomeres or even by the separation of the inner cell mass into two regions within the same blastocyst.

This seems to be exactly what is happening in roughly 0.25 percent of human births. About 33 percent of identical twins have two complete and separate placentas, indicating that separation occurred before the formation of the trophoblast tissue. The remaining identical twins share a common placenta, suggesting that the split occurs within the inner cell mass after the trophoblast has formed.

This ability to produce an entire embryo from cells that would have normally contributed to only a portion of it is called REGULATION. We shall discuss this phenomenon in greater detail in Chapter 7. Regulation is also seen in the ability of two or more early embryos to form one allophenic mouse rather than twins, triplets, or a multiheaded monster. There is even evidence (Chapelle et al., 1974; Mayr et al., 1979) that allophenic regulation can occur in humans. These allophenic in-

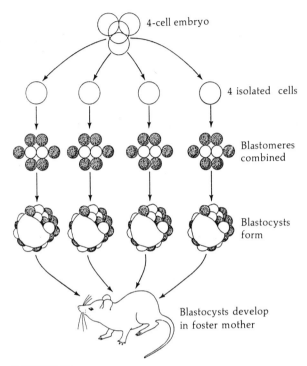

FIGURE 27

Protocol demonstrating the ability of each cell of a 4-cell mouse embryo to become a mouse. After removal of the zona, each of the four cells is isolated and allowed to divide once. The cells are then surrounded by blastomeres from another strain of mouse (differing in coat color and other markers) and cultured to the blastocyst stage. The blastocysts are transferred to a pseudopregnant foster mother. (After Kelly, 1977.)

dividuals have two genetically different cell types (XX and XY) within the same body, each with its own set of genetically defined characteristics. The simplest explanation for the existence of such a phenomenon is that these individuals resulted from the aggregation of two embryos, one male, and one female, which were developing at the same time. If this explanation is correct, two fraternal twins fused to create a single composite individual.

Meroblastic cleavage types

As mentioned earlier in this chapter, yolk concentration plays an important role in cell cleavage. Nowhere is this more apparent than in the meroblastic cleavage types. Here, the large concentrations of yolk prohibit cleavage in all but a small portion of the egg cytoplasm. In DIS-COIDAL cleavage, cell division is limited to a small disc of yolk-free cytoplasm atop a mound of yolk: in SUPERFICIAL cleavage, the centrally located yolk permits cleavage only along the peripheral rim of the oocyte.

Discoidal cleavage

Discoidal cleavage is characteristic of fishes, birds, and reptiles. Figure 28 shows the cleavage of the avian egg. The bulk of the oocyte is taken over by the yolk, allowing cleavage to occur only in the BLASTODISC at the animal pole of the egg. Because these cleavages do not extend into the yolky cytoplasm, the early cleavage-stage cells are actually continuous at their bases. After a single-layered BLASTODERM is formed, equatorial cleavages divide this layer into a tissue 3–4 cell layers thick. Between the blastoderm and the yolk is a space called the SUBGERMINAL CAVITY. At this stage, two distinct regions of the blastodisc can be identified: the AREA PELLUCIDA, composed of the cells above the subgerminal cavity; and the AREA OPACA, consisting of the darker cells at the margin of the blastodisc and yolk.

By the time the hen has laid the egg, the blastoderm contains some 60,000 cells. Some of these cells are shed into the subgerminal cavity to form a second layer (Figure 29). Thus, soon after laying, the chick egg contains two layers of cells: the upper EPIBLAST and the lower HYPO-BLAST. Between them lies the blastocoel.

The yolky eggs of fishes develop similarly, cell division occurring only in the animal pole blastodisc. Scanning electron micrographs (Figure 30) of fish egg cleavage show beautifully the incomplete nature of discoidal cleavage.

FIGURE 28
Discoidal cleavage in a chick egg seen from the animal pole. The cleavage furrows do not penetrate the yolk, and a blastoderm consisting of a single layer of cells is produced.

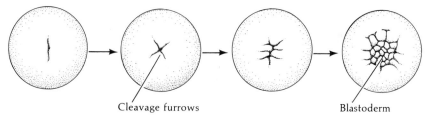

Cleavage furrows   Blastoderm

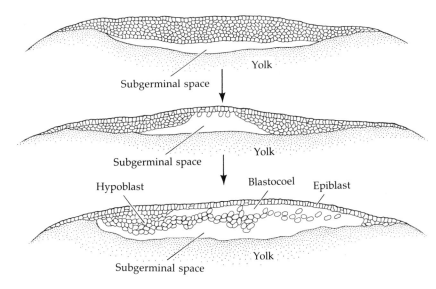

FIGURE 29
Formation of the hypoblast in the avian egg. (After Torrey, 1962.)

FIGURE 30
Discoidal cleavage in a zebra fish, creating a cellular region above the dense yolk. BD signifies the blastodisc region. (From Beams and Kessel, 1976; photographs courtesy of the authors.)

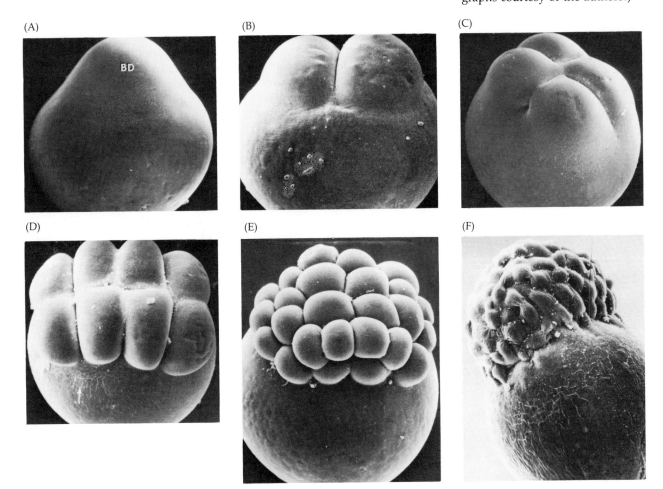

Superficial cleavage

Insect eggs undergo superficial cleavage, wherein a large mass of centrally located yolk confines cleavage to the cytoplasmic rim of the egg. One of the fascinating details of this cleavage type is that the cells do not form until after the nuclei have divided. Cleavage of an insect egg is shown in Figure 31. The zygote nucleus undergoes several mitotic divisions within the central portion of the egg. In *Drosophila*, 256 nuclei are produced. The nuclei, called ENERGIDS, then migrate to the periphery of the egg where the mitoses continue at a prodigious rate. The embryo is now called a SYNCYTIAL BLASTODERM, meaning that all the cleavage nuclei are contained within a common cytoplasm. No cell membranes exist other that those of the egg itself. Those nuclei migrating to the posterior pole of the egg soon become enveloped by new cell membranes to form the pole cells of the embryo. These cells give rise to the germ cells of the adult. Thus, one of the first events of insect development is to separate the future germ cells from the rest of the embryo.

After the pole cells have been formed, the oocyte membrane folds inward between the nuclei, eventually partitioning off each nucleus into a single cell (Figure 32). This creates the CELLULAR BLASTODERM, with all the cells arranged in a single-layered jacket around the yolky core of the egg. In *Drosophila* this layer consists of some 5000 cells and is formed within 3 hours of fertilization. These cells are not scattered equally or

FIGURE 31

Superficial cleavage in a *Drosophila* embryo. The numeral above each embryo corresponds to the number of minutes after deposition of the egg; the numeral at the bottom indicates the number of nuclei (energids) present. Pole cells (which will form the germ cells) are seen at the 512-nuclei stage even though the cellular blastoderm does not form until nearly 3 hours later. (Modified from Zalokar and Erk, 1976.)

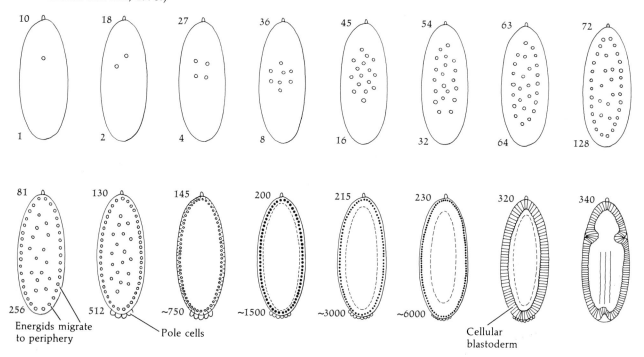

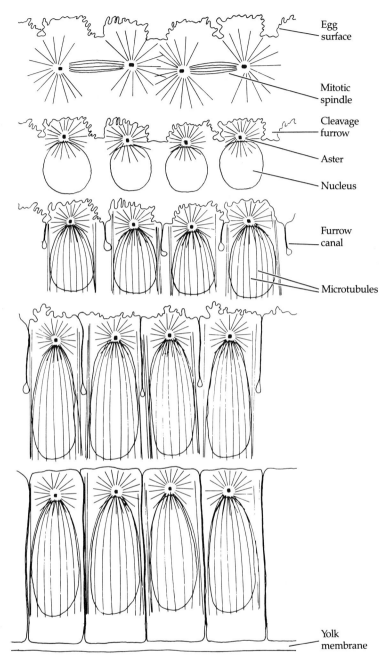

FIGURE 32
Schematic diagram of nuclear elongation and cellularization in *Drosophila* blastoderm. (After Fullilove and Jacobson, 1971.)

Egg surface

Mitotic spindle

Cleavage furrow

Aster

Nucleus

Furrow canal

Microtubules

Yolk membrane

randomly throughout the blastoderm. Instead, there is a concentration of cells along the ventral surface, marking the place where the germ layers will form. This region is called the GERM ANLAGEN and will give rise to all the cells of the insect embryo. The other blastoderm cells become the extraembryonic membranes of the developing insect.

TABLE 2
Karyokinesis and cytokinesis

Process	Mechanical agent	Major protein composition	Location	Major disruptive drug
Karyokinesis	Mitotic spindle	Tubulin microtubules	Central cytoplasm	Colchicine
Cytokinesis	Contractile ring	Actin microfilaments	Cortical cytoplasm	Cytochalasin B

MECHANISMS OF CLEAVAGE

Cleavage is actually the result of two coordinated processes. The first of these cyclic processes is KARYOKINESIS—the mitotic division of the nucleus. The mechanical agent of this division is the mitotic spindle with its MICROTUBULES composed of tubulin (the same type of protein that makes the sperm flagellum). The second process is CYTOKINESIS—the division of the cell. The mechanical agent of cytokinesis is the CONTRACTILE RING of MICROFILAMENTS made of actin (the same type of protein that extends the sperm acrosomal process). Table 2 presents a comparison of the systems of division. The relationship and coordination between these two systems during cleavage is diagrammed in Figure 33. Here, a sea urchin egg is seen undergoing first cleavage. The mitotic spindle and contractile ring are perpendicular to each other, and the spindle is internal to the contractile ring. The cleavage furrow eventually bisects the plane of mitosis, thereby creating two genetically equivalent blastomeres.

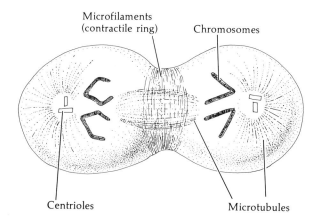

FIGURE 33
Role of microtubules and microfilaments in cell division. In the telophase cell represented here, the chromosomes are being drawn to the centrioles by microtubules while the cytoplasm pinches in through the contraction of microfilaments.

FIGURE 34
Cortical ring of microfilaments (arrows) around the second cleavage furrow of a zebra fish. (From Beams and Kessel, 1976; photograph courtesy of the authors.)

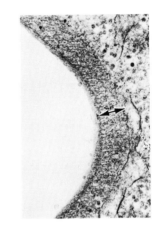

The actin microfilaments are found in the cortex of the egg rather than in the central cytoplasm. Under the electron microscope, the ring of microfilaments can be seen forming a distinct cortical band, 0.1 μm thick (Figure 34). This contractile ring exists only during cleavage and extends 8–10 μm into the center of the egg. It is responsible for exerting the force that splits the zygote into blastomeres, for if it is disrupted, cytokinesis stops. Schroeder (1973) has proposed a model of cleavage wherein the contractile ring splits the egg like an "intercellular purse-string," tightening about the egg as cleavage continues. This tightening of the microfilamentous ring creates the cleavage furrow.

Although karyokinesis and cytokinesis are usually coordinated, they are sometimes separated by natural or experimental conditions. Hiramoto (1965) has shown that cell division can take place even in the absence of the mitotic spindle necessary for nuclear division. He injected sea water or oil droplets into the center of a sea urchin zygote, destroying the mitotic spindle. Even so, the spherical zygote elongated at the usual time and eventually split into two cells (Figure 35). Harvey (1936) found that physically enucleated sea urchin eggs could keep dividing until they formed a blastulalike structure. (Presumably the cells eventually depleted their maternal supply of mRNA and proteins and could not develop further in the absence of new transcription.)

Similarly, karyokinesis can occur in the absence of cell division. One way to cause this state is to treat embryos with the drug cytochalasin B. This drug inhibits the organization of microfilaments in the contractile ring. When the cytochalasin B is added to seawater before the sea urchin zygote has formed its contractile ring, no cleavage furrow forms. When the drug is added after the beginning of the cleavage furrow, the existing contractile ring is disorganized and cleavage stops (Schroeder, 1972). Nuclear division, however, remains unaffected. Lillie (1902) and Whittaker (1979) have shown that even if cleavage is blocked, the nuclei continue to divide and to express their developmentally regulated products at the appropriate time.

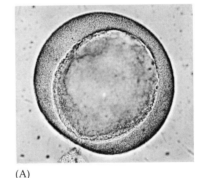

(A)

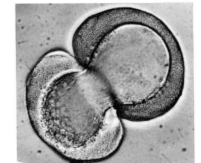

(B)

FIGURE 35
Cytokinesis in the absence of a nucleus. (A) Sea urchin zygote whose nucleus has been replaced with a droplet of seawater. (B) Cytokinesis in a sea urchin egg dividing even though the nucleus has been replaced by an oil droplet. (From Hiramoto, 1965; photographs courtesy of Y. Hiramoto.)

Nuclear division in the absence of cell division is also seen in nature. In arthropods, including insects, the nucleus of the fertilized egg divides repeatedly in the absence of cytokinesis. In *Drosophila*, for instance, hundreds of nuclei are formed within the center of the undivided oocyte cytoplasm. These nuclei then migrate to the surface and along the rim of the peripheral cytoplasm in the egg; only after migration is complete do the nuclei become separated from each other by cell membranes (Figure 29). Thus, we see that in the early stages of normal insect cleavage, cytokinesis and karyokinesis are separate in time.

One of the most intriguing unsolved problems of embryonic cleavage is how cytokinesis and karyokinesis are initiated and coordinated with each other. Although the detailed mechanism is not yet known, three types of evidence suggest that the asters of the mitotic apparatus dictate the location of the cleavage furrow. These asters (Figure 36) are the microtubular "rays" that extend from the poles of the mitotic spindle to the cell periphery. One indication that these asters might control the location of the cleavage furrows is that the number of furrows generated depends upon the number of asters present. Normal cleavage only occurs when a *pair* of asters are present. When there are no asters near the cell periphery, there is no cleavage furrow. E. B. Wilson (1901) showed that when asters are made to disappear (by adding ether to the seawater in which sea urchin embryos are growing), cleavage ceases. When the embryos are then transferred to normal seawater, the asters reappear and cleavage continues. When multiple sperm fertilize an oocyte, three or four asters can form. When there are three asters, three equidistant cleavage furrows appear, and four asters yield four cleavage furrows (see Figure 21 in Chapter 2). It thus appears that asters are the *sine qua non* of cleavage.

FIGURE 36
Diagram of microtubules extending from mitotic poles through the cell to the cell cortex during the first cleavage of a sea urchin egg. (From Wilson, 1896.)

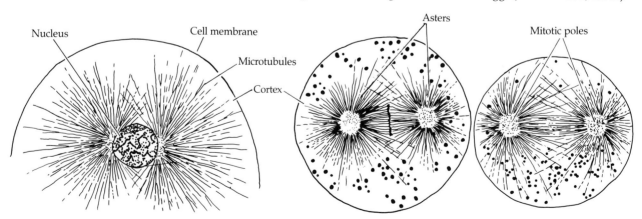

The second type of evidence linking asters with cleavage furrow formation comes from experiments wherein the direction of cleavage is changed by placing the egg under pressure. Pflüger (1884) discovered that when a frog zygote is gently compressed between two glass plates, the direction of the first three cleavages are all perpendicular to the plane of the plates. Both Driesch and Morgan (reviewed in Morgan, 1927) made similar observations with sea urchin embryos. In both cases, the plane of the third cleavage (which is normally parallel to the equator of the oocyte) was displaced by 90 degrees.

These and other experiments showed that pressure causes the elongation of the egg in the plane of the plate and that the mitotic spindles orient themselves along the long axes of the cell. The cleavage furrows always are perpendicular to the spindle, so by changing the location of the mitotic spindle, experimenters can change the location of the cleavage furrow.

Rappaport (1961) has extended this type of experiment by displacing the mitotic spindles to the sides of the cells. In Figure 37, a glass ball has been used to displace the asters from the center of the cell toward the periphery. The cleavage furrow that results extends only as far as the ball and does not appear on the other side. Thus, a binucleate, horseshoe-shaped cell is formed. At the next division, two spindle apparatuses form between four asters, but *three* cleavage furrows are generated! Each arm of the horseshoe has its own mitotic spindle and cleavage furrow as expected, but a *third* furrow appears between the two asters at the top of the arms (arrow in Figure 37C). This demonstrates clearly that if two asters are close enough together, their interactions cause the formation of a cleavage furrow even in the absence of a mitotic spindle between them. Here we see again that cell division can occur without nuclear division as long as the asters are present.

Our last general consideration of embryonic cleavage involves the formation of new cell membranes. Are these membranes newly synthesized or are they merely extensions of the oocyte cell membrane? The answer is probably that both mechanisms contribute to the internal cell membranes. In sea urchin eggs, the zygote's cell membrane is extended into numerous microvilli. During cleavage, these microvilli shrink in size, suggesting that the egg plasma membrane is being "stretched out" as the furrows are formed. However, the surface area of the egg cell membrane is not sufficient to account for all the membrane present toward the end of cleavage. Thus, new membrane is probably synthesized as well.

Evidence that new membrane components may be synthesized during the early cleavage of some species comes from studies done on amphibian embryos with pigmented cortical regions. Figure 38 shows the first cleavage furrow of a pigmented frog zygote. Whereas the original membrane has a pigmented cortical region associated with it, the new membrane is white. This new membrane also has electrical conductance properties different from the original membrane. This evidence, though not conclusive, suggests that in frogs the blastomere

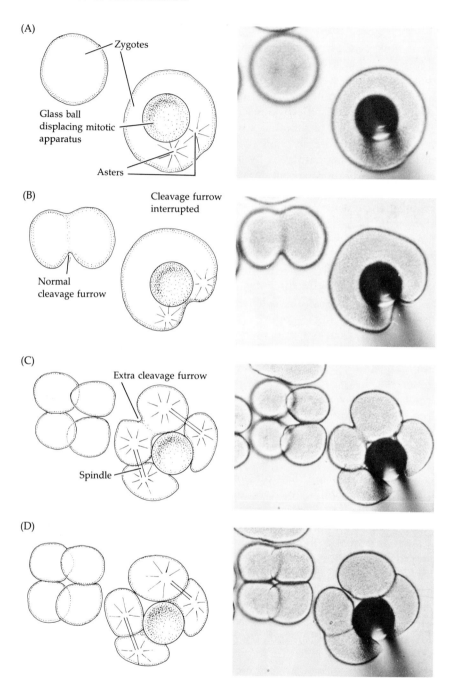

FIGURE 37
Creation of a new cleavage furrow by displacement of the asters. (A, B) By interrupting the cleavage furrow with a glass ball, a horseshoe-shaped cell is created. At the next division (C, D), a new cleavage furrow is created (arrow) even though the chromosomes are not divided between it. (From Rappaport, 1961; photographs courtesy of R. Rappaport.)

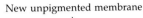

New unpigmented membrane

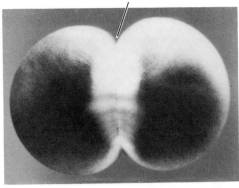

FIGURE 38
Formation of new membranes in the first cleavage of a *Xenopus* egg. Old membrane has pigment granules associated with it. The new membrane appears clear as it does not have these associated granules. (From de Laat and Bluemink, 1974; photograph courtesy of S.W. de Laat.)

membranes may be predominantly newly synthesized, even at the first cleavage division. Although a great deal has been learned about synthesis of new membranes, how such new membranes become connected with the older plasma membranes during furrow formation remains largely unknown. (For a review, see Laat and Bluemink, 1977.)

SUMMARY

Cleavage is the collection of processes by which the fertilized egg is converted into a multicellular structure. The specific type of cleavage depends upon the evolutionary history of the species and on the ability of the egg to provide the nutritional requirements of the embryo. Some animals need large quantities of yolk to support development to the time when feeding begins, whereas other animals form a feeding larval stage and do not require large amounts of yolk in the egg cytoplasm. In mammals, a peculiar type of cleavage forms the beginnings of the placenta, which will supply nutrients to the embryo. Yet we have seen that underneath the incredible diversity of cleavage types, there is an underlying unity of function and mechanism. In all cases, karyokinesis and cytokinesis must be coordinated and the egg divided into cellular regions. The nucleus:cytoplasm ratio is restored, and developmentally important information is sequestered in various cell regions. In the next chapter we shall see how these cleavage-stage blastomeres move about and interact with each other to lay down the framework of the body.

Literature Cited

Balinsky, B. I. (1981). *Introduction to Embryology,* Fifth Edition. Saunders, Philadelphia.

Beams, H. W. and Kessel, R. G. (1976). Cytokinesis: A comparative study of cytoplasmic division in animal cells. *Am. Sci.* 64: 279–290.

Biggelaar, J. A. M. van der and Guerrier, P. (1979). Dorsoventral polarity and mesentoblast determination as concomitant results of cellular interactions in the mollusk *Patella vulgata. Dev. Biol.* 68: 462–471.

Boycott, A. E., Diver, C., Garstang, S. L. and Turner, F. M. (1930). The inheritance of sinestrality in *Limnaea peregra* (Mollusca: Pulmonata). *Philos. Trans. R. Soc. Lon.* [B] 219: 51–131.

Carlson, B. M. (1981). *Patten's Foundations of Embryology.* McGraw-Hill, New York.

Chappelle, A. de la, Schroder, J., Rantanen, P., Thomasson, B., Niemi, M., Tiilikainen, A., Sanger, R. and Robson, E. E. (1974). Early fusion of two human embryos? *Ann. Hum. Genet.* 38: 63–75.

Crampton, H. E. (1894). Reversal of cleavage in a sinistral gastropod. *Ann. N.Y. Acad. Sci.* 8: 167–170.

Dan, K. (1960). Cytoembryology of echinoderms and amphibia. *Int. Rev. Cytol.* 9: 321–367.

Dan-Sohkawa, M. and Fujisawa, H. (1980). Cell dynamics of the blastulation process in the starfish, *Asterina pectinifera. Dev. Biol.* 77: 328–339.

Dulcibella, T. (1977). Surface changes of the developing trophoblast cell. In M. H. Johnson (ed.), *Development in Mammals 1.* Elsevier North-Holland, New York, pp. 5–30.

Fullilove, S. L. and Jacobson, A. G. (1971). Nuclear elongation and cytokinesis in *Drosophila montana. Dev. Biol.* 26: 560–577.

Gardiner, R. C. and Rossant, J. (1976). Determination during embryogenesis in mammals. *Ciba Found. Symp.* 40: 5–18.

Giudice, A. (1973). *Developmental Biology of the Sea Urchin Embryo.* Academic, New York.

Graham, C. F. and Kelly, S. J. (1977). Interactions between embryonic cells during early development of the mouse. In M. Karkinen-Jääskeläinen, L. Saxén and L. Weiss (eds.), *Cell Interactions in Differentiation.* Academic, New York, pp. 45–57.

Gulyas, B. J. (1975). A reexamination of the cleavage patterns in eutherian mammalian eggs: Rotation of the blastomere pairs during second cleavage in the rabbit. *J. Exp. Zool.* 193: 235–248.

Hara, K. (1977). The cleavage pattern of the axolotl egg studied by cinematography and cell counting. *Wilhelm Roux Arch. Entwicklungsmech. Org.* 181: 73–87.

Harvey, E. G. (1936). Parthenogenetic merogony or cleavage without nuclei in *Arbacia punctulata. Biol. Bull.* 71: 101–121.

Hillman, N., Sherman, H. I. and Graham, C. F. (1972). The effects of spatial arrangement on cell determination during mouse development. *J. Embryol. Exp. Morphol.* 28: 263–278.

Hiramoto, Y. (1965). Further studies on cell division without mitotic apparatus in sea urchin eggs. *J. Cell Biol.* 25: 161–167.

Hörstadius, S. (1939). The mechanics of sea urchin development, studied by operative methods. *Biol. Rev.* 14: 132–179.

Hörstadius, S. (1973). *Experimental Embryology of Echinoderms.* Clarendon Press, Oxford.

Hyafil, F., Babinet, C. and Jacob, F. (1981). Cell–cell interactions in early embryogenesis: A molecular approach to the role of calcium. *Cell* 26: 447–454.

Ikeda, M. (1965). Behaviour of sulfhydryl groups of sea urchin eggs under the blockage of cell division by UV and heat shock. *Exp. Cell Res.* 40: 282–291.

Johnson, M. H., Chakraborty, J., Handyside, A. H., Willison, K. and Stern, P. (1979). The effect of prolonged decompaction on the development of the preimplantation mouse embryo. *J. Embryol. Exp. Morphol.* 54: 241–261.

Kalt, M. R. (1971). The relationship between cleavage and blastocoel formation in *Xenopus laevis.* I. Light microscopic observations. *J. Embryol. Exp. Morphol.* 26: 37–49.

Kelly, S. J. (1977). Studies of the developmental potential of 4- and 8-cell stage mouse blastomeres. *J. Exp. Zool.* 200: 365–376.

Kemler, R., Babinet, C., Eisen, H., and Jacob, F. (1977). Surface antigen in early differentiation. *Proc. Natl. Acad. Sci. USA* 74: 4449–4452.

Laat, S. W. de and Bluemink, J. G. (1974). New membrane formation during cytokinesis in normal and cytochalasin B-treated eggs of *Xenopus laevis.* II. Electrophysical observations. *J. Cell Biol.* 60: 529–540.

Laat, S. W. de and Bluemink, J. G. (1977). Plasma membrane assembly as related to cell division. In G. Poste and G. L. Nicolson (eds.), *The Synthesis, Assembly, and Turnover of Cell Surface Components.* Elsevier North-Holland, New York, pp. 403–461.

Laat, S. W. de, Tertoelen, L. G. J., Dorresteijn, A. W. C. and Biggelaar, J. A. M. van der (1980). Intercellular communication patterns are involved in cell determination in early muscular development. *Nature* 287: 546–548.

Lillie, F. R. (1902). Differentiation without cleavage in the egg of the annelid *Chaetopterus pergamentaceous*. *Wilhelm Roux Arch. Entwicklungsmech. Org.* 14: 477–499.

Lutz, B. (1947). Trends towards non-aquatic and direct development in frogs. *Copeia* 4: 242–252.

Markert, C. L. and Petters, R. M. (1978). Manufactured hexaparental mice show that adults are derived from three embryonic cells. *Science* 202: 56–58.

Mayr, W. R., Pausch, V. and Schnedl, W. (1979). Human chimaera detectable only by investigation of her progeny. *Nature* 277: 210–211.

Mintz, B. (1970). Clonal expression in allophenic mice. *Symp. Int. Soc. Cell Biol.* 9: 15.

Morgan, T. H. (1927). *Experimental Embryology.* Columbia University Press, New York.

Mulnard, J. G. (1967). Analyse microcinématographique du développement de l'oeuf de souris du stade II au blastocyste. *Arch. Biol. (Liege)* 78: 107–138.

Newport, J. and Kirschner, M. (1982). A major developmental transformation in early *Xenopus* embryos. II. Control of the onset of transcription. *Cell* 30: 687–696.

Nieuwkoop, P. D. (1973). The "organization center" of the amphibian embryo. Its origin, spatial organization, and morphogenetic action. *Adv. Morphog.* 10: 1–39.

Peyrieras, N., Hyafil, F., Louvard, D., Ploegh, H. L. and Jacob, F. (1983). Uvomorulin: A non-integral membrane protein of early mouse embryo. *Proc. Natl. Acad. Sci. USA* 80: 6274–6277.

Pflüger, E. (1884). Über die Einwirkung der Schwerkraft und anderer Bedingungen auf die Richtung der Zeiltheilung, *Arch. Ges. Physiol.* 3: 4.

Pratt, H. P. M., Ziomek, C. A., Reeve, W. J. D. and Johnson, M. H. (1982). Compaction of the mouse embryo: An analysis of its components. *J. Embryol. Exp. Morphol.* 70: 113–132.

Rappaport, R. (1961). Experiments concerning cleavage stimulus in sand dollar eggs. *J. Exp. Zool.* 148: 81–89.

Rugh, R. (1967). *The Mouse.* Burgess, Minneapolis.

Sakai, H. (1966). Studies on sulfhydryl groups during cell division of sea urchin eggs. VIII. Some properties of mitotic apparatus proteins. *Biochim. Biophys. Acta* 112: 132–145.

Sakai, H. (1968). Contractile properties of protein threads from sea urchin eggs in relation to cell division. *Int. Rev. Cytol.* 23: 89–112.

Saunders, J. W., Jr. (1982). *Developmental Biology.* Mac-millan, New York.

Schroeder, T. E. (1972). The contractile ring. II. Determining its brief existence, volumetric changes, and vital role in cleaving *Arbacia* eggs. *J. Cell Biol.* 53: 419–434.

Schroeder, T. E. (1973). Cell constriction: Contractile role of microfilaments in division and development. *Am. Zool.* 13: 687–696.

Seidel, F. (1952). Die Entwicklungspotenzen einer isolierten Blastomere des Zweizellenstadiums im Säugetierei. *Naturwissenschaften* 39: 355–356.

Sturtevant, M. H. (1923). Inheritance of direction of coiling in *Limnaea*. *Science* 58: 269–270.

Sutherland, A. E. and Calarco-Gillam, P. G. (1983). Analysis of compaction in the preimplantation mouse embryo. *Dev. Biol.* 100: 327–338.

Sze, L. C. (1953). Changes in the amount of deoxyribonucleic acid in the development of *Rana pipiens*. *J. Exp. Zool.* 122: 577–601.

Tarkowski, A. K. and Wróblewska, J. (1967). Development of blastomeres of mouse eggs isolated at the 4- and 8-cell stage. *J. Embryol. Exp. Morphol.* 18: 155–180.

Torrey, T. (1962). *Morphogenesis of the Vertebrates.* Wiley, New York.

Tuchmann-Duplessis, H., David, G., and Haegel, P. (1972). *Illustrated Human Embryology. Vol. I.* Springer-Verlag, New York.

Whittaker, J. R. (1979). Cytoplasmic determinants of tissue differentiation in the ascidian egg. In S. Subtelny and I. R. Konigsberg (eds.), *Determinants of Spatial Organization*. Academic, New York, pp. 29–51.

Wilson, E. B. (1896). *The Cell in Development and Inheritance.* Macmillan, New York.

Wilson, E. B. (1898). Cell lineage and ancestral reminiscence. *Biol. Lect. Woods Hole*, pp. 21–42.

Wilson, E. B. (1901). Experiments in cytology. II. Some phenomena of fertilization and cell division in etherized eggs. III. The effect on cleavage of artificial obliteration of the first cleavage furrow. *Wilhelm Roux Arch. Entwicklungsmech. Org.* 13: 353–395.

Wolpert, L. and Gustafson, T. (1961). Studies in the cellular basis of morphogenesis of the sea urchin embryo: Development of the skeletal pattern. *Exp. Cell Res.* 25: 311–325.

Wolpert, L. and Mercer, E. H. (1963). An electron microscope study of the development of the blastula of the sea urchin embryo and its radial polarity. *Exp. Cell Res.* 30: 280–300.

Zalokar, M. and Erk, I. (1976). Division and migration of nuclei during early embryogenesis of *Drosophila melanogaster*. *J. Microsc. Biol.* 25: 97–106.

CHAPTER 4

Gastrulation: reorganizing the embryonic cells

*My dear fellow . . . life is infinitely stranger
than anything which the mind of man could invent.
We would not dare to conceive the things which are
really mere commonplaces of existence.*
—A. CONAN DOYLE (1891)

*I would on first setting out, inform the reader
that there is a much greater number of miracles
and natural secrets in the frog than anyone hath
ever before thought of or discovered.*
—J. SWAMMERDAM (1737)

Introduction: General features of gastrulation

Gastrulation is the process of highly integrated cell and tissue migrations whereby the cells of the blastula are drastically rearranged. The blastula consists of numerous cells, the positions of which were established during cleavage. During gastrulation, these cells are given new positions and new neighbors, and the multilayered body plan of the organism is established. The cells that will form the endodermal and mesodermal organs are brought inside the embryo while the cells capable of forming the skin and nervous system are extended over the outside surface. Thus, the three germ layers—outer ectoderm, inner endoderm, and interstitial mesoderm—are first produced during gastrulation. In addition, the stage is set for the interactions of these newly positioned tissues.

The movements of gastrulation involve the entire embryo, and cell migrations in one part of the gastrulating organism must be intimately coordinated with other movements occurring simultaneously. Whereas the patterns of gastrulation vary enormously throughout the animal kingdom, there are relatively few mechanisms involved. Gastrulation usually involves combinations of the following types of movements.

- *Epiboly*. The movement of epithelial sheets (usually of ectodermal cells), which spread as a unit, rather than individually, to enclose the deeper layers of the embryo.

- *Invagination*. The infolding of a region of cells, much like the indenting of a soft rubber ball when poked.

- *Involution*. The inturning of an expanding outer layer so that it spreads over the internal surface of the remaining external cells.

- *Ingression*. The migration of individual cells from the surface layers into the interior of the embryo.

- *Delamination*. The splitting of one cellular sheet into two more or less parallel sheets.

Gastrulation affects the entire embryo, providing the embryologist with a spectacle of orchestrated cell movement second to none in animal development. It also provides the opportunity to analyze the mechanisms by which the various embryonic regions are interrelated. As we look at the phenomena of gastrulation in different types of embryos, we should keep in mind the following questions (Trinkaus, 1969).

1. *What is the unit of migration activity?* Is the migration dependent on the movement of individual cells, or are the cells part of a migrating sheet? Remarkable as it first seems, regional migrational properties may be totally controlled by cytoplasmic factors independent of cellulariza-

tion. F. R. Lillie (1902) was able to parthenogenetically activate eggs of the annelid *Chaetopterus* and suppress their cleavage. Many events of early development occurred even in the absence of cells. The cytoplasm of the zygote separated into defined regions, and cilia differentiated in the appropriate parts of the egg. Moreover, the outermost clear cytoplasm migrated down over the vegetal regions in a manner specifically reminiscent of the epiboly of animal hemisphere cells during normal development. This occurred at precisely the time that epiboly would have taken place during gastrulation. Thus, epiboly may be (at least in some respects) independent of the cells that form that migrating region.

2. *Is the spreading or folding of a cell sheet due to factors within the sheet or to external forces stretching or distorting it?* This is essential to know if we are to understand how the various cell movements of gastrulation are integrated. For instance, do involuting cells pull the epibolizing cells down toward them or are the two movements separate?

3. *Is there active spreading of the whole tissue or does the leading edge expand and drag the rest of a cell sheet passively along?*

4. *Are changes in cell shape and motility during gastrulation the consequence of changes in cell surface properties, such as adhesiveness to the substrate or to other cells?*

5. *How are these changes manifest on the subcellular level?* Are rearrangements in the arrays of microtubules or microfilaments necessary for cells to migrate?

Keeping these questions in mind, we shall proceed to observe the various patterns of gastrulation found in echinoderms, amphibians, birds, and mammals.

Sea urchin gastrulation

The sea urchin blastula consists of a single layer of 1000 to 2000 cells. These cells, derived from different regions of the zygote, have different sizes and properties. Figure 1 shows the fates of the various regions of the zygote as it develops through cleavage and gastrulation to the PLUTEUS LARVA[1] characteristic of sea urchins. Here, the fate of each cell layer can be seen through their movements during gastrulation.

Ingression of primary mesenchyme

Approximately 24 hours after the blastula hatches from its fertilization membrane, the vegetal side of the spherical blastula begins to flatten. At the center of this flat VEGETAL PLATE, small cells begin to show pulsating movements on their inner surfaces, extending and contracting long, thin (30 μm x 5 μm) processes called FILOPODIA. These cells then

---

[1]*Pluteus* is the Latin word for "easel" or "shed." Indeed, these larvae do not look unlike an A-frame house.

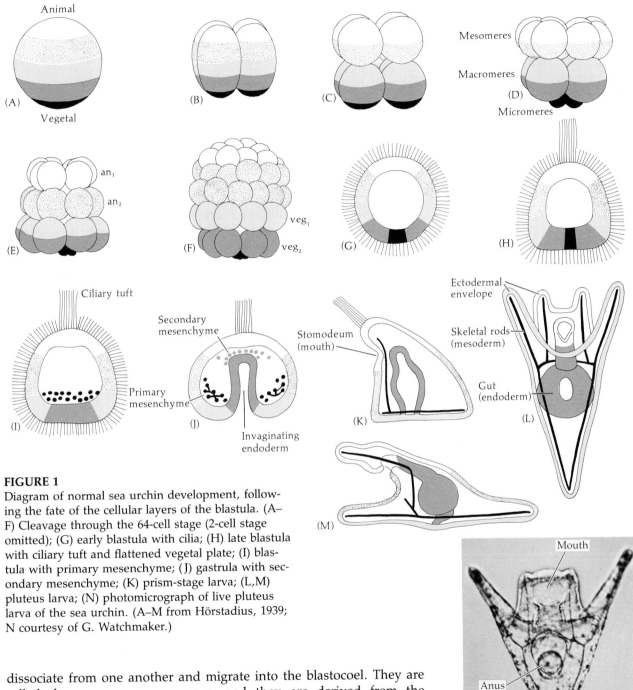

FIGURE 1

Diagram of normal sea urchin development, following the fate of the cellular layers of the blastula. (A–F) Cleavage through the 64-cell stage (2-cell stage omitted); (G) early blastula with cilia; (H) late blastula with ciliary tuft and flattened vegetal plate; (I) blastula with primary mesenchyme; (J) gastrula with secondary mesenchyme; (K) prism-stage larva; (L,M) pluteus larva; (N) photomicrograph of live pluteus larva of the sea urchin. (A–M from Hörstadius, 1939; N courtesy of G. Watchmaker.)

dissociate from one another and migrate into the blastocoel. They are called the PRIMARY MESENCHYME, and they are derived from the micromeres (Figure 2).

Gustafson and Wolpert (1961) have used time-lapse films to follow the microscopic movements of these cells within the blastocoel. At first the mesenchymal cells appear to move randomly along the inner blas-

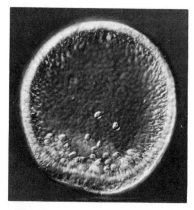

FIGURE 2
Mesenchyme blastula of the sea urchin *S. purpuratus*. (Courtesy of M. A. Harkey.)

tocoel surface, actively making and breaking filopodial connections to the wall of the blastocoel. Eventually, these cells become localized within the ventral region of the blastocoel, where it is believed their adherence is strongest. Here the primary mesenchymal cells will fuse into a SYNCYTIAL CABLE (later referred to as a skeletal rod), which will form the axis of the calcium carbonate skeleton of the pluteus larva (Figure 3).

Both cytoplasmic and cell-surface events are crucial to the ingression and migration of the primary mesenchyme cells. Gibbins and co-workers (1969) have shown that the orientation of microtubules is extremely important in the formation and migration of the primary mesenchyme. Early blastula cells have their microtubules asymmetrically distributed, paralleling the long axis of the cell (Figure 4). The microtubules lose this orientation and become randomly arranged as the primary mesenchyme cells take on an amoeboid shape and lose contact with the blastoderm. The microtubules enter the filopodial regions of the migrating cells and are later found in the syncytial cables.

The importance of microtubules to primary mesenchyme detach-

FIGURE 3
Formation of skeletal rods by mesenchymal cells of the sea urchin *S. purpuratus*. (A) Mesenchymal cells align together and fuse to lay down the matrix of the spicule. (B) Phase-contrast photomicrograph of isolated mesenchymal cells clustering together. (C) Same field as (B), but seen by polarized light to show spicules forming within the clusters. (From Harkey and Whiteley, 1980; photographs courtesy of the authors.)

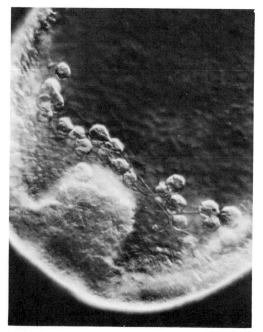

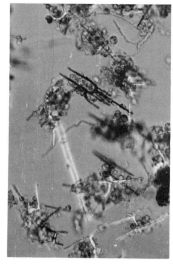

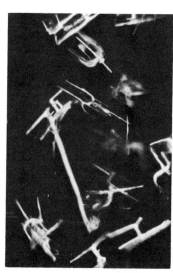

(A)   (B)   (C)

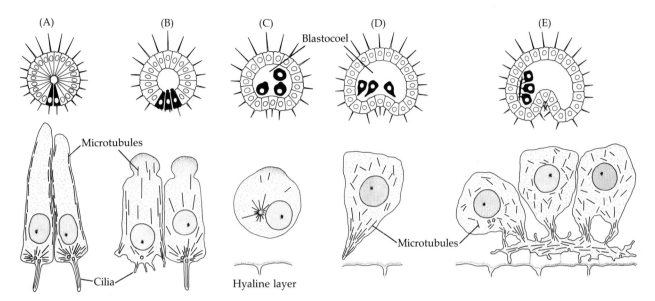

FIGURE 4
Diagrammatic representation of the ingression of primary mesenchymal cells (black; detailed in lower drawings) and the reorientation of microtubules. (A) Early blastula with microtubules parallel to long axes and cilia produced at the periphery. (B) Late blastula with cilia being withdrawn and the rounding up of cell shape. (C) Mesenchyme blastula where micromeres detach from hyaline layer and enter blastocoel. (D) Early gastrula with mobile cells. (E) Formation of syncytium prior to depositing skeletel matrix. (After Gibbins et al., 1969.)

ment was shown by various treatments that disrupt microtubule assembly or disassembly. When colchicine or hydrostatic pressure disrupts the microtubular arrays of early gastrula cells, the mesenchymal cells lose their filopodial extensions and all migration ceases (Tilney and Gibbins, 1969). Moreover, deuterium oxide (an isotopic form of water) was found to stabilize the microtubules, fixing them in place. This, too, stops mesenchymal migration as microtubules cannot depolymerize and form again. Thus, the internal architecture of the cell is extremely important for the development and migration of the primary mesenchyme.

The cell surface is also crucial for the proper migration of mesenchymal cells. Sugiyama (1972) found that sulfated glycoproteins develop on the cell surface of the vegetal plate cells that are destined to form the primary mesenchyme. When the synthesis (or the sulfation) of these glycoproteins is inhibited, the mesenchymal cells have altered cell surfaces and fail to migrate once they have entered the blastocoel (Karp and Solursh, 1974; Figure 5). Lau and Lennarz (1982) have shown that new mRNA molecules for glycoproteins are translated at the beginning of gastrulation.

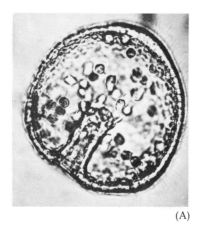

(A)

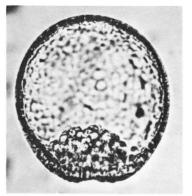

(B)

FIGURE 5
Effect of sulfate deprivation on primary mesenchyme movement in the sea urchin *Lytechinus*. (A) Normal gastrula. (B) Abnormal gastrula formed when embryos are grown in sulfate-free seawater. (From Karp and Solursh, 1974; photographs courtesy of M. Solursh.)

First stage of archenteron invagination

As the ring of primary mesenchymal cells forms in the vegetal region of the blastocoel, important changes are occurring in the cells that remain at the vegetal plate. These cells remain bound to each other and to the hyaline layer of the egg and move to fill the gaps caused by the ingression of the mesenchyme; therefore, the vegetal plate flattens further. One next sees that the vegetal plate bends inward and extends about one-third of the way into the blastocoel (Figure 6). Then invagination suddenly ceases. The invaginated region is called the ARCHEN-TERON (primitive gut), and the opening of the archenteron at the vegetal region is called the BLASTOPORE.

The first stage of vegetal plate invagination appears to be due to the forces within the vegetal plate itself and has nothing to do with pressure from the cells above them. In 1939, Moore and Burt cut echinoderm embryos into animal and vegetal halves. The vegetal halves began to invaginate even in the absence of the animal hemisphere! The invagination was only one-fourth to one-third the distance to the animal pole, about the same as in the first stage of normal invagination (Figure 7). As the folding deepened, the rims of the vegetal cells rolled upward and fused, forming a miniature gastrula.

Moore's suggestion (1941) that differences in cell cohesiveness were important in causing this invagination received support from the cinemicrophotography of Gustafson and Wolpert (1961). Their films showed that the columnar cells of the vegetal plate lose their contact with adjacent cells at their inner borders. The inner parts of each cell round up while the outer borders remain in contact with other cells and with the hyaline layer. Gustafson and Wolpert have shown that this change

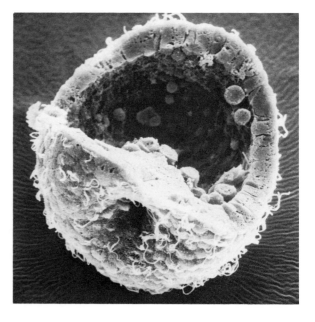

FIGURE 6
Invagination of the vegetal plate in sea urchin gastrulation. Scanning elecron micrograph of *S. purpuratus* early gastrula shows the invagination of the vegetal plate to form the archenteron and shows the primary mesenchymal cells migrating inside the blastocoel. (Photograph courtesy of W. J. Humphreys.)

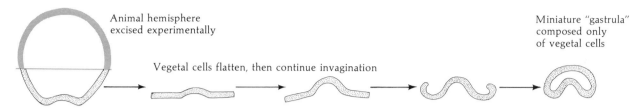

Animal hemisphere
excised experimentally

Miniature "gastrula"
composed only
of vegetal cells

Vegetal cells flatten, then continue invagination

in cell arrangement causes the cell sheet to increase its surface area. They speculate that if the vegetal plate cells were tightly connected by their contacts on their inner surfaces, the curvature would be inward rather than outward.

Second stage of archenteron invagination

The invagination of the vegetal cells occurs in two discrete stages. After a brief pause, the second phase of archenteron formation begins. The cells at the tip of the archenteron release a SECONDARY MESENCHYME that remains near the place where it is formed (Figure 8). Filopodia are

FIGURE 7
Invagination of the vegetal plate in the absence of the animal hemisphere. Vegetal regions of the starfish *Patiria* were isolated; they succeeded in invaginating without attachment of the animal hemisphere. (Modified from Moore and Burt, 1939.)

FIGURE 8
Midgastrula stage of the sea urchin *Lytechinus pictus,* showing filopodial extensions of secondary mesenchyme extending from the archenteron tip to the blastocoel wall. (A) Mesenchymal cells extending filopodia from the tip of the archenteron. (B) Filopodial cables connecting the blastocoel wall to the archenteron tip. The tension of the cables can be seen as they pull on the blastocoel wall at the point of attachment. (Photographs courtesy of C. Ettensohn.)

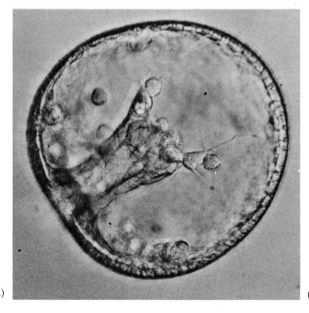

(A)

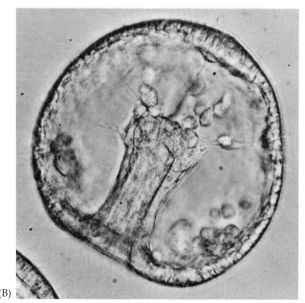

(B)

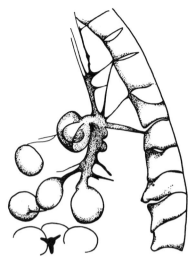

FIGURE 9
Diagram of the attachment of mesenchymal cells to the blastocoel wall by filopodia. (From Gustafson, 1963.)

extended from these cells and extend through the blastocoel fluid to contact the inner surface of the blastocoel wall. The filopodia attach to the wall at the junctions between the cells (Figure 9) and then shorten. Several lines of evidence suggest that the shortening of the filopodia actually *pulls* the archenteron up into the animal hemisphere. Dan and Okazaki (1956) demonstrated that the blastocoel wall is often pulled inward at the point where the filopodia attach and that the archenteron will be drawn closer to the side of the blastula where more filopodial attachments are made. Gustafson and Wolpert (1967) have calculated that the energy needed to move the invaginating cells is on the order of 10^{-2} dynes, well within the ability of a group of mesenchymal cells. Moreover, if filopodial activity is suppressed (by low-calcium seawater or by proteolytic enzymes), the second phase of invagination is inhibited.

As the tip of the archenteron approaches the gastrula wall in the animal hemisphere, the secondary mesenchyme cells disperse into the blastocoel, where they proliferate to form the mesodermal organs. Where the archenteron contacts the wall, a mouth is eventually formed. The mouth fuses with the archenteron to create a continuous digestive tube. Thus, as is characteristic for deuterostomes, the blastopore marks the position of the anus.

Sea urchin gastrulation combines the ingression of the mesenchymal cells, the independent invagination of the vegetal plate cells, and the further invagination of the archenteron, which is mediated by the secondary mesenchymal cells.

Amphibian gastrulation

Although amphibian gastrulation has been intensively studied for the past century, most of our theories concerning the mechanisms of these developmental movements have been revised over the past decade. Thus, the study of amphibian gastrulation is both one of the oldest and one of the newest areas of interest in experimental embryology.

Cell movements during amphibian gastrulation—an overview

Amphibian blastulae are faced with the same task as their echinoderm counterparts, namely, to bring inside those areas destined to form the endodermal organs, to surround the embryo with those cells capable of forming the ectoderm, and to place the mesodermal cells in the proper places between them. The movements whereby this is accomplished can be visualized by the technique of vital dye staining. W. Vogt (1929) saturated agar chips with dyes, such as neutral red or Nile Blue sulfate, that would stain but not damage the embryonic cells. These stained agar chips were pressed to the surface of the blastula, where

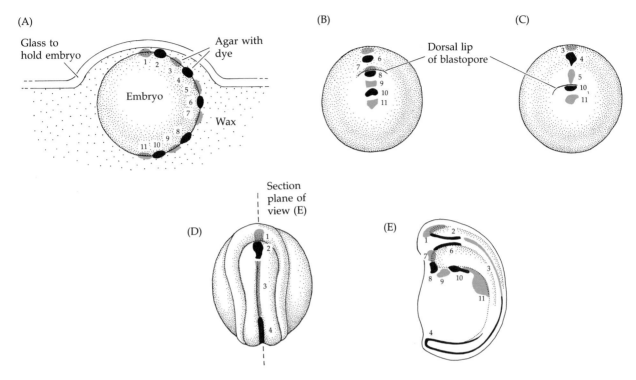

FIGURE 10
Vital dye staining of amphibian embryos. (A) Vogt's method for marking specific cells of the embryo surface with vital dyes. (B–D) Surface views of stain on successively later embryos. (E) Newt embryo dissected in medial plane to show stained cells in the interior. (After Vogt, 1929.)

some of the dye would be transferred onto the contacted cells (Figure 10). The movements of each group of stained cells were followed throughout gastrulation, and the results were summarized in FATE MAPS. Although these fate maps provided a good approximation of the outcome of each cellular region, they could only give information concerning the cells at the surface of the embryo. But the amphibian blastula has both deep *and* superficial layers, so these early fate maps were not complete, and there was a question as to what cells produced the mesoderm. Vital dye studies by Løvtrup (1975; Landström and Løvtrup, 1979) and by Keller (1976, 1978) have modified the older fate maps to show the different fates of the superficial and deep cells in certain regions (Figure 11). The most important difference involves the place-

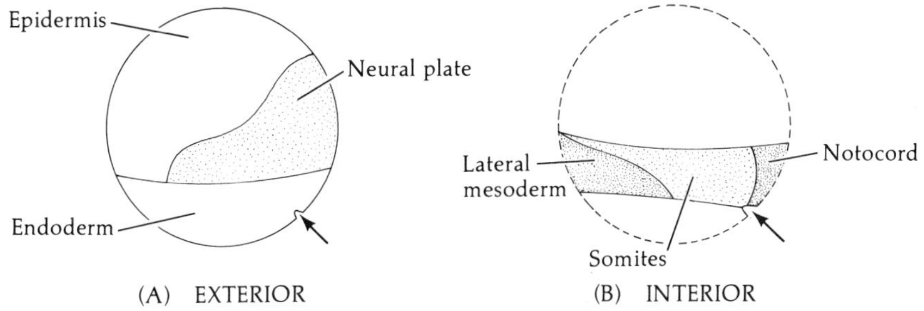

FIGURE 11
Fate map of the embryo of the frog *Xenopus laevis*. Revised fate map for the external (A) and interior (B) cells of the blastula, indicating that most of the mesodermal derivatives are formed from the interior cells. The point at which the dorsal blastopore lip forms is indicated by an arrow. (From Keller, 1975, 1976.)

Epidermis

Neural plate

Endoderm

(A)   EXTERIOR

Lateral mesoderm

Notochord

Somites

(B)   INTERIOR

ment of the mesodermal precursors. According to Vogt's fate maps, the precursors of the mesoderm and notochord exist at different locations on the embryonic surface. The revised fate maps indicate that they are both found in the equatorial region, but at different levels. Vogt himself was hesitant about assigning his positions to the mesodermal precursors (Landström and Løvtrup, 1979), and cellular isolation experiments appear to confirm the revised map, as does Keller's vital dye mapping of frog gastrulation.

Gastrulation is initiated at the future dorsal side of the embryo, just below the equator in the region of the gray crescent (Figure 12). Here the local endodermal cells invaginate to form a slitlike blastopore. These

FIGURE 12
Outline of gastrulation in a frog. (A) Blastula. (B) Gastrulation begins as cells move inward to form the dorsal lip of the blastopore. (C) Involution of cells through the dorsal lip and under the roof of the blastocoel creates the archenteron and displaces the blastocoel. (D,E) Cells involute through ventral and lateral blastopore lips as well as through the dorsal lip; the ectodermal precursors migrate over the vegetal hemisphere. The yolk plug represents the only endoderm visible on the surface. (F) Gastrulation continues until the entire embryo is surrounded by ectoderm, the endoderm has been internalized, and the mesodermal cells are brought between them. (Modified from Rugh, 1951.)

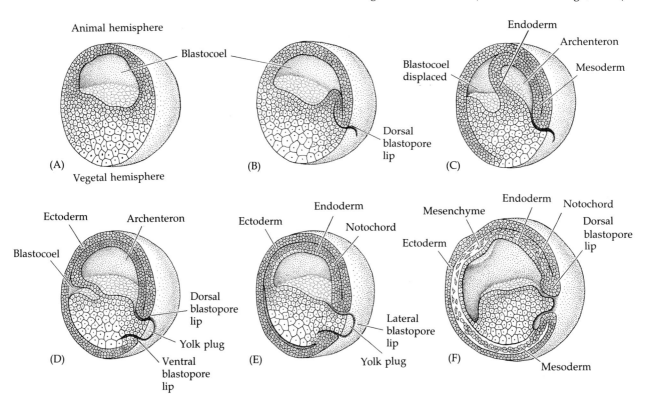

cells change their shape dramatically. The main body of the cell is displaced toward the inside of the embryo while it maintains contact with the outside surface by way of a slender cytoplasmic strand (Figure 13). These BOTTLE CELLS line the initial archenteron. Thus, as in the gastrulating sea urchin, an invagination of cells initiates archenteron formation. However, unlike the sea urchin, gastrulation does not begin at the most vegetal region, but in the MARGINAL ZONE near the equator of the blastula. Here the endodermal cells are not as large or as yolky as the most vegetal blastomeres.

The next phase of gastrulation involves the involution of the marginal zone cells while the animal cells undergo epiboly and converge at the blastopore. When the migrating cells reach the lip of the blastopore, they turn inward and travel along the inner surface of the outer cell sheets. Thus, the cells constituting the lip of the blastopore are constantly changing. The first cells to compose the dorsal lip are the endodermal cells that invaginated to form the leading edge of the archenteron. These cells later become the pharyngeal cells of the foregut. As these first cells pass into the interior of the embryo, the blastopore lip becomes composed of involuting cells that are the precursors of the head mesoderm. The next cells involuting over the dorsal lip of the blastopore are called the CHORDAMESODERM cells. These cells will form the NOTOCHORD, a transient mesodermal "backbone" that is essential for initiating the differentiation of the nervous system (see Chapters 5 and 7).

As the new cells enter the embryo, the blastocoel is displaced to the side opposite the dorsal blastopore lip. Meanwhile, the blastopore widens as the animal hemisphere cells converge at the blastopore lip. The widening blastopore develops lateral and ventral lips through which more of the mesodermal and endodermal precursor cells pass. Eventually, the lips merge to form a ring around the large endodermal cells that remain exposed on the surface (Figure 14). This remaining patch of endoderm is called the YOLK PLUG; and it, too, is eventually surrounded by the epibolizing ectoderm. At this point, all the endo-

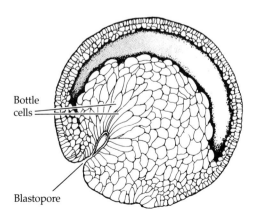

Bottle
cells

Blastopore

FIGURE 13
Diagram of cells seen in a section of gastrulating amphibian embryo, showing the extension of the bottle cells from the blastopore. (After Holtfreter, 1943.)

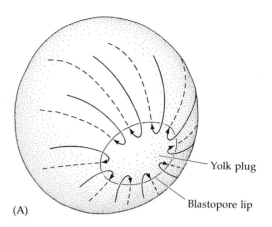

FIGURE 14
Epiboly of the ectoderm. (A) Morphogenic movements of the cells migrating into the blastopore and then under the surface. (B) Changes in the region around the blastopore, as the dorsal, lateral, and ventral lips are formed in succession. When the ventral lip completes the circle, the endoderm becomes progressively internalized. (From Balinsky, 1975.)

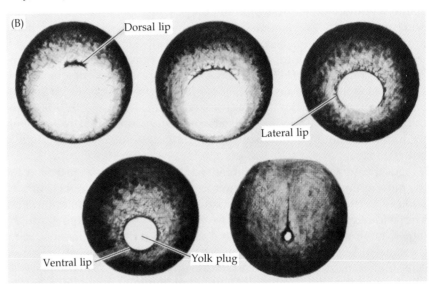

dermal precursors have been brought into the interior of the embryo, the ectoderm has encircled the surface, and the mesoderm has been brought between them.

End of the blastula stage and positioning of the blastopore

Having seen the general features of amphibian gastrulation, we can now look at each step in detail. We start at the time when the blastula prepares itself for the initiation of gastrulation. The first obvious preparatory event is the change in a cell cycle of the blastomeres. Rapid, synchronous cleavage in the amphibian egg lasts for about the first 12 divisions. Then, at midblastula, the cleavage rhythms slow down and lose synchrony. This change is caused by the acquisition of a variable interphase period between mitosis and the time DNA synthesis begins (Hara, 1977; Kirschner and Gerhart, 1981). Shortly thereafter, RNA synthesis begins for the first time since meiosis, and changes in cell motility are seen. The timing of this midblastula transition appears to be dependent on some nucleus–cytoplasm interaction. Signoret and Lefresne (1973) found that haploid embryos undergo one or more extra cleavage cycles before undergoing these midblastula changes, and polyploid embryos undergo this transition at appropriately earlier stages (Kirschner and Gerhart, 1981). The timing of this event appears to be due to a cytoplasmic factor that gets depleted as the amount of chromatin increases. The blastomeres at the late blastula stage are therefore a much more complex population than in earlier cleavage. The cells are now able to synthesize their own RNA, to make contacts with their neighboring cells, and to move within the embryo.

Shortly after this transition, the first evidence of gastrulation is seen. The blastopore forms in the gray crescent area at the margin between the animal and vegetal hemispheres. How is this initial step controlled, and what is the relationship between the gray crescent and the cells creating the dorsal blastopore lip? As we discussed earlier, the entry of the sperm at fertilization causes a radical displacement of cytoplasm within the egg. The shift of cortical cytoplasm in relation to the inner cytoplasm creates a gray crescent 180 degrees away from the point of sperm entry. This gray crescent marks the future dorsal side of the embryo, and the point of sperm entry marks the ventral surface of the embryo. Spemann showed that embryos deprived of the gray crescent region fail to form the dorsal blastopore lip and do not develop further. In 1924, Hans Spemann and Hilde Mangold demonstrated that when the cells constituting the dorsal blastopore lip are transplanted to the *ventral* margin of another salamander gastrula, gastrulation occurs from two positions: One site of gastrulation is from the host's own dorsal blastopore lip, and the other site is initiated by the transplanted lip cells. (These experiments are among the most important in embryology and will be detailed in Chapter 7.) Because the dorsal blastopore lip occupies roughly the same area as the former gray crescent, it was thought that this blastopore lip acquires its capacity to initiate gastrulation from some special cytoplasmic component of the gray crescent region.

What appeared to be conclusive proof of this hypothesis came from the cortical transplantation experiments of Curtis (1960, 1962). Curtis took gray crescent cortical cytoplasm (150 x 150 x 3 μm) from early cleavage stage embryos and grafted it onto the ventral margins of host embryos undergoing their first cleavage. All embryos receiving gray crescent grafts develop two sites of gastrulation and, eventually, two separate embryonic axes in a "Siamese twin" arrangement. Thus, it seemed logical to conclude that the gray crescent cortex contains or organizes whatever factors are required to initiate gastrulation.

124

These experiments eclipsed an alternative hypothesis holding that the factors responsible for blastopore localization are present in the *internal* cytoplasm of the egg. Born (1885) and Pasteels (1948) showed that the position of the blastopore lip could be changed by rotating the egg after the gray crescent had already formed, and some investigators (Schulze, 1894; Penners and Schleip, 1928) were able to produce embryos with two blastopores by rotating the egg in a certain direction at a certain stage in its development.

Kirschner and Gerhart (1981; Kirschner et al., 1980) have repeated and extended these experiments and have concluded that the factors responsible for the initiation of amphibian gastrulation lie in the internal cytoplasm rather than in the cortex. Although these factors usually come to lie close to the gray crescent cortex, the gray crescent and the dorsal lip of the blastopore can be dissociated experimentally. By placing eggs in a solution of 5 percent Ficoll (a large polysaccharide), they were able to collapse the perivitelline space and prevent free rotation of the eggs. Otherwise, the eggs would rotate back to their original position. When fertilized *Xenopus* eggs were taken before first cleavage and were rotated 90 degrees so that the point of sperm entry faced upward, 90 percent of the embryos developed blastopores on the *same* side as the point of sperm entrance (instead of directly opposite it, as is usual). The position of the dorsal lip of the blastopore could readily be changed until about 65 minutes after fertilization, about 20 minutes after the formation of the gray crescent (Figure 15).

Kirschner and Gerhart then reexamined Curtis' experiments. They had reason to believe that they could reproduce the results of Curtis' cortical grafting experiments by merely rotating the fertilized eggs. First,

FIGURE 15
Dissociation of dorsal blastopore lip from the point opposite sperm entry. Eggs were fertilized, dejellied, and marked with a dye immediately below the point of sperm entry. In controls, the unrotated egg forms a blastopore 180 degrees opposite the point of sperm entry. The sperm entry point is later found on the ventral surface of the tadpole. However, if eggs are rotated at 50 minutes after fertilization for 10–60 minutes, such that the sperm entry point is brought to the animal pole, the blastopore will form on the same side as sperm entry. In embryos, the sperm entry point is then found in dorsal structures of the tadpole. The site of dorsal blastopore lip is thought to be coincident with the region where animal and vegetal cytoplasms are brought together by sperm-mediated rearrangement during fertilization (see Figure 35 in Chapter 2). Rotating the egg overrides these displacements as the heavy vegetal cytoplasm slips down to the bottom of the egg, displacing other cytoplasm upwards. The vegetal and animal cytoplasms now interact at a new place, causing the blastopore lip to form at that site. (After Kirschner and Gerhart, 1981.)

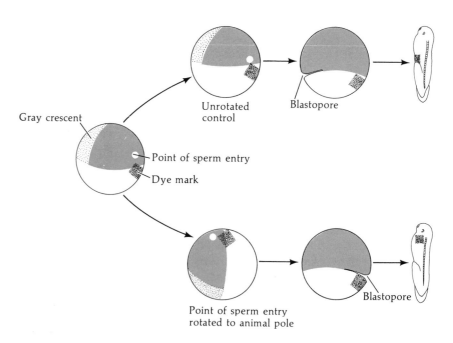

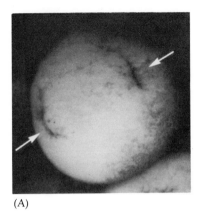

(A)

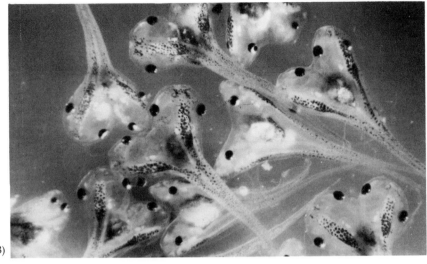

(B)

FIGURE 16
Twin blastopores produced by rotating dejellied *Xenopus* eggs ventral-side up at the time of first cleavage. (A) Two blastopores are instructed to form: the original one (opposite the point of sperm entry) and the new one created by the displacement of cytoplasmic material. (B) Those eggs develop two complete axes, which form twin tadpoles, joined ventrally. (Photographs courtesy of J. Gerhart.)

the eggs in Curtis' experiments had been rotated so that the ventral side could be visualized under a microscope, that is, the point of sperm entry had been placed upward. Second, the host embryos had been at the early stages of the first cleavage, exactly the time that other investigators had found was optimal to produce two sites of gastrulation. Thus, there existed the possibility that gastrulation could be initiated at two sites merely by rotating the egg. This was precisely what they found. A substantial percentage of embryos developed two blastopore lips merely by rotating the eggs during their first cleavage (Figure 16).

Gimlich and Gerhart (1984) then performed another type of transplantation experiment which confirmed the hypothesis that the factor(s) that initiate gastrulation originally lay in the deep cytoplasm rather than in the gray crescent cortex. They demonstrated that the three most dorsal vegetal blastomeres of 64-cell *Xenopus* embryos are able to induce the formation of the dorsal lip of the blastopore and of a complete dorsal axis in UV-irradiated recipients (which otherwise would have failed to properly initiate gastrulation) (Figure 17A). Moreover, these three blastomeres, which underlie the prospective dorsal lip region, can also induce a secondary invagination and axis when transplanted into the ventral side of an unirradiated 64-cell embryo (Figure 17B). Thus, this small cluster of vegetal blastomeres is responsible for allowing their adjacent marginal cells to invaginate and form the dorsal axis of the embryo.

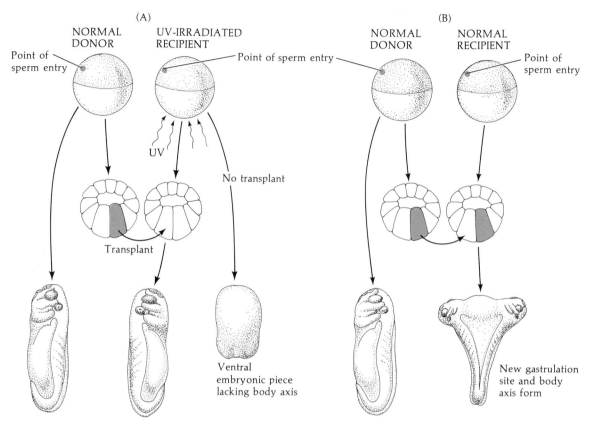

FIGURE 17
Transplantation experiments demonstrating that the vegetal cells underlying the prospective dorsal blastopore lip regions are responsible for causing the initiation of gastrulation. (A) Rescue of irradiated embryos by transplanting the vegetal blastomeres of the most dorsal segment (gray) of a 64-cell embryo into a cavity made by the removal of a similar number of vegetal cells. An irradiated zygote without this transplant fails to undergo normal gastrulation. (B) Formation of new gastrulation site and body axis by the transplantation of the most dorsal vegetal cells of a 64-cell embryo into the ventralmost vegetal region of another 64-cell embryo. (After Gimlich and Gerhart, 1984.)

It appears, then, that internal rearrangements of the cytoplasm, probably those normally occasioned by the entry of the sperm, are responsible for causing the asymmetric distribution of subcellular factors. This asymmetry creates in the egg a dorsal–ventral distinction that ultimately directs the position of the blastopore. Moreover, there appears to be a progressive accumulation of whatever factor (or factors) is needed to form the dorsal blastopore lip. Malacinski and his colleagues (1980) find no dorsal lip-forming activity in the dorsal region of the fertilized egg and early cleavage stages of the amphibian embryo. However, such activity is found there during the mid- and late blastula stages and can be found in the dorsal vegetal cells of the 64-cell frog embryo. Thus, in these species, those factors responsible for blastopore formation do not appear to be localized in the region where the blastopore is formed until shortly before gastrulation is initiated. The gray crescent exerts a bias on the placement of these morphogenic compounds such that during normal development the blastopore-initiating factors will eventually become concentrated in the vegetal cells immediately below the gray crescent. These vegetal cells then interact with the gray crescent cells to enable them to initiate gastrulation and form the dorsal body axis.

Cell movements and construction of the archenteron

Gastrulation is initiated when a group of marginal endoderm cells on the dorsal surface of the blastula sinks into the embryo. This inward movement of cells is accomplished by characteristic changes in cell–cell attachment not unlike those seen in the vegetal plate of sea urchin embryos. The inner borders of these cells lose their intercellular connections while the other border regions remain tightly bound to their neighbors. The outer portions then elongate in a direction perpendicular to the cell surface. Figure 18 shows the formation of these bottle cells in the early amphibian gastrula.

Because these cells remain in the deepest portion of the blastopore groove and appear to lead the small archenteron into the embryo, it has long been thought that the bottle cells have a major role in forming and extending the archenteron. Rhumbler (1902) suggested that the bottle cells actively migrated into the center of the gastrula as the result of lower surface tension there. They would then pull the neighboring cells into the groove as well, thus initiating the cell movement into the embryo.

In 1943–1944, the outstanding embryologist Johannes Holtfreter

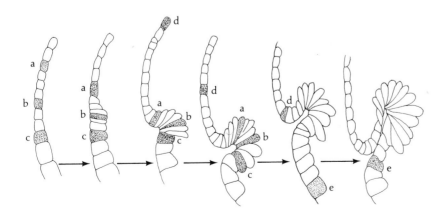

FIGURE 18
Movement of specific cells into the dorsal blastopore lip of *Xenopus laevis*. Time-lapse cinemicrography enables each individual cell to be followed as gastrulation is initiated. (After Keller, 1978.)

produced evidence that this was indeed the case. First Holtfreter isolated the early blastopore lip cells and cultured them on glass. The rounded ends of the cells attached to the glass led the movement of the remainder of the cells. Even more convincing were Holtfreter's recombination experiments in which the marginal zone cells (which would give rise to the dorsal blastopore lip) were combined with inner endoderm tissue. When the dorsal marginal zone cells were excised and placed on inner endoderm tissue, the blastopore cell precursors formed bottle cells and sank below the surface of the inner endoderm (Figure 19). Moreover, as they sank, they created a depression reminiscent of the early blastopore. Thus, Holtfreter claimed that the ability to invaginate into the inner endoderm was an innate property of the dorsal marginal zone cells.

More recently R.E. Keller has shown that while the gray crescent cells do have an intrinsic ability to invaginate into endoderm, the bottle cells may have a more passive role than previously thought. Keller (1981) found that once the bottle cells are formed, they have little to do with extending the archenteron and that partial or complete removal of these cells does not prevent the involution of adjacent cells into the blastopore. Gastrulation can occur without these cells. The major factor in the movement of cells into the embryo appears to be the involution of the *subsurface* marginal cells rather than the superficial ones. It appears that these subsurface, DEEP MARGINAL CELLS turn inward and migrate toward the animal pole along the inside surfaces of the remaining deep cells (Figure 20), and that the superficial layer forms the lining of the archenteron merely by being attached to the actively migrating deep cells. The bottle cells aid in forming the initial blastopore groove, but even this movement is dependent upon their attachment to the underlying deep cells. Removal of the bottle cells does not affect the involution of the deep or superficial marginal cells to the embryo. However, replacement of the dorsal marginal zone deep cells with animal region cells (which do not normally undergo involution) stops archenteron formation.

Figure 21 attempts to integrate various recent studies (Keller, 1978,

FIGURE 19
A graft of amphibian cells from the dorsal blastopore lip region sinks into a layer of endodermal cells and forms a blastopore groove. (After Holtfreter, 1944.)

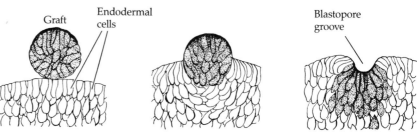

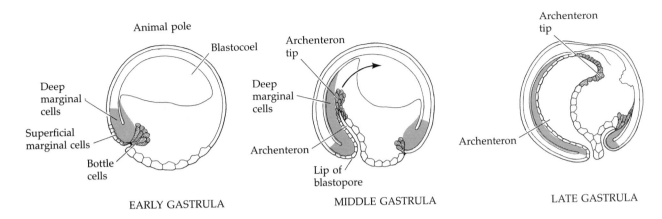

FIGURE 20

Spatial rearrangement of the deep marginal cells to the bottle cells during gastrulation of the frog *Xenopus*. The bottle cells (stippled) originate from superficial marginal cells and become the archenteron tip. The involuting cells that form the mesoderm (shaded) are derived from the internal (deep) marginal cells. (After Keller, 1981.)

FIGURE 21

Integrative model of cell movements during gastrulation. (A) Early gastrulation is characterized by the interdigitation of the marginal deep layers and by involution. (B,C) In later gastrulae, the deep marginal cells flatten and the formerly superficial cells form the wall of the archenteron. Bottle cells are darkly stippled. (After Keller, 1981.)

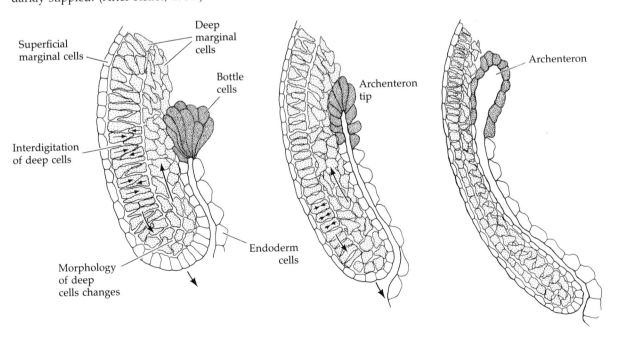

1980, 1981; Keller and Schoenwolf, 1977) into a model of amphibian archenteron formation. According to this model, during the initial stages of gastrulation (Figure 21A), the marginal zone extends toward the vegetal pole by interdigitating several layers of deep cells to form one thin, deep layer. At the same time, the superficial cells spread out by dividing and flattening. Upon reaching the blastopore lip, the deep cells change their morphology and form a mesodermal stream that migrates along the inner surface of the superficial cells toward the animal pole. During the later stages of gastrulation (Figure 21B), the columnar deep cells flatten and spread out as the marginal zone continues to move ventrally (forming the enlarged blastopore about the yolk plug). The mesodermal stream continues to migrate and the overlying layer of superficial cells (including the bottle cells) is passively pulled toward the animal pole, thereby forming the endodermal roof of the archenteron. Thus, although the bottle cells may be responsible for creating the initial groove, the motivating force for this appears to come from the deep layer of the marginal cells. Furthermore, this deep layer of cells appears to be responsible for the continued migration of cells into the embryo.

Epiboly of the ectoderm

While involution is occurring at the blastopore lips, the ectodermal precursors expand over the entire embryo. Keller (1980) and Keller and Schoenwolf (1977) have used scanning electron microscopy to observe the changes in both the superficial and deep cells of the animal and marginal regions. The major mechanism of epiboly in amphibian gastrulation appears to be an increase in cell number (through division) coupled with a concurrent integration of several deep layers into one (Figure 22). During early gastrulation, three rounds of cell division increase the volume of the deep cells by 45 percent. At the same time, the complete integration of the numerous deep cells into one layer is affected. The most superficial layer expands by cell division and flattening. The spreading of cells in the dorsal and ventral marginal zones appears to proceed by the same mechanism, although changes in cell shape appear to play a greater role than in the animal region. The result of these expansions is the epiboly of the superficial and deep cells of the animal and marginal regions over the surface of the egg. Most of the marginal region cells, as previously mentioned, involute to join the mesodermal cell stream within the embryo.

Amphibian gastrulation, then, is the orchestration of several events. The first indication of gastrulation involves the local invagination of endodermal bottle cells in the marginal zone at a precisely defined time and place. Next, the involution of marginal cells through the blastopore lip begins the formation of the archenteron, which is eventually extended around the embryo. At the same time, the ectodermal precursor cells expand vegetally by cell division and by the integration of previ-

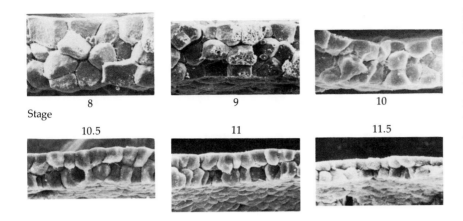

Stage 8   9   10

10.5   11   11.5

FIGURE 22
Scanning electron micrographs of the blastocoel roof showing the changes in cell shape and arrangement. Stages 8 and 9 are blastulae; stages 10 through 11½ represent progressively later gastrulae. (From Keller, 1980; photographs courtesy of R. E. Keller.)

ously independent cell layers. The result of these cell movements is the proper positioning of the three germ layers in preparation for their differentiation into the body organs.

Gastrulation in birds

Overview of avian gastrulation

Cleavage in avian embryos creates a blastodisc above an enormous volume of yolk. This inert underlying yolk mass imposes severe restraints on cell movements. Thus, gastrulation appears at first to be very different from that of a sea urchin or a frog. We shall soon see, though, that there are numerous similarities between avian gastrulation and the gastrulations already studied. Moreover, we shall see that mammalian embryos—which do not have yolk—retain the same type of gastrulation movements as the yolky embryos of birds and reptiles.

The central cells of the avian blastodisc are separated from the yolk by a subgerminal cavity and appear to be clear—hence, the center of the blastodisc is called the area pellucida. In contrast, the cells at the margin of the area pellucida appear opaque because of their contact with the yolk. They form the area opaca (Figure 23). Whereas most of the cells remain at the surface, forming the EPIBLAST, certain cells migrate individually into the subgerminal cavity to form a layer called the PRIMARY HYPOBLAST (Figure 24). Shortly thereafter, a sheet of cells from the posterior margin of the blastoderm ("Koller's crescent") migrate anteriorly to join the primary hypoblast, thereby forming the SECONDARY HYPOBLAST. The two-layered blastoderm (epiblast and hypoblast) is joined together at the margin of the area opaca, and the resulting space

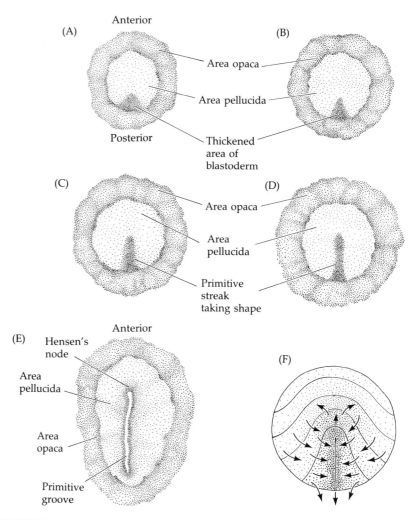

FIGURE 23
Cell movements forming the primitive streak of the chick embryo. Dorsal view of chick blastoderm at (A) 3–4 hours, (B) 5–6 hours, (C) 7–8 hours, (D) 10–12 hours, (E) 15–16 hours. A summary of these cell movements is shown in (F). (Adapted from several sources, especially Spratt, 1946.)

between them is a blastocoel. Thus, the structure of the avian blastodisc is not dissimilar to that of the amphibian or echinoderm blastula.

The fate map for the avian embryo is restricted to the epiblast. That is to say, the hypoblast does not contribute any cells to the developing embryo. Rather, the hypoblast cells form portions of the external membranes that help to nourish and protect the embryo. All three germ layers of the embryo proper (plus a considerable amount of extra-embryonic membrane) are formed from the epiblastic cells. The fate

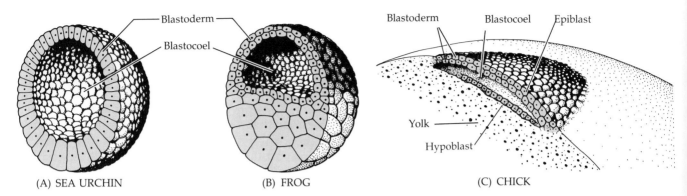

(A) SEA URCHIN   (B) FROG   (C) CHICK

map of the chick epiblast prior to gastrulation is shown in Figure 25. The presumptive notochordal cells are centrally located between the anterior presumptive ectoderm and the posterior presumptive endoderm.

The major structure characteristic of avian, reptilian, and mammalian gastrulation is the PRIMITIVE STREAK. This streak is first seen as a thickening of the cell sheet at the central posterior end of the zona

FIGURE 24
Diagrams of sections comparing the blastulae of (A) echinoderms and *Amphioxus*, (B) amphibians, and (C) birds. (After Carlson, 1981.)

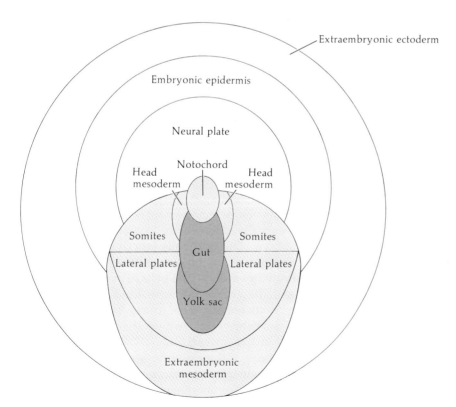

FIGURE 25
Fate map of an avian embryo immediately prior to gastrulation. The primitive streak has not formed yet, but it will eventually extend to the region of the notochord. Presumptive ectoderm is white, presumptive mesoderm is light gray, and presumptive endoderm is dark gray. (After Balinsky, 1975.)

pellucida (Figure 23). This thickening is caused by the migration of cells from the lateral region of the posterior epiblast toward the center. As the thickening narrows, it moves anteriorly and constricts to form the definitive primitive streak. This streak extends 60 to 75 percent the length of the area pellucida and marks the anterior–posterior axis of the embryo. As the cells converge to form the primitive streak, a depression forms within the streak. This depression is called the PRIMITIVE GROOVE, and it serves as a blastopore through which the migrating cells pass into the blastocoel. Thus, the primitive streak becomes analogous to the amphibian blastopore. At the anterior end of the primitive streak is a regional thickening of cells called the PRIMITIVE KNOT, or HENSEN'S NODE. The center of this node contains a funnel-shaped depression through which cells can pass into the blastocoel. Hensen's node is the functional equivalent of the dorsal lip of the amphibian blastopore.

As soon as the primitive streak is formed, blastoderm cells begin to migrate over the lips of the primitive streak and into the blastocoel (Figure 26). Like the amphibian blastopore, the primitive streak is a continually changing cell population. Those cells migrating through Hensen's node pass down into the blastocoel and migrate anteriorly, forming head mesoderm and notochord; and those cells passing through the lateral portions of the primitive streak give rise to the majority of endodermal and mesodermal tissues. Unlike the migration of cell sheets into the amphibian blastocoel, the cells entering the inside of the avian embryo do so as individuals. Rather than forming a tightly organized sheet of cells, the ingressing population creates a loosely connected MESENCHYME. Moreover, there is no true archenteron formed in the avian gastrula.

As the cells enter the primitive streak, the streak elongates toward the future head region. At the same time, the secondary hypoblastic cells are continuing to migrate anteriorly from the posterior margin of the blastoderm. The elongation of the primitive streak appears to be coextensive with the anterior migration of these secondary hypoblast cells.

The first cells to migrate through the primitive streak are those destined to become the foregut. This situation is again similar to what we have seen in amphibians. Once inside the blastocoel, these cells migrate anteriorly and eventually displace the hypoblast cells in the anterior portion of the embryo. The next cells entering the blastocoel through Hensen's node also move anteriorly, but they do not move as far ventrally as the presumptive endodermal cells. These cells remain between the endoderm and the epiblast to form the head mesoderm and the chordamesoderm (notochordal) cells. These first cells into the blastocoel have all moved anteriorly, pushing up the anterior midline region of the epiblast to form the HEAD PROCESS. Meanwhile, cells continue migrating inward through the primitive streak. As they enter the blastocoel, these cells separate into two streams. One stream moves deeper and joins the hypoblast along its midline, displacing the hypoblast cells to the sides. These deep-moving cells give rise to all the

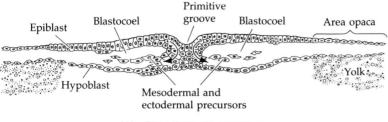

(A)   TRANSVERSE SECTION

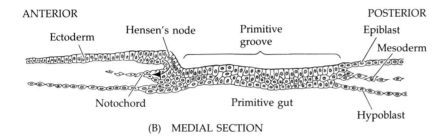

(B)   MEDIAL SECTION

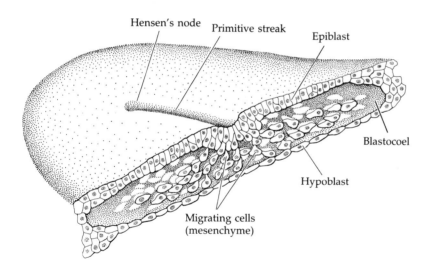

(C)   THREE-DIMENSIONAL VIEW

FIGURE 26

Migration of endodermal and mesodermal cells through the primitive streak.
(A) Diagram of transverse cross-section through a 17-hour embryo, illustrating
the lateral movement of endodermal and mesodermal cells passing into the
blastocoel. (B) Diagram of a median section through the same embryo, showing that those cells migrating through Hensen's node condense to form the
notochord ("head process"). (C) Stereogram of a gastrulating chick embryo,
showing the relationship of the primitive streak, the migrating cells, and the
two original layers of the blastoderm. (A and B modified from Carlson, 1981;
C after Balinsky, 1975.)

endodermal organs of the embryo as well as to most of the extra-embryonic membranes (the hypoblast forms the rest). The second migrating stream spreads throughout the blastocoel as a loose sheet, roughly midway between the hypoblast and the epiblast. This sheet generates the mesodermal portions of the embryo and extraembryonic membranes. By 22 hours of incubation, most of the presumptive endodermal cells are in the interior of the embryo, although presumptive mesodermal cells continue to migrate inward for a longer time.

Now a second phase of gastrulation begins. While the mesodermal ingression continues, the primitive streak starts to regress, moving Hensen's node from near the center of the area pellucida to a more posterior position (Figure 27). It leaves in its wake the dorsal axis of the embryo, the head process. As the node moves further posteriorly, the remaining (posterior) portion of the notochord is laid down. Finally, the node regresses to its most posterior position, eventually forming the anal region in true deuterostome fashion. By this point, the epiblast is composed entirely of presumptive ectodermal cells.

As a consequence of this two-step gastrulation process, avian (and mammalian) embryos exhibit a distinct anterior-to-posterior gradient of developmental maturity. While cells of the posterior portions of the embryo are undergoing gastrulation, cells at the anterior end are already starting to form organs. For the next several days, the anterior end of the embryo is seen to be more advanced in its development than the posterior end.

While the presumptive mesodermal and endodermal cells were moving inward, the ectodermal precursors were busily surrounding the yolk by epiboly. The enclosure of the yolk by the ectoderm (again reminiscent of the epiboly of amphibian ectoderm) is a Herculean task that takes the greater part of 4 days to complete and that involves the

FIGURE 27
Regression of the primitive streak, leaving notochord in its wake. Various points of the streak were followed after it achieved its maximum length. Time represents hours after achieving maximum length. (After Spratt, 1947.)

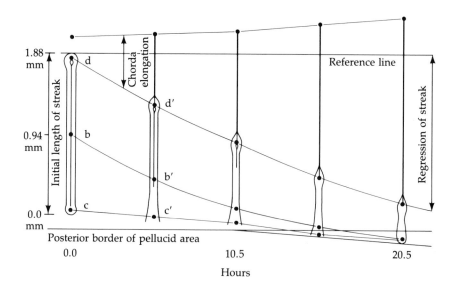

continuous production of new cellular material at the expense of the yolk and the migration of the presumptive ectodermal cells along the underside of the vitelline membrane. Thus, as avian gastrulation draws to a close, the ectoderm has surrounded the yolk, the endoderm has replaced the hypoblast, and the mesoderm has positioned itself between these two regions.

Mechanisms of avian gastrulation

Role of the hypoblast. Although the hypoblast does not contribute any cells to the adult chick (Rosenquist, 1972), it is essential for its proper development. In 1932, C. H. Waddington demonstrated that this lower layer influenced the orientation of the chick embryonic axis. He separated the two cellular layers of the chick blastodisc soon after the primitive streak had formed. After bringing the two layers together again such that their original long axes were now nearly perpendicular to each other, he saw that the primitive streak bent according to the long axis of the hypoblast (Figure 28). Azar and Eyal-Giladi (1979, 1981) have extended these observations to show that the hypoblast *induces* the formation of the primitive streak. First, removal of the hypoblast stops all further development until the remaining epiblast can regenerate a new lower layer. Second, when the population of hypoblast cells derived from the posterior margin is eliminated, no primitive streak is formed. Rather, the area pellucida develops into a mass of disorganized, mesodermlike tissue. Thus, it appears that the secondary hypoblast directs the formation and the directionality of the primitive streak.

Cell movement within the blastocoel. As in amphibian gastrulation, cells passing through the blastopore lip construct their apical ends to become "bottle cells" (Figure 29). Once these cells are released from the primitive streak and enter the blastocoel, they flatten and enter a "stream" of independently migrating cells, each advancing by the extension and contraction of pseudopods. Extracellular polysaccharides may play an important role in this migration. One such complex polysaccharide is *hyaluronic acid*, a linear polymer of glucuronic acid and *N*-acetylglucosamine. This compound is made by the ectodermal cells and accumulates in the blastocoel, where it coats the surfaces of the incoming cells. Fisher and Solursh (1977) have shown that when this material is digested (by injecting the enzyme hyaluronidase into the blastocoel), the mesenchymal cells clump together and fail to migrate properly. Several studies have shown that hyaluronic acid is important in keeping individually migrating mesenchymal cells separated from one another. Moreover, hyaluronic acid is first seen to accumulate at precisely the time when the first cells enter the blastocoel. Hyaluronic acid may be able to keep the cells separated by its ability to expand in water. Once

FIGURE 28
Rotation of embryonic axis of chick embryo, caused by rotating the hypoblast with respect to the developing primitive streak at 20–24 hours incubation. The primitive streak has regressed significantly at this stage. (From Azar and Eyal-Giladi, 1979; photograph courtesy of H. Eyal-Giladi.)

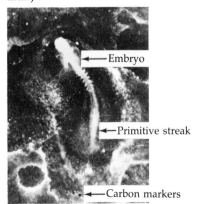

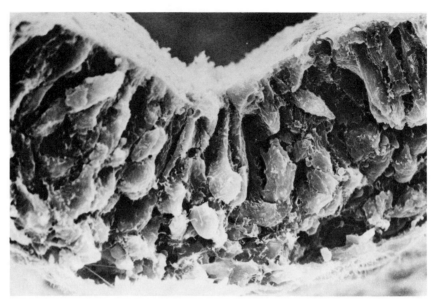

FIGURE 29
Scanning electron micrograph of the primitive streak region of a gastrulating embryo. Epiblast cells passing into the blastocoel extend their apical ends to become "bottle cells." (From Solursh and Revel, 1978; photograph courtesy of M. Solursh.)

in an aqueous environment, this polymer can expand to over 1000 times its original volume. Therefore, hyaluronic acid might be an important factor in keeping the mesenchymal cells dispersed during their migration and thereby ensuring that this migration continues.

Epiboly of the ectoderm. During gastrulation, ectodermal precursor cells expand outward to encircle the yolk. These cells are joined to each other by tight junctions and travel as a unit rather than as individual cells. In the chick egg, the upper surface of the area opaca is seen to adhere tightly to the lower surface of the vitelline membrane and to spread along this inner surface. In culture, the same behavior is seen. New (1959) demonstrated that isolated blastoderm will spread normally on isolated vitelline membrane; and Spratt (1963) demonstrated that this spreading will not occur on other substrates. This result suggests that the vitelline membrane is essential for the spreading of the cell sheet. Interestingly, only the marginal cells (i.e., the cells of the area opaca), attach firmly to the vitelline surface. Most of the blastoderm cells adhere loosely, if at all. These marginal cells are inherently different from the other blastoderm cells, as these cells can extend enormous (500 μm) processes onto the vitelline membrane. These processes are believed to be the locomoter apparatus of the marginal cells.

There are several lines of evidence indicating that the marginal area opaca cells are the agents of ectodermal epiboly. First, the blastoderm spreads only when the margins are expanding. Second, when the marginal cells are cut away from the rest of the blastoderm, they continue to expand alone. Thus, it appears that the ectodermal precursor cells are being moved by the actively migrating cells of the area opaca (Schlesinger, 1958). There also exists a specific relationship between the cell membranes of the marginal cells and the lower surface of the vitelline membrane. New (1959) showed that when the blastoderm is placed on a vitelline membrane upside down (deep layers in contact with the vitelline membrane), the blastoderm curls inward so that the marginal cells of the upper layer are once again contacting the vitelline surface (Figure 30). Thus, there appears to be specific recognition between the marginal cells of the area opaca and the inner surface of the vitelline membrane. This recognition appears essential for the epibolic migration of the ectoderm to enclose the yolk.

Gastrulation in mammals

Overview of mammalian gastrulation

Birds and mammals are both descendants of reptilian species. Therefore, it is not surprising to find that mammalian development parallels that of reptiles and birds. What is surprising is that the gastrulation movements of reptilian and avian embryos, which evolved as an adaptation to yolky eggs, are retained even in the absence of large amounts of yolk in the mammalian embryo. The inner cell mass can be

FIGURE 30
Migratory properties of chick ectodermal precursors. (A) When chick blastoderm is placed on vitelline membrane such that ectodermal precursors contact the vitelline surface, marginal cells migrate, causing the ectoderm to cover the vitelline membrane. (B) When the deep layers are in contact with the vitelline membrane, the blastoderm layer curls such that the cells can adhere to and migrate over the vitelline layer. The result is a closed vesicle. (After New, 1959.)

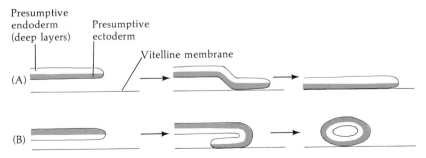

envisioned as sitting atop an imaginary ball of yolk, following instructions that seem more appropriate to its ancestors.

Instead of developing in isolation, most mammals have evolved the remarkable strategy of developing within the mother herself. The mammalian embryo obtains nutrients from its mother and does not use yolk. This evolution has entailed a dramatic restructuring of the maternal anatomy (such as the expansion of the oviduct to form the uterus) as well as the development of a fetal organ capable of absorbing maternal nutrients. This fetal organ—the PLACENTA—is derived primarily from the embryonic trophoblast cells, supplemented with mesodermal cells derived from the inner cell mass.

The origins of the various early mammalian tissues are summarized in Figure 31. The first segregation of cells within the inner cell mass involves the formation of the HYPOBLAST (sometimes called the "primitive endoderm") layer (Figure 32). These cells separate from the inner cell mass to line the blastocoel cavity, where they give rise to the YOLK SAC ENDODERM. As in avian embryos, these cells do not produce any part of the newborn organism. The remaining inner cell mass tissue above the hypoblast is now referred to as the epiblast. The epiblast cells are split by small clefts that eventually coalesce to divide the epiblast into two layers, one of which will produce the EMBRYONIC EPIBLAST while the other forms the lining of the AMNION (Figure 33). Once the lining of the amnion is completed, it fills with a secretion called AMNIOTIC FLUID, which serves as a "shock absorber" to the developing embryo while preventing its dessication.

The embryonic epiblast is believed to contain all the cells that will generate the actual embryo. It is similar to the avian epiblast. At the posterior margin of the embryonic epiblast, a localized thickening oc-

FIGURE 31

Scheme illustrating the derivation of tissues in human and rhesus monkey embryos. (After Luckett, 1978.)

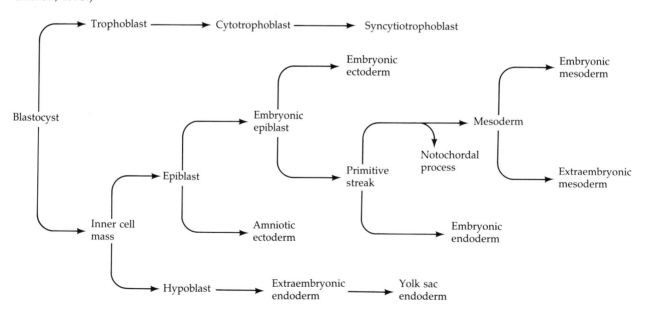

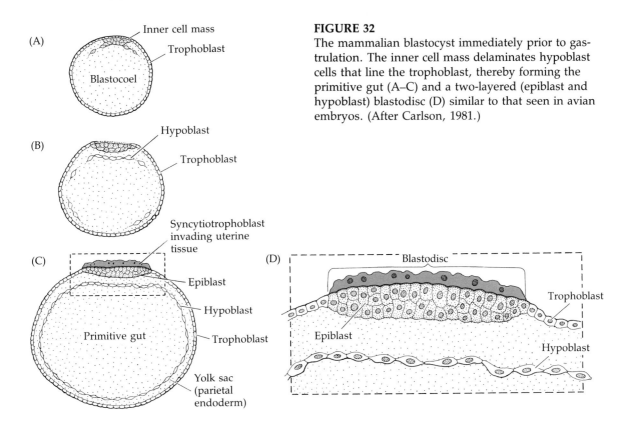

FIGURE 32

The mammalian blastocyst immediately prior to gastrulation. The inner cell mass delaminates hypoblast cells that line the trophoblast, thereby forming the primitive gut (A–C) and a two-layered (epiblast and hypoblast) blastodisc (D) similar to that seen in avian embryos. (After Carlson, 1981.)

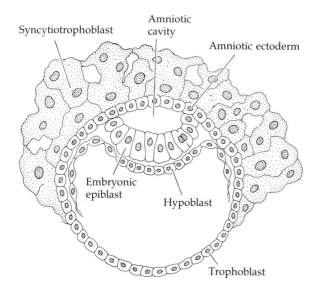

FIGURE 33

Formation of the amnion in human embryos. The hypoblast is complete near the site of the inner cell mass, and the trophoblast cells are dividing to form the syncytiotrophoblast, which will invade the uterus. Meanwhile, the epiblast has split into the amniotic ectoderm and the embryonic epiblast. All subsequent development of the embryo will focus on the embryonic epiblast. (After Carlson, 1981.)

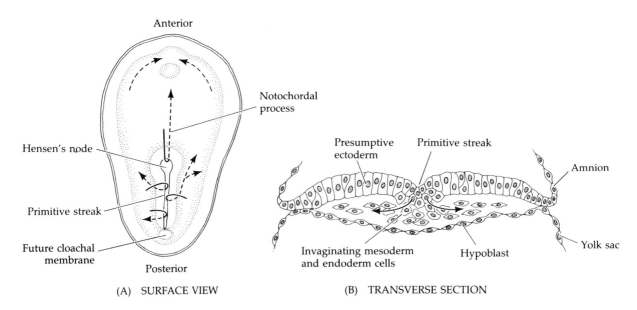

(A)   SURFACE VIEW   (B)   TRANSVERSE SECTION

FIGURE 34

Cell movements during mammalian gastrulation. (A) Schematic diagram of the dorsal surface of the embryonic epiblast. As in chick embryos, cells migrating through Hensen's node travel anteriorly (cephalad) to form the notochord, while the remaining cells traveling through the streak migrate laterally to become the mesoderm and endoderm precursors. (Dotted lines indicate internal migrations.) (B) Transverse section of the embryo. (After Langman, 1981.)

curs, producing a primitive streak through which the endodermal and mesodermal precursors migrate (Figure 34). As in avian embryos, the cells migrating between the hypoblast and epiblast layers appear to be separated by hyaluronic acid, which is first synthesized at the time of primitive streak formation (Solursh and Morriss, 1977).

While the embryonic epiblast is undergoing cell movements reminiscent of those seen in reptile or bird gastrulation, the extraembryonic cells are making the distinctly mammalian tissues that enable the fetus to survive within the maternal uterus. Although the initial trophoblastic cells appear normal, they give rise to a population of cells wherein nuclear division occurs in the absence of cytokinesis. The first type of cell constitutes a layer called the CYTOTROPHOBLAST, whereas the second type of cell forms the SYNCYTIOTROPHOBLAST. This latter tissue invades the uterine lining, embedding the embryo within the uterus. The uterus, in turn, sends blood vessels into this area, where they eventually contact the syncytiotrophoblast. Shortly thereafter, mesodermal tissue extends outward from the gastrulating embryo (Figure 35). Luckett (1978) has shown that this tissue had migrated through the primitive streak but becomes extraembryonic rather than embryonic mesoderm and joins the trophoblastic extensions. This extraembryonic mesoderm gives rise

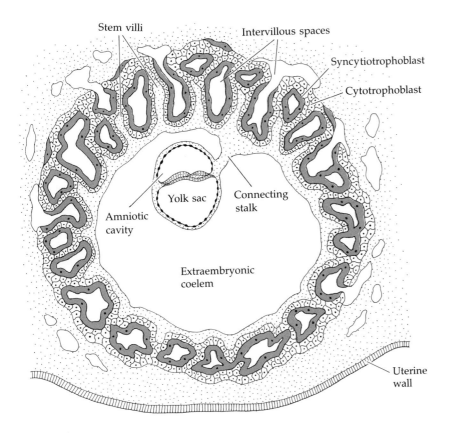

FIGURE 35
Diagram of gastrulating human embryo at the end of the third week of gestation. The amniotic cavity (from the amniotic ectoderm) and yolk sac (from the hypoblast) have formed, the notochord elevates the epiblast in the future anterior region of the embryo, and the trophoblast cells forming the placenta are coming into contact with the blood vessels of the uterus. The embryo is connected to the trophoblast by a connecting stalk of extraembryonic mesoderm that will shortly carry the fetal blood vessels to the placenta. (After Langman, 1981.)

to the blood vessels that carry the nutrients from the mother to the embryo. The narrow connecting stalk of extraembryonic mesoderm that connects the embryo to the trophoblast eventually forms the vessels of the UMBILICAL CORD. The fully developed organ, consisting of trophoblast tissue and blood vessel-containing mesoderm, is called the CHORION, and the chorion fuses with the uterine wall to create the placenta. The chorion may integrate such that the closely apposed fetal and maternal tissues can be readily separated (as in the CONTACT PLACENTA of the pig), or it may be so intimately integrated that the two tissues cannot be separated without damage to both the mother and the developing fetus (as in the DECIDUOUS PLACENTA of most mammals, including humans).

SIDELIGHTS & SPECULATIONS

The roles of the chorion

Two roles of the extraembryonic trophoblast-derived tissues have already been alluded to. The first is the role of the trophoblast itself in the IMPLANTATION of the blastocyst onto the uterine wall. The second function of these tissues is seen after the blood vessels have formed, namely, the transport of soluble substances between the fetus and the mother. Figure 36 shows the relationship between the embryonic and extraembryonic structures of a 6-week human fetus. Here the fetus is encased in the amnion and is further

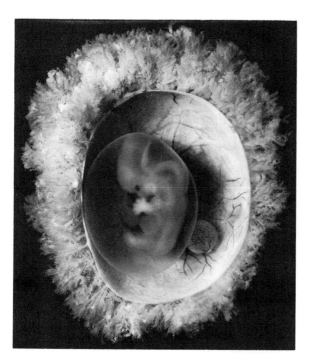

FIGURE 36
Human embryo and placenta after 40 days gestation. The embryo lies within the amnion and its blood vessels can be seen extending into the chorionic villi. The sphere to the right of the embryo is the yolk sac. (Photograph from the Carnegie Institution of Washington; courtesy of C. F. Reather.)

shielded by the chorion. The blood vessels extending to and from the chorion are readily observable, as are the villi that project from the outer surface of the chorion. These villi contain the blood vessels and allow the chorion to have a large area exposed to the maternal blood. Thus, although fetal and maternal circulatory systems never normally merge, diffusion of soluble substances can occur through the villi (Figure 37). In this manner, the mother provides the fetus with nutrients and oxygen, and the fetus sends its waste products (mainly carbon dioxide and urea) into the maternal circulation.

But that is not all the fetus sends to the mother. The third function of the chorion is hormonal. The syncytiotrophoblast produces two hormones that are essential for mammalian development. First, it produces CHORIONIC GONADOTROPIN, a peptide hormone that is capable of causing other cells in the placenta (and in the maternal ovary) to produce PROGESTERONE. Progesterone is the steroid hormone that keeps the uterine wall thick and full of blood vessels. In primates, the ovaries can be removed after the first third of pregnancy without harm to the developing fetus because the chorion itself is able to produce the steroid needed to maintain pregnancy (Zander and Münstermann, 1956). Placental progesterone is also used by the fetal adrenal gland as a substrate for the production of biologically important corticosteroid hormones. The third hormone produced by the chorion is CHORIONIC SOMATOMAMMOTROPIN (often called placental lactogen). This hormone is responsible for promoting maternal breast development during pregnancy, enabling milk production later on.

Recent studies have indicated that the chorion may have yet another function, namely, the protection of the fetus against the immune response of the mother. Each person with a normal immune system can recognize and reject foreign cells within its body. This is reflected by the rejection of skin and organ grafts from genetically different individuals. The glycoproteins responsible for this rejection are called the MAJOR HISTOCOMPATIBILITY ANTIGENS, and these are likely to differ from individual to individual. A human child expresses major histocompatibility antigens from both the mother and the father, and a mother is capable of rejecting her offspring's skin or organs because they contain paternally derived antigens. How, then, can a human fetus remain 9 months within the body of its mother? Why doesn't the mother immunologi-

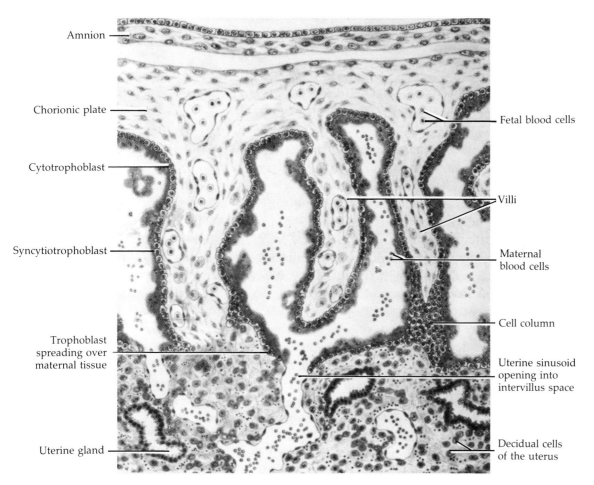

Amnion

Chorionic plate

Cytotrophoblast

Syncytiotrophoblast

Trophoblast
spreading over
maternal tissue

Uterine gland

Fetal blood cells

Villi

Maternal
blood cells

Cell column

Uterine sinusoid
opening into
intervillus space

Decidual cells
of the uterus

FIGURE 37
Schematic representation of the relationship of the
chorionic villi to the maternal blood and uterus.
(Modified from Hill, 1931.)

cally reject her fetus as she would an organ from that
child?

It is known that the fetus does express paternal
antigens early in development and that these antigens
are found on the syncytiotrophoblast cells. Thus, the
paternal antigens appear to be perfectly positioned for
the mother to mount an immune response against
them. One hypothesis for why the mother's immune
system does not reject the fetus is that the chorion
produces substances that block the immune response.
One study supporting this hypothesis will be discussed here. In 1979, McIntire and Faulk showed that
the cell membranes of the human syncytiotrophoblast
contain a compound that inhibits certain cells of the

immune system from dividing. The cell-mediated immune response that leads to graft rejection is mediated
through a type of cell called the T LYMPHOCYTE, or
simply, T CELLS. These T cells can divide rapidly when
they encounter another cell that contains different major histocompatibility antigens. They then help other
T cells destroy the foreign cell or tissue. The T cell
response is measured by the MIXED LYMPHOCYTE RE-
ACTION, in which cell proliferation is measured after
two lymphocyte populations are combined. When person W's lymphocytes are combined with another sample of W's lymphocytes (which have been irradiated
to stop their possible proliferation), the original lymphocytes do not respond by dividing (Figure 38). However, when W's lymphocytes are combined with irradiated lymphocytes from person J, they divide rapidly,
incorporating radioactive precursors into DNA. This
reaction can be inhibited by adding an extract of syncytiotrophoblast cell membranes. Extracts of other or-

FIGURE 38

Inhibition of mixed lymphocyte reaction by material extracted from the placenta. (A) When lymphocytes from two individuals are placed together, they multiply upon recognizing foreign antigens. They will therefore incorporate large amounts of radioactive thymidine into new DNA. (B) This is not the case when lymphocytes from the same individual are brought together. (C) When the lymphocyte combination in (A) is brought together in the presence of trophoblast membrane proteins, their multiplication is inhibited. (Modified from McIntire and Faulk, 1979.)

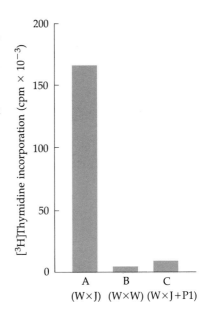

gans prepared in the same way fail to inhibit this specific type of cell division. Thus, the cell-mediated immune response appears to be inhibited by a substance found in the syncytiotrophoblast cell membrane. This is probably only one of many ways that the chorion protects the embryo against the immunological assault of its mother. The chorion must be recognized, then, as a complex organ with several functions, all of which enable yolk-deficient mammalian embryos to develop within the mother.

In gastrulation, then, we see an incredibly well coordinated series of cell movements whereby the cleavage-stage blastomeres are rearranged and begin to interact with new neighbors. Moreover, although there are differences between the gastrulation movements of sea urchin, amphibian, bird, and mammalian embryos, there exist certain mechanisms common to them all. Each group has the problem of bringing the mesodermal and endodermal precursor cells inside the body while surrounding the embryo with ectodermal precursors. Given the different amounts and distributions of yolk, as well as other environmental considerations, each type of organism has evolved a way of accomplishing this goal. The stage is now set for the formation of the first organs.

Literature cited

Azar, Y. and Eyal-Giladi, H. (1979). Marginal zone cells—the primitive streak inducing component of the primary hypoblast in the chick. *J. Embryol. Exp. Morphol.* 52: 79–88.

Azar, Y. and Eyal-Giladi, H. (1981). Interaction of epiblast and hypoblast in the formation of the primitive streak and the embryonic axis in chick, as revealed by hypoblast-rotation experiments. *J. Embryol. Exp. Morphol.* 61: 133–141.

Balinsky, B. I. (1975). *Introduction to Embryology,* Fourth Edition. Saunders, Philadelphia.

Born, G. (1885). Ueber den Einfluss der Schwere auf das Froschei. *Arch. Mikrosk. Anat.* 24: 475–545.

Carlson, B. M. (1981). *Patten's Foundations of Embryology,* Fourth Edition. McGraw-Hill, New York.

Curtis, A. S. G. (1960). Cortical grafting in *Xenopus laevis. J. Embryol. Exp. Morphol.* 8: 163–173.

Curtis, A. S. G. (1962). Morphogenic interactions before gastrulation in the amphibian *Xenopus laevis:* The cortical field. *J. Embryol. Exp. Morphol.* 10: 410–422.

Dan, K. and Okazaki, K. (1956). Cyto-embryological studies of sea urchins. III. Role of the secondary mesenchymal cells in the formation of the primitive gut in sea urchin larvae. *Biol. Bull.* 110: 29–42.

Fisher, M. and Solursh, M. (1977). Glycosaminoglycan localization and role in maintenance of tissue spaces in the early chick embryo. *J. Embryol. Exp. Morphol.* 42: 195–207.

Gibbins, J. R., Tilney, L. G. and Porter, K. R. (1969). Microtubules in the formation and development of the primary mesenchyme of *Arbacia punctulata. Dev. Biol.* 18: 523–539.

Gimlich, R. L. and Gerhart, J. C. (1984). Early cellular interactions promote embryonic axis formation in *Xenopus laevis. Dev. Biol.* 104: 117–130.

Gustafson, T. (1963). Cellular mechanisms in the morphogenesis of the sea urchin embryo: Cell contacts within the ectoderm and between mesenchyme and ectoderm cells. *Exp. Cell Res.* 32: 570–589.

Gustafson, T. and Wolpert, L. (1961). Studies on the cellular basis of morphogenesis in sea urchin embryos: Directed movements of primary mesenchyme cells in normal and vegetalized larvae. *Exp. Cell Res.* 24: 64–79.

Gustafson, T. and Wolpert, L. (1967). Cellular movement and contact in sea urchin morphogenesis. *Biol. Rev.* 42: 442–498.

Hara, K. (1977). The cleavage pattern of the axolotl egg studied by biocinematography and cell counting. *Wilhelm Roux Arch. Dev. Biol.* 181: 73–87.

Harkey, M. A. and Whitely, A. M. (1980). Isolation, culture and differentiation of echinoid primary mesenchyme cells. *Wilhelm Roux Arch. Dev. Biol.* 189: 111–122.

Hill, J. P. (1932). The developmental history of primates. *Philos. Trans. Roy. Soc. Lond. [B]* 221: 45–178.

Holtfreter, J. (1943). A study of the mechanics of gastrulation: Part I. *J. Exp. Zool.* 94: 261–318.

Holtfreter, J. (1944). A study of the mechanics of gastrulation: Part II. *J. Exp. Zool.* 95: 171–212.

Hörstadius, S. (1939). The mechanics of sea urchin development, studied by operative methods. *Biol. Rev.* 14: 132–179.

Karp, G. C. and Solursh, M. (1974). Acid mucopolysaccharide metabolism, the cell surface, and primary mesenchyme cell activity in the sea urchin embryo. *Dev. Biol.* 41: 110–123.

Keller, R. E. (1975). Vital dye mapping of the gastrula and neurula of *Xenopus laevis.* I. Prospective areas and morphogenetic movements of the superficial layer. *Dev. Biol.* 42: 222–241.

Keller, R. E. (1976). Vital dye mapping of the gastrula and neurula of *Xenopus laevis.* II. Prospective areas and morphogenetic movements in the deep layer. *Dev. Biol.* 51: 118–137.

Keller, R. E. (1978). Time-lapse cinemicrographic analysis of superficial cell behaviour during and prior to gastrulation in *Xenopus laevis. J. Morphol.* 157: 223–248.

Keller, R. E. (1980). The cellular basis of epiboly: An SEM study of deep cell rearrangement during gastrulation of *Xenopus laevis. J. Embryol. Exp. Morphol.* 60: 201–234.

Keller, R. E. (1981). An experimental analysis of the role of bottle cells and the deep marginal zone in the gastrulation of *Xenopus laevis. J. Exp. Zool.* 216: 81–101.

Keller, R. E. and Schoenwolf, G. C. (1977). An SEM study of cellular morphology, contact, and arrangement as related to gastrulation in *Xenopus laevis. Wilhelm Roux Arch. Dev. Biol.* 182: 165–186.

Kirschner, M. W. and Gerhart, J. C. (1981). Spatial and temporal changes in the amphibian egg. *Bioscience* 31: 381–388.

Kirschner, M. W., Gerhart, J. C., Hara, K. and Ubbels, G. (1980). Initiation of cell cycle and establishment of bilateral symmetry in *Xenopus* eggs. *Symp. Soc. Dev. Biol.* 38: 187–215.

Landström, U. and Løvtrup, S. (1979). Fate maps and cell differentiation in the amphibian embryo—An experimental study. *J. Embryol. Exp. Morphol.* 54: 113–130.

Langman, J. (1981). *Medical Embryology,* Fourth Edition. William and Wilkins, Baltimore.

Lau, J. T. Y. and Lennarz, W. J. (1982). Developmental regulation of the glycosylation of proteins in sea urchin embryos. *J. Cell Biol.* 95: 158.

Lillie, F. R. (1902). Differentiation without cleavage in the egg of the annelid *Chaetopterus pergamentaceous. Wilhelm Roux Arch. Entwicklungsmech. Org.* 14: 477–499.

Løvtrup, S. (1975). Fate maps and gastrulation in amphibia: A critique of current views. *Can. J. Zool.* 53: 473–479.

Luckett, W. P. (1978). Origin and differentiation of the yolk sac and extraembryonic mesoderm in presomite human and rhesus monkey embryos. *Am. J. Anat.* 152: 59–98.

Malacinski, G. M., Chung, H. M. and Asashima, M. (1980). The association of primary embryonic organizer activity with the future dorsal side of amphibian eggs and early embryos. *Dev. Biol.* 77: 449–462.

McIntire, J. A. and Faulk, W. P. (1979). Trophoblast modulation of maternal allogeneic recognition. *Proc. Natl. Acad. Sci. USA* 76: 4029–4032.

Moore, A. R. (1941). On the mechanism of gastrulation in *Dendraster excentricus. J. Exp. Zool.* 87: 101–111.

Moore, A. R. and Burt, A. S. (1939). On the locus and nature of the forces causing gastrulation in the embryos of *Dendraster excentricus. J. Exp. Zool.* 82: 159–171.

New, D. A. T. (1959). Adhesive properties and expansion of the chick blastoderm. *J. Embryol. Exp. Morphol.* 7: 146–164.

Pasteels, J. (1948). Les bases de la morphogènese chez les vertebres anamniotes au function de la structure de l'oeuf. *Folia Biotheoret. (Leiden)* 3: 83–108.

Penners, A. and Schleip, W. (1928). Die Entwicklung der Schultzeschen Doppelbildungen aus dem Ei von *Rana fusca*. Teil V. und VI. *Z. Wissen. Zool.* 131: 1–156.

Rhumbler, L. (1902). Zur Mechanik des Gastrulationvorganges insbesundere der Invagination. Eine Entwicklungsmechanische Studie. *Wilhelm Roux Arch. Entwicklungsmech. Org.* 14: 401–476.

Rosenquist, G. C. (1972). Endoderm movements in the chick embryo between the short streak and head process stages. *J. Exp. Zool.* 180: 95–104.

Rugh, R. (1951). *The Frog: Its Reproduction and Development*. Blakiston, Philadelphia.

Schlesinger, A. G. (1958). The structural significance of the avian yolk in embryogensis. *J. Exp. Zool.* 138: 223–258.

Schulze, O. (1894). Die Kunstliche Erzeugung von Doppelbildungen bei Froschlarven mit Hilfe abnormer Gravitationwirkung. *Wilhelm Roux Arch. Entwicklungsmech. Org.* 1: 269–305.

Signoret, J. and Lefresne, J. (1973). Contribution à l'étude de la segmentation de l'oeuf d'Axolotl. II. Influence de modifications du noyau et du cytoplasme sur les modalités, de la segmentation. *Ann. Embryol. Morphogen.* 6: 299–307.

Solursh, M. and Morriss, G. M. (1977). Glycosaminoglycan synthesis in rat embryos during the formation of the primary mesenchyme and neural folds. *Dev. Biol.* 57: 75–86.

Solursh, M. and Revel, J. P. (1978). A scanning electron microscope study of cell shape and cell appendages in the primitive streak region of the rat and chick embryo. *Differentiation* 11: 185–190.

Spemann, H. and Mangold, H. P. (1924). Induction of embryogenic primordia by the implantation of organizer from different species. In B. Willier and J. M. Oppenheimer (eds.), *Foundations of Experimental Embryology*. Hafner, New York, pp. 144–184.

Spratt, N. T., Jr. (1946). Formation of the primitive streak in the explanted chick blastoderm marked with carbon particles. *J. Exp. Zool.* 103: 259–304.

Spratt, N. T., Jr. (1947). Regression and shortening of the primitive streak in the explanted chick blastoderm. *J. Exp. Zool.* 104: 69–100.

Spratt, N. T., Jr. (1963). Role of the substratum, supracellular continuity, and differential growth in morphogenetic cell movements. *Dev. Biol.* 7: 51–63.

Sugiyama, K. (1972). Occurrence of mucopolysaccharides in the early development of the sea urchin embryo and its role in gastrulation. *Dev. Growth Differ.* 14: 62–73.

Tilney, L. G. and Gibbins, J. R. (1969). Microtubules in the formation and development of the primary mesenchyme in *Arbacia punctulata*. II. An experimental analysis of their role in the development and maintenance of cell shape. *J. Cell Biol.* 41: 227–250.

Trinkaus, J. P. (1969). *Cells into Organs: The Forces That Shape the Embryo*. Prentice-Hall, New York.

Vogt, W. (1929). Gestaltungsanalyse am Amphibienkeim mit örtlicher Vitalfärbung. II. Teil Gastrulation und Mesodermbildung bei Urodelen und Anuren. *Wilhelm Roux Arch. Entwicklungsmech. Org.* 120: 384–706.

Waddington, C. H. (1932). Experiments in the development of chick and duck embryos cultivated *in vitro*. *Philos. Trans. Roy. Soc. Lona. [Biol.]* 13: 221.

Zander, J. and Münstermann, A. M. von (1956). Progesteron in menschlichem Blut und Geweben. III. *Klin. Wschr.* 34: 944–953.

Early vertebrate development: neurulation and ectoderm

For the real amazement, if you wish to be amazed, is this process. You start out as a single cell derived from the coupling of a sperm and an egg; this divides in two, then four, then eight, and so on, and at a certain stage there emerges a single cell which has as all its progeny the human brain. The mere existence of such a cell should be one of the great astonishments of the earth. People ought to be walking around all day, all through their waking hours calling to each other in endless wonderment, talking of nothing except that cell.
—LEWIS THOMAS (1979)

The phenomenon of life itself negates the boundaries that customarily divide our disciplines and fields.
—HANS JONAS (1966)

The vertebrate pattern of development

In 1828, Karl Ernst von Baer, the foremost embryologist of his day,[1] exclaimed, "I have two small embryos preserved in alcohol, that I forgot to label. At present I am unable to determine the genus to which they belong. They may be lizards, small birds, or even mammals." Figure 1 allows us to appreciate his quandary and it illustrates von Baer's four general laws of embryology. From his detailed study of chick development and his comparison of those embryos to embryos of other vertebrates, von Baer derived four generalizations, illustrated here with some vertebrate examples.

1. *The general features of a large group of animals appear earlier in the embryo than do the specialized features.* All developing vertebrates (fishes, reptiles, amphibians, birds, and mammals) appear very similar during certain early developmental stages. It is only later in development that the special features of class, order, and finally species emerge (Figure 1). All vertebrate embryos have gill arches, notochords, spinal chords, and pronephric kidneys.

2. *Less general characters are developed from the more general, until finally the most specialized appear.* Certain fishes use a pronephric type of kidney as adults. This type of kidney can be found in the embryos of all other types of vertebrates. However, in these more complex species, it never becomes functional but contributes to the subsequent development of more specialized types of kidneys. All vertebrates initially have the same type of skin. Only later does skin develop into one type of scales in fishes, another type of scales in reptiles, feathers in birds, and hair, claws, and nails in mammals. Similarly, the early development of the limb is essentially the same in all vertebrates. Only later do the differences between fins, legs, wings, and arms become apparent.

3. *Each embryo of a given species, instead of passing through the adult stages of other animals, departs more and more from them.* The visceral clefts of embryonic birds and mammals do not resemble the gill slits of adult fishes in detail. Rather, they resemble those visceral clefts of *embryonic* fishes and other *embryonic* vertebrates. Whereas fishes preserve and elaborate these clefts into true gill slits, mammals convert them into structures such as the eustachian tubes (between the ear and mouth).

4. *Therefore, the early embryo of a higher animal is never like (the adult of) a lower animal, but only like its early embryo.*

Thus, von Baer saw that different groups of animals shared certain common features during early embryonic development and that these features became more and more characteristic of the species as development proceeded. Human embryos never pass through a stage equivalent to an adult fish or bird; rather, human embryos initially share characteristics in common with fish and avian embryos. Later, the mam-

---

[1]K. E. von Baer discovered the notochord, the mammalian egg, and the human egg, as well as making the conceptual advances described here.

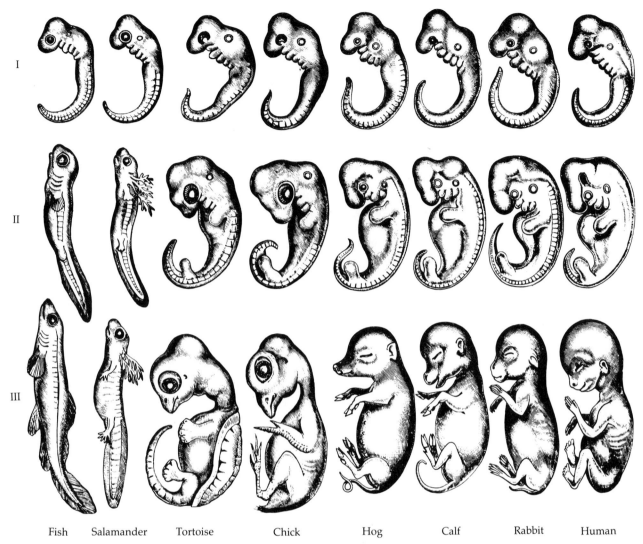

I

II

III

Fish   Salamander   Tortoise   Chick   Hog   Calf   Rabbit   Human

FIGURE 1
Illustration of von Baer's law. Early vertebrate embryos exhibit features common to the entire subphylum. As development progresses, embryos become recognizable as members of their class, their order, their family, and finally their species. (From Romanes, 1901.)

malian and other embryos diverge, none of them passing through the stages of the other.[2]

These conclusions were extremely important for Charles Darwin because evolutionary classification depended upon finding those similarities that demonstrated that certain diverse animals arose from a common ancestor. "Community of embryonic structure," wrote Darwin

[2]This work was largely ignored after Ernst Haeckel made a synthesis of German romanticism and Darwinism wherein "ontogeny [the development of an individual] recapitulated phylogeny [the evolutionary history of the species]." Here, the various stages of the human embryo were seen to correspond with the adult stages of "lower" organisms. Even though this view was scientifically discredited before it was even proposed, Haeckel had a gift for showmanship and his theory "explained" human progress. It caught on like wildfire throughout biology and the social sciences before it was shown to be based on false premises (see Gould, 1977).

in *Origin of Species* (1859), "reveals community of descent." In fact, Darwin believed that embryology was the strongest foundation for his evolutionary theory. Von Baer, of course, was not an evolutionist when he wrote his four laws in 1828. In fact, he never became one. Von Baer believed that he had found the divine plan upon which all the organisms in a group such as the vertebrates developed. Because earlier evolutionary theories had envisioned a single, nonbranching series of transformations, von Baer's observations were often used against evolution. Darwin, however, recognized that von Baer's work supported an evolutionary hypothesis in which a common ancestor could radiate into several different types of organisms by heritable modifications of embryonic development. The reason human embryos, fish embryos, and chick embryos all had visceral clefts was because they had a common ancestor whose embryo had such visceral clefts. Moreover, the reason visceral clefts gave rise to different structures in different groups of organisms (gills in fishes, eustachian tubes in mammals) was that the ancestral plan had been modified through the action of natural selection.

Von Baer also recognized that there was a common pattern to vertebrate development. The three germ layers gave rise to different organs, and their derivation was constant whether the organism was a fish, a frog, or a chick. ECTODERM formed skin and nerves; ENDODERM formed respiratory and digestive tubes; and MESODERM formed connective tissue, blood cells, heart, the urogenital system, and parts of most of the internal organs. In this chapter we shall follow the early development of ectoderm, focusing on formation of the nervous system. In the next chapter we shall follow the early development of endodermal and mesodermal organs.

Neurulation

In vertebrates, gastrulation creates an embryo having an internal endodermal layer, an intermediate mesodermal layer, and an external ectoderm. In addition, a cord of mesodermal cells, the notochord, lies directly beneath the most dorsal portion of the ectoderm. The interaction between the notochord and its superadjacent ectoderm is one of the most important interactions of all development, for the notochord directs the ectoderm to form the hollow NEURAL TUBE, which will differentiate into the brain and spinal cord. Thus, we begin a new phase of development—ORGANOGENESIS, the creation of tissues and organs. The action by which the notochord instructs the ectoderm to become neural tube is called PRIMARY EMBRYONIC INDUCTION, and the cellular response by which the flat layer of ectodermal cells is transformed into a hollow tube is called NEURULATION. The events of neurulation are diagrammed in Figure 2. Here, the original ectoderm is divided into three sets of cells: (1) the internally positioned neural tube, (2) the epidermis of the skin, and (3) the neural crest cells, which migrate from the region that had connected the neural tube and epidermal tissues.

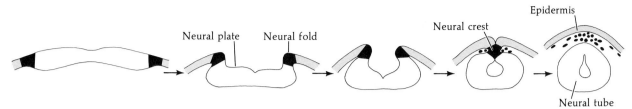

An embryo undergoing such changes is called a NEURULA. The mechanism of primary embryonic induction will be detailed in Chapter 7. In this chapter, we are concerned with the response by the various ectodermal tissues.

The process of neurulation in frog embryos is depicted in Figure 3.

FIGURE 3

Three views of neurulation in a frog embryo, showing early (left), middle (center), and late (right) neurulae in each case. (A) Transverse section through the center of the embryo. (B) The same sequence looking down on the dorsal surface of the whole embryo. (C) Sagittal section through the median plane of the embryo. (After Balinsky, 1975.)

FIGURE 2

Diagrammatic representation of neural tube formation in amphibians and amniotes. The ectodermal cells are represented either as precursors of the neural crest (black) or as precursors of the epidermis (gray). The ectoderm folds in at the most dorsal point, forming an outer epidermis and an inner neural tube connected by neural crest cells. (After Balinsky, 1975.)

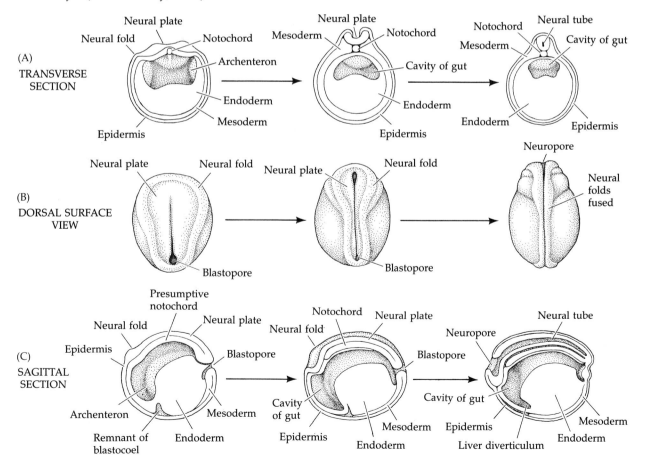

The mechanisms of neural tube formation appear to be similar in amphibians, reptiles, birds, and mammals (Gallera, 1971), so we will be considering various groups as we proceed through our survey.[3] The first indication that a region of ectoderm is destined to become neural tissue is a change in cell shape (Figure 4). Midline ectodermal cells become elongated, whereas those cells destined to form the epidermis become more flattened. The elongation of dorsal ectodermal cells causes these prospective neural regions to rise above the surrounding ectoderm, thus creating the NEURAL PLATE. As much as 50 percent of the ectoderm is included in this plate. Shortly thereafter, the edges of the neural plate thicken and move upward to form the NEURAL FOLDS, while a U-shaped NEURAL GROOVE appears in the center of the plate, dividing the future right and left sides of the embryo (Figures 3 and 5). The

[3]Among the vertebrates, fishes generate their neural tube in a different manner. Fish neural tubes do not form from an infolding of the overlying ectoderm, but rather a solid cord of cells is induced to sink into the embryo. This cord subsequently hollows out to form the neural tube. In mammals, only the most posterior portion of the neural tube forms by cavitation.

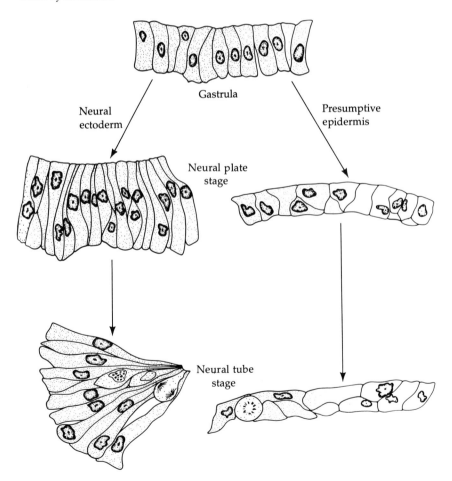

FIGURE 4
Schematic diagram of the shape changes during neurulation in the salamander. At the end of gastrulation, the ectoderm is a uniform epithelium. During neurulation, neural epithelial cells elongate to form the neural plate and then constrict at their apices to form the neural tube. Presumptive epidermal cells flatten throughout neurulation. (After Burnside, 1971.)

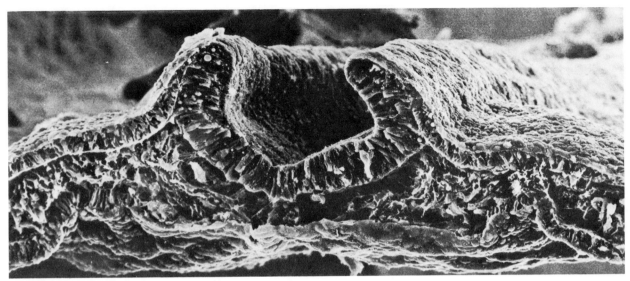

FIGURE 5
Scanning electron micrograph of neural tube formation in the chick embryo.
Elongated neural epithelial cells form a tube as the flattened epidermal cells
are brought to the midline of the embryo. (Photograph courtesy of K. W.
Tosney.)

neural folds migrate toward the midline of the embryo, eventually
fusing to form the neural tube beneath the overlying ectoderm. The
cells at the junction between the outer ectoderm and the neural tube
become the NEURAL CREST cells. These crest cells will migrate through
the embryo and will give rise to several cell populations, including
pigment cells and the cells of the peripheral nervous system.

The neural tube and the origins
of the central nervous system

The formation of the neural tube does not occur simultaneously
throughout the ectoderm. This is best seen in those vertebrates (such
as birds and mammals) whose body axis is elongated prior to neurulation. Figure 6 depicts neurulation in a 24-hr chick embryo. Neurulation
in the cephalic (head) region is well advanced while the caudal (tail)
region of the embryo is still undergoing gastrulation. Regionalization
of the neural tube also occurs as a result of changes in the shape of the
tube: In the cephalic end (where the brain will form), the tube is broad,
and a series of swellings and constrictions emerge and define the various brain compartments. Posterior to the head region, however, a
simple tube that tapers off toward the tail is formed. The two open
ends of the neural tube are called the ANTERIOR NEUROPORE and the

FIGURE 6
Stereogram of a 24-hour chick embryo. Cephalic portions are finishing neurulation while the caudal portions are still undergoing gastrulation. (From Patten, 1971; after Huettner, 1949.)

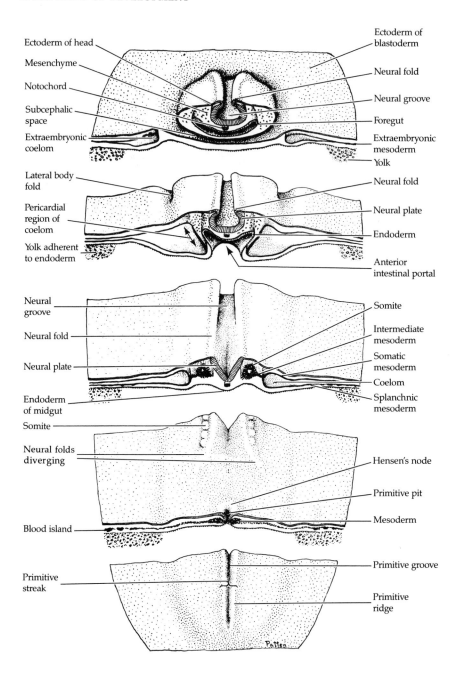

POSTERIOR NEUROPORE. In mammals, these pores allow amniotic fluid to flow through the neural tube for a time (Figure 7). Failure to close the human posterior neuropore (at day 27) results in SPINA BIFIDA, the severity of which depends upon how much of the spinal cord remains open. However, failure to close the anterior neuropore results in a lethal condition, ANENCEPHALY (Figure 8). Here, the forebrain tissue becomes

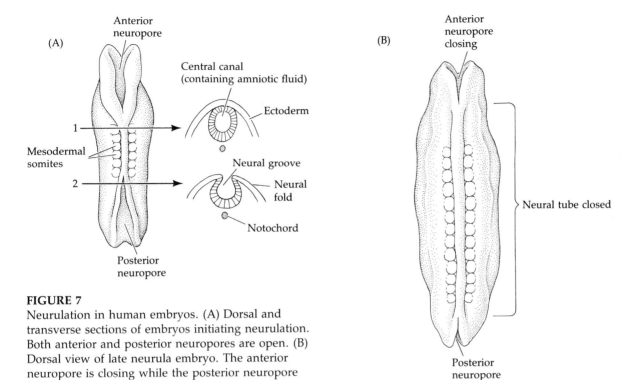

FIGURE 7
Neurulation in human embryos. (A) Dorsal and transverse sections of embryos initiating neurulation. Both anterior and posterior neuropores are open. (B) Dorsal view of late neurula embryo. The anterior neuropore is closing while the posterior neuropore remains open.

exposed to the surface and degenerates. Fetal brain development ceases, and the vault of the skull fails to form. This abnormality is not that uncommon in humans, occurring in about 0.1 percent of all pregnancies; neural tube closure defects can now be detected during pregnancy by various physical and chemical tests.

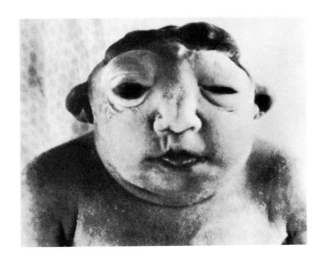

FIGURE 8
Head of anencephalic human stillborn fetus, delivered at 13 weeks gestation. (From Warkany, 1971.)

Mechanism of neural tube formation

Neural tube formation is intimately linked to changes in cell shape, and microtubules and microfilaments are both involved in these changes that create the neural tube. Ectodermal cells elongate as the randomly arranged microtubules of these cells align themselves parallel to the lengthening axis (Figure 9). This stage of neural tube formation can be blocked by colchicine, an inhibitor of microtubule polymerization (Burnside, 1973). A second change in cell shape involves the apical constriction of cells to form a cylinder; this change is directed by a ring of contractile microfilaments encircling the apical margins of the cells. The contraction of these microfilaments produces a "purse-string" effect, constricting the apical end of each cell. Burnside (1971) and Karfunkel (1972) have shown that when embryos are cultured in the presence of cytochalasin B, the neuroectodermal cells can elongate but cannot constrict to form the neural folds.

An interdisciplinary effort between biologists and mathematicians has succeeded in showing that changes in elongation and constriction may be all that are needed to change a flat sheet into a hollow tube

(A)
Microfilaments

Microtubules

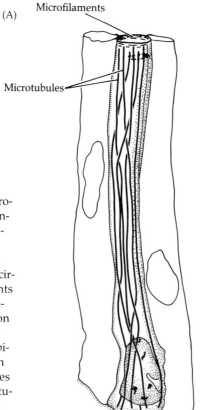

(B)

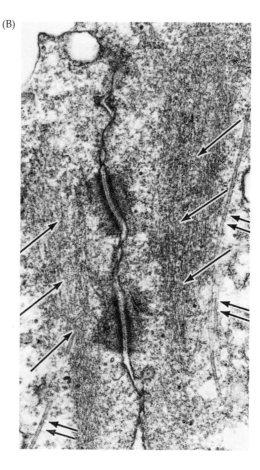

FIGURE 9
(A) Schematic diagram of microtubule and microfilament orientation in neural plate cells. Microtubules are seen aligned parallel to the long axis of the cell; microfilaments (black) encircle the apex. The microfilaments are often seen to attach to desmosomes, specific structures on the cell membrane where two cells come into contact. (B) Apical microfilament bundles seen near desmosomes on both sides of the cell-cell contacts. Microtubules (double arrows) are also seen. (From Burnside, 1971; photograph courtesy of B. Burnside.)

10 μm

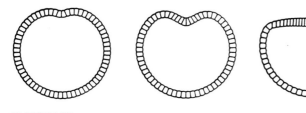
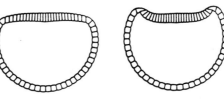
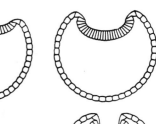
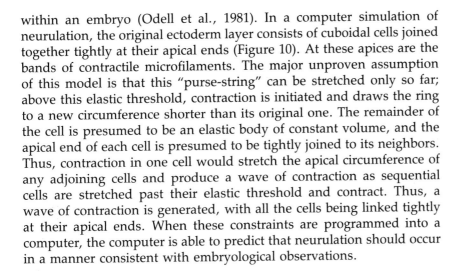
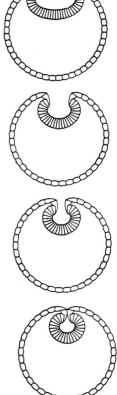

FIGURE 10
Computer-generated model of neural tube formation in amphibians. (From Odell et al., 1981.)

within an embryo (Odell et al., 1981). In a computer simulation of neurulation, the original ectoderm layer consists of cuboidal cells joined together tightly at their apical ends (Figure 10). At these apices are the bands of contractile microfilaments. The major unproven assumption of this model is that this "purse-string" can be stretched only so far; above this elastic threshold, contraction is initiated and draws the ring to a new circumference shorter than its original one. The remainder of the cell is presumed to be an elastic body of constant volume, and the apical end of each cell is presumed to be tightly joined to its neighbors. Thus, contraction in one cell would stretch the apical circumference of any adjoining cells and produce a wave of contraction as sequential cells are stretched past their elastic threshold and contract. Thus, a wave of contraction is generated, with all the cells being linked tightly at their apical ends. When these constraints are programmed into a computer, the computer is able to predict that neurulation should occur in a manner consistent with embryological observations.

Differentiation of the neural tube

Formation of brain regions.   The differentiation of the neural tube into the various regions of the central nervous system occurs simultaneously in three different ways. On the gross anatomical level, the neural tube bulges and constricts to form the chambers of the brain and the spinal cord. At the tissue level, the wall of the neural tube becomes rearranged in various ways to form the different functional regions of the brain and the spinal cord. Finally, on the cellular level, the neuroepithelial cells themselves differentiate into the numerous different types of neurons present in the body. The early development of most vertebrate brains is similar, but because the human brain is probably the most interesting organ in the animal kingdom, we shall concentrate on the development that is supposed to make *Homo* sapient.

The early mammalian neural tube is a straight structure. However, even before the posterior portion of the tube has formed, the most anterior portion of the tube is undergoing drastic changes. In this an-

FIGURE 11
Early brain development in a 4-week human embryo. (A) Lateral view. (B) Extended diagram to illustrate the bulges of the neural tube. (After Langman, 1969.)

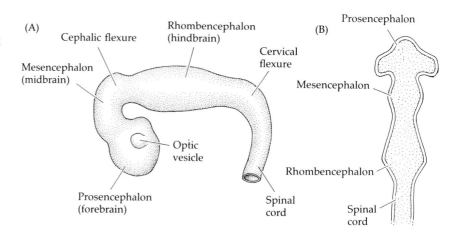

terior region, the neural tube balloons into three primary vesicles (Figure 11): forebrain (PROSENCEPHALON), midbrain (MESENCEPHALON), and hindbrain (RHOMBENCEPHALON). By the time the posterior end of the neural tube closes, secondary bulges—the OPTIC VESICLES—have extended laterally from each side of the developing forebrain. Moreover, the future brain has bent so that the creases demark the boundaries of the brain cavities. The major creases are the CEPHALIC FLEXURE and the CERVICAL FLEXURE.

The forebrain becomes subdivided into the anterior TELENCEPHALON and the more caudal DIENCEPHALON (Figure 12). The telencephalon will eventually form the CEREBRAL HEMISPHERES, and the diencephalon will form the thalamic and hypothalamic brain regions as well as the region that receives neural input from the eyes. The rhombencephalon becomes subdivided into a posterior MYELENCEPHALON and a more anterior METENCEPHALON. The myelencephalon eventually becomes the MEDULLA OBLONGATA, the nerves of which regulate respiratory, gastrointestinal, and cardiovascular movements; the metencephalon gives rise

FIGURE 12
Further human brain development. Lateral view (A) and midline extended diagram (B) of a 6-week human brain showing secondary bulges of the neural tube. (After Langman, 1969.)

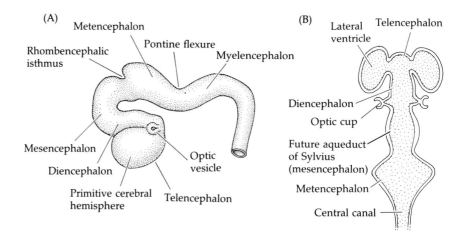

to the CEREBELLUM, the part of the brain responsible for coordinating movements, posture, and balance. The mesencephalon does not subdivide, and its lumen eventually becomes the cerebral aqueduct. The lineages of the brain are illustrated in Figure 13.

Tissue architecture of the central nervous system.   The original neural tube is composed of a germinal neuroepithelium, one cell layer thick. This is a rapidly dividing cell population. Sauer (1935) and others have shown that all of these cells are continuous from the lumen of the neural tube to the edges but that the nuclei of these cells are at different heights, thereby giving the superficial impression that the neural tube has numerous cell layers. The position of the nucleus between the lumen and the edge of the tube depends on the stage of the cell's cycle (Figure 14). DNA synthesis (S phase) occurs as the nucleus is at the

FIGURE 13
Schema of early human brain development in terms of regional specialization.

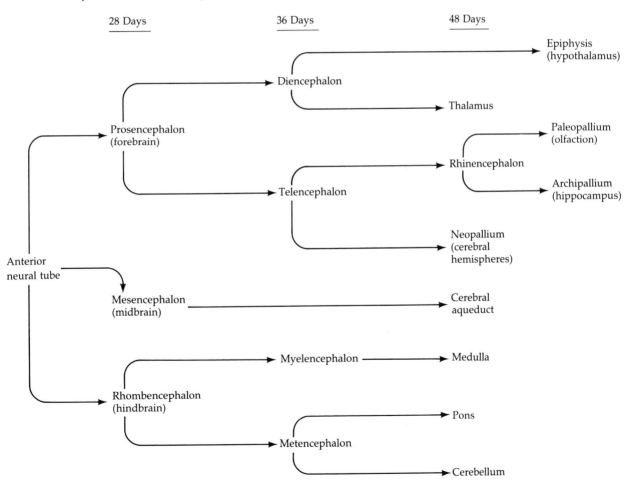

Stage of cell cycle

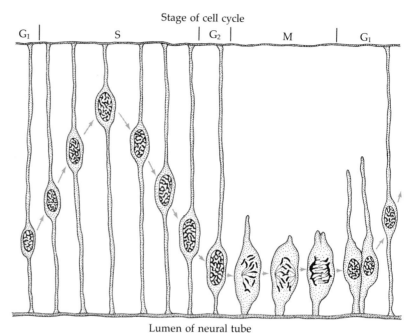

Lumen of neural tube

FIGURE 14

Section of the neural tube of a chick embryo showing the position of a single nucleus in a neuroepithelial cell as a function of the cell cycle. Mitotic cells are to be found near the center of the neural tube, adjacent to the lumen. (Adapted from Sauer, 1935.)

edge of the tube, and the nucleus migrates luminally as mitosis proceeds. Mitosis occurs at the lumen of the tube. During early development, 100 percent of the neural tube cells will incorporate radioactive thymidine into DNA (Fujita, 1964). Shortly thereafter, certain cells stop incorporating these DNA precursors, thereby indicating that they are no longer participating in DNA synthesis and mitosis. These are the young neuronal and glial (supporting) cells that migrate to the periphery of the neural tube to differentiate (Fujita, 1966; Jacobson, 1968). Subsequent neural differentiation is dependent upon the position these NEUROBLASTS occupy once outside the region of dividing cells (Jacobson, 1978; Letourneau, 1977).

As the cells adjacent to the lumen continue to divide, the migrating cells form a second layer around the original neural tube. This layer gets progressively thicker as more cells are added to it from the germinal epithelium. This new layer is called the MANTLE ZONE (Figure 15A), and the germinal epithelium is now called the EPENDYMA. The mantle zone cells differentiate into both neurons and glia. The neurons make connections among themselves and send forth axons away from the lumen, thereby creating a cell-poor MARGINAL ZONE. Eventually, glial cells cover each of these axons in a myelin sheath, giving them a whitish appear-

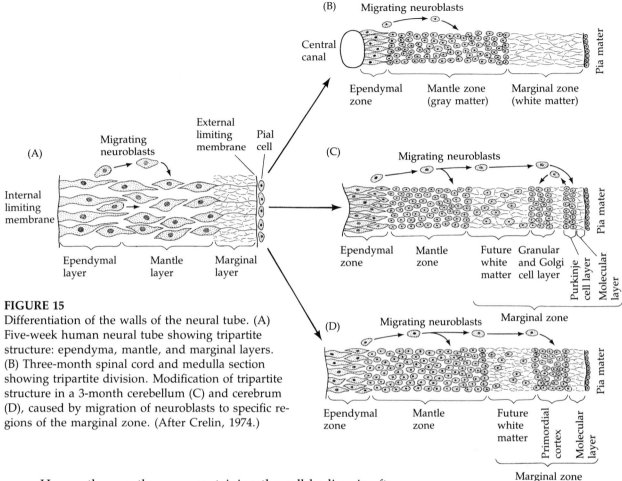

FIGURE 15
Differentiation of the walls of the neural tube. (A)
Five-week human neural tube showing tripartite
structure: ependyma, mantle, and marginal layers.
(B) Three-month spinal cord and medulla section
showing tripartite division. Modification of tripartite
structure in a 3-month cerebellum (C) and cerebrum
(D), caused by migration of neuroblasts to specific regions of the marginal zone. (After Crelin, 1974.)

ance. Hence, the mantle zone, containing the cell bodies, is often referred to as the GRAY MATTER; and the axonal, marginal layer is often
called the WHITE MATTER.

In the spinal cord and medulla, this basic three-zone pattern of
ependymal, mantle, and marginal layers is retained throughout development. The gray matter (mantle) gradually becomes an H-shaped
structure surrounded by white matter; and both become encased in
connective tissue. During its maturation, a longitudinal groove—the
SULCUS LIMITANS—divides the tube into dorsal and ventral halves. The
dorsal half receives sensory neurons and the ventral half is involved in
effecting the motor functions.

In the brain, however, cell migration, differential growth, and selective cell death produce modifications of this three-zone pattern, especially in the cerebellum and cerebrum (Figure 15B, C, and D). Within
the gray matter of the cerebellum, the neuroblasts often cluster together
into groups called nuclei. Each nucleus works as a functional unit. Other
gray matter neuroblasts migrate *through* the white matter. Some of these
cells will form a new germinal zone near the outer boundary of the

neural tube. The neuroblasts formed by these germinal cells migrate back into the developing white matter to produce a region of granular neurons and glial cells. The cells remaining in the new germinal region are the precursors of the PURKINJE NEURONS. Once the layer of granular neurons is established, the precursors of the Purkinje neurons migrate down as well. These differentiate into one of the most complex of all neuronal types. Each Purkinje neuron has an enormous DENDRITIC APPARATUS, which spreads like a fan above a bulblike soma (Figure 16a). A typical Purkinje cell may form as many as 100,000 synapses with other neurons, more than any other neuron studied. Each Purkinje neuron emits a slender AXON, which connects to a cell in a cerebellar nucleus in the gray matter.

The development of spatial organization, then, is critical for the proper functioning of the cerebellum. All impulses eventually regulate the activity of the Purkinje cells, which are the only output neurons of the cerebellum cortex. For this to happen, the proper cells must differentiate at the appropriate place and time.

Some insight into the mechanism of spatial ordering has come from the analysis of neurological mutations in mice. Over 100 mutations are known to affect the arrangement of cerebellar neurons. The defect in *staggerer* mice appears to reside in the Purkinje cells. These cells are smaller than normal and have stunted dendrites with very few spines (Figure 16B). The granule cells originate on schedule, migrate normally, but then die. Sidman (1974) and Sotelo and Changeaux (1974) have suggested that the granule cells die because they are dependent upon connections with the Purkinje dendrites. The resulting mice are small, have tremors, and exhibit a characteristic staggering walk. A similar mutation is *weaver*. Here, too, the Purkinje cell dendritic tree is grossly stunted, but this defect is believed to be secondary to a deficiency of a type of glial cell called the Bergmann glia. The mutant mice contain less of these glial cells than do normal mice, and those glia that do exist are morphologically abnormal. Microscopic studies of normal and mutant cerebella suggest that the processes of Bergmann glial cells serve to guide Purkinje cell migration (Figure 17). Without this migratory guide, there is no distinct Purkinje layer and thus few Purkinje–granule neuron connections (Rakic and Sidman, 1973).

In the cerebrum, the three-zone arrangement is also modified. Certain neuroblasts from the mantle zone migrate through the white matter to generate a second zone of neurons. This new layer is called the NEOPALLIAL CORTEX. The cortex eventually stratifies into six layers of cell bodies, and the final differentiation of these neopallial neurons is not complete until the middle of childhood.

(A)

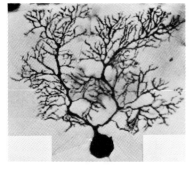

(B)

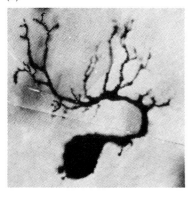

FIGURE 16
Purkinje neurons. (A) Normal mouse Purkinje neuron. (B) Stunted Purkinje neuron in the *staggerer* mutation of the mouse. This mutant has impaired cerebellar function, leading to its characteristic walk. (From Berry et al., 1980; courtesy of M. Berry.)

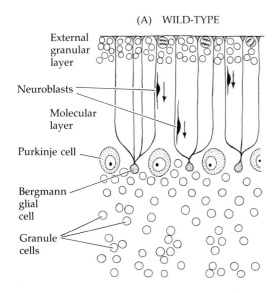

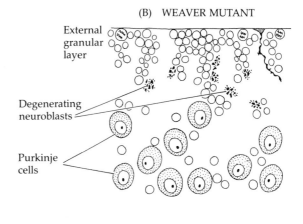

FIGURE 17

Failure of cell migration in the *weaver* mutant cerebellum. In normal mice (A), neuroblasts migrate inward from the external granular layer (as diagrammed in Figure 15) along the processes emanating from the Bergmann glial cells. In the *weaver* mutant (B), these glial cells are absent, and the neuroblasts do not migrate back into the wall of the neural tube. Instead, most of them degenerate. The illustration has been simplified by omitting other cell types and the Purkinje cell axons and dendrites. (After Rakic and Sidman, 1973.)

The evolution of cerebral development

SIDELIGHTS & SPECULATIONS

The evolution of the cerebral cortex is among the most spectacular stories of vertebrate anatomy. The cerebral lobes originated as paired outgrowths of the forebrain; primarily for olfactory perception (as in fishes). Amphibians and reptiles overlaid this primitive olfactory-oriented PALEOPALLIUM with ARCHIPALLIUM and CORPUS STRIATUM (Figure 18). The archipallium appears to be associated with "emotional" types of behavior, whereas the corpus striatum is concerned with automatic "instinctual" responses. In birds, this corpus striatum is greatly developed. In advanced reptiles and mammals, the NEOPALLIAL neurons develop. It is the growth of the neopallium that characterizes mammalian and especially human evolution. Here, the archipallium has been pushed internally to form the hip-

pocampus (which is involved in sexual and aggressive behaviors), and the corpus callosum has become a relay center for certain involuntary reactions. The neopallium has become the seat of learning, memory, and intelligence. Birds lack any neopallium, and those species capable of learning have developed a HYPERSTRIATUM tissue, where these abilities reside (Romer, 1976).

The development of the human neopallium continues for a remarkably long time. In fact, the human brain continues to develop at fetal rates even after birth (Holt et al., 1975). Portmann (1941, 1945) has suggested that, compared to other primates, human gestation is very much too short. By comparing human development at birth to that of other primates, Portmann determined that human gestation "should" be 21 months

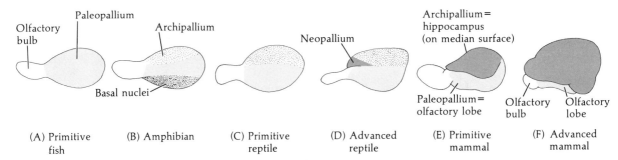

(A) Primitive fish   (B) Amphibian   (C) Primitive reptile   (D) Advanced reptile   (E) Primitive mammal   (F) Advanced mammal

FIGURE 18
Evolution of the vertebrate brain, showing the progressive differentiation of the cerebral cortex. (A) In its most primitive form, the cerebrum is solely for olfaction (smell). (B) In amphibians, the dorsal and ventral portions of the cerebrum become differentiated into the archipallium (hippocampus) and the basal nuclei (corpus striatum). (C) In reptiles, the basal nuclei are found deeper within the cerebrum, and a new group of neurons, the neopallium, is seen on the cortex of advanced reptiles (D). With the advent of mammals (E, F), the expanding neopallial neurons displace the archipallium, eventually dominating the cerebral cortex. (From Romer, 1976.)

long. Moreover, he and Gould (1977) have speculated that we actually do have such a long gestation. During our first year of life, we are essentially extrauterine fetuses.

The reason for this state is that our brain keeps growing at its enormous fetal rate. No woman could deliver a 21-month-old infant because the head would not pass through the birth canal. Birth must occur while the head is still small enough to fit through the mother's pelvis. Thus, humans give birth at the end of 9 months rather than at the end of 21 months. It is during this time, moreover, that we are first exposed to the world. Our developing nervous system takes advantage of the incredible stimulation given it during its first year.

Neuronal types. The human brain consists of over 100 billion nerve cells (neurons) associated with over a trillion supporting cells called GLIA. Both types of cells are believed to be derived from the neural tube (Figure 19). Those cells that remain as integral components of the neural tube become EPENDYMAL CELLS. If these cells migrate from the tube very early, they form ASTROGLIAL CELLS or OLIGONDENDROGLIAL CELLS. Those cells that migrate later from the germinal epithelium become NEUROBLASTS. These neuroblasts differentiate into various types of neurons, depending upon the portion of the brain that they reside in. Some develop only a few cytoplasmic regions where other cells can relay electrical impulses, whereas other neurons develop extensive areas for cellular interaction. These fine extensions of the cell that are used to pick up electrical impulses are called DENDRITES (Figure 20). There are very few dendrites on cortical neurons at birth, but one of the amazing things about the first year of life is the increase in the number of such receptive regions in the cortical neurons. During this initial year of life, each cortical neuron develops enough dendrites to accommodate as

FIGURE 19
Summary diagram of some of the differentiated cell types of the nervous system and their derivation. (Modified from Crelin, 1974.)

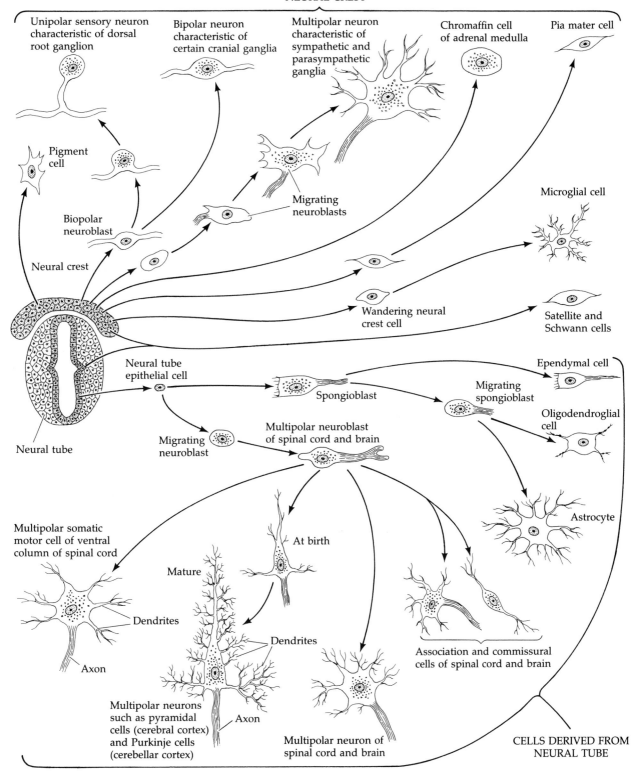

Unipolar sensory neuron characteristic of dorsal root ganglion

Bipolar neuron characteristic of certain cranial ganglia

Multipolar neuron characteristic of sympathetic and parasympathetic ganglia

Chromaffin cell of adrenal medulla

Pia mater cell

Pigment cell

Biopolar neuroblast

Neural crest

Migrating neuroblasts

Microglial cell

Wandering neural crest cell

Satellite and Schwann cells

Neural tube epithelial cell

Spongioblast

Migrating spongioblast

Ependymal cell

Oligodendroglial cell

Neural tube

Migrating neuroblast

Multipolar neuroblast of spinal cord and brain

Astrocyte

Multipolar somatic motor cell of ventral column of spinal cord

At birth

Mature

Dendrites

Dendrites

Axon

Association and commissural cells of spinal cord and brain

Multipolar neurons such as pyramidal cells (cerebral cortex) and Purkinje cells (cerebellar cortex)

Axon

Multipolar neuron of spinal cord and brain

CELLS DERIVED FROM NEURAL TUBE

167

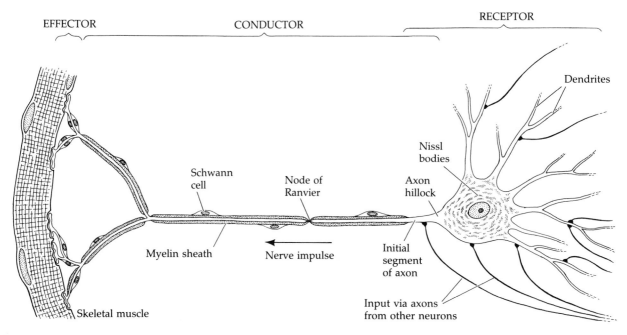

EFFECTOR   CONDUCTOR   RECEPTOR

Dendrites

Schwann cell

Node of Ranvier

Nissl bodies

Axon hillock

Myelin sheath   Nerve impulse

Initial segment of axon

Skeletal muscle

Input via axons from other neurons

FIGURE 20

Diagrammatic representation of a large motor neuron. Impulses received by the dendrites and the stimulated neuron can transmit electrical impulses through the axon (which may be 2–3 feet long) to its target tissue. The myelin sheath, which provides insulation for the axon, is formed by adjacent Schwann cells. (After Bloom and Fawcett, 1975.)

many as 100,000 connections with other neurons. The average cortical neuron connects with 10,000 other neural cells. This pattern of neural development is essential for the human cortex to function as the center for learning, reasoning, and memory, to develop the capacity for symbolic expression, and to produce voluntary responses to interpreted stimuli.

Another feature of neural development is the AXON. Whereas the dendrites are often numerous and do not extend far from the nerve cell body (or SOMA), axons may span several feet. Thus, the rhythms of the heart are controlled by nerves whose axons are located as far away as the medulla oblongata. The pain receptors on one's big toe have a long way to transmit their message to the spinal cord. One of the fundamental concepts of neurobiology is that the axon is a continuous extension of the nerve cell body. At the turn of the last century, there were still numerous competing theories of axon formation. Schwann, one of the founders of the cell theory, believed that numerous neural cells linked themselves together in a chain to form an axon. Hensen (the discoverer of the node) thought that the axon formed around preexisting cytoplasmic threads between the cells. Wilhelm His (1886) and S. Ramón y Cajal (1890) postulated that the axon was indeed an outgrowth (albeit an extremely large one) of the nerve soma.

In 1907, Ross Harrison demonstrated the validity of the outgrowth theory in an elegant experiment that founded both the science of developmental neurobiology and the technique of tissue culture. Harrison isolated a portion of neural tube from a 3-mm frog tadpole. At this stage, shortly after the closure of the neural tube, there is no visible

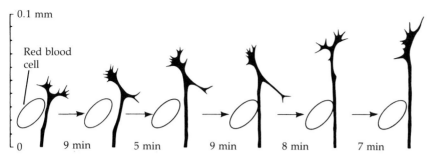

differentiation of axons. He placed these neuroblasts in a drop of frog lymph on a coverslip and inverted the coverslip over a depression slide. Harrison could thus watch what was happening within this "hanging drop." What Harrison saw (Figure 21) was that the axons emerged as outgrowths from the neuroblasts, elongating at about 56 μm/hr.

Such nerve outgrowth is led by the tip of the axon, which is called the GROWTH CONE (Figure 22). This cone does not proceed in a straight line but rather "feels" its way along the substrate. The growth cones move by the elongation and contraction of pointed filopodia called MICROSPIKES. These microspike filopodia contain microfilaments, which are oriented in an array parallel to the long axis of the axon. (You will no doubt recall a similar situation existing in the filopodial microfilaments of secondary mesenchymal cells in echinoderms.) Treating neurons with cytochalasin B will inhibit their further advance (Yamada et al., 1971). Within the axon itself, structural support is provided by microtubules and the axon will disassociate if placed in a solution of colchicine (Figure 23). Thus, the developing neuron retains the same

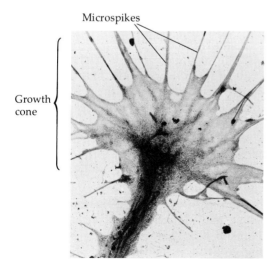

FIGURE 22
The growth cone of the axon. Transmission electron micrograph of growth cone region showing microspikes. (From Letourneau, 1979; photograph courtesy of P. C. Letourneau.)

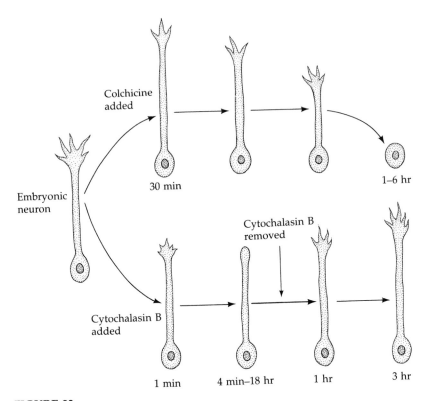

FIGURE 23
Effects of cytochalasin B and colchicine on axon elongation in cultured embryonic neurons. Colchicine causes the collapse of the axon, an event that correlates with the disruption of the microtubules. Cytochalasin B causes the reversible depolymerization of microfilaments, thereby leading to the disappearance of the microspikes. When cytochalasin B is removed from the medium, the growth cone reappears.

features that we have already noted in neural tube formation, namely, elongation by microtubules and apical shape changes by microfilaments. In most migrating cells, the exploratory filopodia will attach to the substrate and pull the rest of the cell over to it. This, of course, would be difficult given the length of the axon. Thus, the contractile apparatus in the axonal filopodium appears to be solely exploratory; and elongation is accomplished by cytoplasmic movement through the axon.

Neurons transmit electrical impulses from one region to another. These impulses usually go from the dendrites to the nerve soma and ultimately to the axon. To insulate the axon and to increase the velocity of its conduction, a type of glial cell called the oligodendrocyte wraps itself around the developing axon and secretes myelin (Figure 24). (In the peripheral nervous system, Schwann glial cells accomplish this myelination.) This myelin sheath is essential for proper neural function, and demyelination of nerve fibers is associated with several lethal diseases. The axon must also be specialized for secreting a specific NEU-ROTRANSMITTER SUBSTANCE across the small gaps (synaptic clefts) that separate the axon of one cell from its target, the soma or dendrites of a receiving neuron (or a receptor on the effector organ). Thus, the developing neuron must also develop the specific enzymes required for the production and degradation of its neurotransmitter substance, and the development of neurons involves both structural and molecular differentiation.

FIGURE 24
Myelination in the central and peripheral nervous systems. (A) In the peripheral nervous system, Schwann cells wrap themselves around the axon; in the central nervous system, myelination is accomplished by oligodendroglial cells. (B) The mechanism of this wrapping entails the production of an enormous membrane complex. (C) Micrograph of an axon enveloped by the myelin membrane of a Schwann cell. (Photograph courtesy of C. S. Raine.)

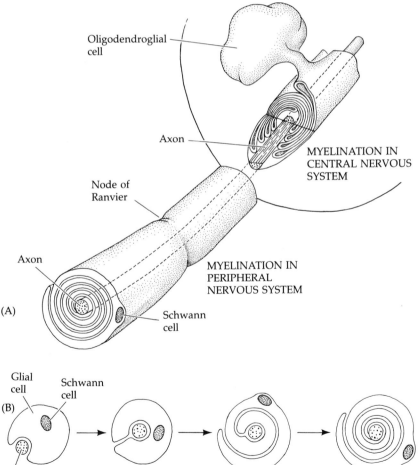

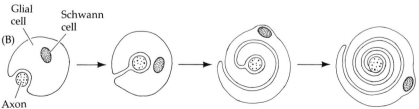

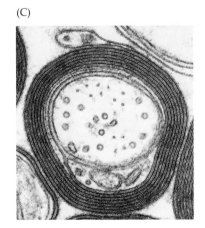

Development of the eye

Dynamics of optic development.   An individual gains knowledge of its environment through its sensory organs. In this section, we will focus on the eye because this organ, perhaps more than any other organ in the body, must develop with precision and perfect coordination of all its components.

The story of optic development begins at the wall of the diencephalon. In humans, optic development begins as the wall of the 22-day embryonic diencephalon bulges out laterally from the neural tube. This differential growth produces the OPTIC VESICLES, which are connected to the diencephalon by the OPTIC STALK. Subsequently, these vesicles contact the surface ectoderm and induce this ectoderm to form the LENSES (Figure 25).[4] In this induction, the optic vesicles are said to induce the overlying ectoderm to form the lens precursors. If isolated optic vesicles are implanted adjacent to any region of head ectoderm,

[4]The inductions forming the eye will be detailed in Chapters 7 and 16.

FIGURE 25
Development of the eye. (A) Optic vesicle from the brain contacts the overlying ectoderm. (B–C) Overlying ectoderm differentiates into lens cells as the optic vesicle folds in on itself. (D) Optic vesicle becomes the neural and pigmented retina as the lens is internalized. (E) The lens induces the overlying ectoderm to become the cornea as the optic stalk develops to carry impulses from the eye to the brain.

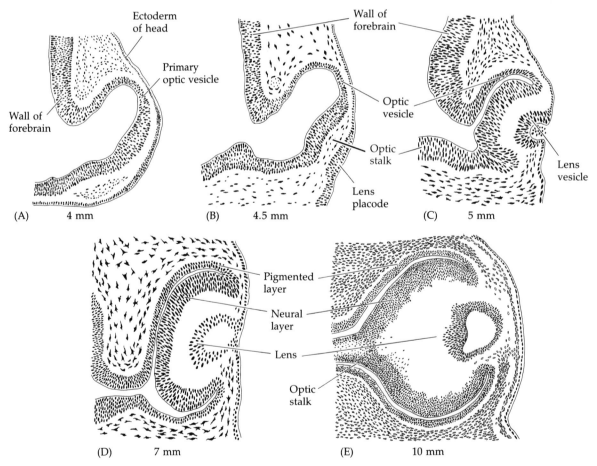

that ectoderm will form lens tissue instead of epidermal cells, whereas in the absence of optic vesicle contact, no lens forms. The necessity for close contact between the optic vesicles and the surface ectoderm is also seen in both experimental cases and in certain mutants. For example, in the mouse mutant *eyeless*, the optic vesicles fail to contact the surface and eye formation ceases.

As the optic vesicle induces the formation of the LENS PLACODE, the lens placode reciprocates and causes changes in the optic vesicle. The vesicle invaginates to form a double-walled OPTIC CUP (Figure 26). As the invagination continues, the connection between the optic cup and the brain is reduced to a narrow slit. At the same time, the two layers of the optic cup begin to differentiate in different directions. The outer cup becomes thinner and develops melanin-containing granules. This layer ultimately becomes the PIGMENTED RETINA. The cells of the inner layer elongate and begin to develop light-sensitive rods and cones. This layer becomes the NEURAL RETINA. The axons from the neural retina

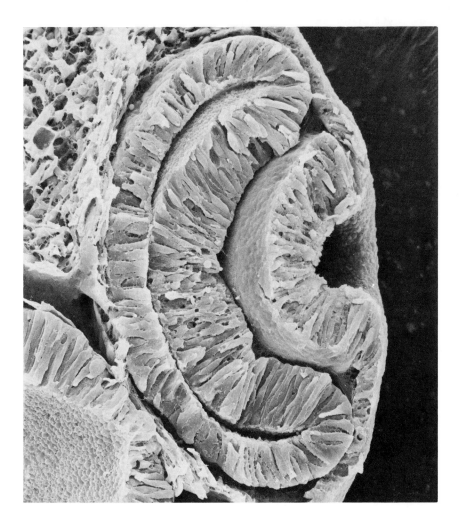

FIGURE 26
Scanning electron micrograph of the formation of the optic cup and lens placode in a chick. (From Hilfer and Yang, 1980; courtesy of S. R. Hilfer.)

meet at the base of the eye and travel down the OPTIC STALK. This stalk is then called the OPTIC NERVE.

Neural retina differentiation.   Like the cerebral and cerebellar walls, the neural retina develops into a layered array of different neuronal types (Figure 27). These layers include the light- and color-sensitive cells, the cell bodies of the optic nerve axons, and the bipolar neurons, which transmit the electrical stimulus from the sensory cells to the cell bodies of the optic nerve. In addition, there are numerous cells that serve to maintain the integrity of the retina.

In the early stages of retinal development, cell division occurs primarily at the margin of the optic cup (the opposite of neural tube cell divisions). Thus, division occurs on the outer surface of the neural layer. These cells migrate into the deeper regions of the optic cup, eventually filling the cup with neuroblastic cells. Then, in some unknown fashion, most of the neuroblasts die, leaving only bands at certain depths within the optic cup. Differentiation of these neuroblasts begins at the innermost layer of the retina, the ganglionic cells of the optic nerve, and proceeds in stepwise fashion until the sensory apparatuses of the rods and cones are formed.

The axons of the ganglionic cells form the optic nerve. Meanwhile, the dendrites of these neurons contact the neuroblasts of the INNER NUCLEAR LAYER, causing them to differentiate into the bipolar neurons of the retina. The OUTER NUCLEAR LAYER, which contains the nuclei of the photoreceptive neurons, is the last to differentiate. The axons of these photoreceptor cells synapse with the dendrites of the bipolar neurons. As they differentiate, the cell bodies of these outer neurons produce a bud of cytoplasm that contains several specialized organelles, which elongate the bud and which adjust the size and shape of the photoreactive regions (Detwiler, 1932). The cell membrane of these cells folds back upon itself to form sacs upon which the photoreceptive pigments are placed (Figure 28). Light induces these pigments to undergo chemical changes that result in a release of electrons, and the electrical impulse so generated is transmitted to the brain through the optic nerve.

Lens and cornea differentation.   During its continued development into a lens, the lens placode rounds up and contacts the new overlying ectoderm (Figure 29). The lens placode then induces the ectoderm to form the transparent CORNEA. Although we are primarily interested in the ectodermal components of the eye, the mesoderm plays an interesting indirect role in the differentiation of the cornea. Intraocular fluid pressure is necessary for the correct curvature of the cornea so that light can be focused upon the retina. The importance of such ocular pressure can be demonstrated experimentally; the cornea will not develop its characteristic curve when a small glass tube is inserted through the wall of a developing chick eye to drain away intraocular fluids (Coulombre, 1956, 1965). Intraocular pressure is sustained by a ring of scleral bones

FIGURE 27
Schematic representation of the organization of the neural retina in a 25-week human fetus. (From Langman, 1969.)

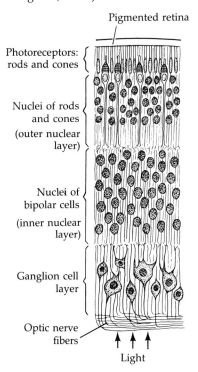

Pigmented retina

Photoreceptors: rods and cones

Nuclei of rods and cones (outer nuclear layer)

Nuclei of bipolar cells (inner nuclear layer)

Ganglion cell layer

Optic nerve fibers

Light

(A)

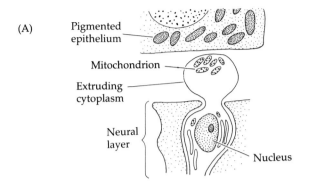

Pigmented epithelium

Mitochondrion

Extruding cytoplasm

Neural layer

Nucleus

FIGURE 28
Differentiation of the photoreceptive apparatus in a rod or cone cell. (A) Cells extrude a bulb of cytoplasm into the space between the neural and pigmented layers of the retina. (B) The bulb expands and forms a cytoplasmic region and a region of stacked membranes, supported by a nonmotile cilium (axial filament). (C) The membranes become internalized and the photoreceptive pigments become localized to these regions. (After Coulombre, 1965.)

(B)

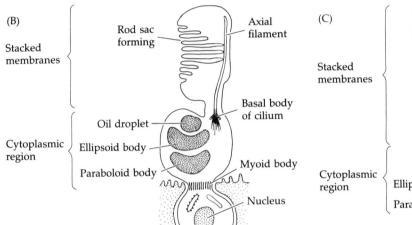

Stacked membranes

Rod sac forming

Axial filament

Cytoplasmic region

Oil droplet

Ellipsoid body

Paraboloid body

Basal body of cilium

Myoid body

Nucleus

(C)

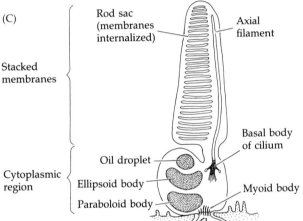

Rod sac (membranes internalized)

Axial filament

Stacked membranes

Cytoplasmic region

Oil droplet

Ellipsoid body

Paraboloid body

Basal body of cilium

Myoid body

(the mesodermal structures alluded to above), which act as inelastic restraints. Without these bones, the intraocular pressure could not be maintained, and the cornea would not be properly formed.

The differentiation of the lens tissue into a transparent membrane capable of directing light onto the retina involves changes in cell structure and shape as well as synthesis of lens-specific proteins called CRYSTALLINS. These crystallins are synthesized as cell shape changes occur, thereby causing the lens vesicle to become the definitive lens. The cells at the inner portion of the lens vesicle elongate and, under the influence of the neural retina, produce the lens fibers. As these fibers continue to grow, they synthesize crystallins, which eventually fill up the cell and cause the extrusion of the nucleus. The crystallin-synthesizing fibers continue to grow and eventually fill the space between the two layers of the lens vesicle. The anterior cells of the lens vesicle constitute a germinal epithelium, which keeps dividing. These dividing cells move toward the equator of the vesicle, and as they pass through the equatorial region, they too begin to elongate (Figure 30). Thus, the lens contains three regions: an anterior zone of dividing epithelial cells, an equatorial zone of cellular elongation, and a posterior

FIGURE 29
Scanning electron micrograph of
the rounded and internalized
lens placode in chick embryo.
(Courtesy of K. W. Tosney.)

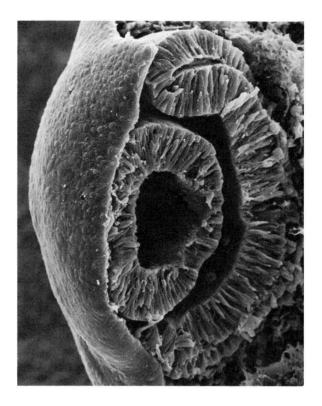

and central zone of crystallin-containing fiber cells. This arrangement
persists throughout the lifetime of the animal as fibers are continuously
being laid down. In the adult chicken, the differentiation from an epithelial cell to a lens fiber takes 2 years (Papaconstantinou, 1967). The
details of lens and cornea formation are discussed in Chapter 16.

The muscles that control the width of the lens are known as the
IRIS muscles. These muscles are unlike any other muscles in the body.
Not only do they contain pigment granules (which give an individual's
eyes their characteristic color), but these muscles are derived from the
ectoderm. Specifically, the iris muscles are formed from folds of the
pigmented retina that fail to develop photoreceptors. Whereas most
muscles are derived from the mesoderm, the muscles of the iris come
from the neural ectoderm cells.

FIGURE 30
Differentiation of the lens cells. (A) Lens vesicle as shown in Figure 29. (B)
Elongation of the interior cells, producing lens fibers. (C) Lens filled with
crystallin-synthesizing cells. (D) New lens cells derived from anterior lens epi-
thelium. (E) As the lens grows, new fibers differentiate. (After Paton and
Craig, 1974.)

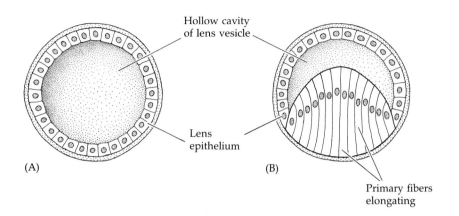

Hollow cavity
of lens vesicle

Lens
epithelium

(A)   (B)

Primary fibers
elongating

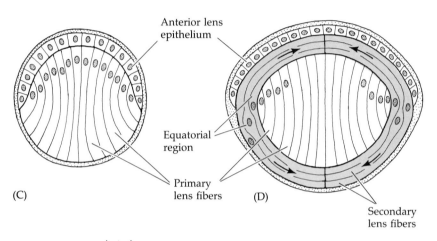

Anterior lens
epithelium

Equatorial
region

Primary
lens fibers

(C)   (D)

Secondary
lens fibers

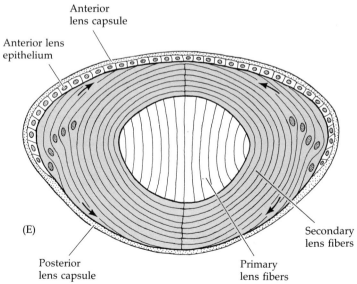

Anterior
lens capsule

Anterior lens
epithelium

(E)

Posterior
lens capsule

Primary
lens fibers

Secondary
lens fibers

TABLE 1
Major neural crest derivatives

Pigment cells	Sensory nervous system	Autonomic nervous system	Skeletal and connective tissue	Endocrine
TRUNK CREST (INCLUDING CERVICAL CREST)				
Melanocytes Xanthophores (erythrophores) Iridophores (guanophores) in dermis, epidermis, and epidermal derivatives	Spinal ganglia Some contributions to vagal (X) root ganglia	Sympathetic Superior cervical ganglion Prevertebral ganglia Paravertebral ganglia Adrenal medulla Parasympathetic Remak's ganglion Pelvic plexus Visceral and enteric ganglia Some supportive cells Glia (oligodendroglia) Schwann sheath cells Some contribution to meninges	Mesenchyme of dorsal fin in amphibia Walls of aortic arches Connective tissue of parathyroid	Adrenal medulla Calcitonin- producing cells Type I cells of carotid body Parafollicle cells of thyroid
CRANIAL CREST				
Small, belated contribution	Trigeminal (V) Facial (VII) root Glossopharyngeal (IX) root (superior ganglia) Vagal (X) root (jugular ganglia) Supportive cells	Parasypathetic ganglia Ciliary Ethmoidal Sphenopalatine Submandibular Intrinsic ganglia of viscera	Most visceral cartilages Trabeculae craniae (ant.) Contributes cells to posterior trabeculae, basal plate, para- chordal cartilages Odontoblasts Head mesenchyme (membrane bones)	

Sources: Weston (1970); Bockman and Kirby (1984).

The neural crest and its derivatives

Although derived from the ectoderm, the neural crest has sometimes been called the "fourth germ layer" because of its importance. The neural crest cells give rise to a bewildering number of differentiated cell types including (1) the neurons and supporting glial cells of the sympathetic and parasympathetic nervous systems, (2) the central portion of the adrenal gland, (3) the pigment-containing cells of the epidermis,

(A)

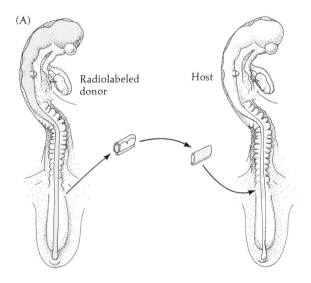

(B)

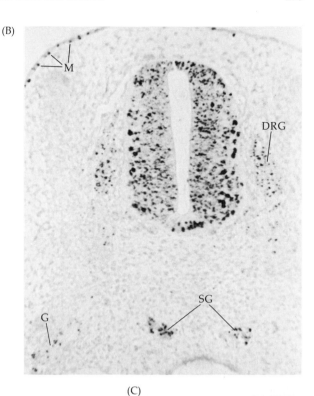

FIGURE 31

(A) Grafting technique for mapping neural crest cells. A piece of the dorsal axis is excised from a donor embryo; the neural tube and its associated crest is isolated and implanted into a host embryo whose neural tube and crest have been excised. If the donor crest cells are radiolabeled (with tritiated thymidine) or differ genetically (such as being from a different species or strain), they can be followed in the host embryo. (B) Autoradiograph showing locations of neural crest cells that have migrated from radioactive donor neural crest to form melanoblasts (M), sympathetic ganglia (SG), dorsal root ganglia (DRG), and glial cells (G). (C) Chick resulting from the transplantation of a neural crest region from a pigmented strain of chicken into the crest region of an unpigmented strain. The crest cells that gave rise to pigment were able to migrate into the wing skin. (From Weston, 1963; photographs courtesy of J. Weston.)

(C)

and (4) skeletal and connective tissue components of the head (Table 1). The fate of the neural crest cells depends on where the cells migrate.

As shown in Figure 2, the neural crest is a transient structure, the cells of which disperse soon after the neural tube closes. By grafting a portion of the chick neural tube and its associated crest from radioactively or genetically marked embryos onto other embryos (Weston, 1963; Thiery et al., 1982), investigators have discovered four major pathways of neural crest migration (Figures 31 and 32). The first migration takes place into the region between two consecutive somites. These cells reach the area around the aorta and aggregate to form the SYMPATHETIC GANGLIA (which transmit neural impulses to target cells when stimulated by spinal cord neurons). Some of these cells (at specific regions of the body) migrate to form the epinephrine-secreting cells of the ADRENAL

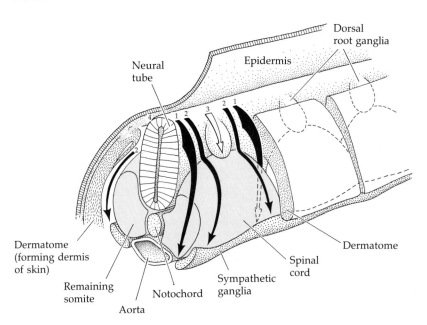

FIGURE 32

Paths of neural crest cell migration in the trunk of an embryo. Path 1, between adjacent somites, results in the formation of the sympathetic ganglia and adrenal medullary cells. In other regions of the trunk, neural crest cells form the parasympathetic ganglia. Those cells taking path 2 also contribute to the sympathetic ganglia. Neural crest cells migrating in path 3 become the neurons of the dorsal root ganglia, and those neural crest cells migrating beneath the ectoderm (4) give rise to the pigment cells. (From Le Douarin et al., 1984.)

MEDULLA. Sympathetic ganglia cells are also formed by neural crest cells migrating over the SOMITES (blocks of mesodermal tissue that will generate cartilage, muscles, and dermis). The third pathway that opens up for neural crest cells takes them between the somite and the neural tube. These cells become localized about the neural tube and aggregate to form the DORSAL ROOT GANGLIA (clusters of neurons that relay sensory information to the spinal cord).

The fourth major pathway of neural crest cell migration follows a DORSOLATERAL route beneath the embryonic ectoderm. Some of the cells of this pathway differentiate into pigment cells (in mammals, the MELANOCYTES). These cells travel from the central dorsal region to the most ventral, eventually terminating in the belly skin of the organism. In several mutations that affect neural crest migration, there is a white "belly spot," a finding indicating that the pigment cells did not arrive there (Figure 33).

Thus, neural crest cells do not migrate randomly through the body; rather, they follow precise pathways. Although the nature of these pathways is still a major question, the direction of neural crest migration appears to be controlled by the substrate over which they travel. In

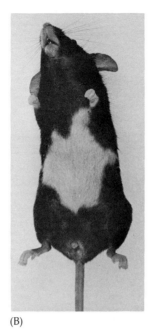

(A)   (B)

FIGURE 33
Pigment cell migration. The dorsal (A) and ventral (B) sides of a mouse heterozygous for the mutation *White*. Here the melanoblasts do not migrate completely around the mouse; a white spot (lacking pigment cells) remains in the center of the belly. In those mice homozygous for this mutation, the body is white, but the melanin pigment that forms in the retina can be seen in their eyes. (Photograph courtesy of The Jackson Laboratory.)

1941, Paul Weiss speculated that SELECTIVE CONTACT GUIDANCE may be the only condition needed to direct neural crest cells to their appropriate destination. In this model, temporary linkages would form between specific molecules of the cell surface and complementary molecules on the substrate. More recent evidence suggests that this is probably the case. First, when the neural tube and its associated crest are inverted, the crest cells continue to stream out. However, they now move dorsally instead of ventrally, thereby indicating that they keep their original orientation with regard to the neural tube. Moreover, when neural crest cells or their derivatives are placed (either by transplantation or injection) at various points within the embryo, they will migrate along the normal neural crest pathways. Other embryonic cells will not orient themselves in this manner and will stay where they are placed (Erickson et al., 1980; Bronner-Fraser and Cohen, 1980). Therefore, the neural crest cells are able to recognize certain pathways in the embryo and migrate along them.

Recognition of the proper road may be due to that road's being marked by the protein FIBRONECTIN. Rosavio and his co-workers (1983) have found that migratory neural crest cells bind specifically to this molecule. If neural crest cells were placed on a strip of fibronectin, the crest cells move only on that strip, and they move away from higher concentrations of cells (Figure 34). This mimics the behavior of neural crest cells in vivo, which also travel on thin fibronectin-rich tracks away from regions of high cell density (Mayer et al., 1981). Moreover, when neural crest cells stop migration in vivo, they accumulate in regions lacking fibronectin (Newgreen and Thiery, 1980). Thus, recognition of

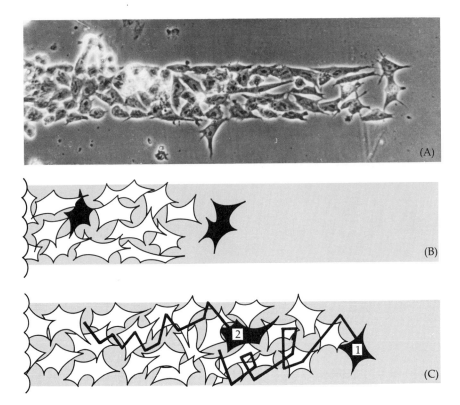

FIGURE 34

Migration of neural crest cells along a fibronectin substrate. Neural tubes and their associated crests were removed from the embryo and placed in a petri dish on a line of fibronectin. (A) Crest cells migrating from a fixed point (at left) in a nonrandom fashion along the fibronectin line. (B, C) Tracks of representative cells as they traveled on a fibronectin strip. The undulating line at the left represents the neural tube and crest that was the source of the migrating cells. (From Rosavio et al., 1983; photograph courtesy of J. P. Thiery.)

fibronectin probably plays an important role in neural crest cell migration.

Although the preceding experiments suggest that the neural crest cells actively migrate along a fibronectin-rich surface, experiments by Bronner-Fraser (1982) suggest that the migration is more passive. She found that dead neural crest cells and even polystyrene beads will follow the correct routes and accumulate in the correct site as long as they are placed in an appropriate position at the appropriate time. However, if the polystyrene beads are coated with fibronectin, the beads fail to move. Thus, neural crest cells may be transported over certain regions of the embryo, so long as they lack fibronectin on their cell surface. Loss of cell-surface fibronectin, moreover, has been correlated with the beginning of neural crest migration.

The mechanism for neural crest migration

The migration of neural crest cells is probably mediated by the physical and chemical nature of the embryonic environment. One such feature is the somitic tissue upon which the ventral pathway cells follow. Meier (1979, 1981) has found that when neural crest cells enter a region where mesodermal somites have already organized (Chapter 6) the neural crest cells were influenced to aggregate into ganglia. Furthermore, the surface of the somites probably undergoes temporal changes. Weston and Butler (1966) showed that when older neural crests were placed into younger environments, the older cells were able to migrate into all the areas normally colonized by the crest cells. However, when younger crests were placed into older environ-

ments, most cells were restricted to form dorsal root ganglia. This observation suggests that some alterations had occurred in the environment over which these cells travel.

Another component of neural crest migration involves the formation of spaces into which the cells can migrate. It is thought that hyaluronic acid causes the formation of cell-free spaces, which can then be invaded by the crest cells (Pratt et al., 1975; Solursh et al., 1979). This is a similar model to that proposed for the invasion of the avian blastocoel by mesodermal precursors, and Meier (1981) has found that hyaluronic acid does indeed accumulate in those regions over which the crest cells travel.

Pluripotentiality of neural crest cells

One of the most exciting features of neural crest cells is their PLURI-POTENTIALITY. A single neural crest cell can differentiate into different cell types depending upon its location within the embryo. For example, the parasympathetic neurons formed by the cervical neural crest cells produce acetylcholine as their neurotransmitter. They are therefore CHOLINERGIC neurons. The sympathetic neurons formed by the thoracic neural crest cells produce norepinephrine. They are ADRENERGIC neurons. However, when cervical and thoracic neural crests are reciprocally transplanted, the former thoracic crest is found to produce the cholinergic neurons of the parasympathetic ganglia and the former cervical crest forms adrenergic neurons in the sympathetic ganglia (Le Douarin et al., 1975). Thus, the thoracic crest cells are capable of developing into cholinergic neurons when they are placed into the neck, and the cervical crest cells are capable of becoming adrenergic neurons when they are placed in the trunk. Kahn and co-workers (1980) found that premigratory neural crest cells from both the thoracic and the cervical regions had the enzymes capable of synthesizing both acetylcholine and norepinephrine.

In addition to producing the peripheral neural structures, the neural crest is also responsible for the production of all the melanin-containing cells in the organism (with the exception of the pigmented retina). In a series of classic experiments, Rawles and others transplanted the neural tube and crest from a pigmented strain of chicken into the neural tube of an albino chick embryo. The result (Figure 31C) was a white chicken with a specific region of colored feathers. It should be noted that the production of norepinephrine and epinephrine (sympathetic neurons and adrenal medulla) and melanin (melanocytes) both involve the hydroxylation of tyrosine and the subsequent oxidation of the product.

From the preceding discussion, it would appear that all neural crest cells are originally identical in their potencies. This, however, is not the case when one investigates the ability to form the head cartilage. Only the cranial neural crest seems able to produce these cells, and the thoracic crest cannot substitute. Moreover, the head neural crest, when transplanted into the trunk region, will participate in forming trunk cartilage that normally does not arise from neural crest components. However, the head and trunk neural crest cells are extremely good examples of single cell types that can give rise to many different types of differentiated tissues.

The epidermis and the origin of cutaneous structures

The cells covering the embryo after neurulation form the presumptive epidermis. Originally, this tissue is one cell layer thick, but in most vertebrates this shortly becomes a two-layered structure. The outer layer gives rise to the PERIDERM, a temporary covering that will be shed once the bottom layer differentiates to form a true epidermis. The inner layer, called the BASAL LAYER (STRATUM GERMINATIVUM), will give rise to all the cells of the epidermis (Figure 35). The stratum germinativum first divides to give rise to an outer population of cells that constitutes the SPINOUS LAYER. These two epidermal layers are referred to as the MALPIGHIAN LAYER. The cells of the malpighian layer divide to produce the GRANULAR LAYER of the epidermis, so called because the cells are characterized by granules of the protein KERATIN. Unlike the malpighian layer, the cells of the granular layer do not divide. Rather, they begin to differentiate into skin cells (KERATINOCYTES). The keratin granules become more prominent as the cells of the granular layer age and migrate outward. Here, they form the HORNY LAYER (STRATUM CORNEUM), in which the cells have become flattened sacs of keratin protein. The nuclei have been pushed to one edge of the cell. Shortly after birth, the cells of the horny layer are shed and are replaced by new cells coming up from the granular layer. Throughout life, the dead keratinized cells of the horny layer are being shed (we humans lose about 1.5 grams each day[5]) and are replaced by new cells, the source of which is the mitotic cells of the malpighian layer. The pigment cells from the

[5]Most of this skin becomes "house dust" atop furniture and floors. Should you doubt this, burn some of this dust. It will smell just like singed skin.

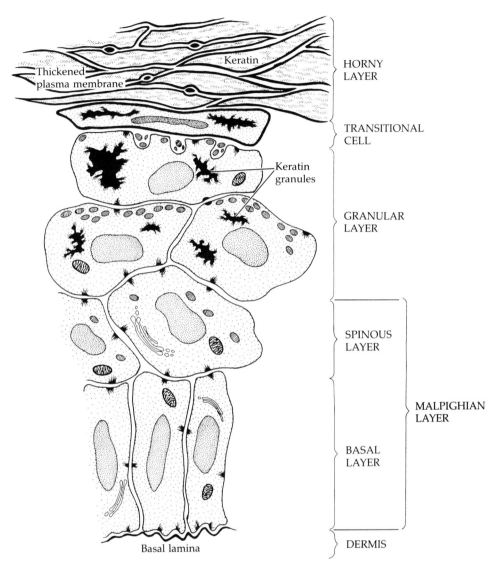

FIGURE 35
Schematic diagram of the layers of the epidermis. The basal cells are mitotically active, whereas the fully keratinized cells characteristic of external skin
are dead and shed off. (From Montagna and Parakkal, 1974.)

neural crest also reside in the malpighian layer, where they transfer
their pigment sacks (melanosomes) to the developing keratinocyte. In
adult skin, a cell born in the malpighian layer takes roughly 2 weeks to
reach the stratum corneum. In individuals with psoriasis, a disease
characterized by the exfoliation of enormous amounts of epidermal
cells, the time required is only 2 days (Weinstein and van Scott, 1965;
Halprin, 1972).

Cutaneous appendages

The epidermis alone does not make the skin. As we shall see in the next chapter, a region of mesoderm underlies the epidermis to form the DERMIS of the skin. This layer consists of loose connective tissue embedded in an elastic glycoprotein matrix. The dermis and epidermis interact at specific sites to create the cutaneous appendages—hair, scales, or feathers (depending on the species), sweat glands, and apocrine glands.

The first indication that a hair follicle will form at a particular place is an aggregation of cells in the basal layer of the epidermis. This occurs at different times and different places in the embryo. The basal cells elongate and divide into the dermis, forming the hair follicle (Figure 36). At this stage, two epithelial swellings begin to grow. The cells of the upper bulge will form the SEBACEOUS GLANDS, which produce an oily secretion, SEBUM. In many mammals, including humans, the sebum mixes with the desquamated peridermal cells to form the whitish VERNIX CASEOSA, which surrounds the fetus at birth. The lower swelling forms the HAIR BUD. The hair bud becomes split by upwelling mesenchyme. The germinal epidermis directly above this mesenchyme proliferates to form the keratinized shaft of hair.

The first hairs in the human embryo are of a thin, closely spaced type called LANUGO. This type of hair is usually shed before birth and is replaced (at least in part, by new follicles) by the short and silky VELLUS. Vellus remains on many parts of the human body usually considered hairless, such as the forehead and eyelids. In other areas of the body, vellus gives way to the "terminal" hair. During a person's life, some of the follicles that produced vellus can later form terminal hair, and still later revert to vellus production. The armpits of infants, for instance, have follicles producing vellus until adolescence. At that time, terminal shafts are generated. Conversely, in normal masculine "baldness," the scalp follicles revert back to producing unpigmented and very fine vellus (Montagna and Parakkal, 1974). The placement and pattern of hair, feathers, and scales involves the interactions of the dermis and the epidermis, and these will be discussed in more detail in Chapters 16 and 17.

The germinative cells of the epidermis also give rise to the sweat glands and (what are embryologically modified sweat glands) the mammary glands. Like the formation of the hair bulb, these glands are formed by the ingrowth of the germinative basal epidermal cells in the dermis. The epithelial–mesenchymal interactions necessary for mammary gland development will be addressed in Chapter 18.

In this chapter we have followed the differentiation of the embryonic ectoderm into a wide variety of tissues. We have seen, then, that the ectoderm produces three sets of cells during neurulation: (1) the neural tube, which gives rise to the neurons, glia, and ependymal cells of the system; (2) the neural crest cells, which give rise to the peripheral nervous systems, pigment cells, adrenal medulla, and certain areas of

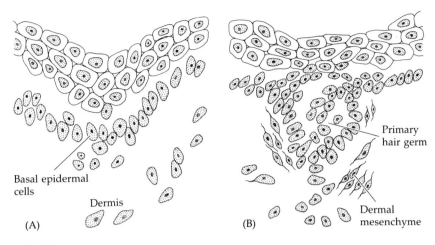

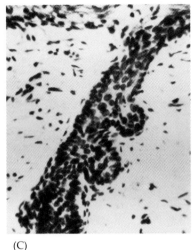

(C)

FIGURE 36
Development of the hair follicles in fetal human skin. (A) Basal epidermal cells become columnar and bulge slightly into the dermis. (B) Epidermal cells continue to proliferate, and dermal mesenchyme cells collect at the base of the primary hair germ. (C) Elongated hair germ. The uppermost bulge will develop into an apocrine sweat gland, while the central bud forms the sebaceous gland. The lowest bulb differentiates into the hair bud. (Photograph courtesy of W. Montagna.)

head cartilage; and (3) the epidermis of the skin, which contributes to the formation of cutaneous structures such as hair, feathers, scales, and sweat and sebaceous glands, as well as forming the outer protective covering of our body. We also observed how the interactions of epidermal cells are involved in generating the various tissues of the eye.

In later chapters we shall discuss in more detail the induction of the neural tube, the coordinated development of the eye, and the manner by which the neurons are directed to travel to specific sites, thus enabling the development of reflexes and behaviors. Meanwhile, we shall see how the endoderm and the mesoderm layers initiate the formation of their organ systems.

Literature cited

Balinsky, B. I. (1975). *Introduction to Embryology,* Fourth Edition. Saunders, Philadelphia.

Berry, M., McConnell, P. and Sievers, J. (1980). Dendritic growth and the control of neuronal form. *Curr. Top. Dev. Biol.* 15: 67–101.

Bloom, W. and Fawcett, D. W. (1975). *Textbook of Histology,* Tenth Edition. Saunders, Philadelphia.

Bockman, D. E. and Kirby, M. L. (1984). Dependence of thymus development on derivatives of the neural crest. *Science* 223: 498–500.

Bronner-Fraser, M. (1982). Distribution of latex beads and retinal pigment epithelial cells along the ventral neural crest pathway. *Dev. Biol.* 91: 50–63.

Bronner-Fraser, M. and Cohen, A. M. (1980). Analysis of the neural crest ventral pathway using injected tracer cells. *Dev. Biol.* 77: 130–141.

Burnside, B. (1971). Microtubules and microfilaments in newt neurulation. *Dev. Biol.* 26: 416–441.

Burnside, B. (1973). Microtubules and microfilaments in amphibian neurulation. *Am. Zool.* 13: 989–1006.

Cohen, A. M. and Konigsberg, I. R. (1975). A clonal approach to the problem of neural crest determination. *Dev. Biol.* 46: 262–251.

Coulombre, A. J. (1956). The role of intraocular pressure in the development of the chick eye. I. Control of eye size. *J. Exp. Zool.* 133: 211–225.

Coulombre, A. J. (1965). The eye. In R. DeHaan and H. Ursprung (eds.), *Organogenesis.* Holt, Rinehart & Winston, New York, pp. 217–251.

Crelin, E. S. (1974). Development of the central nervous system: A logical approach to neuroanatomy. *Ciba Clin. Symp.* 26(2): 2–32.

Detwiler, S. R. (1932). Experimental observations upon the developing retina. *J. Comp. Neurol.* 55: 473–492.

Erickson, C. A., Tosney, K. W. and Weston, J. A. (1980). Analysis of migrating behaviour of neural crest and fibroblastic cells in embryonic tissues. *Dev. Biol.* 77: 142–156.

Fujita, S. (1964). Analysis of neuron differentiation in the central nervous system by tritiated thymidine autoradiography. *J. Comp. Neurol.* 122: 311–328.

Fujita, S. (1966). Application of light and electron microscopy to the study of the cytogenesis of the forebrain. In R. Hassler and H. Stephen (eds.), *Evolution of the Forebrain.* Plenum, New York, pp. 180–196.

Gallera, J. (1971). Primary induction in birds. *Adv. Morphol.* 9: 149–180.

Gould, S. J. (1977). *Ontogeny and Phylogeny.* Belknap, Cambridge, MA.

Halprin, K. M. (1972). Epidermal "turnover time"—a re-examination. *J. Invest. Dermatol.* 86: 14–19.

Harrison, R. G. (1907). Observations on the living developing nerve fiber. *Anat. Rec.* 1: 116–118.

Harrison, R. G. (1910). The outgrowth of a nerve fiber as a mode of protoplasmic movement. *J. Exp. Zool.* 9: 787–846.

Hilfer, S. R. and Yang, J.-J. W. (1980). Accumulation of CPC-precipitable material at apical cell surfaces during formation of the optic cup. *Anat. Rec.* 197: 423–433.

His, W. (1886). Zur Geschichte des menschlichen Rückenmarks und der Nervenwurzeln. *Ges. d. Wissensch.* BD 13, S. 477.

Holt, A. B., Cheek, D. B., Mellitz, E. D. and Hill, D. E. (1975). Brain size and the relation of the primate to the non-primate. In D. B. Cheek (ed.), *Fetal and Postnatal Cellular Growth: Hormones and Nutrition.* Wiley, New York, pp. 23–44.

Jacobson, M. (1968). Cessation of DNA synthesis in retinal ganglion cells correlated with the time of spec- ification of their central connections. *Dev. Biol.* 17: 219–232.

Jacobson, M. (1978). *Developmental Neurobiology.* Plenum, New York.

Kahn, C. R., Coyle, J. T. and Cohen, A. M. (1980). Head and trunk neural crest *in vitro*: Autonomic neuron differentiation. *Dev. Biol.* 77: 340–348.

Karfunkel, P. (1972). The activity of microtubules and microfilaments in neurulation in the chick. *J. Exp. Zool.* 181: 289–302.

Langman, J. (1969). *Medical Embryology,* Third Edition. Williams & Wilkins, Baltimore.

LeDouarin, N. M., Renaud, D., Teillet, M. A. and Le-Douarin, G. H. (1975). Cholinergic differentiation of presumptive adrenergic neuroblasts in interspecific chimeras after heterotopic transplantation. *Proc. Natl. Acad. Sci. USA* 72: 728–732.

LeDouarin, N. M., Cochard, P., Vincent, M., Duband, J. L., Tucker, G. C., Teillet, M.-A. and Thiery, J.-P. (1984). Nuclear, cytoplasmic, and membrane markers to follow neural crest cell migration. A comparative study. In R. L. Trelstad (ed.), *The Role of the Extracellular Matrix in Development (Dev. Biol. Soc. Symp. 42).* Alan R. Liss, New York, pp. 373–398.

Letourneau, P. C. (1977). Regulation of neuronal morphogenesis by cell–substratum adhesion. *Soc. Neurosci. Symp.* 2: 67–81.

Letourneau, P. C. (1979). Cell substratum adhesion of neurite growth cones, and its role in neurite elongation. *Exp. Cell Res.* 124: 127–138.

Mayer, B. W., Jr., Hay, E. D. and Hynes, R. O. (1981). Immunocytochemical localization of fibronectin in embryonic chick trunk and area vasculosa. *Dev. Biol.* 82: 267–286.

Meier, S. (1979). Development of the chick embryo mesoblast: Formation of the embryonic axis and establishment of the metameric pattern. *Dev. Biol.* 73: 25–45.

Meier, S. (1981). Development of the chick embryo mesoblast: Morphogenesis of the prechordal plate and cranial segments. *Dev. Biol.* 83: 49–61.

Montagna, W. and Parakkal, P. F. (1974). The piliary apparatus. In W. Montagna (ed.), *The Structure and Formation of Skin.* Academic, New York, pp. 172–258.

Newgreen, D. F. and Thiery, J. P. (1980). Fibronectin in early avian embryos: Synthesis and distribution along the migratory pathways. *Cell Tissue Res.* 211: 269–291.

Odell, G. M., Oster, G., Alberch, P. and Burnside, B. (1981). The mechanical basis of morphogenesis. I. Epithelial folding and invagination. *Dev. Biol.* 85: 446–462.

Papaconstantinou, J. (1967). Molecular aspects of lens

cell differentiation. *Science* 156: 338–346.

Paton, D. and Craig, J. A. (1974). Cataracts: Development, diagnosis, and management. *Ciba Clin. Symp.* 26(3): 2–32.

Patten, B. M. (1971). *Early Embryology of the Chick*, Fifth Edition. McGraw-Hill, New York.

Piatigorsky, J. (1981). Lens differentiation in vertebrates: A review of cellular and molecular features. *Differentiation* 19: 134–153.

Pierce, M., Turley, E. A. and Roth, S. (1980). Cell surface glycosyltransferase activities. *Int. Rev. Cytol.* 65: 1–47.

Portmann, A. (1941). Die Tragzeiten der Primaten und die Dauer der Schwangerschaft beim Menschen: Ein Problem der vergleichen Biologie. *Rev. Suisse Zool.* 48: 511–518.

Portmann, A. (1945). Die Ontogenese des Menschen als Problem der Evolutionsforschung. *Verh. Schweiz. Naturf. Ges.* 125: 44–53.

Pratt, R. M., Larsen, M. A. and Johnston, M. C. (1975). Migration of cranial neural crest cells in a cell-free hyaluronate-rich matrix. *Dev. Biol.* 44: 298–305.

Rakic, P. and Sidman, R. L. (1973). Weaver mutant mouse cerebellum: Defective neuronal migration secondary to abnormality of Bergmann glia. *Proc. Natl. Acad. Sci. USA* 70: 240–244.

Ramón y Cajal, S. (1890). Sur l'origene et les ramifications des fibres neuveuses de la moelle embryonnaire. *Anat. Anz.* 5: 111–119.

Romanes, G. J. (1901). *Darwin and After Darwin*. Open Court Publishing Co., London.

Romer, A. S. (1976). *The Vertebrate Body*, Fifth Edition. Saunders, Philadelphia.

Rosavio, R. A., Delouvée, A., Yamada, K. M., Timpl, R. and Thiery, J. P. (1983). Neural crest cell migration: Requirements for exogenous fibronectin and high cell density. *J. Cell Biol.* 96: 462–473.

Sauer, F. C. (1935). Mitosis in the neural tube. *J. Comp. Neurol.* 62: 377–405.

Sidman, R. L. (1974). Cell–cell recognition in the central nervous system. In F. O. Schmitt and F. G. Worden (eds.), *The Neurosciences: Third Study Program*. MIT Press, Cambridge, MA, pp. 743–758.

Soleto, C. and Changeaux, J. P. (1974). Transsynaptic degeneration "en cascade" in the cerebellar cortex of staggerer mutant mice. *Brain Res.* 67: 519–526.

Solursh, M., Fisher, M. and Singley, C. T. (1979). The synthesis of hyaluronic acid by ectoderm during early organogenesis in the chick embryo. *Differentiation* 14: 77–85.

Thiery, J. P., Duband, J. L. and Delouvée, A. (1982). Pathways and mechanisms of avian trunk neural crest cell migration and localization. *Dev. Biol.* 93: 324–343.

Turley, E. A. and Roth, S. (1979). The spontaneous glycosylation of glycosaminoglycan substratum by adherent fibroblasts. *Cell* 17: 109–115.

von Baer, K. E. (1828). Entwicklungsgeschichte der Thiere: Beobachtung und Reflexion. Bornträger, Konigsberg.

Warkany, J. (1971). *Congenital Malformations*. Year Book Medical Publishers, Chicago.

Weinstein, G. D. and van Scott, E. J. (1965). Turnover times of normal and psoriatic epidermis. *J. Invest. Dermatol.* 45: 257–262.

Weiss, P. (1941). The mechanics of nerve growth. *Third Growth Symp., Growth* 5 [Suppl.]: 163–203.

Weston, J. (1963). A radiographic analysis of the migration and localization of trunk neural crest cells in the chick. *Dev. Biol.* 6: 274–310.

Weston, J. (1970). The migration and differentiation of neural crest cells. *Adv. Morphog.* 8: 41–114.

Weston, J.A. and Butler, S. L. (1966). Temporal factors affecting localization of neural crest cells in the chicken embryo. *Dev. Biol.* 14: 246–266.

Yamada, K. M., Spooner, B. S. and Wessells, N. K. (1971). Ultrastructure and function of growth cones and axons of cultured nerve cells. *J. Cell Biol.* 49: 614–635.

Early vertebrate development: mesoderm and endoderm

Of physiology from top to toe I sing,
Not physiognomy alone or brain alone is worthy for the Muse,
I say the form complete is worthier far,
The Female equally with the Male I sing.
—WALT WHITMAN (1867)

The greatest progressive minds of embryology
have not searched for hypotheses; they have
looked at embryos.
—J. M. OPPENHEIMER (1955)

Introduction

In the last chapter we followed the various tissues formed by developing ectoderm. In this chapter, we shall follow the early development of the mesodermal and endodermal germ layers. Endoderm will be seen to form the lining of the digestive and respiratory tubes with their associated organs; mesoderm will be seen to generate all the organs between the ectodermal wall and the endodermal tissues.

MESODERM

The mesoderm of a neurula-stage embryo can be divided into five regions (Figure 1). The first region is the CHORDAMESODERM. This tissue forms the notochord, a transient organ the major functions of which include inducing the formation of the neural tube and establishing the body axis. The second region is the DORSAL MESODERM. The term *dorsal* refers to the observation that this position will be in the back of the embryo, along the spine. Located on both sides of the neural tube, this region will produce many of the connective tissues of the body—bone, muscles, cartilage, and dermis. The INTERMEDIATE MESODERM forms the urinary system and genital ducts, and we will discuss this region in detail in later chapters. Further away from the notochord, the LATERAL MESODERM will give rise to the heart, blood vessels, and blood cells of the circulatory system as well as to the lining of the body cavities and all the mesodermal components of the limbs except the muscles. It will also form the extraembryonic membranes of the embryo. Lastly, the HEAD MESODERM will form the muscles of the face.

Dorsal mesoderm: Differentiation of somites

One of the major tasks of gastrulation is to position the endoderm deep within the embryo and to sandwich the mesodermal cells between the ectodermal and endodermal layers. As shown in Figure 2, the formation of mesodermal and endodermal organs is not subsequent to neural tube formation, but occurs synchronously. Those mesodermal cells of the chick that are not involved in notochord formation have migrated laterally to form thick bands running longitudinally along each side of the notochord and neural tube. These bands are called PARAXIAL MESODERM. As the primitive streak regresses and the neural folds begin to gather at the center of the embryo, the paraxial mesoderm separates into triangular blocks of cells called SOMITES. The first somites appear in the anterior portion of the embryo, and new somites are formed posteriorly

192

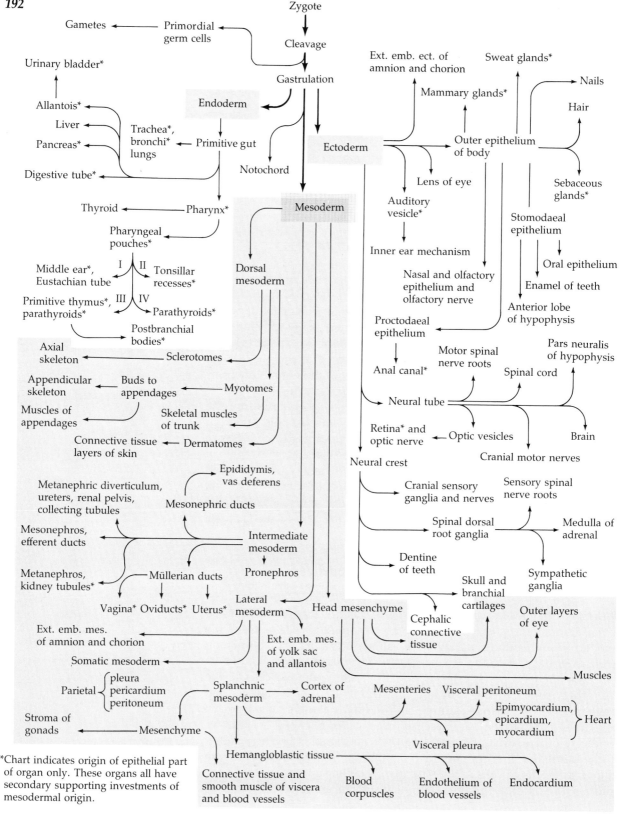

*Chart indicates origin of epithelial part of organ only. These organs all have secondary supporting investments of mesodermal origin.

FIGURE 1
Chart depicting the lineage of the specialized parts of the body through the three primary germ layers. The germ cells are represented as being formed separately from the three somatic germ layers because, although the germ cell precursors are located in the presumptive endoderm or mesoderm, they are probably a cell type unto themselves. (Modified from Carlson, 1981.)

at regular intervals (Figure 2D; Figure 3). Because embryos develop at different rates when incubated at slightly different temperatures, the number of somites present is usually the best indication as to how far development has proceeded. The mechanism for somite formation is not well established, but recent studies in chicks have shown that the cells of the paraxial mesoderm may be organized into clumps of cells

FIGURE 2
The progressive development of the chick embryo, focusing on the mesodermal component. (A) Primitive streak region showing migrating mesodermal and endodermal precursors. (B) Formation of the notochord and paraxial mesoderm. (C, D) Differentiation of the somites, coelom, and aortae (which will eventually come together). (A–C) are from 24-hour embryos; (D) is from a 48-hour embryo.

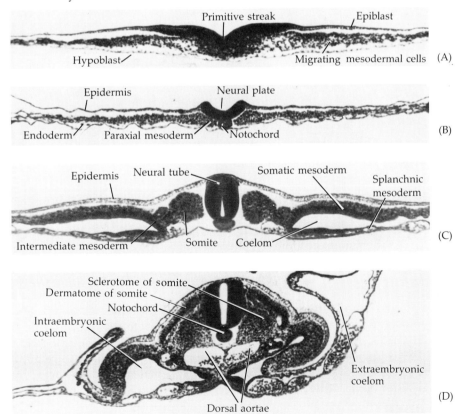

FIGURE 3
Scanning electron micrograph of neural tube and somites. Paraxial mesoderm on bottom right-hand side has not yet been segmented into somitic regions. Neural crest cells can be seen migrating ventrally from the roof of the neural tube. (Courtesy of K. W. Tosney.)

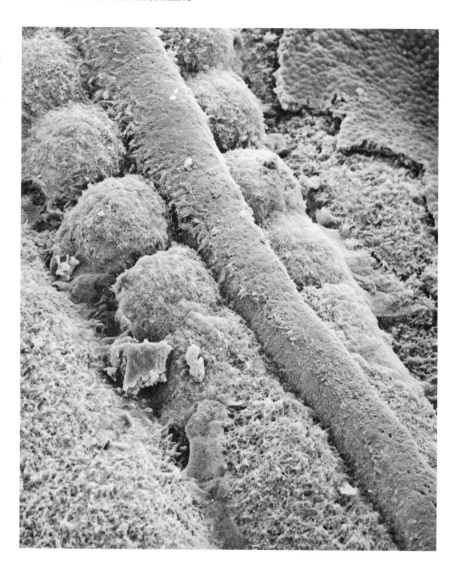

called SOMITOMERES (Meier, 1979) and that the regressing primitive streak may exert a shearing force, which separates these regions at preexisting weak points (Lipton and Jacobson, 1974).

As the somite becomes a coherent entity, its cells become rounded, and the outer cells of the somites become attached to each other by tight junctions. Like the notochord and the neural tube, the somite becomes covered with a BASAL LAMINA (Figure 4) consisting of collagen and glycosaminoglycans (GAG). In fact, GAG from the neural tube and notochord appear to induce the somites to secrete their own GAG. If the GAGs are enzymatically removed from the surface of the notochord, GAG synthesis in the somites will stop until the notochord once more has elaborated GAG on its own surface. Whether through the action of GAG or some other factor secreted by the notochord or neural tube,

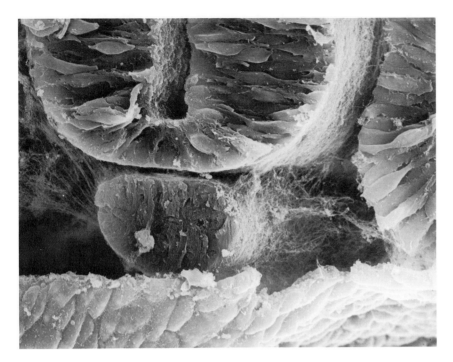

FIGURE 4
Scanning electron micrograph of the neural tube-notochord-somite borders. The webbing of collagen and glycosaminoglycan complexes connecting these structures is important in cell migration and the intercellular reactions occurring between them. (Courtesy of K. W. Tosney.)

the *ventral* cells of the somite (those cells located opposite the back) undergo mitosis, lose their round epithelial characteristics, and become mesenchymal cells again. The portion of the somite that gives rise to these cells is called the SCLEROTOME and the mesenchymal cells ultimately become the CHONDROCYTES (Figures 2 and 5). Chrondrocytes are responsible for secreting the special type of collagen and the special type of GAG (chondroitin sulfate) characteristic of cartilage. These particular chondrocytes will be responsible for constructing the axial skeleton (vertebrae, ribs, and so forth).

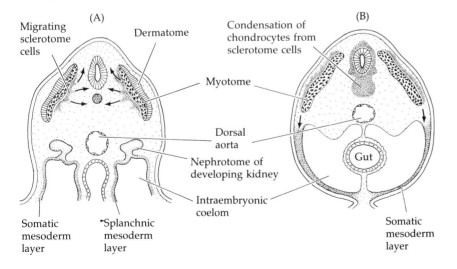

FIGURE 5
Diagram of a transverse section through the trunk of an early 4-week (A) and late 4-week (B) human embryo. In (A), the sclerotome cells begin to migrate away from the myotome and dermatome. By the end of the fourth week (B), the sclerotome cells are condensing to form the vertebra, the dermatome is starting to form the dermis, and myotome cells are migrating ventrally down the walls of the embryo. (Modified from Langman, 1969.)

Once the cells of the sclerotome have migrated away from the somite, the remaining epithelial cells there separate to form a bilayered tube (Figures 2 and 5). The dorsal layer is called the DERMATOME, and it contributes to the mesenchymal connective tissue of the skin—the dermis. The inner layer of cells is called the MYOTOME, and these cells give rise to the striated muscles of both the body wall and the limbs (Chevallier et al., 1977).

Myogenesis: Differentiation of skeletal muscle

A skeletal muscle cell is an extremely large elongated cell that contains many nuclei. In the mid-1960s it was debated whether each of these cells (often called MYOTUBES) was derived from the fusion of several mononucleate muscle precursor cells (MYOBLASTS) or from a single myoblast that underwent nuclear division without cytokinesis. Evidence from two independent sources demonstrates that the myotube is derived from the fusion of several mononucleate myoblasts.

One of the most significant advances in the study of muscle development came when Konigsberg (1963) demonstrated that myoblasts could differentiate into myotubes in culture. The observations of such fusion in vitro have enabled scientists to study the mechanism of this fusion and to relate it to the differentiation of the muscle cell. Cells from embryonic regions containing myoblasts are separated from each other by digesting the extracellular lamina with trypsin. These cells are then placed in collagen-coated petri dishes containing nutrients and serum (Figure 6). By 48 hours, the large majority of the cells are rapidly dividing myoblasts. Shortly thereafter, these myoblasts cease making DNA and fuse with their neighbors to produce extended myotubes (Konigsberg, 1963), which begin synthesizing muscle-specific proteins. DNA synthesis and nuclear division were not seen in the multinucleated myotubes.

The fusion of myoblasts to produce myotubes has been divided into two separate processes. The first stage involves cell migration, recognition, and alignment. When 11-day chick embryo pectoral muscle (which still contains a large amount of mononucleate myoblasts) is placed into culture, the isolated cells divide and migrate. Many of the migrating cells eventually line up to form chains of mononucleate cells. These cells eventually will fuse to form the myotube. Nameroff and Munar (1976) have been able to separate this recognition phenomenon (the making of such chains) from the fusion event by treating the cultured cells with phospholipase C. This enzyme prohibits membrane fusion but permits recognition. Thus, chains are formed but no fusion takes place. These experiments also suggest that the recognition event is responsible for stopping DNA synthesis in the myoblast. Radioactive thymidine is incorporated into the DNA of the migrating myoblasts but is not incorporated into those myoblasts within the chains (Figure 7).

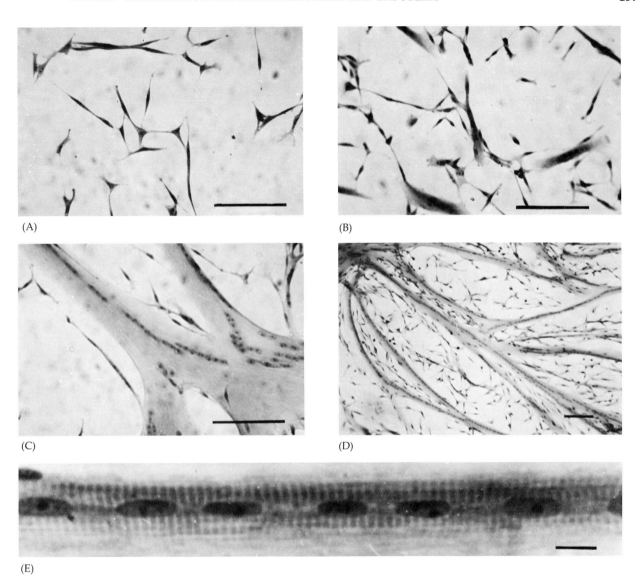

FIGURE 6

Photomicrographs of quail myoblasts differentiating in culture. All cultures were inoculated, fixed, stained, and photographed on specific days. (A) Day 2 culture consisting of unicellular myoblasts (bar = 0.1 mm). (B) Day 3 culture in which fusion of myoblasts is first seen. (C) Day 5 culture showing the increase in the number of nuclei seen within multinucleated myotubes. (D) Lower magnification of Day 5 culture showing the extent of cell fusion. (E) Higher magnification of multinucleated myotube demonstrating the characteristic cross-striations (bar = 0.01 mm). (From Buckley and Konigsberg, 1974; photographs courtesy of I. R. Konigsberg.)

FIGURE 7
Autoradiograph of chick myoblast cells treated with phospholipase C in culture and then exposed to radioactive thymidine. Unattached myoblasts still divide and incorporate the radioactive thymidine into their DNA. The incorporated radioactive nucleotides cause the silver grains of the photographic emulsion to darken when the emulsion is developed. Lined up (but not yet fused) cells (arrows) do not incorporate the label. (From Nameroff and Munar, 1976; photograph courtesy of M. Nameroff.)

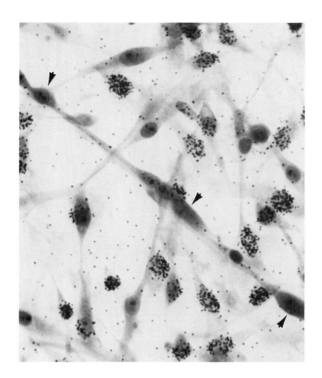

Thus, cell–cell recognition may withdraw the myoblasts from the cell cycle.

The second stage of myogenesis is cell fusion itself. Both the recognition event and the fusion event are dependent upon calcium ions, but in fusion, the calcium ions must get into the myoblasts. Thus, fusion can be activated by calcium ionophores, such as A23187, which carry calcium ions across cell membranes (Shainberg et al., 1969; David et al., 1981). Kalderon and Gilula (1979) have shown that calcium-mediated membrane fusion occurs at those regions of the myoblast membranes that are enriched for phospholipids. This would explain why phospholipase C is such a potent inhibitor of cell fusion.

Although cell–cell recognition and fusion are essential for myogenesis, species specificity is not. Myoblasts will fuse only with other myoblasts, but the myoblasts need not be of the same species. Yaffe and Feldman (1965) have shown that embryonic rat and chick myoblasts will readily fuse to form hybrid myotubes.

The second type of evidence for myoblast fusion came from allophenic mice. These mice can be formed from the fusion of two early embryos, which regulate to produce a single mouse having two distinct cell populations (Figure 25 in Chapter 3). Beatrice Mintz and W. W. Baker (1967) fused mouse embryos that produced different types of the enzyme isocitrate dehydrogenase. This enzyme, found in all cells, is composed of two identical subunits. Thus, if myotubes are formed from one cell whose nuclei divide without cytokinesis, one would expect to find two distinct forms of the enzyme, that is, the two parental forms, in the allophenic mouse (Figure 8). On the other hand, if myotubes are formed by fusion between cells, one would expect to find muscle cells expressing not only the two parental types of enzymes (AA and BB) but also a third class composed of a subunit from each of the parental

FIGURE 8
Diagram illustrating the two possible mechanisms of skeletal muscle formation and how to distinguish between them. Allophenic mice are made from the fusion of mouse embryos from two different strains, each strain making a different form of the enzyme isocitrate dehydrogenase. This enzyme is composed of two subunits such that one strain of mouse makes AA isocitrate dehydrogenase and the other makes BB. (A) If the enzymes are made in a single cell, or in multinucleate cells made from the division of a single cell, the enzyme will be purely AA or BB. (B) If there are two different nuclei in the same cell, however, one might code for B subunits while the other might code for A, with the result that some of the enzyme will be a hybrid molecule, AB. Electrophoresis can separate these three types of molecules. The presence of the AB molecule in skeletal muscle cells (but not other cell types) confirms the fusion model. (After Mintz and Baker, 1967.)

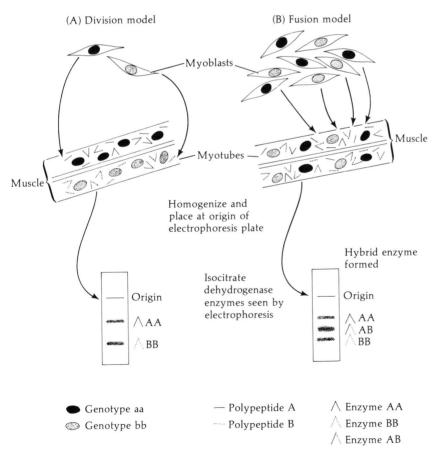

types (AB). The different forms of isocitrate dehydrogenase can be separated by their mobility in an electric field. The results clearly demonstrated that although only the two parental types of enzyme were present in all the other tissues of the allophenic mice, the hybrid (AB) enzyme was present in extracts of skeletal muscle tissue. Thus, the myotubes must have been formed from the fusion of numerous myoblasts.

There is little disagreement that myoblast fusion and the synthesis of muscle-specific proteins are separate events that usually occur at the same time. Once the myoblasts stop dividing, tropomyosin, myosin heavy chain, and the muscle-specific forms of actin and creatine kinase are all seen to develop and the myoblasts are seen to fuse. For example, creatine kinase is a dimeric protein. Myoblasts only make creatine kinase composed of B subunits (BB). At the time of fusion, the cultured myotubes begin expressing a different, muscle-specific (M) form of the enzyme, thereby creating MM dimers (Figure 9A). Myoblast fusion can

FIGURE 9
New enzyme synthesis concomitant with myoblast fusion. (A) Decline of nonmuscle (B-type) creatine kinase subunits and concomitant replacement with muscle-specific (M-type) creatine kinase subunits as cultured embryonic chick myoblasts begin fusing. Fusion was seen to begin after one day of culture, and by day 2, 38 percent of the nuclei were in myotubes. (B) Synthesis of muscle-specific contractile proteins when cultures of dividing myoblasts are shifted to a medium that promotes fusion. Cultures were labeled with [^{35}S]methionine, which is incorporated into proteins, and the synthesis of specific proteins was monitored. (A after Lough and Bischoff, 1977; B from Devlin and Emerson, 1978.)

(A)

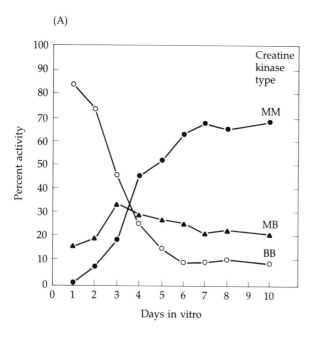

(B)

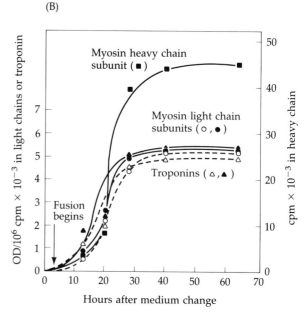

be stimulated by changing the tissue culture medium from one that supports proliferation to one that lacks certain growth factors. In the limited medium, myoblasts cease to divide and will fuse within a short period of time. Devlin and Emerson (1978) have used this technique to synchronize cell fusion and look at the synthesis of myosin, actin, and tropomyosin (Figure 9B). At first, none of these proteins are seen. At the time of fusion, however, these structural proteins rapidly appeared. Southerland and Konigsberg (1983) have shown that once DNA synthesis has stopped, the muscle-specific proteins increase normally even if the fusion event is inhibited by low calcium concentrations. It appears then that once the cells stop dividing, they acquire the capacities to fuse and to make muscle-specific proteins.

Although there is agreement that enormous increases in the synthesis of muscle-specific proteins occurs at the time of fusion, it is not clear yet whether such control is exerted at the transcriptional or translational levels, or both. Some investigators (see Chapter 14) have reported finding the mRNAs from muscle-specific proteins within prefusion myoblasts. If this is so, then fusion would act as a signal allowing the *translation* of these preexisting messages. Other studies (Devlin and Emerson, 1979; Daubas et al., 1981) find no evidence for contractile protein messages within the dividing myoblasts. They see these proteins arising from newly transcribed messages. In any event, the fusion of the myoblasts and the accumulation of muscle-specific proteins appears to be coordinated by the cessation of DNA synthesis in the proliferating myoblasts.

SIDELIGHTS & SPECULATIONS

Differentiation and the cell cycle

Rapidly proliferating cells are said to be in the CELL CYCLE. After each division, there is (except in very young embryos) a time gap (G_1) before the synthesis of DNA begins. The cell is metabolically active during this part of the cycle and many of the DNA precursors are made at this time. The period of DNA synthesis (S) involves the replication of the chromosomal DNA and is followed by another time gap (G_2), after which mitosis (M) occurs. There are many variations on the cell cycle from tissue to tissue. Sea urchin blastomeres, for instance, replicate their DNA in the telophase of the preceding division, thus eliminating G_1 entirely.

Most differentiated cells rarely divide. While this is obvious in such an extreme case as the mature red blood cell, division is also suppressed in nucleated cells such as epidermal cells, fat cells, skeletal muscle cells, and cartilage cells. Differentiated cells are sometimes said to be in G_0, an extended G_1 after their last division.

If there is indeed a great dichotomy between division and differentiation, then one of the most important cellular "decisions" is that which takes a cell out of the cell cycle and causes it to differentiate. Howard Holtzer and his colleagues have speculated that the cell is "reprogrammed" for differentiation during its last mitosis. Thus, this special QUANTAL MITOSIS is essential for differentiation. Myoblasts, then, can be expected to undergo repeated divisions until they undergo a change that forbids further mitosis. After such a change, they differentiate (Bischoff and Holtzer, 1969; Holtzer et al., 1976). Abbott and Holtzer (1965) show that this quantal mitosis may be essential in the development of muscle, cartilage, and blood cells.

Holtzer's hypothesis claims that a round of DNA synthesis is indispensable for a cell's commitment to differentiate. More recent studies of mammary gland development (Smith and Vonderhaar, 1981) show that hormone-stimulated mammary epithelial cells differentiate only if the DNA is replicated in the presence of the hormones. In this case, a final "quantal" mitosis does not have to occur, and a majority of the differentiating cells are believed to be polyploid. Therefore, the events responsible for initiating differentiation may require DNA synthesis but not necessarily cell division. Similarly, Southerland and Konigsberg (1983) have argued that it is not myoblast fusion that initiates the burst of muscle-specific protein synthesis but the removal of myoblasts from the cell cycle. When they blocked myoblasts from fusing by putting them into calcium-deficient medium, the postmitotic, fusion-blocked myoblasts developed the muscle-specific proteins at the same rate as the fusing cells.

Although this hypothesis may help explain differentiation in some tissues, it does not appear to be universally applicable. Heart muscle cells, for example, undergo numerous cell divisions even after they have formed their contractile filaments and begin to function. The heart grows for months or years after it has begun beating, and this growth is due primarily to the division of contractile cells. No dichotomy between division and differentiation seems to exist here (Polinger, 1973; Przybylski and Chlebowski, 1972). Similarly, pigment cells are seen to divide while laden with their differentiated product. Other studies (O'Neill and Stockdale, 1972) have suggested that limb-derived myoblasts can undergo further divisions if fusion is prevented after what would normally have been their quantal mitosis. The mechanisms by which some cells are able to switch between division and differentiation while other types of cells are able to do both simultaneously remains enigmatic and is the subject of many investigations.

Osteogenesis: Development of bones

Some of the most obvious structures derived from the somitic mesoderm are the bones. In this chapter we can only begin to outline the mechanisms of bone formation, and students wishing further details are invited to consult histology textbooks that devote entire chapters to this topic. There are two major modes of bone formation (OSTEOGENESIS), and both involve the transformation of a preexisting connective tissue into bone tissue. When primitive connective tissue is converted into bone, this is called INTRAMEMBRANOUS OSSIFICATION. When bone formation occurs in preexisting cartilage, it is called ENDOCHONDRAL OSSIFICATION. Intramembranous ossification is the characteristic way in which the flat bones of the skull are formed. Loose mesenchymal cells proliferate and condense into compact nodes. Some of these cells develop into capillaries and others change their shape to become OSTEOBLASTS, cells capable of secreting the bone matrix. The secreted network of collagen-mucopolysaccharide matrix is able to bind calcium salts, which are brought to the region through the capillaries. In this way, the matrix becomes calcified. In most cases, the osteoblasts are separated from the region of calcification by a layer of the prebone (osteoid) matrix they secrete. Occasionally, though, osteoblasts become trapped in the calcified matrix and become OSTEOCYTES, bone cells. As calcification proceeds, the bony spicules radiate out from the center where ossification began (Figure 10). Furthermore, the entire region of calcified spicules becomes surrounded by compact mesenchymal cells that form

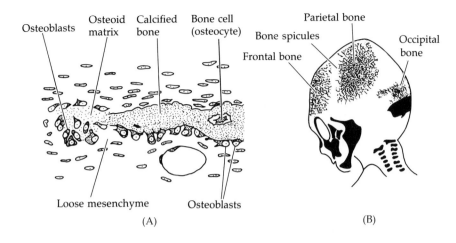

(A)   (B)

FIGURE 10
Schematic diagram of membranous ossification. (A) Mesenchymal condensation produces osteoblasts that deposit osteoid matrix. These osteoblasts become arrayed along the calcified region of the matrix. The osteoblasts that are trapped within the bone matrix become osteocytes. (B) Spread of bone spicules from the primary ossification site in the flat skull bones of a 3-month-old human embryo. The bones shown in black are formed by endochondral ossification. (Modified from Langman, 1969.)

the PERIOSTEUM. The cells on the inner surface of the periosteum also become osteoblasts and deposit bone matrix parallel to that of the existing spicules. In this manner, many layers of bone are formed.

Endochondral ossification involves the transformation of cartilage tissue into bone. The cartilage tissue is a model for the bone that follows. The bones of the vertebral column, the pelvis, and the extremities are first formed of cartilage and are later changed into bone. In humans, the long bones of the 7-week embryo are represented as cartilaginous tissue surrounded by a sheath of dense mesenchyme. This sheath becomes the PERIOSTEUM (Figure 11). A capillary from the periosteum then

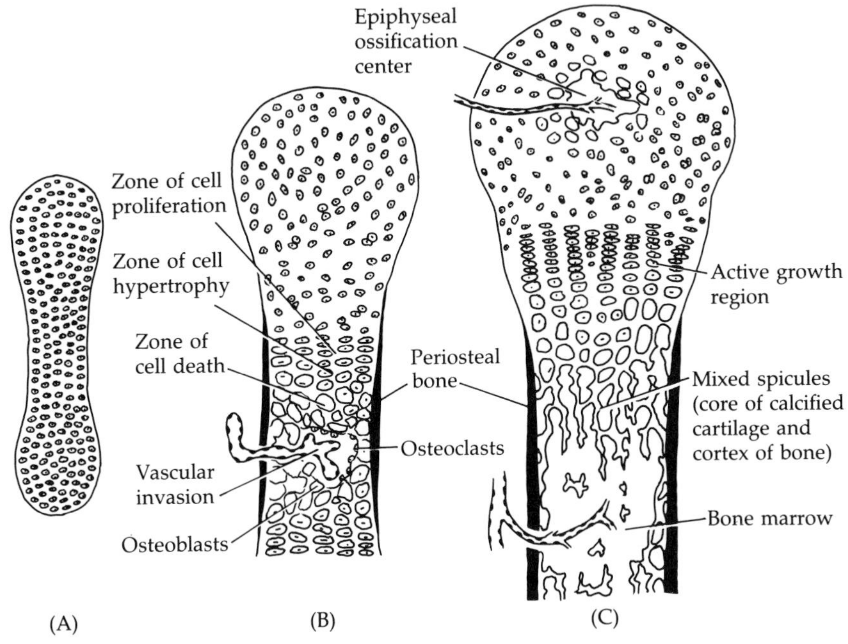

(A)   (B)   (C)

FIGURE 11
Schematic diagram of endochondral ossification. (A) Cartilage "model" of bone. (B) Entry of blood vessel into center of cartilagenous shaft, causing regions of cell proliferation, cell hypertrophy, and cell death. (C) Formation of a central bone marrow and secondary ossification center in the epiphysis. (Modified from Langman, 1969.)

invades the center of the previously avascular cartilage shaft. The cartilage reacts by forming a zone of rapid cell proliferation, a region of cell enlargement, and a region of cell death. In the region of cell death, osteoblasts, entering from the periosteum, align themselves on the cartilaginous strands and begin secreting osteoid matrix. Calcification soon occurs, creating bone where there had been cartilage. In long bones, osteogenesis begins in the shaft and gradually progresses toward the ends of the cartilaginous tissues. At birth, the shafts of human long bones are calcified, but the ends, known as epiphyses, are still cartilaginous. Shortly after birth, secondary ossification centers arise in the epiphyses, as blood vessels enter the cartilage in those regions. However, a cartilaginous plate continues to exist between the calcified tissues of the epiphyses and the shaft. This EPIPHYSEAL PLATE plays an important role in human growth after birth, for it proliferates in response to growth hormone (Figure 12). Hormones are also responsible for the cessation of growth, as high amounts of either estrogen or testosterone cause the fusion of the epiphyseal and shaft ossification centers, thereby resulting in the loss of the epiphyseal plate. This EPIPHYSEAL FUSION occurs during adolescence, about 2 years earlier for girls than for boys.

As new bone material is added peripherally, there is a hollowing out of the internal region to form the bone marrow. This destruction of bone tissue is thought to be due to OSTEOCLASTS, multinucleated cells that enter the bone through the blood (Kahn and Simmons, 1975). Osteoclasts probably act by dissolving the matrix upon which the calcium phosphate is crystallized. The blood vessels also impart the blood-forming cells, which will reside in the marrow for the duration of the organism's life.

Lateral plate mesoderm

Not all of the mesodermal mantle is organized into somites. Adjacent to somitic mesoderm is the intermediate mesodermal region. This cord of mesodermal cells develops into the pronephric tubule, which is the precursor of kidney and genital ducts. The development of these organ systems will be discussed in detail in Chapters 16 and 19, respectively. Further laterally, we come to the LATERAL PLATE MESODERM. These plates exist on both sides of the embryo and become split into the dorsal SOMATIC (or PARIETAL) LAYER, which underlies the ectoderm, and a ventral SPLANCHNIC (or VISCERAL) LAYER, which overlies the endoderm (Figure 2D). Between these layers is the body cavity—the COELOM—which stretches from the future neck region to the posterior of the body. During later development, mesodermal folds extend from the mesoderm, dividing the coelom into separate cavities. In mammals, the coelom is subdivided into the pleural, pericardial, and peritoneal spaces, enveloping the thorax, heart, and abdomen, respectively. The mechanism for creating the mesodermal somites and body linings has changed

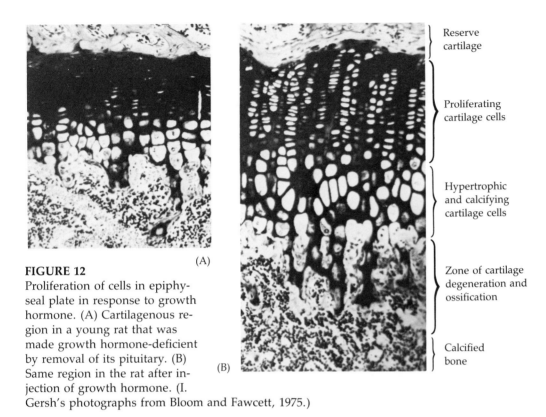

Reserve cartilage

Proliferating cartilage cells

Hypertrophic and calcifying cartilage cells

Zone of cartilage degeneration and ossification

Calcified bone

(A)

(B)

FIGURE 12
Proliferation of cells in epiphyseal plate in response to growth hormone. (A) Cartilagenous region in a young rat that was made growth hormone-deficient by removal of its pituitary. (B) Same region in the rat after injection of growth hormone. (I. Gersh's photographs from Bloom and Fawcett, 1975.)

little throughout vertebrate evolution, and the development of chick mesoderm can thus be compared with similar stages of frog embryos (Figure 13).

Formation of extraembryonic membranes

In amniote vertebrates (reptiles, birds, and mammals), embryonic development has taken a new direction. Reptiles evolved a mechanism for laying eggs on dry land, thus freeing them to explore niches that were not close to ponds. To accomplish this, the embryo produced four sets of EXTRAEMBRYONIC MEMBRANES to mediate between it and the environment, and even though most mammals have evolved placentas instead of shells, the basic pattern of extraembryonic membranes remains the same. In developing reptiles, birds, and mammals, there initially is no distinction between embryonic and extraembryonic domains. However, as the body of the embryo takes shape, the border epithelia create a series of BODY FOLDS, which surround the developing embryo, isolating it from the yolk, and delineating which areas are to be embryonic.

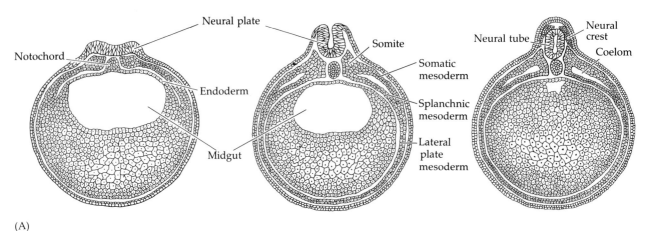

(A)

FIGURE 13
Comparison of mesodermal development in frog and chick embryos. (A) Neurula stage frog embryos showing progressive development of the mesoderm and coelom. (B) Transverse section of a chick embryo torn and brought together. Without the enormous yolk mass, it resembles the amphibian neurula at a similar stage. (A after Rugh, 1951; B adapted from Patten, 1951.)

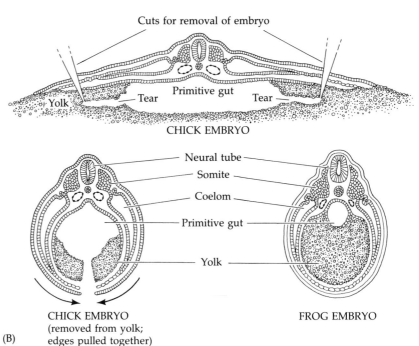

(B)

The membranous folds are formed by the extension of ectodermal and endodermal epithelium underlaid with mesoderm. The combination of ectoderm and mesoderm is often referred to as the SOMATOPLEURE and forms the amnion and chorion membranes; the combination of endoderm and mesoderm—the SPLANCHNOPLEURE—forms the yolk sac and allantois. The endodermal or ectodermal tissue acts as the functioning epithelial cells; and the mesoderm generates the essential blood supply to and from this epithelium. A schematic of the finished membranes is seen in Figure 24 in Chapter 1. The formation of these folds can be followed in Figure 14.

The first problem of a land-dwelling egg is dessication. Embryonic cells would quickly dry out if they were not in an aqueous environment. This environment is supplied by the AMNION. These cells of this membrane secrete and absorb amniotic fluid. Thus, embryogenesis still occurs in water. This evolutionary advance is so significant and characteristic that reptiles, birds, and mammals are grouped together as the amniote vertebrates.

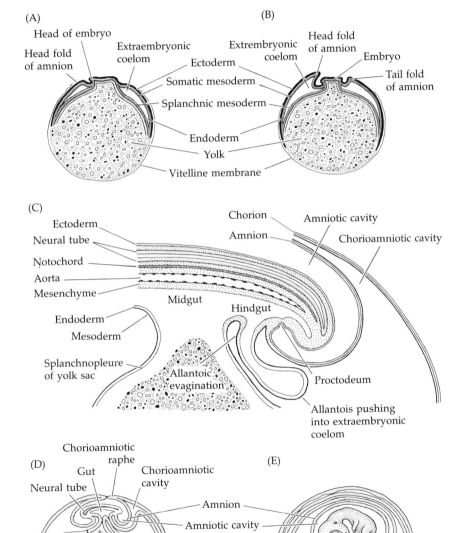

FIGURE 14
Schematic drawings of the extraembryonic membranes of the chick. The embryo is cut longitudinally and the albumen and shell coatings are not shown. (A) 2-day embryo. (B) 3-day embryo. (C) Detailed schematic diagram of the caudal (hind) region of the chick embryo showing the formation of the allantois. (D) 5-day embryo. (E) 9-day embryo. (After Carlson, 1981.)

The second problem of a land-dwelling egg is gas exchange. This exchange is provided for by the CHORION, the outermost extraembryonic membrane. In birds and reptiles, this membrane adheres to the shell, allowing the exchange of gases between the egg and the environment. In mammals, as we have seen, the chorion has evolved into the placenta, which has many other functions besides respiration.

The ALLANTOIS acts to store urinary wastes and to mediate gas exchange. In reptiles and birds, the allantois becomes a large sac, as there is no other way to keep toxic by-products of metabolism from the developing embryo. The mesodermal layer of allantoic membrane often will reach and fuse with the mesodermal layer of the chorion to create the CHORIOLLANTOIC MEMBRANE. This extremely vascular envelope characterizes the chick embryo and is responsible for transporting calcium from the eggshell into the embryo for bone production. In mammals, the size of the allantois depends upon how well nitrogenous wastes can be removed by the chorionic placenta. In humans, the allantois is a vestigial sac, whereas in pigs it is a large and important organ.

The YOLK SAC is the first extraembryonic membrane to be formed as it mediates nutrition in developing birds and reptiles. It is derived from endodermal cells that grow over the yolk to enclose it. The yolk sac is connected to the midgut by an open tube, the YOLK DUCT, so that the walls of the yolk sac and the walls of the gut are continuous. The blood vessels within the mesoderm of the splanchnopleure serve to take the nutrients from the yolk into the body, for the yolk is not taken directly into the body through the yolk duct. Rather, the endodermal cells digest the protein into soluble amino acids that can then be passed on to the blood vessels surrounding the yolk sac. In this way, the four extraembryonic membranes enable the egg to develop on land.

Heart and circulatory system

The circulatory system is one of the great achievements of the lateral plate mesoderm. Consisting of a heart, blood cells, and an intricate system of blood vessels, the circulatory system provides nourishment to the developing vertebrate embryo. The circulatory system is the first functional unit in the developing embryo, and the heart is the first functional organ.

The heart

In vertebrate embryos, the heart develops in an anterior position, within the splanchnic mesoderm of the neck; only later does it move to the chest. In amphibians, the two presumptive heart-forming regions are initially found at the most anterior position of the mesodermal mantle. While the embryo is undergoing neurulation, these two regions are brought together in the ventral region of the embryo such that they form a common PERICARDIAL CAVITY. In birds and mammals, the heart also develops from the fusion of paired primordia, but the fusion of these two rudiments does not occur until much later in development. In amniote vertebrates, the embryo is a flattened disc, and the lateral mesoderm does not circle completely around the yolk. Thus, the two heart-forming primordia undergo significant differentiation independently of each other. The presumptive heart cells of birds and mammals form a double-walled tube consisting of an inner ENDOCARDIUM and an outer EPIMYOCARDIUM. The endocardium will form the inner lining of the heart and the outer layer will form the layer of heart muscles, which will pump for the lifetime of the organism.

As neurulation proceeds, the foregut is closed by the inward folding of the splanchnic mesoderm (Figure 15). This brings the two tubes together, eventually uniting the epimyocardium into a single tube. The two endocardia lie within a common chamber for a short while; but these will also fuse together. At this time, the originally paired coelomic chambers unite to form the body cavity in which the heart resides. The bilateral origin of the heart can be demonstrated by surgically preventing the merger of the lateral plate mesoderm (Graeper, 1907; DeHaan, 1959). This results in a condition called CARDIA BIFIDA, wherein separate hearts form on each side of the body (Figure 16). The next step in heart formation is the fusion of the endocardial tubes to form a single pumping chamber (Figure 15C and D). This occurs at around 29 hours in chick development or at 3 weeks of human gestation. The unfused posterior portions of the endocardium become the opening of the VITELLINE VEINS into the heart (Figure 17). These veins will carry nutrients from the yolk sac into the SINUS VENOSUS. The blood then passes through a valvelike flap into the atrial region of the heart. Contractions of the TRUNCUS ARTERIOSUS speed the blood into the AORTA.

Pulsations of the heart begin while the paired primordia are still fusing. The pacemaker of this contraction is the sinus venosus. The contractions begin here and a wave of muscle contraction is then propagated up the tubular heart. In this way, the heart can pump blood even before its intricate system of valves has been completed. Heart muscle cells have their own inherent ability to contract—and isolated heart cells from 7-day rat or chick embryos will still continue to beat in petri dishes (Harary and Farley, 1963; DeHaan, 1967). In the embryo, these contractions become regulated by electrical stimuli from the medulla oblongata via the vagus nerve; and by 4 days, the electrocardiogram of a chick embryo approximates that of an adult.

Formation of blood vessels

The developing embryo has needs that are different from those of the adult organism, and its circulatory system reflects those differences. First, food is absorbed, not through the intestine, but from either the

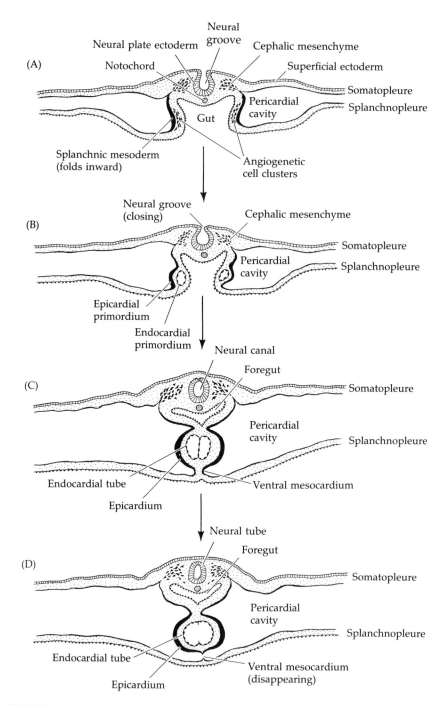

FIGURE 15

Transverse sections through the heart-forming region of the chick embryo at (A) 25 hours; (B) 26 hours; (C) 28 hours; and (D) 29 hours. (After Carlson, 1981.)

yolk or the placenta. Second, respiration is not conducted through the gills or lungs but through the chorionic or allantoic membranes. Thus, major embryonic blood vessels are constructed to serve these extraembryonic structures.

In addition to fulfilling the physiological needs of the embryo, the construction of the blood vessels also is dictated by evolutionary history. The mammalian embryo will extend blood vessels to the yolk sac even though there is no yolk therein. Moreover, the blood leaving the heart loops over the foregut to form the dorsally located aorta. These six pairs of AORTIC ARCHES arch over the pharynx (Figure 18). In primitive fishes, these arches persist and allow the gills to oxygenate the blood. In the adult bird or mammal, where lungs oxygenate the blood, such a system makes little sense. Yet in mammalian and avian embryos, all six pairs of aortic arches are formed, but the system eventually becomes simplified into a single aortic arch. Thus, our embryonic condition reflects

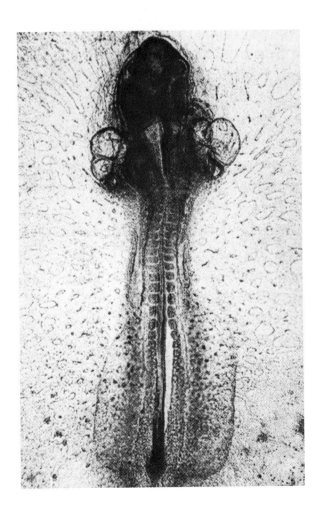

FIGURE 16
Cardia bifida in chick embryo caused by inhibiting the two heart primordia from fusing. (From DeHaan, 1969; courtesy of R. L. DeHaan.)

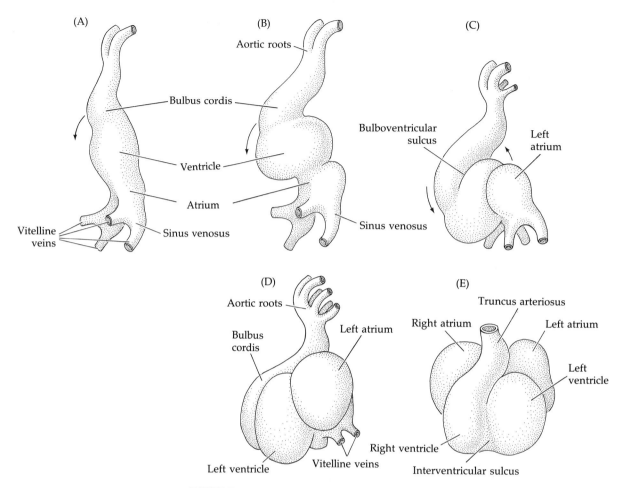

FIGURE 17

Formation of the heart chambers from a simple tube during the third week of human development. Views A–D show the developing heart from the left; E is a frontal view. Note that there are two aortic roots and that these branch out to form the aortic arches (Figure 18). (After Langman, 1981.)

our evolutionary history even though our physiology does not require such a structure.

The major embryonic blood vessels are those that obtain nutrients and bring them to the body and that transport gases to and from the sites of respiratory exchange. In those vertebrates with yolk, the VITEL-LINE (OMPHALOMESENTERIC) VEINS are formed by the aggregation of splanchnic mesodermal cells (BLOOD ISLANDS) that line the yolk sac (Figure 19). These cords hollow out into double-walled tubes analogous to the double tube of the heart. The inner wall becomes the flat EN-DOTHELIAL CELL lining of the vessel and the outer cells become the smooth muscle. Between these layers is a basal lamina containing a type of collagen specific for blood vessels. It has been suggested (Mur-

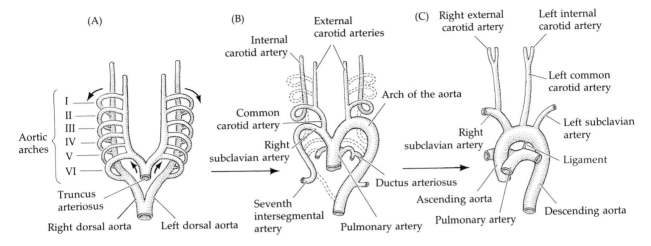

FIGURE 18
The aortic arches of the human embryo. (A) Originally, the truncus arteriosus pumps blood into the aorta, which branches on either side of the foregut. The six aortic arches take blood from the ventral aorta and allow it to flow into the dorsal aorta. (B) The arches begin to disintegrate or become modified; the dotted lines indicate degenerating structures. (C) Eventually, the remaining arches are modified and the adult artery system is formed. (After Langman, 1981.)

phy and Carlson, 1978) that this basal lamina initiates the differentiation of the cell types in the vessel. The central cells of the blood islands differentiate into the embryonic blood cells. As the blood islands grow, they eventually merge to form the capillaries of the two vitelline veins, which bring the food and blood cells to the newly formed heart. The embryonic circulatory system to and from the embryo and yolk sac is shown in Figure 20.

FIGURE 19
Angiogenesis. Blood vessel formation is first seen in the wall of the yolk sac where undifferentiated mesenchyme (A) condenses to form angiogenetic cell clusters (B). The centers of these clusters form the blood cells and the outside of the clusters develop into blood vessel endothelial cells (C). (After Langman, 1981.)

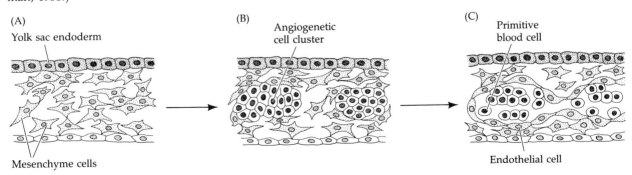

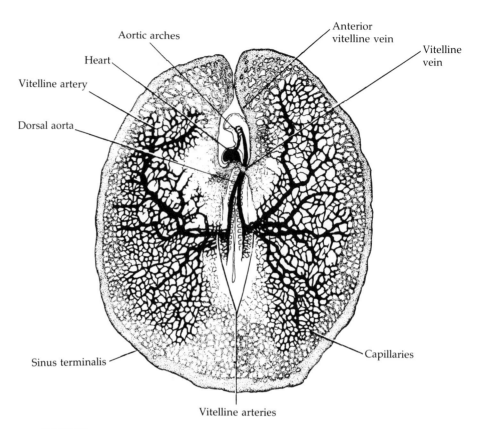

Aortic arches

Heart

Vitelline artery

Dorsal aorta

Anterior vitelline vein

Vitelline vein

Capillaries

Sinus terminalis

Vitelline arteries

FIGURE 20
Diagram of the circulatory system of a 44-hour chick embryo. This ventral view shows arteries in black; the veins are stippled. The sinus terminalis is the limit of blood circulation and is where new blood cells are being generated. (From Carlson, 1981.)

In mammalian embryos, food and oxygen is obtained from the placenta. Thus, although the mammalian embryo has vessels analogous to the vitelline veins, the main supply of food and oxygen comes from the UMBILICAL VEIN, which unites the embryo with the placenta (Figure 21). This vein, which takes the oxygenated and food-laden blood back into the embryo, is derived from what would be the right vitelline vein in birds. The UMBILICAL ARTERY, carrying wastes to the placenta, is derived from what would have become the allantoic artery of the chick. It extends from the caudal portion of the aorta and proceeds along the allantois and then out to the placenta.

After entering the embryonic mammalian heart, the blood is then pumped into a series of aortic arches, which encircle the pharynx to bring the blood dorsally. In mammals, the left member of the fourth pair of aortic arches is the only one serving to reach the aorta. The right member of this pair has become the root of the subclavian artery. The third aortic arches have been modified to form the internal common

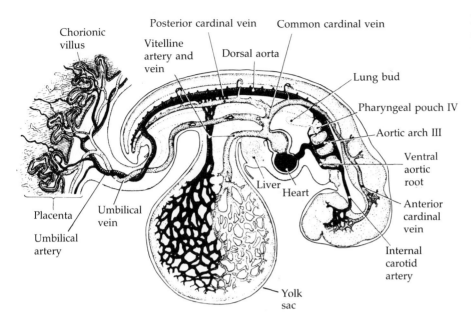

FIGURE 21
Circulatory system of a 4-week human embryo. Although at this stage all the major blood vessels are paired left and right, only the right vessels are shown. Arteries are shown in black. (From Carlson, 1981.)

carotid arteries, which supply blood to the brain and head. The sixth arch is modified to form the pulmonary artery; and the first, second, and fifth arches degenerate. The aorta and pulmonary artery, therefore, have a common opening to the heart for much of their development. Eventually, partitions form within the truncus arteriosus to create two different vessels (Figure 22). Only when the first breath of the newborn animal indicates that the lungs are ready to handle the oxygenation of the blood does the heart become modified to pump blood separately to the pulmonary artery.

FIGURE 22
Schematic diagrams illustrating the development of the truncus arteriosus into the aorta and pulmonary arteries. (After Kramer, 1942.)

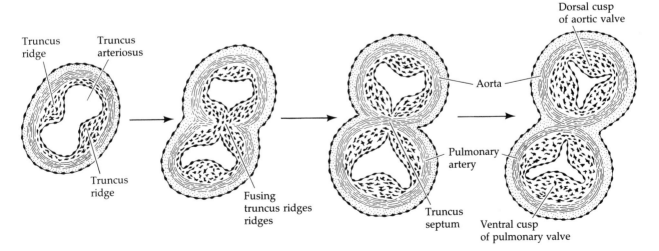

SIDELIGHTS & SPECULATIONS

Redirecting blood flow at birth

During mammalian development, the placenta rather than the lungs functions as the respiratory organ. At birth, when the newborn is no longer attached to the placenta, blood must be immediately shunted into the lungs. During fetal development, an opening—the DUCTUS ARTERIOSUS—diverts the passage of blood from the pulmonary artery into the aorta (and thus to the placenta). Because blood does not return from the pul-

monary vein in the fetus, the developing mammal has to have some other way of getting blood into the left ventricle to be pumped. This is accomplished by the FORAMEN OVALE, a hole in the septum separating the right and left atria. Blood can enter the right atrium, pass through the foramen to the left atrium, and then enter the left ventricle.

When the first breath is drawn, the oxygen in the blood causes the muscles surrounding the ductus arteriosus to close the opening. As the blood pressure in the left side of the heart rises, it shuts a flap over the foramen ovale, thereby separating the pulmonary and systemic circulation. Thus, when breathing begins, the respiratory circulation is shunted from the placenta to the lungs.

One of the most interesting details of the circulatory system is that the capillary network—the region where gas and nutrient exchange actually take place in each tissue—develops independently within the tissues themselves. The capillaries do not arise as smaller and smaller extensions of the major vessels growing from the heart. Rather, the mesoderm of each organ contains cells, called ANGIOBLASTS, which arrange themselves to form the capillary vessels. Thus, each organ forms its own capillary network, which eventually links to the extensions of the major blood vessels.

SIDELIGHTS & SPECULATIONS

Tumor-induced angiogenesis

Judah Folkman (1974) has estimated that as many as 350 billion mitoses occur in each human everyday. With each cell division comes the potential that the resulting cells will be malignant. Yet very few tumors actually do develop in any individual. Folkman has suggested that cells capable of forming tumors develop at a certain frequency but that a large majority never become able to form observable tumors. The reason is that solid tumors, like any other rapidly dividing tissue, need oxygen and nutrients to survive. Without a blood supply, potential tumors either die or remain

dormant. Such "microtumors" remain a stable cell population wherein dying cells are replaced by new cells.

The critical point at which this node of tumorous cells becomes a rapidly growing tumor occurs when the pocket of cells becomes vascularized. Such exponential tumor growth is seen in Figure 23. The microtumor can expand to 16,000 times its volume in 2 weeks after vascularization. Without the blood supply, no growth is seen (Folkman, 1974; Ausprunk and Folkman, 1977). To accomplish this vascularization, the original microtumor elaborates a substance called tumor angiogenesis factor (TAF). This factor has not been purified but appears to have an extremely low mass (250 daltons) (Weiss et al., 1979). Tumor angiogenesis factor causes the mitosis of endothelial cells and their formation into blood vessels in the direction of the tumor. This can be demonstrated by implanting a piece

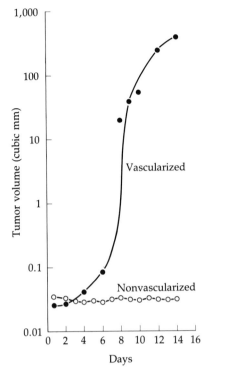

FIGURE 23
Exponential growth of a tumor after it receives blood vessels compared with that of a non-vascular tumor. Tumor spheroids were placed in the anterior of rabbit eyes; some tumors became vascularized (by capillaries growing from the iris) while others did not. (Modified from Folkman, 1974.)

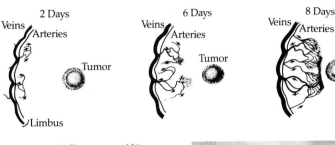

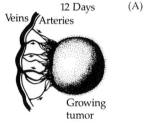

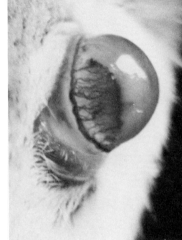

FIGURE 24
New blood vessel growth to the site of a tumor in the cornea of an albino mouse. (A) Sequence of events leading to the vascularization of a mammary adenocarcinoma tumor on days 2, 6, 8 and 12. The veins and arteries of the limbus surrounding the cornea both provide vessels. (B) Photograph of living cornea of an albino mouse, with new vessels being constructed to enter the tumor graft. (A from Muthukkaruppan and Auerbach, 1979; B, photograph courtesy of R. Auerbach.)

of tumor tissue within the layers of a rabbit or mouse cornea. The cornea is not vascularized, but it is surrounded by a vascular LIMBUS. Tumor tissue induces the formation of blood vessels to come toward the tumor (Figure 24). Most other adult tissues will not induce these vessels to form. (The exceptions are antigen-stimulated lymphocytes and macrophages.) Once the blood vessels enter the tumor, the tumor cells undergo explosive growth, eventually bursting the eye.

More recently a natural inhibitor of TAF has been isolated from cartilage. (Cartilage is one of the few tissues that is not extensively vascularized. Tumors of cartilage likewise have few blood vessels.) This inhibitor has been found to stop angiogenesis in the tumor-implanted cornea and an infusion of the inhibitor has stopped tumor expansion in the eye of rabbits (Langer et al., 1980). Thus, tumors can elaborate a substance that causes the development of blood vessels in the adult animal.

Development of blood cells

Pluripotential stem cells and hematopoietic microenvironments

Most vertebrate tissues are composed of differentiated cells that no longer divide. Myoblasts, for instance, are a rapidly dividing cell population until they develop into myotubes. Some tissues, such as epidermis, intestinal epithelium, and blood cells, however, retain an "embryonic" cell population within themselves, such that their cellular composition is always changing, even in adult animals. This is most evident in the case of the mammalian red blood cell. This cell lacks a nucleus and has a life span of only 120 days in circulation. A normal person will lose and replace 3×10^{11} red blood cells every day (Hay, 1966). The continuous formation of new red blood cells (as well as all the other types of blood cells) is accomplished in the bone marrow by the HEMATOPOIETIC ("blood-forming") STEM CELLS.

Stem cells are an intriguing and little-understood phenomenon. Yet our lives depend upon them. A stem cell is a cell that is capable of extensive proliferation and that can generate more stem cells ("self-renewal") as well as more differentiated progeny (Siminovitch et al., 1963). Thus, a single stem cell can generate a clone containing millions of differentiated cells as well as a few stem cells. The notion of a stem cell is depicted in Figure 25.

In mammals and birds, there appears to be a common PLURIPOTENTIAL HEMATOPOIETIC STEM CELL, which can give rise to red blood cells (erythrocytes), white blood cells (granulocytes), macrophages, platelets, and immunocompetent cells (lymphocytes). The existence of such stem cells was shown by Till and McCulloch (1961), who injected bone mar-

FIGURE 25
Model of the dynamics of stem cell proliferation and differentiation.

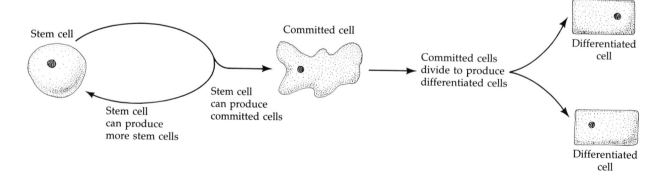

Stem cell

Committed cell

Committed cells divide to produce differentiated cells

Differentiated cell

Stem cell can produce committed cells

Stem cell can produce more stem cells

Differentiated cell

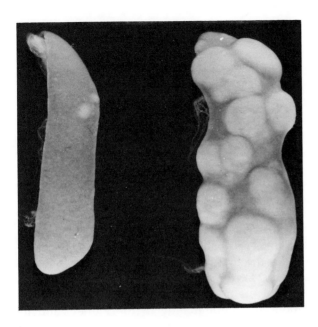

FIGURE 26
Isolated blood-forming colonies. When bone marrow containing hematopoietic stem cells is injected into irradiated mice, discrete colonies of blood cells are seen on the surface of the spleen of that mouse. The spleen on the right has more of these colonies than does the spleen on the left. (From Till, 1981; photograph courtesy of J. E. Till.)

row cells into lethally irradiated mice of the same genetic strain as the marrow donors. Some of these cells produced discrete nodules on the spleens of the host animals (Figure 26). Microscopic studies showed these nodules to be composed of erythrocyte, granulocyte, and platelet precursors. Thus, a single cell from the bone marrow was capable of forming many of the different blood cell types. The cell responsible was called the CFU-S, the colony forming unit of the spleen. Further studies used chromosomal markers to prove that the different types of cells within the colony were formed from the same CFU-S. Here, marrow cells were irradiated so that very few survived. Many of those that did survive had abnormal chromosomes, which could be detected microscopically. When such irradiated CFU-S were injected into an irradiated mouse, each cell of a spleen colony, be it granulocyte or erythrocyte precursor, had the same chromosomal anomaly (Becker et al., 1963). It is important to the stem cell concept that the stem cell be able to form more stem cells in addition to its differentiated cell types. This has indeed been found to be the case. When spleen colonies derived from a single CFU-S are resuspended and injected into other mice, many spleen colonies are seen to emerge (Jurásková and Tkadleček, 1965; Humphries et al., 1979). Thus, we see that a single marrow cell can form numerous different cell types and can also undergo self-renewal. We can therefore consider the CFU-S to be a pluripotential hematopoietic stem cell.

The preceding data indicate that although the CFU-S can generate many of the blood cell types, it is not capable of generating lymphocytes. However, Abramson and her colleagues (1977a, b) have shown that the CFU-S and the lymphocytes are both derived from a different type of stem cell. When they injected irradiated bone marrow cells into mice having a hereditary deficiency of blood-forming cells, they found the same chromosomal abnormalities in both the spleen colonies and the circulating lymphocytes. Figure 27 is a model that summarizes several studies. The first pluripotential hematopoietic stem cell is the CFU-M,L (myeloid and lymphoid colony forming unit). This gives rise to the CFU-S (blood cells) and the CFU-L (lymphocytes). The CFU-S and CFU-L are pluripotent stem cells in that they can give rise to numerous cell types. The immediate progeny of the CFU-S, however, are *committed* stem cells. They can produce only one type of cell in addition to renewing themselves. The BFU-E (burst-forming unit-erythroid), for instance, is formed from the CFU-S, and it can form only one cell type in addition to itself. This new cell is the CFU-E (colony forming unit-erythroid), which is capable of responding to the hormone ERYTHROPOIETIN to produce the first recognizable differentiated erythrocyte, the PROERYTHROBLAST. Erythropoietin is a glycoprotein that rapidly induces the synthesis of the mRNA for globin (Krantz and Goldwasser, 1965). It is produced predominantly in the kidney, and its synthesis is responsive to environmental conditions. If the level of blood oxygen falls, erythropoietin production is increased, thus leading to more red blood cells being produced. As the red blood cell matures, it becomes an ERYTHROBLAST, synthesizing enormous amounts of hemoglobin. Eventually, the mammalian erythroblast expels its nucleus, becoming a RETICULOCYTE. Reticulocytes can no longer synthesize globin mRNA but can still translate existing messages into globin. The final stage of differentiation is the ERYTHROCYTE stage. Here, no division, RNA synthesis, or protein synthesis takes place. The cell leaves the bone marrow to undertake its role delivering oxygen to the bodily tissues.

The CFU-L cells are thought to develop in a similar fashion, namely, the pluripotential stem cell generates another type of stem cell that is committed to a certain type of differentiation. Thus, there is a committed stem cell for platelets and a committed stem cell for granulocytes and macrophages. The determination of stem cell fate is thought to be controlled by the interactions of the stem cell with the cells of its immediate environment. In the stroma of the spleen, the stem cells are predominantly committed to erythroid development. In the bone marrow, granulocyte development predominates. That short-range interactions yield these results was shown by Wolf and Trentin (1968). These investigators placed plugs of bone marrow into the spleen and then injected stem cells. Those colonies in the spleen were predominantly erythroid whereas those forming in the marrow plugs were predominantly granulocytic. In fact, those colonies that straddled the borders were predominantly erythroid in the spleen and granulocytic in the marrow. The regions of determination are referred to as HEMATOPOIETIC INDUCTIVE MICROENVIRONMENTS (HIM).

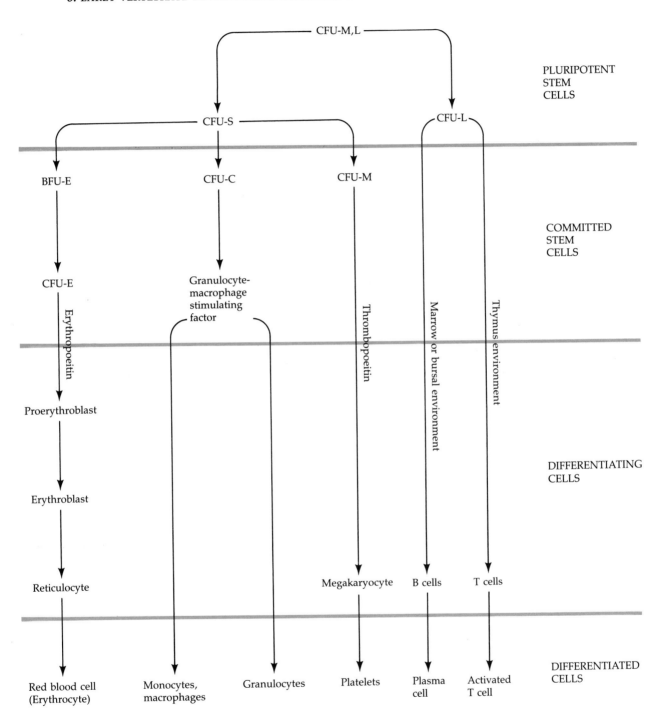

FIGURE 27
A model for the origin of mammalian blood and lymphoid cells. (After Quesenberry and Levitt, 1979.)

Sites of hematopoiesis

In most mammals, the major source of blood cells in the adult is the bone marrow. (In some species, like the mouse, the spleen also participates.) However, numerous experiments have shown that the hematopoietic stem cells do not originate in the marrow or spleen, but rather, they migrate there from other regions of the embryo. In fact, the mammalian hematopoietic stem cell probably comes from *outside* the embryo proper, that is, from the yolk sac. When Moore and his coworkers removed the yolk sacs from presomite mouse embryos, no hematopoietic development took place in the embryo. Moreover, in

FIGURE 28

The successive sites of blood formation in the embryonic mouse. Arrows indicate potential migratory paths of pluripotential hemopoietic stem cells (SC). The comparative size of the red blood cells produced in each site is shown at the right. (After Russell, 1979.)

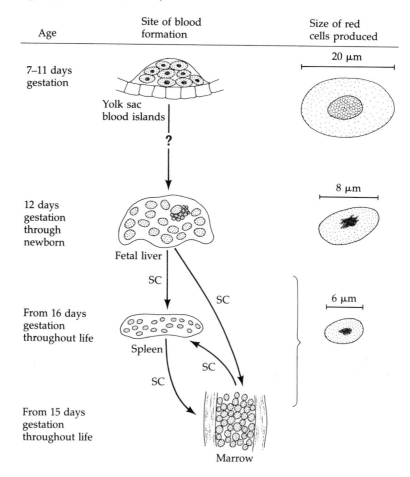

cultures of the isolated yolk sac, erythropoiesis and CFU-S production were seen. Moore and Metcalf (1970) demonstrated that the peak of CFU-S activity in the yolk sac was at 10 days. In contrast, the embryonic liver begins to show the presence of stem cells only after 11 days. Thus, it was postulated that the stem cells migrate from the yolk sac to the embryonic liver and then to the spleen and bone marrow. Embryonic yolk sac cells have also been seen to carry out lymphocytic functions before any stem cell activity is present within the embryo proper (Hofman and Globerson, 1973; Dahl et al., 1980). Thus, the original mammalian pluripotential stem cell is thought to arise within the yolk sac and migrate first to the liver and then to the adult hematopoietic organs (Figure 28).

A similar situation was believed to occur in avian embryos, but birds appear to have two separate sources of hematopoietic stem cells. The first is, as in mammalian embryos, located in the yolk sac. These cells, however, appear to be a transient population. The major source of stem cells in birds seems to be local blood islands, which form within the embryo proper. This was discovered in a series of elegant experiments by Dieterlen-Lievre, who grafted the blastoderm of chickens onto the yolk of the Japanese quail (Figure 29). Chick cells can be distinguished readily from quail cells because the quail cell nucleus stains much more darkly, providing a permanent marker for distinguishing the two cell types. Using these "yolk sac chimeras," Dieterlen-Lievre and Martin (1981) have shown that the yolk sac stem cells do not contribute cells to the adult animal but that the true stem cells are formed within nodes of mesoderm that line the mesentery and the major blood vessels. In mammals, the yolk sac is still thought to be the source of all hematopoietic stem cells. However, recent studies (Kubai and Auerbach, 1983) have suggested that these yolk sac blood precursors may be derived from stem cells lining the abdominal cavity.

ENDODERM

Pharynx

The function of embryonic endoderm is to construct the linings of two tubes within the body. The first tube, extending throughout the length of the body, is the digestive tube. Buds from this tube form the liver, gall bladder, and pancreas. The second tube is the respiratory tube. This tube eventually bifurcates into two lungs. The digestive and respiratory tubes share a common chamber in the anterior region of the embryo, and this region is called the PHARYNX. Epithelial outpockets of the pharynx give rise to the tonsils, thyroid, thymus, and parathyroid glands.

The respiratory and digestive tubes are both derived from the primitive gut (Figure 30). As the endoderm folds in toward the center of the embryo, the foregut and hindgut regions are formed. At first, the oral end is blocked by a region of ectoderm called the ORAL (or STOMODEAL) PLATE. Eventually (about 22 days in human embryos), the stomodeal

FIGURE 29
Blood cell mapping by chick–quail chimeras. (A) Photograph of a "yolk sac chimera" where the blastoderm of a quail was put on the yolk sac of a chick. (B) Photograph of chick and quail cells in thymus of a chimeric animal showing the difference in the nuclear staining. The lymphoid cells are all chick whereas the structural cells of the thymus are of quail origin. (C) Aorta of a 3-day chick embryo showing the cells (arrows) that give rise to the hemopoietic stem cells. If this region of cells is taken from quail embryos and placed into chick embryos, the chick embryos have quail blood. (From Martin and Dieterlen-Lievre, 1978; Dieterlen-Lievre and Martin, 1981; photographs courtesy of F. Dieterlen-Lievre.)

(A)

(B)

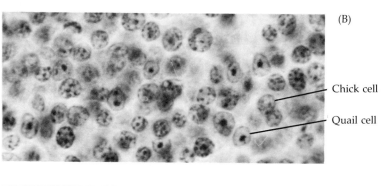

Chick cell

Quail cell

(C)

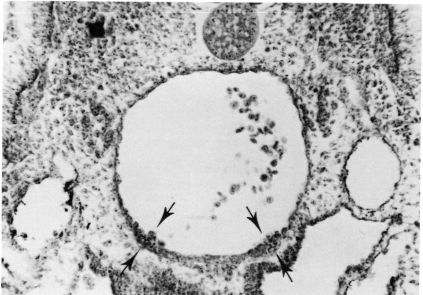

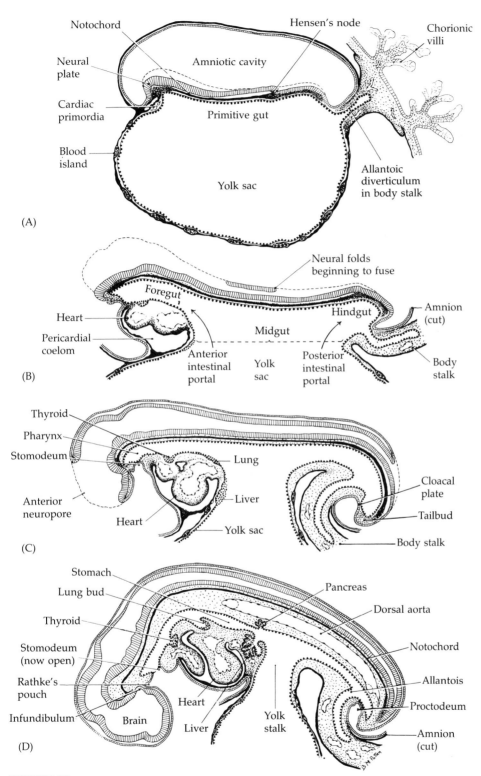

FIGURE 30
Formation of the human digestive system, depicted at about (A) 16 days, (B) 18 days, (C) 22 days, (D) 28 days. (From Carlson, 1981.)

plate breaks, thereby creating the oral opening of the digestive tube. The opening itself is lined by ectodermal cells, and this creates an interesting situation, for the oral plate ectoderm is in contact with the brain ectoderm, which has curved around toward the ventral portion of the embryo. The two ectodermal regions mutually interact with each other. The roof of the oral region forms Rathke's pouch and becomes the glandular part of the pituitary gland. The neural tissue on the floor of the diencephalon gives rise to the infundibular process, which becomes the neural portion of the pituitary. Thus, the pituitary gland has a dual origin; and this dual nature of the pituitary is reflected in its adult functions.

The endodermal portion of the digestive and respiratory tubes begins in the pharynx. Here, the mammalian embryo produces four pairs of PHARYNGEAL POUCHES. The furrows between these pouches are sometimes called the GILL CLEFTS, as they resemble those structures found in fish embryos (Figure 31). However, instead of generating gills, mammalian pharyngeal pouches have been modified for the terrestrial environment. The first pharyngeal pouches become the auditory cavities of the middle ear and the associated eustachian tubes. The second pair of pouches gives rise to the wall of the tonsils. The thymus is derived from the third pair of pharyngeal pouches. This gland will direct the differentiation of lymphocytes during later stages of development. One pair of parathyroid glands is also derived from the third pair of pharyngeal pouches, and the other pair is derived from the fourth. In addition to these paired pouches, a small diverticulum is formed between the second pharyngeal pouches on the floor of the pharynx. This pocket of endoderm and mesenchyme will bud off from the pharynx and migrate down the neck to become the thyroid gland.

The digestive tube and its derivatives

Posterior to the pharynx, the digestive tube constricts to form the esophagus, which is followed in sequence by the stomach, small intestine,

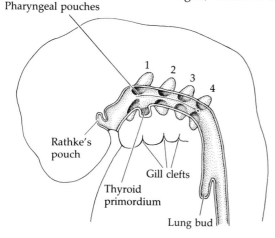

FIGURE 31
The pharyngeal pouches and gill clefts of a 5-week human embryo.

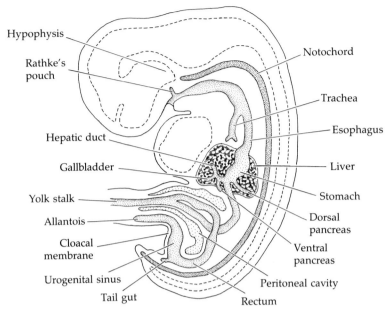

FIGURE 32
Development of the digestive system in a 6-week human embryo. The stomach region of the digestive tube has begun to dilate and the pancreas is represented by two buds which will eventually fuse. (After Langman, 1981.)

and large intestine. The endodermal cells generate only the lining of the digestive tube and its glands, for mesenchymal cells will surround this tube to provide the muscles for peristalsis.

Figure 32 shows that the stomach develops as a dilated region close to the pharynx. More caudally, the intestines develop, and the connection between the intestine and yolk sac is eventually severed. At the caudal end of the intestine, a depression forms where the endoderm meets the overlying ectoderm. Here, a thin CLOACAL MEMBRANE separates the two tissues. It eventually ruptures, forming the opening that will become the anus.

Liver, pancreas, and gallbladder

Endoderm also forms the lining of three accessory organs that develop immediately caudal to the stomach. The HEPATIC DIVERTICULUM is the tube of endoderm that extends out from the foregut into the surrounding mesenchyme. The mesenchyme induces the endoderm to proliferate, to branch, and to form the glandular epithelium of the liver. A portion of the hepatic diverticulum (that region closest to the digestive tube) continues to function as the drainage duct of the liver, and a diverticulum from this duct produces the gallbladder (Figure 33).

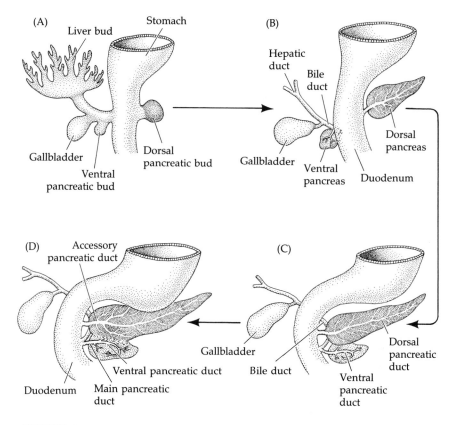

FIGURE 33

Pancreatic development in humans. At 30 days (A), the ventral pancreatic bud is close to the liver primordium, but by 35 days (B) it begins migrating posteriorly and comes into contact with the dorsal pancreatic bud during the sixth week of development (C). In most individuals, the dorsal pancreatic bud loses its duct into the duodenum. However, in about 10 percent of the population, the dual duct system persists (D). (After Langman, 1969.)

The pancreas develops from the fusion of distinct dorsal and ventral diverticula. Both of these primordia arise from the endoderm immediately caudal to the stomach; and as they grow, they come closer together and eventually fuse. In humans, only the ventral duct survives to carry digestive enzymes into the intestine. In other species (such as the dog), both the dorsal and ventral ducts empty into the intestine.

The respiratory tube

The lungs can also be seen as a derivative of the digestive tube, even though they serve no role in digestion. In the center of the pharyngeal floor, between the fourth pair of pharyngeal pouches, the LARYNGOTRA-CHEAL GROOVE extends ventrally (Figure 34). This groove then bifurcates

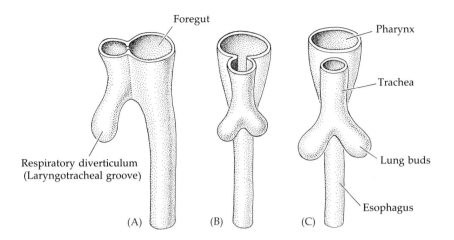

FIGURE 34
The partitioning of the foregut into the esophagus and respiratory diverticulum during the third and fourth weeks of human gestation. (A) Lateral view, end of week 3; (B, C) Ventral views, week 4.

into two major branches. The laryngotracheal endoderm becomes the lining of the trachea, the two bronchi, and the air sacs (alveoli) of the lungs. As we will see in a later chapter, the branching of this endodermal tube depends upon interactions with the different types of mesodermal cells in its path.

The lungs are an evolutionary novelty; and they are among the last of the mammalian organs to fully differentiate. The lungs must be able to draw in oxygen at the baby's first breath. To accomplish this, alveolar cells secrete a SURFACTANT into the fluid bathing the lungs. This surfactant, consisting of phospholipids such as sphingomyelin and lecithin, is secreted very late in gestation, and it usually reaches physiologically useful levels around week 34 of gestation. These compounds enable the alveolar cells to touch one another without sticking together. Thus, infants born prematurely often have difficulty breathing and have to be placed in respirators until their surfactant-producing cells mature.

SIDELIGHTS &
SPECULATIONS

Teratocarcinoma

A mammalian germ cell or early blastomere contains within it all the information needed for subsequent development. What would happen if such a cell became malignant? The result is a TERATOCARCINOMA. This type of tumor sometimes arises spontaneously within the ovary or testis. Teratocarcinomas can also be produced experimentally by merely implanting a mammalian blastocyst in a site other than the uterus (for example, inside the connective tissue capsule that surrounds a kidney). Whether spontaneous or experimentally produced, a teratocarcinoma contains an undifferentiated stem cell population that has biochemical and developmental properties remarkably similar to those cells of the inner cell mass (Graham, 1977). Moreover, these stem cells (called EMBRYONAL CARCINOMA cells) not only divide but can also differentiate into a wide variety of tissues including gut and respiratory epithelia, muscle, nerve, cartilage, and bone (Fig-

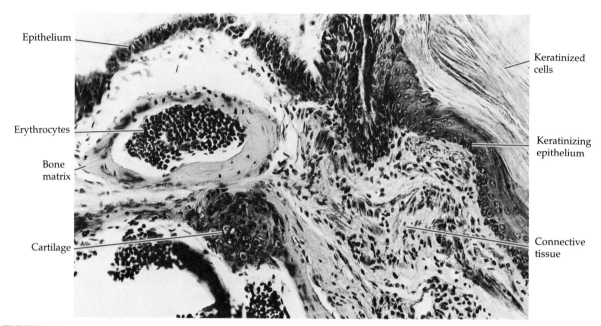

Epithelium

Erythrocytes

Bone matrix

Cartilage

Keratinized cells

Keratinizing epithelium

Connective tissue

FIGURE 35

Photomicrograph of a solid teratocarcinoma showing numerous differentiated cell types. (From Gardner, 1982; photograph from C. Graham, courtesy of R. L. Gardner.)

ure 35). These differentiated cells no longer divide and are therefore not malignant. Thus, such tumors can give rise to most of the tissue types in the body.

Embryonal carcinomas can be transferred from mouse to mouse and can be induced to form EMBRYOID BODIES by injecting these cells into the peritoneal cavity

of a mouse. These bodies contain a core of embryonal carcinoma cells surrounded by a cell type that looks extremely similar to the early endoderm of the mouse. Indeed, such embryoid bodies can encyst themselves,

FIGURE 36

Embryoid bodies. (A) Phase-contrast photomicrograph of living embryoid bodies in culture. Arrows point to regions where the two different (inner and outer) cell types can be seen. (B) Section of such an embryoid body stained with hematoxylin and eosin. The core stem cells are seen to be surrounded by a cell type very reminiscent of early embryonic endoderm. (Courtesy of G. Martin.)

(A)

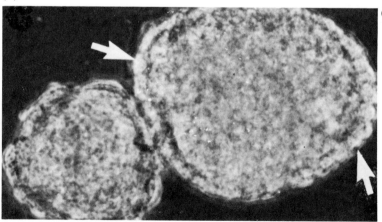

(B)

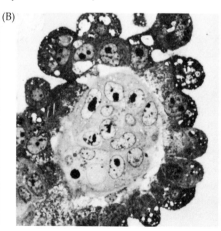

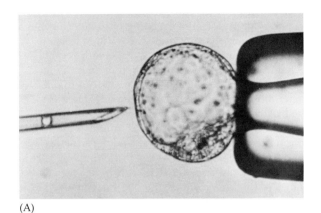

(A)

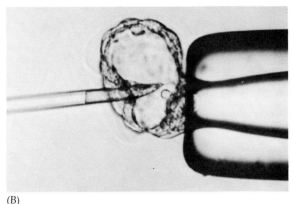

(B)

(C)

FIGURE 37

Insertion of embryonal carcinoma cell into a mouse blastocyst. (A) Teratocarcinoma stem cell in micropipette approaches blastocyst held fast by suction pipette. (B) Injection of stem cell onto the inner cell mass. (C) Mice produced when teratocarcinoma stem cell from a black strain of mice integrated into the inner cell mass of a white strain of mice. Black fur can be seen on heads and backs of mice, indicating that at least some of the pigment cells were of tumor origin. (A and B courtesy of K. Illmensee; C from Papaioannou, 1979; photograph courtesy of V. E. Papaioannou.)

in which case amorphous types of differentiated cells appear (including hematopoietic colonies) in the "endoderm" (Figure 36). Adamson and co-workers (1977) showed that this "endoderm" is also capable of synthesizing α-fetoprotein, a protein usually synthesized by normal mammalian endoderm cells. Thus, the embryonal carcinoma cells mimic early mammalian development. However, the tumor they form is characterized by random, haphazard development.

Although these tumors are probably derived from germ cells, the biochemistry, morphology, and cell surface behavior of embryonal carcinoma cells resemble those of inner cell mass blastomeres. In 1981, Stewart and Mintz formed a mouse from cells derived in part from a teratocarcinoma stem cell! Stem cells that had arisen in a teratocarcinoma of an agouti (yellow-tipped) strain of mice were cultured for several cell generations and were seen to maintain the characteristic chromosome complement of the parental mouse. Individual

stem cells were injected into the blastocysts of black mice (Figure 37). The blastocysts were then transferred to the uterus of a foster mother and live mice were born. Some of these mice had coats of two colors, indicating that the tumor cell had integrated itself into the embryo. Moreover, when mated to a mouse carrying an appropriate marker, the chimeric mouse was able to generate mice having some of the phenotypes of the tumor "parent." The malignant embryonal carcinoma cell had produced many, if not all, types of normal somatic cells and even had produced normal, functional germ cells! When mice having a tumor cell for one parent were mated together, the resultant litter contained mice that were homozygous for a large number of genes from the tumor cell (Figure 38).

Because the cultivation of early mammalian embryos is a difficult task, researchers are looking at embryonal carcinoma cells as a unique way to analyze early mammalian development. Moreover, the ability to mutate embryonal carcinoma cells in culture and inject know mutant cells into blastocysts also provides the possibility of creating mouse mutants to order.

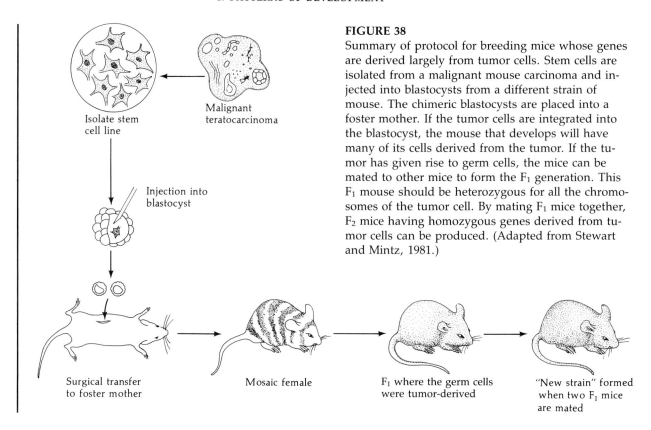

FIGURE 38

Summary of protocol for breeding mice whose genes are derived largely from tumor cells. Stem cells are isolated from a malignant mouse carcinoma and injected into blastocysts from a different strain of mouse. The chimeric blastocysts are placed into a foster mother. If the tumor cells are integrated into the blastocyst, the mouse that develops will have many of its cells derived from the tumor. If the tumor has given rise to germ cells, the mice can be mated to other mice to form the F_1 generation. This F_1 mouse should be heterozygous for all the chromosomes of the tumor cell. By mating F_1 mice together, F_2 mice having homozygous genes derived from tumor cells can be produced. (Adapted from Stewart and Mintz, 1981.)

Isolate stem cell line

Malignant teratocarcinoma

Injection into blastocyst

Surgical transfer to foster mother

Mosaic female

F_1 where the germ cells were tumor-derived

"New strain" formed when two F_1 mice are mated

Here we shall conclude our survey of the major features of vertebrate development. We shall now attend to the *mechanisms* that enable this development to take place. In Part II we shall focus upon the ways in which cells differentiate; and in Part III we shall be concerned with how cells interact with one another to form tissues and organs.

Literature cited

Abbott, J. and Holtzer, H. (1965). Critical number of mitoses and the differentiation of chondroblasts and myoblasts. *Anat. Rec.* 151: 439–447.

Abramson, S., Miller, R. G. and Phillips, R. A. (1977). The identification in adult bone marrow of pluripotent and restricted stem cells of the myeloid and lymphoid systems. *J. Exp. Med.* 145: 1567–1579.

Adamson, E., Evans, M. J. and Magrane, G. G. (1977). Biochemical markers of the progress of differentiation in cloned teratocarcinoma cell lines. *Eur. J. Biochem.* 79: 607–615.

Ausprunk, D. H. and Folkman, J. (1977). Migration and proliferation of endothelial cells in preformed and newly formed blood vessels during tumor angiogenesis. *Microvasc. Res.* 14: 53–65.

Balinsky, I. (1981). *Introduction to Embryology,* Fifth Edition. Saunders, Philadelphia.

Becker, A. J., McCulloch, E. A. and Till, J. E. (1963). Cytological demonstration of the clonal nature of spleen cells derived from transplanted mouse marrow cells. *Nature* 197: 452–454.

Bischoff, R. and Holtzer, H. (1969). Mitosis and the processes of differentiation of myogenic cells *in vitro. J. Cell Biol.* 41: 188–200.

Bloom, W. and Fawcett, D. W. (1975). *Textbook of Histology,* Tenth Edition. Saunders, Philadelphia.

Buckley, P. A. and Konigsberg, I. R. (1974). Myogenic fusion and the postmitotic gap. *Dev. Biol.* 37: 193–212.

Carlson, B. M. (1981). *Patten's Foundations of Embryology.* McGraw-Hill, New York.

Chevallier, A., Kieny, M., Mauger, A. and Sengel, P. (1977). Developmental fate of the somitic mesoderm in the chick embryo. In D. A. Ede, J. R. Hinchliffe and M. Balls (eds.), *Vertebrate Limb and Somite Morphogenesis.* Cambridge University Press, Cambridge, pp. 421–432.

Dahl, C. A., Kahan, B. W. and Auerbach, R. (1980). Studies with mouse embryonic yolk sac cells: An approach to understanding the patterns of ontogeny of cell-mediated immune functions. In J. D. Horton (ed.), *Development and Differentiation of Vertebrate Lymphocytes.* Elsevier North-Holland, Amsterdam, pp. 241–253.

Daubas, P., Caput, D., Buckingham, M. and Gros, F. (1981). A comparison between the synthesis of the contractile proteins and the accumulation of their translatable mRNAs during calf myoblast differentiation. *Dev. Biol.* 84: 133–143.

Davis, J. D., See, W. M. and Higginbotham, C. A. (1981). Fusion of chick embryo skeletal myoblasts: Role of calcium influx preceding membrane union. *Dev. Biol.* 82: 297–307.

DeHaan, R. (1959). *Cardia bifida* and the development of pacemaker function in the early chicken heart. *Dev. Biol.* 1: 586–602.

DeHaan, R. L. (1967). Regulation of spontaneous activity and growth of embryonic chick heart cells in tissue culture. *Dev. Biol.* 16: 216–249.

Devlin, R. B. and Emerson, C. P., Jr. (1978). Coordinate regulation of contractile protein synthesis during myoblast differentiation. *Cell* 13: 599–611.

Devlin, R. B. and Emerson, C. P., Jr. (1979). Coordinate accumulation of contractile protein mRNAs during myoblast differentiation. *Dev. Biol.* 69: 202–216.

Dieterlen-Lievre, F. and Martin, C. (1981). Diffuse intraembryonic hemopoiesis in normal and chimeric avian development. *Dev. Biol.* 88: 180–191.

Folkman, J. (1974). Tumor angiogenesis. *Adv. Cancer Res.* 19: 331–358.

Gardner, R. L. (1982). Manipulation of development. In C. R. Austin and R. V. Short (eds.), *Embryonic and Fetal Development,* Cambridge University Press, Cambridge, pp. 159–180.

Graham, C. F. (1977). Teratocarcinoma cells and normal mouse embryogenesis. In M. I. Sherman (ed.), *Concepts in Mammalian Embryogenesis.* MIT Press, Cambridge, MA, pp. 315–394.

Gräper, L. (1907). Untersuchungen über die Herzbildung der Vögel. *Wilhelm Roux Arch. Entwicklungsmech. Org.* 24: 375–410.

Harary, I. and Farley, B. (1963). *In vitro* studies on single beating rat heart cells. II. Intercellular communication. *Exp. Cell Res.* 29: 466–474.

Hay, E. (1966). *Regeneration.* Holt, Rinehart & Winston, New York.

Hofman, F. and Globerson, A. (1973). Graft-versus-host response induced *in vitro* by mouse yolk sac cells. *Eur. J. Immunol.* 3: 179–181.

Holtzer, H., Rubinstein, N., Fellini, S., Yeoh, G., Chi, J., Birbaum, J. and Okayama, M. (1975). Lineages, quantal cell cycles, and the generation of cell diversity. *Q. Rev. Biophys.* 8: 523–557.

Humphries, R. K., Jacky, P. B., Dill, F. J., Eaves, A. C. and Eaves, C. J. (1979). CFUs in individual erythroid colonies derived *in vitro* from adult mature mouse marrow. *Nature* 279: 718–720.

Jurśšková, V. and Tkadleček, L. (1965). Character of primary and secondary colonies of haematopoiesis in the spleen of irradiated mice. *Nature* 206: 951–952.

Kahn, A. J. and Simmons, D. J. (1975). Investigation of cell lineage in bone using a chimaera of chick and quail embryonic tissue. *Nature* 258: 325–327.

Kalderon, N. and Gilula, N. B. (1979). Membrane events involved in myoblast fusion. *J. Cell Biol.* 81: 411–425.

Konigsberg, I. R. (1963). Clonal analysis of myogenesis. *Science* 140: 1273–1284.

Kramer, T. C. (1942). The partitioning of the truncus and conus and the formation of the membranous portion of the intraventricular septum in the human heart. *Am. J. Anat.* 71: 343–370.

Krantz, S. B. and Goldwasser, E. (1965). On the mechanism of erythropoietin induced differentiation. II. The effect on RNA synthesis. *Biochim. Biophys. Acta.* 103: 325–332.

Kubai, L. and Auerbach, R. (1983). A new source of embryonic lymphocytes in the mouse. *Nature* 301: 154–156.

Langer, R., Conn, H., Vacantai, J., Haudenschild, C. and Folkman, J. (1980). Control of tumor growth in animals by infusion of an angiogenesis inhibitor. *Proc. Natl. Acad. Sci. USA* 77: 4331–4335.

Langman, J. (1969). *Medical Embryology,* Third Edition. Williams & Wilkins, Baltimore.

Lipton, B. H. and Jacobson, A. G. (1974). Analysis of normal somite development. *Dev. Biol.* 38: 73–90.

Lough, J. and Bischoff, R. (1977). Differentiation of creatine phosphokinase during myogenesis: Quantitative fractionation of isozymes. *Dev. Biol.* 57: 330–344.

Martin, C., Beaupain, D. and Dieterlen-Lievre, F. (1978). Developmental relationships between vitelline and intraembryonic haemopoiesis studied in avian yolk sac chimeras. *Cell Differ.* 7: 115–130.

Meier, S. (1979). Development of the chick mesoblast: Formation of the embryonic axis and establishment of the metameric pattern. *Dev. Biol.* 73: 25–45.

Mintz, B. and Baker, W. W. (1967). Normal mammalian muscle differentiation and gene control of isocitrate dehydrogenase synthesis. *Proc. Natl. Acad. Sci. USA* 58: 592–598.

Moore, M. A. S. and Metcalfe, D. (1970). Ontogeny of the haemopoietic system: Yolk sac origin of *in vivo* and *in vitro* colony forming cells in the developing mouse embryo. *Br. J. Haematol.* 18: 279–296.

Murphy, M. E. and Carlson, E. C. (1978). Ultrastructural study of developing extracellular matrix in vitelline blood vessels of the early chick embryo. *Am. J. Anat.* 151: 345–375.

Muthukkaruppan, V. R. and Auerbach, R. (1979). Angiogenesis in the mouse cornea. *Science* 205: 1416–1418.

Nameroff, M. and Munar, E. (1976). Inhibition of cellular differentiation by phospholipase C. II. Separation of fusion and recognition among myogenic cells. *Dev. Biol.* 49: 288–293.

O'Neill, M. C. and Stockdale, F. E. (1972). Differentiation without cell division in cultured skeletal muscle. *Dev. Biol.* 29: 410–418.

Papahadjopoulos, D. (1978). Calcium induced phase changes and fusion in natural and model membranes. In G. Poste and G. C. Nicolson (eds.), *Membrane Fusion*. Elsevier North-Holland, Amsterdam, pp. 766–790.

Papaioannou, V. (1979). Interactions between mouse embryos and teratocarcinomas. In N. LeDouarin (ed.), *Cell Lineage, Stem Cells, and Cell Determination*, Elsevier North-Holland, New York, pp. 141–155.

Patten, B. M. (1951). *Early Embryology of the Chick*, Fourth Edition. McGraw-Hill, New York.

Polinger, I. S. (1973). Growth and DNA synthesis in embryonic chick heart cells, *in vivo* and *in vitro*. *Exp. Cell Res.* 76: 253–262.

Przybylski, R. J. and Chlebowski, J. S. (1972). DNA synthesis, mitosis, and fusion of myocardial cells. *J. Morphol.* 137: 417–432.

Quesenberry, P. and Levitt, L. (1979). Hematopoietic stem cells. *N. Engl. J. Med.* 301: 755–760.

Rugh, R. (1951). *The Frog: Its Reproduction and Development*. Blakiston, Philadelphia.

Russell, E. S. (1979). Hereditary anemias of the mouse: A review for geneticists. *Adv. Genet.* 20: 357–459.

Shainberg, A., Yagil, G. and Yaffe, D. (1969). Control of myogenesis *in vitro* by Ca^{++} concentration in nutritional medium. *Exp. Cell Res.* 58: 163–167.

Siminovitch, L., McCulloch, E. A. and Till, J. E. (1963). The distribution of colony-forming cells among spleen colonies. *J. Cell. Comp. Physiol.* 62: 327–336.

Smith, G. H. and Vonderhaar, B. K. (1981). Functional differentiation in mouse mammary gland epithelium is attained through DNA synthesis, inconsequent of mitosis. *Dev. Biol.* 88: 167–179.

Stewart, T. A. and Mintz, B. (1981). Successive generations of mice produced from an established culture line of euploid teratocarcinoma cells. *Proc. Natl. Acad. Sci. USA* 78: 6314–6318.

Sutherland, W. M. and Konigsberg, I. R. (1983). CPK accumulation in fusion-blocked quail myocytes. *Dev. Biol.* 99: 287–297.

Till, J. E. (1981). Cellular diversity in the blood-forming system. *Am. Sci.* 69: 522–527.

Till, J. E. and McCulloch, E. A. (1961). A direct measurement of the radiation sensitivity of normal mouse bone marrow cells. *Rad. Res.* 14: 213–222.

Weiss, J. B., Brown, R. A., Kumar, S. and Phillips, P. (1979). An angiogenic factor isolated from tumors: A potent low molecular weight compound. *Br. J. Cancer* 40: 493–496.

Wolf, N. S. and Trentin, J. J. (1968). Hemopoietic colony studies. V. Effect of hemopoietic organ stroma on differentiation of pluripotent stem cells. *J. Exp. Med.* 127: 205–214.

Yablonka, Z. and Yaffe, D. (1977). Synthesis of myosin light chains and accumulation of mRNA coding for light chain-like polypeptides in differentiating muscle cultures. *Differentiation* 8: 133–143.

Yaffe, D. and Feldman, M. (1965). The formation of hybrid multinucleated muscle fibres from myoblasts of different genetic origin. *Dev. Biol.* 11: 300–317.

Mechanisms of cellular differentiation

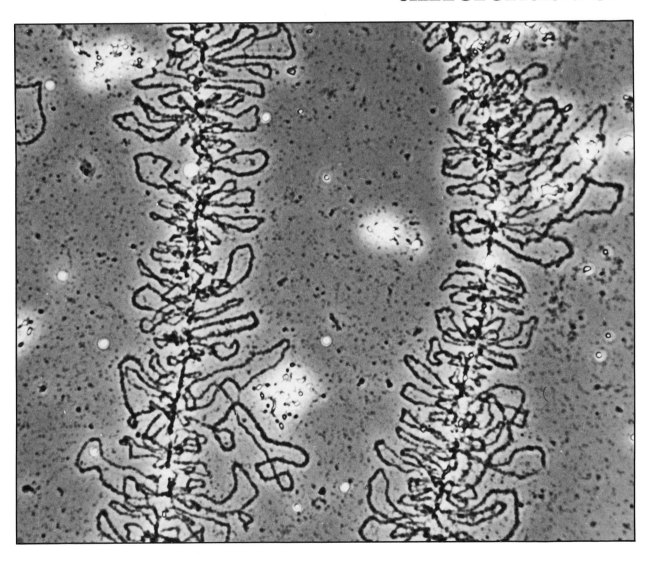

Progressive determination

Tom is now a great man of science . . . and knows everything about everything except why a hen's egg don't turn into a crocodile.
—CHARLES KINGSLEY (1863)

Spemann once remarked that his whole life-work consisted of placing the amphibian embryo into increasingly embarrassing situations.
—CURT STERN (1955)

Introduction

Each metazoan organism is a complex assortment of specialized cell types. For example, the red and white blood cells differ not only from each other but also from heart cells, which propel them through the body. They also differ from the outstretched neurons, which conduct neural impulses from the brain to the heart, and from the glandular cells, which secrete hormones into the blood. Table 1 presents a very incomplete list of specialized cell types, their characteristic products, and their functions.

The development of specialized cell types from the single fertilized egg is called DIFFERENTIATION. This overt change in cellular biochemistry and function is preceded by a process called DETERMINATION, wherein the fate of the cell becomes committed. We know of two major ways by which this determination takes place. The first is EMBRYONIC INDUCTION, which involves cells or tissues interacting with one another to determine the fates of one or both of the participants. The second mechanism involves the CYTOPLASMIC SEGREGATION of determinative molecules during embryonic cleavage. In this situation, the cleavage planes separate qualitatively different regions of the zygote cytoplasm. As we shall see, both mechanisms are used in the development of any organism. This chapter will focus on those experiments demonstrating embryonic induction; and the next chapter will cover determination by cytoplasmic segregation.

Because each of these cell types arose from the same original cell—the fertilized egg—one can ask, *"What are the mechanisms by which the zygote can give rise to the diverse cell types of the body?"* The invention of the microscope allowed this crucial question to be asked in a meaningful manner, and it elicited two major opposing hypotheses: preformation and epigenesis.

Preformation and epigenesis

Any explanation of the differentiation of the various bodily cells from the fertilized egg has to explain (1) the constant morphology of each species (i.e., that chickens beget only chickens, not crocodiles), and (2) the diversity among the bodily parts of each organism. Indeed, one of the major characteristics of development is that each species reproduces its characteristic developmental pattern. Development involves the expression of the inherited properties of the species.

In the seventeenth century, a union of development and inheritance was achieved in the hypothesis of PREFORMATIONISM. According to this view, all the organs of the adult were prefigured in miniature within the sperm or (more usually) the ovum. Thus, organisms were not seen to be "developed"; rather they were "unrolled." This hypothesis had

TABLE 1
Some differentiated cell types and their major products

Cell type	Differentiated cell product	Specialized function
Keratinocyte (skin cell)	Keratin	Protection against abrasion, dessication
Erythrocyte (red blood cell)	Hemoglobin	Transport of oxygen
Lens cell	Crystallins	Transmission of light
B lymphocyte	Immunoglobulins	Antibody synthesis
T lymphocyte	Cell surface antigens (Thy 1, Lyt series)	Destruction of foreign cells; regulation of immune response
Melanocyte	Melanin	Pigment production
Pancreatic islet cells	Insulin	Regulation of carbohydrate metabolism
Leydig cell (♂)	Testosterone	Male sexual characteristics
Chondrocyte (cartilage cell)	Chondroitin sulfate; type II collagen	Tendons and ligaments
Osteoblast (bone-forming cell)	Bone matrix	Skeletal support
Myocyte (muscle cell)	Muscle actin and myosin	Contraction
Hepatocyte (liver cell)	Serum albumin; numerous enzymes	Production of serum proteins and numerous enzymatic functions
Neurons	Neurotransmitters (acetylcholine, epinephrine, etc.)	Transmission of electrical impulses
Tubule cell (♀) of hen oviduct	Ovalbumin	Egg white proteins for nutrition and protection of embryo
Follicle cell (♀) of insect oviduct	Chorion proteins	Eggshell proteins for protection of embryo

the backing of both science and philosophy (see Gould, 1977). First, because all the organs were prefigured, embryonic development merely required the growth of existing structures, not the formation of new ones. No extra mysterious force was needed for embryonic development. Second, just as the adult organism was prefigured in the germ cells, another generation already existed in a prefigured state within the germ cells of the first prefigured generation. This corollary, called EMBOÎTMENT (encapsulation), assured that the species would always remain constant. Although certain microscopists claimed to see fully formed human miniatures within the sperm or egg, the major proponents of this hypothesis—Albrecht von Haller and Charles Bonnet—knew that organ systems develop at different rates and that embryonic structures need not be in the same place as those in the newborn.

The preformationists had no cell theory to provide a lower limit to the size of their preformed organisms, nor did they view mankind's tenure on earth as potentially immortal. Rather, said Bonnet (1764), "Nature works as small as it wishes," and the human species existed in that finite time spanning Creation and Resurrection. This was in accord with the best science of its time, conforming to the French mathematician–philosopher René Descartes' principle of the infinite divisibility of a mechanical nature initiated, but not interfered with, by God.

Preformation was a conservative theory, emphasizing the lack of change between generations. Its principal failure was its inability to account for the variations known by the limited genetic evidence of the time. It was known, for instance, that matings between white and black parents produced children of intermediate skin color, an impossibility if inheritance and development were solely through either the sperm or the egg. In more controlled experiments, the German botanist Joseph Kölreuter (1766) had produced hybrid tobacco plants having the characteristics of both species. Moreover, by mating the hybrid to either the male or female parent, Kölreuter was able to "revert" the hybrid back to one or the other parental types after several generations. Thus, inheritance seemed to arise from a mixture of parental components. In addition, preformationism could not explain the generation of "monstrosities" and such deviations as hexadactylism (six fingers per hand) when both parents were normal.

There developed, then, an alternative hypothesis: EPIGENESIS. This view of development, having philosophical roots as far back as Aristotle, was revived by a German embryologist working in St. Petersburg, Kaspar Friedrich Wolff. By carefully observing the development of chicken embryos, Wolff demonstrated that the embryonic parts developed from tissues having no counterpart in the adult organism. The heart and blood vessels (which, according to preformationism, had to be present from the beginning in order to ensure embryonic *growth*) could be seen to develop anew in each embryo. Similarly, the intestinal tube was seen arising by the folding of an originally flat tissue. This latter observation was explicitly detailed by Wolff, who proclaimed (1767), "When the formation of the intestine in this manner has been duly weighed, almost no doubt can remain, I believe, of the truth of epigenesis." However, to create an organism anew each generation, Wolff had to postulate an unknown force, the *vis essentialis* ("essential force"), which, acting like gravity or magnetism, would organize embryonic development.

Thus, preformationism best explained the continuity between generations, whereas epigenesis best explained the variation and the direct observations of organ formation. A reconciliation of sorts was attempted by the German philosopher Immanuel Kant (1724–1804) and his colleague, biologist Johann Friedrich Blumenbach (1752–1840). Attempting to construct a scientific theory of racial descent, Blumenbach postulated a mechanical goal-directed force called the *Bildungstrieb* ("development force"). Such a force, he said, was not theoretical but could be shown to exist by experimentation. A *Hydra*, when cut, will regenerate its amputated parts from the rearrangement of existing elements. Thus, some purposive organizing force could be observed in operation, and this force was a property of the organism itself. This *Bildungstrieb* was thought to be inherited through the germ cells. Thus, development could proceed epigenetically through a predetermined force inherent in the matter of the embryo (Cassirer, 1950; Lenoir, 1980). Moreover, such a force was susceptible to change, and the left-handed variant of snail coiling was used as an example of such modifications in the organizing force. In this hypothesis, where epigenetic development is directed by preformed instructions, we are not far from the modern view that "the complete description of the organism is already written in the egg" (Brenner, 1979). However, until the rediscovery of Mendel's work at the beginning of the twentieth century, there was no consistent genetic theory in which to place such ideas of inherited variation, and each scientist was free to speculate on the mechanisms by which developmental patterns were inherited.

Foremost among these speculators was Charles Darwin. His "provisional hypothesis," called PANGENESIS, claimed that each somatic cell contained particles that migrated back into the sex cells to provide for the transmission of that cell's characteristics. These particles ("pangenes") were susceptible to change by environmental factors, thereby providing the basis for inherited variation. In this theory (and in numerous variations of it), the germ cells were clearly influenced by the somatic cells. For decades, biologists argued about the validity of such ideas, but the first testable alternative was proposed by one of the great synthesizers of biology, August Weismann (1834–1914).

August Weismann: The germ-plasm theory

In 1883, Weismann began proposing a theory that integrated such diverse biological phenomena as heredity, development, regeneration, sexual reproduction, and evolution by natural selection. Weismann was a brilliant chemist and physician who had turned his attention to the problem of embryonic development and metamorphosis. In 1863, when a severe eye disorder curtailed his microscopy, he used his time to ponder the ways by which the germ cells produced their differentiated progeny. This quest freed Weismann from the strict Darwinian type of explanation (i.e., the use of embryology to support theories of evolutionary descent) and set embryology on a new, physiological course that emphasized experimental manipulation over comparative observation.

The cytology of the 1870s had given Weismann important new information concerning sexual generation. Hermann Fol and Richard Hertwig had independently observed the union of egg and sperm and their pronuclei. Researchers such as van Beneden and Strasburger had demonstrated that each somatic nucleus contained a defined number of chromosomes depending on the species and that this number was halved during germ cell maturation and restored during fertilization. Using these scant data, Weismann proposed a mechanical model of cellular differentiation, the GERM-PLASM THEORY.

First, the sperm and egg were posited to provide equal chromosomal contributions, both quantitatively and qualitatively, to the new organism. Second, chromosomes carried the inherited potentials of this new organism and were the basis for the continuity between generations. [Embryologists were thinking in these terms some 15 years before the rediscovery of Mendel's work. Weismann (1892, 1893) also speculated that these nuclear determinants of inheritance functioned by elaborating substances that became active in the cytoplasm!] However, not all the determinants on the chromosomes went into every cell of the embryo; for instead of dividing equally, the chromosomes divided in such a way that different nuclear determinants entered different cells. Weismann, a former military physician, likened these determinants to specialized army brigades. While the fertilized egg would carry the full complement of all determinants, certain cells would retain the "blood-forming" brigades while other cells would retain the "muscle-forming" determinants. Only in the nuclei of those cells destined to become gametes (the germ cells) were all types of determinants retained. The nuclei of all other cells had only a fraction of the original determinant types.

Thus, Weismann's hypothesis proposed the continuity of the germ plasm and the diversity of the somatic lines. Differentiation was due to the "segregation of the nuclear determinants" into the various cell types and was accomplished by "the architecture of the germ plasm," that is, the germ cell nucleus. The chromosomes, while appearing equal in all cells, would be unequal in their qualities. The germ cell line was totally *independent* of the somatic cells. Hence, there could be no inheritance of characteristics acquired by the somatic cells. Weismann obtained support for this model by cutting off the tails of newborn mice for nineteen generations. The mice of each succeeding generation had normal length tails, thereby indicating that the germ line was insulated from the insults to the somatic tissue.[1]

---

[1]According to pangenesis, one should have obtained mice with shorter tails. The most convincing evidence for pangenesis was the claim by physiologists Brown-Sequard and Westphal that experimentally induced epilepsy in guinea pigs could be transmitted to the next generation. Other anecdotal evidence for the inheritance of acquired traits was rife

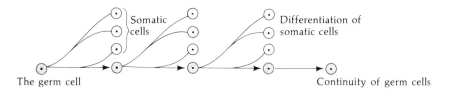

The germ cell — Somatic cells — Differentiation of somatic cells — Continuity of germ cells

FIGURE 1
Diagram illustrating Weismann's theory of inheritance. The germ cell gives rise to the differentiating somatic cells of the body (white), as well as to new germ cells (shaded). (From Wilson, 1896.)

Weismann's germ plasm theory is depicted in Figure 1. It emphasizes the continuity and immortality of the germ line as contrasted with the temporary nature of the adult organism, showing, as physiologist Michael Foster noted, that "the animal body is in reality a vehicle for ova." E. B. Wilson, who claimed that his remarkable textbook *The Cell in Development and Inheritance* (1896) grew out of Weismann's hypothesis, also saw the implications of Weismann's scheme:

> The death of the individual involves no breach of continuity in the series of cell divisions by which the life of the race flows onwards. The individual dies, it is true, but the germ-cells live on, carrying with them, as it were, the traditions of the race from which they have sprung, and handing them on to their descendants.

Wilhelm Roux: Mosaic development

Weismann had intuited that the chromosomes were the bearers of the inherited information for development. More importantly, though, he had proposed a hypothesis that could be tested immediately. Weismann had claimed that when the first cleavage division separated the future right half of the embryo from the future left half, there would be a separation of "right" determinants from "left" determinants in the resulting blastomeres. This assertion was tested by Wilhem Roux, a young German embryologist and a former student of Ernst Haeckel. Roux, too, had been an ardent Darwinian (his first major work in embryology proposed a competition for survival between embryonic cells), but, like Weismann, he felt that development needed to be studied analytically. In 1883, the same year that Weismann first proposed the germ plasm theory, Roux proposed that nuclear divisions need not be equal but may apportion hereditary determinants qualitatively. Five years later, Roux published the results of a series of experiments in which he took

---

(for a critical review, see Thomson, 1907). Weismann's mouse mutilation experiments were confirmed by Bos (15 generations), Cope (11 generations), and Mantezza and Rosenthal (11 generations). Recently, Thomas Jukes, commenting on these results, quoted Hamlet's intuition that "There's a divinity that shapes our ends, rough-hew them how we will." In any event, the case for pangenesis was severely weakened. Weismann said that Darwin's provisional hypothesis "has been one of those indirect roads which science has been compelled to travel in order to arrive at the truth."

2- and 4-cell frog embryos and destroyed some of the cells of each embryo with a hot needle. Weismann's hypothesis would predict the formation of right or left half-embryos. Roux obtained half-morulae, just as Weismann had predicted (Figure 2). These developed into half-tadpoles having a complete right or left side, containing one medullary fold, one ear pit, and so on. He therefore concluded that the frog embryo was a mosaic of self-differentiating parts and that it was likely that each cell received a specific set of determinants and differentiated accordingly. With this series of experiments, Roux inaugurated his program of developmental mechanics (*Entwicklungsmechanik*), the physiological approach to embryology. No longer, insisted Roux, would embryology merely be the servant of evolutionary studies. Rather, embryology would assume its role as an independent experimental science.

Hans Driesch: Regulative development

Nobody appreciated this experimental approach to embryology more than another former student of Haeckel's, Hans Driesch. Driesch's goal was to reduce embryology to the laws of physics and mathematics, and his investigations began similarly to those of Roux. Roux's experiments were, technically, *defect* studies that answered the question of how the remaining blastomeres of an embryo would develop when part of them were destroyed. Driesch (1892) sought to extend these researches by performing *isolation* experiments. Here, sea urchin blastomeres were separated from each other by vigorous shaking (or, later, by placing them in calcium-free seawater). To Driesch's surprise, each of the blastomeres from a 2-cell embryo developed into a complete larva. Similarly,

FIGURE 2
Roux's attempt to show "mosaic" development. Destroying one cell of a 2-cell frog embryo results in the development of only one-half of the embryo.

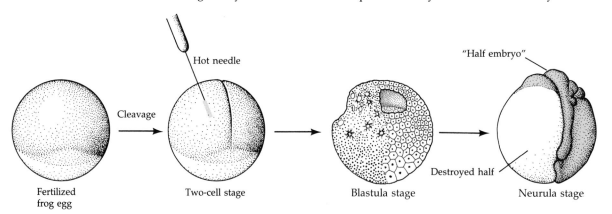

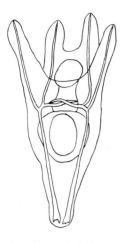

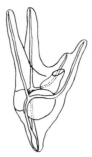
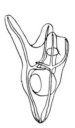
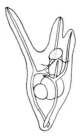
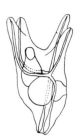

(B)   Plutei developed from single cells of 4-cell embryo

FIGURE 3
Driesch's demonstration of "regulative" development. (A) A normal pluteus larva. (B) Smaller, but normal, plutei that each developed from one blastomere of a dissected 4-cell embryo. (All larvae are drawn to the same scale.) Note that the four larvae derived in this way (B) are not identical, despite their ability to generate all the necessary cell types. (After Hörstadius and Wolsky, 1936.)

(A)   Normal pluteus larva

when Driesch separated the blastomeres from 4- and 8-cell embryos, some of the cells produced entire pluteus larvae (Figure 3). Here was a result drastically different from that predicted by Weismann's or Roux's hypotheses. Rather than self-differentiating into its future embryonic part, each blastomere could regulate its development so as to produce a complete organism. This phenomenon, called REGULATIVE DEVELOPMENT, also was demonstrated by another experiment by Driesch.

In sea urchin eggs, the first two cleavage planes are longitudinal, passing through the animal and vegetal poles, whereas the third division is equatorial, dividing the embryo into four upper and four lower cells (see Figure 3 in Chapter 3). Driesch (1893) changed the direction of the third cleavage by gently compressing the early embryos between two glass plates, causing the third division to be longitudinal like the preceding two cleavages. After he released the pressure, the fourth division was equatorial. This procedure reshuffled the nuclei such that a nucleus that normally would be found in the region destined to form endoderm was now in the presumptive ectoderm region. Nuclei that would have normally produced dorsal structures were now found in the ventral cells (Figure 4). If the segregation of nuclear determinants had occurred, the resulting embryo should be strangely disordered. However, Driesch obtained perfectly normal larvae from these embryos.

The consequences of these experiments were momentous both for embryology and for Driesch personally. First, Driesch had demonstrated that the "prospective potency" of an isolated blastomere (those cell types it was possible for it to form) was greater than its "prospective fate" (the cell types it would normally give rise to an unaltered course of its development). According to Weismann and Roux, the prospective

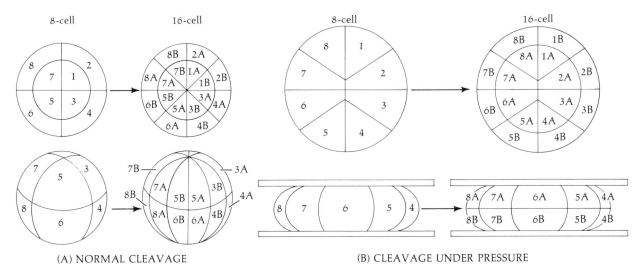

(A) NORMAL CLEAVAGE   (B) CLEAVAGE UNDER PRESSURE

FIGURE 4
Diagram of Driesch's pressure-plate experiment to alter the distribution of nuclei. (A) Normal cleavage from 8- to 16-cell sea urchin embryos seen from the animal pole (upper sequence) and from the side (lower sequence). (B) Abnormal cleavage planes formed under pressure, as seen from the animal pole and from the side. (After Huxley and de Beer, 1934.)

potency and fate of a blastomere should be identical. Second, Driesch concluded that the sea urchin embryo was a "harmonious equipotential system." It was equipotential in that each blastomere was equivalent and could give rise to an entire embryo. It was harmonious because all these potentially independent parts functioned together to form a single organism. Third, the fate of a nucleus depended solely on its location in the embryo. Driesch (1894) hypothesized a series of events wherein development proceeded by the interactions of the nucleus and cytoplasm:

> Insofar as it contains a nucleus, every cell, during ontogenesis, carries the totality of all primordia; insofar as it contains a specific cytoplasmic cell body, it is specifically enabled by this to respond to specific effects only. . . . When nuclear material is activated, then, under its guidance, the cytoplasm of its cell that had first influenced the nucleus is in turn changed, and thus the basis is established for a new elementary process, which itself is not only the result but also a cause.

This strikingly modern concept of nuclear–cytoplasmic interaction and nuclear equivalence was too much for Driesch. He could no longer envision the embryo as a physical machine because it could be subdivided into parts that each were capable of reforming the entire organism. In other words, Driesch had come to believe that development could not be explained by physical forces. He was driven to invoke a vital force, *entelechy* ("internal goal-directed force"), to explain how development proceeds. Essentially, the embryo was imbued with an internal psyche and wisdom to accomplish its goals despite the obstacles embryologists placed in its path. Feeling outsmarted by the embryo, Driesch renounced the study of developmental physiology and became a professor of philosophy, proclaiming vitalism until his death in 1941.

TABLE 2
Experimental procedures and results of Roux and Driesch

Investigator	Organism	Type of Experiment	Conclusion	Interpretation concerning potency and fate
Roux (1888)	Frogs (*Rana fusca* and *Rana esculenta*)	Defect	Mosaic (self-differentiating) development	Prospective potency equals prospective fate
Driesch (1892)	Sea urchin (*Echinus microtuberculatus*)	Isolation	Regulative development	Prospective potency is greater than prospective fate
Driesch (1893)	Sea urchin (*Echinus* and *Paracentrotus*)	Recombination	Regulative development	Prospective potency is greater than prospective fate

The differences between Roux's experiments and those of Driesch are summarized in Table 2. The difference between isolation and defect experiments and the importance of the interactions provided by the destroyed blastomeres were highlighted in 1910 when J. F. McClendon showed that isolated frog blastomeres behaved just like the separated sea urchin cells. Therefore, the mosaic-like development of frog blastomeres in Roux's study was an artifact of the defect experiment. Something in or on the dead blastomere still informed the live cells that it existed. We have also seen that early mammalian blastomeres have a regulative type of development. As we saw in Chapter 3, each isolated blastomere of a mouse inner cell mass is capable of generating an entire fertile mouse. The ability of two or more early mouse or rat embryos to fuse into one normal embryo (Figure 25 in Chapter 3) and the phenomenon of identical twins also attest to the regulative ability of mammalian blastomeres. Therefore, even though Weismann and Roux pioneered the study of developmental physiology, their proposition that differentiation is caused by the segregation of nuclear determinants was shortly shown to be incorrect.

Sven Hörstadius: Potency and oocyte gradients

But Driesch was not 100 percent correct either. As we shall see in the next chapter, there are numerous animals that do develop largely as a mosaic of self-differentiating parts. More importantly, though, even the sea urchin embryo is not a collection of completely equipotential cells. In a series of experiments from 1928 to 1935, Swedish biologist Sven Hörstadius separated various layers of early sea urchin embryos with

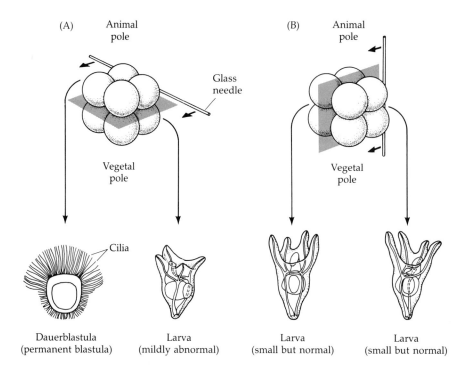

FIGURE 5

Early asymmetry in the sea urchin embryo. (A) When the four animal blastomeres are separated from the four vegetal blastomeres and each half is allowed to develop, the animal cells form a ciliated *dauerblastula* and the vegetal
cells form a larva with an expanded gut. (B) When the 8-cell embryo is split
so that each half contains animal and vegetal cells, small, normal-appearing
larvae develop.

fine glass needles and observed their subsequent development (Hörstadius, 1939). When the 8-cell embryo was divided longitudinally
through the animal and vegetal poles, both halves produced pluteus
larvae, just as Driesch had foretold. However, when embryos at the
same stage were split equatorially (separating animal and vegetal poles),
neither part developed into a complete larva (Figure 5). Rather, the
animal half became a hollow ball of ciliated epidermal cells, and the
vegetal half developed into a slightly abnormal embryo with an expanded gut. Hörstadius was able to duplicate these results by cutting
the unfertilized sea urchin eggs in half and fertilizing the halves separately. In sea urchins, egg fragments (MEROGONES) can divide and develop even if they have only a haploid nucleus. If a sperm enters the
half that lacks the haploid egg nucleus, the merogone will still develop
(Figure 6). When the egg had been split longitudinally, normal embryos
could be formed from either half of the egg. However, when the oocyte
had been equatorially cut, fertilization produced either the ciliated animal ball or the expanded vegetal gut. Therefore, even in sea urchin

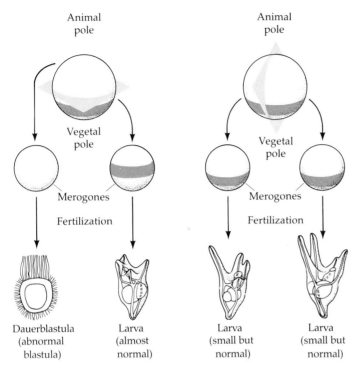

FIGURE 6
Asymmetry in the sea urchin egg. When Hörstadius divided the sea urchin egg into animal and vegetal halves (merogones) and allowed the halves to be fertilized by sperm, the animal half developed into a ciliated dauerblastula and the vegetal half produced a pluteus with an expanded gut. When the egg was split such that both merogones contained animal and vegetal cytoplasm, small, normal-appearing plutei developed.

embryos, there appears to be some degree of mosaicism, at least along the animal–vegetal axis.

These observations led Hörstadius to perform some of the most exciting experiments in the history of embryology. First, Hörstadius (1935) traced the normal development of each of the six tiers of cells of the 32-cell sea urchin embryo. As shown in Figure 7A, the animal cells and the first vegetal layer normally produce ectoderm; the second vegetal layer gives rise to endoderm; and the micromeres generate the mesodermal structures.

Next, Hörstadius removed the fertilization membrane from the 32-cell embryos, separated the tiers with fine glass needles, and recombined them in various ways. The isolated animal hemisphere, alone, became a DAUERBLASTULA of ciliated ectoderm cells (Figure 7B). Such an embryo was called "animalized." Recombining an isolated animal hemisphere with the veg$_1$ tier (Figure 7C), Hörstadius noted that the resulting larva was less animalized. Ciliary development was suppressed and a portion of the gut was formed. However, when the

FIGURE 7
Hörstadius' demonstation of a "vegetalizing" gradient. (A) Fate of each cell layer of the 64-cell sea urchin embryos through blastula to pluteus stage. The different cell layers are marked as in Figure 1 in Chapter 4. (B) Fate of the isolated animal half. (C) Recombination of animal half plus the veg$_1$ tier of cells. (D) Recombination of animal half plus the veg$_2$ tier of cells. (E) Recombination of animal half plus the micromeres. (Modified from Hörstadius, 1939.)

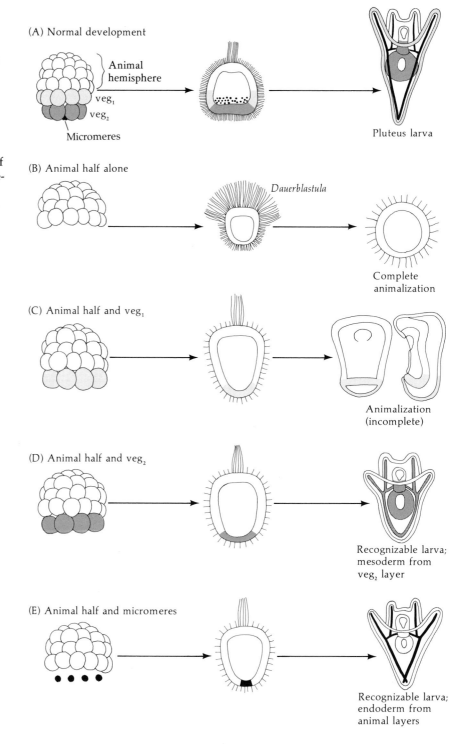

(A) Normal development

Animal hemisphere

veg$_1$

veg$_2$

Micromeres

Pluteus larva

(B) Animal half alone

Dauerblastula

Complete animalization

(C) Animal half and veg$_1$

Animalization (incomplete)

(D) Animal half and veg$_2$

Recognizable larva; mesoderm from veg$_2$ layer

(E) Animal half and micromeres

Recognizable larva; endoderm from animal layers

animal hemisphere was combined with the veg$_2$ tier (Figure 7D), a normal-looking pluteus larva developed. In this combination, the veg$_2$ cells, which normally form only the endoderm, now formed both endoderm and mesoderm! Similarly, when the animal half was recombined with just the micromeres (Figure 7E), a small, normal-looking pluteus was formed, but in this case the endoderm was completely derived from the animal cells. Here, the gut was formed by cells that would normally have created the ciliated ectoderm. These experiments showed that the animal cells had the genetic potential to be gut cells even at the 32-cell stage. More importantly, it also suggested that the ability to suppress "animalization" was localized as a GRADIENT. The micromeres were stronger "vegetalizers" than the veg_2 layer; but the veg_2 layer, in turn, was stronger than the veg_1 tier.

Hörstadius' next series of experiments suggested the existence of an animalizing factor that also formed a gradient. Micromeres were combined with each of the tiers of the 32-cell embryo. An isolated an_1 tier (Figure 8A) would produce an ectodermal dauerblastula. As each successive micromere was added, a more complete embryo formed. The combination of just the an_1 tier with four micromeres resulted in the production of a normal pluteus. The an_2 tier formed a normal pluteus with only two micromeres (Figure 8B); and four micromeres added the an_2 tier caused an abnormally expanded endoderm. The veg_1 layer would also produce a dauerblastula when alone. However, even one micromere caused a severe expansion of gut tissue (Figure 8C) and the isolated veg_2 tier showed this tendency to "vegetalize" without the aid

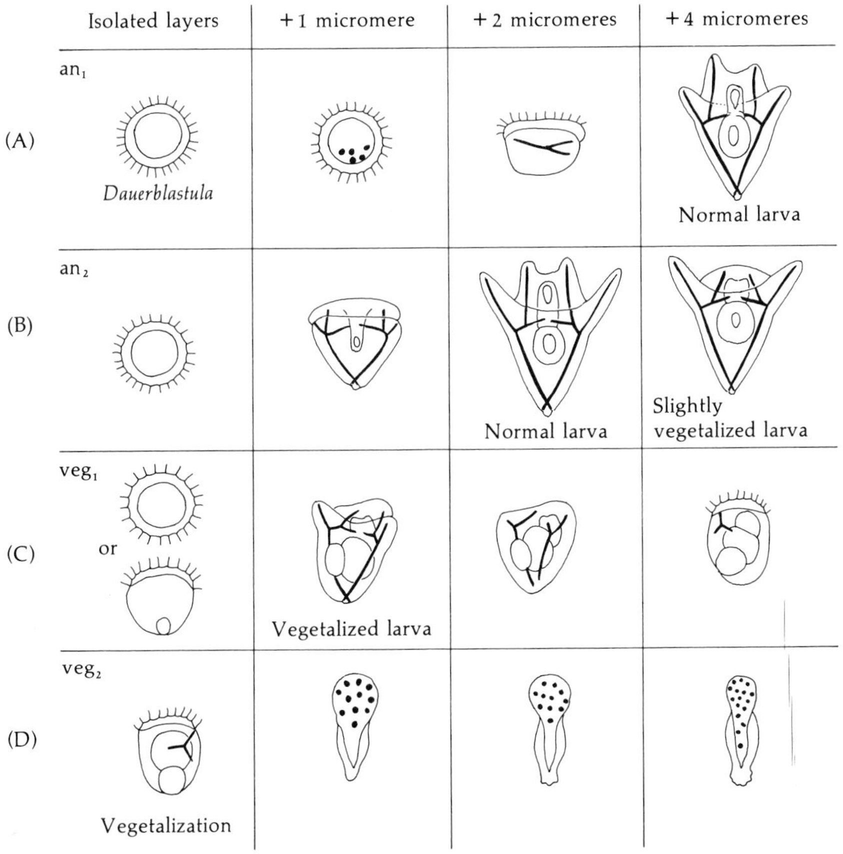

FIGURE 8
Hörstadius' demonstration of an "animalizing" gradient. Each tier of a 32-cell sea urchin embryo is isolated and then recombined with 0, 1, 2, or 4 micromeres. It takes 4 micromeres to form a normal pluteus larva from cells derived from the an_1 tier (A), but only 2 are needed to produce a pluteus from the an_2 tier (B). The veg_1 layer produces a vegetalized larva with only one micromere (C), and the veg_2 layer lacks enough animalizing properties to form a normal pluteus (D). (Modified from Hörstadius, 1935.)

of any micromeres (Figure 8D). Thus, there appeared to be a gradient of "animalization" proceeding in strength from an_1 to the micromeres.

The most logical explanation of these results is that of two opposed gradients: a vegetalizing gradient with its maximum at the vegetal pole and an animalizing gradient with its maximal activity at the animal region. Exactly such a system of dual gradients had been proposed by Hörstadius' adviser, J. Runnström, in 1929. Because the relative ratios of substances are important for development, the recombination of the two poles reestablishes all the intermediate positions. Similarly, the intermediate cells still have a maximum and minimum for both gradients and can thus give rise to an entire pluteus as well (Figure 9). This dual-gradient model has been extremely useful in explaining other recombination experiments. For example, when micromeres are transplanted from the vegetal pole to a region near the center of a 32-cell embryo, the implanted micromeres invaginate into the host blastocoel, thereby causing the formation of a well-developed secondary archenteron (Figure 10). When the micromeres are added to the animal pole, a very small secondary gut forms. So far neither the vegetal factor nor

FIGURE 9
Pluteus larva formation from the extreme and intermediate cells of a 32-cell sea urchin embryo. (A) Micromeres added to mesomere (an_1 + an_2) layer. (B) The an_2 plus the veg_1 and veg_2 layers, without the two extremes. Both combinations form pluteus larvae. Below each figure is the hypothetical gradient. (After Czihak, 1971.)

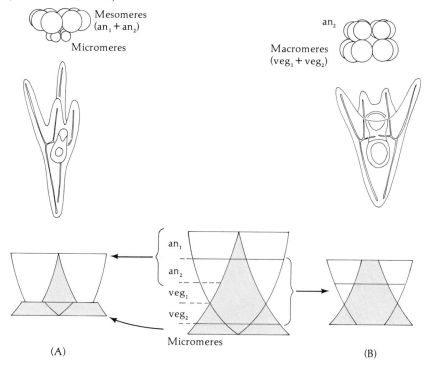

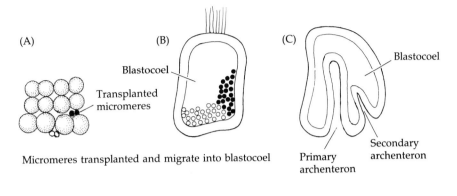

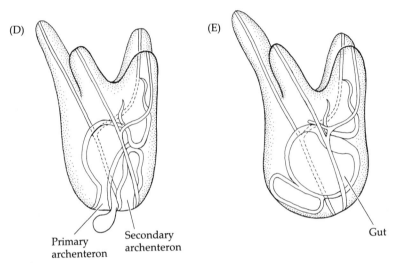

(A) Micromeres transplanted and migrate into blastocoel

Primary archenteron

Secondary archenteron

Primary archenteron

Secondary archenteron

Gut

FIGURE 10
(A) Micromeres (black) are transplanted from the vegetal pole to a region between the animal and vegetal halves of a 32-cell embryo. (B) The micromeres invaginate into the blastocoel. (C) A secondary archenteron forms and eventually (D) a pluteus larva that has two archenterons is formed. (E) These separate archenterons later fuse into a large gut. (After Hörstadius, 1935.)

the animal factor has been isolated. Protein inhibitors (heavy metals, NaSCN, Evans Blue dye) appear to diminish the vegetalizing gradient and thus animalize the embryo. Respiratory inhibitors (CO, KCN, NaN$_3$, Li) appear to vegetalize the embryo. Thus, von Ubisch (1929) was able to get a normal pluteus from an isolated animal hemisphere grown in dilute lithium chloride. The lithium appeared to weaken the animal portion of the gradient within the animal cells, leading to balanced development.

We have, then, a model for regulative development based on relative concentration gradients within the oocyte. When animal cells are combined with micromeres only, a normal pluteus is formed, with its gut cells coming from animal cells that would normally have formed the ectoderm. Thus, the potency is still greater than the fate. However, the animal cells alone would have given rise only to an animalized blastula. The ability for one cell to form the entire larva is gone. Here we see the restriction of potency. Any cell with a particular ratio of animal and vegetal substances will produce a certain type of cell. Thus, animal pole cells, with a high animal factor to vegetal factor ratio, would normally produce ectodermal tissue. However, if recombined with a strong source of the vegetal factor, certain animal cells will produce gut tissue. When the embryo is separated so that each half has a complete animal and vegetal gradient (i.e., along the animal–vegetal axis), a complete larva forms. The blastomeres will stay regulative so long as they are not completely formed from animal or vegetal cytoplasm. It is not surprising, then, that after the 32-cell stage, most of the individual blastomeres can no longer give rise to complete larvae (Morgan, 1895), and even the micromeres of the 16-cell stage are incapable of doing so (Hagström and Lonning, 1965). Even in an embryo that undergoes regulative development, there comes a time when the potencies of its cells are restricted.

Hans Spemann:
Progressive determination of embryonic cells

In the previous section, we saw evidence for regulative development. We noted that the two major aspects of regulation—(1) that an isolated blastomere has a potency greater than its normal embryonic fate, and (2) that rearranged blastomeres develop according to their new locations—hold true during the early stages of sea urchin cleavage. Eventually, however, the blastomeres become committed to certain fates. Hörstadius was able to relate this restriction in potency to the plane of cleavage, as blastomeres could regulate only so long as they had sufficient material from both the animal and vegetal parts of the egg. In 1918, Hans Spemann of the University of Freiburg discovered that a similar situation existed in the salamander egg. The experiments by which he and his colleagues analyzed this phenomenon over the next 20 years form the basis for much of our knowledge of embryonic physiology and won for Spemann a Nobel Prize in 1935.

Spemann, like Roux and Driesch, sought to test Weismann's hypothesis, and by an ingenious method he demonstrated that early newt blastomeres had identical nuclei, each capable of producing an entire larva. Shortly after fertilizing a newt egg, Spemann took a baby's hair and lassoed the zygote in the plane where the first cleavage furrow was expected. He would then partially constrict the egg, causing all the nuclear divisions to remain on one side of the constriction. Eventually, often as late as the 16-cell stage, a nucleus would escape across the constriction into the nonnucleated side. Cleavage then began on this side, too, whereupon the lasso was tightened until the two halves were completely separated. Twin larvae developed, one slightly later than the other (Figure 11). Therefore, Spemann concluded that early amphibian nuclei were genetically identical and that each was capable of giving rise to an entire organism. In this respect, amphibian blastomeres were similar to those of the sea urchin.

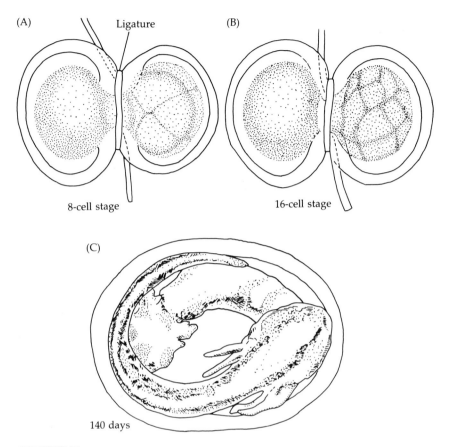

FIGURE 11

Spemann's demonstration of nuclear equivalence in newt cleavage. (A) When the fertilized egg of the newt *Triturus taeniatus* was constricted by a ligature, the nucleus was restricted to one half of the embryo. The cleavage on that side of the embryo reached the 8-cell stage while the other side remained undivided. (B) At the 16-cell stage, a single nucleus entered the as yet undivided half and the ligature was constricted to complete the separation of the two halves. (C) After 140 days, each side had developed into a normal embryo. (From Spemann, 1938.)

However, when Spemann performed the same experiment perpendicularly to the plane of the first cleavage (separating the future dorsal and ventral regions rather than right and left sides), he obtained a different result altogether! The nuclei would continue to divide on both sides of the constriction, but only one side would give rise to a normal larva. The other side would produce an unorganized tissue mass, which Spemann called the *Bauchstuck*—the belly piece. This belly piece contained gut cells of endodermal origin but had no skeletal (mesodermal) or neural (ectodermal) components (Figure 12).

FIGURE 12

Asymmetry in an amphibian egg. When the plane of the first cleavage divides the egg such that each blastomere gets one half of the gray crescent, each experimentally separated cell develops into a normal embryo. However, when one of the two blastomeres receives the entire gray crescent, it alone forms a normal embryo.

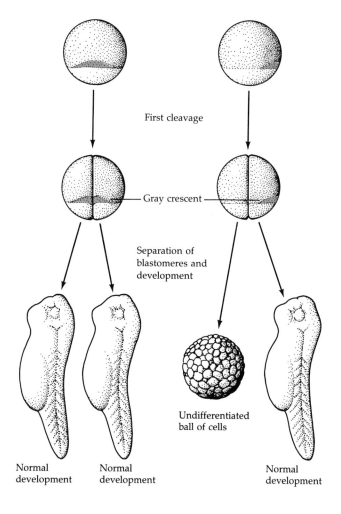

First cleavage

Gray crescent

Separation of blastomeres and development

Undifferentiated ball of cells

Normal development   Normal development   Normal development

Why should these two experiments give different results? Could it be that when the egg was divided perpendicularly to the first cleavage plane, some cytoplasmic substance was not equally distributed to the two halves? Fortunately, the salamander egg was a good place to look for answers. As we have seen in Chapter 2, there are dramatic movements in the cortical cytoplasm following the fertilization of amphibian eggs. These movements expose a gray, crescent-shaped area of cytoplasm in the region directly opposite the point of sperm entry. Moreover, the first cleavage plane normally bisects this region equally into the two blastomeres. If these cells are then separated, two complete larvae develop. However, should this cleavage plan be aberrant (either in the rare natural event or in an experiment in which an investigator constricts a hair-loop lasso perpendicular to the normal cleavage plane), the gray crescent material passes into only one of the two blastomeres. Spemann found that when these two blastomeres are separated, only the blastomere containing the gray crescent developed normally.

It appears, then, that gray crescent cytoplasm is essential for proper embryonic development. But how does it function? What role does it play in normal development? The most important clue came from the fate map of this area of the egg, for it showed that the gray crescent region gave rise to the cells that initiate gastrulation. These cells form the dorsal lip of the blastopore. As was shown in Chapter 4, the cells of the dorsal blastopore lip are somehow "programmed" to invaginate into the blastula, thus initiating gastrulation and the formation of the archenteron. Because all future amphibian development depends on the interaction of cells moved about by gastrulation, Spemann speculated that the importance of gray crescent material lay in its ability to initiate gastrulation and that crucial developmental changes occurred during gastrulation.

In 1918, Spemann demonstrated that enormous changes in cell potency did indeed take place during gastrulation. He found that the cells of the *early* gastrula were uncommitted with respect to their eventual differentiation, but the fates of the *late* gastrula cells were fixed. Spemann exchanged tissues between the early gastrulae of two differently pigmented species of newts (Figure 13). When a region of pro-

FIGURE 13

Determination of ectoderm during newt gastrulation. Presumptive neural ectoderm from one newt embryo is transplanted into a region in another embryo that normally becomes epidermis. (A) When the transfer is done in early gastrula, the presumptive neural tissue develops into epidermis and only one neural plate is seen. (B) When the same experiment is performed on late gastrula tissues, the presumptive neural cells form neural tissue, thereby causing two neural regions to form on the host. (Modified from Saxén and Toivonen, 1962.)

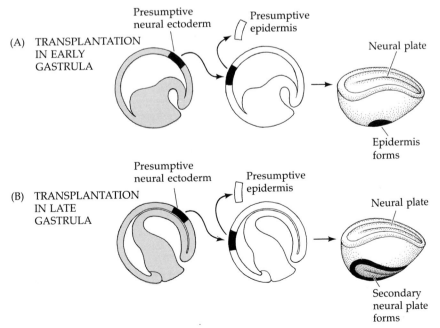

spective epidermal cells was transplanted to an area where the neural plate formed, the cells gave rise to neural tissue. When prospective neural plate cells were transplanted to the region fated to become belly skin, these cells became epidermal (Table 3). Thus, early newt gastrula cells were not yet committed to a specific type of differentiation. Their prospective potencies were still greater than their prospective fates. These cells were said to exhibit DEPENDENT DEVELOPMENT as their ultimate fates depended on their location in the embryo. However, when the same heteroplastic (interspecies) transplantation experiments were performed on *late* gastrulae, Spemann obtained completely different results. Rather than regulating their differentiation in accordance with their new location, the transplanted cells exhibited INDEPENDENT DEVELOPMENT. Their prospective fate was fixed, and the cells developed independently of their new embryonic location. Specifically, prospective neural cells now developed into brain tissue even when placed in the region of prospective epidermis; and prospective epidermis formed skin, even in the region of the prospective neural tube. Thus, within the time separating early and late gastrulae, the potencies of these groups of cells had become restricted to their eventual paths of differentiation. These cells are now said to be DETERMINED. Determination refers to the cell's commitment to eventually differentiate into a specific cell type and no other. Therefore, in the late gastrula, those animal pole cells on the same side as the blastopore lip are now committed (determined) to produce neural tissue in whatever location they might be placed (including a petri dish). They can no longer regulate their differentiation into other cell types. It should be noted that the criteria for determination are completely operational. There are no obvious changes

TABLE 3
Results of tissue transplantation during early and late gastrula stages in the newt

Donor region	Host region	Differentiation of donor tissue	Conclusion
EARLY GASTRULA			
Prospective neurons	Prospective epidermis	Epidermis	Dependent differentiation
Prospective epidermis	Prospective neurons	Neurons	Dependent differentiation
LATE GASTRULA			
Prospective neurons	Prospective epidermis	Neurons	Independent differentiation (Determined)
Prospective epidermis	Prospective neurons	Epidermis	Independent differentiation (Determined)

occurring in the cells and no overt differentiation can yet be seen. The molecular basis of determination remains one of the major unsolved puzzles of development.

Hans Spemann and Hilde Mangold: Primary embryonic induction

The most spectacular transplantation experiments were published by Hans Spemann and Hilde Mangold in 1924. They showed that the dorsal lip of the blastopore is the only self-differentiating region in the early gastrula and that it indeed initiated gastrulation and embryogenesis in the surrounding tissue. In these experiments, Spemann and Mangold used differently pigmented embryos from two species of newt, the darkly pigmented *Triturus taeniatus* and the nonpigmented (clear) *Triturus cristatus*. When Spemann and Mangold prepared heteroplastic transplants, they were able to readily identify host and donor tissues on the basis of color. The dorsal blastopore lips of early *T. cristatus* gastrulae were removed and implanted into the regions of early *T. taeniatus* gastrula fated to become ventral epidermis (Figure 14). Unlike the other early gastrulae tissues, which developed according to their new location, the donor blastopore lip did not become belly skin. Rather, it invaginated just as it would normally have done (showing self-determination) and disappeared beneath the yolky ventral cells. The light-colored donor tissue then continued to self-differentiate into the chordamesoderm and other mesodermal structures that constituted the original fate of that blastopore tissue. As the axis was formed, host cells began to participate in the production of the new embryo, becoming organs that they never would normally have formed. Thus, a somite could be seen containing both colorless (donor) and pigmented (host) tissue. Even more spectacularly, the donor chordamesoderm was able to interact with the overlying ectodermal cells to give rise to a complete neural plate! Eventually, a secondary embryo formed face-to-face with its host (Figure 14).

The process by which one embryonic region interacts with a second region to cause the latter tissue to differentiate in a direction it otherwise would not is called INDUCTION. Because there are numerous inductions during embryonic development, the first key interaction in which the dorsal mesoderm induces ectoderm to differentiate into neural structures is called PRIMARY EMBRYONIC DEVELOPMENT. Therefore, Spemann referred to the dorsal lip cells as the ORGANIZER. It is now known (thanks largely to Spemann and his students) that the interaction of the chordamesoderm and ectoderm is not sufficient to "organize" the entire embryo. Rather, it is the first in a series of inductive events. We also know that the dorsal blastopore lip is active in organizing secondary embryos in *Amphioxus*, cyclostomes, teleosts, and a variety of amphibians. The anterior portion of the primitive streak (that is, Hensen's node, the region initiating gastrulation in birds and mammals) is simi-

FIGURE 14

Self-differentiation of the dorsal blastopore lip tissue. Dorsal blastopore lip from early gastrula (A) is transplanted into another early gastrula (B) in the region that normally becomes ventral epidermis. Tissue invaginates and forms a second archenteron and then a second embryonic axis. Both donor (black) and host (white) tissues are seen in this new neural tube, notochord, and somites. (C) Eventually, a second embryo forms that is joined to host.

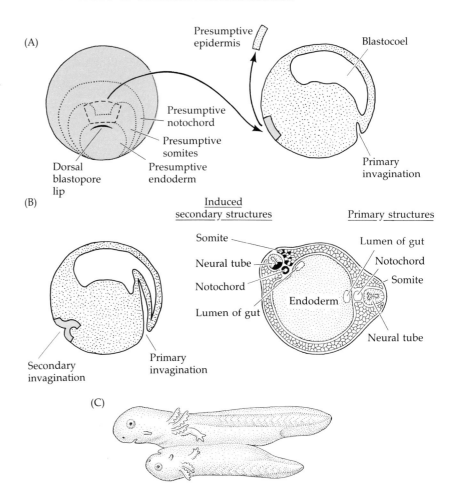

larly effective in organizing secondary embryos in these classes of vertebrates (Figure 15).

Regional specificity of induction

One of the most fascinating phenomena in primary embryonic induction is the regional specificity of the neural structures that are produced. Forebrain (archencephalic), hindbrain (deuterencephalic), and spinocaudal regions of the neural tube must all be properly organized in an anterior to posterior direction. Thus, the chordamesoderm tissue of the archenteron roof not only induces the neural tube, but it also induces the specific region of the neural tube. This region-specific induction was shown by Otto Mangold (1933) in a series of elegant experiments wherein various regions of the *Triturus* archenteron roof were transplanted into early gastrula embryos (Figure 16). After removing the

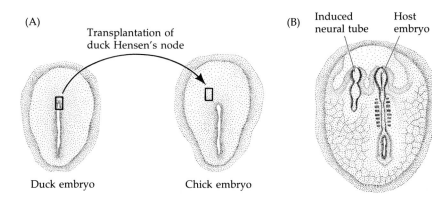

(A) Transplantation of duck Hensen's node

Duck embryo   Chick embryo

(B) Induced neural tube   Host embryo

FIGURE 15

Induction of a new embryonic axis by Hensen's node. (A) Diagram of experiment by which Hensen's node tissue is removed from a duck embryo and implanted within a host chick embryo. (B) Accessory neural tube induced at the graft site.

FIGURE 16

Regional specificity of induction can be demonstrated by implanting different regions (shown in dark gray) of the archenteron roof into early *Triturus* gastrulae. The resulting animals have secondary parts. (A) Head with balancers; (B) head with eyes and forebrain; (C) posterior part of head, deuterencephalon, and otic vesicles; and (D) trunk–tail segment. (After Mangold, 1933.)

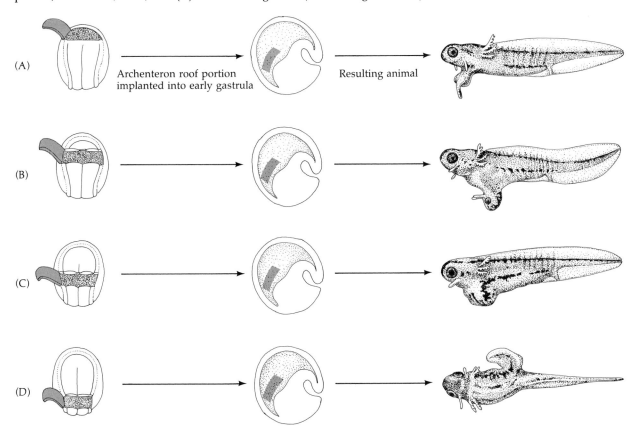

(A) Archenteron roof portion implanted into early gastrula   Resulting animal

(B)

(C)

(D)

superadjacent neural plate, four successive sections of the archenteron roof were excised from embryos that had just completed gastrulation and were placed into the blastocoels of early gastrulae. The most anterior portion of the archenteron roof gave rise primarily to suckers and portions of the oral apparatus (Figure 16A); the next most anterior section induced the formation of various head structures, including nose, eyes, suckers, and otic vesicles (Figure 16B). The third section induced the otic (hearing organ) vesicles without any other head organs (Figure 16C); and the most posterior segment induced the formation of trunk and tail structures (Figure 16D).

To further study this phenomenon, Holtfreter (1936) made a blastopore "sandwich," enclosing the dorsal blastopore lip between slices of undifferentiated ectoderm (Figure 17). The blastopore lips from the early gastrula stages induced the most archencephalic structures, whereas the blastopore lips of later embryos caused the differentiation of the more posterior neural elements. It is currently theorized that regional specification is caused by the interaction of two substances secreted by the cells of the chordamesoderm. A high concentration of one yields forebrain development, whereas a high concentration of the other produces the formation of spinal cord and trunk structures. Mixtures of the two substances elicit the midbrain and hindbrain regions.

FIGURE 17
Holtfreter's "sandwich" technique. Young or old dorsal blastopore lips are placed between two layers of ectoderm. The age of the dorsal lip determines what structures are induced.

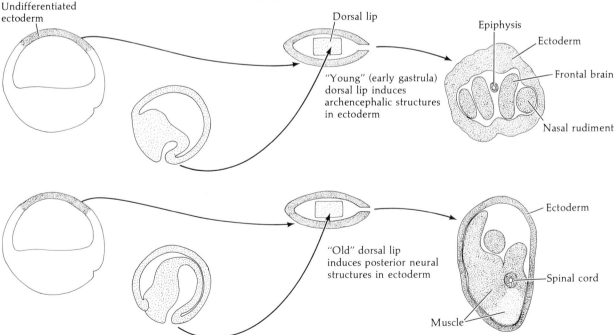

Evidence for this model came from studies involving artificial tissue-specific inducers. Guinea pig bone marrow, for instance, was found to induce mesodermal structures only. Pellets of guinea pig liver, however, could only cause the induction of forebrain structures. Toivonen and Saxén (1955) implanted these inducers together within the blastocoel of the same early gastrula. Whereas the liver would only have induced forebrain and the bone marrow would have induced only mesoderm, the two together induced all the normal forebrain, hindbrain, spinal cord, and trunk mesoderm. Thus, primary embryonic induction may also be due to a double gradient (Figure 18). Further evidence for a double gradient model comes from the isolation of forebrain- and trunk-inducing activity from a preparation of chick embryo extract that had been able to induce hindbrain structure preferentially (Tiedmann, 1967). When the hindbrain-inducing substance was separated by passing it through a column, two activities were isolated: one that induced mesodermal structures and one that induced the forebrain (Figure 18B).

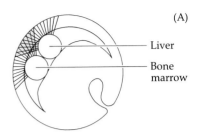

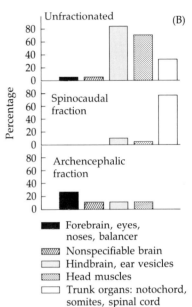

FIGURE 18
Evidence for a double gradient of inducers in the archenteron roof. (A) Simultaneous implantation of an archencephalic inducer (guinea pig liver) and a mesodermal inducer (guinea pig bone marrow pellets) into young newt gastrulae produces all of the dorsal regions of the brain and somites. (B) Separation of a midbrain (deuterencephalic)-inducing extract of chick embryos into an archencephalic-inducing fraction and a spinocaudal-inducing fraction.

SIDELIGHTS &
SPECULATIONS

The mechanism of primary embryonic induction

There are actually two major processes involved in primary embryonic induction. The first involves the capacity of certain cells in the late blastula to obtain the precocious ability to invaginate (see Nieuwkoop, 1969), whereas the second process concerns the ability of the chordamesoderm cells to induce the neural plate from the overlying ectoderm. We have already seen (Chapter 4) that the invaginating and organizing ability is not present in the gray crescent cytoplasm of the fertilized egg but that it arises there shortly before gastrulation. Again, the progressive nature of this type of determination should be noted. We do not know the molecular basis for either event.

The progressive development of inducing capacity

The ability to induce the formation of the axial structures is not found preformed in some particular region of the egg cytoplasm. Rather, it is likely that such determinants lie throughout the egg but are activated by contact with other regions of the cytoplasm. Thus, the beginnings of gastrulation and induction are to be found in the cytoplasmic rearrangements that attend fertilization (Figure 35 in Chapter 2). There are several lines of evidence that lead to this model (Gerhart et al., 1981). First, treating the eggs to prevent cytoplasmic rearrangements will produce normal-looking late blastulae; but the treated embryos will gastrulate symmetrically (instead of starting with a dorsal lip) and will fail to induce neural structures. By exposing fertilized eggs briefly to cold shock, pressure, or ultraviolet irradiation (all of which will destroy the microtubules of the aster), the embryos develop into tissue resembling the "belly pieces" of Spemann's nongastrulating embryos. Conversely, when microtubules are stabilized beyond their normal duration, the embryos develop excessive axial structures, sometimes having extra eyes. (Thus, it may be possible to "overdo" the activation or localization of these determinants.) Second, if the cytoplasmic regions are artificially mixed (such as by centrifuging the embryos at an angle to the dorsal–ventral axis or by rotating them after fertilization), the embryo will form a new dorsal lip and

FIGURE 19

Speculative sequence of events leading to the induction of the body axis in frogs. Fertilization of a radially symmetric egg causes displacement of cortical and inner cytoplasms, as described in Chapter 2. Deep cytoplasm adhering to the cortex on the gray crescent side activates "dorsal" determinants (shown in gray) in the vegetal hemisphere on that side. These dorsalizing determinants are included in the cytoplasm of the vegetal cells during the next 12 cleavages. Sometime during cleavage, the dorsal vegetal cells induce adjacent animal cells to become prospective chordamesoderm. During gastrulation, these chordamesoderm cells migrate up the blastocoel wall to induce the neural plate. (After Gerhart et al., 1983.)

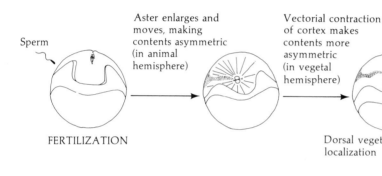

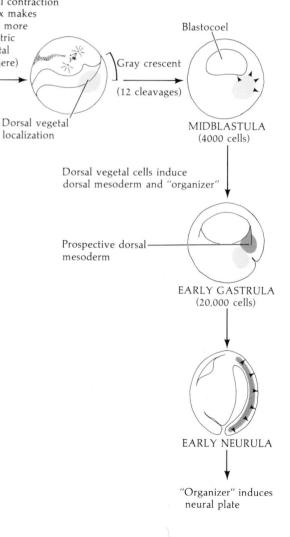

develop a second embryonic axis. In this manner, "Siamese twins" can be readily created (Figure 16 in Chapter 4). Third, Nieuwkoop has removed the equatorial cells of the midblastula frog and salamander embryo and has shown that neither of the remaining animal or vegetal hemispheres could produce mesodermal cells when separately cultured. Only the equatorial cells had this potency. However, when the cap and base portions were recombined, the healed embryo could form mesodermal structures from the cells of the animal hemisphere. Several of these embryos formed normal body axes, implying that the "organizer" region of dorsal mesoderm had been formed when animal hemisphere cells had been in contact with the vegetal hemisphere cells. Moreover, there was a distinct polarity to this induction because different fragments of vegetal hemisphere produced different inductions: The fragment adjacent to the excised gray crescent induced the animal hemisphere cells above it to become dorsal mesoderm (just as the "organizer" had). Ventral fragments produced ventral structures (such as kidney tubules) and the most ventral fragments induced ventral mesoderm. At midblastula, the dorsal vegetal cells possess a pattern of inductive potencies wherein they can induce the responsive cells of the animal hemisphere. Thus, these vegetal cells are the ones carrying the activated determinants. Moreover, the vegetal cells on the gray crescent side—where the cytoplasmic rearrangements of fertilization were most extensive—are responsible for inducing the cells that will constitute the "organizer" chordamesoderm. The vegetal cells accomplishing all this do not become part of the axial structures but rather give rise to the ventral endoderm of the gut. Thus, it is proposed (Gerhart et al., 1981; Gimlich and Gerhart, 1984) that vegetal cells induce their overlying equatorial cells to become mesodermally determined long before gastrulation and that these prospective dorsal mesoderm cells migrate through the blastocoel to induce the neural plate (Figure 19; Figure 17 in Chapter 4).

The primary inducer as a diffusible molecule

Although Spemann thought that searching for an organizer molecule was folly, his students pressed forward. They discovered that the identification of the active inducer was not going to be an easy task. The problem was a lack of specificity; it seemed that a huge variety of things could induce a neural plate. These compounds included turpentine, formaldehyde, and methylene blue dye, as well as dead archenteron and an assortment of adult tissues from several phyla. On the basis of such eclectic induction, Holtfreter (1948) suggested that the real inducer lay within the ectoderm itself and that it was released by mild cytolysis. Anything causing such sublethal damage would suffice, even a mildly alkaline or hypertonic medium. Barth and Barth (1969) have proposed a modification of this view, wherein the various compounds all act by re-

leasing bound sodium ions within the ectodermal cells. At this moment, a purified inducer molecule has not been isolated.

Whatever is responsible for the induction does not necessarily demand physical contact between the ectoderm and chordamesoderm. Following up on a chance observation, Niu and Twitty (1953) showed that whatever caused the embryonic induction was diffusible. They cultured blastopore lips or chordamesoderm for a week in vitro. When they removed the inducer and placed embryonic ectoderm tissue in this "conditioned" medium, the tissue differentiated into neural cells, pigment cells, and mesodermal precursor cells. The predominant cell type depended upon the age of the inducer, the earlier notochords causing the most neural cell and pigment cell differentiation. No differentiation occurred when the medium remained unconditioned by inducer tissue. Thus, induction does not need physical contact between the inducer and the responding tissues. More recently, Finnish embryologist Sulo Toivonen (1979) demonstrated that induction could occur when the dorsal blastopore lip and the responding ectodermal tissue were placed on opposite sides of a membrane having a pore size of 0.5 μm. Induction was found to occur even though no cell processes were observed to traverse the membrane. This suggests that cell–cell contact is not necessary for primary induction to occur.

Thus, there is still debate as to the components and mechanism of primary embryonic induction, one of the oldest areas of developmental biology. Surveying the field in 1927, Spemann remarked,

> What has been achieved is but the first step; we still stand in the presence of riddles, but not without hope of solving them. And riddles with the hope of solution—what more can a scientist desire?

The challenge still remains.

Competence and secondary induction

Any system of embryonic induction has at least two components: a tissue capable of producing the inducing stimulus, and a tissue capable of receiving and responding to it. So far, we have been looking at the specificity of production; now we must look at the specificity of responding ectodermal cells. This ability to respond in a specific manner to a given stimulus is called COMPETENCE. We have already seen that in the early gastrula, an implanted blastopore lip can induce a new neural plate and embryonic axis just about anywhere in the embryo where it can meet ectoderm. However, with increasing embryonic age the ectoderm loses this ability to respond, and the implantation of a dorsal blastopore lip beneath the prospective epidermis of a neurula stage embryo will not cause it to form a new neural plate.

Although the late neurula ectoderm is no longer competent to respond to chordamesoderm, it has acquired new competences. For instance, it is now competent to respond to contact from the optic vesicle (derived from the forebrain) to become lens. Hindbrain can similarly induce the adjacent ectodermal area to form the otic (ear) vesicles. This ectoderm, therefore, has acquired the ability to respond to SECONDARY INDUCERS.

It will be useful to discuss one of these secondary induction schemes here. (Such inductions are among the most important events in development and will be discussed in detail in Chapter 16.) If Spemann had never performed the preceding experiments demonstrating primary induction, his fame would still be secure because of his earlier analysis of the tissue interactions during eye development. These secondary inductions are illustrated in Figure 20. The formation of the lens, as previously mentioned, starts when the bulge of the forebrain (optic vesicle) contacts the overlying ectoderm of the mid-late neurula. After this contact, the wall of the optic vesicle invaginates to form a two-layered neural structure, the optic cup. This optic cup is responsible for the induction of the overlying ectoderm into the lens placode. If a barrier is placed between the optic cup and the overlying ectoderm, no lens placode forms. The optic cup differentiates into the pigmented and neural portions of the retina. The lens vesicle, however, is both an inducing and a responding tissue. It induces the new overlying ectoderm to become cornea, while it responds to the inductive stimulation from the neural retina and differentiates into the definitive lens.

FIGURE 20
Cascade of secondary inductions in the eye region of the brain.

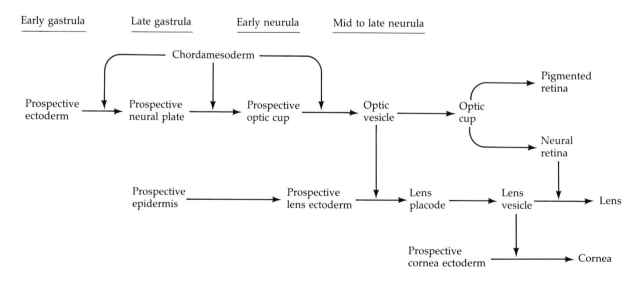

Thus, in both primary and secondary inductions, there is restriction in the potency of ectodermal cells. For the primary interactions, the change occurs during gastrulation; for the secondary inductive interactions, the determination takes place later. Here, then, we find determination to be dependent on interactions between groups of cells. However, because the basis for such commitments can be traced back to the material of the egg cytoplasm, it should not be surprising to find, as we shall in the next chapter, organisms in which the determination is not progressive, but immediate. Here we leave those embryos whose blastomeres interact to determine their respective fates; we next observe embryos whose early blastomeres are determined by cytoplasmic segregation.

Literature cited

Barth, L. G. and Barth, L. J. (1969). The sodium dependence of embryonic induction. *Dev. Biol.* 20: 236–262.

Brenner, S. (1979). Cited in H. F. Judson (1979). *The Eighth Day of Creation*. Simon and Schuster, New York, p. 219.

Bonnet, C. (1764). *Contemplation de la Nature*. Marc-Michel Ray, Amsterdam.

Cassirer, E. (1950). Developmental mechanics and the problem of cause in biology. In E. Cassirer (ed.), *The Problem of Knowledge*. Yale University Press, New Haven.

Czihak, G. (1971). Echinoids. In G. Reverberi (ed.), *Experimental Embryology of Marine and Fresh-water Invertebrates*. Elsevier North-Holland, Amsterdam, pp. 363–506.

Driesch, H. (1892). The potency of the first two cleavage cells in echinoderm development. Experimental production of partial and double formations. In B. H. Willier and J. M. Oppenheimer (eds.), *Foundations of Experimental Embryology*. Hafner, New York.

Driesch, H. (1893). Zur Verlagerung der Blastomeren des Echinideneies. *Anat. Anz.* 8: 348–357.

Driesch, H. (1894). *Analytische Theorie de organischen Entwicklung*. W. Engelmann, Leipzig.

Gerhart, J., Ubbels, G., Black, S., Hara, K. and Kirschner, M. (1981). A reinvestigation of the role of the gray crescent in axis formation in *Xenopus laevis*. *Nature* 292: 511–516.

Gerhart, J., Black, S., Gimlich, R. and Scharf, S. (1983). Control of polarity in the amphibian egg. In W. R. Jeffery and R. A. Raff (eds.), *Time, Space and Pattern in Embryonic Development, MBL Lectures in Biology 2*. Alan R. Liss, New York, pp. 261–296.

Gould, S. J. (1977). *Ontogeny and Phylogeny*. Belknap Press, Cambridge, MA.

Hagstrom, B. E. and Lonning, S. (1965). Studies in cleavage and development of isolated sea urchin blastomeres. *Sarsia* 18: 1–9.

Haller, A. von (1758). *Sur la Formation du Coeur dans le Poulet, Mem. II*. M. M. Bousquet and Co., Lausanne.

Holtfreter, J. (1936). Regional induktionen in Xenoplastisch zusammengesetzten explantaten. *Wilhelm Roux Arch. Entwicklungsmech. Org.* 134: 466–561.

Holtfreter, J. (1948). Concepts on the mechanism of embryonic induction and its relation to parthenogenesis and malignancy. *Symp. Soc. Exp. Biol.* 2: 17–48.

Hörstadius, S. (1928). Über die Determination des Keimes bei Echinodermen. *Acta Zool.* 9: 1–191.

Hörstadius, S. (1935). Über die Determination im Verlaufe der Eiasche bei Seeigeln. *Publ. Staz. Zool. Napoli* 14: 251–479.

Hörstadius, S. (1939). The mechanics of sea urchin development studied by operative methods. *Biol. Rev.* 14: 132–179.

Hörstadius, S. and Wolsky, A. (1936). Studien über die Determination der Bilateralsymmetrie des jungen Seeigelkeimes. *Wilhelm Roux Arch. Entwicklungsmech. Org.* 135: 69–113.

Huxley, J. S. and deBeer, G. R. (1934). *Elements of Experimental Embryology*. Cambridge University Press, Cambridge.

Kölreuter, J. G. (1761–1766). *Vorläufige Nachricht von einigen das Geschlecht der Pflanzen betreffenden Versuchen und Beobachtungen, nebst Fortsetzungen 1,2, und 3*. Leipzig.

Lenoir, T. (1980). Kant, Blumenbach, and vital materialism in German biology. *Isis* 71: 77–108.

Mangold, O. (1933). Über die Induktionsfahigkeit der verschiedenen Bezirke der Neurula von Urodelen. *Naturwissenschaften* 21: 761–766.

McClendon, J. F. (1910). The development of isolated blastomeres of the frog's egg. *Am. J. Anat.* 10: 425–430.

Morgan, T. H. (1895). Studies on the "partial" larvae of *Sphaerechinus. Wilhelm Roux Arch. Entwicklungsmech. Org.* 2: 81–126.

Nieuwkoop, P. D. (1960). The formation of the mesoderm in urodele amphibians. I. Induction by the endoderm. *Wilhelm Roux Arch. Entwicklungsmech. Org.* 162: 341–373.

Niu, M. C. and Twitty, V. C. (1953). The differentiation of gastrula ectoderm in medium conditioned by axial mesoderm. *Proc. Natl. Acad. Sci. USA* 39: 985–989.

Roux, W. (1888). Contributions to the developmental mechanics of the embryo. On the artificial production of half-embryos by destruction of one of the first two blastomeres and the later development (postgeneration) of the missing half of the body. In B. H. Willier and J. M. Oppenheimer (eds.), *Foundations of Experimental Embryology*. Hafner, New York, pp. 2–37.

Saxén, L. and Toivonen, S. (1962). *Embryonic Induction.* Prentice-Hall, Englewood Cliffs, NJ.

Spemann, H. (1918). Über die Determination der ersten Organanlagen des Amphibienembryo. *Wilhelm Roux Arch. Entwicklungsmech. Org.* 43: 448–555.

Spemann, H. (1938). *Embryonic Development and Induction.* Yale University Press, New Haven.

Spemann, H. and Mangold, H. (1924). Induction of embryonic primordia by implantation of organizers from a different species. In B. H. Willier and J. M. Oppenheimer (eds.), *Foundations of Experimental Embryology*. Hafner, New York, pp. 144–184.

Thomson, J. A. (1907). *Heredity.* Putnam, New York.

Tiedmann, H. (1967). Biochemical aspect of primary induction and determination. In R. Weber (ed.), *The Biochemistry of Animal Development*, Vol. 2. Academic, New York, pp. 3–55.

Toivonen, S. (1979). Transmission problems in primary induction. *Differentiation* 15: 177–181.

Toivonen S. and Saxén, L. (1955). The simultaneous inducing action of liver and bone marrow of the guinea pig in implantation and explantation experiments with embryos of *Triturus. Exp. Cell Res. Suppl.* 3: 346–357.

Ubisch, L. von (1919). Über die Determination der larvalen Organe und der Imaginalanlage bei Seeigeln. *Wilhelm Roux Arch. Entwicklungsmech. Org.* 117: 80–122.

Weismann, A. (1892). *Essays on Heredity and Kindred Biological Problems.* Translated by E. B. Poulton, S. Schoenland and A. E. Shipley. Clarendon, Oxford.

Weismann, A. (1893). *The Germ-Plasm: A Theory of Heredity.* Translated by W. Newton Parker and H. Ronnfeld. Walter Scott, Ltd., London.

Wilson, E. B. (1896). *The Cell in Development and Inheritance.* Macmillan, New York.

Wolff, K. F. (1767). De formatione intestinorum praecipue. *Novi Commentarii Academine Scientarum Imperlialis Petropolitanae.*

Determination by cytoplasmic specification

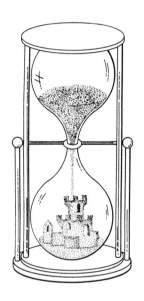

*Though the substance of a cell may appear homogeneous under
the most powerful microscope, excepting for the fine granular
matter suspended in it, it is quite possible, indeed certain, that it
may contain,* already formed and individualized, *various
kinds of physiological molecules. The visible process of
segregation is only the sequel of a differentiation already
established, and not visible.*
—E. R. LANKESTER (1877)

Cytoplasmic specification: Mosaic development

We have seen that during the development of sea urchins and amphibians, blastomeres eventually become determined through their interactions with other cells. In this "regulative" development, cytoplasmic differences between the blastomeres are not great enough to prevent a blastomere from giving rise to some cell type for which it is not usually determined. There are some organisms, however, where the cytoplasmic localization of determinative materials in the zygote is such that the first few blastomeres are qualitatively different. Therefore, if the blastomeres of such an early embryo are isolated, each will give rise to only a portion of an embryo. Such embryos cannot regulate for the missing parts, and the potency of each blastomere is equal to its prospective fate. This type of determination, where there is a qualitative separation of organ-forming materials to the early blastomeres, is often called MOSAIC determination. The terms *regulative* and *mosaic*, however, should not be considered as two separate categories. Rather, they are poles of a continuum wherein some species are more mosaic or more regulative than others. In this chapter, we shall discuss the more mosaic organisms.

Cytoplasmic specification in tunicate embryos

Roux, you may recall, believed he had shown such immediate self-determination in frog embryos, but his results derived from an inadequate experimental procedure. The first person to show truly mosaic development was the Frenchman, Laurent Chabry. Like Roux, Driesch, and Spemann, Chabry tested the potency of early blastomeres. Unlike his German contemporaries, however, he was not the slightest bit interested in testing theories of differentiation. Rather, Chabry's research was motivated by medical concerns, namely, TERATOLOGY—the study of developmental errors that lead to malformed infants. In 1886, he set out to artificially produce specific embryonic malformations.

He accomplished this by lancing one of the first two blastomeres of tunicate embryos, and he obtained the same result that Roux did. The surviving blastomere continued to develop as if it were part of the whole embryo, producing a half-tadpole.[1] Unlike frog cells, however, *isolated* tunicate blastomeres also behave mosaically, the blastomere isolated from a 2- or 4-cell embryo giving rise to only those structures that it would normally form in situ.

[1]The published results of this work appeared in 1887, the same year as Roux's public presentation of his experiment. Neither had known of the other's work. However, while Roux's paper (and its erroneous conclusion) was widely quoted, Chabry's medical paper was largely ignored and might have been altogether forgotten had Driesch not referred to it in his isolation experiments (see Churchill, 1973).

More recent studies have shown that the tunicate embryo does indeed approximate "a mosaic of self-differentiated parts" constructed from information stored in the oocyte cytoplasm. These studies have been helped considerably by the eggs of certain tunicate species that segregate their cytoplasm into a series of colored regions immediately after fertilization. In 1905, E. G. Conklin described how these colored plasms became apportioned into various blastomeres. The first cleavage separates the egg into right and left mirror images. From then on, each cell division on one side parallels a similar division on the other. By following the fate of *each* blastomere of the tunicate *Styela partita*, Conklin came to the astonishing conclusion that each colored plasm delineates a specific embryonic fate (Figure 1). The yellow cytoplasmic crescent gave rise to the muscle cells; the gray equatorial crescent produced the notochord and the neural tube; the clear animal cytoplasm became the larval epidermis; and the yolky gray vegetal region gave rise to the larval gut.

G. Reverberi and A. Minganti (1946) analyzed tunicate determination in a series of isolation experiments and they observed the self-differentiation of each isolated blastomere and the remaining embryo. The results of one of these experiments is shown in Figure 2. When the

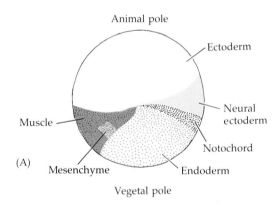

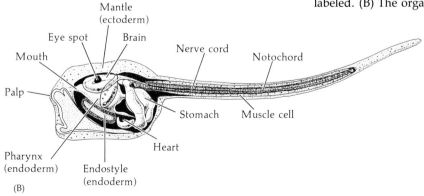

FIGURE 1
Segregation of cytoplasmic determinants at fertilization. (A) Cytoplasmic regions of the fertilized tunicate egg (*Styela partita*) with their prospective fates labeled. (B) The organs of the tunicate larva.

FIGURE 2
Mosaic determination in tunicates. The four dissociated blastomere couples of the 8-cell embryo develop as indicated, each forming separate structures. (After Reverberi and Minganti, 1946.)

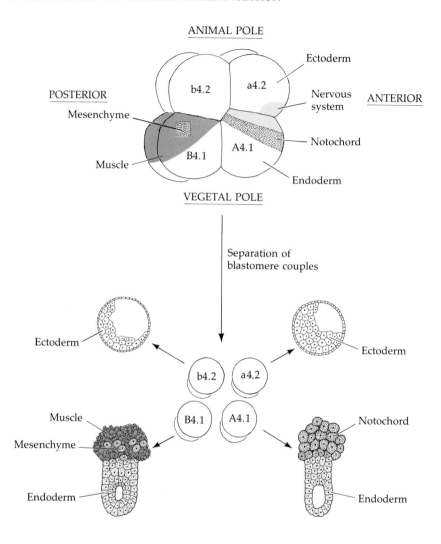

8-cell embryo is separated into its four doublets (the right and left sides being equivalent), mosaic self-determination is the rule. The animal posterior pair of blastomeres gives rise to the ectoderm; the vegetal posterior pair produces endoderm, mesenchyme, and muscle tissue, just as expected from the fate map. Neural development, however, is an exception. The nerve-producing cells are generated from both the animal and vegetal anterior quadrants, yet neither produces them alone. When these anterior pairs are reunited, though, the brain and palp tissues arise. Even in an embryo as strictly determined as in tunicates, some inductive interactions take place between blastomeres. In fact, Ortolani (1959) has shown that this region of ectoderm is not determined for "neuralness" until the 64-cell stage, right before gastrulation. Thus, although most tissues are determined immediately, certain tissues in these embryos do have a progressive determination.

In 1973, J. R. Whittaker provided dramatic biochemical confirmation of the cytoplasmic segregation of tissue determinants. Whittaker stained cells for the presence or absence of the enzyme ACETYLCHOLINESTERASE. This enzyme is found only in the larval muscle tissue and is involved in allowing muscles to respond to repeated nerve impulses. From the cell lineage studies of Conklin and others (Figures 2 and 3), it was known that only one pair of blastomeres (posterior vegetal; B4.1) in the

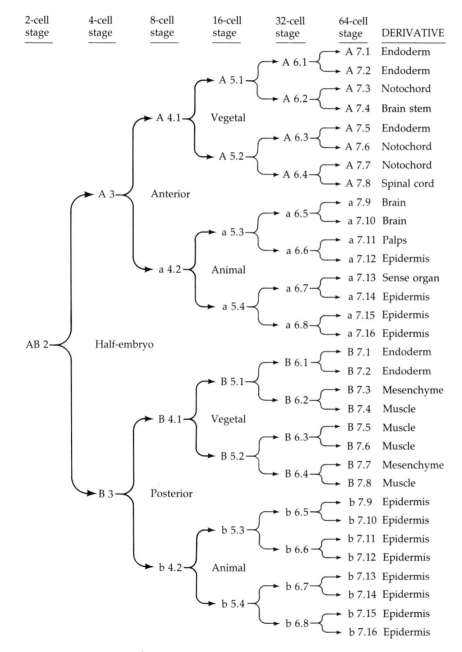

FIGURE 3
Lineage fate map of embryonic tunicate development. Because both right and left halves develop identically, only one half of the embryo is represented here. (From Whittaker, 1979).

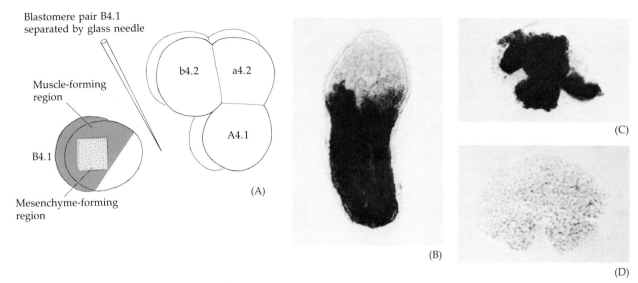

FIGURE 4

Acetylcholinesterase development in the progeny of the muscle lineage blastomeres (B4.1) isolated at the 8-cell stage. (A) Diagram of the isolation procedure. (B) Localization of acetylcholinesterase in the tail muscles of the intact tunicate larva. Enzyme presence is demonstrated in the progeny of the B4.1 blastomere pair (C), but not in the remaining 6/8 of the embryo (D) when incubated for the length of time it takes to normally form a larva. (From Whittaker et al., 1977; photographs courtesy of J. R. Whittaker.)

8-cell embryo was capable of producing muscle tissue. When Whittaker removed these two cells and placed them in isolation, they produced muscle tissue that stained positively for the presence of acetylcholinesterase. The remaining 6/8 embryo produced a muscleless larva without any detectable amount of acetylcholinesterase activity (Figure 4).[2]

Whittaker then took various stage tunicate embryos and arrested further cleavage by treating them with cytochalasin B. This drug binds to microfilaments, thus preventing cytokinesis (cell division) while allowing nuclear division to occur normally. Thus, all further development occurs within the population of cells present at the time when cytochalasin was added. After the embryos had their further cleavages blocked, they were allowed to develop and then were stained for the presence of acetylcholinesterase. The results were striking. A comparison of these results (Figure 5) with the cell lineage chart (Figure 3) shows a direct concordance. The ability to produce muscle cells was originally present in both of the 2-cell blastomeres. However, by the 4-cell stage, the ability to produce acetylcholinesterase-synthesizing cells

[2]Deno and co-workers (1984) have shown that in at least some species of tunicates, the b4.2 pair may also make acetylcholinesterase. However, this is seen only in whole embryos and not in isolated blastomeres. It is possible that some induction may be causing muscle formation in these ascidians.

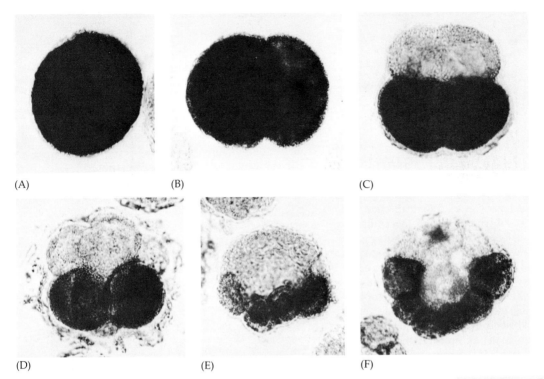

(A) (B) (C)

(D) (E) (F) (G)

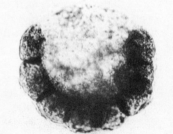

(H)

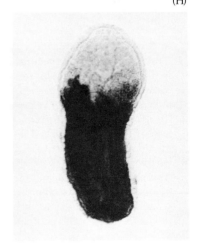

FIGURE 5
Acetylcholinesterase localization in tunicate embryos whose development has been arrested at various times by cytochalasin B. (A) 1-cell; (B) 2-cell; (C) 4-cell; (D) 8-cell; (E) 16-cell; (F) 32-cell; (G) 64-cell; (H) tailbud muscle localization in 9-hour control embryo. (From Whittaker, 1973a; photographs courtesy of J. R. Whittaker.)

is limited to the vegetal blastomeres. In the 8-cell embryo, only the two posterior vegetal blastomeres can give rise to such cells. These are the cells known to form the tail muscle of the tunicate larva.

The cells whose descendants were shown to synthesize acetylcholinesterase are precisely those cells destined to produce muscle. Moreover, the production of that enzyme in the isolated cells occurs at exactly the time when it appears in normal embryos (Satoh, 1979). T. C. Tung and colleagues (1977) also have shown that when larval nuclei are transplanted into enucleated tunicate egg fragments, the newly formed cells show the structures typical of those cells determined by the egg cytoplasmic regions, not the source of the nuclei. We can conclude, then, that certain determinants that exist in the cytoplasm cause the formation of certain tissues. These MORPHOGENIC DETERMINANTS (or MORPHOGENS) appear to work by selectively activating (or inactivating) specific genes. Thus, the determination of the blastomeres and the activation of certain genes are controlled by the spatial localization of morphogenic determinants within the egg cytoplasm.

SIDELIGHTS & SPECULATIONS

Intracellular localization and movements of morphogenic determinants

Whereas the various inclusions of the fertilized egg (including yolk, pigment, and soluble proteins) can easily be displaced by centrifugation, such displacement does not usually affect embryogenesis (reviewed in Morgan, 1927). It appears, then, that either the determinants are too small to be moved by centrifugation or that they are somehow anchored within the egg. The lack of diffusion manifest in the cytoplasmic localization of these determinants weighs against the first possibility. Most likely, the determinants are attached to insoluble material, probably the cytoskeletal framework of the cell. This infrastructure of filaments and tubules is particularly prominent in the oocyte cortex, but it also extends throughout the cell. Cervera et al. (1981) have reported that most cellular RNA in cultured cells is associated with the cytoskeletal framework. Thus, the cytoskeleton might be a means of specifically localizing cytoplasmic determinants.

The cytoskeleton can be isolated by extracting cells with nonionic detergents such as Triton X-100. The detergent solubilizes lipids, tRNA, and monoribosomes. The remaining cytoskeleton contains microtu-

bules, microfilaments, intermediate filaments, and roughly 200 proteins, including one that is capable of binding the 5′ cap of mRNAs (Zumbe et al., 1982; Moon et al., 1983). In the tunicates *Styela* and *Boltenia*, the muscle-forming "yellow crescent" is characterized by an actin-containing cytoskeletal domain. This domain is originally coextensive with the unfertilized egg. However, after fertilization, the actin microfilaments contract and become segregated into those blastomeres fated to form muscle cells, taking the yellow pigment granules with them (Jeffery and Meier, 1983). Figure 6 shows that the cytoskeletal framework contains the yellow pigment granules and that these granules are given their intracellular localization by the movements of the oocyte cytoplasm during fertilization. The cytoskeleton may be the anchor for localizing the morphogenic factors determining embryonic cell fate.

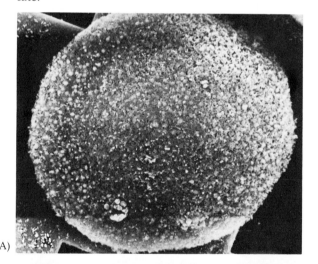

(A)

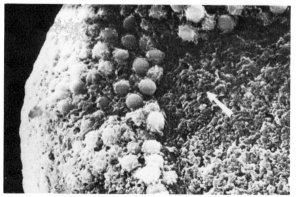

(B)

FIGURE 6
Scanning electron micrographs of tunicate eggs undergoing the segregation of morphogenic determinants. The eggs have been extracted with Triton X-100 detergent to solubilize the membrane and to allow the observation of cytoskeletal components. (A) Unfertilized *Styela* egg; yellow pigment granules and the cytoskeleton can be seen around the entire surface. (B) A newly-fertilized *Boltenia* zygote segregating its yellow cytoplasm. The region containing the yellow pigment granules is elevated and is seen to consist of a plasma membrane lamina and a deeper network of filaments. The presumed direction of the pigment migration is shown by the arrow. (From Jeffrey and Meier, 1983; photographs courtesy of S. Meier.)

Cytoplasmic localization in mollusc embryos

The "mosaic" type of differentiation is widespread throughout the animal kingdom, especially in organisms such as ctenophores (comb jellies), annelids, nematodes, and molluscs, which initiate gastrulation at the future anterior end after only a few cell divisions. Molluscs provide some of the most impressive examples of "mosaic" development and of the phenomenon of CYTOPLASMIC LOCALIZATION, wherein the morphogenic determinants are segregated to specific regions of the oocyte.

E. B. Wilson, the outstanding embryologist at Columbia University, isolated early blastomeres from embryos of the mollusc *Patella coerulea* and compared their development to the same cells that had been left within the embryos. Figure 7 shows one set of results published by Wilson in 1904. Not only did the isolated blastomeres follow their normal developmental fates (in this case, to produce the ciliated trochoblast cells), but they completed the normal number of cell divisions at precisely the same time as those cells remaining within the embryo. Their cleavages were in the correct orientation, and the derived cells

FIGURE 7
Differentiation of trochoblast cells in the normal embryo of the mollusc *Patella* (A–C) and of trochoblast cells isolated and cultured in vitro (D–G). (A) 16-cell stage seen from the side; the presumptive trochoblast cells are shaded. (B) 48-cell stage. (C) Ciliated larval stage, seen from the animal pole. Cilia are seen on trochoblast cells. (D) Isolated trochoblast cell. (E,F) Results of first and second divisions in culture. (G) Ciliated product of F; even in isolated culture, cells become ciliated at the correct time. (After Wilson, 1904.)

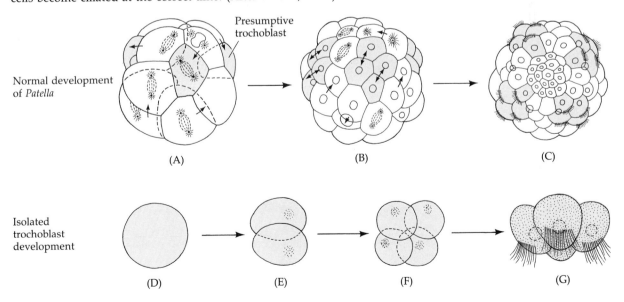

became ciliated at the appropriate time. Wilson (1904) concluded from these experiments that

> the history of these cells gives indubitable evidence that they possess within themselves all the factors that determine the form and rhythm of cleavage, and the characteristic and complex differentiation that they undergo wholly independently of their relation to the remainder of the embryo.

In his next experiment, Wilson was able to demonstrate that such development was predicated upon the segregation of specific morphogenic determinants into specific blastomeres. Certain spirally cleaving embryos (mostly in the mollusc and annelid phyla) extrude a bulb of cytoplasm immediately before first cleavage (Figure 8). This is called the POLAR LOBE. In certain species of the snails, the region uniting the polar lobe to the rest of the egg becomes a fine tube. The first cleavage splits the zygote asymmetrically such that the polar lobe is connected only to the CD blastomere. In several species, nearly one-third of the total cytoplasmic volume is present in these anucleate lobes, giving them the appearance of another cell. This three-lobed structure is therefore often referred to as the TREFOIL STAGE embryo (Figure 9). The CD blastomere then absorbs the polar lobe material but extrudes it again prior to second cleavage. After this division, the polar lobe is attached only to the D blastomere, which absorbs its material. Thereafter, no polar lobe is formed.

Wilson showed that if one removes the polar lobe at the trefoil stage, the remaining cells divide normally. However, instead of pro-

FIGURE 8
Cleavage in the mollusc *Dentalium* showing the extrusion and reincorporation of the polar lobe. (From Wilson, 1904.)

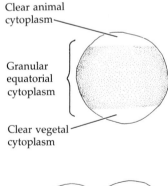

Clear animal cytoplasm

Granular equatorial cytoplasm

Clear vegetal cytoplasm

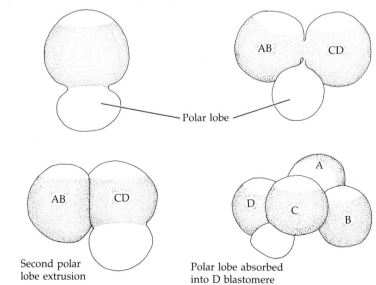

Polar lobe

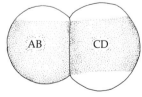

Polar lobe absorbed into CD blastomere

Second polar lobe extrusion

Polar lobe absorbed into D blastomere

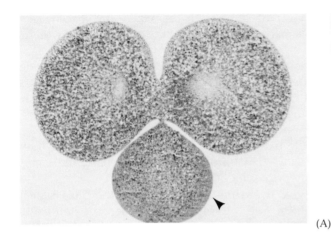

(A)

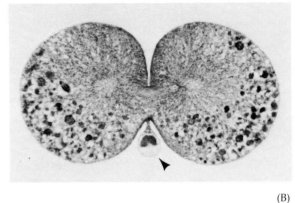

(B)

FIGURE 9
Polar lobes of molluscs. Sections through the first cleavage eggs of *Dentalium* (A) and *Bithynia* (B). Cells were stained with hematoxylin-eosin. (C) Scanning electron micrograph of the extension of the polar lobe in the uncleaved egg of *Buccinum undatum*. The surface ridges are confined to the polar lobe region. (Photographs courtesy of M. R. Dohmen.)

(C)

ducing a normal trochophore (snail) larva, they produced an incomplete larva, wholly lacking its mesodermal derivatives—muscles, mouth, shell gland,[3] and foot. Moreover, Wilson demonstrated that the same type of abnormal embryo can be produced by removing the D blastomere from the 4-cell embryo. Therefore, Wilson concluded that the polar lobe cytoplasm contained the mesodermal determinants and that these determinants give the D blastomere its mesoderm-forming capacity. Wilson also showed that the localization of the mesodermal determinant is established shortly after fertilization, thus demonstrating that a specific cytoplasmic region of the egg, destined for inclusion into the D blas-

[3]The shell gland is an ectodermal organ formed through induction by the mesodermal cells. Without the mesoderm, there are no cells present to induce the competent ectoderm. Again, we see some limited induction within a "mosaic" embryo.

tomere, contains whatever "factors" are necessary for the special cleavage rhythms of the D blastomere and for the differentiation of the mesoderm.

A. C. Clement has extended Wilson's work by analyzing the development of the snail *Ilyanassa obsoleta*. Clement (1968) centrifuged the trefoil-stage embryos, causing the fluid part of the cytoplasm to flow back into the CD blastomere. However, when he then cut off the polar lobe, the resulting embryos *still* lacked their mesodermal derivatives. Therefore, the diffusible part of the cytoplasm does not contain these morphogenic determinants. They probably reside in the nonfluid cortical cytoplasm or in the cytoskeleton.

Clement also analyzed the further development of the D blastomere in order to observe the further appropriation of these determinants. The development of the D blastomere is illustrated in Figure 10. This macromere, having received the contents of the polar lobe, is larger than the other three. When one removes the D blastomere *or* its first or second macromere derivatives (D, 1D, or 2D), one obtains an incomplete larva, lacking heart, intestine, velum (the ciliated border of the larva), shell gland, eyes, and foot. When one removes the 3D blastomere (*after* the division of the 2D cell to form the 3d blastomeres), one obtains an almost normal embryo, having eyes, feet, velum, and some shell gland, but no heart or intestine (Figure 11). Therefore, some of the

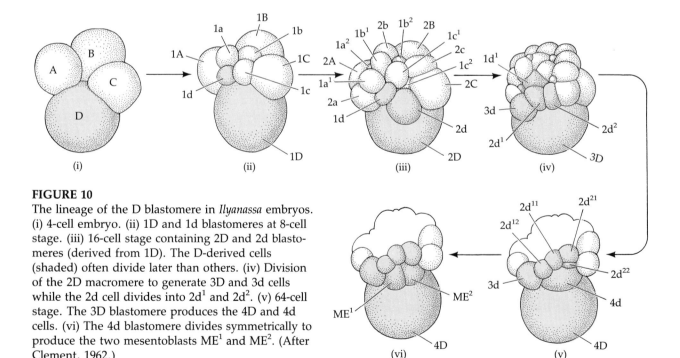

FIGURE 10

The lineage of the D blastomere in *Ilyanassa* embryos. (i) 4-cell embryo. (ii) 1D and 1d blastomeres at 8-cell stage. (iii) 16-cell stage containing 2D and 2d blastomeres (derived from 1D). The D-derived cells (shaded) often divide later than others. (iv) Division of the 2D macromere to generate 3D and 3d cells while the 2d cell divides into $2d^1$ and $2d^2$. (v) 64-cell stage. The 3D blastomere produces the 4D and 4d cells. (vi) The 4d blastomere divides symmetrically to produce the two mesentoblasts ME^1 and ME^2. (After Clement, 1962.)

morphogenic determinants originally present in the D blastomere were apportioned in the 3d cell. After the 4d cell is given off (by the division of the 3D blastomere), removal of the D derivative (the 4D cell) produces no qualitative difference in development. In fact, all the essential determinants for heart and intestine formation are now in the 4d blastomere and removal of *that* cell results in a deficient larva. The 4d blastomere is responsible for forming (at its next division) the two MESENTOBLASTS, the cells that give rise to both the mesodermal (heart) and endodermal (intestine) organs.

Thus, Wilson and Clement have demonstrated that the nondiffusible polar lobe cytoplasm is extremely important for normal mollusc development because

1. It contains the determinants for the proper rhythm and cleavage orientation of the D blastomere.
2. It contains certain determinants (those entering the 4d blastomere and hence leading to the mesentoblasts) for mesodermal and intestinal differentiation.
3. It is responsible for permitting the inductive interactions (through the material entering the 3d blastomere) leading to the formation of the shell gland and eye.

Cytoplasmic localization of germ cell determinants

Cytoplasmically localized determinants are found throughout the animal kingdom. The most frequently observed determinants are those responsible for the determination of germ cell precursors, that is, those cells that give rise to the gametes. Even in many embryos in which other aspects of early development are regulative, those cells containing a certain region of egg cytoplasm are destined to become germ cell precursors. We shall presently discuss the germ cell determinants in three different groups of animals: nematodes, insects, and frogs.

Germ cell determination in nematodes

Theodor Boveri (1862–1915) was the first person to look at an organism's chromosomes throughout its development. In so doing, he discovered a fascinating feature in the development of the roundworm *Ascaris megalocephala*. This nematode has only two chromosomes per haploid cell, thus allowing detailed observations of the individual chromosomes. The cleavage plane of the first embryonic division is unusual in that it is equatorial, separating the animal from the vegetal side of the zygote (Figure 12A). More bizarre, however, is the behavior of the chromosomes in the subsequent division of these first two blastomeres. The ends of the chromosomes in the animal-derived blastomere fragment

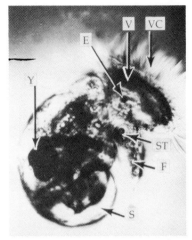

(A)

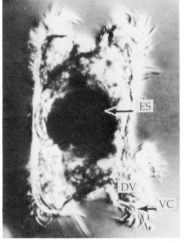

(B)

FIGURE 11
(A) Normal veliger larva. (B) Aberrant larva, typical of those produced when the polar lobe or D blastomere is removed. E, eye; F, foot; S, shell; St, statocyst (balancing organ); V, velum; VC, velar cilia; Y, residual yolk; ES, everted stomodeum; DV, disorganized velum. (From Newrock and Raff, 1975; courtesy of K. Newrock.)

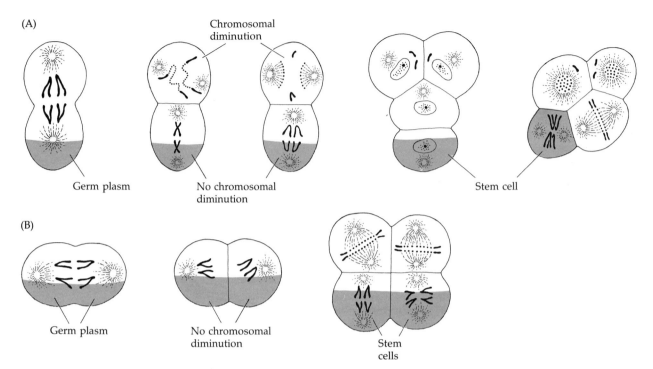

FIGURE 12
Distribution of germ plasm (shaded) during the cleavage of normal (A) and
centrifuged (B) zygotes of *Ascaris*. The germ plasm is normally conserved
in the most vegetal blastomere as shown by the lack of chromosomal diminution in that particular cell. Thus, at the 4-cell stage, the embryo has one stem
cell for its gametes. When the first cleavage is displaced 90 degrees by centrifugation, both resulting cells have vegetal germ plasm and neither cell undergoes chromosomal diminution. After second cleavage, these two cells give
rise to germinal stem cells. (After Waddington, 1966.)

into dozens of pieces just before the cleavage of this blastomere. This
is called CHROMOSOME DIMINUTION because only a portion of the original
chromosome survives. Numerous genes are lost in these cells when the
chromosome fragments are not included in the newly formed nuclei
(Tobler et al., 1972). Meanwhile, in the vegetal blastomere, the chromosomes remain normal. During second division, the animal cell splits
meridionally while the vegetal cell divides equatorially. Both vegetally
derived cells have normal chromosomes. However, the chromosomes
of the more animally located of these two vegetal blastomeres fragment
before third division. Thus, at the 4-cell stage, only one cell—the most
vegetal—contains a full set of genes. At successive cleavages, somatic
nuclei are given off from this line, until at the 16-cell stage, there are
only two cells that have undiminished chromosomes. One of these two
blastomeres gives rise to the germ cells; the other blastomere eventually
undergoes chromosome diminution and forms somatic cells. The chro-

mosomes are kept intact only in those cells destined to form the germ line. If this were not the case, the genetic information would degenerate from one generation to the next. The cells that have undergone chromosome diminution generate the somatic cells.

Boveri has been called the last of the great "observers" of embryology and the first of the great experimenters. (Both Spemann and Wilson dedicated their major books to him.) Not content with observing the retention of the full chromosome complement solely by the germ cell precursors, he set out to test whether a specific region of cytoplasm protects the nuclei within it from diminution. If so, any nucleus happening to reside in this region should be protected. Boveri (1910) tested this by centrifuging the *Ascaris* eggs shortly before their first cleavage. This treatment shifted the orientation of the mitotic spindle (just as Driesch had done in sea urchin embryos by using glass plates). When the spindle forms perpendicularly to its normal orientation, both resulting blastomeres should contain some of the vegetal cytoplasm (Figure 12B). Indeed, Boveri found that after the first division neither nucleus underwent chromosomal diminution. However, the next division was equatorial along the animal–vegetal axis. Here the resulting "animal" blastomeres both underwent diminution whereas the two vegetal cells did not. Boveri concluded that the vegetal cytoplasm contained a factor (or factors) that protected nuclei from chromosomal diminution and determined them to be germ cells.

Germ cell determination in insects

Certain insect eggs also contain a germ plasm that appears to act very similarly to the one observed in *Ascaris*. In the midge *Wachtiella persicariae*, most nuclei lose 32 of their original 40 chromosomes! However, two undiminished nuclei are found at the posterior pole of the egg and do not divide for a period of time (Figure 13). These two nuclei eventually give rise to the germ cells.[4] When nuclei are prevented by a ligature from migrating into the posterior pole region, every nucleus undergoes diminution, and the resulting midge is sterile. When the ligature is loosened and a diminished nucleus enters the posterior pole, germ cells are made but never differentiate into functional gametes (Geyer-Duszynska, 1959). Kunz and co-workers (1970) have shown that the eliminated chromatin contains genes that are active during germ cell production.

The germinal cytoplasm of insects is different from that of any other cytoplasm in the egg. Hegner (1911) found that when he removed or destroyed this region of beetle eggs before pole cell formation had

---

[4]This and the *Ascaris* example sound like perfect evidence for Weismann's hypothesis of the segregation of nuclear determinants. These cases of chromsomal diminution and elimination are exceptions to the general rule (Chapters 9 and 10) that the nuclei of differentiated cells retain unused genes. Moreover, there is no evidence that different somatic cells in *Wachtiella* or *Ascaris* retain different parts of the genome.

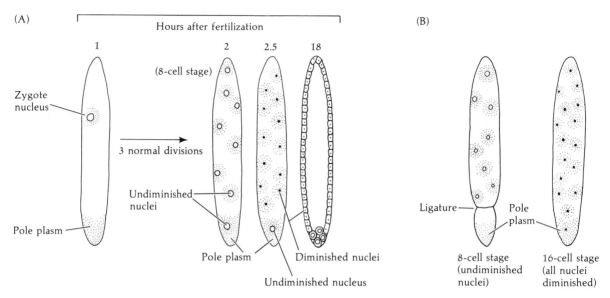

FIGURE 13

Segregation of the germ line and chromosome diminution in *Wachtellia*. (A) Chromosomal diminution depends on location. The first three divisions are normal. Thereafter all nuclei except the nucleus at the posterior pole lose 32 of their 40 chromosomes. (B) When the zygote is ligated such that no nucleus enters the pole region at the 8-cell stage, all nuclei undergo diminution at the 16-cell stage, even if the connection to the pole plasm is reestablished. (Adapted from Geyer-Duszynska, 1959.)

FIGURE 14

Electron micrograph of polar granules from particulate fraction from *Drosophila* pole cells. (Courtesy of A. P. Mahowald.)

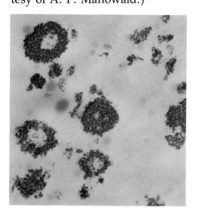

occurred the resulting embryos had no germ cells and were sterile. This posterior POLE PLASM is conveniently marked with POLAR GRANULES (Figure 14). Although there is not much direct evidence for their role in germ cell determination, their constant association with the pole plasm and the POLE CELLS derived from it makes them a convenient marker for this region. This region of pole cells is easily identifiable under a scanning electron microscope (Figure 15).

Recent work on pole cell cytoplasm has focused primarily on *Drosophila* embryos. For the first two hours after fertilization, the *Drosophila* embryo develops as a SYNCYTIUM. Here the nuclei divide without any corresponding cellular division, eventually forming a BLASTODERM layer containing some 3500 nuclei. Each nucleus is eventually enclosed in a cellular membrane, creating the cellular blastoderm. Transplantation experiments (Zalokar, 1971; Illmensee, 1968, 1972) have shown that the SYNCYTIAL NUCLEI are each equivalent and totipotent. The cells of the CELLULAR BLASTODERM, however, are precisely determined. A similar conclusion was reached by Schubiger and Wood (1977), who ligated *Drosophila* eggs during various stages of their development. *Drosophila* nuclei do not undergo diminution, and any syncytial-stage nucleus can give rise to germ cells or to somatic cells. Their determination depends upon the region of the egg into which they migrate. One of these regions is the posterior pole plasm. Again, this pole plasm appears to contain the morphogenic determinants for germ cell production.

Geigy (1931) showed that irradiating the pole plasm with ultraviolet light produced sterile flies; and Okada and co-workers (1974) extended this line of experimentation by showing that the addition of pole plasm from unirradiated donor embryos could cure the sterility of irradiated eggs (Figure 16). No other part of the cytoplasm could accomplish this reversal of sterility.

The autonomy of this cytoplasmic region and its ability to determine

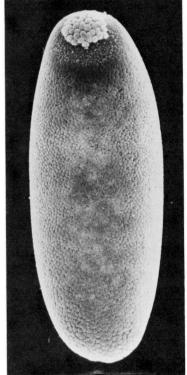

FIGURE 16
Ability of pole plasm to correct radiation-induced sterility. (A) Technique of pole plasm transplantation from unirradiated donor to irradiated host. (B) Longitudinal sections of the posterior portion of the *Drosophila* embryo fixed at the completion of cleavage. (i) Normal embryo with complete blastoderm and pole cells. (ii) Embryo that was irradiated during early cleavage. Blastoderm has formed, but pole cells are absent. (iii) Embryo irradiated during early cleavage but subsequently injected with pole plasm from normal embryos. Polar bodies and blastoderm are both seen. (From Okada et al., 1974.)

(A)

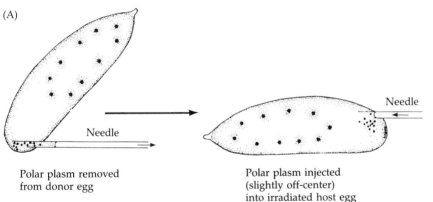

Polar plasm removed
from donor egg

Polar plasm injected
(slightly off-center)
into irradiated host egg

Needle

FIGURE 15
Scanning electron micrograph of the *Drosophila* embryo just prior to the completion of cleavage. The pole cells can be seen at the upper end of this picture. (Photograph courtesy of A. P. Mahowald.)

(B)

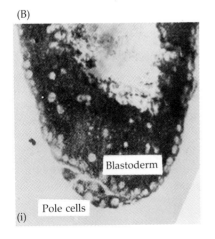

(i) Blastoderm / Pole cells

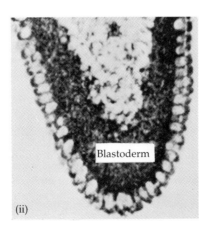

(ii) Blastoderm

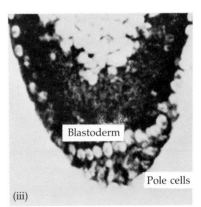

(iii) Blastoderm / Pole cells

any *Drosophila* nucleus was shown in 1974 by the ingenious experiments of Karl Illmensee and Anthony Mahowald. In these experiments (outlined in Figure 17), an incredibly small amount (5–100 picoliters) of anucleate pole plasm was transferred from wild-type *Drosophila* eggs into the *anterior* pole of genetically marked eggs, prior to their cellularization. (These donor eggs carried the chromosomal mutations *multiple wing hair* and *ebony*.) After the cellular blastoderm had formed, the cells in the anterior pole of the donor embryo resembled normal posterior pole cells, having incorporated the polar granules and developed a typical pole cell morphology. In order to test whether or not these cells had become functioning germ cell precursors, Mahowald and Illmensee transplanted these anterior cells into the posterior region of cleaving embryos containing their own genetically marked pole cells.[5] These new host embryos were marked by different mutations (recessives *yellow*, *white*, and *singed*). When these embryos developed, they all became flies bearing the mutations *yellow*, *white*, and *singed*. These flies were mated to other flies carrying the same mutations. In most cases (88/92), these matings produced individuals identical to both parents. However, in four cases, there emerged wild-type progeny, indicating that some of the germ cells in these flies derived from the transplanted cells; the germ plasm from one embryo was able to cause the anterior nuclei of another embryo to develop into functioning germ cells! This technique has also proved useful in showing when this determinant is localized. Illmensee and his colleagues (1976) found that the germ cell determinant is able to function before fertilization and gets localized in the developing oocyte at about the same time the yolk reaches the posterior of the egg.

Nature has also provided a confirmation of the importance of both pole plasm and the polar granules. Female *Drosophila* homozygous for the *grandchildless* mutation produce normal, but sterile, offspring [*GG* (male) × *gg* (female) → *Gg* (sterile)]. Mahowald and colleagues (1979) have shown that when such females are mated with normal males, the nuclei of the resulting embryos never migrate into the pole plasm of the egg. No pole cells are formed and the resulting adults have no primordial germ cells to produce gametes. Another maternal-effect mutation—*agametic*—causes the absence of germ cells in about

---

[5]The reason for doing this was so that these cells may get incorporated into the developing gonads. There is evidence from other organisms that the pole plasm also contains determinants for the proper *migration* of germ cells (Züst and Dixon, 1977; Ikenishi and Kotani, 1979).

FIGURE 17
Ability of germ plasm to determine fate of cells which contain it. Pole plasm from wild-type *Drosophila* eggs (white) is transplanted into the anterior (but not the posterior) pole of genetically marked (mutant) embryos (dark gray). The cells in the anterior pole resemble the normal germ cell precursors seen at the posterior pole. These anterior pole cells are then transplanted into host posterior regions (marked with different mutations; light gray) so that their

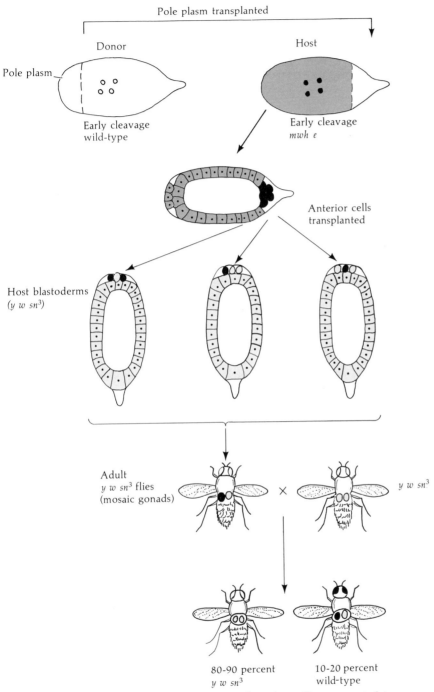

descendants might migrate to the gonads. When these flies are mated to other flies with the second series of mutations (which did not have transplanted cells), some of their progeny were wild-type, indicating their germ cells came from a nonparental strain of fly. (Modified from Illmensee and Mahowald, 1974.)

half the gonads of offspring derived from homozygous female flies. Here, the normal number of pole cells form, but the polar granules degenerate shortly after fertilization (Engstrom et al., 1982). Transplantation experiments demonstrate that the defect is in the polar cytoplasm and not in the ovarian environment. Thus, we now have fairly strong evidence that the polar granules are directly concerned with germ cell production.

In *Drosophila*, these polar granules have been isolated (Waring et al., 1978) and appear to be composed of both protein and RNA. The protein (there only seems to be one major protein in the granules) has a molecular weight of 95,000; it also has basic properties. It is synthesized anew during oogenesis, so the polar granules are not inherited directly through the egg cytoplasm. The RNA component is present in the oocyte polar granules but is no longer seen once the pole cells are formed (Mahowald, 1971a, b). Other changes happen to these granules, too. Before fertilization, the dense, membranous granules cluster around the mitochondria and then disperse before the nuclei reach the pole. The pole cells eventually incorporate these granules. When the pole cells migrate toward the gonadal ridge, the granules further decondense into filaments called NUAGE. This material clusters about the nuclear envelope. Similar changes occur in amphibian polar granules, and nuage material is seen to attach to nuclear membrane-associated mitochondria. This behavior is seen throughout the animal kingdom (Eddy, 1975) and is probably important in gametogenesis.

Germ cell determination in amphibians

Cytoplasmic localization of germ cell determinants has also been observed in vertebrate embryos. Bounoure (1934) showed that the ventral region of fertilized frog eggs contains a material with staining properties similar to that of *Drosophila* pole plasm (Figure 18). He was able to trace this cortical cytoplasm into those few cells in the presumptive endoderm that would normally migrate into the genital ridge. Blackler (1962) showed that these cells are the primordial germ cell precursors. Blackler removed the endodermal region from the neurula of one frog and inserted it into the neurula endoderm of another (Figure 19). The donor frogs were genetically marked in that their cells contained only one nucleolus instead of the usual two. The host frogs were normal (i.e., two nucleoli per nucleus). After the operation, the donor frogs were sterile, indicating that the neurula could not regulate for the production of germ cells and that the correct region had been excised. The host frogs, however, were fertile. Moreover, when these frogs were mated to normal adults, the offspring were a mixture of one-nucleolus and two-nucleoli frogs. Therefore, some of the gametes came from the endodermal region of the *donor* embryo. (If none came from the donor, one would get 100 percent two-nucleoli frogs. If every germ cell were derived from the donor, one would expect a 1:1 ratio, as is shown in

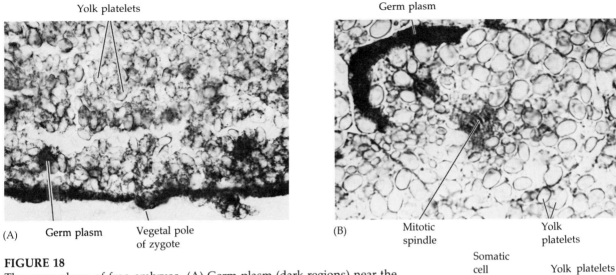

Yolk platelets

Germ plasm

(A)   Germ plasm   Vegetal pole of zygote

(B)   Mitotic spindle   Yolk platelets

FIGURE 18

The germ plasm of frog embryos. (A) Germ plasm (dark regions) near the ventral pole of a newly-fertilized zygote. (B) Germ plasm-containing cell in endodermal region of blastula in mitotic anaphase. Note the germ plasm entering into only one of the two yolk-laden daughter cells. (C) Primordial germ cell and somatic cells near the floor of the blastocoel in early gastrula. (Courtesy of A. Blackler.)

Somatic cell   Yolk platelets

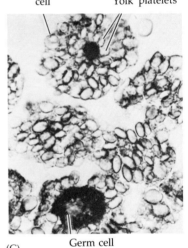

(C)   Germ cell

Figure 19B). In several experiments, this latter figure was actually obtained.)

With his associates, Bounoure (1939) showed that when ultraviolet light is applied to the ventral poles of fertilized frog eggs, the resulting adults are sterile but otherwise normal. In 1966, L. Dennis Smith was able to extend these studies by irradiating various regions of the frog embryo with ultraviolet light. He found that ultraviolet irradiation of the animal hemisphere did not affect normal development, whereas irradiation of the vegetal pole produced tadpoles that lacked germ cells. No large primordial germ cells were seen to enter the genital ridge. Smith showed the importance of this region of cytoplasm by transferring the cytoplasm from normal zygotes into the ventral region of irradiated zygotes. When the animal region cytoplasm was added to the irradiated ventral regions, there was no effect, and the resulting tadpoles lacked the large germ cells. However, when vegetal pole cytoplasm was added, primordial germ cells were ultimately seen in the genital ridge. More recent studies (Züst and Dixon, 1977) have shown that primordial germ cells do arrive in the genital ridge of tadpoles whose germ plasm was irradiated at the 2- or 4-cell stage. However, there are significantly fewer of these cells, they are only about one-tenth the size of normal primordial germ cells, they have aberrantly shaped nuclei, and they enter the genital ridge much later than their counterparts from nonirradiated eggs. So, like the *Drosophila* pole plasm, frog zygotes contain a deter-

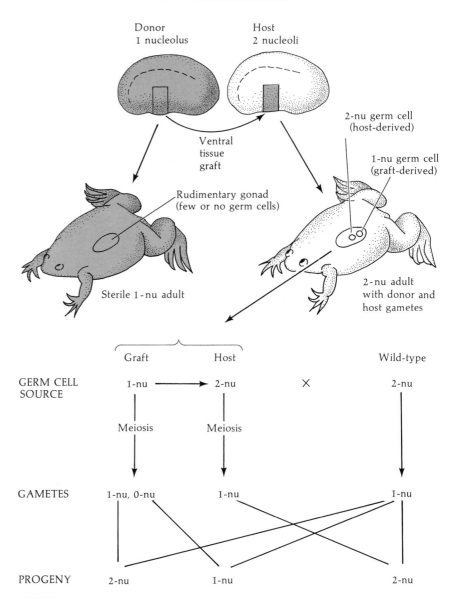

FIGURE 19

Demonstration of primordial germ cells in early tadpole endoderm. A block of ventral tissue from a strain of frog having one nucleolus per cell is transplanted into the analogous region in a host tadpole having two nucleoli per cell. The donor tadpole metamorphoses into a sterile adult, while the host tadpole develops into a frog with both host and donor gametes. This can be seen by crossing that frog with a normal (two-nucleoli) adult; the results of such a cross are shown in the graph. Some of the offspring have only one nucleolus per cell, indicating that they received one nucleolus from one parent (the two-nucleoli frog) and no nucleolus from the other (the two-nucleoli frog with one-nucleolus germ cells). (After Blackler, 1966.)

minant for germ cell formation and migration that is sensitive to ultraviolet irradiation and that can be transferred through the fluid portion of the cytoplasm.

The nature of cytoplasmic determinants

The ability to transfer functional germ plasm from prefertilization *Drosophila* eggs and from recently fertilized frog zygotes demonstrates that such factors must be placed in the oocyte during gametogenesis. Nevertheless, there are several ways that developmental information could be stored. Although the muscle determinant of tunicates has not been identified as yet, one of its actions appears to be the selective activation of certain genes. Actinomycin D is a drug that prevents the transcription of RNA from its DNA template. If tunicate embryos are grown in actinomycin D, no acetylcholinesterase activity emerges. Moreover, when RNA is purified from different stages of *Ciona* embryos and injected into frog oocytes, the oocytes will translate the newly inserted messages. By this technique, Meedel and Whittaker (1983) found that there is no detectable mRNA for acetylcholinesterase until the gastrula stage. Therefore, new RNA synthesis has to occur for the production of this enzyme. This is not the case, however, when one looks at the synthesis of intestinal alkaline phosphatase in tunicate embryos. This enzyme is synthesized by the larval endoderm and becomes segregated into those cells that form the gut (Figure 20). Actinomycin D has no effect on the synthesis of this enzyme, although inhibitors of translation prevent its expression (Whittaker, 1977). Therefore, mRNA for intestinal alkaline phosphatase must already exist in the oocyte cytoplasm. Somehow, the mRNA itself must get sequestered into the appropriate blastomere.

The polar lobe of snail embryos may preferentially harbor certain mRNA sequences of regulators of their translation. The small polar lobes of certain snails appear to be heavily endowed with RNA (Figure 21); and Brandhorst and Newrock (1981) have shown large quantitative differences between the proteins synthesized in normal and lobeless embryos of *Ilyanassa* when transcription has been eliminated by actinomycin D. Collier and McCarthy (1981) have suggested that the polar lobe contains regulators of both transcription and translation, because the removal of the polar lobe from actinomycin-treated embryos actually

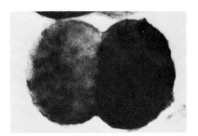

(A)

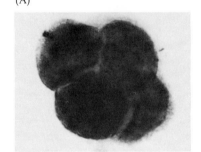

(B)

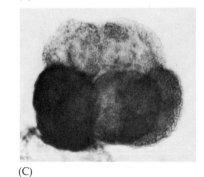

(C)

FIGURE 20
Alkaline phosphatase localization in cytochalasin B-arrested *Ciona* embryos. (A) 2-cell; (B) 4-cell; (C) 8-cell; (D) 16-cell. Staining reactions 14–16 hours after fertilization reveal localization of alkaline phosphatase development. (From Whittaker, 1977; photographs courtesy of J. R. Whittaker.)

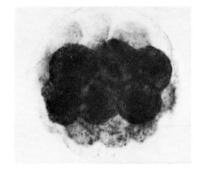

(D)

FIGURE 21
First cleavage embryo of *Bithynia tentaculata* stained with methyl green-pyronin for RNA. (This can be compared with Figure 9B, which was stained with hematoxylin-eosin.) The polar lobe is heavily stained, indicating a large RNA concentration. (From Dohmen and Verdonk, 1974; photograph courtesy of M. R. Dohmen.)

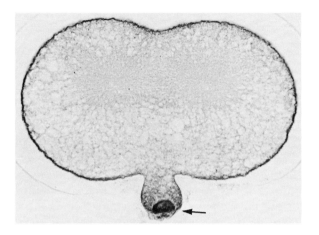

causes the increased translation of certain messages. Nor is such sequestering of messages confined to mosaic embryos. Rodgers and Gross (1978) have grown sea urchin embryos in actinomycin D and then have separated the blastomeres. They find that the mRNA sequences in the micromeres differ from those in the mesomeres.

Another example of localized mRNAs serving as morphogenic regulators comes from studies of insect eggs. In insects, the anterior–posterior axis is defined during oogenesis. In the anterior end, a head region is formed, followed by three thoracic segments. Eight abdominal segments form in the posterior half of the egg, and the germ cell pole cytoplasm marks the posterior end. When the anterior portion of the egg of the midge *Smittia* is exposed to ultraviolet light of a wavelength capable of inactivating RNA (265 and 285 nm), the resulting embryo lacks head and thorax and instead develops a mirror-image duplication of the posterior end (without the germ cells). The double-abdomen embryos made by this technique (Kalthoff and Sander, 1968) are shown in Figure 22.

In an experiment that confirms RNA as the anterior determinant, Kandler-Singer and Kalthoff (1976) submerged *Smittia* embryos in solutions of different enzymes and punctured the embryo in specific regions. Double abdomens resulted when ribonuclease (an enzyme used to digest RNA) was permitted to enter the anterior end. Other enzymes admitted into the anterior end did not cause this abnormality, nor did ribonuclease effect this change if it entered other regions of the embryo. Thus, it appears that a specific type of RNA is a morphogenic determinant for the anterior region of certain insect eggs.

The dorsal-ventral axis of insects may also be specified by maternally stored RNA. In *Drosophila*, there exists a maternal effect mutant called *snake*. Females homozygous for this mutation produce eggs that develop into embryos that are completely dorsal. Such embryos lack the structures and cuticle typical of the larval belly. These eggs can be "rescued" by the injection of a small amount of cytoplasm from wild-

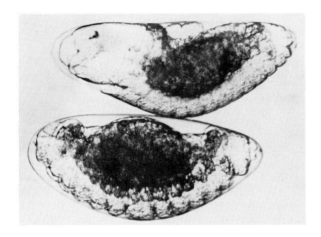

FIGURE 22
Normal and irradiated embryos of the midge *Smittia*. The normal embryo (top) shows a head on the left and abdominal segments on the right. The irradiated embryo has no head region, but has abdominal segments at both ends. (From Kalthoff, 1969; photograph courtesy of K. Kalthoff.)

type eggs. The active component of this wild-type cytoplasm has been shown to be stored mRNA, since when the mRNA is extracted from the wild-type egg and injected into early cleavage stage mutant embryos, the ventral structures and cuticle develop normally (Anderson and Nüsslein-Volhard, 1984). Here, then, is a situation in which a maternal gene is coding for an mRNA necessary for the dorso-ventral differentiation. In the absence of this product, all the cells develop dorsally, and when the RNA is supplied to deficient embryos their normal development is restored. In 1936 embryologist E. E. Just criticized those geneticists who sought to explain development by looking at specific mutations affecting eye color, bristle number, and wing shape. He said that he wasn't interested in the development of the bristles on a fly's back; rather, he wanted to know how the fly embryo made the back itself. Fifty years later, embryologists and geneticists are finally meeting to answer that question.

Summary

We have evidence, then, for two major mechanisms of embryonic determination. In one type, the determination of a cell's fate is due to the portion of egg cytoplasm that it acquires during cleavage. This cell differentiates independently of other cells. Organisms that utilize such a mechanism tend toward a "mosaic" type of development. The second type of determination is caused by the interaction of cells later in development. Here, cells are dependent upon their new positions in the embryo, and organisms whose cells are so determined tend toward a "regulative" type of development. Thus, a cell of the most animal tier of an early sea urchin embryo will normally give rise to an ectodermal cell; but if it is combined with micromeres, it will produce gut tissue characteristic of the endoderm. It is important to note that both mech-

FIGURE 23
The continuum of mosaic and regulative development. Different organisms have different amounts of regulative potential and mosaic determination. (David Kirk, personal communication.)

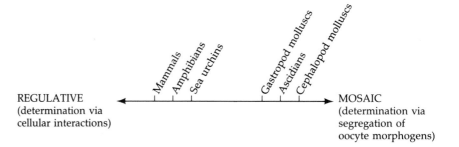

anisms are used during the development of most any organism and that there is a continuum from mosaic to regulative development (Figure 23). "Mosaic" embryos such as tunicates and snails still use cellular interactions to form certain types of tissues, and "regulative" embryos such as frogs have obvious regions of cytoplasm that contain morphogenic determinants such as polar granules. It is also important to note that both "mosaic" and "regulative" styles of determination must be traced back to the heterogeneously distributed materials in the zygote cytoplasm. The mechanisms by which morphogenic determinants control gene regulation and the ways in which these morphogens become localized within the egg are still to be determined.

Literature cited

Anderson, K. V. and Nüsslein-Volhard, C. (1984). Information for the dorsal-ventral pattern of the *Drosophila* embryo is stored as maternal mRNA. *Nature* 311: 223–227.

Blackler, A. W. (1962). Transfer of primordial germ cells between two subspecies of *Xenopus laevis. J. Embryol. Exp. Morphol.* 10: 641–651.

Blackler, A. W. (1966). The role of a "germinal plasm" in the formation of primordial germ cells in *Rana pipiens. Dev. Biol.* 14: 330–347.

Bounoure, L. (1934). Recherches sur la lignée germinale chez la grenouille rousse aux premiers stades du développement. *Ann. Sci. Natl. 10e wer.* 17: 67–248.

Bounoure, L. (1939). *L'origine des cellules reproductries et le probleme de la lignée germinale.* Gauthier-Villars, Paris.

Boveri, T. (1910). Über die Teilung centrifugierter Eier von *Ascaris megalocephala. Wilhelm Roux Arch. Entwicklungsmech. Org.* 30: 101–125.

Brandhorst, B. P. and Newrock, K. M. (1981). Post-transcriptional regulation of protein synthesis in *Ilyanassa* embryos and isolated polar lobes. *Dev. Biol.* 83: 250–254.

Cervera, M., Dreyfuss, G. and Penman, S. (1981). Messenger RNA is translated when associated with the cytoskeletal framework in normal and VSV-infected HeLa cells. *Cell* 23: 113–120.

Churchill, F. B. (1973). Chabry, Roux, and the experimental method in nineteenth century embryology. In R. N. Giere and R. S. Westfall (eds.), *Foundations of Scientific Method, The Nineteenth Century.* Indiana University Press, Bloomington, pp. 161–205.

Clement, A. C. (1962). Development of *Ilyanassa* following removal of the D macromere at successive cleavage stages. *J. Exp. Zool.* 149: 193–215.

Clement, A. C. (1968). Development of the vegetal half of the *Ilyanassa* egg after removal of most of the yolk by centrifugal force, compared with the development of animal halves of similar visible composition. *Dev. Biol.* 17: 165–186.

Collier, J. R. and McCarthy, M. E. (1981). Regulation of polypeptide synthesis during early embryogenesis of *Ilyanassa obsoleta. Differentiation* 19: 31–46.

Conklin, E. G. (1905a). The organization and cell-lineage of the ascidian egg. *J. Acad. Nat. Sci. Phila.* 13: 1–119.

Conklin, E. G. (1905b). Organ-forming substances in the eggs of ascidians. *Biol. Bull.* 8: 205–230.

Conklin, E. G. (1905c). Mosaic development in ascidian

eggs. *J. Exp. Zool.* 2: 145–223.

Deno, T., Nishida, H. and Satoh, N. (1984). Autonomous muscle cell differentiation in partial ascidian embryos according to the newly verified cell lineages. *Dev. Biol.* 104: 322–328.

Dohmen, M. R. and Verdonk, N. H. (1974). The structure of a morphogenetic cytoplasm present in the polar lobe of *Bithynia tentaculata* (Gastropoda, Prosobranchia). *J. Embryol. Exp. Morphol.* 31: 423–433.

Eddy, E. M. (1975). Germ plasm and the differentiation of the germ line. *Int. Rev. Cytol.* 43: 229–280.

Engstrom, L., Caulton, J. H., Underwood, E. M. and Mahowald, A. P. (1982). Developmental lesions in the agametic mutant of *Drosophila melanogaster*. *Dev. Biol.* 91: 163–170.

Geigy, R. (1931). Action de l'ultra-violet sur le pôle germinal dans l'oeuf de *Drosophila melanogaster* (castration et mutabilite). *Rev. Suisse Zool.* 38: 187–288.

Geyer-Duszynska, I. (1959). Experimental research on chromosome diminution in *Cecidomiidae* (Diptera). *J. Exp. Zool.* 141: 381–441.

Hegner, R. W. (1911). Experiments with chrysomelid beetles. III. The effects of killing parts of the eggs of *Leptinotarsa decemlineata*. *Biol. Bull.* 20: 237–251.

Ikenishi, K. and Kotani, M. (1979). Ultraviolet effects on presumptive primordial germ cells (pPGCs) in *Xenopus laevis* after the cleavage stage. *Dev. Biol.* 69: 237–246.

Illmensee, K. (1968). Transplantation of embryonic nuclei into unfertilized eggs of *Drosophila melanogaster*. *Nature* 219: 1268–1269.

Illmensee, K. (1972). Developmental potencies of nuclei from cleavage, preblastoderm, and syncytial blastoderm transplanted into unfertilized eggs of *Drosophila melanogaster*. *Wilhelm Roux Arch. Entwicklungsmech. Org.* 170: 267–298.

Illmensee, K. and Mahowald, A. P. (1974). Transplantation of posterior polar plasm in *Drosophila*. Induction of germ cells at the anterior pole of the egg. *Proc. Natl. Acad. Sci. USA* 71: 1016–1020.

Illmensee, K., Mahowald, A. P. and Loomis, M. R. (1976). The ontogeny of germ plasm during oogenesis in *Drosophila*. *Dev. Biol.* 49: 40–65.

Jeffery, W. R. and Meier, S. (1983). A yellow crescent cytoskeletal domain in ascidian eggs and its role in early development. *Dev. Biol.* 96: 125–143.

Just, E. E. (1936). Quoted in Harrison, R. G. (1937). Embryology and its relations. *Science* 85: 369–374.

Kalthoff, K. (1969). Der Einfluss verscheidener Versuchsparameter auf die Häufigkeit der Missbildung "Doppelabdomen" in UV-bestrahlten Eiern von *Smittia* sp. (Diptera, Chironomidae). *Zool. Anz. Suppl.* 33: 59–65.

Kalthoff, K. and Sander, K. (1968). Der Entwicklungsgang der Missbildung "Doppelabdomen" im partiell UV-bestrahlten Ei von *Smittia parthenogenetica* (Dipt., Chironomidae). *Wilhelm Roux Arch. Entwicklungsmech. Org.* 161: 129–146.

Kandler-Singer, I. and Kalthoff, K. (1976). RNase sensitivity of an anterior morphogenetic determinant in an insect egg (*Smitta* sp., Chironomidae, Diptera). *Proc. Natl. Acad. Sci. USA* 73: 3739–3743.

Kunz, W., Trepte, H. H., and Bier, K. (1970). On the function of the germ line chromosomes in the oogenesis of *Wachtiella persiariae* (Cecidomyiidae). *Chromosoma* 30: 180–192.

Mahowald, A. P. (1971a). Polar granules of *Drosophila*. III. The continuity of polar granules during the life cycle of *Drosophila*. *J. Exp. Zool.* 176: 329–343.

Mahowald, A. P. (1971b). Polar granules of *Drosophila*. IV. Cytochemical studies showing loss of RNA from polar granules during early stages of embryogenesis. *J. Exp. Zool.* 176: 345–352.

Mahowald, A. P., Caulton, J. H. and Gehring, W. J. (1979). Ultrastructural studies of oocytes and embryos derived from female flies carrying the *grandchildless* mutation in *Drosophila subobscura*. *Dev. Biol.* 69: 118–132.

Meedel, T. H. and Whittaker, J. R. (1983). Development of transcriptionally active mRNA for larval muscle acetylcholinesterase during ascidian embryogenesis. *Proc. Natl. Acad. Sci. USA* 80: 4761–4765.

Moon, R. T., Nicosia, R. F., Olsen, C., Hille, M. and Jeffery, W. R. (1983). The cytoskeletal framework of sea urchin eggs and embryos: Developmental changes in the association of messenger RNA. *Dev. Biol.* 95: 447–458.

Morgan, T. H. (1927). *Experimental Embryology*. Columbia University Press, New York.

Newrock, K. M. and Raff, R. A. (1975). Polar lobe specific regulation of translation in embryos of *Ilyanassa obsoleta*. *Dev. Biol.* 42: 242–261.

Okada, M., Kleinman, I. A. and Schneiderman, H. A. (1974). Restoration of fertility in sterilized *Drosophila* eggs by transplantation of polar cytoplasm. *Dev. Biol.* 37: 43–54.

Ortolani, G. (1959). Ricerche sulla induzione del sistema nervoso nelle larve delle Ascidie. *Boll. Zool.* 26: 341–348. (Quoted in Reverberi, 1971, p. 539).

Reverberi, G. (1971). *Experimental Embryology of Marine and Freshwater Invertebrates*. Elsevier, New York.

Reverberi, G. and Minganti, A. (1946). Fenomeni di evocazione nello sviluppo dell'uovo di Ascidie. Risultati dell'indagine sperimentale sull'uovo di *Asciadiella aspersa* e di *Ascidia malaca* allo stadio di 8 blastomeri. *Pubbl. Staz. Zool. Napoli* 20: 199–252. (Quoted in Reverberi, 1971, p. 537).

Rodgers, W. H. and Gross, P. R. (1978). Inhomogeneous distribution of egg RNA sequences in the early embryo. *Cell* 14: 279–288.

Satoh, N. (1979). On the "clock" mechanism determining the time of tissue-specific enzyme development during ascidian embryogenesis. I. Acetylcholinesterase development in cleavage-arrested embryos. *J. Embryol. Exp. Morphol.* 54: 131–139.

Schubiger, G. and Wood, W. J. (1977). Determination during early embryogenesis in *Drosophila melanogaster. Am. Zool.* 17: 565–576.

Smith, L. D. (1966). The role of a "germinal plasm" in the formation of primordial germ cells in *Rana pipiens. Dev. Biol.* 14: 330–347.

Tobler, H., Smith, K. D. and Ursprung, H. (1972). Molecular aspects of chromatin elimination in *Ascaris lumbricoides. Dev. Biol.* 27: 190–203.

Tung, T. C., Wu, S. C., Yeh, Y. F., Li, K. S. and Hsu, M. C. (1977). Cell differentiation in ascidians studied by nuclear transplantation. *Scientia Sinica* 20: 222–233.

Waddington, C. H. (1966). *Principles of Development and Differentiation.* Macmillan, New York.

Waring, G. L., Allis, C. D. and Mahowald, A. P. (1978). Isolation of polar granules and the identification of polar granule-specific protein. *Dev. Biol.* 66: 197–206.

Whittaker, J. R. (1973a). Segregation during ascidian embryogenesis of egg cytoplasmic information for tissue-specific enzyme development. *Proc. Natl. Acad. Sci. USA* 70: 2096–2100.

Whittaker, J. R. (1973b). Evidence for the localization of an RNA template for endodermal alkaline phosphatase in an ascidian egg. *Biol. Bull.* 145: 459–460.

Whittaker, J. R. (1977). Segregation during cleavage of a factor determining endodermal alkaline phosphatase development in ascidian embryos. *J. Exp. Zool.* 202: 139–153.

Whittaker, J. R., Ortolani, G. and Farinella-Ferruzza, N. (1977). Autonomy of acetylcholinesterase differentiation in muscle lineage cells in asicidian embryos. *Dev. Biol.* 55: 196–200.

Wilson, E. B. (1904). Experimental studies on germinal localization. I. The germ regions in the egg of *Dentalium.* II. Experiments on the cleavage-mosaic in *Patella* and *Dentalium. J. Exp. Zool.* 1: 1–72.

Zalokar, M. (1971). Transplantation of nuclei in *Drosophila melanogaster. Proc. Natl. Acad. Sci. USA* 68: 1539–1541.

Zumbe, A., Stahli, C. and Trachsel, H. (1982). Association of a M_r 50,000 cap-binding protein with the cytoskeleton in baby hamster kidney cells. *Proc. Natl. Acad. Sci. USA* 79: 2927–2931.

Zust, B. and Dixon, K. E. (1977). Events in the germ cell lineage after entry of the primordial germ cells into the genital ridges in normal and UV-irradiated *Xenopus laevis. J. Embryol. Exp. Morphol.* 41: 33–46.

Genomic equivalence and differential gene expression: embryological investigations

Heredity is effected by the transmission of a nuclear preformation which in the course of development finds expression in a process of cytoplasmic epigenesis.
—E. B. WILSON (1925)

Two cells are differentiated with respect from one another if, while they harbor the same genome, the pattern of proteins they synthesize is different.
—F. JACOB AND J. MONOD (1963)

Introduction

Developmental genetics is the study of how the inherited potential of the fertilized egg becomes expressed during the life of an organism. Observing the developing embryo, it becomes apparent that different cell types are expressing different genes. Hemoglobin, for instance, characterizes the red blood cell, whereas keratin is found only in the epidermal cells of our skin. The cells of the neural retina are capable of transmitting electrical impulses across large distances, whereas the adjacent cells of the pigmented retina are darkened by melanin granules and lack electrical conductivity. Yet each of these cell types arose from the mitotic divisions of the same fertilized egg. Each should contain the same nuclear information. Development, then, involves the differential expression of specific genes at specific places and times. The problem of developmental genetics then becomes: How is the genetic information regulated such that cells become different?

The central hypothesis of developmental genetics has been that cell differentiation occurs in the absence of genetic alteration. Thus, within any organism, each somatic cell is postulated to contain the identical complement of genes. Each different cell type would then utilize different genes from their common inherited repertoire. The basis for this hypothesis of DIFFERENTIAL GENE EXPRESSION comes from both genetics and embryology. In this chapter we shall look at those studies seeking to determine whether or not the genome has undergone irreversible changes during development.

Genomic equivalence

Certain large, nondividing cells of larval flies such as *Drosophila* and *Chironomus* contain POLYTENE CHROMOSOMES. Such chromosomes undergo DNA replication in the absence of mitosis and therefore contain 512, 1024, or even more parallel DNA double helices instead of just one (Figures 1 and 2). These chromosomes never undergo mitosis and are visible under the light microscope, where they are seen to exhibit characteristic banding patterns. In *Drosophila* approximately 5150 individual bands have been counted in the haploid genome. In some tissues, thick bands are seen, which, upon stretching, resolve into two or more thinner bands. Several genetic studies (see Judd and Young, 1973) have suggested a correlation between the number of these bands, called CHROMOMERES, and the number of genes in the fly (Swanson et al., 1981). Beermann (1952) demonstrated that these chromosomes and their banding patterns were constant throughout the larval organism (Figure 3) and that no loss of any chromosomal region could be seen when different cell types were compared. When it became possible to study the individual chromosomes of vertebrate nuclei, Tjio and Puck (1958)

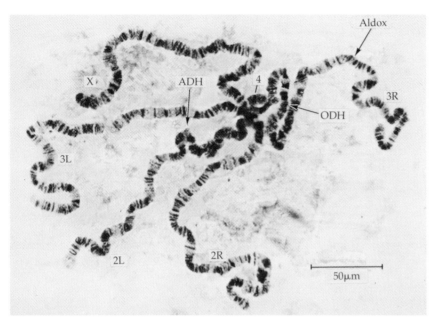

FIGURE 1

Polytene chromosomes from the salivary gland cells of *Drosophila melanaogaster*. The four chromosomes are connected at their centromere regions, forming a dense chromocenter. The structural genes for alcohol dehydrogenase (ADH), aldehyde oxidase (aldox), and octanol dehydrogenase (ODH) have been mapped to the assigned positions on these chromosomes. (From Ursprung et al., 1968; photograph courtesy of H. Ursprung.)

FIGURE 2

Electron microscopic views of the band (dark) and interband (light) regions of *Drosophila* polytene chromosomes. The bands are highly condensed in comparison to the interband chromatin. The figures represent different degrees of stretching so that the fine structure of the bands may be seen. (From Burkholder, 1976; photographs courtesy of G. D. Burkholder.)

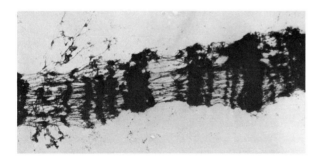

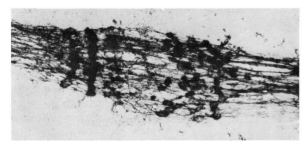

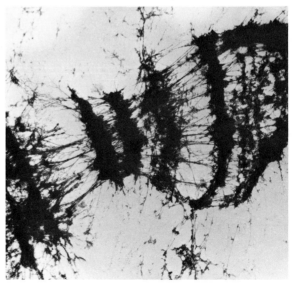

FIGURE 3

A region of the polytene chromosome set of the midge *Chironomus tentans*. Note the constancy of band number in different tissues. (From Beerman, 1952.)

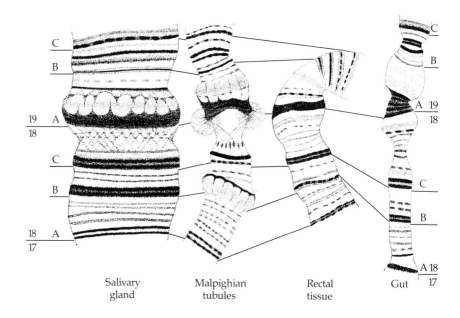

Salivary gland   Malpighian tubules   Rectal tissue   Gut

also found chromosome constancy throughout the different tissues of the adult organism. As we shall see later, various studies have demonstrated that when DNA was extracted from several somatic tissues, they each had very similar compositions and properties.

The second support for the hypothesis of genomic equivalence came from embryology. Driesch and Spemann clearly demonstrated that the nuclei of early blastomeres of sea urchins and salamanders were TOTI- POTENT, that is, capable of generating every type of differentiated cell. In both sea urchin and salamander embryos, a cell that would normally have produced only a small fraction of the embryo was found to be capable of generating the entire organism. The nuclei of such cells would have had to retain those genes for the products of all the other cell types. Likewise, Spemann showed that early salamander gastrula cells could change their prospective fates when transplanted into another area of the embryo. Could these embryological observations be extrapolated to cells that had already made a commitment to differentiate in a certain direction? Does a cell type that has become differentiated or determined still retain other potencies? Two lines of evidence lead to the conclusion that this is indeed the case.

Transdetermination

Upon hatching, a *Drosophila* larva has two distinct cell populations. About 10,000 cells form the larval tissue. Most of these cells have polytene chromosomes, and they grow by expanding to about 150 times their original volume. In addition, about 1000 nonpolytene diploid cells occur in clusters throughout the larva. These clusters of undifferentiated cells are called IMAGINAL DISCS (from the Latin *imago*, meaning "adult"), and they divide throughout the larval growth period. During metamorphosis, the hormone ECDYSONE triggers enormous changes throughout the organism (see Chapter 18). The larval cells degenerate while the imaginal disc cells are signaled to differentiate into the organs of the adult fly. Figure 4 shows the location of the *Drosophila* imaginal discs and the structures into which they develop.

Larval imaginal disc cells are determined. For example, an eye disc can be removed from one larva and implanted into the abdomen of a second larva. After metamorphosis, the fly developing from the second larva will have an extra eye in its abdomen. If a portion of the eye disc is transplanted, only a portion of the eye will develop. Transplantation of a disc or disc fragment into a *larva*, then, provides an excellent system to test what structure that disc or disc segment is determined to produce at metamorphosis. When discs are transplanted into *adult* flies, however, no differentiation takes place. Rather, the imaginal disc cells continue to proliferate. These proliferating cells can be continually cultured by transplanting them from adult fly to adult fly. At the same time, they can also be tested for their state of determination by removing pieces of the growing discs and placing them back into metamorphosing larvae (Figure 5).

Ernst Hadorn and his co-workers used these discs to demonstrate that a cell can change its developmental commitment (Hadorn, 1968). Usually fragments of a disc determined to give rise to antennae would continue to produce antennal structures every time they were tested, even after several serial transplantations in adult flies. However, an occasional disk would surprise the scientists. Instead of monotonously

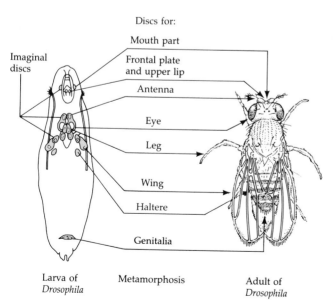

Discs for:

Mouth part

Frontal plate
and upper lip

Imaginal
discs

Antenna

Eye

Leg

Wing

Haltere

Genitalia

Larva of
Drosophila

Metamorphosis

Adult of
Drosophila

FIGURE 4
The locations and developmental fates of the imaginal discs in *Drosophila melanogaster*.

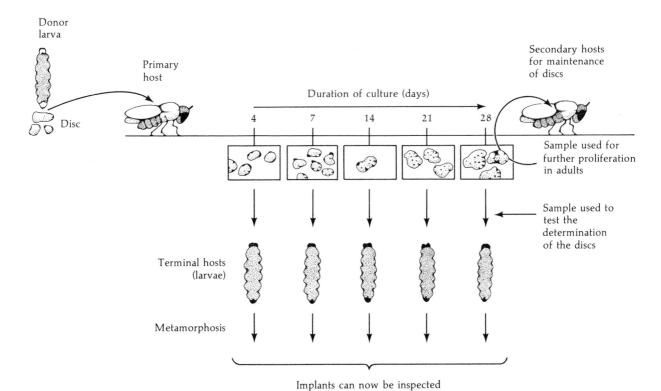

FIGURE 5

Scheme for testing the potency of imaginal discs. Discs can be cut and placed in adult flies, where they divide. If they are removed from the adults after various periods of incubation and are transplanted into normal larvae, they will become adult structures after metamorphosis. (After Markert and Ursprung, 1971.)

producing antennal structures, portions of the antenna disc would form parts of the leg, mouth, or wing. This is called TRANSDETERMINATION. Instead of developing the "proper" organ, the imaginal cells develop into some other part of the adult fly. For example, a disc normally determined to develop into an antenna can yield structures appropriate for the fly leg (Figure 6). Moreover, like the original state of determination, the transdetermined state is relatively stable and is inherited by the disc cells over many generations of cell division.

Transdetermination happens more frequently after several passages through adult flies and occurs preferentially in certain directions (Figure 7). A wing disc, for instance, can generate thorax structures, but thorax discs have not been seen to differentiate into parts of the wing. Genital discs can transdetermine to give rise to antennae or legs, but no other disc type has been observed to produce genital structures. Although the cause of this directionality remains unsolved, it is clear that determined cells can give rise to cell types other than their normal fate.

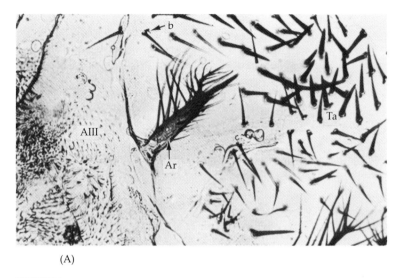

(A)

(B)

FIGURE 6
Transdetermination between antenna and leg structures. (A) Trans-determination in an antennal disc transplanted as shown in Figure 5. In addition to normal antennal structures (AIII, the third antennal segment; Ar, aristae), there are leg structures such as tarsal bristles (Ta) and their associated bracts (b). (B) Head of an adult fly carrying the mutation *Antennapedia*. In this mutant, the antennae are almost completely transformed into normal legs. Such mutations, where one structure is transformed into another, are called homeotic mutations. Although the mechanism of transde-termination in transplanted discs probably differs from that of homeotic mutations, they both demonstrate the change in fate of the representative discs. This will be discussed in more detail in Chapter 17. (A from Gehring, 1969; courtesy of W. J. Gehring. Photograph in B courtesy of J. Haynie.)

Imaginal disc cells, therefore, have retained the genes for the specific products of some other differentiated cell types.

Metaplasia

The study of salamander eye regeneration has demonstrated that even adult *differentiated* cells can retain their potential to produce other cell types. Here, removal of the neural retina promotes its regeneration from the pigmented retina, and new lens can be formed from the cells of the dorsal iris. This latter type of regeneration (called "Wolffian

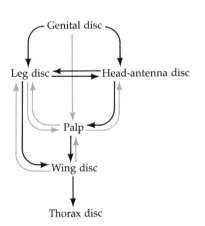

FIGURE 7
Routes of transdetermination in imaginal discs. The black arrows represent frequently observed changes; the gray arrows show more rare events. Those conversions not depicted by arrows are extremely rare events.

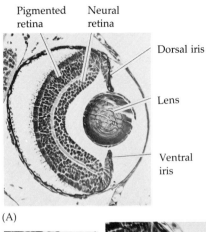

Pigmented retina　Neural retina

Dorsal iris

Lens

Ventral iris

(A)

FIGURE 8
Wolffian regeneration of the newt lens from the dorsal margin of the iris. (A) Normal unoperated eye of the larval stage newt *Notophthalmus viridiscens*. (B–G) Regeneration of lens, seen respectively on days 5, 7, 9, 16, 18 and 30. The new lens is complete at day 30. (From Reyer, 1954; photographs courtesy of R. W. Reyer.)

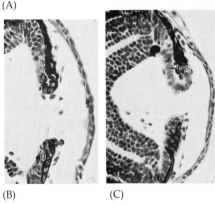

(B)　　　　　(C)

(D)

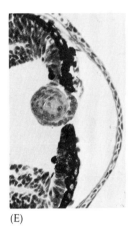

(E)

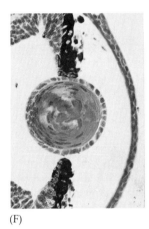

(F)

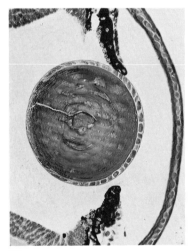

(G)

regeneration" after its first observer) has been intensively studied by Tuneo Yamada and his colleagues (Yamada, 1966; Dumont and Yamada, 1972). They find that after the removal of lens, a series of events lead to the production of a new lens by the iris (Figure 8).

1. The nuclei of the iris cells change their shape.
2. The dorsal iris cells begin to produce enormous amounts of ribosomes.
3. The DNA of these cells begins to replicate and cell divisions soon follow.
4. These cells dedifferentiate. They throw out their melanosomes (the differentiated products that give the eye its characteristic color), and these melanosomes are ingested by macrophages that have entered the wounded site.
5. Dorsal iris cells continue to divide, forming dedifferentiated tissue in the region of the removed lens.
6. These dedifferentiated iris cells now start synthesizing the differentiated products of lens cells, the crystallin proteins. These proteins are made in the same order as in normal lens development.
7. Once a new lens has formed, the dorsal side of the iris ceases mitosis.

These events are not the normal route by which the lens is formed. As you will recall, in embryogenesis, the lens develops from a layer of epithelial cells induced by the underlying neural ectoderm. The formation of the lens by the differentiated cells of the iris represents METAPLASIA, the transformation of one differentiated cell type into another. Therefore, the evidence of genetics and developmental biology points to the hypothesis of differential gene expression from genetically identical nuclei.

Amphibian cloning: Restriction of cell potency

The ultimate test of whether or not the nucleus from a differentiated cell has undergone any irreversible functional restriction would be to have that nucleus generate every other type of differentiated cell in the body. In 1936, Hans Spemann suggested a "somewhat fantastical" experiment for determining whether various cellular genomes were indeed identical. One would have to implant a nucleus from some differentiated cell into a host egg whose own nucleus had been removed. If each nucleus were identical to the zygote nucleus, this new nucleus would be capable of directing the entire development of the organism. Before such an experiment could be done, however, three techniques had to be perfected: (1) a method for enucleating host eggs without destroying them; (2) a method for isolating intact donor nuclei; and (3) a method for transferring such nuclei into the egg without damaging either the nucleus or the oocyte.

These techniques were developed by Robert Briggs and Thomas King. First, they combined the enucleation of the egg with its parthenogenetic activation. When an oocyte from the leopard frog (*Rana pipiens*) is pricked with a clean glass needle, the egg undergoes all the cytological and biochemical changes associated with fertilization. The cortical granules burst, internal cytoplasmic rearrangements of fertilization occur, and meiosis occurs near the animal pole of the cell. This meiotic spindle can be easily located as the pigment granules at the animal pole move away from it, and puncturing the oocyte at this site will cause the spindle and its chromosomes to flow outside the egg (Figure 9). The host egg is now considered to be both activated (in that the fertilization reactions necessary to initiate development have been completed) and enucleated. The transfer of nuclei into the eggs is accomplished by disrupting donor cells and transferring the released nucleus into the oocyte through a micropipette. Some cytoplasm accompanies the nucleus to its new home, but the ratio of donor to recipient cytoplasm is only $1:10^5$, and the donor cytoplasm does not seem to affect the outcome of the experiments.

In 1952, Briggs and King demonstrated that blastula cell nuclei could direct the development of complete tadpoles when transferred into the oocyte cytoplasm. Spemann had shown that blastula cells alone were not yet determined, so their nuclei were already known to be pluripo-

FIGURE 9
Procedure for transplanting blastula nuclei into activated enucleated *Rana pipiens* eggs. The relative dimensions of the meiotic spindle have been exaggerated to show the technique. (After King, 1966.)

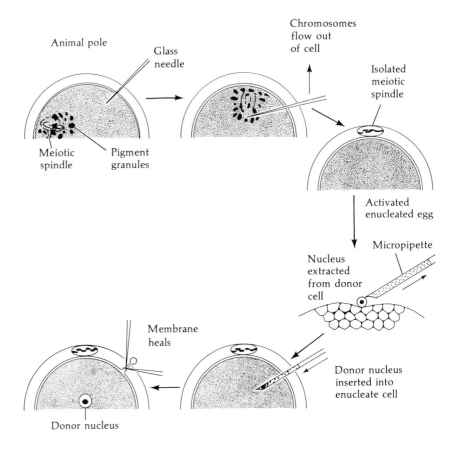

FIGURE 10
This handsome specimen of *Rana pipiens* was derived from the transplantation of a blastula nucleus into an activated enucleated egg. (Photograph courtesy of M. DiBerardino and N. Hoffner.)

tent. Therefore, if the nuclear transfer system worked, such blastula nuclei might be able to promote complete development. They did. Sixty percent of all transferred nuclei were able to direct the development of the oocytes into functional, swimming tadpoles, and all these tadpoles were diploid (suggesting that the nucleus was from the donor cells). Thus, the nuclear transfer system worked and could be used to study nuclear potency (Figure 10).

What happens when nuclei from more advanced stages are transferred into activated enucleated oocytes? The results of King and Briggs (1956) are outlined in Figure 11. Whereas most *blastula* nuclei could produce entire tadpoles, there was a dramatic decrease in the ability of nuclei from later stages to direct development to the tadpole stage. When nuclei from the *somatic cells* of tailbud-stage tadpoles were used as donors, no nucleus was able to direct normal development. However, *germ cell* nuclei from tailbud-stage tadpoles (which eventually will give rise to a complete organism after fertilization) *were* capable of directing normal development in 40 percent of the blastulae that developed (Smith, 1956). Thus, somatic cells appear to lose their ability to direct complete development as they become determined and differentiated.

This decrease in nuclear potency was shown to be stable and tissue

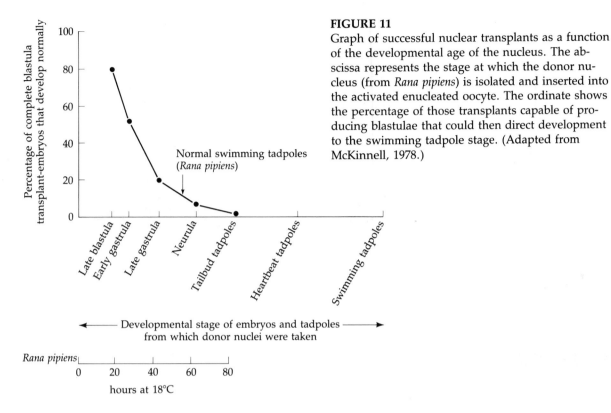

FIGURE 11

Graph of successful nuclear transplants as a function of the developmental age of the nucleus. The abscissa represents the stage at which the donor nucleus (from *Rana pipiens*) is isolated and inserted into the activated enucleated oocyte. The ordinate shows the percentage of those transplants capable of producing blastulae that could then direct development to the swimming tadpole stage. (Adapted from McKinnell, 1978.)

specific. Endoderm nuclei from late gastrula embryos were amplified by serial transplantation (Figure 12). Here, one nucleus is transferred into an enucleated oocyte and allowed to produce thousands of identical blastula nuclei. These blastula nuclei can then be transferred into more enucleated oocytes. In this way, more copies of the original nucleus are prepared and the potential of that nucleus can be evaluated. This technique is called NUCLEAR CLONING. In the first transfer prior to serial transplantation, there was wide variation in the stages to which individual nuclei could direct development. Some nuclei could direct development all the way through the swimming tadpole stage whereas other nuclei would abort development at gastrulation. Although King and Briggs found this "normal" variation *between* their clones of endoderm nuclei, they found little variation *within* these clones (Figure 13). Often, the stage of arrest was similar for all the embryos produced from a single, cloned endoderm nucleus. This was the case through several clonal generations. Moreover, when aberrant larvae were produced, they were all aberrant in the same way. They had endodermal structures (notably, gut) but were lacking several mesodermal or ectodermal derivatives. The endoderm nuclei appear to be good at forming endoderm but are restricted in their ability to form ectoderm or mesoderm. DiBerardino and King (1967) found a similar loss of potency in nuclei from

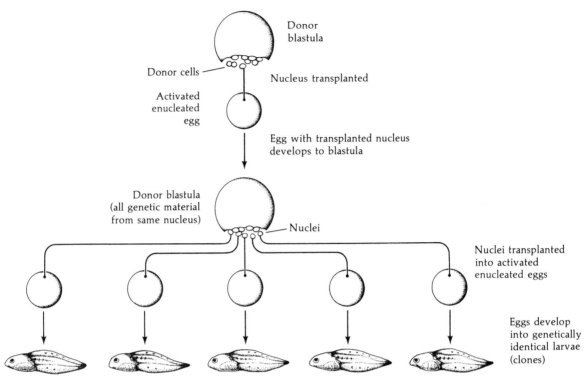

FIGURE 12
Testing potency by serial nuclear transplantation. Nuclei of donor cells from a single blastula are placed into activated, enucleated eggs. The blastula produced by such a transplant is used as a source of nuclei for a second generation of transplants, a process called nuclear cloning.

ectodermal cells. Here the aberrant tadpoles had excellent neural differentiation but lacked endodermal structures. Thus, the progressive restriction of nuclear potency during development appears to be the general rule.

Amphibian cloning: Exceptions to restriction

However, there are other explanations for the limited potency of differentiated cell nuclei. When transferring a nucleus of a differentiated cell

FIGURE 13
Serial transplantation of endodermal nuclei from *Rana pipiens* gastrula cells. Nuclei from cells in the endodermal region of a frog gastrula are transplanted into activated enucleated oocytes. The blastulae are used as sources of new clones, and secondary clones are generated from some of the blastulae thus formed. Comparisons can be made within the clones and between them. (After King and Briggs, 1956.)

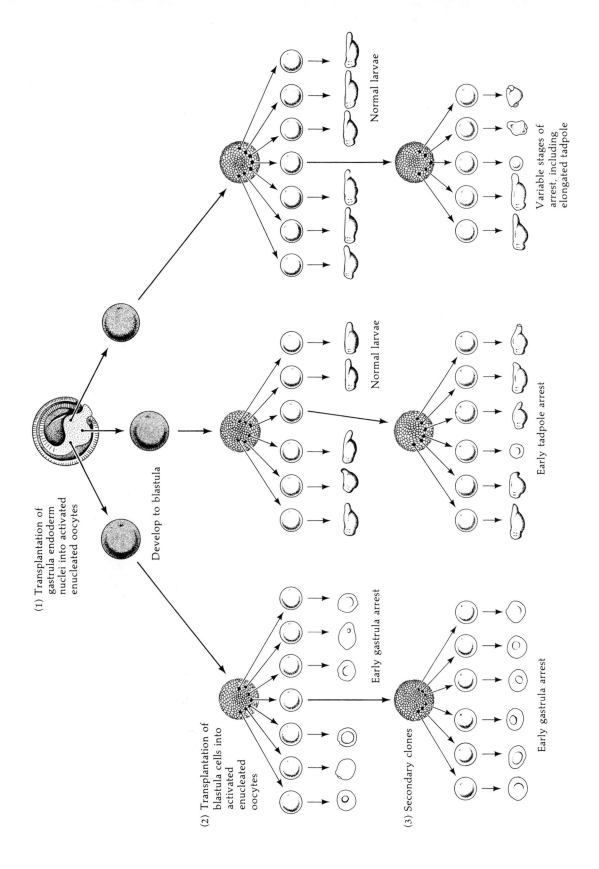

(1) Transplantation of gastrula endoderm nuclei into activated enucleated oocytes

Develop to blastula

Normal larvae

Variable stages of arrest, including elongated tadpole

Normal larvae

Early tadpole arrest

(2) Transplantation of blastula cells into activated enucleated oocytes

Early gastrula arrest

(3) Secondary clones

Early gastrula arrest

into oocyte cytoplasm, one is asking the nucleus to revert back to physiological conditions that it is not used to. The cleavage nuclei of frogs divide at a rapid rate, whereas some differentiated cell nuclei divide rarely if at all. Failure to replicate DNA rapidly can lead to chromosomal breakage, and such chromosomal abnormalities have been seen in many cells of the cloned tadpoles. John Gurdon and his colleagues, using slightly different methods of nuclear transplantation, have obtained results suggesting that many of the nuclei of differentiated cells remain totipotent.

A major difference between the experiments of Gurdon and those of Briggs and King concerns the organism being studied. Gurdon isolated nuclei from *Xenopus laevis*, the South African clawed frog. *Xenopus* (Figure 14) is a much more primitive frog than *Rana*, lacking the eyelids, tympanum (ear), and even the tongue so characteristic of more recently evolved species of frogs. *Xenopus* also has different developmental properties. Unlike the leopard frog, adult *Xenopus* can regenerate lost limbs; and early development in *Xenopus* is about three times as rapid as early development in *Rana pipiens*. This means that whereas *R. pipiens* takes 80 hours to reach tailbud-stage tadpoles, *Xenopus* accomplishes the same amount of development in only 26 hours. Thus, nuclei from *Xenopus* tailbud endoderm are as young as early gastrula nuclei in *Rana* (McKinnell, 1978).

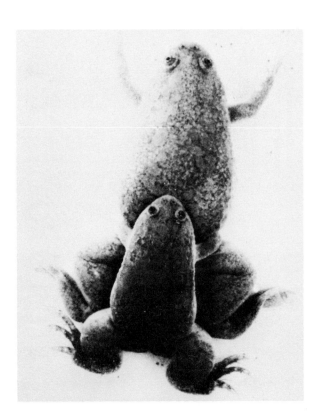

FIGURE 14
Mating pair of *Xenopus laevis*. The smaller male grasps the female and fertilizes the newly-shed egg externally. (From Deuchar, 1975.)

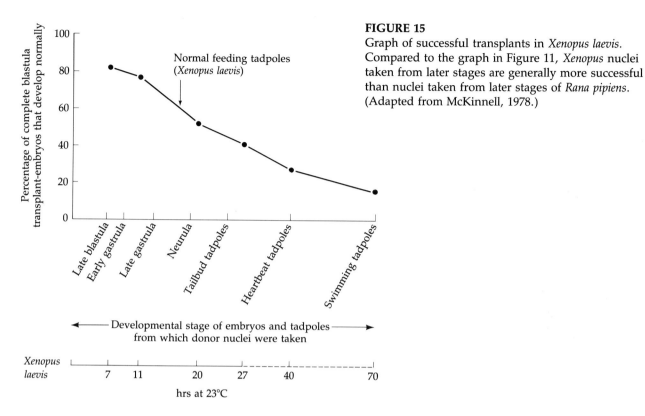

FIGURE 15
Graph of successful transplants in *Xenopus laevis*. Compared to the graph in Figure 11, *Xenopus* nuclei taken from later stages are generally more successful than nuclei taken from later stages of *Rana pipiens*. (Adapted from McKinnell, 1978.)

Gurdon, too, found a progressive loss of potency with increased development (Figure 15). The exceptions to this rule, however, proved very interesting. Gurdon had transferred the intestinal endoderm of feeding *Xenopus* tadpoles to activated enucleated eggs. These donor nuclei contained a genetic marker (one nucleolus per cell instead of the usual two) that could distinguish them from host nuclei. Out of 726 transfers, only 10 nuclei promoted development of feeding tadpoles. Serial transplantation (placing an intestinal nucleus into an egg; and when the egg has become a blastula, transferring the nuclei of the blastula cells into several more eggs) increased the yield to 7 percent (Gurdon, 1962). In some instances nuclei from intestinal epithelial cells were capable of generating all the cell lineages—neurons, blood cells, nerves, and so forth—of a living tadpole. Moreover, seven of these tadpoles (from two original nuclei) metamorphosed into fertile adult frogs (Gurdon and Uehlinger, 1966). These nuclei were totipotent (Figure 16).

King and his colleagues, however, criticized these experiments, pointing out that (1) not enough precautions were taken to make certain that primordial germ cells—which migrate through and often stay in the gut—were not used as sources of nuclei, and (2) the intestinal epithelial cell of such a young tadpole may not qualify as a truly differentiated cell type. Such cells of feeding tadpoles still contain yolk

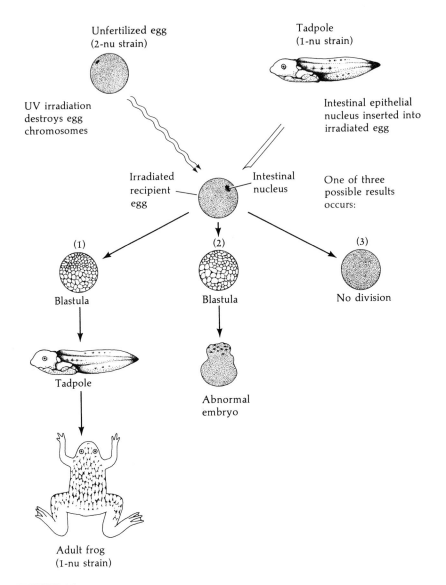

FIGURE 16
Procedure used to obtain mature frogs from the intestinal nuclei of *Xenopus* tadpoles. The wild-type egg (2-nu) is irradiated to destroy the maternal chromosomes, and an intestinal nucleus from a 1-nu tadpole is inserted. In some cases there is no division; in some cases, the embryo is arrested in development; but in other cases, an entire new frog is formed an has a 1-nu genotype.

platelets (DiBerardino and King, 1967; McKinnell, 1978; Briggs, 1979).

To circumvent these criticisms, Gurdon and his colleagues have cultured epithelial cells from adult frog foot webbing. These cells were found to be differentiated in that each of them contained keratin, the characteristic protein of adult skin cells. When nuclei from these cells were transferred into activated, enucleated *Xenopus* oocytes, none of the first generation transfers progressed further than neurulation. By serial transplantation, however, numerous tadpoles were generated, but these tadpoles all died prior to feeding (Gurdon et al., 1975). A similar developmental arrest was reported by Wabl and colleagues (1975), who attempted to produce entire frogs from lymphocyte nuclei.

One can look at these amphibian cloning experiments in two ways. First, one can recognize a general restriction of potency concomitant with development. This restriction is genetically determined and characteristic of the donor nucleus type. Second, one can readily see that the differentiated cell genome is remarkably potent in its ability to produce all the cell types of the amphibian tadpole. In other words, even if there is debate over the *totipotency* of such nuclei, there is little doubt that they are extremely *pluripotent*. Certainly many unused genes in skin or lymphocytes can be reactivated to produce the nerves, stomach, or heart of a swimming tadpole.

FIGURE 17
Procedure for transferring nuclei into an activated enucleated mammalian egg. Single-cell embryo incubated in colcemid and cytochalasin is held on a suction pipette. The enucleation pipette pierces the zona pellucida and sucks up adjacent cell membrane and the area of the cell containing the pronuclei. The enucleation pipette is then withdrawn (A) and the pronuclei-containing cytoplasm is removed from the egg. The cell membrane is not broken and the continuity of the membrane-bound cytoplasm is indicated by the arrow. (B) The cell membrane forms a vesicle around the pronuclear cytoplasm. (C) This vesicle is mixed with Sendai virus and is inserted into the perivitteline space between the zona and another enucleated egg. (D) The Sendai virus mediates the fusion of the enucleated egg and the membrane-bound pronuclei, allowing the pronuclei (arrow) to enter the cell. The polar body, seen beside the suction pipette, does not enter the fusion reaction. (From McGrath and Solter, 1983; courtesy of the authors.)

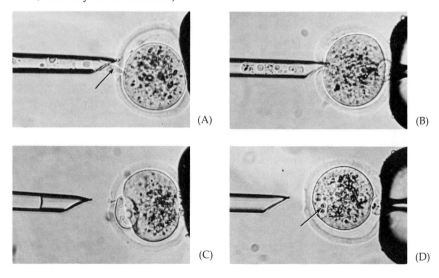

(A)

(B)

(C)

(D)

SIDELIGHTS & SPECULATIONS

Cloning

One can see (as we have also seen while discussing the work of Roux and Driesch) that the details of experimental technique can greatly influence the answers one obtains to a given question. By slight modifications of the cloning procedure, even late stage *Rana pipiens* nuclei can develop into normal larva. Sally Hennen has shown that the developmental success of donor nuclei can be increased by treating such nuclei with spermine and by cooling the egg to give the nucleus time to adapt to the egg cytoplasm (Hennen, 1970). By so treating *Rana pipiens* tailbud-stage endoderm nuclei, she obtained from 60 percent of those nuclei normal larvae that could support development through the blastula stage (compared to 0 percent in the untreated controls). Muggleton-Harris and Pezzella (1972) have used this technique on adult frog lens cell nuclei and claim to have generated an entire frog from such donors. This report used no genetic markers and has not been confirmed (see McKinnell, 1978). In such an experiment it is essential to have some proof that the nucleus was derived from the donor cell because some parthenogenetically activated oocytes can indeed develop into mature frogs (Miyada, 1960). It is obviously important to repeat such an experiment, as it would be the first example of cloning an adult organism from a differentiated adult cell. The experiments on cultured skin cell transplants also used spermine but did not result in complete development to an adult frog.

Cloning humans from previously differentiated cells seems to be the goal of several newspaper writers and novelists (novelists tend to clone politically important persons such as Hitler or Kennedy; periodical writers extend this technique to athletes and movie stars). It should be obvious from the above that cloning a fully developed individual from differentiated cells is a formidable undertaking, and the results are not conclusive, even in amphibians. To date, the best evidence is still the work of Gurdon. Thus, at best we can expect less than 2 percent chance of success if all were equal between amphibian and human eggs.

But all is not equal. Besides the ethical questions, the human organism is much harder to work with. Whereas the amphibian female produces hundreds of eggs simultaneously, human females produce only a few ripe ova each month. Moreover, there may be significant differences in the ability of oocyte cytoplasm to be directed from a nucleus of a more advanced stage cell.

Nuclear transplantation has quite recently been achieved in mice by removing the two pronuclei of the zygote and then replacing them with pronuclei from other zygotes (McGrath and Solter, 1983). This is done by the procedure shown in Figure 17. One-cell embryos are first incubated in cytochalasin and colchicine to relax the microfilaments and microtubules of the cytoskeleton. While the embryo is held in place by a suction pipette, the enucleation pipette penetrates the zona pellucida. The enucleation pipette does not pierce the cell membrane. Rather, it is positioned adjacent to the pronuclei and the region containing the pronuclei is sucked into the pipette (A). The enucleation pipette is then withdrawn, and the bud of pronuclei-containing cytoplasm separates from the egg. This cytoplasm is contained in a plasma membrane (B). The pipette containing the membrane-bounded pronuclei is moved to a drop containing inactivated Sendai virus, whose viral coat can cause membrane fusion. After aspirating some Sendai virus, the pipette is moved to a zygote whose pronuclei have been similarly removed. The zona is penetrated, and the membrane-bounded pronuclei are deposited within the perivitelline space, between the zona and the egg cell membrane (C). The embryo is then incubated at 37°, thereby allowing the membranes to fuse (D). Thus, the two donor pronuclei are found to reside within the host cytoplasm. When these embryos are cultured for 5 days, they form a blastocyst that can be implanted into the uterus of a pseudopregnant female adult. The resulting mice display the phenotype of the donor nucleus.

Whereas over 90 percent of enucleated mouse zygotes receiving pronuclei from other zygotes develop successfully to the blastocyst stage, not a single embryo (out of 81) developed even this far when nuclei from 4-cell embryos were transferred into the enucleated zygotes. Similarly, nuclei from 8-cell embryos and the inner cell mass would not support preimplantation development (McGrath and Solter, 1984). Unlike sea urchins and amphibians, the nuclei of early mammalian blastomeres (known to be totipotent) do not support full development. It is probable that the blastomere nuclei cannot function properly in zygote cytoplasm. Thus, the cloning of Elvis Presley (as reported in the *National Examiner* in September 1984) by simple nuclear transfer is not biologically possible.

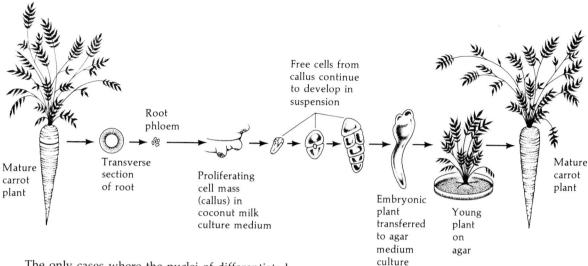

FIGURE 18
Steward's experimental procedure demonstrating the totipotency of carrot phloem cells.

The only cases where the nuclei of differentiated cells of adult organisms are readily seen to be capable of directing the development of another adult organism occur in plants. This ability has been demonstrated dramatically in cells from carrots and tobacco. In 1958, F. C. Steward and his colleagues established a procedure by which the differentiated tissue of carrot roots could give rise to an entire new plant (Figure 18). Small pieces of the phloem tissue were isolated from the carrot and rotated in large flasks containing coconut milk. This fluid (which is actually the liquid endosperm of the coconut seed) contains the factors and nutrients necessary for plant growth and the hormones required for plant differentiation. Under these conditions, the tissues proliferate and form a disorganized mass of tissue called a CALLUS. Continued rotation causes sheering of individual cells away from the callus and into suspension. These individual cells give rise to rootlike nodules of cells that continue to grow as long as they remain in suspension. When these nodules are placed into a medium solidified with agar, the rest of the plant is able to develop, ultimately forming a complete, fertile carrot plant (Steward et al., 1964; Steward, 1970).

Although the process from single cell to flowering plant cannot be observed under the rotating culture conditions, Vasil and Hildebrandt (1965) have been able to follow these events by isolating single tobacco cells and watching their development either directly or by time-lapse cinematography. As with carrot cells, the tobacco cells were able to give rise to a plantlet that eventually could flower and seed.

We therefore have an example of nuclear totipotency. All the genes needed to produce an entire plant exist within the nucleus of a differentiated cell. But plants develop differently than do animals; the vegetative propagation of plants by cuttings (i.e., portions of the plant that, when nourished, regenerate the missing parts) is common agricultural practice. Moreover, in contrast to animals (in which germ cells are set off as a distinctive cell lineage early in development), plants normally derive their gametes from somatic cells. It is not overly surprising, then, that a single plant cell can differentiate into other cell types and form a genetically identical clone (from the Greek *klon*, meaning "twig").

So far, we can conclude that differential gene loss is not the cause of differentiation. The nuclei of differentiated cells contain most, if not all, the genes of the zygote; and these genes can be expressed under the appropriate conditions. The process of differentiation, then, involves the selective expression of different portions of a common genetic repertoire. With the advent of molecular biology and gene cloning techniques, the problems of genome constancy and differential gene expression have been reinvestigated. The next chapter extends this discussion into contemporary molecular biology.

Literature cited

Beermann, W. (1952). Chromomeren Konstanz und spezifische modifikationen der Chromosomenstrucktur in der Entwicklung und Organdifferenzierung von *Chironomus tentans*. *Chromosoma* 5: 139–198.

Blackler, A. W. and Fischberg, M. (1961). Transfer of primordial germ cells in *Xenopus laevis*. *J. Embryol. Exp. Morphol.* 9: 634–641.

Briggs, R. (1979). Genetics of cell type determination. *Int. Rev. Cytol. Suppl.* 9: 107–127.

Briggs, R. and King, T. J. (1952). Transplantation of living nuclei from blastula cells into enucleated frogs' eggs. *Proc. Natl. Acad. Sci. USA* 38: 455–463.

Burkholder, G. D. (1976). Whole mount electron microscopy of polytene chromosome from *Drosophila melanogaster*. *Can. J. Genet. Cytol.* 18: 67–77.

Deuchar, E. M. (1975). *Xenopus: The South African Clawed Frog*. Wiley-Interscience, New York.

DiBerardino, M. A. and King, T. J. (1967). Development and cellular differentiation of neural nuclear transplants of known karyotypes. *Dev. Biol.* 15: 102–128.

Dumont, J. N. and Yamada, T. (1972). Dedifferentiation of iris epithelial cells. *Dev. Biol.* 29: 385–401.

Fischberg, M. and Blackler, A. W. (1961). How cells specialize. *Sci. Am.* 205(3): 124–170.

Fristrom, J. W., Raikow, R., Petri, W. and Stewart, D. (1969). *In vitro* evagination and RNA synthesis in imaginal discs of *Drosophila melanogaster*. In E. W. Hanly (ed.), *Problems in Biology: RNA in Development*. University of Utah Press, Salt Lake City, pp. 381–402.

Gehring, W. J. (1969). Problems of cell differentiation in *Drosophila*. In E. W. Hanly (ed.), *Problems in Biology: RNA in Development*. University of Utah Press, Salt Lake City, pp. 231–244.

Gurdon, J. B. (1962). The developmental capacity of nuclei taken from intestinal epithelial cells of feeding tadpoles. *J. Embryol. Exp. Morphol.* 10: 622–640.

Gurdon, J. B. (1968). Transplanted nuclei and cell differentiation. *Sci. Am.* 219(6): 24–35.

Gurdon, J. B. (1977). Egg cytoplasm and gene control in development. *Proc. Roy. Soc. Lond. [Biol.]* 198: 211–247.

Gurdon, J. B. and Uehlinger, V. (1966). "Fertile" intestinal nuclei. *Nature* 210: 1240–1241.

Gurdon, J. B., Laskey, R. A. and Reeves, O. R. (1975). The developmental capacity of nuclei transplanted from keratinized cells of adult frogs. *J. Embryol. Exp. Morphol.* 34: 93–112.

Hadorn, E. (1968). Transdetermination in cells. *Sci. Am.* 219(5): 110–120.

Ham, R. G. and Veomett, M. J. (1980). *Mechanisms of Development*. Mosby, St. Louis.

Hennen, S. (1970). Influence of spermine and reduced temperature on the ability of transplanted nuclei to promote normal development in eggs of *Rana pipiens*. *Proc. Natl. Acad. Sci. USA* 66: 630–637.

Judd, B. H. and Young, M. W. (1973). An examination of the one cistron–one chromomere concept. *Cold Spring Harbor Symp. Quant. Biol.* 38: 573–579.

King, T. J. (1966). Nuclear transplantation in amphibia. *Methods Cell Physiol.* 2: 1–36.

King, T. J. and Briggs, R. (1956). Serial transplantation of embryonic nuclei. *Cold Spring Harbor Symp. Quant. Biol.* 21: 271–289.

Markert, C. L. and Ursprung, H. (1971). *Developmental Genetics*. Prentice-Hall, Englewood Cliffs, NJ.

McGrath, J. and Solter, D. (1983). Nuclear transplantation in the mouse embryo by microsurgery and cell fusion. *Science* 220: 1300–1302.

McGrath, J. and Solter, D. (1984). Inability of mouse blastomere nuclei transferred to enucleated zygotes to support development in vitro. *Science* 226: 1317–1319.

McKinnell, R. G. (1978). *Cloning: Nuclear Transplantation in Amphibia*. University of Minnesota Press, Minneapolis.

Miyada, S. (1960). Studies in haploid frogs. *J. Sci. Hiroshima Univ. Ser. B. Div. 1*, 19: 1–56.

Muggleton-Harris, A. L. and Pezzella, K. (1971). The ability of the lens cell nucleus to promote embryonic development through to metamorphosis, and its implications to ophthalmic gerontology. *Exp. Gerontol.* 7: 427–431.

Reyer, R. W. (1954). Regeneration in the lens in the

amphibian eye. *Q. Rev. Biol.* 29: 1–46.

Smith, L. D. (1956). Transplantation of the nuclei of primordial germ cells into enucleated eggs of *Rana pipiens. Proc. Natl. Acad. Sci. USA* 54: 101–107.

Steward, F. C. (1970). From cultured cells to whole plants: The induction and control of their growth and morphogenesis. *Proc. Roy. Soc. Lond. [Biol.]* 175: 1–30.

Steward, F. C., Mapes, M. O., Kent, A. E. and Holsten, R. D. (1964). Growth and development of cultured plant cells. *Science* 143: 20–27.

Swanson, C. P., Merz, T. and Young, W. J. (1981). *Cytogenetics: The Chromosome in Division, Inheritance, and Evolution,* Second Edition. Prentice-Hall, Englewood Cliffs, NJ.

Tjio, J. H. and Puck, T. T. (1958). The somatic chromosomes of man. *Proc. Natl. Acad. Sci. USA* 44: 1229–1237.

Ursprung, H., Smith, K. D., Sofer, W. H. and Sullivan, D. T. (1968). Assay systems for the study of gene function. *Science* 160: 1075–1081.

Vasil, V. and Hildebrandt, A. C. (1965). Differentiation of tobacco plants from single isolated cells in microcultures. *Science* 150: 889–892.

Wabl, M. R., Brun, R. B. and DuPasquier, L. (1975). Lymphocytes of the toad *Xenopus laevis* have the gene set for promoting tadpole development. *Science* 190: 1310–1312.

Yamada, T. (1966). Control of tissue specificity: The pattern of cellular synthetic activities in tissue transformation. *Am. Zool.* 6: 21–31.

CHAPTER

Genomic equivalance and differential gene expression: molecular investigations

Seek simplicity and distrust it.
—ALFRED NORTH WHITEHEAD (1919)

For it is not cell nuclei, not even individual chromosomes, but certain parts of certain chromosomes from certain cells that must be isolated and collected in enormous quantities for analysis; that would be the precondition for placing the chemist in such a position as would allow him to analyze (the hereditary material) more minutely than the morphologists.
—THEODOR BOVERI (1904)

Introduction

Embryologists have asked: Is the set of genes found in each different cell of an organism the same as those that were found in the zygote? Molecular biologists frame the same problem: Is the DNA of each cell the same despite the different proteins made by each cell type? The problem remains the same, but the techniques used to answer it have become more sophisticated. Instead of analyzing the generation of organs or embryos, we can now look at individual sequences of DNA and monitor whether or not these genes are present in a cell and whether or not they are actively transcribing RNA. In order to further explore this question, then, we must become familiar with the techniques of molecular biology.

Molecular biology techniques:
Nucleic acid hybridization and gene cloning

Our modern understanding of eukaryotic genes and their RNA products is derived largely from experiments involving NUCLEIC ACID HYBRIDIZATION. This technique involves the re-annealing of single-stranded nucleic acids such that isolated complementary strands will pair to form double-stranded regions. Therefore, whenever DNA is cut into small pieces and each piece dissociated into two single strands (i.e., denatured) each strand should find and stick to its complementary partner, given sufficient time. The conditions of this renaturation must be such that all specific binding between complementary strands is maintained while nonspecific matchings are dissociated. This is usually accomplished by varying the temperature of the ionic conditions in the solution in which such renaturation or hybridization is taking place (Wetnur and Davidson, 1968).

Similarly, RNA that was synthesized from any region of DNA would be expected to bind to the strand from which it was transcribed. Thus, RNA is expected to hybridize specifically with the gene that codes for it. In order to measure this hybridization, one of the nucleic acid strands is usually radiolabeled. One such experiment will illustrate how this technique can be used. Unlabeled DNA from a frog liver can be denatured (its strands being separated in alkali) and immobilized on nitrocellulose filter paper. Radiolabeled RNA can then be prepared by feeding radioactive RNA precursors either to another frog or to frog cells growing in culture. One can then isolate ribosomes and extract from them a radioactive 28 S ribosomal RNA (rRNA), which can be added to the DNA-containing filters. We would expect the radioactive RNA to bind only to the DNA pieces containing the genes for the 28 S ribosomal RNA. The amount of binding can be quantified by measuring the radioactivity of these filters after washing.

The procedure and results for exactly such an experiment are shown in Figure 1. Here, Wallace and Birnstiel (1966) isolated the DNA from normal tadpoles with two nucleoli per cell, from homozygous mutant tadpoles that lacked nucleoli altogether, and from heterozygous tadpoles with only one nucleolus per cell. The isolated DNA from each

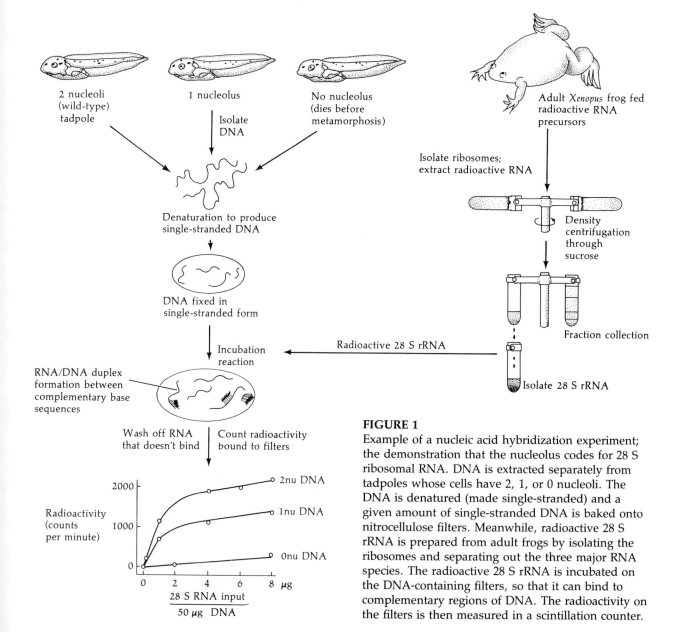

FIGURE 1

Example of a nucleic acid hybridization experiment; the demonstration that the nucleolus codes for 28 S ribosomal RNA. DNA is extracted separately from tadpoles whose cells have 2, 1, or 0 nucleoli. The DNA is denatured (made single-stranded) and a given amount of single-stranded DNA is baked onto nitrocellulose filters. Meanwhile, radioactive 28 S rRNA is prepared from adult frogs by isolating the ribosomes and separating out the three major RNA species. The radioactive 28 S rRNA is incubated on the DNA-containing filters, so that it can bind to complementary regions of DNA. The radioactivity on the filters is then measured in a scintillation counter.

group was denatured. Fifty micrograms of DNA were then placed on each of dozens of filter papers, such that one group of filters contained denatured DNA from the wild-type tadpoles, one group of filters had DNA from the mutant tadpoles, and one group of filters had DNA from the heterozygous tadpoles. Then each group of filters had various amounts of radioactive 28 S ribosomal RNA hybridized to them. Some filters received small amounts of the radioactive rRNA, other filters received more. It was found that the radioactive rRNA bound well to the normal tadpole DNA, eventually saturating all the available DNA. No specific binding was seen to occur to the DNA from the tadpoles lacking nucleoli; and the DNA from the heterozygous tadpoles bound only one-half as much as their normal counterparts. Thus, it was shown that the genes for 28 S ribosomal RNA are transcribed from the nucleolar region of the tadpole genome.

One technical problem of nucleic acid hybridization is that it is not always possible to get enough radioactive precursors into an RNA molecule. This problem has been circumvented by isolating the RNA and making a complementary DNA (cDNA) copy in the presence of radioactive precursors. This can be done in a test tube containing the RNA, a short stretch of DNA, radioactive DNA precursors, and the viral enzyme REVERSE TRANSCRIPTASE. This enzyme is capable of making DNA from an RNA template (Figure 2). Because the DNA is synthesized in vitro, one need not worry about the dilution of the radioactive precursor. Furthermore, such a cDNA can hybridize with both the gene that produced the RNA (albeit the other strand) and the RNA itself. It is therefore extremely useful in detecting small amounts of specific RNAs.

More recently nucleic acid hybridization techniques have enabled developmental biologists to isolate individual genes through a technique called GENE CLONING. If one wanted to isolate a single human gene, for instance, the problem would be enormous. The human genome has

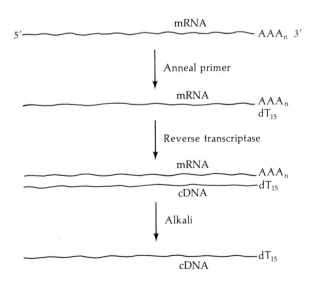

FIGURE 2
Method for preparing complementary DNA (cDNA). Most messenger RNA (mRNA) contains a long stretch of adenosine residues (AAA_n) at the 3' end of the message (to be discussed in Chapter 11); therefore, investigators anneal "primer" consisting of 15 deoxythymidine residues (dT_{15}) to the 3' end of the message. Reverse transcriptase then transcribes a complementary DNA strand, starting at the dT_{15} primer. The cDNA can be isolated by raising the pH of the solution, thereby denaturing the double-stranded hybrid and cleaving the RNA.

enough DNA to code for some 2 million genes of about 1500 base pairs each. Ordinary procedures would not allow a single gene to be resolved from the rest. Gene cloning, however, allows the purification of specific regions of DNA as well as their amplification into an enormous number of copies.

The first step in this process involves cutting the nuclear DNA into various pieces. This is done by incubating the DNA with a RESTRICTION ENDONUCLEASE (more commonly called a "restriction enzyme"). These endonucleases are bacterial enzymes that recognize specific sequences of DNA (see Table 1) and cleave the DNA at these sites (Nathans and Smith, 1975). For example, when human DNA is incubated with the enzyme *Eco*RI (from *Escherichia coli* strain RY), the DNA is cleaved at every site where the sequence GAATTC occurs. The result would be the production of variously sized pieces of DNA all ending with G on one end and AATTC on the other (Figure 3).

The next step is to incorporate these DNA fragments into CLONING VECTORS. These vectors are circular DNA molecules that replicate in bacterial cells independently of the bacterial chromosome. Usually, drug-resistant plasmids are used, although specially prepared bacteriophages have also been useful (especially for cloning large DNA fragments). Such a plasmid can be constructed so as to have only one *Eco*RI-sensitive site. This plasmid can be opened by incubating it with that

TABLE 1
Commonly used restriction enzymes

Enzyme	Derivation	Recognition and cleavage site[a]
*Eco*RI	*Escherichia coli*	G▼A A T T C C T T A A▲G
*Bam*HI	*Bacillus amyloliquifaciens*	G▼G A T C C C C T A G▲G
*Hind*III	*Haemophilus influenzae*	A▼A G C T T T T C G A▲A
*Sal*I	*Streptomyces albus*	G▼T C G A C C A G C T▲G
*Sma*I	*Serratia marcescens*	C C C▼G G G G G G▲C C C
*Hha*I	*Haemophilus haemolyticus*	G C G▼C C▲G C G
*Hae*III	*Haemophilus aegyptius*	G G▼C C C C▲G G
*Alu*I	*Arthrobacter luteus*	A G▼C T T C▲G A

[a]All restriction enzyme recognition sites have a center of symmetry. The double-stranded sequence read in one direction is identical to the sequence read backward in the other direction.

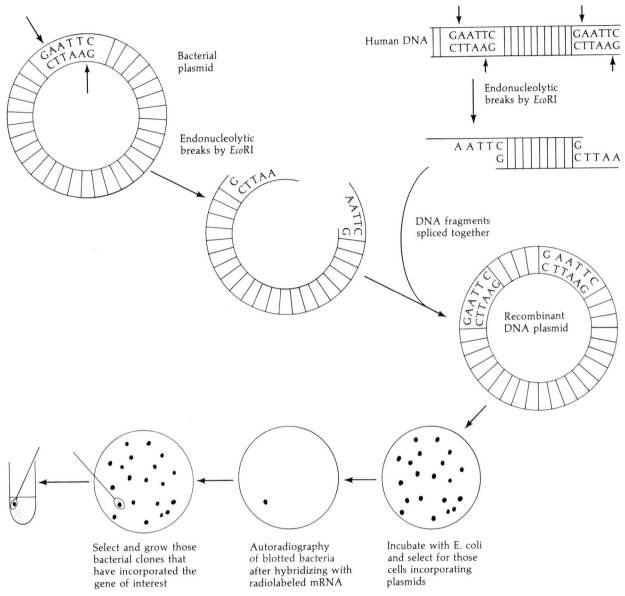

Human DNA

Endonucleolytic
breaks by *Eco*RI

Bacterial
plasmid

Endonucleolytic
breaks by *Eco*RI

DNA fragments
spliced together

Recombinant
DNA plasmid

Select and grow those
bacterial clones that
have incorporated the
gene of interest

Autoradiography
of blotted bacteria
after hybridizing with
radiolabeled mRNA

Incubate with E. coli
and select for those
cells incorporating
plasmids

FIGURE 3
Diagram of the general protocol
for cloning DNA, using as an example the cloning of a human
sequence into a plasmid with
one *Eco*RI-sensitive site.

restriction enzyme. After opening, it can be mixed with fragmented
DNA. In numerous cases, the DNA pieces will become incorporated
into these vectors (since their ends are complementary to the vector's
open ends) and the pieces can be joined covalently by placing them in
a solution containing the enzyme DNA LIGASE. This whole process would
yield bacterial plasmids that each contain a single piece of human DNA.
These are called RECOMBINANT PLASMIDS, or usually RECOMBINANT DNA
(Cohen et al., 1973; Blattner et al., 1978).

These recombinant plasmids are incubated with *E. coli* cells that are susceptible to a specific antibiotic. The bacteria are grown on agar dishes in the presence of that drug so that only those bacteria that incorporate a plasmid carrying the drug-resistance genes will survive. Therefore, the only bacteria producing colonies will be those that incorporate a recombinant plasmid. These colonies are allowed to grow; the agar plate is then blotted with a nitrocellulose filter. Some cells remain on the agar and others are picked up by the paper. When cells sticking to the paper are lysed, their DNA remains bound to the paper in the place where the cell burst.

Next, the DNA strands are separated while they are still on the paper; and the paper is incubated in a solution containing the radioactive RNA (or its cDNA copy) of the gene one wishes to clone. If a plasmid contains that gene, its DNA should be on the paper and only that DNA should be able to bind the radioactive RNA or cDNA. Therefore, only those areas will be radioactive. The radioactivity of these regions is detected by AUTORADIOGRAPHY. Sensitive X-ray film is placed over the treated paper. The high-energy electrons emitted by the radioactive RNA sensitize the silver grains in the film, causing them to turn dark when the film is developed, and a black spot is produced over each colony containing the appropriate recombinant plasmid carrying that particular gene (Figure 3). This colony is then isolated and grown. Such techniques enable one to produce trillions of bacteria, each containing hundreds of recombinant molecules.

These recombinant plasmids can be separated from the *E. coli* chromosome by centrifugation; and incubating the recombinants in *Eco*RI releases the human DNA fragment that contains the gene. This fragment can also be separated from the plasmid DNA. Thus, the investigator has picograms of purified DNA sequences containing a specific gene. [Although this procedure sounds very logical and easy, the numbers of colonies that have to be screened is often astronomical. The number of random fragments that must be cloned in order to obtain the gene you want gets larger with the increasing complexity[1] of the organism's genome. To be 99 percent certain of cloning the desired gene from *E. coli*, one needs to screen some 1500 colonies. With mammals, this number increases to 800,000; and rather than growing bacteria in standard petri dishes, the colonies are grown on cafeteria trays (Blattner et al., 1978; Slightom et at., 1980). To select against "useless" bacteria that picked up plasmids not containing any of the foreign gene fragments, most plasmids are constructed so that the restriction enzyme site lies within another antibiotic-resistant gene, so that any plasmid incorporating new DNA will lose that drug resistance. By replicating colonies onto normal agar and onto agar containing each of the two antibiotics, only those bacteria which took up recombinant plasmids (that is, growing in the first antibiotic, but not in the second) need be screened further.]

[1]Complexity is the term referring to the number of different types of genes within a nucleus.

Differential gene expression

With the advent of such biochemical techniques, the hypothesis of differential gene expression became testable on the molecular level. This hypothesis can be divided into three testable postulates.

1. Every cell nucleus contains the complete genome as established in the fertilized egg. In molecular terms, the DNAs of all differentiated cells are identical.
2. The unused genes in differentiated cells are not destroyed or mutated, and they retain the potential for being expressed.
3. Only a small percentage of the genome is being expressed in each cell, and a portion of the RNA synthesized is specific for that cell type.

We have already analyzed some of the genetic and embryological evidence for genomic equivalence. We now can ask whether biochemical studies support this view. Such evidence has been obtained from several studies involving nucleic acid hybridization.

The first large-scale molecular analysis to test the identity of DNA throughout the organism was done by Brian McCarthy and B.H. Hoyer in 1964. By supplying a mouse tumor cell with radioactive precursors, they were able to extract radioactive mouse DNA. The DNA was denatured into its two strands and added to a plate of denatured mouse embryonic DNA that had been immobilized in agar (Figure 4). After a certain time, the agar was washed, and the renatured DNA was measured. This amount of radioactivity was defined as 100 percent binding. McCarthy and Hoyer then repeated this procedure but added various amounts of nonradioactive DNA as well. If the nonradioactive DNA had sequences in common with the labeled DNA, one would expect to see competition; and the amount of radioactive DNA bound should decrease as more nonradioactive DNA is added. Figure 5 shows their results. No competition was observed when *bacterial* DNA was used as a possible competitor, a finding indicating that bacterial DNA had little or no sequences in common with the radioactive mouse DNA. When DNAs from various mouse organs were tested, however, competition was indeed observed. Moreover, the rate and extent of competition was identical for the DNAs from all mouse sources, thereby indicating the identity (within experimental error) of all the DNA. It is now known that the DNA being measured represented only those DNA sequences that were present in multiple copies and not those of the protein-synthesizing genes (see Chapter 11); but subsequent investigations have demonstrated the identity of single-copy genes as well (Davidson, 1976).

Nucleic acid hybridization can be modified for studying specific genes. By a technique called IN SITU HYBRIDIZATION, genes for specific differentiated cell proteins can be found on the chromosomes of cells not synthesizing these products. Developed by Mary Lou Pardue and

(A) HYBRIDIZATION PROCEDURE

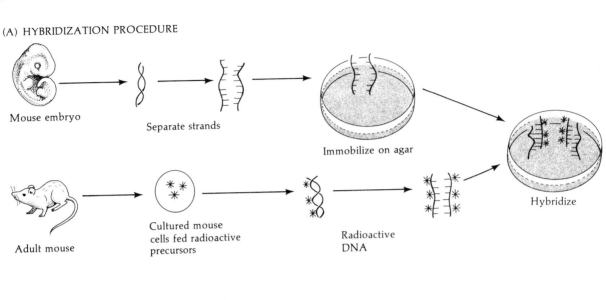

Mouse embryo

Separate strands

Immobilize on agar

Hybridize

Adult mouse

Cultured mouse
cells fed radioactive
precursors

Radioactive
DNA

(B) COMPETITION-HYBRIDIZATION PROCEDURE

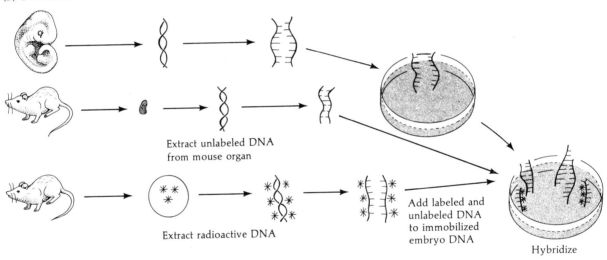

Extract unlabeled DNA
from mouse organ

Extract radioactive DNA

Add labeled and
unlabeled DNA
to immobilized
embryo DNA

Hybridize

(C) MEASURE COMPETITION

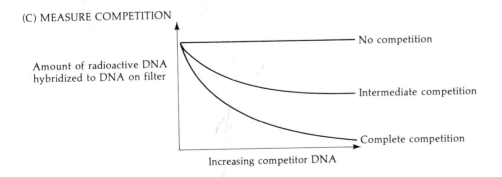

Amount of radioactive DNA
hybridized to DNA on filter

No competition

Intermediate competition

Complete competition

Increasing competitor DNA

FIGURE 4

Protocol for determining the identity of DNA sequences from two organs.
(A) Binding of radioactive DNA from cultured mouse cells to denatured and
immobilized DNA. (B) Competition–hybridization procedure wherein the
reaction shown in (A) is modified by the addition of competing nonradioactive DNA from specific organs or tissues. (C) Possible results of such a
competition–hybridization procedure.

Joseph Gall (1970), in situ hybridization involves denaturing polytene
chromosomes in such a way that the strands of the DNA helices are
separated while the chromosomes are still on a microscope slide. Radioactive RNA or cDNA can then be placed on the slide and washed
off after an appropriate incubation time. During that incubation, the
RNA is able to bind to the regions of DNA that encoded it. The slides
are then covered with a transparent photographic emulsion. The radioactivity from any bound RNA will sensitize the silver grains of the
emulsion, causing a black dot to form over the chromosomal band once
the emulsion is developed. Therefore, an RNA for a specific product
should be able to reveal the band from which it was synthesized.

   Figure 6 shows an autoradiograph localizing the genes for a specific
yolk protein of *Drosophila*. *Drosophila* genes were cloned and screened
with partially purified yolk-protein mRNA. Of those clones binding the
yolk-protein mRNA, several were found to contain DNA capable of
directing the synthesis of a yolk protein. These clones were grown and
the yolk-protein genes were isolated. A radioactive cDNA was made
from the mRNA of one of these clones and hybridized with preparations
of polytene salivary gland chromosomes. As illustrated in the figure,

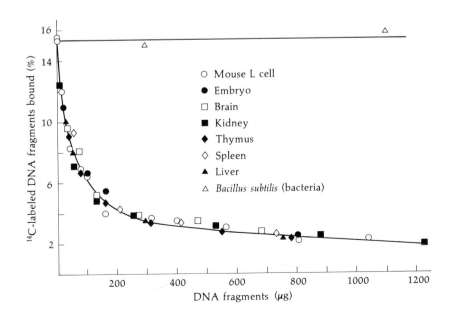

FIGURE 5

Competition of unlabeled DNA
from mouse organ fragments for
sites on immobilized mouse
DNA that could bind radioactive
DNA from mouse cell cultures.
Competitor DNA sources are indicated by the various symbols
in the key. (From McCarthy and
Hoyer, 1964.)

FIGURE 6
In situ hybridization of yolk protein cDNA to the polytene chromosomes of the larval *Drosophila* salivary gland. The dark grains show where radioactive yolk protein cDNA (made from yolk protein mRNA) bound to the chromosomes. (From Barnett et al., 1980; photograph courtesy of P. C. Wensink.)

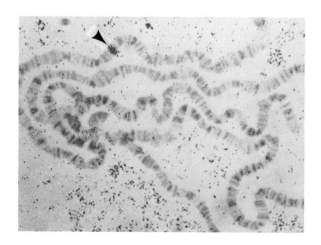

this cDNA binds to one specific band. However, salivary glands do not produce this protein. The only cells synthesizing yolk proteins are fat body cells in the adult female fly. Thus, a gene that is active solely in the adult fat body cells is found to exist in larval salivary gland chromosomes.

It has also been possible to show that unused genes of differentiated cells can be activated under certain conditions and that they can produce proteins specific for other cell types. In mammals, the clearest evidence for the reactivation of unused genes comes from the laboratory of Mary Weiss (Peterson and Weiss, 1972; Brown and Weiss, 1975), who fused together differentiated cells of various kinds. Cell fusion can occur naturally (as in muscle development) or can be mediated by agents such as inactivated Sendai virus (mouse measles) or polyethylene glycol. Such fusion creates a situation in which two nuclei reside in a common cytoplasm (Figure 7). If conditions allow, the nuclei of such cells will enter mitosis at the same time and eventually form a hybrid nucleus containing the chromosomes of both parental cell types. In most instances, when two different types of cells are fused together, the resulting hybrid lacks the differentiated traits that characterize the parental cells. By fusing rat liver tumor cells to mouse fibroblasts, Weiss was able to isolate hybrids having two sets of liver chromosomes to one fibroblast set. These cells retained their ability to make rat liver-specific proteins such as albumin, aldolase, and tyrosine aminotransferase (TAT). More surprisingly, they also synthesized *mouse* albumin, mouse aldolase, and mouse TAT—three proteins that no fibroblast ever synthesizes. The mouse fibroblast had retained liver-specific genes in a form that was capable of expression under certain circumstances. This appears to be a general rule in animal development: No irreversible genetic changes occur during cellular differentiation. Yet, the techniques of molecular biology have also uncovered an intriguing exception to this rule: the differentiation of the plasma cell.

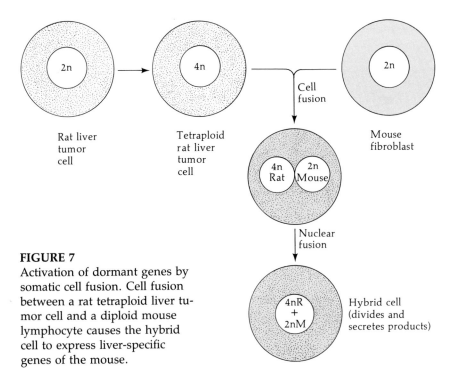

FIGURE 7

Activation of dormant genes by somatic cell fusion. Cell fusion between a rat tetraploid liver tumor cell and a diploid mouse lymphocyte causes the hybrid cell to express liver-specific genes of the mouse.

Changes in lymphocyte genes

In most cases, unused genes appear to be present and potentially functional in differentiated cells. The genetic repertoire is the same in all these tissues. There is one cell type in mammals, however, in which the genome of each cell is different from that of almost any other cell in the body. This cell type is represented by B LYMPHOCYTE, the cell that synthesizes antibody molecules.

Antibodies are produced when a foreign substance—the ANTIGEN—comes into contact with B LYMPHOCYTES, which reside in the lymph nodes and spleen. Even before contact with an antigen, each of these resting B lymphocytes makes antibody molecules. However, they do not secrete them. Rather, the antibody molecules are inserted into their cell membranes. Each B lymphocyte makes an antibody that recognizes one and only one antigenic shape. Therefore, an antibody recognizing the protein shell of polio virus would not be expected to recognize cholera toxin, *E. coli* membranes, or influenza virus. Once the membrane-bound antibody binds the antigen, the B lymphocyte divides numerous times and differentiates into an antibody-secreting PLASMA CELL (Figures 8 and 9). (The mechanism by which this differentiation occurs will be detailed in Chapter 16.) Only those B lymphocytes having the ability to bind specific antigens are stimulated to multiply and secrete antibodies. According to this model, called the CLONAL SELECTION THEORY (Burnett, 1959), each B lymphocyte acquires its specificity

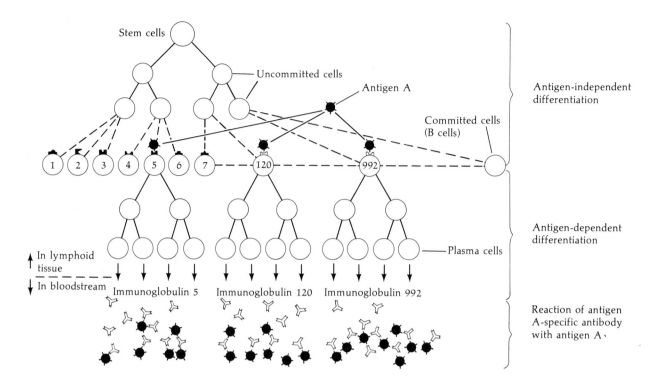

↑ In lymphoid tissue
↓ In bloodstream

Immunoglobulin 5   Immunoglobulin 120   Immunoglobulin 992

FIGURE 8
Clonal selection model of antibody formation. B cells obtain their cell-surface specificity before contact with antigen. Only one type of antibody is placed in the membrane. B cells whose antibodies are capable of binding the antigen to their cell surfaces proliferate to produce clones of plasma cells that secrete antibody with the same specificity as the cell-surface antibody that bound the antigen. (Adapted from Edelman, 1970.)

FIGURE 9
(A) B cell and (B) plasma cell. Note the expansion of the rough endoplasmic reticulum as the B cell becomes the antibody-secreting plasma cell. (Photographs courtesy of L. Weiss.)

(A)

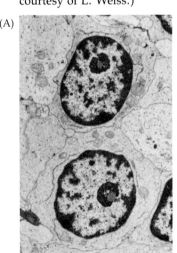

(B)

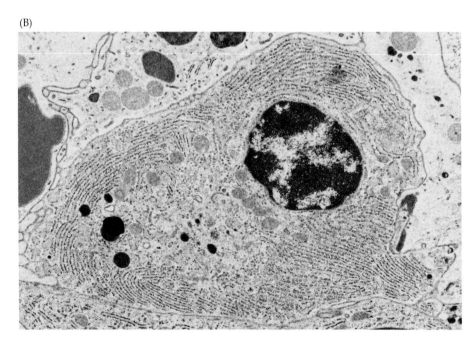

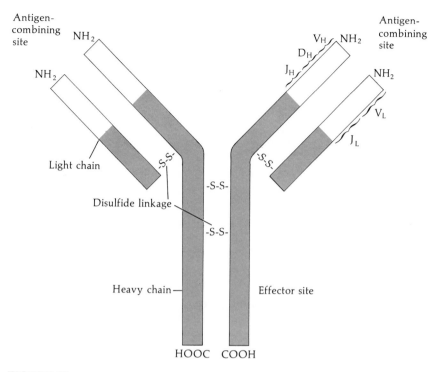

FIGURE 10

The structure of a typical immunoglobulin (antibody) protein. Two identical heavy chains are connected by disulfide linkages. The antigen-combining site is composed of the variable regions (white) of the heavy and light chains, whereas the effector site of the antibody (whether it agglutinates antigens, binds to macrophages, or enters into mucous secretions) is determined by the amino acid sequence of the heavy chain constant (gray) region.

before contact with the antigen. That is, out of the millions of possible antibodies it can make, each B lymphocyte "chooses" one of them and places antibodies of that specificity on its cell surface.

The mechanism of this selection of antibody specificity involves the creation of new genes during B lymphocyte differentiation. The antibody protein on the cell surface consists of two pairs of polypeptide subunits (Figure 10). There are two identical heavy chains and two identical light chains. The chains are linked together by disulfide bonds. The specificity of the immunoglobulin molecule (i.e., whether it will bind to a polio virus, an *E. coli* cell, or some other molecule) is determined by the amino acid sequence of the VARIABLE REGION. This region is made of the amino-terminal ends of both the heavy and the light chains. The variable regions of the immunoglobulin molecules are attached to CONSTANT REGIONS, which give the antibody its effector properties. In the case of cell-surface immunoglobulin molecules, the heavy chain constant region anchors the protein into the cell membrane.

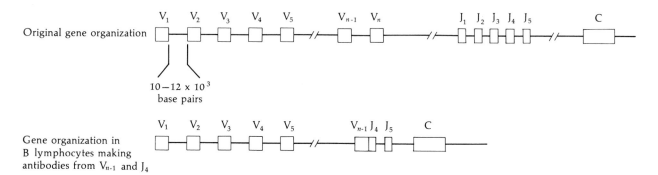

FIGURE 11

Rearrangement of the light-chain genes during B lymphocyte development. Prior to antigenic stimulation, one of the 200 or more V gene segments combines with one of the five J gene segments and moves closer to the constant (C) gene segment.

Creation of antibody light chain genes. Lymphocyte heavy and light chain genes are organized in segments. Light chain genes contain three segments (Figure 11). The first gene segment is the variable (V) region. It contains some 200 different sequences, all of which code for the first 97 amino acids of an antibody. The second segment, the J region, consists of four or five possible DNA sequences for the last 15–17 residues of the variable portion of the antibody. The third segment is the constant (C) region.

In lymphocyte development, one of the 200 V regions and one of these five J regions combine to form the variable portion of the antibody gene. This is done by moving a V region sequence to a J region sequence, eliminating the intervening DNA.

This gene rearrangement was first shown by Nobumichi Hozumi and Susumu Tonegawa. Hozumi and Tonegawa (1976) isolated DNA from a mouse embryo and from a light chain-secreting B cell tumor.[2] They then digested these two DNAs separately with the restriction enzyme *Bam*HI, which cleaved the DNA wherever it located the sequence GGATCC. The result was a series of random-sized fragments of DNA (the size of each fragment being determined by the length of the DNA molecule between two cleavage sites).

These DNA fragments were placed at one end of a gelatinous slab and an electric current was passed through the gel (Figure 12). The DNA migrated to the positive electrode, the smaller fragments moving faster than the larger ones.[3] (This separation of compounds by their

[2]Tumors were used because they produce an enormous amount of one specific immunoglobulin (and the mRNA for that immunoglobulin).

[3]Given the same charge to mass ratio, smaller fragments obtain a faster velocity than larger ones when propelled by the same energy. This is a function of the kinetic energy formula ($E = mv^2/2$); the velocity becomes inversely proportional to the square root of the mass.

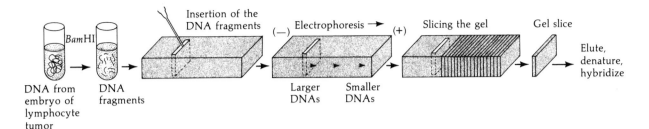

FIGURE 12
Electrophoretic separation of DNA fragments based on their size. DNA cut by restriction enzyme *Bam*HI is placed in a slot in a gel. In an electric current, the smaller pieces move faster than the larger ones. After separating the fragments according to their sizes, the gel can be sliced and the DNA fragments eluted for hybridization.

migration in an electric field is called ELECTROPHORESIS.) After separating the DNA fragments by their size, the gel (containing the DNA spread throughout it) was cut into several pieces, each containing pieces of DNA of a certain size.[4] The DNA from each slice of the gel was eluted and denatured. Part of this DNA was treated with radioactive RNA coding for the entire light chain isolated from the original B cell tumor. The other part was hybridized with radioactive RNA coding only for the C region of the light chain. The DNA from the embryo bound the light chain mRNA in two slices. The DNA of the first slice had a molecular weight (MW) of about 6 million, and the molecular weight of the DNA in the second band was 3.9 million. When the mouse embryo DNA was reacted with the light chain C region mRNA, only the 6 million MW DNA bound the RNA. Thus, in the mouse embryo, the C region was coded within DNA fragments having a molecular weight of 6 million (between *Bam*HI sites) and the V region was coded within a region of 3.9 million (Figure 13).

The lymphocyte tumor DNA, however, gave a very different result. The only lymphocyte DNA that bound the light chain mRNA had a molecular weight of 2.4 million. Moreover, it bound the light chain C region mRNA segment as well. Both the C and the V regions were found to be coded on the same fragment of DNA! The simplest explanation (and one confirmed by numerous other laboratories and methods; see Brack et al., 1978; Bernard et al., 1978; Seidman et al., 1979) was that two gene fragments, one coding for the light chain C region and one coding for a specific light chain V region, had fused together *to form a new gene*. A new gene had been created during the development

[4]The size of the DNA in each piece was determined by running DNAs of known sizes in an adjacent column.

(A) EXPERIMENT PROTOCOL

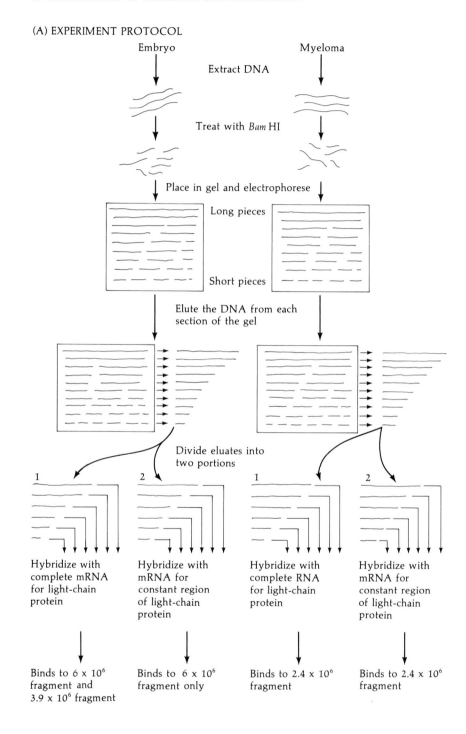

(B) RESULTS

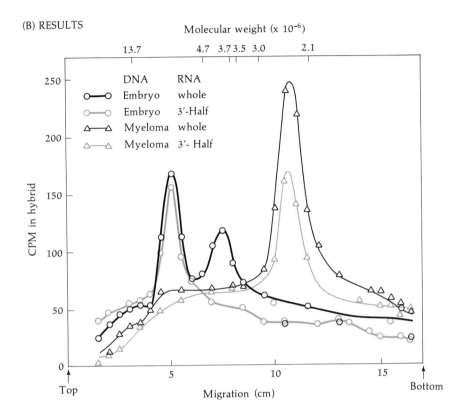

FIGURE 13
Protocol and results of the Hozumi and Tonegawa experiment. DNA from mouse embryo cells and B cell tumors (myelomas) were separately digested in *Bam*HI, electrophoresed, separated and eluted from their gels. After being made single-stranded, each eluted DNA sample was hybridized to (1) mRNA coding for the V and C regions of the immunoglobulin light chain and (2) a fragmented mRNA encoding only the constant region of that light chain protein. In the embryonic DNA, the variable and constant regions of the light chain protein were found on two different pieces of DNA (the constant region being on a piece having a molecular weight of 3.9×10^6, the variable region being on a DNA fragment of 6×10^6. In the lymphocyte tumor, the constant and variable regions are together on a single DNA fragment having a molecular weight of 2.4×10^6. (Graph adapted from Hozumi and Tonegawa, 1976.)

of the lymphocyte. Their model for such gene synthesis is presented in Figure 14.

Creation of antibody heavy chain genes. The heavy chain genes of antibodies contain even more segments than the light chain genes. Heavy chain gene segments include a V region (200 different sequences for the first 97 amino acids), a D region (10 to 15 different sequences of 3–14 amino acids) and a J region (four sequences for the last 15–17

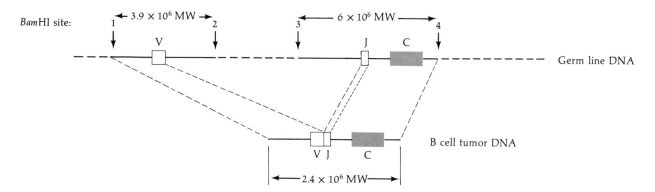

FIGURE 14
Model of the changes in DNA between embryonic cells and the B lymphocyte, according to the data of Hozumi and Tonegawa (1976).

amino acids of the V region). The next region is the C region. The heavy chain V region is formed by adjoining one V sequence and one D sequence to one J sequence (Figure 15A and B). This V region sequence is now adjacent to the first C region of the heavy chain—the C region for antibodies that can be inserted into the plasma membrane. Thus, an antibody molecule is formed from two genes created during the antigen-independent stage of B lymphocyte development. This antibody is placed in the cell membrane as an antigen receptor.

Class switching in the heavy chain gene. Upon stimulation by antigen, the B cell divides and differentiates into an antibody-secreting plasma cell. At first, the antibodies produced by these cells contain the same C region as before. However, as antibody synthesis continues, the C region can change. The first C region is called a mu constant (Cμ) region. The second C region—that of the secreted antibody—can remain mu (albeit with a modification that allows secretion), but it can also be a gamma, an epsilon, or an alpha constant region. Thus, a single heavy chain V region can be seen first on a mu constant region and later on, say, a gamma constant region. This is called CLASS SWITCHING. (The heavy chain CLASS of an antibody determines its mode of operation. Mu and gamma chains promote lysis, agglutination, or macrophage digestion of an antigen. Epsilon chains induce inflammatory responses, and alpha chains allow the antibodies to be secreted into mucus, tears, sweat, and milk without being digested.) Class switching is accomplished by taking the entire variable region gene segment (VDJ) and translocating it from in front of the mu constant region to a place in front of the gamma, alpha, or epsilon constant region (Figure 15C, D). This process results in the deletion of the mu constant region gene segment (Davis et al., 1980; Cory et al., 1980; Rabbits et al., 1980; Yaoita and Honjo, 1980).

Thus, the genome of a plasma cell is markedly different from that

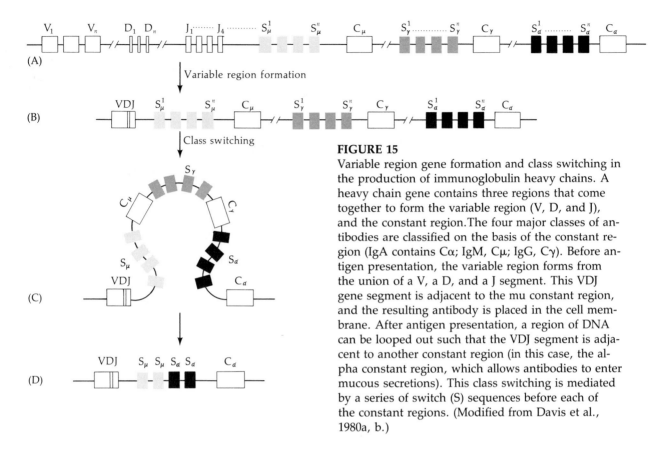

FIGURE 15

Variable region gene formation and class switching in the production of immunoglobulin heavy chains. A heavy chain gene contains three regions that come together to form the variable region (V, D, and J), and the constant region. The four major classes of antibodies are classified on the basis of the constant region (IgA contains Cα; IgM, Cμ; IgG, Cγ). Before antigen presentation, the variable region forms from the union of a V, a D, and a J segment. This VDJ gene segment is adjacent to the mu constant region, and the resulting antibody is placed in the cell membrane. After antigen presentation, a region of DNA can be looped out such that the VDJ segment is adjacent to another constant region (in this case, the alpha constant region, which allows antibodies to enter mucous secretions). This class switching is mediated by a series of switch (S) sequences before each of the constant regions. (Modified from Davis et al., 1980a, b.)

of any other cell. First, it has created a variable region gene sequence by bringing together different DNA segments. In all other organs, these DNA segments are apart; in B lymphocytes and plasma cells they have been brought together. Second, in many plasma cells, a portion of the genome (namely, the DNA of the mu heavy chain constant region) has been eliminated from the nucleus. A specific part of the genetic repertoire has been lost in the development of the plasma cell; new genes are created while other genes are destroyed.

What can we conclude from the preceding discussion? Present evidence suggests that the nuclei of differentiated cells do indeed retain most of their specific genetic information in a form that can be expressed under appropriate conditions. However, it is clear that some loss of genetic material accompanies at least one form of cellular differentiation. As yet, we have no way of knowing how many other cases of irreversible genetic change there may be during animal development, but our present knowledge of lymphocyte gene rearrangements suggests that irreversible genetic loss is a result, not a cause, of cellular differentiation.

SIDELIGHTS & SPECULATIONS

Gene alterations

FIGURE 16

Maize kernels exhibiting pigmented regions produced by cells in which a transposon has left the region of the pigment-forming genes. The cells within the white areas fail to produce pigment because the transposon is present and continues to inactivate the pigment-forming gene. (From Peterson, 1977; photographs courtesy of R. A. Peterson.)

Plasma cells have a genome deficient in certain DNA sequences needed for differentiation. It is possible, then, that this might be the mechanism for the restriction of nuclear potency, especially with regard to amphibian metamorphosis. This is only one form of genomic instability. Other examples are known. The tropical parasite *Trypanosoma brusei* alters its genome to produce different cell surface glycoproteins (thereby escaping the immune system). A similar mechanism appears to be responsible for the alteration of mating types in yeast (Hoeijmakers et al., 1980; Haber at al., 1980; Strathern et al., 1979). Here, each haploid cell contains both mating-type genes. The determination of mating type depends upon which of them is in a particular position on the chromosome. The positions are switched from one division to the next. Thus, in one generation, the *a* mating-type gene is in the "on" locus, whereas in the next generation, the α gene is in that position while the *a* gene waits nearby. In each of the three known cases, the genes involved are responsible for cell surface recognition phenomena. Such phenomena are found throughout vertebrate development (Part III), so it would not be too surprising if such irreversible genomic change were responsible for the loss of totipotency. In addition to the recombinational events described earlier, antibody diversity can also be generated by somatic mutation. Crews and colleagues (1981) have demonstrated that a significant number of point mutations occur during the differentiation of the B cell to a plasma cell. Another type of mutation comes from the insertion of TRANSPOSABLE ELEMENTS (or TRANSPOSONS) into a gene. Transposons are migratory pieces of DNA that can integrate throughout the genome. When such a sequence interrupts a structural gene, the gene is inactivated. The best-known examples are in bacteria, maize, and *Drosophila*. In maize, a transposon inserted in or near the gene for kernal pigment will produce a colorless kernal (McClintock, 1952; Peterson, 1980). However, upon removing itself from the gene, pigment synthesis is restored. The result is a variegated phenotype (Figure 16). In *Drosophila*, the *white-apricot* mutation is caused by the insertion of a transposable element into the region of the *white* gene (Green, 1980; Gehring and Paro, 1980). So far, such changes in the positions of transposons are not known to direct any developmental process. However, transposons do synthesize mRNA (Rubin et al., 1980), and it is possible that such DNA fragments are able to change the development of infected cells (Shimotohno and Temin, 1980).

It is clear, then, that the genome is a dynamic entity and not a perfectly stable structure. Although the changeability of the nucleus came as a surprise to many biologists, it was predicted by the founder of the gene theory, Thomas Hunt Morgan. Morgan, writing in 1927, acknowledged that "the most common genetic assumption is that the genes remain the same throughout this time [of development]." He reasoned that

the basic constitution of the gene remains always the same, the postulated addition or changes in the gene being of the same order as those that take place in the protoplasm. If the latter can change its differentiation in a new environment without losing its fundamental properties, why not the genes also? This question is clearly beyond the range of present evidence, but as a possibility, it need not be rejected. The answer, for or against such an assumption, will have to wait until evidence can be obtained by experimental investigation.

Differential RNA synthesis

The third postulate—that only a small portion of the genome is active in making tissue-specific products—has also been tested in flies and mice. Cytoplasmic RNA from insects and vertebrates hybridizes with less than 10 percent of the possible DNA sites available, but evidence shows that this RNA contains tissue-specific sequences. Becker (1959), studying *Drosophila*, and Beermann (1952), studying the larval gall midge *Chironomus*, found that there were regions of the chromosomes that were "puffed out." These puffs appeared in different places on the chromosomes in different tissues, and their appearances changed with the development of these cells (Figure 17). Furthermore, certain puffs could be induced or inhibited by certain physiological changes caused by heat or hormones (Clever, 1966; Ashburner, 1972a,b).

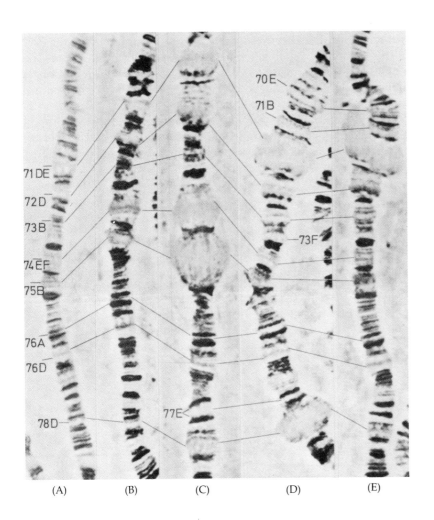

(A)   (B)   (C)   (D)   (E)

FIGURE 17
Puffing sequence of a portion of chromosome 3 in the larval *Drosophila melanogaster* salivary gland. (A, B) 110-hour larva; (C) 115-hour larva; (D, E) prepupal stage (4 hours apart). Note the puffing and regression of bands 74EF and 75B. Other bands (71DE, 78D) puff later and most did not puff at all during this time. (Courtesy of M. Ashburner.)

These puffs are now known to represent the spinning out of the polytene chromosome into a more loosely compacted arrangement (Figure 18); Beermann also presented the first evidence that these puffs were sites of active messenger RNA synthesis (Beermann, 1961). He found two interbreeding species of *Chironomus* that differed: one produced a major salivary protein, whereas the other did not (Figure 19). The producers had a large puff (Balbiani ring) at a certain band in the polytene chromosomes in cells of the larval salivary gland; that puff was absent in the nonproducers. When producer was mated with nonproducer, the hybrid larva produced an intermediate amount of salivary protein. When two hybrid flies were mated, the ability to produce salivary protein segregated in proper Mendelian fashion (1 high producer:2 intermediate producers:1 nonproducer). Moreover, whereas producers were found to have two puffs (one on each homologous chromosome), the intermediate producers had only one puff. Beermann concluded that the genetic information needed for the synthesis of this salivary protein was present in this distal chromosome band and that the synthesis of that product depended upon that band becoming a puff region.

Further proof that chromosomal puffs are making messenger RNA comes from studies of Balbiani ring 2 puffs in *Chironomus tentans*. Because of its exceptionally large size, Balbiani ring 2 (BR2) can be isolated by microdissection. Its products can be analyzed by electrophoresis and autoradiography (Lambert and Daneholt, 1975). Figure 20 shows the isolation of BR2 from chromosome 4 of *C. tentans*. The nuclear sap and cytoplasm of these salivary gland cells can also be collected to monitor the fate of the Balbiani ring RNA. Balbiani ring 2 transcription was demonstrated by giving isolated salivary glands radioactive RNA pre-

(A)

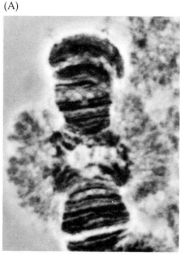

FIGURE 18
Proximal end of chromosome 4 from the salivary gland of *Chironomus pallidivitatus*, showing the enormous puff, BR2. (A) Phase-contrast photomicrograph of stained preparation showing extended puff on the polytene chromosome. (B) Diagrammatic representation of the BR2 region undergoing puffing. (A from Grossbach, 1973; photograph courtesy of U. Grossbach. B from Beermann, 1963.)

(B)

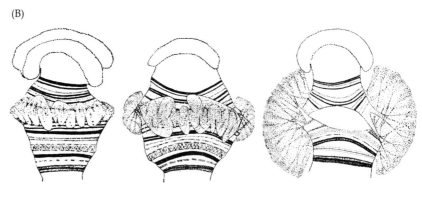

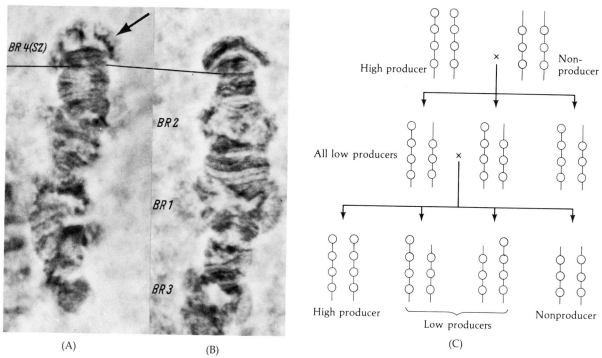

(A)   (B)   (C)

FIGURE 19

Correlation of puffing patterns with specialized functions in salivary gland cells of *Chironomus pallidivitatus*. (A) Chromosome 4 from a salivary gland cell producing granular secretion and showing additional Balbiani ring [BR4(SZ)]. (B) Chromosome 4 from a clear salivary cell, showing only BR1, 2, and 3. (C) Genetic evidence that the synthesis of a major salivary protein is dependent on the formation of BR4(SZ) puffs. Larvae with high amounts of granular secretions have salivary gland cells with BR4(SZ) puffs on both chromosomes 4, whereas those larvae with no such secretions have no such puffs. Intermediate producers have only one puffed BR4(SZ) region in each salivary cell making the secretion. (B from Beerman, 1961; photographs courtesy of W. Beermann.)

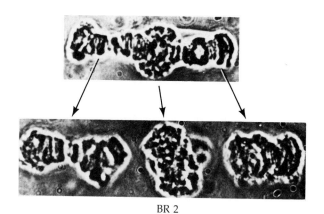

BR 2

FIGURE 20

Isolation of BR2 region of *Chironomus tentans* by micromanipulation. The intact chromosome 4 can be divided into three regions, one containing BR2. (From Lambert and Daneholt, 1975; photograph courtesy of B. Lambert.)

cursors. Radioactive RNA could then be extracted from the BR2 portion of the dissected chromosome (Lambert, 1972). This RNA was found to sediment at 75 S, thereby indicating an exceptionally large RNA of about 50,000 bases. This 75 S RNA hybridized specifically to the BR2 region of the chromosome, thereby showing that the puffed DNA and no other locus had been actively transcribing it (Figure 21). Unlike most RNAs, the 75 S BR2 RNA was not seen to decrease its size noticeably when it passed through the nucleus into the cytoplasm. This made it especially valuable because, if it were coding for a protein, the large number of ribosomes binding to it would form a complex with an extraordinarily high molecular weight. Therefore, salivary glands were again given radioactive RNA precursors and the cytoplasmic polysome complexes were sedimented through a sucrose gradient. Some of the polysomes had very high sedimentation values, a result suggesting the presence of an RNA capable of binding large numbers of ribosomes. This was indeed seen to be the case (Figure 22). The mRNA from these large polysomes was extracted (by dispersing the ribosome with EDTA) and 75 S RNA was obtained. This RNA hybridized specifically to the BR2 region of the chromosome (Wieslander and Daneholt, 1977). Thus, an RNA transcribed from a specific band of DNA which puffs in the larval salivary gland is seen later to be making proteins on cytoplasmic polysomes. Therefore, the puffs on salivary chromosomes are actively making messenger RNA.

More recently, Bonner and Pardue (1977) provided molecular evidence that at least some puffs reflect the developmentally regulated transcription of particular genes. Knowing the classic observation that ecdysone causes certain puffs to appear while diminishing others (Figure 23A), they grew *Drosophila* salivary glands in medium containing radioactive precursors of RNA. Some of these cultures contained ecdysone, whereas others did not. The RNA from these cultures was

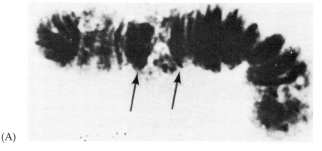

(A)

FIGURE 21
Transcription from the BR2 region of chromosome 4 of *Chironomus tentans* salivary gland cells. (A) Toluidine blue-stained chromosome preparation. Arrows show point at which the chromosome was cut for in vitro RNA synthesis. (B) In situ autoradiograph after hybridizing BR2 RNA to the chromosome preparation. (From Lambert, 1972; photographs courtesy of B. Lambert.)

(B)

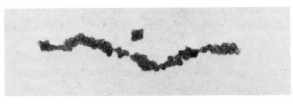

FIGURE 22

Electron micrographs of rapidly sedimenting polysomes derived from the BR2 region of *C. tentans* chromosome 4. The polysomes synthesize an extremely large secretory protein specific for the salivary gland. (From Daneholt et al., 1976; photographs courtesy of B. Daneholt.)

FIGURE 23

Ecdysone-induced puffs in cultured salivary gland cells of *D. melanogaster*. The chromosomal region here is the same as that shown in Figure 17. (A) Puffing cycle induced by ecdysone. (i) Unincubated control. (ii–v) Ecdysone-stimulated chromosomes at 25 minutes, 1 hour, 2 hours, and 4 hours, respectively. (B) In situ hybridization of radioactive RNA showing that chromosomes from cultured salivary glands transcribed RNA from specific regions when stimulated by ecdysone. (A courtesy of M. Ashburner. B after Bonner and Pardue, 1977.)

(A)

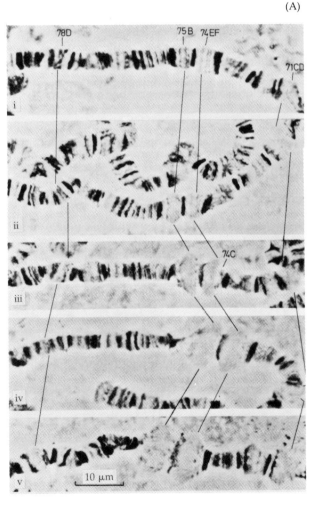

(B)

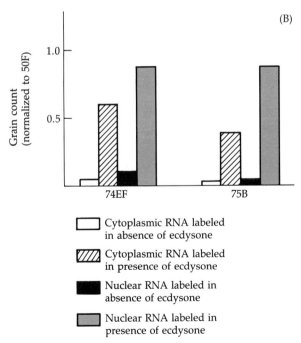

isolated and bound to salivary gland chromosomes by in situ hybridization. Hybridization of RNA to ecdysone-induced puffs was detected only if the RNA was labeled in the presence of ecdysone (Figure 23B). Hybridization to ecdysone-repressed puffs occurred best when the RNA came from those glands cultured without ecdysone. Moreover, the RNA from ecdysone-stimulated imaginal discs did not hybridize to the puffs of ecdysone-stimulated salivary glands. It appears, therefore, that these puffs are tissue specific, that they change with alteration in cellular activity, and that they are synthesizing tissue-specific RNA.

The use of cloned DNA fragments has significantly advanced our understanding of tissue-specific mRNAs. Darnell and his colleagues (Derman et al., 1981) isolated mRNA from mouse liver, made cDNA copies of this liver mRNA, and inserted these cDNA pieces into plasmids (Figure 24). If different tissues contain different populations of mRNA, certain of these recombinant plasmids should contain DNA sequences that can be recognized by liver mRNA but not by brain or kidney RNA populations. This, indeed, was the case. The recombinant plasmids were inserted into bacteria, and the individual clones were grown and blotted onto nitrocellulose filters. The colonies were lysed and the DNA affixed to the filters. The filters were then incubated with radioactive mRNA from liver cells or from brain cells and analyzed by autoradiography. The autoradiograms demonstrated that the liver mRNA bound to several clones that were not bound by the brain mRNA.

Messenger RNA showing stage and tissue specificity has more recently been visualized by a modification of in situ hybridization. The

FIGURE 24

Protocol for demonstrating organ-specific mRNAs. The cDNA from liver mRNA was made double stranded and the ends of these strands were modified, allowing their insertion into plasmids. The bacterial colonies containing the recombinant genes were lysed and hybridized with radioactive mRNA from the liver and other organs such as the brain.

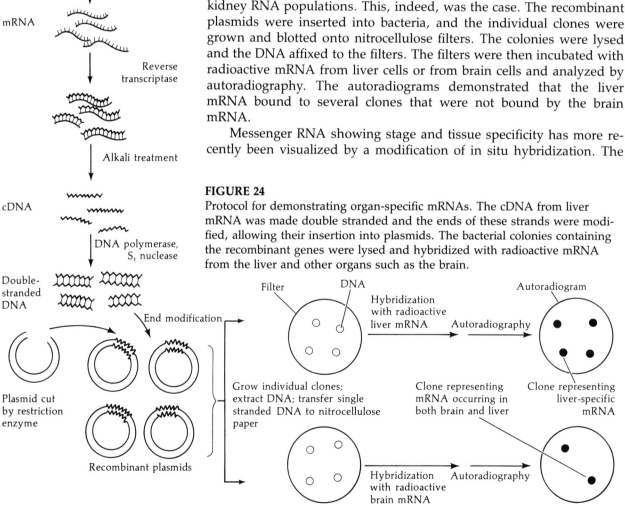

DNA of sea urchins was digested with a restriction enzyme and the pieces were inserted into plasmids. (The resulting collection of recombinant plasmids containing sea urchin genes is called the sea urchin DNA "library".) Some of these clones were found to contain DNA that hybridized specifically with mRNA isolated from the ectodermal regions of sea urchin plutei. These ectoderm-specific plasmids were then grown and made radioactive. This radioactive DNA was denatured into single strands and placed upon sea urchin embryos that had been fixed for in situ hybridization (Lynn et al., 1983). Figure 25 shows that the radioactive DNA from these plasmids recognizes RNA only in those gastrula cells that will give rise to the dorsal ectoderm of the pluteus. It also recognizes the dorsal ectoderm pluteus larva itself. Because these mRNAs are encoded by the nucleus and increase greatly at the blastula stage to synthesize a set of ectodermal acidic proteins, these studies show that certain messages are produced only by specific cells at specific times during development.

We can conclude from the studies presented in this chapter that (1) most differentiated cell types retain unused genes in a form that allows their expression under appropriate conditions; (2) there exists at least one case—the plasma cell—where loss of genetic material is associated with cellular differentiation (yet even here, the loss of specific DNA sequences is the result of, and not the cause of, differentiation); (3) there exist mRNAs whose synthesis is developmentally regulated. That is to say, the mRNA is produced only by certain cells at certain stages of development. (This does not imply that all mRNAs are so regulated or that all development is regulated by differential gene transcription.)

(A) (B)

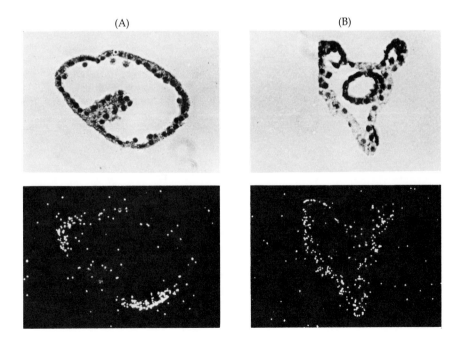

FIGURE 25
Visualization of ectoderm-specific mRNA in the sea urchin *S. purpuratus* by in situ hybridization. The upper row shows sections of midgastrula (A) and pluteus (B) larvae seen by phase-contrast microscopy. The lower row shows the same sections seen with dark-field microscopy and hybridized with radioactive cDNA made to an ectoderm-specific message. The hybridization (shown by white dots) is only to the RNA in those cells that are or become the dorsal ectoderm. (From Lynn et al., 1983; photographs courtesy of R. and L. Angerer.)

We have, then, the paradigm of differential gene expression: Differentiated cells contain the same genes but regulate their expression such that different cells make different proteins. The remaining chapters of this section will investigate the mechanisms by which differential gene expression takes place.

Literature cited

Ashburner, M. (1972a). Patterns of puffing activity in the salivary glands of *Drosophila*. VI. Induction by ecdysone in salivary glands of *D. melanogaster* cultured *in vitro*. *Chromosoma* 38: 255–281.

Ashburner, M. (1972b). Puffing patterns in *Drosophila melanogaster* and related species. In W. Beerman (ed.), *Developmental Studies on Giant Chromosomes*, Springer-Verlag, Berlin, pp. 101–151.

Barnett, T., Pachl, C., Gergen, J. P. and Wensink, P. C. (1980). The isolation and characterization of *Drosophila* yolk protein genes. *Cell* 21: 729–738.

Becker, H. J. (1959). Die Puffs der Speicheldrüsenchromosomen von *Drosophila melanogaster*. I. Beobachtungen zum Verhalten des Puffmusters im Normalstamm und bei zwei Mutanten, giant- und lethal-giant larvae. *Chromosoma* 10: 654–678.

Beermann, W. (1952). Chromomerenkonstanz und spezifische Modifikationen der Chromosomenstruktur in der Entwicklung und Organdifferenzierung van *Chironomus tentans*. *Chromosoma* 5: 139–198.

Beermann, W. (1961). Ein Balbiani-ring als Locus einer Speicheldrüsen-Mutation. *Chromosoma* 12: 1–25.

Beermann, W. (1963). Cytological aspects of information transfer in cellular differentiation. *Am. Zool.* 3: 23–28.

Bernard, O., Hozumi, N. and Tonegawa, S. (1978). Sequences of mouse immunoglobulin light chain genes before and after somatic change. *Cell* 15: 1133–1144.

Blattner, F. R., Blechl, A. E., Denniston-Thompson, K., Faber, H. E., Richards, J. E., Slightom, J. L., Tucker, P. W. and Smithies, O. (1978). Cloning human fetal γ globin and mouse α type globin DNA: Preparation and screening of shotgun collections. *Science* 202: 1279–1283.

Bonner, J. T. and Pardue, M. L. (1977). Ecdysone-stimulated RNA synthesis in salivary glands of *Drosophila melanogaster*: Assay by *in situ* hybridization. *Cell* 12: 219–225.

Brack, C., Hirama, M., Lenhard-Schuller, R. and Tonegawa, S. (1978). A complete immunoglobulin gene is created by somatic recombination. *Cell* 15: 1–14.

Brown, J. E. and Weiss, M. C. (1975). Activation of production of mouse liver enzymes in rat hepatoma–mouse lymphoid hybrids. *Cell* 6: 481–493.

Burnett, F. M. (1959). *The Clonal Selection Theory of Immunity*. Vanderbilt Press, Nashville.

Clever, U. (1966). Induction and repression of a puff in *Chironomus tentans*. *Dev. Biol.* 14: 421–438.

Cohen, S. N., Chang, A. C. Y., Boyer, H. W. and Helling, R. B. (1973). Construction of biologically functional bacterial plasmids *in vitro*. *Proc. Natl. Acad. Sci. USA* 70: 3240–3244.

Cory, S., Jackson, J. and Adams, J. M. (1980). Deletions in the constant region locus can account for switches in immunoglobulin heavy chain expression. *Nature* 285: 450–456.

Crews, S., Griffin, J., Huang, H., Calame, K. and Hood, L. (1981). A single V_H gene segment encodes the immune response to phosphorylcholine: Somatic mutation is correlated with the class of the antibody. *Cell* 25: 59–66.

Daneholt, B., Case, S. T., Hyde, J., Nelson L. and Wieslander, L. (1976). Production and fate of Balbiani ring products. *Prog. Nucleic Acid Res.* 19: 319–334.

Davis, M. M., Calame, K., Early, P. W., Livant, D. L., Joho, R., Weissman, I. L. and Hood, I. (1980). An immunoglobulin heavy chain gene is formed by at least two recombinational events. *Nature* 283: 733–739.

Davis, M. M., Kim, S. K. and Hood, L. (1980). Immunoglobulin class switching: Developmentally regulated DNA rearrangements during differentiation. *Cell* 22: 1–2.

Derman, E., Krauter, K., Walling, L., Weinberger, C., Ray, M. and Darnell, J. E., Jr. (1981). Transcriptional control of the production of liver-specific mRNAs. *Cell* 23: 731–739.

Edelman, G. M. (1970). The structure and function of antibodies. *Sci. Am.* 223(2): 34–42.

Gehring, W. J. and Paro, R. (1980). Isolation of a hybrid

plasmid with homologous sequences to a transposing element of *Drosophila melanogaster*. *Cell* 10: 897–904.

Green, M. M. (1980). Transposable elements in *Drosophila* and other diptera. *Annu. Rev. Genet.* 14: 109–120.

Grossbach, U. (1973). Chromosome puffs and gene expressions in polytene cells. *Cold Spring Harbor Symp. Quant. Biol.* 38: 619–627.

Haber, J. E., Rogers, D. T. and McCusker, J. H. (1980). Homothallic conversions of yeast mating types occur by intrachromosomal recombination. *Cell* 22: 277–289.

Hoeijmakers, J. H. J., Frasch, A. C. C., Bernards, A., Borst, P. and Cross, G. A. M. (1980). Novel expression-linked copies of the genes for variant surface antigens in trypanosomes. *Nature* 284: 78–80.

Hozumi, N. and Tonegawa, S. (1976). Evidence for somatic rearrangement of immunoglobulin genes coding for variable and constant regions. *Proc. Natl. Acad. Sci. USA* 73: 3628–3632.

Lambert, B. (1972). Repeated DNA sequences in a Balbiani ring. *J. Mol. Biol.* 72: 65–75.

Lambert, B. and Daneholt, B. (1975). Microanalysis of RNA from defined cellular components. *Methods Cell Biol.* 10: 17–47.

Lynn, D. A., Angerer, L. M., Bruskin, A. M., Levin, W. H. and Angerer, R. C. (1983). Localization of a family of mRNAs in a single cell type and its precursors in sea urchin embryos. *Proc. Natl. Acad. Sci. USA* 80: 2656–2660.

McCarthy, B. J. and Hoyer, B. H. (1964). Identity of DNA and diversity of messenger RNA molecules in normal mouse tissues. *Proc. Natl. Acad. Sci. USA* 52: 915–922.

McClintock, B. (1951). Chromosome organization and genic expression. *Cold Spring Harbor Symp. Quant. Biol.* 16: 13–47.

Morgan, T. H. (1927). *Experimental Embryology.* Columbia University Press, New York.

Nathans, D. and Smith, H. O. (1975). Restriction endonucleases in the analysis and restructuring of DNA molecules. *Annu. Rev. Biochem.* 44: 273–293

Pardue, M. L. and Gall, J. G. (1970). Chromosomal localization of mouse satellite DNA. *Science* 168: 1356–1358.

Peterson, J. A. and Weiss, M. C. (1972). Expression of differentiated functions in hepatoma cell hybrids: Induction of mouse albumin production in rat hepatoma–mouse fibroblast hybrids. *Proc. Natl. Acad. Sci. USA* 69: 571–575.

Peterson, R. A. (1977). The position hypothesis for controlling elements in maize. In A. I. Bukhari, J. A. Shapiro and S. L. Adhya (eds.), *DNA Insertion Elements, Plasmids, and Episomes.* Cold Spring Harbor Laboratory, Cold Spring Harbor, NY, p. 429.

Peterson, R. A. (1980). Instability among the components of a regulatory element transposon in maize. *Cold Spring Harbor Symp. Quant. Biol.* 45: 447–455.

Rabbits, T. H., Forster, A., Dunnick, W. and Bentley, D. L. (1980). The role of gene deletion in the immunoglobulin heavy chain switch. *Nature* 283: 351–356.

Rubin, G. M., Brosein, W. J., Dunsmuir, P., Flavell, A. J., Lavis, R., Strobel, E., Toole, J. J. and Young, E. (1980). *Copia*-like transposable elements in *Drosophila* genome. *Cold Spring Harbor Symp. Quant. Biol.* 45: 619–628.

Seidman, J. G., Max, E. E. and Leder, P. (1979). A κ-immunoglobulin gene is formed by site-specific recombination without further somatic mutation. *Nature* 280: 370-375.

Shimotohno, K., and Temin, H. M. (1980). Evolution of retroviruses from cellular moveable genetic elements. *Cold Spring Harbor Symp. Quant. Biol.* 45: 719–730.

Slightom, J. L., Blechl, A. E. and Smithies, O. (1980). Human fetal γ^G and γ^A globin genes: Complete nucleotide sequences suggest that DNA can be exchanged between these duplicated genes. *Cell* 21: 627–638.

Strathern, J. N., Newton, C. S., Herskowitz, I. and Hicks, J. B. (1979). Isolation of a circular derivative of yeast chromosome. III. Implications for the mechanism of mating type interconversion. *Cell* 18: 309–319.

Wallace, H. and Birnstiel, M. L. (1966). Ribosomal cistrons and the nucleolar organizer. *Biochim. Biophys. Acta* 114: 296–310.

Weislander, L. and Daneholt, B. (1977). Demonstration of Balbiani ring RNA sequence in polysomes. *J. Cell Biol.* 73: 260–264.

Wetmur, J. G. and Davidson, N. (1968). Kinetics of renaturation of DNA. *J. Mol. Biol.* 31: 349–370.

Yaoita, Y. and Honjo, T. (1980). Deletion of immunoglobulin heavy chain genes from expressed allelic chromosome. *Nature* 286: 850–853.

CHAPTER

Transcriptional regulation of gene expression: the nature of eukaryotic genes

We have entered the cell, the mansion of our birth, and have started the inventory of our acquired wealth.
—ALBERT CLAUDE (1974)

Introduction

The development of a eukaryotic organism depends upon the differential expression of specific genes at specific places during specific times. Therefore, the problem of differentiation becomes one of differential gene expression. In any cell at any time, only a certain subset of the inherited genetic repertoire is being expressed. There are numerous ways to regulate gene expression, and controls have evolved at the levels of transcription, RNA processing, translation, and protein modification.

In this chapter, we shall discuss the regulation of gene expression at the transcriptional level. We shall first study the physical structure of chromatin and the chemical nature of the gene, detailing the conditions necessary for transcription to take place. In Chapter 12, we shall document cases in which development is seen to be regulated by differential gene transcription.

Structure of the eukaryotic genome

When we compare the eukaryotic genome with that of bacteria, we find that eukaryotes have an enormous amount of DNA. The *Escherichia coli* genome, for instance, contains some 4.5×10^6 base pairs of DNA, whereas the human genome contains about 2.8×10^9 base pairs. Thus, for every bacterial gene, there are some 600 human DNA sequences. This is very surprising because it is estimated that bacteria synthesize most of the known eukaryotic enzymes (Britten and Davidson, 1971). Moreover, as animals have become more complex, this DNA content has tended to become increasingly large (Figure 1). There is a great deal of diversity within each group, but the minimum genome size has obviously increased greatly over the amount that is sufficient to direct the processes of bacteria. What is the function of all this DNA?

DNA renaturation experiments show that much of this extra DNA is highly repetitive. Some of the sequences are present thousands or even millions of times in the genome. Such DNA would not be expected to encode enzymes or other proteins. In these experiments, the DNA of an organism is separated into its component strands and the strands are allowed to renature after they find their respective complementary sequences. When bacteriophage DNA is sheared into small pieces and heated to 90°C, the double helices separate into single strands. When the solution of single-stranded DNA is then cooled to 65°C, the strands are able to reassociate, once they meet up with complementary sequences. At any given time, the solution can be poured over a column of hydroxyapatite (a form of calcium phosphate), which absorbs the double-stranded DNA and lets the single strands pass through. The double helices can be eluted from the hydroxyapatite with high concen-

FIGURE 1
DNA content of the haploid genome observed in each group of animals. The bars show the ranges of the haploid genome sizes. The animal groups have been organized according to the number of different cell types described for representative members of their respective groups. (Modified from Raff and Kaufman, 1983.)

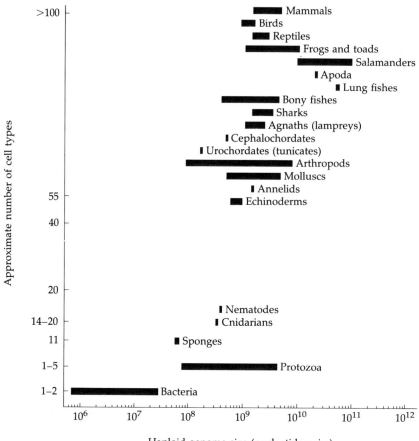

trations of sodium phosphate. By this procedure, one can measure the percentage of reassociated DNA at any given time (Britten and Kohne, 1968).

Because two complementary molecules have to come together in solution, the kinetics of renaturation follow that for a second-order reaction. The rate at which two complementary chains reassociate varies directly with their concentration and the time allowed for reaction. The curves obtained from reassociating prokaryotic DNA are shown in Figure 2. Notice that the ordinate is inverted (starting at 0 percent reassociation) and that the abscissa is a function of the original DNA concentration (C_0) multiplied by time (t). Hence, they are often referred to as "C_0t CURVES." The reason for using both concentration and time on this axis can be seen in the following manner. If one places 1 mg of DNA into solution (DNA that contains many copies of a single, short sequence and its complement), each sequence should find its mate fairly rapidly. However, if the original concentration (C_0) is cut in half, it should take twice as much time (t) for complementary strands to find each other. Thus, the extent of reassociation should be a product of C_0

by t. Now consider what would happen if that 1 mg of DNA were to contain a thousand different sequences instead of just one. Although the total DNA concentration would be the same as in the first experiment, the concentration of any one sequence (and its complement) would be 1000 times lower, so it would take 1000 times longer for complementary strands to bump into one another in the second sample than it would take in the first sample. This aspect of reassociation has a very important consequence. The half-C_0t value—the C_0t value at which half the DNA has reannealed—is directly proportional to the COMPLEXITY of the genome, that is, to the number of different nucleotide sequences present. Bacteriophage T4, which has a genome of 1.5×10^5 base pairs, has a half-C_0t value around 0.3 mole $\times$ sec/liter, whereas *E. coli*, with a genome containing 4.5×10^6 base pairs, has a half C_0t value 30 times greater (Figure 2). Because T4 has 30 times less genetic information, each gene will be present 30 times more often in a milligram of T4 DNA than in a milligram of *E. coli* DNA.

When prokaryotic DNA is tested, simple curves are generated because each gene is represented in the genome only once (or at most, very few times) and has as good a chance at finding its mate as any other sequence. With eukaryotic DNA, however, a more complicated curve is generated (Figure 3). Here one finds some sequences reassociating at a C_0t of 5×10^3, whereas other sequences reassociate much earlier, at 5×10^{-3}. This means that some sequences of the eukaryotic

FIGURE 2
Reassociation of DNA from various sources. The curve represents the normal kinetics expected when each gene is present at the same frequency in the genome. Because the renaturing reaction demands that two complementary sequences come together, the rate of the reaction can be described as $dC/dt = -kC^2$, where C is the concentration of single-stranded sequences at time t and k is the reassociation constant. For any given time, the ratio of single-stranded DNA present to that found initially can be found by integrating the formula to get $C/C_0 = 1/(1 + kC_0t)$, where C_0 is the initial concentration. (After Britten and Kohne, 1968.)

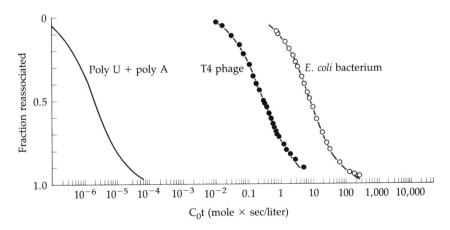

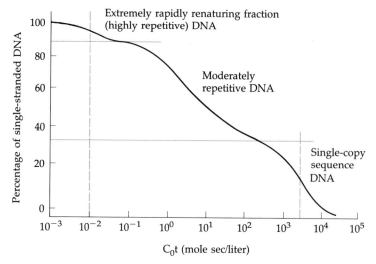

FIGURE 3

Reassociation of DNA from a typical eukaryotic organism. The curve shows wide ranges of gene representation within the genome. The single-copy sequences show the expected reassociation kinetics of DNA sequences represented equally in each nucleus. (After Hood et al., 1975.)

genome are present in a million more copies than certain other sequences. In general, the C_0t curves of eukaryotic organisms are able to identify three sets of DNA.

1. The first group to reassociate is called the EXTREMELY RAPIDLY RENATURING FRACTION (ERRF). This DNA is not used for protein synthesis and does not seem to have any role in differentiation. It is usually found in the centromeric region of the chromosome (Figure 4) and

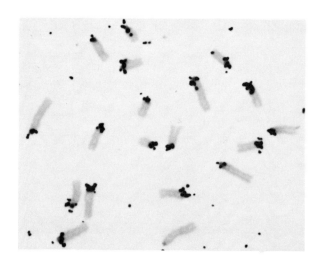

FIGURE 4

Localization of ERRF DNA by in situ hybridization of radioactive mouse ERRF DNA to mouse cell chromosomes. The ERRF was found to bind to the chromosome around the centromere. (From Pardue and Gall, 1970; photograph courtesy of M. L. Pardue.)

probably serves to keep chromosomes intact during mitosis. The ERRF represents approximately 10 percent of the mouse genome (*Mus musculus*) and can constitute as much as 30 percent of the genome in some organisms. The ERRF DNA is not known to be transcribed in somatic cell nuclei.

2. The last set of DNA sequences to reassociate is that containing the SINGLE-COPY SEQUENCES. Most enzymes and structural proteins are coded by the DNA in this fraction, and this is the region where the classic genes have been defined. We know from Mendelian genetics that most proteins are coded by only two genes per diploid cell, although certain proteins (such as hemoglobins, actins, tubulins, and histones) are coded by small families of closely related genes.

3. The MODERATELY REPETITIVE DNA may be the most interesting from the standpoint of developmental gene regulation. These genes are represented about 10^3 to 10^5 times per nucleus and represent about 15 percent of the mammalian genome. Rather than being located at any one place (like the ERRF), the moderatively repetitive DNA is interspersed among the single-copy sequences. This was first shown by double renaturation experiments (Davidson et al., 1973a,b). DNA was sheared to 3700-base pair fragments and denatured. The single strands were then allowed to reassociate at a C_0t value sufficiently low to enable the moderately repetitive sequences (but not the single-copy sequences) to find their complementary strands. Those DNAs containing double-stranded regions were isolated on hydroxyapatite and eluted. The eluted DNAs were denatured again and sheared into lengths of DNA about 450 base pairs long. When these were reassociated to the same low C_0t value, only 45 percent formed double-stranded structures (Figure 5). Therefore, moderately repetitive and single-copy sequences must have existed next to each other and were separated when the fragment size became smaller.

This interspersion of moderately repetitive and single-copy sequences has been demonstrated by electron microscopy of long DNA fragments reassociated at low C_0t values (Chamberlin et al., 1975). One finds that for a given piece of DNA there are regions of double-strandedness that alternate with single-stranded regions (Figure 6). The double-stranded regions are the repetitive sequences that have found partners that were originally located elsewhere in the genome. The single-stranded areas represent unique sequences that are located adjacent to the equivalent repetitive sequences. Recent studies on nuclear RNA have shown that moderately repetitive DNA can be found in the intervening sequences between the coding regions of the genes (see later), a finding indicating that moderately repetitive sequences are often transcribed but not translated (Ryffel et al., 1981). There is also evidence in sea urchins, slime molds, and rat brains that specific repeat sequences are adjacent to or within genes that get expressed together (Constantini et al., 1978; Kimmel and Firtel, 1979; Zucker and Lodish, 1981). We shall see in Chapter 13 that this may be an important clue to how gene expression is controlled during development.

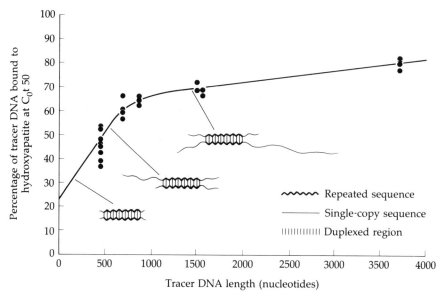

FIGURE 5

The interspersion of single-copy and moderately repetitive DNA sequences, shown by the effect of fragment length on the amount of DNA retained by hydroxyapatite. DNA is sheared and annealed at a C_0t value that will allow moderately repetitive DNA to hybridize. Regions of single-copy DNA that would not be bound to the hydroxyapatite column can be stuck to it by adjacent repetitive DNA that becomes double-stranded at the selected C_0t value. As fragment size decreases, the moderately repetitive DNA is separated from the single-copy sequences. The diagrams represent electron micrographs of the DNA reassociated at that length. (From Davidson et al., 1973a.)

FIGURE 6

Interspersion of single-copy and moderately repetitive DNA shown by electron microscopy. *Xenopus* DNA was sheared to 2400 base pairs, denatured, and run over a hydroxyapatite column after incubating to C_0t 10. The bound fraction was eluted from the column and observed under the electron microscope. An interpretive diagram is shown in Figure 5. The bar represents the length of 1000 base pairs. (From Davidson et al., 1973b; photographs courtesy of E. H. Davidson.)

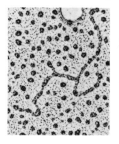
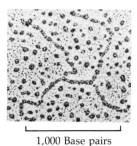
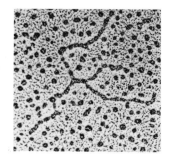

1,000 Base pairs

Structure of eukaryotic chromatin: Nucleosomes

In addition to having a nuclear membrane that separates the events of transcription from those of translation, eukaryotic cells are also characterized by the presence of nucleoprotein chromosomes. Eukaryotic chromosomes contain as much protein (by weight) as nucleic acid, and this DNA–protein complex is called CHROMATIN. The most abundant of the chromatin proteins are the HISTONES. These five proteins—H1, H2A, H2B, H3, and H4—are highly basic polypeptides containing a large proportion of the positively charged amino acids lysine and arginine. Another histone, H5, is found in a few cell types (notably avian and amphibian erythrocytes) in which the DNA is extremely tightly packed. In these cells, H5 replaces H1.

The five major histones are present throughout the animal, plant, and protist kingdoms and have changed very little during the course of evolution. Such conservation usually implies that the proteins must be playing an extremely important role in all the cells studied; and for histones, the major function is the packaging of DNA into specific coiled structures called NUCLEOSOMES. The nucleosome is the basic unit of chromatin structure; it is composed of a spherical histone octamer (H2A, H2B, H3, H4)$_2$ wrapped about with two loops of approximately 140 base pairs of DNA (Figure 7). Between each nucleosome is another 60

FIGURE 7
(A) Model of nucleosome structure. About 140 base pairs of DNA encircle the histone octamer (containing histones H2A, H2B, H3, and H4), and about 60 base pairs of DNA link the nucleosome together. (B) Model for the arrangement of nucleosomes into the highly compacted, solenoidal chromatin structure. (A after Wolfe, 1983; B after Stewart and Hunt, 1982.)

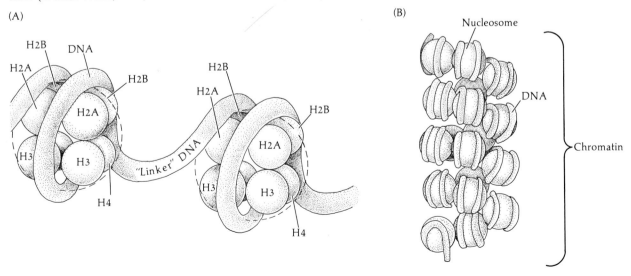

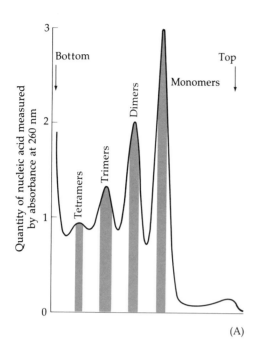

(A)

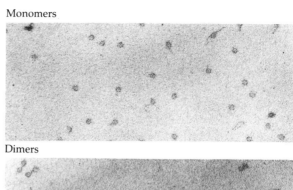

Monomers

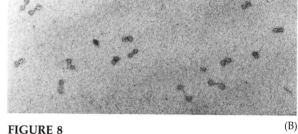

Dimers

(B)

FIGURE 8

Characterization of nucleosome multimers. (A) Sucrose density gradient separating discrete groups of particles formed by treating rat liver chromatin with micrococcal DNase. (B) Electron micrographs of the four sucrose gradient peaks, showing nucleosome monomers, dimers, trimers, and tetramers. (C) The particles of chromatin in each of the peaks were isolated, and the DNA was extracted from the chroma-

or so base pairs of DNA; these "linker" nucleotides can be covered by H1. This arrangement was suggested by the cleavage pattern of chromatin when it is subjected to small amounts of a deoxyribonuclease (DNase) derived from a certain bacterium (*Micrococcus*). This treatment cleaves the DNA at several sites, and the resulting chromatin fragments can be separated from each other by centrifugation through a sucrose gradient (Figure 8). When the DNA from the treated chromatin is extracted and run on gels, one obtains a clearly defined pattern. Instead of showing random-sized pieces of DNA, the DNA appears to be cleaved into pieces containing multiples of 200 base pairs (Hewish and Burgoyne, 1973; Noll, 1974). When these fragments of DNase-treated chromatin were observed under the electron microscope, the fragment containing 200 base pairs of DNA was seen as a spherical body with a small tail. The 400-base pair sequence is found in two connected bodies, and the 600-base pair sequence has three connected spherical units (Finch et al., 1975). Thus, it was concluded that the basic chromatin subunit contains about 200 base pairs of DNA and that micrococcal DNase preferentially cleaves the DNA between nucleosomes.

The role of histones in forming this structure was demonstrated by Kornberg and Thomas (1974), who were able to synthesize such chromatin subunits in the test tube. Mixing histones H2A, H2B, H3, and

Trimers

Tetramers

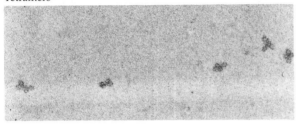

(B)

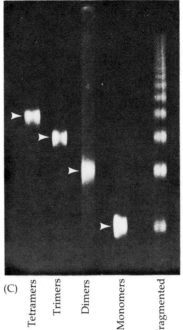

tin and analyzed by electrophoresis. Discrete integral jumps of 200 base pairs are seen. The right-hand side of the gel shows DNA extracted after DNase digestion and before sedimentation on sucrose. (B from Finch et al., 1975.)

(C) Tetramers Trimers Dimers Monomers Unfragmented

H4 together merely created tetramers of H2A and H2B and tetramers of H3 and H4. However, when DNA was added to the histone mixture, the nucleic acids and histones aggregated to form a nucleosome structure with the same physical and chemical properties as normal chromatin.

Chromatin can thus be visualized as a string of nucleosome beads linked by 10–100 base pairs of DNA. The histone octamer consists of two molecules each of H2A, H2B, H3, and H4 plus about 140 base pairs of DNA that is wrapped around the protein bead. Histone H1 is not present on the nucleosomes but can be bound to the linker DNA between them, depending on the physiological state of the cell. These nucleosomes are packed into tight structures themselves, and the H1 may be involved in the winding of nucleosomes about each other, especially during the preparation for cell division (Figure 9). Moreover, in at least one instance (Schlissel and Brown, 1984), the H1-dependent conformation of nucleosomes appears to inhibit the transcription of specific genes in somatic cells.

Regulation of gene accessibility on the nucleosome

The homogeneity of histones and the ubiquity of nucleosomes throughout all eukaryotes suggest that the nucleosomes do not, of themselves,

FIGURE 9
Role of H1 in compacting chromatin. (A) Chicken liver chromatin observed in the electron microscope. The beads represent the nucleosomes. (B) The same chromatin after the removal of histone H1 by salt elution. The chromatin has become far less compacted. The bar indicates 0.25 μm. (From Oudet et al., 1975; photograph courtesy of P. Chambon.)

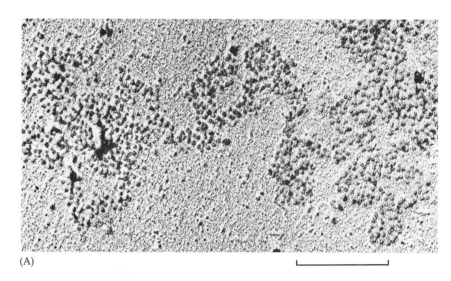

(A)

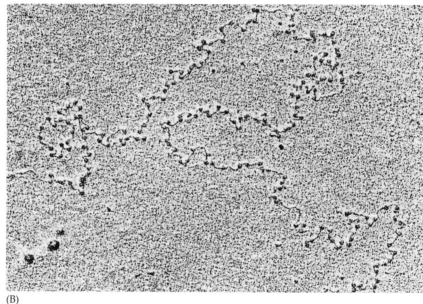

(B)

create the conditions for differential gene transcription. How, then, might the nucleosomes be modified to enable specific genes to be transcribed? One such modification may merely involve regulating the number of nucleosomes in a given region of the genome. Theoretically, the absence of nucleosomes would ensure that all the DNA was accessible to RNA polymerase and that transcription could be efficiently accomplished. In the milkweed bug *Oncopeltus fasciatus*, the portion of the genome containing the ribosomal RNA genes is almost totally devoid of nucleosomes and the DNA in this region is very actively transcribed. No other region of the *Oncopeltus* genome is so free of nucleosomes, a finding suggesting that this mechanism may be peculiar to their ribosomal RNA genes (Foe et al., 1976). Although this finding has been confirmed in *Oncopeltus* and in some other species (Labhart and Koller, 1982; Figure 10), the ribosomal genes of many other species can be rapidly transcribed despite being wrapped about nucleosomes. Most active single-copy genes, such as the globin gene of developing erythrocytes, are also found in nucleosomes; so the absence of nucleosomes is not necessary for transcription.

If quantitative differences in nucleosome number are not the rule in gene regulation, might there be differences in the nucleosomes themselves? There is now an excellent correlation between the transcriptional accessibility of specific genes and specific nonhistone chromosomal proteins, which may bind to the nucleosome to make these genes accessible. The major evidence for this comes from studies concerning the DNase susceptibility of specific genes in different cells. If a gene is to be transcribed, it must first be capable of being recognized by RNA polymerase. The accessibility of the gene can be detected by treating the chromatin of a tissue with small amounts of DNase I (a type of DNase different from that of *Micrococcus*). The DNA of the treated chromatin is extracted and mixed with radioactive cDNA for a particular gene (Figure 11).

FIGURE 10
Active and inactive regions of the *Xenopus* genome. Isolated *Xenopus* chromatin was treated with RNase to remove nascent RNA strands and was fixed for electron microscopy. Two regions of chromatin were seen: on the right, region P, in which the large protein molecules (believed to be RNA polymerase) were present; and, on the left, region N, in which nucleosomes without polymerase were seen. (From Labhart and Koller, 1982; photograph courtesy of T. Koller.)

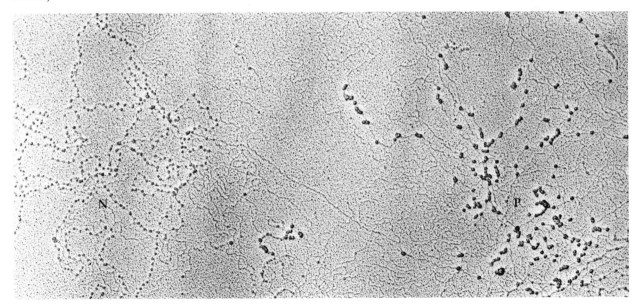

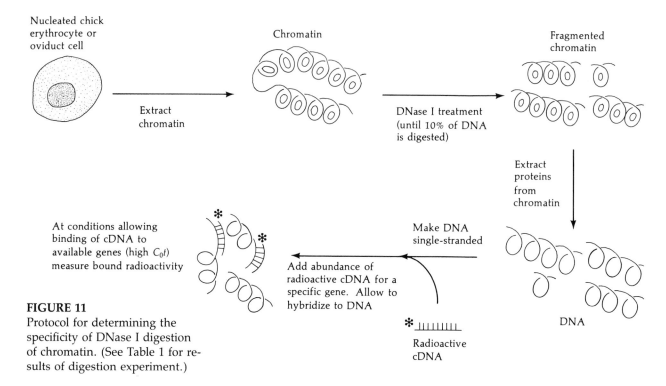

FIGURE 11
Protocol for determining the specificity of DNase I digestion of chromatin. (See Table 1 for results of digestion experiment.)

This binding test is based on the assumption that DNase I digests only those genes actively transcribing RNA. All other genes have been protected by chromatin proteins and after extraction are available to bind with radiolabeled cDNA. When the cDNA finds sequences to bind to, then the gene has been protected from digestion by chromatin proteins—and it probably was not accessible to RNA polymerase either. However, when the cDNA probe does not find sequences to bind to, then the gene has been exposed to the DNase—and probably was accessible to RNA polymerase, too.

The DNase I sensitivity of a given gene was found to be dependent upon the cell type in which it resides (Table 1; Weintraub and Groudine, 1976). When chromatin from developing chick *red blood cells* was treated with DNase I and its DNA extracted and mixed with radioactive globin cDNA, the globin cDNA found little with which to bind. The globin genes in the chromatin had been digested by the small amounts of DNase I. However, treating *brain* cell chromatin with the same amounts of DNase I did not destroy the globin genes. Therefore, the globin gene was accessible to outside enzymes in developing red blood cell chromatin but not in brain cell chromatin.

Similarly, the ovalbumin gene is susceptible to DNase I digestion in oviduct chromatin but not in red blood cell chromatin. When the chromatin is treated with DNase I and the DNA is extracted, ovalbumin cDNA is able to find sequences in the erythrocyte preparation but not in the DNA from the treated oviduct chromatin. Here, too, we have

TABLE 1
Binding studies with DNase I-treated DNAs

Source of DNase I-treated chromatin	Radioactive cDNA probe	Percentage of maximum binding of radioactive cDNA to DNA extracted from treated chromatin
Chick red blood cell DNA (no DNase treatment)	Globin cDNA	94
Chick brain cell chromatin	Globin cDNA	90–100
Chick fibroblast chromatin	Globin cDNA	90–100
Chick red blood cell chromatin	Globin cDNA	25
Chick red blood cell chromatin	Ovalbumin cDNA	90–100

Source: Weintraub and Groudine (1976).

clear correlation of differential gene regulation and chromatin structure.

When chromatin is digested by DNase I, there is a preferential release of two nonhistone chromatin proteins: HMG (high mobility group) 14 and HMG 17. These low-molecular-weight proteins can be extracted from chromatin with 0.35 M NaCl. When these proteins are removed from the chromatin, the differential susceptibility of specific genes to DNase I treatment is simultaneously removed. Thus, when the HMG proteins are extracted from red blood cell chromatin, the globin genes are no longer specifically susceptible to DNase I digestion (Figure 12). However, when these proteins are added back to such

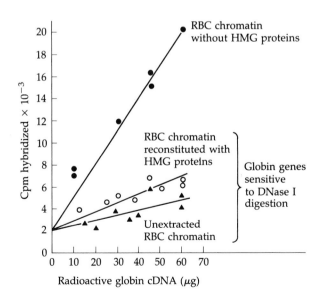

FIGURE 12
Reconstruction of DNase I sensitivity of the globin gene. The HMG proteins were removed from chick red blood cells (RBC) chromatin. One fraction was maintained without HMG proteins and the other fraction had them restored. DNase I was added to both groups and to unextracted RBC chromatin such that 10 percent of the DNA was digested. Radioactive globin cDNA bound to complementary sequences much better when the chromatin lacked HMG proteins. (Data from Weisbrod and Weintraub, 1979.)

depleted chromatin, the differential sensitivity of the globin genes is restored (Weisbrod and Weintraub, 1979; Gazit et al., 1980). The ability of these proteins to restore sensitivities is still based on the chromatin. HMG 14 and HMG 17 isolated from the brain will restore the DNase I sensitivity of globin genes in red blood cells but not in brain cells. Thus, two nonhistone chromatin proteins appear necessary (but not in themselves sufficient) to allow the transcription of specific genes in different tissues.

SIDELIGHTS & SPECULATIONS

Transcriptional regulation by the nuclear matrix

Chromatin does not float freely in the nucleus. Rather, it is attached to the NUCLEAR MATRIX, a protein scaffolding that permeates the nucleus. It is known that DNA synthesis occurs on this matrix (Infante et al., 1973; Robinson et al., 1982) and it is possible that RNA synthesis also occurs here. Three sets of observations suggest that transcription occurs only with the help of the nuclear matrix. First, essentially all the newly transcribed RNA is associated with this matrix (Herman et al., 1978), and this attachment is mediated by specific matrix proteins (Van Eekelen and Van Venrooij, 1981). Second, genes are associated with the nuclear matrix in a tissue-specific manner. The ovalbumin gene is preferentially associated with the matrix in chick oviduct cells but not in chicken liver or blood cells. The globin genes, however, are not associated with the matrix of oviduct nuclei (Robinson et al., 1982). Third, a "transcription complex" has been localized to the nuclear matrix (Jackson et al., 1981). These nuclear matrix particles are capable of synthesizing RNA from DNA templates. Jackson and co-workers speculate that the RNA polymerase is located on the nuclear matrix and that the DNA passes through it, rather than the polymerase traveling along the DNA template. This arrangement would have the advantage of placing all the component enzymes needed for transcription in a concentrated area and would provide a structure upon which the heavily condensed chromatin could relax. It also has the advantage of immediately associating the newly synthesized RNA with a solid structure so that it can be processed and transported out of the nucleus. It is doubtful that such a rapid process is dependent on the random collision of freely floating enzymes with their nucleic acid substrates. Therefore, in addition to the aforementioned changes in chromatin structure, we must also consider the possibility that the accessibility of the gene for RNA polymerase is controlled by the attachment of the chromatin to the nuclear matrix.

Structure of eukaryotic genes: RNA polymerase recognition sites

The accessibility of the gene in the chromatin is only the first step in transcription. Next, RNA polymerase must recognize a specific sequence of DNA that enables it to initiate transcription. In bacteria, the RNA polymerase recognizes the 7-base pair sequence 5'-TAT(pu-

TABLE 2
General classification of eukaryotic RNA polymerases

Property	RNA polymerase I	RNA polymerase II	RNA polymerase III
Cellular location	Nucleolar	Nucleoplasmic	Nucleoplasmic
Proportion of cellular activity (%)	50–70	20–40	10
Elution on DEAE-Sephadex at $(NH_4)_2SO_4$	$\sim0.1\ M$	$\sim0.2\ M$	$\sim.2$–$0.3\ M$
α-Amanitin concentration (μg/ml) needed for 50% inhibition:			
Mammalian polymerases	>400	0.025	20
Insect polymerases	—	0.03	>1000
Yeast polymerases	600	1.0	>2000

Source: Lewin (1980).

rine)AT(purine)-3′, often referred to as the PRIBNOW BOX. This sequence serves as the PROMOTER for bacterial transcription.

In eukaryotic cells, the situation is more complex. There are three different RNA polymerases in eukaryotic nuclei, and they each have different properties and affinities (Rutter et al., 1976). RNA POLYMERASE I is found in the nucleolar region of the nucleus and is responsible for transcribing large ribosomal RNAs. RNA POLYMERASE II transcribes messenger RNA precursors; and RNA POLYMERASE III transcribes small RNAs such as transfer RNA, 5 S ribosomal RNA, and certain small, moderately repetitive DNA sequences. These polymerases have different sensitivities to ionic conditions and drugs; Table 2 lists their properties.

RNA polymerase II recognition

The gene sequences recognized by these polymerases differ enormously. These recognition sequences have been determined by experiments testing the specific transcription from cloned DNA. Cloned genes can be accurately transcribed when they are placed into the nuclei of frog oocytes or when they are incubated with purified RNA polymerase II in the presence of nucleotides and cell supernatant fluid (Wasylyk et al., 1980). After the transcription of a gene is confirmed, one uses restriction enzymes to make specific deletions in the gene or in the regions surrounding it. One can then see whether such a modified gene will still be accurately transcribed. Several genes have been so analyzed, and at least two sites are found to be necessary for transcrip-

TABLE 3
RNA polymerase II binding and stabilizing sites

Gene	"CAAT" region[a]		"TATA" region[a]	
Histone				
Sea urchin H2A	GGA C AAT T G	(-85)	T AT A AAA	(-34)
Sea urchin H2B	GAC C AAT GA	(-92)	T AT A AAA	(-26)
Sea urchin H3	GAC C AAT C A	(-75)	T AT A AAT	(-30)
Drosophila H2A	AGT C AATT C		T AT A AAT	
Drosophila H3	CGT C AAAT G		T AT A AGT	
Globin				
Mouse α	AGC C AAT GA	(-88)	CAT AT AA	(-29)
Human α2	AGC C AAT GA	(-70)	CAT A AAC	(-28)
Collagen				
Chick α2 type 1	GCC C ATT GC	(-78)	T AT A AAT	
Insulin				
Human	GGC C AGG CG	(-73)	T AT A AAG	(-29)
Rat I	GGC C AAA CG	(-78)	T AT A AAG	(-30)
Ovalbumin	GGT C AAA CT	(-74)	T AT AT AT	(-31)
Conalbumin	GGA C AAA CA	(-81)	T AT A AAA	(-30)
Silk fibroin	GT AC AAA T A	(-93)	T AT A AAA	(-29)

Source: Efstratiadis et al. (1980); Vogeli et al. (1981).
[a]Numbers in parentheses correspond to the position upstream from the point where transcription is initiated.

tion by RNA polymerase II (Table 3). The first site, called the GOLDBERG-HOGNESS BOX, is usually TATA (but sometimes CATA) and is located about 30 base pairs "upstream" from the site where transcription is initiated (Benoist et al., 1980; Efstratiadis et al., 1980). The second site is called the "CAAT box" and is a set of nucleotides, CCAT or CAAT, about 70–80 base pairs in front of the site where transcription is initiated. The Goldberg-Hogness box is thought to specify the correct 5' end of the RNA precursor. The CAAT box probably stabilizes the polymerase binding. These promoter regions have the potential for alternate schemes of base pairing, and this flexibility may be important in regulating when these genes are transcribed. These "promoter" structures for a chicken collagen gene are shown in Figure 13. In addition to the TATA and CAAT boxes, there may be other regions farther upstream that are needed for accurate transcription of certain genes by RNA polymerase II (Grosschedl and Birnstiel, 1980).[1]

---

[1]Most of the RNA polymerase II-dependent genes studied to date are genes whose products (hemoglobin, ovalbumin and collagen, for example) characterize a particular cell. This is so because such genes make enormous amounts of a particular message

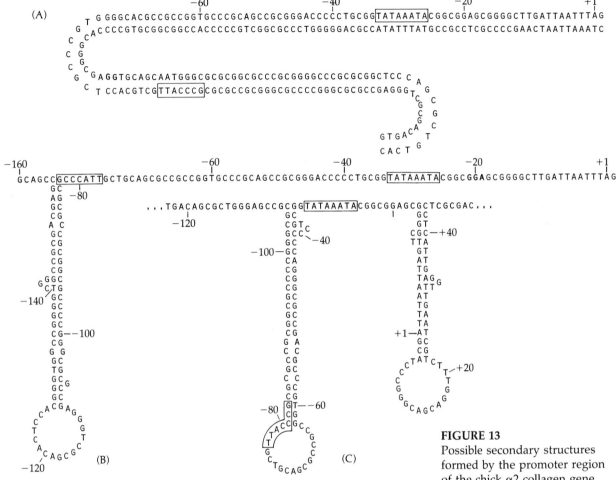

FIGURE 13
Possible secondary structures formed by the promoter region of the chick α2 collagen gene. The TATA and CAAT sequences are enclosed in boxes; +1 represents the initiation site of RNA synthesis. (A) Standard base pairing; (B, C) Stable "hairpin loop" structures containing single-stranded DNA regions. Only one strand is represented, but the complementary strand would form a similar loop. (After Vogeli et al., 1981.)

Purified RNA polymerase, however, cannot recognize the eukaryotic promoter without other factors present in the nucleus. Without these factors, transcription is initiated randomly. The protein that binds the RNA polymerase II to its promoter has been purified; it binds specifically to those gene regions containing the TATA box (Davison et al., 1983). Dynan and Tjian (1983) have also purified factors essential for the proper initiation of transcription by RNA polymerase II; one of

which can be readily isolated and then used to probe for cloned genes. These genes are developmentally regulated in that their expression occurs in specific cell types at specific times. Those genes which are not developmentally regulated (whose products occur in most cells at most times) may be controlled in a different manner and may have a different structure. Such genes need not have either the TATA or CAAT sequences (Melton et al., 1984; Reynolds et al., 1984).

these factors may control transcriptional specificity by directing the polymerase to a certain subset of promoters. This factor (isolated from cultured human tumor cells) will produce a 40-fold stimulation of transcription from some promoters while inhibiting the transcription of other genes.

SIDELIGHTS & SPECULATIONS

Enhancers

In addition to the promoter sequences at the 5' end of the gene, there may exist other sequences necessary for the initiation of transcription by RNA polymerase II. These regions of the gene are called ENHANCERS and they can activate transcription either "upstream" or "downstream" from the promoter sequence. These enhancer sequences were first discovered in viruses, but recent evidence has shown that enhancer sequences exist in the immunoglobulin genes as well. Gillies and his co-workers (1983) placed a cloned immunoglobulin heavy chain gene into cultured B lymphocyte tumor cells that had lost the ability to make their own heavy chain. This can be done by putting the cloned gene in a solution of calcium phosphate. The calcium phosphate will crystallize about the gene and will be phagocytized into the cell. In a certain small percentage of cells, such a gene is incorporated into the nuclear chromosomes. These myeloma cells were then able to synthesize the heavy chain encoded by the incorporated gene. However, if they added the same gene, but without a small region of the intron between the variable and the Cμ regions, they observed very little transcription of the gene. There was a region within the intron necessary for transcription.

This would explain why one does not see random transcription from all the promoter sequences in the V region genes, for the promoter would have to be brought near the enhancer in order to function. During class switching, the enhancer region remains by the VDJ piece (Figure 14A). Enhancers might also explain tissue-specific transcription because the cloned immunoglobulin genes were *not* transcribed when inserted into the nuclei of cells other than B lymphocytes (Gillies et al., 1983; Banerji et al., 1983). Moreover, when the enhancer region of the immunoglobulin

heavy chain is inserted into a cloned gene for hemoglobin, it stimulates the transcription of that hemoglobin gene over 100-fold when it is inserted into a myeloma cell. Thus, we see a specific region of DNA that stimulates transcription of nearby genes in a cell-specific fashion.

It is possible that hormone-responsive sites on DNA may also act as enhancers once the hormone or hormone–receptor protein complex has bound to them. In the mouse mammary tumor virus, there is a series of DNA sequences that can bind glucocorticoid hormone complexes. The genes adjacent to these sites are quiescent except when activated by these hormones, and a nonresponsive gene can be made glucocorticoid-responsive by placing these DNA sequences adjacent to it (Figure 14B; Chandler et al.,

FIGURE 14
Proposed enhancer regions. (A) Immunoglobulin heavy chain enhancer region appears to involve sequences between the J region gene segments and the switch sequences (Sμ) preceding Cμ. If deleted, transcription is greatly diminished. The 5' promoter precedes each of the V region gene segments and is originally very distant from the enhancer. The VDJ gene rearrangement brings the promoter near the enhancer and allows transcription to take place. During class switching, the enhancer stays with the VDJ segments as they are placed near a new constant region. (B) A recombinant virus containing the glucocorticoid-responsive enhancer of the mouse mammary tumor virus and the thymidine kinase gene of herpes simplex virus can integrate into the genome of a cell lacking thymidine kinase gene. Upon treatment with glucocorticoids, the recombinant gene transcribes viral thymidine kinase. P is the promoter region. (A modified from Marx, 1983; B modified from Chandler et al., 1983.)

1983). Moreover, these sequences can be a distance of 500 base pairs in front of or after the promoter sequence. The hormone complex appears to bind to four regions of DNA containing the sequence 5'-TGTTCT-3'. This sequence shows a strong homology to sequences of DNA in a region around 250 base pairs away from the human metallothionine IIA gene, which is also induced by glucocorticoid hormones (Scheidereit et al., 1983). These data support the notion that certain regions of DNA can bind hormone–receptor protein complexes and in some manner activate the promoter to initiate transcription.

(A)

Germ line gene: No transcription

Rearranged gene: Transcription of immunoglobulin

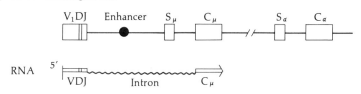

Class switched gene: Transcription of new class of immunoglobulin

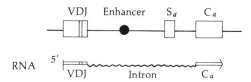

(B)   Hormone-responsive viral gene

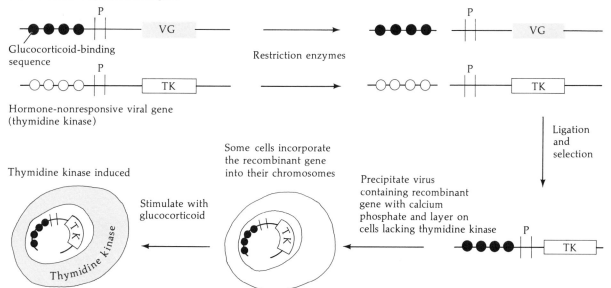

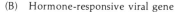

RNA polymerase III recognition

The recognition site for RNA polymerase III surprised everybody. Whereas the RNA polymerase II recognition site was upstream from the gene (in the region before the 5′ end of the RNA coding sequence), the recognition site for RNA polymerase III is located *within* the gene. Accurate transcription still occurred even after the 5′ and 3′ flanking regions from the *Xenopus* 5 S rRNA gene were enzymatically removed (Figure 15). Only when the nucleotides from positions 50 to 83 were removed did transcriptions cease (Sakonju et al., 1980; Bogenhagen et al., 1980). Moreover, there appears to be a conserved region within the genes transcribed by RNA polymerase III. The sequence AGCAGGGT is found between positions 55 and 62 of the *Xenopus* 5 S rRNA gene, and similar sequences (within two base pairs) are found within *Bombyx mori* (moth) tyrosine tRNA and adenovirus VA-I genes. Like the binding of RNA polymerase II, the binding of RNA polymerase III to its control region appears to be mediated by special proteins. When isolated 5 S ribosomal RNA genes of *Xenopus laevis* are incubated with RNA polymerase III, accurate transcription does not take place. However, faithful transcription does occur when 5 S rRNA *chromatin* is used, or when the 5 S rRNA genes are incubated in a nuclear extract. There is a 38,500-d protein that binds to the control region of the 5 S rRNA gene and serves to direct the RNA polymerase III to bind (Ng et al., 1979; Engelke et al., 1980). Without this protein, the 5 S rRNA gene is inactive. Moreover, this protein is specific for 5 S rRNA genes and is developmentally regulated. It is extremely low in mature oocytes (where there is no 5 S rRNA produced), but it is abundant when large amounts of 5 S rRNA are synthesized, as in early oogenesis.

FIGURE 15
Deletion map of a 5 S rRNA gene of *Xenopus laevis*. The heavy horizontal lines represent the deleted portions of DNA in plasmids containing the 5 S rRNA gene. A plus mark adjacent to a line indicates that the remaining plasmid DNA still supported accurate transcription of 5 S rRNA. A negative sign indicates a failure to be transcribed accurately. A control region within the gene, roughly 30 base pairs in length, is defined by these deletions. (Modified from Brown, 1981.)

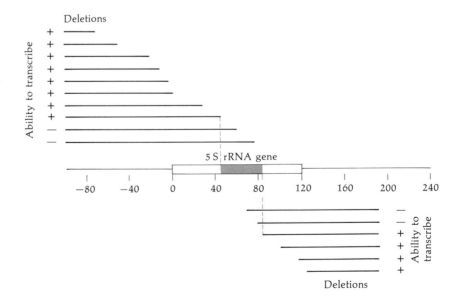

RNA polymerase I recognition

RNA polymerase I is responsible for synthesis of large ribosomal RNAs, and there appears to be a major promoter region and a minor promoter region in these ribosomal RNA genes. When cloned ribosomal RNA genes are injected into the nuclei of growing frog oocytes (which are transcribing such genes at an enormous rate), the cloned genes are efficiently transcribed. When various segments of these clones are deleted and the defective clones are injected into the oocyte nuclei, the promoter was found to consist of a single site, from −7 to +6, spanning the site where transcription is initiated (+1). However, when the conditions were not as optimal (as in in vitro transcription systems), a larger region, from −140 to +6, was required (Sollner-Webb et al., 1983). It is thought that the smaller region directs the polymerase to initiate transcription at the +1 site, whereas the larger region behind it collects polymerase molecules or their cofactors to ensure rapid transcription of the gene.

Thus, the recognition sites for RNA polymerases differ greatly, there is a division of labor for the different types of RNA polymerases, and the polymerases can be differently regulated during development.

SIDELIGHTS & SPECULATIONS

Promoter structure and differentiation

Earlier in this chapter, we saw that in chromatin there must be at least two functional entities that are necessary for transcription. First, there is an elutable fraction that is not tissue specific but that confers accessibility upon certain genes. This fraction consists of the HMG 14 and HMG 17 proteins. The second factor is nonelutable and instructs the HMG proteins to globin genes in red blood cells and to ovalbumin genes in chick oviduct tubules. More recent evidence has suggested that such instruction might be encoded in the promoter structure itself and that the promoter structure may cause changes in the placement of the chromatin proteins.

Whereas transcriptionally active genes are sensitive to small amounts of DNase I, the 5′ ends of such genes (i.e., the portion containing the promoter) are *hypersensitive*. Thus, when a very small amount of DNase I is added to chromatin for a very short duration, only the active promoter sites are cleaved (Elgin,

1981). It has been shown that these hypersensitive sites are also digested by another enzyme, S1 nuclease (Larsen and Weintraub, 1982). S1 only attacks single-stranded nucleic acids, so it appears that active promoter regions contain single-stranded DNA. Looking again at Figure 13 we see that single-stranded regions of DNA can be generated at the promoter region. The promoter sites have several alternative configurations, but it is thought that usually the DNA is forced into the double helix by the nucleosomes. However, nucleosomes must be replaced during cell division, and so it is possible that altered conformational forms can be assumed at this time. Depending on the cellular conditions, a certain promoter could remain in its standard configuration or it could form hairpin loops. Once these loops form, they could be recognized by nonspecific proteins, and nucleosomes would not bind there (thus making the sites hypersensitive to DNase I). Moreover, the changes caused by the loops could orient the way in which other nucleosomes bind to the gene, making it accessible for RNA polymerase to transcribe from it. It is known that nucleosomes bind to HMG 14 and HMG 17 (Mardian et al., 1980; Weisbrod et al., 1980) and that the capacity to bind these proteins increases when genes become potentially active. All nucleosomes, therefore, may not be alike in their constituents.

Such speculations have recently gained support from studies of viral genes which become integrated into cellular DNA. A recent study by Zaret and Yamamoto (1984) probes the mechanism for enhancer activity by studying mouse mammary tumor virus (MTV) whose genes are regulated by glucocorticoid hormones. They find that a specific region of this viral DNA binds the glucocorticoid hormone-receptor complex and acts as a hormone-dependent enhancer sequence. Moreover, before the addition of the hormone (to cells containing the virus), this enhancer sequence shows no special DNase I sensitivity. After administering the hormone, a discrete DNase I-hypersensitive site develops in this region. The formation of the hypersensitive site coincides with the initiation of viral gene transcription; when the hormone is withdrawn, both the hypersensitive site and viral gene transcription disappear. Zaret and Yamamoto speculate that the interaction between the glucocorticoid receptor complex and the enhancer DNA alters the chromatin configuration to facilitate transcription from the nearby promoter.

DNA methylation

Another mechanism for changing the gene structure (and, hence, chromatin structure) is DNA METHYLATION. Methyl groups have been found on certain cytosine residues of DNA such that in mammals about 5 percent of the cytosines are converted to 5-methylcytosine. Recent studies have suggested that methylation may control transcriptional activity of specific genes in specific cells. In developing human and chick red blood cells, the DNA involved in globin synthesis is totally or nearly completely unmethylated, whereas the same genes are highly methylated in cells that do not produce globin (Figure 16A). Fetal liver cells that produce hemoglobin in early development have unmethylated genes for fetal hemoglobin. These genes are methylated in the adult tissue (van der Ploeg and Flavell, 1980; Groudine and Weintraub, 1981).

A further correlation between hypomethylation and globin gene activity comes from experiments on cloned globin genes (Busslinger et al., 1983). When human fetal globin genes are cloned and added to cells by coprecipitation with calcium phosphate, the cells ingest the DNA and, in many cases, incorporate the DNA into the nucleus. In such cases, the cloned globin gene is transcribed. By protecting certain regions of the cloned globin genes from methylation before adding it to the cells, it is possible to create clones in which the globin genes are all the same, but their methylation patterns differ. Figure 16B depicts the results. A completely unmethylated gene is transcribed, whereas a completely methylated gene (methyl groups on every C residue) is not transcribed. Using partially methylated clones, it was shown that methylation in the 5' region of the globin gene (nucleotides -760 to $+100$) prevents transcription. So, methylation in the 5' end of a gene might play a direct role in the regulation of gene expression.

Organ-specific methylation patterns are also seen in the ovalbumin gene; the gene is unmethylated in the oviduct cells but is methylated

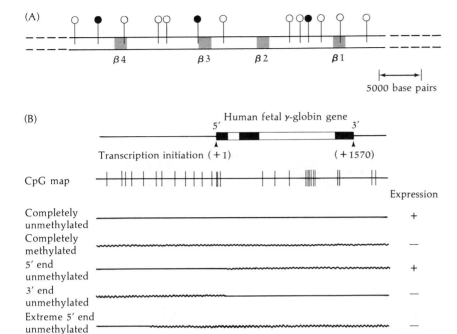

FIGURE 16
Summary of methylation in the hemoglobin gene. (A) Methylation in rabbit β globin genes. Each gene of this cluster is represented by a black box. The positions of each CCGG sequence are noted by stemmed circles. The dark circles indicate sites that are totally methylated only in cells not transcribing globin. (B) Transcription from a cloned human fetal γ-globin gene controlled by methylation. The top diagram depicts the cloned globin gene and its 5′ and 3′ flanking regions. Exons are black, introns are white, and the flanking regions are represented by thin lines. Beneath it is a map of the possible sites of DNA methylation, that is, those regions where the dinucleotide CpG is found. Beneath those maps are diagrams of the methylation patterns of five clones used in this study. The straight line indicates unmethylated regions; the wavy line indicates methylated segments. To the right is a summary of whether or not they were expressed. (A after Shen and Maniatis, 1980; B after Busslinger et al., 1983.)

in other chick tissues (Mandel and Chambon, 1979). Demethylation is seen to accompany class switching in immunoglobulin synthesis (Rogers and Wall, 1981) and is seen to correlate with the ability of murine lymphocytes to produce the metal binding protein metallothionine I (Comprere and Palmiter, 1981). Thus, the lack of DNA methylation correlates well with the tissue-specific expression of certain genes.

The correlation between demethylation and gene activation may have important medical consequences. In the disease β-thalassemia, the adult β-hemoglobin protein is not synthesized in normal amounts. Ley and his co-workers (1982) have proposed treating this disease by reactivating the fetal (γ) hemoglobin gene. To accomplish this, they have given three patients 5-AZACYTIDINE, an analogue of cytidine that can be incorporated into DNA but that cannot be methylated. For the first 2 weeks after this treatment, the patients made large numbers of blood cells containing the fetal hemoglobin chain. Moreover, the methylation analysis on their red blood cell precursors demonstrated that the sites near the γ-globin genes were undermethylated. Even though this may be a risky medical procedure (we do not know yet what other "silent" genes may be activated in other organs, nor do we know the toxicity or tumorigenicity of 5-azacytidine), this study does provide us with a tantalizing clue to the mechanism of gene expression.

We still do not know the mechanisms by which promoter site conformation changes, methylation, and HMG 14 and HMG 17 binding are related. There is probably more than one mechanism involved (especially as specific methylation differences are not seen in *Drosophila* tissues), but it is probable that conformational changes in the DNA lead to the differential regulation of gene transcription.

SIDELIGHTS & SPECULATIONS

Z-DNA and nucleosome structure

One possible link between methylation and the spacing of nucleosomes in chromatin is the existence, in the genome, of Z-DNA. Most nuclear DNA is the right-handed, double-helical B-DNA described by Watson and Crick. However, Alexander Rich and his colleagues discovered that certain CG-rich sequences, such as CGCGCG, preferentially formed *left-handed* double helices (Wang et al., 1979). The physical properties of this left-handed structure differ from those of standard DNA. The left-handed helix has 12 base pairs per turn (rather than 10) and forms a zigzag conformation (hence, the name Z-DNA). Z-DNA also lacks the major groove that allows the binding of certain molecules to B-DNA (Figure 17).

This left-handed Z-DNA has been found in mammalian and *Drosophila* genomes. Using antibodies to Z-DNA, Rich and his colleagues have demonstrated that Z-DNA is present throughout the polytene chromosomes of *Drosophila* (Figure 18). Here they are seen to

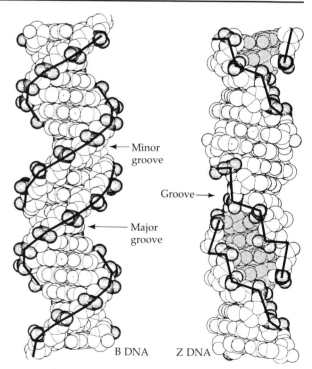

FIGURE 17
Comparison of right-handed (B) and left-handed (Z) DNA showing differences in the two helical structures. (From Wang et al., 1979; courtesy of A. Rich.)

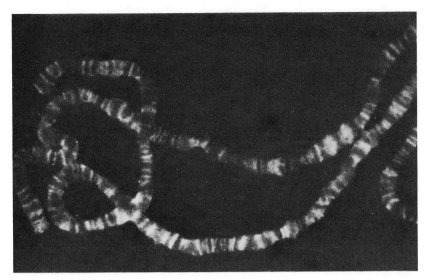

FIGURE 18
Polytene *Drosophila* salivary chromosomes stained with fluorescence-labeled antibodies to Z DNA. Numerous interband regions are seen to bind the antibody. (From Nordheim et al., 1981; photograph courtesy of A. Rich.)

be restricted to the regions between bands. These *interband* regions are thought to control the expression of genes within the bands (Nordheim et al., 1981).

It had previously been thought that the transition from B-DNA to Z-DNA could only occur under high salt conditions. However, Behe and Felsenfeld (1981) demonstrated that Z-DNA will form naturally under physiological salt concentrations if the cytosine residues of an alternating GC polymer are methylated. Thus, methylation of cytosine may enable the B-DNA to enter the Z configuration. Moreover, whereas B-DNA promotes nucleosome formation from histones, Z-DNA actively prevents nucleosome formation (Nickol et al., 1982). Histones will indeed bind to Z-DNA, but they will not associate into the nucleosome octamer. These results suggest that methylation may cause the formation of Z-DNA, and this new configuration will disrupt the normal formation of nucleosomes along the chromatin.

Weintraub (1984) has recently advanced a hypothesis which attempts to unite these various mechanisms of gene activation. He observed that histone H1 behaved differently depending upon whether the DNA sequence was part of an actively transcribing gene or an inactive one. The nucleosomes of nontranscribed genes appear to be linked together into solenoidal clusters such that they remain together even if their DNA is fragmented by micrococcal DNase treatment. Nucleosomes from actively transcribing genes, however, are readily split apart by similar DNase digestion. Weintraub speculates that H1-dependent supranucleosome structures are responsible for suppressing gene activity, and that nucleosomes normally form these inactive clusters unless some mechanism prevents this association. This mechanism can be the binding of HMG 14 or 17, the creation of DNase I-hypersensitive sites, histone modification, Z-DNA, or the binding of a sequence-specific DNA binding protein. All these mechanisms may have a final result—the inhibition of nucleosome aggregation—that permits the expression of the gene.

Structure of eukaryotic genes: Exons and introns

One of the most surprising discoveries of the 1970s—a decade full of remarkable discoveries—was the structure of the eukaryotic gene. Bacterial genetics has taught us that the protein product is colinear with the gene that encodes it; that the 5' end of the message creates the amino terminal of the protein and that each codon is translated into an amino acid until the carboxyl terminus of the protein is reached. In 1977, this was seen not to be the case in certain eukaryotic messages. Since then, most of the eukaryotic genes sequenced have been found to be split. In other words, the 5' and 3' ends of eukaryotic messenger RNA come from noncontiguous regions on the chromosome. Between the regions of DNA coding for the proteins (EXONS) are intervening sequences (INTRONS), which have nothing whatever to do with the amino acid sequence of the protein.[2] The structure of the human β-hemoglobin gene (Lawn et al., 1980) is shown in Figure 19. This gene consists of the following elements:

1. A CAAT BOX: In the case of this hemoglobin gene, a CCAAT sequence is located 77 base pairs upstream from the initiation point of transcription.
2. A GOLDBERG-HOGNESS BOX: Most hemoglobins use CATA for RNA polymerase II recognition. This sequence is located 32 base pairs before the initiation point of transcription.
3. The sequence ACATTTG, where *transcription* is initiated. This is often called the CAP SEQUENCE because it represents the 5' end of the RNA, which will receive a "cap" of modified nucleotides soon after it is transcribed (as we shall see later).
4. The *ATG codon* for the initiation of *translation*. This codon is located 50 base pairs after the initiation point of transcription. The intervening 50 nucleotide pairs are called the LEADER SEQUENCE.
5. An EXON containing 90 base pairs coding for amino acids 1–30 of human β-hemoglobin.

[2]There are some notable exceptions (such as the genes for histones), which lack the intervening sequences. Any hypothesis concerning their function must take these exceptions into account.

FIGURE 19
Nucleotide sequences of the human β globin gene. (A) Schematic representation of the locations of the promoter region, transcriptional initiation (cap) site, leader sequence, exons and introns of the β globin gene. Exons are in black, and the numbers flanking them indicate the amino acid positions they code for in β globin. (B) The nucleotide sequence of the β globin gene, shown from the 5' to the 3' end of the RNA. The CAAT and Goldberg-Hogness sequences are boxed, as are the translation initiation and termination codes ATG and TAA. The large capital letters correspond to exons, and the amino acids for which they code are abbreviated above them. The small capital letters are the bases of the intervening sequences. The codons represented by capital letters after the translation terminator are in the globin mRNA but are not translated into proteins. Within this group is the sequence thought to be needed for polyadenylation. A G in the first intron (arrow) is mutated to A in one form of β[+]-thalassemia. (Sequence from Lawn et al., 1980.)

(A)

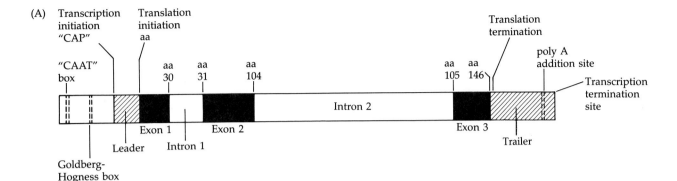

(B)

```
ccctgtggagccacaccctagggttgg ccaat ctactcccaggagcagggagggcaggagccagggctggg cataaaa

gtcagggcagagccatctattgcttACATTTGCTTCTGACACAACTGTGTTCACTAGCAACCTCAAACAGACACC ATG

ValHisLeuThrProGluGluLysSerAlaValThrAlaLeuTrpGlyLysValAsnValAspGluValGlyGlyGlu
GTGCACCTGACTCCTGAGGAGAAGTCTGCCGTTACTGCCCTGTGGGGCAAGGTGAACGTGGATGAAGTTGGTGGTGAG

AlaLeuGlyArg
GCCCTGGGCAGGTTGGTATCAAGGTTACAAGACAGGTTTAAGGAGACCAATAGAAACTGGGCATGTGGAGACAGAGAAG

                                                    ↓
ACTCTTGGGTTTCTGATAGGCACTGACTCTCTCTGCCTATTGGTCTATTTTTCCCACCCTTAGGCTGCTGGTGGTCTAC
                                                        LeuLeuValValTyr

ProTrpThrGlnArgPhePheGluSerPheGlyAspLeuSerThrProAspAlaValMetGlyAsnProLysValLys
CCTTGGACCCAGAGGTTCTTTGAGTCCTTTGGGGATCTGTCCACTCCTGATGCTGTTATGGGCAACCCTAAGGTGAAG

AlaHisGlyLysLysValLeuGlyAlaPheSerAspGlyLeuAlaHisLeuAspAsnLeuLysGlyThrPheAlaThr
GCTCATGGCAAGAAAGTGCTCGGTGCCTTTAGTGATGGCCTGGCTCACCTGGACAACCTCAAGGGCACCTTTGCCACA

LeuSerGluLeuHisCysAspLysLeuHisValAspProGluAsnPheArg
CTGAGTGAGCTGCACTGTGACAAGCTGCACGTGGATCCTGAGAACTTCAGGGGTGAGTCTATGGGACCCTTGATGTTTT

CTTTCCCCTTCTTTTCTATGGTTAAGTTCATGTCATAGGAAGGGGAGAAGTAACAGGGTACAGTTTAGAATGGGAAAC

AGACGAATGATTGCATCAGTGTGGAAGTCTCAGGATCGTTTTAGTTTCTTTTATTTGCTGTTCATAACAATTGTTTTC

TTTTGTTTAATTCTTGCTTTCTTTTTTTTTCTTCTCCGCAATTTTTACTATTATACTTAATGCCTTAACATTGTGTAT

AACAAAAGGAAATATCTCTGAGATACATTAAGTAACTTAAAAAAAAACTTTACACAGTCTGCCTAGTACATTACTATT

TGGAATATATGTGTGCTTATTTGCATATTCATAATCTCCCTACTTTATTTTCTTTTATTTTTAATTGATACATAATCA

TTATACATATTTATGGGTTAAAGTGTAATGTTTTAATATGTGTACACATATTGACCAAATCAGGGTAATTTTGCATT

TGTAATTTTAAAAAATGCTTTCTTCTTTTAATATACTTTTTTGTTTATCTTATTTCTAATACTTTCCCTAATCTCTTT

CTTTCAGGGCAATAATGATACAATGTATCATGCCTCTTTGCACCATTCTAAAGAATAACAGTGATAATTTCTGGGTTA

AGGCAATAGCAATATTTCTGCATATAAATATTTCTGCATATAAATTGTAACTGATGTAAGAGGTTTCATATTGCTAA

TAGCAGCTACAATCCAGCTACCATTCTGCTTTTATTTTATGGTTGGGATAAGGCTGGATTATTCTGAGTCCAAGCTAG
                                                LeuLeuGlyAsnValLeuValCysValLeuAla
GCCCTTTTGCTAATCATGTTCATACCTCTTATCTTCCTCCCACAGCTCCTGGGCAACGTGCTGGTCTGTGTGCTGGCC

HisHisPheGlyLysGluPheThrProProValGlnAlaAlaTyrGlnLysValValAlaGlyValAlaAsnAlaLeu
CATCACTTTGGCAAAGAATTCACCCCACCAGTGCAGGCTGCCTATCAGAAAGTGGTGGCTGGTGTGGCTAATGCCCTG

AlaHisLysTyrHis
GCCCACAAGTATCAC TAA GCTCGCTTTCTTGCTGTCCAATTTCTATTAAAGGTTCCTTTGTTCCCTAAGTCCAACTAC

TAAACTGGGGGATATTATGAAGGGCCTTGAGCATCTGGATTCTGCCTAATAAAAAACATTTATTTTCATTGCaatgat

gtatttaaattatttctgaatatttactaaaaaaggggaatgtgggaggtcagtgcatttaaaacataaagaaatgatg

agctgttcaaaccttgggaaaatacactatatcttaaactccatgaaagaaggtgaggctgcaaccagctaatgcaca

ttggcaacagcccctgatgcctatgccttattcatccctcagaaaaggattcttgtagaggcttgatttgcaggttaa

agttttgctatgctgtattttacattacttattgtttttagctgtcctcatgaatgtcttttcactacccatttgctta

tcctgcatctctctcagccttgact
```

6. An INTRON containing 130 base pairs having no coding sequences for hemoglobin.
7. An EXON containing 222 base pairs coding for amino acids 31–104.
8. A large INTRON—850 base pairs—having nothing to do with the hemoglobin protein structure.
9. An EXON containing 126 base pairs coding for amino acids 105–146.
10. A TRANSLATION TERMINATION CODON, TAA.
11. A 3′ TRANSCRIBED TRAILER REGION, which does not get translated into protein. This region includes the sequence AATAAA, which is needed to place a "tail" of some 200–300 adenylate residues on the RNA transcript. Transcription continues slightly beyond the AATAAA sequence before being terminated.

The original nuclear RNA transcript for such a gene contains the capping sequence, the leader sequence, the exons, the introns, and the 3′ nontranslated region (Figure 20). In addition, both its ends become modified. A CAP consisting of methylated guanosine is placed on the 5′ end of the RNA in opposite polarity to the RNA itself. Thus, whereas all the bases in the message precursor are linked 5′ to 3′, the cap structure is linked 5′ to 5′. This means that there is no free 5′ phosphate group on the nuclear RNA (Figure 21). Messenger RNA molecules are

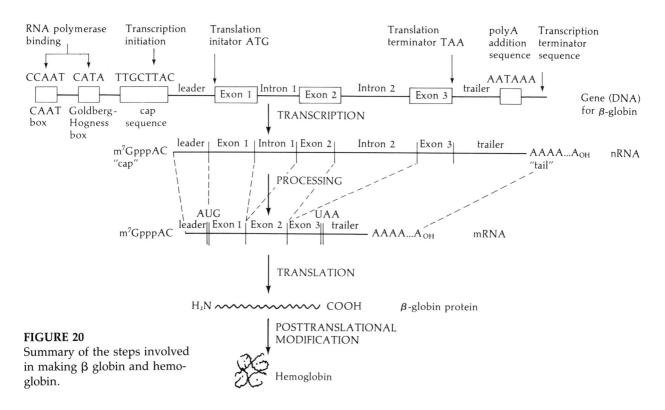

FIGURE 20
Summary of the steps involved in making β globin and hemoglobin.

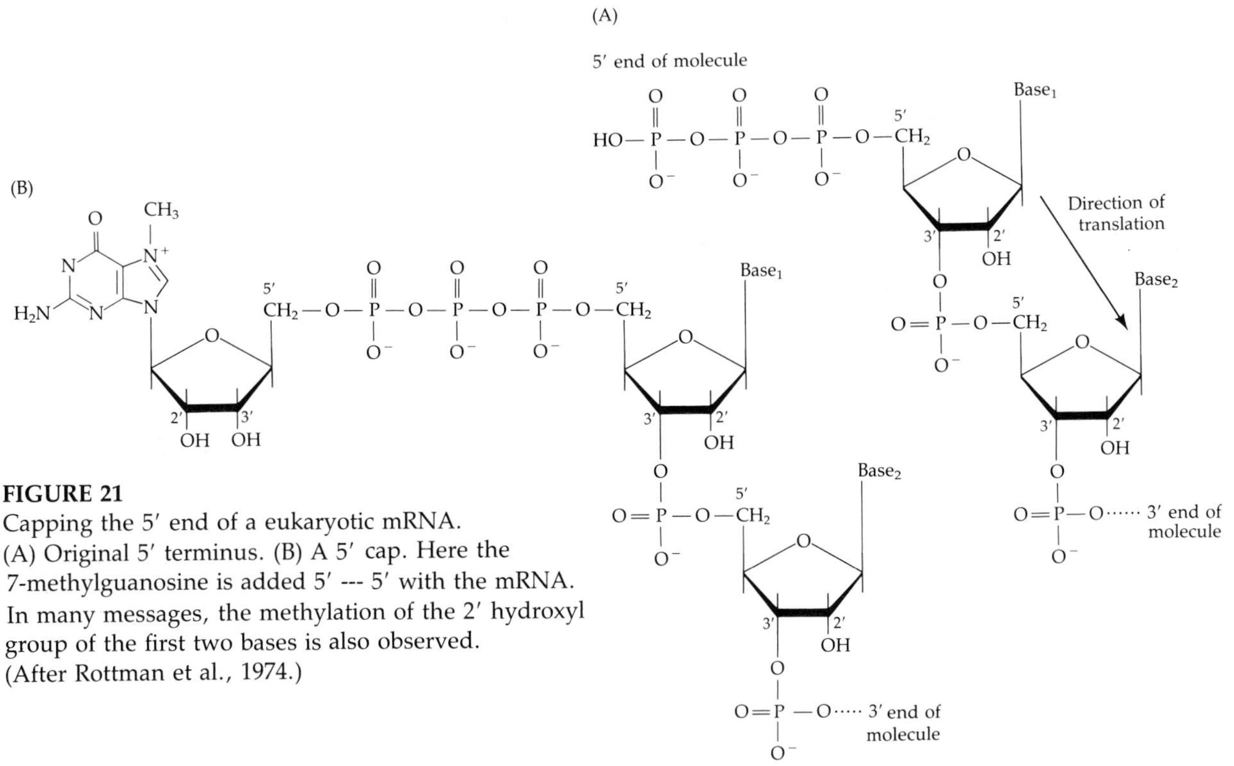

FIGURE 21

Capping the 5' end of a eukaryotic mRNA.
(A) Original 5' terminus. (B) A 5' cap. Here the
7-methylguanosine is added 5' --- 5' with the mRNA.
In many messages, the methylation of the 2' hydroxyl
group of the first two bases is also observed.
(After Rottman et al., 1974.)

likewise "capped," although it is not certain that the mRNA cap is the original one it received in the nucleus. The 5' cap is also known to be necessary for the binding of mRNA to the ribosome (Shatkin, 1976).

The 3' terminus is usually modified in the nucleus by having roughly 200 adenylate residues added on as a tail. These adenylic acid residues are put together enzymatically and are added to the transcript. They are not part of the gene sequence. Both the 5' and 3' modifications may protect the RNA from exonucleases (Sheiness and Darnell, 1973; Gedamu and Dixon, 1978), thereby stabilizing the message and its precursor. These poly(A) tails get progressively shorter as the mRNA ages. Newly synthesized globin message in mouse and rabbit cells has about 150 adenylate residues, whereas older messages have 100, 60, or 40 extra adenylates (Merkel et al., 1975; Nokin et al., 1976). Marbaix and co-workers (1975) have shown that globin message lacking poly(A) is degraded rapidly when injected into *Xenopus* oocytes. Globin mRNA with its poly(A) tails lasts over 20 hours.

But the mRNA is useless so long as it remains cloistered in the nucleus. Unlike prokaryotic cells, in which transcription and translation are intimately connected, eukaryotic RNA must first get out of the nucleus to be translated. In the past few years, evidence has accumulated to show that the introns are responsible for getting the mRNA coding sequences out into the cytoplasm.

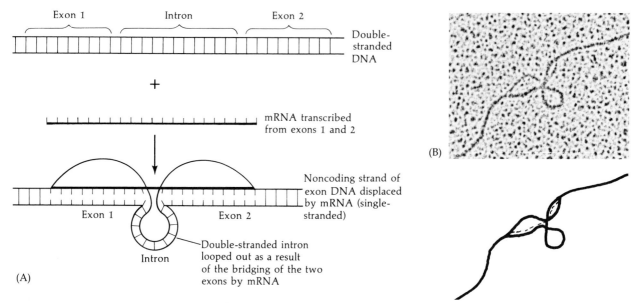

Exon 1   Intron   Exon 2

Double-
stranded
DNA

+

mRNA transcribed
from exons 1 and 2

Noncoding strand of
exon DNA displaced
by mRNA (single-
stranded)

Exon 1   Exon 2

Double-stranded intron
looped out as a result
of the bridging of the two
exons by mRNA

Intron

(A)

(B)

FIGURE 22

Electron microscopic demonstration of introns by replacement loop (R-loop) mapping. A mouse β globin gene is mixed in formamide with its own mRNA. The formamide conditions favor RNA–DNA binding over DNA–DNA binding. When the RNA replaces the homologous DNA strands, the displaced DNA forms a single-stranded R-loop. In addition, the intron between the two coding sequences is forced to form a double-stranded loop. (A) Schematic diagram illustrating the principle of R-loop mapping. (B) An electron micrograph of the interaction of mouse β globin and its message. In the accompanying interpretive diagram, the solid black lines are DNA and the dashed line represents the β globin message. (Electron micrograph from Tiemeier et al., 1978.)

The existence of intron regions in the nuclear precursors of mRNA was shown by electron microscopy (Figure 22). Complementary DNA (made by reverse transcriptase) to the β globin message binds to specific regions of nuclear RNA precursor, looping out the introns. Several types of evidence show that these introns are necessary for the messenger RNA sequences to escape the nucleus.

First, artificial viruses can be constructed such that part of the globin gene is inserted into them. The viruses are then allowed to infect the cells and transcribe RNA in the nucleus. When the virus contains an intron (either from the globin gene or from the viral gene), stable globin message sequences can be detected in the cytoplasm. However, when neither the virus nor the globin gene provides an intron, no globin message sequences accumulate in the cytoplasm (Hamer and Leder, 1979).

Second, as mentioned earlier, certain individuals have the disease β⁺-THALASSEMIA, which is characterized by the underproduction of β globin chains. These people do not lack the genes for β globin, and transcription of the RNA precursor from these genes appears to be normal. Moreover, the mRNA from such patients gets translated properly into the amino acid sequence expected for human β globin. The defect is in the *amount* of β globin message. Therefore, the disease appears to be at the level of RNA processing. Maquat and colleagues (1980) showed this to be the case by a "pulse–chase" experiment. Developing red blood cells were isolated from the patients' bone marrow and given a 12-minute "pulse" of radioactive nucleotides. After 12 minutes, transcription was stopped with actinomycin D. At various intervals, RNA was harvested from the cells, subjected to electrophoresis to separate RNA molecules by size, and then hybridized to β-globin cDNA. In normal individuals, a 1900-base RNA is seen soon after labeling. Soon thereafter, the cDNA binds to RNAs containing 1550, 1150, 960, and 880 bases. In a few more minutes, 85 percent of the RNA is seen as mRNA. In the β⁺-thalassemic cells, there is a buildup of the intermediates. Very few message-size RNAs exist by 30 minutes. These intermediates are probably digested within the nucleus and very few get into the cytoplasm (Maquat et al., 1980; Kantor et al., 1980). Thus, β⁺-thalassemia appears to be a defect in the processing of the β globin message precursor. When the β globin genes from these individuals were sequenced, they were found to contain a mutation in the first intervening sequence (Spritz et al., 1981; Figure 19).

Mechanism for RNA splicing

Comparison of the DNA sequences of many different genes revealed certain similarities at the intron–exon junctions. The base sequence of the intron usually begins with GT, and the base sequence of the intron usually ends with AG (Table 4). In fact, the "consensus sequence" for 5′ ends is AG/GT (i.e., most exons end in AG and the intron begins with GT) and the 3′ sequence of most introns is TT*N*CAG (where *N* can be any nucleotide). These sites appear to be very important for intron processing. Again, consider β-thalassemia. In the β globin genes of these individuals, there is a single base pair mutation. However, the mutation is in the first *intron*, not in the coding sequence. In this intron, the sequence TTGGTCT is mutated to TTAGTCT. However, the usual 3′ splicing site for this intron is TTAG/GCT. The mutation makes an artificial splicing site, and in so doing creates the possibility for erroneous splicing (Spritz et al., 1981). In fact, the thalassemic β globin precursor splices at the wrong junction over 90 percent of the time (Busslinger et al., 1981).

Although the actual splicing mechanism is not yet known, it probably involves a fascinating group of polynucleotides called SMALL NUCLEAR RNAs. These snRNAs consist of only 90–200 bases and are extremely stable within the nucleus. The structure of the 5′ end of one of these snRNAs, U1a, is shown in Figure 23. What is of interest is that

TABLE 4
Nucleotide sequences at the junctions between coding regions and intervening sequences[a]

Organism	Gene and intron	Coding sequence	Intervening sequence	Coding sequence
Chicken	Ovalbumin 1	... TCAAAAG	GTAGGC ... TGCTCTAG	ACAACTC ...
	Ovalbumin 2	... AAATAAG	GTGAGC ... AATTACAG	GTTGTTC ...
	Ovalbumin 3	... AGCTCAG	GTACAG ... GTATTCAG	TGTGGCA ...
	Ovalbumin 4	... CCTGCCA	GTAAGT ... CTTTACAG	GAATACT ...
	Ovalbumin 5	... ACAAATG	GTAAGT ... TCTTAAAG	GAATTAT ...
	Ovalbumin 6	... GACTGAG	GTATAT ... TGCTCTAG	CAAGAAA ...
	Ovalbumin 7	... TGAGCAG	GTATAT ... CCTTGCAG	CTTGAGA ...
Mouse	β^{maj} globin 1	... TGGGCAG	GTGAGC ... CTTTTTAG	GCTGCTG ...
Rabbit	β globin 1	... TGGGCAG	GTGAGC ... CTTTTTAG	GCTGCTG ...
Mouse	β^{maj} globin 2	... CTTCAGG	GTGAGT ... TCCCACAG	CTCCTG ...
Rabbit	β globin 2	... CTTCAGG	GTGAGT ... TCCTACAG	CTCCTG ...
Mouse	α globin 2	... CTTCAAG	GTATGC ...	
	VγII immunoglobulin 1	... TGCTCAG	GTCAGC ... GTTTGCAG	GAGCCA ...
	Cγ1 immunoglobulin 2	... TGTACAG	GTAAGT ... ATCCTTAG	TCCAGA ...
Rat	Insulin I & II–1	... CAAGCAG	GTACTC ... TCTTCCAG	GTCATTG ...
	Insulin II–2	... CCACAAG	GTAAGC ... CCTGGCAG	TGCACA ...
Bombyx mori	Silk fibroin	... TCTGCAG	GTGAGT ... TGTTTCAG	TATGTCG ...
Consensus sequence		AG	GT TTNCAG	

Source: Lewin (1980).

[a]The DNA sequence homologous to the RNA is shown such that the junctions between exons and introns are aligned. The exact point of splicing cannot be assigned as the last bases in the first coding region are often repeated in the right end of the intron. At the bottom of the table, the consensus sequence common to most exon/intron boundaries (AG/GT) is identified, as is the consensus sequence for most intron/exon junctions (TTNCAG).

it contains regions that are complementary to most consensus sequences found at the exon–intron borders. When the sequence of one of these snRNAs is compared to part of a message precursor sequence, a possible mechanism instantly suggests itself (Figure 23). Here we see that U-1a can be used to bridge the 5' and 3' ends of the intron, thus bringing the exon regions into contact (Lerner et al., 1980; Rogers and Wall, 1980).[3] The intron would be excised by some enzyme that recognizes such a remarkable double-stranded structure. Thus, the coding sequences are brought together. Once the sites have been so aligned, splicing is thought to have two further steps. The first step involves cleaving the RNA; and the second involves putting the new ends back

[3]It has recently been found that the removal of the 5' end from U1a snRNA abolishes the splicing activity of nuclear homogenates (Krämer et al., 1984).

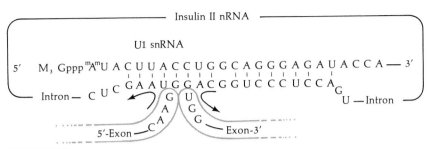

FIGURE 23
Possible binding site of the 5' end of U1 small nuclear RNA to the intron-exon boundaries. In this example—the rat insulin II gene—the 5' end of the U1 RNA (seen to be capped) forms a stable duplex with the first 5 bases of the intron and with 11 of the last 13 bases of the intron, bringing together the two coding regions. (After Rogers and Wall, 1980.)

together. The enzymes involved in cleavage and ligation of yeast transfer RNA have been isolated from growing yeast cells. The cleaving enzyme has been localized to membranes; it cuts the precursor tRNA at two precise locations, thereby removing the intron. The two exons are ligated by a different enzyme, a soluble RNA ligase (Peebles et al., 1983; Greer et al., 1983). The identity of splice junction sites throughout plant and animal RNAs and the ability of *Xenopus* or human cell extracts to efficiently splice yeast tRNA precursor indicate that the mechanism of RNA splicing is highly conserved. So the structure of the intron, while not providing any information for protein structure, is necessary to move the message into the cytoplasm. The evolutionary change resulting in the nuclear membrane occasioned a change in the structure of the genes.

Summary

Within the past decade, our conception of eukaryotic transcription has broadened enormously. At the moment, we have in our hands the pieces of one or more puzzles. We are not certain that we have all the pieces; neither do we know how many puzzles there are. There may be one major mechanism by which all eukaryotic genes are transcribed, or there may be different types of transcriptional controls for different types of genes. The next few years should see the integration of much of the material presented here—methylation, nucleosome structure and spacing, Z-DNA, DNase sensitivity of promoter regions, enhancer elements, and polymerase binding specificity—into unified pictures of transcription. Moreover, the mechanism(s) by which processing occurs within the nucleus and how this allows passage of RNA into the cytoplasm may also be elucidated soon. As Albert Claude remarked, we have just started the inventory of our acquired wealth.

Literature cited

Banerji, J., Olson, L. and Schaffner, W. (1983). A lymphocyte-specific cellular enhancer is located downstream of the joining region in immunoglobulin heavy chain genes. *Cell* 33: 729–740.

Behe, M. and Felsenfeld, G. (1981). Effects of methylation on a synthetic polynucleotide. The B–Z transition in poly(dG-m⁵dC)·poly(dG-m⁵dC). *Proc. Natl. Acad. Sci. USA* 78: 1619–1623.

Benoist, C., O'Hare, K., Breathnach, R. and Chambon, P. (1980). Ovalbumin gene sequence putative control region. *Nucleic Acid Res.* 8: 127–142.

Bogenhagen, D. F., Sakonju, S. and Brown, D. D. (1980). A control region in the center of the 5S RNA gene directs specific initiation of transcription. II. The 3′ border of the region. *Cell* 19: 27–35.

Britten, R. J. and Davidson, E. H. (1971). Repetitive and nonrepetitive DNA sequences and a speculation on the origins of evolutionary novelty. *Q. Rev. Biol.* 46: 111–133.

Britten, R. J. and Kohne, D. E. (1968). Repeated sequences in DNA. *Science* 161: 529–540.

Brown, D. D. (1981). Gene expression in eukaryotes. *Science* 211: 667–674.

Busslinger, M., Hurst, J. and Flavell, R. A. (1983). DNA methylation and the regulation of globin gene expression. *Cell* 34: 197–206.

Busslinger, M., Moschonas, N. and Flavell, R. A. (1981). β⁺ Thalassemia: Aberrant splicing results from a single point mutation in an intron. *Cell* 27: 289–298.

Chamberlin, M. E., Britten, R. J. and Davidson, E. H. (1975). Sequence organization in *Xenopus* DNA studied by the electron microscope. *J. Mol. Biol.* 96: 317–334.

Chandler, V. L., Maier, B. A. and Yamamoto, K. R. (1983). DNA sequences bound specifically by glucocorticoid receptor *in vitro* render a heterologous promoter hormone responsive *in vivo*. *Cell* 33: 489–499.

Compere, S. J. and Palmiter, R. D. (1981). DNA methylation controls the inducibility of the mouse metallothionein-I gene in lymphoid cells. *Cell* 25: 233–240.

Constantini, F. D., Scheller, R. H., Britten, R. J. and Davidson, E. H. (1978). Repetitive sequence transcripts in the mature sea urchin oocyte. *Cell* 15: 173–187.

Davidson, E. H., Graham, D. E., Newfeld, B. R., Chamberlin, M. E., Amenson, C. S., Hough, B. R. and Britten, R. J. (1973a). Arrangement and characterization of repetitive sequence elements in animal DNAs. *Cold Spring Harbor Symp. Quant. Biol.* 38: 295–301.

Davidson, E. H., Hough, B. R., Amenson, C. S. and Britten, R. J. (1973b). General interspersion of repetitive with nonrepetitive sequence elements in the DNA of *Xenopus*. *J. Mol. Biol.* 77: 1–24.

Davison, B. L., Edgly, J. M., Mulvihill, E. R. and Chambon, P. (1983). Formation of stable preinitiation complexes between eukaryotic class B transcription factors and promoter sequences. *Nature* 301: 680–686.

Dynan, W. S. and Tjian, R. (1983). Isolation of transcription factors that discriminate between different promoters recognized by RNA polymerase II. *Cell* 32: 669–680.

Efstratiadis, A., Posakony, J. W., Maniatis, T., Lawn, R. M., O'Connell, C., Spritz, R. A., DeRiel, J. K., Forget, B. G., Weissman, S. M., Slightom, J. L., Blech, A. E., Smithies, O., Baralle, F. E., Shoulders, C. C. and Proudfoot, N. J. (1980). The structure and evolution of the human β-globin gene family. *Cell* 21: 653–668.

Elgin, S. (1981). DNase I hypersensitive sites of chromatin. *Cell* 27: 413–415.

Engelke, D. R., Ng, S.-Y., Shastry, B. S. and Roeder, R. G. (1980). Specific interaction of a purified transcription factor with an internal control region of 5S RNA genes. *Cell* 19: 717–728.

Finch, J. T., Noll, M. and Kornberg, R. D. (1975). Electron microscopy of defined lengths of chromatin. *Proc. Natl. Acad. Sci. USA* 72: 3320–3322.

Foe, V. E., Wilkinson, L. E. and Laird, C. D. (1976). Comparative organization of active transcription units in *Oncopeltus fasciatus*. *Cell* 9: 131–146.

Gazit, B., Panet, A. and Cedar, H. (1980). Reconstitution of deoxyribonuclease-I sensitive structure on active genes. *Proc. Natl. Acad. Sci. USA* 77: 1787–1790.

Gedamu, L. and Dixon, G. H. (1978). Effect of enzymatic decapping on protamine messenger RNA translation in wheat-germ S-30. *Biochem. Biophys. Res. Commun.* 85: 114–124.

Gillies, S. D., Morrison, S. L., Oi, V. T. and Tonegawa, S. (1983). A tissue-specific transcription enhancer element is located in the major intron of a rearranged immunoglobulin heavy-chain gene. *Cell* 33: 717–728.

Greer, C. L., Peebles, C. L., Gegenheimer, P. and Abelson, J. (1983). Mechanism of action of a yeast RNA ligase in tRNA splicing. *Cell* 32: 537–546.

Grosschedl, R. and Birnstiel, M. L. (1980). Spacer DNA upstream from the TATAATA sequence are essential for promotion of H2A histone gene transcription *in vivo*. *Proc. Natl. Acad. Sci. USA* 77: 7102–7106.

Groudine, M. and Weintraub, H. (1981). Activation of globin genes during chick development. *Cell* 24: 393–401.

Hamer, D. H. and Leder, P. (1979). Splicing and the formation of stable RNA. *Cell* 18: 1299–1302.

Herman, R., Weymouth, L. and Penman, S. (1978). Heterogeneous nuclear RNA–protein fibers in chromatin-depleted nuclei. *J. Cell Biol.* 78: 663–674.

Hewish, D. R. and Burgoyne, L. A. (1973). Chromatin substructure. The digestion of chromatin DNA at regularly spread sites by nuclear DNase. *Biochem. Biophys. Res. Commun.* 52: 504–510.

Hood, L. E., Wilson, J. H. and Wood, W. B. (1975). *The Molecular Biology of Eukaryotic Cells.* Benjamin, New York.

Infante, A. A., Nauta, R., Gilbert, S., Hobart, P. and Firshein, W. (1973). DNA synthesis in developing sea urchins: Role of a DNA–nuclear membrane complex. *Nature New Biol.* 242: 5–8.

Jackson, D. A., McCready, S. J. and Cook, P. R. (1981). RNA is synthesized at the nuclear cage. *Nature* 292: 552–555.

Kantor, J. A., Turner, P. H. and Nienhuis, A. W. (1980). Beta thalassemia: Mutations which affect processing of the β-globin mRNA precursor. *Cell* 21: 149–157.

Kimmel, A. R. and Firtel, R. A. (1979). A family of short interspersed repeat sequences at the 5′ end of a set of *Dictyostelium* single copy mRNAs. *Cell* 16: 787–796.

Kornberg, R. D. and Thomas, J. D. (1974). Chromatin structure: Oligomers of histones. *Science* 184: 865–868.

Krämer, A., Keller, W., Appel, B. and Lührmann, R. (1984). The 5′ terminus of the RNA moiety of U1 small nuclear ribonucleoprotein particles is required for the splicing of messenger RNA precursors. *Cell* 38: 299–307.

Labhart, P. and Koller, T. (1982). Structure of the active nucleolar chromatin of *Xenopus laevis* oocytes. *Cell* 28: 279–292.

Larsen, A. and Weintraub, H. (1982). An altered DNA conformation detected by S1 nuclease occurs at specific regions in active chick globin chromatin. *Cell* 29: 609–622.

Lawn, R. M., Efstratiadis, A., O'Connell, C. and Maniatis, T. (1980). The nucleotide sequence of the human β-globin gene. *Cell* 21: 647–651.

Lerner, M. R., Boyle, J. A., Mount, S. M., Wolin, S. L. and Steitz, J. A. (1980). Are snRNPs involved in splicing? *Nature* 283: 220–224.

Lewin, B. (1980). *Gene Expression 2* (Second Edition), Wiley-Interscience, New York.

Ley, T. J., DeSimone, J., Anagnous, N. P., Keller, G. H., Humphries, R. K., Turner, P. H., Young, N. S., Heller, P. and Nienhuis, A. W. (1982). 5-Azacytidine selectively increased γ-globin synthesis in a patient with β⁺ thalassemia. *N. Engl. J. Med.* 307: 1469–1475.

Mandel, J. L. and Chambon, P. (1979). DNA methylation differences: Organ specific variations in methylation pattern within and around ovalbumin and other chick genes. *Nucleic Acid Res.* 7: 2081–2103.

Maquat, L. E., Kinniburgh, A. J., Beach, L. R., Honig, G. R., Lazerson, J., Ershler, W. B. and Ross, J. (1980). Processing of human β-globin mRNA precursor to mRNA is defective in three patients with β⁺ thalassemias. *Proc. Natl. Acad. Sci. USA* 77: 4287–4291.

Marbaix, G., Huez, G., Burny, A., Cleuter, Y., Hubert, E., Lecleroq, M., Chantrenne, H., Soreq, H., Nudel, U. and Littauer, U. Z. (1975). Absence of polyadenylate segment in globin messenger RNA accelerates its degradation in *Xenopus* oocytes. *Proc. Natl. Acad. Sci. USA* 72: 3065–3067.

Mardian, J. K. W., Paton, A. E., Bunick, G. J. and Olins, D. E. (1980). Nucleosome cores have two specific binding sites for non-histone chromosomal proteins HMG 14 and HMG 17. *Science* 209: 1534–1536.

Marx, J. L. (1983). Immunoglobulin genes have enhancers. *Science* 221: 735–737.

Melton, D., Konecki, D. S., Brennard, J. and Caskey, C. T. (1984). Structure, expression, and mutation of the hypoxanthine phosphoribosyltransferase gene. *Proc. Natl. Acad. Sci. USA* 81: 2147–2151.

Merkel, C. G., Kwan, S.-P. and Lingrel, J. B. (1975). Size of the poly(A) region of newly synthesized globin mRNA. *J. Biol. Chem.* 250: 3725–3728.

Ng, S.-Y., Parker, C. S. and Roeder, R. G. (1979). Transcription of cloned *Xenopus* 5S RNA gene by *Xenopus laevis* RNA polymerase III in reconstituted systems. *Proc. Natl. Acad. Sci. USA* 76: 136–140.

Nickol, J., Behe, M. N. and Felsenfeld, G. (1982). Effect of the B-Z transformation in poly(dG-m⁵dC)·poly(dG-m⁵dC) on nucleosome formation. *Proc. Natl. Acad. Sci. USA* 79: 1771–1775.

Nokin, P., Burny, A., Huez, G. and Marbaix, G. (1976). Globin mRNA from anaemic rabbit spleen. Size of its polyadenylate segment. *Eur. J. Biochem.* 68: 431–436.

Noll, M. (1974). Subunit structure of chromatin. *Nature* 251: 249–251.

Nordheim, A., Pardue, M. L., Lafer, E. M., Moller, A., Stollar, B. D. and Rick, A. (1981). Antibodies to left-handed Z-DNA bind to interband regions of *Dro*-

sophila polytene chromosomes. *Nature* 294: 417–422.

Oudet, P., Gross-Bellard, M. and Chambon, P. (1975). Electron microscope and biochemical evidence that chromatin structure is a repeating unit. *Cell* 4: 281–300.

Pardue, M. L. and Gall, J. F. (1970). Chromosomal localization of mouse satellite DNA. *Science* 168: 1356–1358.

Peebles, C. L., Gegenheimer, P. and Abelson, J. (1983). Precise excision of intervening sequences from precursor tRNAs by a membrane-associated yeast endonuclease. *Cell* 32: 525–536.

Raff, R. A. and Kaufman, T. C. (1983). *Embryos, Genes, and Evolution*. Macmillan, New York.

Reynolds, G. A., Basu, S. K., Osborne, T. F., Chin, D. J., Gil, G., Brown, M. S., Goldstein, J. L. and Luskey, K. L. (1984). HMG CoA reductase: A negatively regulated gene with unusual promoter and 5′ untranslated regions. *Cell* 38: 275–285.

Robinson, S. I., Nelkin, B. D. and Vogelstein, B. (1982). The ovalbumin gene is associated with the nuclear matrix of chicken oviduct cells. *Cell* 28: 99–106.

Rogers, J. and Wall, R. (1981). Immunoglobulin heavy-chain genes: Demethylation accompanies class switching. *Proc. Natl. Acad. Sci. USA* 78: 7497–7501.

Rogers, J. and Wall, R. (1980). A mechanism for RNA splicing. *Proc. Natl. Acad. Sci. USA* 77: 1877–1879.

Rothwell, N. V. (1983). *Understanding Genetics*, Third Edition. Oxford University Press, New York.

Rottman, F. A., Shatkin, A. J. and Perry, R. P. (1974). Sequences containing methylated nucleotides at the 5′ termini of messenger RNAs: Possible implications for processing. *Cell* 3: 197–199.

Rutter, W., Jr., Valenzuela, P., Bell, G. E., Holland, M., Hager, G. L., Degennaro, L. J. and Bishop, R. J. (1976). The role of DNA-dependent RNA polymerase in transcriptive specificity. In E. M. Bradbury and K. Javeherian (eds.), *The Organization and Expression of the Eukaryotic Genome*. Academic, New York, pp. 279–293.

Ryffel, G. U., Muellener, D. B., Wyler, T., Wahli, W. and Weber, R. (1981). Transcription of single-copy vitellogenin gene of *Xenopus* involves expression of middle repetitive DNA. *Nature* 291: 429–431.

Sakonju, S., Bogenhagen, D. F. and Brown, D. D. (1980). A control region in the center of the 5S RNA gene directs specific initiation of transcription. I. The 5′ border of the region. *Cell* 19: 13–25.

Scheidereit, C., Geisse, S., Westphal, H. M. and Beato, M. (1983). The glucocorticoid receptor binds to defined nucleotide sequences near the promoter of mouse mammary tumour virus. *Nature* 304: 749–752.

Schlissel, M. S. and Brown, D. D. (1984). The transcriptional regulation of *Xenopus* 5 S RNA genes in chromatin: The roles of active stable transcription complex and histone H1. *Cell* 37: 903–913.

Shatkin, A. J. (1976). Capping of eucaryotic mRNAs. *Cell* 9: 645–653.

Sheiness, D. and Darnell, J. E. (1973). Polyadenylic segment in mRNA becomes shorter with age. *Nature New Biol.* 241: 265–268.

Shen, C. K. J. and Maniatis, T. (1980). Tissue-specific DNA methylation in a cluster of rabbit β-like globin genes. *Proc. Natl. Acad. Sci. USA* 77: 6634–6638.

Sollner-Webb, B., Wilkinson, J. A. K., Roan, J. and Reeder, R. H. (1983). Nested control regions promote *Xenopus* ribosomal RNA synthesis by RNA polymerase I. *Cell* 35: 199–206.

Spritz, R. A., Jagadeeswaran, P., Choudary, P. V., Biro, P. A., Elder, J. T., deReil, J. K., Manley, J. L., Gefter, M. L., Forget, B. G. and Weissman, S. M. (1981). Base substitution in an intervening sequence of a β⁺-thalassemic human globin gene. *Proc. Natl. Acad. Sci. USA* 78: 2455–2459.

Stewart, A. D. and Hunt, D. M. (1982). *The Genetic Basis of Development*. Wiley, New York.

Tiemeier, D. C., Tilghman, S. M., Polsky, F. I., Seidman, J. G., Leder, A., Edgell, M. H. and Leder, P. (1978). A comparison of two cloned mouse globin genes and their surrounding intervening sequences. *Cell* 14: 237–246.

Van Eekelen, C. A. G. and Van Venrooij, W. J. (1981). HnRNA and its attachment to a nuclear protein matrix. *J. Cell Biol.* 88: 554–563.

Vogeli, G., Ohkubo, H., Sobel, M. E., Yamada, Y., Pastan, I. and Crombrugghe, B. de (1981). Structure of the promoter for chicken α2 type I collagen gene. *Proc. Natl. Acad. Sci. USA* 78: 5334–5338.

Wang, A. H. J., Quigley, G. J., Kolpack, F. J., Crawford, J. L., vanBoom, J. H., van der Marel, G. and Rich, A. (1979). Molecular structure of a left-handed double helical DNA fragment at atomic resolution. *Nature* 282: 680–686.

Wasylyk, B., Kedinger, C., Corden, J., Brison, D. and Chambon, P. (1980). Specific *in vitro* initiation of transcription on conalbumin and ovalbumin genes and comparison with adenovirus 2 early and late genes. *Nature* 285: 367–373.

Weintraub, H. (1984). Histone-H1-dependent chromatin superstructures and the suppression of gene activity. *Cell* 38: 17–27.

Weintraub, H. and Groudine, M. (1976). Chromosomal subunits in active genes have an altered configuration. *Science* 193: 848–856.

Weisbrod, S., Groudine, M. and Weintraub, H. (1980). Interaction of HMG 14 and 17 with actively transcribing genes. *Cell* 19: 289–301.

Weisbrod, S. and Weintraub, H. (1979). Isolation of a subclass of nuclear proteins responsible for conferring a DNase I-sensitive structure on globin chromatin. *Proc. Natl. Acad. Sci. USA* 76: 630–634.

Zaret, K. S. and Yamamoto, K. R. (1984). Reversible and persistent changes in chromatin structure accompany activation of a glucocorticoid-dependent enhancer element. *Cell* 38: 29–38.

Zucker, C. and Lodish, H. F. (1981). Repetitive DNA sequences cotranscribed with developmentally regulated *Dictyostelium discoideum* mRNAs. *Proc. Natl. Acad. Sci. USA* 78: 5386–5390.

CHAPTER

Transcriptional regulation of gene expression:
transcriptional changes
in developing cells

Hence, we cannot categorically deny that perhaps we may be able to grind genes in a mortar and cook them in a beaker after all. Must we geneticists become bacteriologists, physical chemists and physicists simultaneously with being zoologists and botanists? Let us hope so.
—H. J. MULLER (1922)

Introduction

In the previous chapter, we studied the structure of chromatin and how it might be changed in order to effect differential gene expression. Here, we shall investigate cases where developmental processes are controlled at the level of transcription. Three major strategies of differential gene transcription will be described: (1) inactivation of large regions of the genome, (2) amplification of those specific genes needed for the production of specific products in particular cells, and (3) selective transcription (by RNA polymerase II) of just those genes needed to synthesize the cell-specific proteins.

Heterochromatin

Chromatin is grossly divided into two major categories, depending on its state of condensation in interphase nuclei. Most of the chromosomal material decondenses during interphase and loses the dark-staining properties that caused it to be seen as individual chromosomes. This material is called EUCHROMATIN. Some chromosomal regions, however, do not decondense. Rather, they retain their tight coiling and heavy-staining properties throughout interphase. This is HETEROCHROMATIN. Heterochromatin is found throughout the animal and plant kingdoms, and it has several properties, besides staining, that distinguish it from euchromatin. First, heterochromatin is relatively, if not completely, inactive in RNA synthesis; second, heterochromatin is the last DNA to replicate at each cell division; and third, heterochromatin suppresses crossing-over between chromatids during meiosis.

There are two types of heterochromatin: constitutive and facultative. CONSTITUTIVE HETEROCHROMATIN is always seen at the same position on both members of a homologous pair of chromosomes. Usually, it is found at the centromeres, and it is often composed of highly repetitive DNA sequences. FACULTATIVE HETEROCHROMATIN is formed by the condensation of chromatin at specific stages of the organism's life cycle and is usually present on only one of the two members of a homologous chromosome pair. Somewhere in development, specific regions of DNA are rendered transcriptionally inactive by the condensation of their DNA into heterochromatin. This has been found to be a widespread mechanism of gene regulation, and we shall see it here in insects and mammals.

Paternal heterochromatin in meally bugs

One of the most impressive feats of facultative heterochromatization occurs in the male meally bug (*Planococcus citri*) (Figure 1). Females of

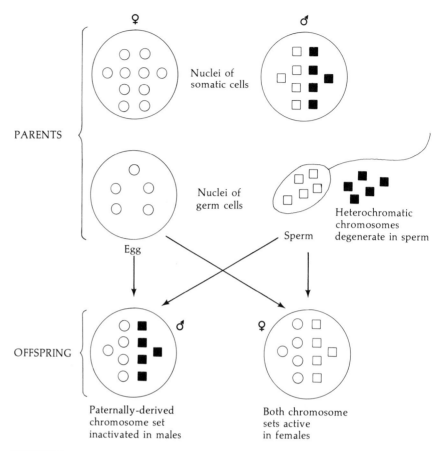

FIGURE 1

Heterochromatization of the paternally derived chromosomes in male mealy
bugs. The male mealy bug becomes functionally haploid and only the euchromatic chromosomes are packaged in the sperm. These chromosomes become heterochromatic if the resulting organism is male.

this species have no facultative heterochromatin. In males, however,
the entire paternally derived haploid chromosome set becomes heterochromatic. Even though this chromosome set was euchromatic in the
father, it becomes heterochromatic when transmitted to the son. The
male *Planococcus*, then, has a euchromatic haploid chromosome set derived from the mother and a heterochromatic haploid set derived from
the father. In male meiosis, the heterochromatic chromosomes disintegrate, leaving only the euchromatic ones to be packaged into sperm. If
the offspring is to be male, these chromosomes will condense.

A result of this arrangement is that all the gene expression in male
coccids comes from maternally derived chromosomes (except for the
few tissues in the males where the heterochromatization is reversed).
This genetic inertness of the paternally derived chromosome set in

males was elegantly demonstrated by Brown and Nelson-Rees (1961), who irradiated male and female coccids and mated them to unirradiated animals. When females were irradiated and mated, there were many deaths in the male and female offspring produced. This showed that the X-rays were capable of inducing dominant lethal mutations. However, when the male parents were irradiated, there were few deaths among the male progeny but many deaths among the daughters; the heterochromatization of the irradiated chromosomes in male offspring prevented the expression of dominant lethal mutations.

Mammalian X chromosome inactivation

In mammals, facultative heterochromatin is seen in the phenomenon of X chromosome inactivation. In 1949, Barr and Bertram discovered a deeply staining nuclear body residing on the nuclear envelope of female cat neurons. This BARR BODY was later demonstrated in numerous female mammals, including humans, and was shown to be an inactive X chromosome (Figure 2). This means that even though female cells have two X chromosomes and male cells have only one, only one X chromosome in the female cells can be transcriptionally active. This equalization is called DOSAGE COMPENSATION.

One of the earliest analyses of X chromosome inactivation was performed by Mary Lyon (1961), who observed the coat color patterns in mice. If a mouse is heterozygous for an autosomal gene controlling hair pigmentation, the mouse resembles one of the two parents or has a color intermediate between the two. In any case, the mouse will be a single color. However, if a female mouse is heterozygous for a pigmentation gene on the X chromosome, a different result is seen (Figure 3). Patches of one parental color are seen alternating with patches of the other parental color. Lyon proposed the following hypotheses to account for these results: (1) In the early development of female mammals, both X chromosomes are active. (2) As development proceeds, one X

FIGURE 2
Nuclei of human oral epithelial cells stained with cresyl violet. (A) Cell from a normal XY male, showing no Barr body. (B) Cell from a normal XX female, showing a single Barr body (arrow). (C) Cell from a female with three X chromosomes. Two Barr bodies can be seen and only one X chromosome is active per cell. (From Moore, 1977.)

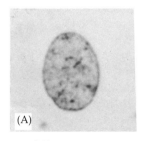
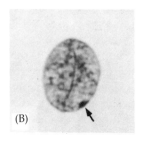
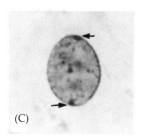

(A)   (B)   (C)

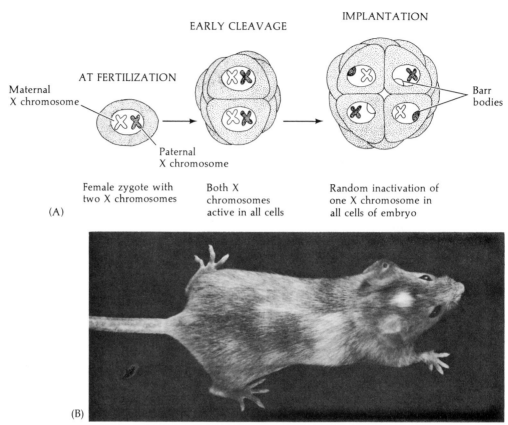

FIGURE 3

X chromosome inactivation in mammals. (A) Schematic diagram illustrating random X chromosome inactivation. The inactivation is believed to occur at about the time of implantation. (B) A female mouse heterozygous for the X-linked coat color gene *dappled*. Distinctly different pigmented regions are seen. (Photograph courtesy of M. F. Lyon.)

chromosome is turned off in each cell. (3) This inactivation is random. In some cells, the paternally derived X chromosome is inactivated; in other cells, the maternally derived X chromosome is shut off. (4) This process is irreversible. Once an X chromosome has been inactivated, the same X chromosome is inactivated in all that cell's progeny. (The areas of pigment in these mice are large patches. It is not a "salt-and-pepper" pattern.) Thus, female mammals are mosaics of two cell types.

Some of the most impressive evidence for this model comes from biochemical studies on clones of human cells. In humans, there is a genetic disease—Lesch-Nyhan syndrome—that is characterized by the lack of the X-linked enzyme hypoxanthine-guanosine phosphoribosyl-transferase (HGPRT). This disease is transmitted through the X chromosome, that is, the males having this mutation in their one X chromosome suffer (and die) from the disease. In females, however, the presence of the mutant HGPRT gene can be masked by the other X chromosome, which carries the wild-type allele. A woman who has sons with this disease is said to be a carrier, having a mutant HGPRT gene on one chromosome and a wild-type HGPRT gene on the other X chromosome. If the Lyon hypothesis is correct, each cell from such a woman should be making either the active or the inactive HGPRT, depending on which X chromosome was active. Barbara Migeon (1971) tested this by taking individual skin cells from a woman heterozygous for the HGPRT gene and placing them in culture. Each of these cells then divided to form a clone of cells. When Migeon stained the clones for the presence of wild-type HGPRT, about half of the clones had the enzyme and the other half did not (Figure 4).

Another X chromosome gene codes for glucose-6-phosphate dehydrogenase (G6PD). There are two common electrophoretic variants of this enzyme: G6PD-A and G6PD-B. Males can be either A or B, but females can be A, B, or AB with regard to their G6PD phenotype. Skin taken from heterozygous women show both G6PD variants (Figure 5). However, when individual cells from these heterozygotes are cloned, the isolated clones express either one of the two possible variants. No clone expressed both (Davidson et al., 1963).

The Lyon hypothesis of X chromosome inactivation provides an excellent account of differential gene inactivation at the level of transcription. Some interesting exceptions to the general rules further show its importance. First, the hypothesis holds true only for *somatic* cells. In female *germ* cells, the inactive X chromosome is reactivated shortly before the cells enter meiosis (Kratzer and Chapman, 1981; Gartler et al., 1980). Thus, in mature oocytes, both X chromosomes are seen to

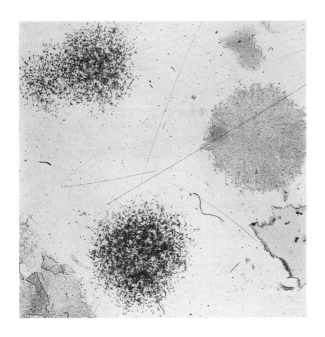

FIGURE 4

Retention of X chromosome inactivation. Approximately 30 cells from a woman heterozygous for HGPRT deficiency were placed into a petri dish and allowed to grow. The cells were stained by autoradiography after incubation in a medium containing radioactive hypoxanthine. Cells with HGPRT incorporate the radiolabeled compound into their RNA and darken the photographic emulsion placed over them. The clones of cells without HGPRT appear lighter because their cells cannot incorporate the radioactive compound. (From Migeon, 1971; photograph courtesy of B. Migeon.)

FIGURE 5

Two populations of cells in human females. Electrophoresis of skin cells from women heterozygous for G6PD indicates that both chromosomes are synthesized, but in different cells. When heterozygous skin cells are cultured (line 3), both types of enzymes are present. However, in each clone of skin cells (lines 4–10), only one form of the enzyme is seen.

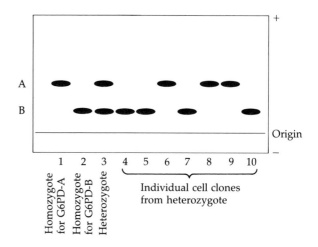

FIGURE 6

Both X chromosomes are active in mammalian oocytes. Electrophoresis of cells from the ovary (lanes 1 and 2) and lung (lane 3) of a 14-week human fetus heterozygous for G6PD. The lung cells express the A (AA) and B (BB) forms of the enzyme, whereas the ovary cells also contain a heterodimer (AB) of G6PD. (The heavy A and B bands of the ovary reflect the fact that the ovarian cells themselves express only the A and B enzymes; the heterodimer is expressed only in the oocytes.) (Photograph courtesy of B. Migeon.)

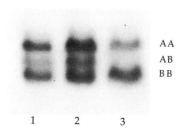

be active. This is shown in Figure 6. G6PD is a dimeric enzyme, so each somatic cell has enzymes constructed of either two A subunits or two B subunits. In a heterozygous female, some cells will have G6PD composed of two A subunits, whereas other cells will have G6PD composed of two B subunits. If both X chromosomes are functional in the same cell, however, one expects to find some G6PD composed of one subunit of A and one subunit of B. This is precisely what is seen when oocytes are tested (Gartler et al., 1973; Migeon and Jelalian, 1977). An AB *heterodimer* is seen, evidence that both X chromosomes have been transcriptionally active in the same cell.

The second class of exceptions concerns the randomness of X chromosome inactivation. Randomness is certainly the rule, but there are some cases where the preferential inactivation of the paternal X chromosome is seen. In the trophoblast tissues of female mice (but not humans), expression of the maternal X predominates—to the extent of an almost complete absence of paternal X expression. In marsupials, the paternally derived X chromosome is also preferentially inactivated (Sharman, 1971; Cooper et al., 1971). We still do not know why this happens, but it may ultimately provide an important key to understanding the mechanics of X chromosome inactivation.

Third, X chromosome inactivation may not extend throughout the entire length of the X chromosome. Most all known X chromosome genes map to the long arm of the X chromosome. However, the gene for steroid sulfatase is known to be located on the short arm of the human X. This locus is not seen to undergo inactivation (Mohandas et al., 1980). Thus, heterochromatization may not extend entirely throughout the X chromosome.

The fourth exception really ends up being the proof of the rule. There are some male mammals with coat color patterns that should not be expected unless the animals exhibit X chromosome inactivation. Male calico and tortoiseshell cats are among these examples. These spotted coat patterns are normally seen in females and are thought to result from the Lyon effect. But rare males exhibit these coat patterns as well. How can this be? It turns out that these cats are XXY. The Y chromosome makes them male (see Chapter 19), but the X chromosomes undergo inactivation, just as in females, such that there is only one active X per cell (Centerwall and Benirschke, 1973). Thus, these cats are seen to have a Barr body and have random X chromosome inactivation. It is clear, then, that one mechanism of transcription-level control of gene regulation is to make a large number of genes heterochromatic and, thus, transcriptionally inert.

SIDELIGHTS & SPECULATIONS

Mechanism of X chromosome inactivation

There are several theories for how one X chromosome becomes inactivated and how this inactivation is passed on to the progeny of that cell. However, the actual inactivation event is very difficult to observe and manipulate. During the earliest stages of mammalian development, both X chromosomes function (Epstein et al., 1978); and the inactivation event occurs soon after implantation when the embryo is composed of very few cells and is very hard to experiment upon. However, certain cells can be grown in large quantities and they will inactivate an X chromosome in culture. These are the teratocarcinoma cells mentioned in Chapter 6. We have already noted that these cells are totipotent and resemble the early inner cell mass blastomeres. This being the case, might we not expect them to inactivate an X chromosome when they differentiate? McBurney and Strutt (1980) have shown this to be the case. Both X chromosomes were active in the teratocarcinoma stem cells derived from female mice. Yet, as these cells differentiated into the array of tissues found in the tumor, one of the X chromosomes of each cell became late replicating and transcriptionally inert.

The mechanism by which all the genes on one X chromosome are coordinately inactivated is still unknown. Whatever the mechanism is, however, it may operate by making the genes of one X chromosome more susceptible to methylation at their 5' ends. Wolf and his co-workers (1984) have data indicating that human X chromosome inactivation can be correlated with the methylation of a cluster of sites at the 5' end of the HGPRT gene. In the *active* X chromosome, there is extensive methylation throughout most of the gene, but the 5' end cluster of cytosines is unmethylated. The methylation of the *inactive* X chromosome varies from cell to cell, but it is characterized by extensive methylation of this 5' cluster. Moreover, the transcriptional activity of the HGPRT gene on the inactive X can be restored by growing cultured cells in 5-azacytidine (Mohandas et al., 1981). When such reverted genes were observed, a loss of methylation was seen at the 5' end. It appears, then, that whatever the signaling mechanism might be, the inhibition of transcription from the inactive X chromosome may be regulated in the same manner as cell type-specific genes.

Amplified genes

Amplification of ribosomal RNA genes

One strategy for transcribing enormous amounts of a particular RNA is to make many more copies of the specific gene. This strategy is used only rarely, and it is mainly seen in the amplification of insect and amphibian ribosomal RNA genes and in the chorion (egg shell) genes of certain insects such as *Drosophila*. In many respects, the ribosome can be considered as a differentiated cell product of the oocyte. This is especially so in amphibians, where the oocyte contains around 200,000 times the number of ribosomes found in most larval cells. Thus, for one cell the number of ribosomes is truly spectacular. In fact, in those zygotes lacking the nucleolar organizing regions that make ribosomal RNA, development can still continue on the stored ribosomes up until the feeding tadpole stage (at which time the organism finally dies from the lack of ribosomes needed for growth).

Synthesis of ribosomal RNA in amphibian oocytes occurs during the prolonged prophase of the first meiotic division. Specifically, rRNA synthesis takes place during the months-long diplotene stage, after pairing and replication of the homologous chromosomes.

The ribosomal RNA genes of the frog *Xenopus laevis* are among the best studied; they are diagrammed in Figure 7. There is a 5' leader sequence followed by the coding sequence for 18 S ribosomal RNA. This is followed by a transcribed spacer region and the coding sequence for the 5.8 S ribosomal RNA gene, another transcribed spacer, and then the gene for the 28 S ribosomal RNA. There are no introns within these genes, and the entire unit is transcribed into a 40 S ribosomal RNA precursor, which is then processed to form the three ribosomal RNAs. (The 5 S ribosomal RNA is made from another region of the genome, and these 5 S genes are present in thousands of copies in the genome and are not amplified.)

The nontranscribed spacer (NTS) DNA between these functional units may serve as a "loading zone" for RNA polymerase I; for although the length of the NTS varies between each transcription unit, each NTS is composed of short repeated DNA sequences. These repeated sequences are capable of binding RNA polymerase I. Thus, the NTS can store RNA polymerase for transcribing the ribosomal RNA precursor (Moss, 1983).

Mechanism of ribosomal gene amplification

There are two homologous regions of large ribosomal RNA genes per diploid frog cell. Each region contains roughly 450 copies of the unit described in the preceding section, separated by nontranscribed spacers. All of the ribosomal gene units on a chromosome are identical, but the nontranscribed spacers vary in length from unit to unit, even on the same chromosome. We would expect, then, that the diplotene oocyte (in which the DNA has already replicated) would contain 4×450, or roughly 1800 genes coding for ribosomal RNA. But even this is not enough! Rather than a few thousand copies, one finds nearly a million ribosomal RNA genes. One sees, not four nucleoli, but thousands (Figure 8). These extra nucleoli reside within the nucleus but are not attached to any chromosome, as nucleoli usually are.

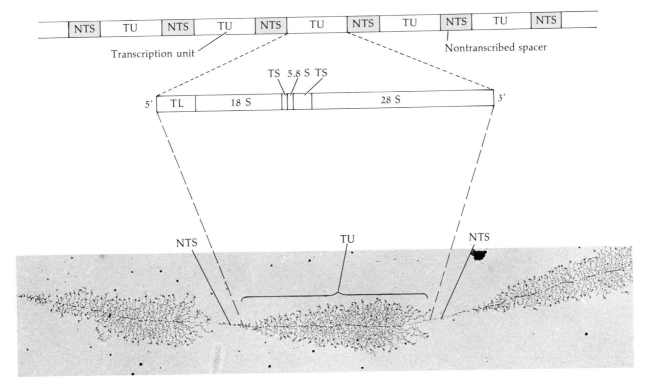

FIGURE 7

Organization of the genes for 18 S, 5.8 S, and 28 S ribosomal RNA in amphibians. Each transcription unit (TU) is flanked by nontranscribed spacer regions (NTS). The transcription unit contains a transcribed leader sequence (TL), the 18 S rRNA gene, a transcribed spacer (TS), the 5.8 S rRNA gene, another transcribed spacer, and the gene for 28 S rRNA. Beneath the diagram is an electron micrograph of tandemly arranged rRNA genes actively transcribing in the newt oocyte. The RNA gets progressively larger as it is transcribed, and dozens of transcripts are made simultaneously. (After Watson, 1976; photograph courtesy of O. L. Miller, Jr.)

By comparing the hybridization of radioactive 28 S ribosomal RNA to oocyte and somatic cell DNA, it was determined that the ribosomal RNA genes were amplified 1500 times over their normal amount. If each nucleolus contained 450 gene copies, we would estimate that there are 6.8×10^5 genes for the large ribosomal RNA precursors in the amphibian oocyte. This amplification can also be seen on cesium chloride gradients. When DNA is fragmented and centrifuged in such gradients, the fragments sediment according to their density. The DNA of *Xenopus laevis* somatic cells has an average buoyant density of 1.699 grams per cubic centimeter. However, if one looks at the density pattern for the sheared DNA from oocytes, one sees a slightly different pattern (Figure 9). A second, or "satellite," peak is seen; this DNA is derived from the amplified ribosomal RNA genes. The base composition of this

FIGURE 8
Nucleus isolated from an oocyte of *Xenopus laevis*. The deeply stained spots represent the extra-chromosomal nucleoli. (From Brown and Dawid, 1968; courtesy of D. D. Brown.)

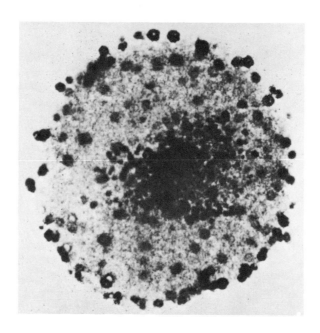

DNA gives it a higher density than the majority of the *Xenopus* DNA. The average DNA of *Xenopus laevis* is 40 percent GC, whereas the ribosomal genes are 67 percent GC. Thus, the ribosomal genes are denser than most. In the somatic tissues, the ribosomal genes make up only 0.06 percent of the genome and do not form a separate, detectable satellite peak. However, when these genes are amplified 1500 times, they make a band of dense DNA easily distinguished from the main band. This DNA will hybridize to ribosomal DNA but not to any other genes. So we see the specific amplification of a gene, thereby allowing the transcription of enormous amounts of RNA (Brown and Dawid, 1968).

The next question concerns how these genes might be amplified. Although we do not know the exact details of the mechanism, we do have some clues as to how it is done. We know, for instance, that the extrachromosomal nucleoli are made during the pachytene stage of meiotic prophase, just before diplotene. Indeed, there is a "pachytene cap," which represents a large mass of extrachromosomal DNA. Thus, the extra genes are made during the pachytene stage of meiosis and are transcribed during the diplotene stage. When the oocyte nucleus

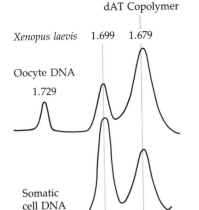

FIGURE 9
Isolation of extrachromosomal (amplified) nucleoli. DNA from somatic cell nuclei (lower) and oocyte nuclei (upper) were isolated; each was mixed with a standard amount of dAT (for aligning the results) and centrifuged in cesium chloride to equilibrium. In both cases, the major band of DNA had a density of 1.699 g/mL. In the oocyte, a "satellite" band of greater density is also seen. (Modified from Brown and Dawid, 1968.)

disintegrates during the first meiotic division, the extra nucleoli are thrown into the cytoplasm and are destroyed.

One of the most interesting clues to the mechanism of amplification is that in any given extrachromosomal nucleolus, the nontranscribed spacer regions are all the same length (Wellauer et al., 1976). In the chromosomal gene clusters of the same cells, however, there is marked heterogeneity of spacer lengths, so the extrachromosomal nucleoli are not merely copies of the entire chromosomal nucleolus. One model that could explain the generation of hundreds of ribosomal gene units separated by identical nontranscribed spacers is the rolling circle model. This replicative mechanism is used by many viruses. In the oocyte, a single ribosomal gene unit would detach in some way from the chromosome and form a circle. The circle would be nicked (one strand broken) and DNA synthesis would begin in both directions from this site. In such a way, identical copies of the ribosomal genes and their spacers can be formed (Figure 10). Electron micrographs of the amplified genes (Hourcade et al., 1973; Rochaix et al., 1974) give support to this model. They reveal closed circles typical of extrachromosomal nucleoli, as well as "lariat" structures believed to be the rolling circle intermediates (Figure 11).

Visualization of transcription from amplified genes

The amplified ribosomal RNA genes of amphibian oocytes were the first genes to be isolated and purified. They were also the first genes whose transcription could actually be seen by electron microscopy. Knowing that these genes would be intensely active in transcribing ribosomal RNA during the diplotene stage of meiosis, Miller and Beatty (1969) used low ionic strength buffers to disperse the chromatic strands for electron microscopy. The circles of DNA unwound to reveal the configuration seen in Figure 7. There is an axial core of DNA upon

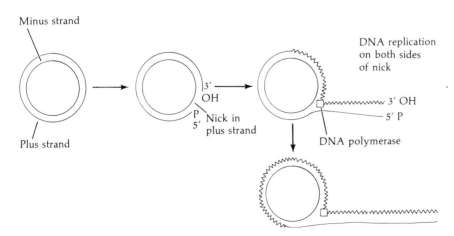

FIGURE 10

The rolling circle model of replication by which many identical units can be synthesized. The circle would include one ribosomal RNA transcription unit and one nontranscribed spacer. By this method, extrachromosomal nucleoli would be formed having the same length NTS between their genes.

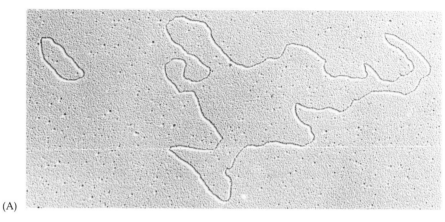

(A)

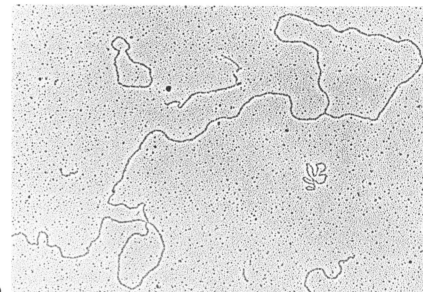

(B)

FIGURE 11
Isolated extrachromosomal rRNA in *Xenopus laevis* oocytes. Electron micrographs find circles (A) and "lariats" (B) suggestive of the rolling circle mechanism of replication. (From Rochaix et al., 1974; courtesy of A. Bird.)

which rRNA is being synthesized. Numerous ribosomal RNA chains can be seen being transcribed from any one unit, and several such units are seen in this picture. In each unit, transcription begins at the small end of the "Christmas tree" and continues until the large ribosomal RNA precursor is formed. At each junction of the DNA with an RNA transcript, there is a molecule of RNA polymerase (type I). The length of the RNA in these large transcripts corresponds to about 7200 bases, roughly the same size as is estimated for the ribosomal precursor. Between the transcribing units of DNA are nontranscribed spacers.

Thus we can get an accurate picture of eukaryotic gene transcription in action.

Drosophila chorion genes

The chorion genes of *Drosophila melanogaster* are also known to be amplified during the synthesis of the chorion proteins of the egg. These proteins are synthesized by the ovarian follicle cells. Prior to chorion gene expression, the entire genome of the follicle cells undergoes extra rounds of DNA synthesis to achieve 16 times the haploid DNA content. After these replications, the chorion protein genes are selectively replicated about tenfold again. This does not happen in tissues other than the ovarian follicle cell (Spradling and Mahowald, 1980). Spradling (1981) showed that this amplification is due to additional rounds of DNA synthesis in specific regions of the genome only. These regions of DNA are characterized by several branches, each containing the amplified chorion genes (Figures 12 and 13). Here, amplification occurs, but the genes remain attached to the chromosomes.

Selective gene transcription

Most genes do not utilize selective gene amplification as a mechanism for large-scale gene expression. In developing red blood cells, where hemoglobin constitutes 98 percent of the protein being synthesized, no selective amplification of the globin genes is seen. In the silk gland of caterpillars producing tremendous amounts of the silk protein fibroin, the entire genome is amplified by polyteny. Even so, there is no selective

FIGURE 12
Amplification of chorion genes within the chromosome. Diagram of local amplification by multiple rounds of DNA synthesis in a region already containing multiple copies of the chorion genes.

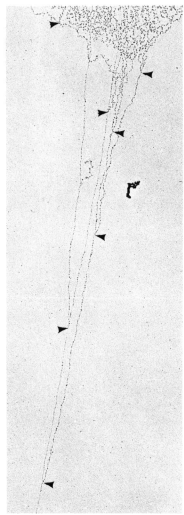

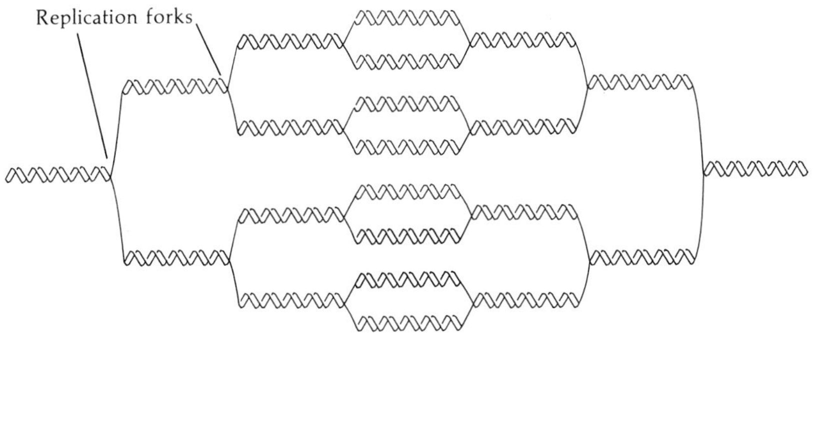

FIGURE 13
Electron micrograph of chorion gene amplification in the follicle cells of *Drosophila*. Arrows point to the sites of DNA replication. (From Osheim and Miller, 1983; photograph courtesy of O. L. Miller, Jr.)

amplification of the fibroin genes. Thus, we expect that certain genes can be selectively transcribed.

Chromosomal puffs and lampbrush chromosomes

The nature of polytene chromosomes and their puffs was discussed in Chapter 10. It should be realized that the data presented there illustrates the transcription-level control of specific genes. When ecdysone is added to salivary glands, certain puffs are produced and others regress. The puffing is mediated by the binding of ecdysone to specific places on the chromosomes. This can be shown by crosslinking the ecdysone to the chromatin so that it stays where it has been bound. Then, rabbit antibodies recognizing ecdysone are added and washed away, and a fluorescence-labeled goat antibody against rabbit immunoglobulins is added. The fluorescent tag will be located wherever a goat antibody has bound to a rabbit antibody. Rabbit antibodies should bind only to ecdysone. In this manner, a fluorescent label is brought to every place on the chromosome where ecdysone has been bound (Gronemeyer and Pongs, 1980). The result (Figure 14) shows that all the ecdysone-sensitive puff sites bind ecdysone.

Ecdysone-sensitive puffs can be grossly divided into three categories. There are those puffs that ecdysone causes to regress; there are those puffs that ecdysone induces rapidly; and there are those puffs that are first seen several hours after ecdysone stimulation. The early puffs are stimulated to form within 4 hours after the addition of ecdysone. Moreover, they appear to synthesize a protein that is necessary

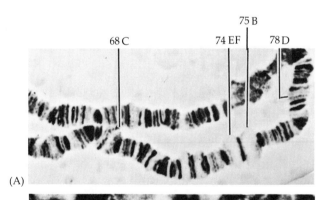

(A)

(B)

FIGURE 14
Localization of ecdysone on the polytene chromosomes of *Drosophila*. After ecdysone is linked in situ on the chromosomes, it can be recognized by fluorescent antibodies. (A) Phase-contrast micrograph of a segment of salivary gland chromosomes. (B) Same segment seen under ultraviolet light, showing immunofluorescence at ecdysone-sensitive areas. (Courtesy of O. Pongs.)

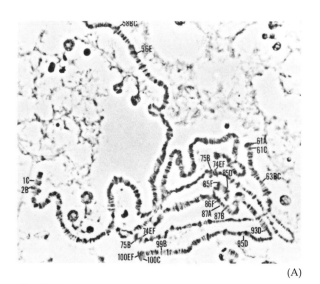
(A)

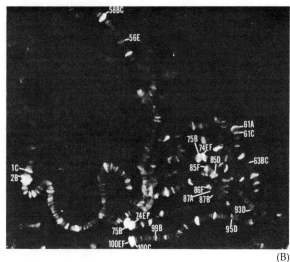
(B)

FIGURE 15

Immunofluorescence localization of a protein that is released upon DNase I digestion of *Drosophila* chromatin. When antibodies are made against that protein and are bound to the polytene chromosomes, they can be recognized by other, fluorescent antibodies. (A) Phase-contrast micrograph. (B) Same region seen under ultraviolet light, showing the labeling of ecdysone-sensitive sites. (Figures 14 and 15 can be compared to Figure 17 in Chapter 10, which shows the normal puffing sequence.) (From Mayfield et al., 1978; photographs courtesy of S. Elgin.)

for the induction of the later puffs. If protein synthesis inhibitors are added to cultured *Drosophila* cells soon after the early puffs have formed, ecdysone will not stimulate the formation of the later puffs. Presumably, the later puffs need both ecdysone and the early puff product (Ashburner et al., 1974).

It is worth noting that ecdysone not only stimulates the puffing of certain regions but also causes the regression of certain existing larval puffs. One of these latter puffs is present at position 68C of the left arm of chromosome 3. This puff site contains the gene for *Sgs3*, a "glue protein" responsible for attaching the pupal case to a solid surface (Korge, 1975). Only in the salivary gland does this region of the chromosome puff out at the last instar. The addition of ecdysone to cultured salivary glands causes the rapid regression of this puff and the cessation of transcription from this gene (Meyerowitz and Hogness, 1982). Thus, the glue can be synthesized when it is needed to adhere the larva to a substrate; but afterward the gene is shut off while metamorphosis occurs.

As in other organisms, susceptibility to DNase I is correlated with transcriptional activity and appears to be mediated by nonhistone chromatin proteins. DNase I digestion of *Drosophila* chromatin releases a 63,000-d protein that will bind to all the ecdysone-sensitive loci of third instar larvae (Figure 15). While the size of the protein is larger than the

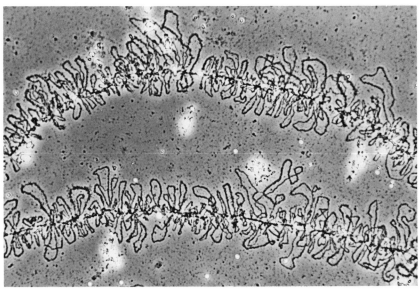

FIGURE 16
Portions of lampbrush chromosomes from the nucleus of a newt oocyte. Regions of DNA can be seen looping out of a central axis. (Courtesy of J. Gall.)

FIGURE 17
Localization of histone genes on a lampbrush chromosome by in situ hybridization. (From Old et al., 1977; courtesy of H. G. Callan.)

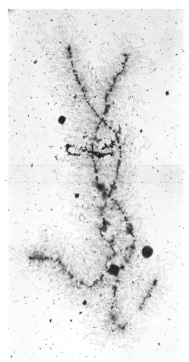

HMG proteins, it may serve the same function in establishing or maintaining the chromatin structure characteristic of active genes (Mayfield et al., 1978).

Puffs represent the unwinding of DNA in specific areas of the larval insect chromosomes. Because the chromosomes are polytene, the thousands of strands produce the characteristic puff appearance. An analogous unwinding of DNA from nonpolytene chromosomes is seen in amphibian oocytes. These structures are the LAMPBRUSH CHROMOSOMES. During the diplotene stage of meiosis, the compacted amphibian chromosomes will stretch out large loops of DNA, retracting them after this stage is completed (Figure 16).

The notion that this unfolding of the chromosomes enables certain genes to be expressed can be demonstrated by in situ hybridization. Preparation of oocyte chromosomes can be prepared, denatured, and incubated with radioactive RNA for a specific protein. After the unbound RNA is washed away, autoradiography can find the precise location of the gene. Figure 17 shows diplotene chromosome I of the newt *Triturus cristatus* after incubation with radioactive histone mRNA. It is obvious that a histone gene (or set of histone genes) is located on one of these loops of the lampbrush chromosome (Old et al., 1977). Electron micrographs of gene transcripts for lampbrush chromosomes look very much like the transcripts seen on the amplified ribosomal RNA genes. The same "Christmas tree" polarity is seen, but usually (the exception being gene families like the histones) only one transcription unit is observed for each loop (Hill and MacGregor, 1980).

Ovalbumin synthesis

The enormous egg of the laying hen is protected by numerous extra layers, which are secreted upon it during its passage through the oviduct (Figure 18). The ovum is gathered by the INFUNDIBULUM of the left oviduct (the right oviduct having atrophied in most birds) and becomes fertilized shortly thereafter. The egg then passes sequentially into the MAGNUM, where ovalbumin and the other egg white proteins are secreted; into the ISTHMUS, where the shell membranes are put on; and finally into the UTERUS, where water, salts, and the calcium carbonate shell are added.

The major product of the tubular cells of the magnum is ovalbumin, and during this time ovalbumin constitutes more than 50 percent of the total cell protein. The structure of the ovalbumin gene is illustrated in Figure 19. It contains eight coding sequences and seven introns. The production of ovalbumin depends on the presence of the sex hormone

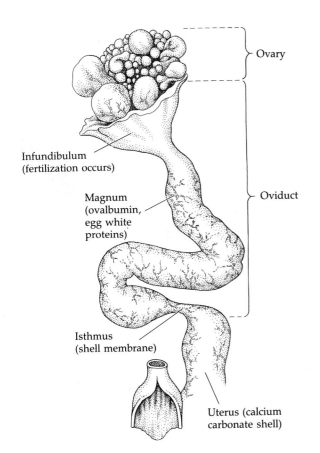

Ovary

Infundibulum
(fertilization occurs)

Magnum
(ovalbumin,
egg white
proteins)

Oviduct

Isthmus
(shell membrane)

Uterus (calcium
carbonate shell)

FIGURE 18
The reproductive system of a mature hen. Numerous layers of secretions coat the egg during its passage along the oviduct.

FIGURE 19
Structure of the ovalbumin gene. This gene contains eight exons (white) and seven introns (black). The sites of transcription initiation and translation initiation [AUG] are separated by the first intron. (Modified from Dugaiczyk et al., 1978.)

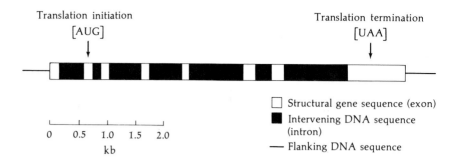

estrogen. Estrogen enters the cell and is picked up by a receptor protein, which translocates it into the nucleus. The estrogen–receptor complex then binds to the chromatin and stimulates transcription.

The estrogen dependence of ovalbumin production is readily seen when young female chicks are injected with estrogen (Palmiter and Schimke, 1973). These injections cause the tubular glands of the magnum to differentiate and to synthesize the major egg white proteins: ovalbumin, conalbumin, ovomucoid, and lysozyme. During 2 weeks of estrogen treatment, ovalbumin levels rise from undetectable amounts to become a majority of newly synthesized cellular proteins (Figure 20). Withdrawal of the estrogen causes a decline in ovalbumin production even though the cells retain their differentiated state. After 2 weeks of estrogen withdrawal, no ovalbumin can be detected.

This only demonstrates that the synthesis of ovalbumin is dependent upon the presence of the hormone. It does not tell us at what level this hormone works. Conceivably, it could either increase transcription, or stabilize the ovalbumin protein or its message, or even aid in the

FIGURE 20
The effect of estrogen on ovalbumin synthesis in the oviduct of 4-day-old chicks. Chicks were given daily injections of estrogen (primary stimulation); after 10 days the injections were stopped. Two weeks after this withdrawal, the injections were resumed (secondary stimulation). (After Palmiter and Schimke, 1973.)

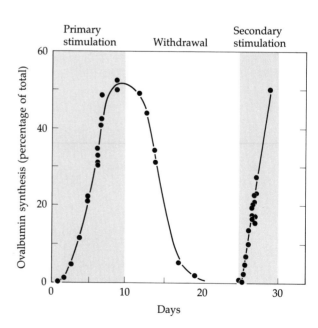

transport of the ovalbumin message through the nuclear membrane. The next step was to find a correlation between estrogen and the presence of ovalbumin mRNA (Harris et al., 1975; McKnight and Palmiter, 1979). To do this, a complementary DNA was made from isolated ovalbumin message. The ovalbumin message was obtained by passing the cytoplasmic mRNA from oviduct cells over a column containing oligo(dT)-cellulose beads (Figure 21). These beads will bind mRNA through the attraction of the deoxythymidine chains for the poly(A) tails of the mRNA. All the other nucleic acids pass through. The mRNA is then eluted from the column and the different mRNAs are separated

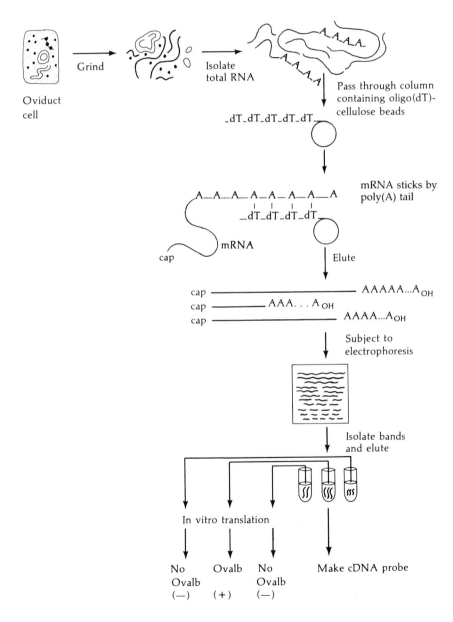

FIGURE 21
Diagram of protocol used to isolate ovalbumin mRNA to make cDNA probe. (See text for details.)

by electrophoresis. The ovalbumin message can be readily isolated (as it is the most prevalent message in the cell), and it is then transcribed in vitro to make certain that it codes for ovalbumin. The message is then used as a template for reverse transcriptase, and a cDNA probe is made. This probe enables one to "fish out" any complementary sequence, and it can recognize a single copy of ovalbumin message per cell. Thus, the rate of hybridization is a measure of how many messages a cell has for ovalbumin. Table 1A shows that the estrogen treatment is responsible for inducing the ovalbumin message. Even more importantly, the cDNA probe has shown that estrogen stimulates the appearance of ovalbumin-coding sequences in the *nucleus* (Roop et al., 1978; Table 1B). Thus, the control of ovalbumin production by the magnum cells is mainly regulated at the level of transcription.

TABLE 1
Transcriptional regulation of ovalbumin

A. INDUCTION OF CYTOPLASMIC OVALBUMIN MESSENGER RNA DURING PRIMARY AND SECONDARY STIMULATION WITH ESTROGEN

Hormonal state of oviduct[a]	Molecules mRNA$_{ov}$ per tubular gland cell[b]
UNSTIMULATED	None detectable
+4 days DES	20,000
+9 days DES	44,000
+18 days DES	48,000
WITHDRAWN FOR 12 DAYS	0-4
+0.5 hour DES	9
+1.0 hours DES	50
+4.0 hours DES	2,300
+8.0 hours DES	5,100
+29.0 hours DES	17,000

B. INDUCTION OF NUCLEAR RNA TRANSCRIPTS CONTAINING OVALBUMIN SEQUENCES

Tissue	hnRNA molecules containing ovalbumin-coding sequences per tubule gland cell nucleus[c]
ESTROGEN-STIMULATED OVIDUCT	3,075
ESTROGEN-WITHDRAWN OVIDUCT	2
+2 hours DES	6
+4 hours DES	71
+8 hours DES	460
+16 hours DES	995
+48 hours DES	3,049

Source (A): Harris et al. (1975).
Source (B): Roop et al. (1978).
[a]Ten-day-old White Leghorn chicks received daily subcutaneous injections of diethylstilbestrol (DES, a synthetic estrogen; 2.5 mg in oil) and were killed at the indicated times. For experiments involving secondary stimulation with estrogen, the chicks were first treated with DES for 10 days, followed by 11 days of withdrawal from hormone. On day 12 of withdrawal, chicks were given one subcutaneous injection of 2.5 mg of DES and oviducts were collected at the indicated time intervals.
[b3]H-labeled cDNA was made to ovalbumin message and was hybridized to total oviduct RNA extracted from chicks treated as indicated. The number of molecules of ovalbumin mRNA was calculated.
[c]Determined by hybridization to cDNA$_{ov}$.

Coordinated gene expression

One of the fundamental phenomena of differentiation is the coordinated expression of unlinked genes. A kidney cell, for instance, must produce numerous tissue-specific enzymes and structural proteins to function as a kidney. Similarly, estrogen not only initiates the structural differentiation of the oviduct magnum tubule cell, but it also causes these cells to secrete large quantities of several egg white proteins: ovalbumins, conalbumin, lysozyme, and ovomucoid.

Although steroids other than estrogen are not able to induce the initial maturation of the gland, progesterone—a steroid closely related to estrogen—is able to induce the appearance of egg white proteins after primary estrogen stimulation (McKnight and Palmiter, 1979). Studies similar to those presented earlier show that the genes coding for ovalbumin, conalbumin, ovomucoid, and egg white proteins X and Y are each

under transcriptional control. Moreover, they are all secondarily inducible by progesterone. It is reasonable, then, to expect that near each of these genes is a DNA sequence that binds the progesterone–receptor protein complex. The actual DNA sequence for binding this complex is determined by fixing double-stranded calf thymus DNA (which is presumed to have many progesterone binding sites) to cellulose filters and adding the radioactive progesterone–receptor protein complex. The complex is allowed to bind to the DNA, and the bound radioactivity is counted. Next, the same assay is run with an excess of some other DNA in solution. If this competitor DNA has the progesterone–receptor complex binding site, it will compete for the radioactive complex and less radioactivity will be

FIGURE 22
Diagram of protocol for determining the site of the progesterone receptor. (See text for details.)

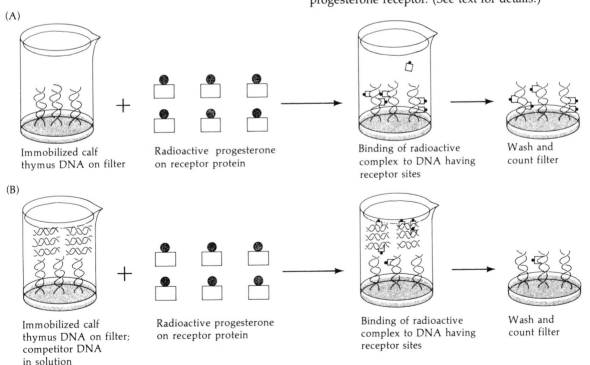

(A)

Immobilized calf thymus DNA on filter

Radioactive progesterone on receptor protein

Binding of radioactive complex to DNA having receptor sites

Wash and count filter

(B)

Immobilized calf thymus DNA on filter; competitor DNA in solution

Radioactive progesterone on receptor protein

Binding of radioactive complex to DNA having receptor sites

Wash and count filter

bound to the calf thymus DNA (Figure 22). Cloned ovalbumin, conalbumin, and ovomucoid genes did compete in this assay, indicating that they have such a binding site. By using various restriction enzyme-derived fragments of these genes, Mulvihill and colleagues (1982) found the progesterone–receptor binding sequence to be 5'-ATCNNATTNTCTGNTTGTA-3'. Each of the egg white protein genes was found to have several sequences homologous to this consensus sequence. These sequences were found in the region immediately upstream from the gene, in the leader sequences, and in the introns. No such sequences have been found near or in any of the non-steroid-inducible genes such as globin. We therefore have a clue as to how a variety of genes can be induced by the same stimulus.

Consensus sequences may also be found by "brute force." By making cDNA probes to brain-specific mRNA, one can identify cloned genes whose products are active only in brain tissue (see Chapter 10). Eleven such "brain genes" have been isolated from the rat DNA clone library. The sequences of each of these genes revealed the presence of an 82-nucleotide consensus sequence in their introns (Sutcliffe et al., 1984). It should be remembered, however, that although all of the brain-specific genes sequenced to date have this sequence, this represents only eleven of the thousands of genes whose mRNA is brain-specific. Sequencing is presently being performed on other groups of coordinately regulated genes in order to determine if specific DNA sequences may be involved, and the results of some of these studies are summarized in Table 2.

TABLE 2
Consensus sequences in coregulated gene expression

Organism	Gene	Regulation	Consensus sequence	Percentage of homology with consensus sequence
Yeast	his1	Induced by amino acid starvation	A${A \atop T}$GTGACTC	89
	his3			78
	his4			89
	trp5			78
Drosophila	hsp70	Induced by heat shock	CTNGAATNTTCAGA	85
	hsp83			92
	hsp26			85
	hsp27			54
	hsp68			69
	hsp23			62
	hsp22			85
Rat	Metallothionein	Glucocorticoid-induced	${T \atop C}$GGTN${A \atop T}$CA${A \atop C}{A \atop T}$NTGT${T \atop G}$CT	95
Chick	Lysozyme			95
Mouse	Mammary tumor virus			95
Rat	11 brain genes	Transcriptionally active in brain tissue	82 nucleotide consensus sequence in introns at 5' end flanking region	~75

Globin gene transcription

One of the best-documented cases for the transcriptional regulation of differentiation is the activation of the globin gene in developing red blood cells. In chick and human embryos, changes in hemoglobin synthesis can be observed during development. In chick erythroblasts, the earliest transcription of hemoglobin sequences can be monitored, and changes can be seen in the chromatin structure when the hemoglobin gene is being transcribed. When the posterior region of the area opaca of a 20- to 23-hour chick embryo is isolated, blood islands are observed to contain the precursors of the red blood cells. Complementary DNA probes to globin messages show that these precursor cells are not yet transcribing globin genes. However, after two more cell divisions (35 hours of development), the cells (now called erythroblasts) are rapidly synthesizing hemoglobin. Somehow during this time the globin genes were turned on.

This transcriptional change is well correlated with alteration of the chromatin structure (Groudine and Weintraub, 1981). First, when erythroblast chromatin was exposed to low concentrations of DNase I, portions of the globin gene were found to be digested. This was not the case when chromatin of the precursor cells were so treated. Second, the methylation pattern of the globin genes differs between the precursor cells and the erythroblasts. The coding regions for the globin proteins are methylated in the precursor cells but are unmethylated in the hemoglobin-producing erythroblasts. Thus, changes in the structure of the chromatin are seen to be correlated with the regulation of globin gene expression at the level of transcription.

Another type of transcription-level regulation occurs later in development. In many species, including chicks and humans, the embryonic or fetal hemoglobin differs from that found in adult red blood cells. A schematic diagram of human hemoglobin types and the genes that code for them is shown in Figure 23. Human EMBRYONIC HEMOGLOBIN consists largely of two zeta (ζ) globin chains, two epsilon (ϵ) globin chains, and four molecules of heme. During the second month of human gestation, ζ and ϵ globin synthesis abruptly ceases, while alpha (α) and gamma (γ) globin synthesis increases (Figure 24). The association of two γ globin chains with two α globin chains produces FETAL HEMOGLOBIN. At three months gestation, the beta (β) globin and delta (δ) globin genes begin to become active, and their products slowly increase while γ globin gradually declines. This switchover is greatly accelerated after birth, and fetal hemoglobin is replaced by adult hemoglobin, $\alpha_2\beta_2$. The normal adult hemoglobin profile is 97 percent $\alpha_2\beta_2$, 2–3 percent $\alpha_2\delta_2$ and one percent $\alpha_2\gamma_2$.

In humans, the ζ and α globin genes are on chromosome 16, but the ϵ, γ, δ and β globin genes are linked together, in order of appearance, on chromosome 11. It appears, then, that there is a mechanism

FIGURE 23
Sequential gene activation in hemoglobin synthesis during development.

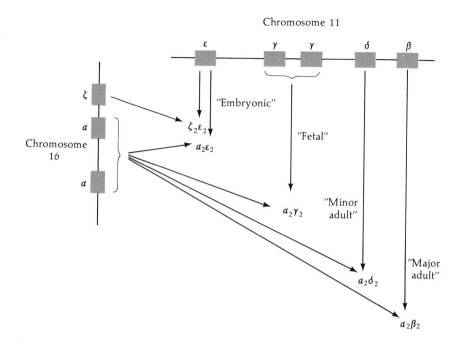

that directs the sequential switching of the chromosome 11 genes from embryonic, to fetal, to adult globins. The mechanism is so far unknown, but one clue comes from certain individuals who, though perfectly healthy, never shut off their fetal hemoglobin synthesis. In such cases, the γ globin genes are still active in the adult and the β globin genes are not turned on. More recent studies (see Jagadeeswaran et al., 1982)

FIGURE 24
Percentages of hemoglobin polypeptide chains as a function of human development. (After Huehns et al., 1964.)

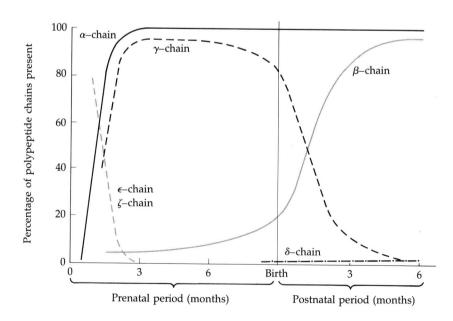

suggest that the nucleotides involved in the switching mechanism exist between the γ and δ genes and may involve repetitive DNA sequences existing in this area. The β globin gene family is one example of a set of related genes in which switching occurs from one member of the family to the other. Other such families include the genes for histones, actins, tubulins, and insect chorion proteins.

Transcription is only the first step in gene expression. In the next chapters, we shall explore other steps where regulation can occur. In these next chapters, it is useful to keep your attention on hemoglobin because we shall find that the globin genes are regulated at all the other levels as well.

Literature cited

Ashburner, M. (1974). Sequential gene activation by ecdysone in polytene chromosomes of *Drosophila melanogaster*. II. Effects of inhibitors of protein synthesis. *Dev. Biol.* 39: 141–157.

Barr, M. L. and Bertram, E. G. (1949). A morphological distinction between neurones of the male and female, and the behavior of the nucleolar satellite during accelerated nucleoprotein synthesis. *Nature* 163: 676.

Brown, D. D. and Dawid, I. B. (1968). Specific gene amplification in oocytes. *Science* 160: 272–280.

Brown, S. W. and Nelson-Rees, W. A. (1961). Radiation analysis of a lecanoid genetic system. *Genetics* 46: 983–1007.

Centerwall, W. R. and Benirschke, K. (1973). Male tortoiseshell and calico (T-C) cats. *J. Hered.* 64: 272–278.

Cooper, D. W., Vandeberg, J. L., Sharmen, G. B. and Poole, W. E. (1971). Phosphoglycerate kinase polymorphism in kangaroos provides further evidence for paternal X inactivation. *Nature New Biol.* 230: 155–157.

Davidson, E. H., Jacobs, H. T. and Britten, R. J. (1983). Very short repeats and coordinate induction of genes. *Nature* 301: 468–470.

Davidson, R. G., Nitowsky, H. M. and Childs, B. (1963). Demonstration of two populations of cells in the human female heterozygous for glucose-6-phosphate dehydrogenase variants. *Proc. Natl. Acad. Sci. USA* 50: 481–485.

Dugaiczyk, A., Woo, S. L. C., Lai, E. C., Mace, M. L., McReynolds, L. and O'Malley, B. W. (1978). Natural ovalbumin gene contains seven intervening sequences. *Nature* 274: 328–333.

Epstein, C. J., Smith, S., Travis, B. and Tucker, G. (1978). Both X chromosomes function before visible X-chromosome inactivation in female mouse embryos. *Nature* 274: 500–502.

Fox, J. L. (1983). Gene splicers contemplate the rat brain. *Science* 222: 828–829.

Gartler, S. M., Liskay, R. M. and Grant, N. (1973). Two functional X chromosomes in human fetal oocytes. *Exp. Cell Res.* 82: 464–466.

Gartler, S. M., Rivest, M. and Cole, R. E. (1980). Cytological evidence for an inactive X-chromosome in murine oogonia. *Cytogenet. Cell Genet.* 28: 203–207.

Gronemeyer, H. and Pongs, O. (1980). Localization of ecdysterone on polytene chromosomes of *Drosophila melanogaster*. *Proc. Natl. Acad. Sci. USA* 77: 2108–2112.

Groudine, M. and Weintraub, H. (1981). Activation of globin genes during chicken development. *Cell* 24: 393–401.

Harris, S. E., Rosen, J. M., Meens, A. R. and O'Malley, B. W. (1975). Use of specific probe for ovalbumin messenger RNA to quantitate estrogen-induced transcripts. *Biochemistry* 14: 2072–2081.

Hill, R. S. and MacGregor, H. C. (1980). The development of lampbrush chromosome-type transcription in the early diplotene oocytes of *Xenopus laevis*: An electron microscope analysis. *J. Cell Sci.* 44: 87–101.

Hourcade, D., Dressler, D. and Wolfson, J. (1973). The amplification of ribosomal RNA genes involves a rolling circle intermediate. *Proc. Natl. Acad. Sci. USA* 70: 2926–2930.

Huehns, E. R., Dance, N., Beaven, G. H., Hecht, F. and Motulsky, A. G. (1964). Human embryonic hemoglobins. *Cold Spring Harbor Symp. Quant. Biol.* 29: 327–333.

Jagadeeswaran, P., Tuan, D., Forget, B. G. and Weissman, S. M. (1982). A gene deletion ending at the midpoint of a repetitive DNA sequence in one form of hereditary persistence of fetal hemoglobin. *Nature* 296: 469–470.

Korge, G. (1975). Chromosome puff activity and protein synthesis in larval salivary glands of *Drosophila melanogaster. Proc. Natl. Acad. Sci. USA* 72: 4550–4554.

Kratzer, P. G. and Chapman, V. M. (1981). X-chromosome reactivation in oocytes of *Mus caroli. Proc. Natl. Acad. Sci. USA* 78: 3093–3097.

Lyon, M. F. (1961). Gene action in the X-chromosome of the mouse (*Mus musculus* L.). *Nature* 190: 372–373.

Mayfield, J. E., Serunian, L. A., Silver, L. M. and Elgin, S. C. R. (1978). A protein released by DNase I digestion of *Drosophila* nuclei is preferentially associated with puffs. *Cell* 14: 539–544.

McBurney, M. W. and Strutt, B. J. (1980). Genetic activity of X chromosomes in pluripotent female teratocarcinoma cells and their differentiated progeny. *Cell* 21: 357–364.

McKnight, G. S. and Palmiter, R. D. (1979). Transcriptional regulation of the ovalbumin and conalbumin genes by steroid hormones in chick oviduct. *J. Biol. Chem.* 254: 9050–9058.

Meyerowitz, E. M. and Hogness, D. S. (1982). Molecular organization of a *Drosophila* puff site that responds to ecdysone. *Cell* 28: 165–176.

Migeon, B. R. (1971). Studies of skin fibroblasts from ten families with HGPRT deficiency, with reference to X-chromosomal inactivation. *Am. J. Human Genet.* 23: 199–209.

Migeon, B. R. and Jelalian, K. (1977). Evidence for two active X-chromosomes in germ cells of female before meiotic entry. *Nature* 269: 242–243.

Miller, D. L., Jr. and Beatty, B. R. (1969). Visualization of nucleolar genes. *Science* 164: 955–957.

Mohandas, T., Sparkes, R. S., Hellkuhl, B., Grzeschik, K. H. and Shapiro, L. J. (1980). Expression of an X-linked gene from an inactive human X chromosome in mouse–human hybrid cells: Further evidence for the non-inactivation of the steroid sulfatase locus in man. *Proc. Natl. Acad. Sci. USA* 77: 6759–6763.

Mohandas, T., Sparkes, R. S. and Shapiro, L. J. (1981). Reactivation of an inactive human X chromosome: Evidence for X inactivation by DNA methylation. *Science* 211: 393–396.

Moore, K. L. (1977). *The Developing Human.* Saunders, Philadelphia.

Moss, T. (1983). A transcriptional function for the repetitive ribosomal spacer in *Xenopus laevis. Nature* 302: 223–228.

Mulvihill, E. R., LePennec, J.-P. and Chambon, P. (1982). Chicken oviduct progesterone receptor: Location of specific region of high-affinity binding in cloned DNA fragments of hormone-responsive genes. *Cell* 24: 393–401.

Old, R. W., Callan, H. G. and Gross, K. W. (1977). Localization of histone gene transcripts in newt lampbrush chromosomes by *in situ* hybridization. *J. Cell Sci.* 27: 57–80.

Osheim, Y. N. and Miller, O. L. (1983). Novel amplification and transcriptional activity of chorion genes in *Drosophila melanogaster* follicle cells. *Cell* 33: 543–653.

Palmiter, R. D. and Schimke, R. T. (1973). Regulation of protein synthesis in chick oviduct. III. Mechanism of ovalbumin "superinduction" by actinomycin D. *J. Biol. Chem.* 248: 1502–1512.

Rochaix, J. D., Bird, A. and Bakken, A. (1974). Ribosomal RNA gene amplification by rolling circles. *J. Mol. Biol.* 87: 473–488.

Roop, D. R., Nordstrom, J. L., Tsai, S. Y., Tsai, M.-J. and O'Malley, B. W. (1978). Transcription of structural and intervening sequences in the ovalbumin gene and identification of potential ovalbumin mRNA precursors. *Cell* 15: 671–685.

Sharman, G. B. (1971). Late DNA replication in the paternally derived X chromosome of female kangaroos. *Nature* 230: 231–232.

Silver, L. M. and Elgin, S. C. R. (1977). Distribution patterns of three subfractions of *Drosophila* nonhistone chromosomal proteins: Possible correlations with gene activity. *Cell* 11: 971–983.

Spradling, A. C. (1981). The organization and amplification of two chromosomal domains containing *Drosophila* chorion genes. *Cell* 27: 193–201.

Spradling, A. C. and Mahowald, A. P. (1980). Amplification of genes for chorion proteins during oogenesis in *Drosophila melanogaster. Proc. Natl. Acad. Sci. USA* 77: 1096–1100.

Sutcliffe, J. G., Milner, R. J., Gottesfeld, J. M. and Lerner, R. A. (1984). Identifier sequences are transcribed specifically in brain. *Nature* 308: 237–241.

Watson, J. D. (1976). *Molecular Biology of the Gene,* Third Edition. Benjamin/Cummings, Menlo Park, CA.

Wellauer, P. K., Dawid, I. B., Brown, D. D. and Reeder, R. H. (1976). The arrangement of length heterogeneity in repeating units of amplified and chromosomal ribosomal DNA from *Xenopus laevis. J. Mol. Biol.* 105: 487–506.

Wolf, S. F., Jolly, D. J., Lunnen, K. D., Friedmann, T. and Migeon, B. R. (1984). Methylation of the HPRT locus on the human X: Implications for X-inactivation. *Proc. Natl. Acad. Sci. USA* 81: 2806–2810.

CHAPTER

Control of development by RNA processing

*It may be in the interpretation and analysis of differentiation
that the new concepts derived from the study of microorganisms
will prove of the greatest value. . . . Eventually, however,
differentiation will have to be studied in differentiated cells.*
—J. MONOD AND F. JACOB (1961)

*Between the conception
And the creation . . .
Between the potency
And the existence
Between the essence
And the descent
Falls the Shadow.*
—T.S. ELIOT (1936)

Introduction

The essence of differentiation is the production of different subsets of proteins in different types of cells. In bacteria, such regulation can be effected at the levels of transcription, translation, and protein degradation. In eukaryotes, however, another possible level for regulation occurs, namely, control at the level of RNA processing. This chapter will present some of the recent evidence suggesting that this type of regulation is crucial for development to occur, and that different cells can process the same transcribed RNA in different ways, thus creating different populations of cytoplasmic messenger RNAs from the same set of nuclear transcripts.

Transcriptional model of developmental regulation

During the 1960s, developmental biologists had a strong bias toward explaining differentiation in terms of transcriptional regulation of gene expression. First, research on polytene chromosomes and their products indicated that transcriptional regulation did occur in insect development. Second, DNA–RNA hybridization studies showed that whereas the DNA of different cell types was identical, the RNA was different (McCarthy and Hoyer, 1964). Third, gene expression could be studied more readily in bacteria than in eukaryotic cells, and this research demonstrated the importance of transcriptional regulation in producing different physiological states in *Escherichia coli*.

Numerous bacterial genes have been shown to be regulated at the level of transcription, but we only need to consider one set of them to realize how powerfully this appears to model differential gene expression in development. The *lac* operon in *E. coli* consists of three structural genes linked together on a common promoter (Figure 1). Between the promoter and the structural genes is a region of DNA called the OPERATOR. This operator is capable of binding a REPRESSOR PROTEIN made from a gene at a different place on the chromosome. Once the repressor protein binds to the operator, RNA polymerase cannot transcribe the three genes. However, when lactose is added to the bacterial environment, these genes are able to be expressed, for the lactose combines with the repressor protein to change its conformation. Once in this altered shape, the repressor cannot bind to the operator sequence. Thus, RNA polymerase is free to transcribe the genes. Both the inducer and the repressor protein were seen to be soluble substances, and the experiment showing the latter is seen in Figure 1D. Here, *E. coli* is made diploid with respect to the *lac* genes and the repressor protein from one chromosome is found to act upon the operon of the other chromosome. Thus, bacteria have a mechanism for differential gene transcription

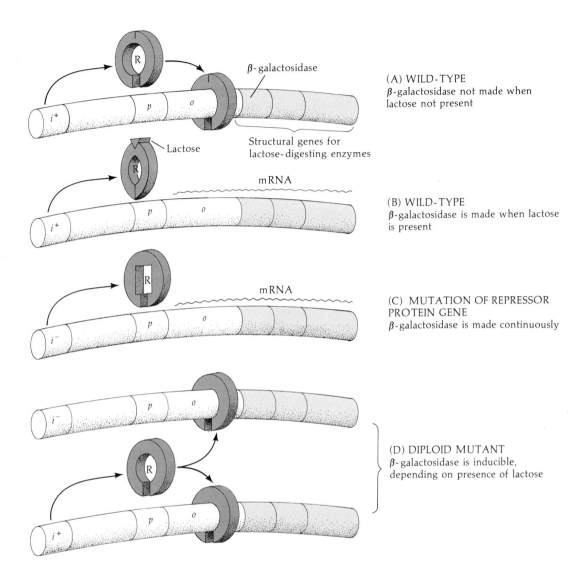

(A) WILD-TYPE
β-galactosidase not made when lactose not present

β-galactosidase

Structural genes for lactose-digesting enzymes

Lactose

mRNA

(B) WILD-TYPE
β-galactosidase is made when lactose is present

mRNA

(C) MUTATION OF REPRESSOR PROTEIN GENE
β-galactosidase is made continuously

(D) DIPLOID MUTANT
β-galactosidase is inducible, depending on presence of lactose

FIGURE 1
Differential gene regulation in the *lac* operon. (A) In its wild-type, inducible state, no β-galactosidase RNA is transcribed unless lactose is present. When no lactose is available, a repressor protein, R, made by gene *i*, binds to the operator site, *o*, inhibiting transcription by RNA polymerase from the promoter, *p*. (B) When lactose is present and combines with the repressor protein, the repressor does not bind to the DNA and transcription ensues. The soluble nature of this repressor is shown in studies on mutant *E. coli*. (C) In certain mutants the *i* gene manufactures a repressor protein that will not bind to the operator. Therefore, β-galactosidase is made continuously. (D) When haploid bacteria cells with an i^- gene are made partially diploid with the wild-type *i* gene (i^+), wild-type repressor is manufactured and is able to make the original β-galactosidase gene inducible.

whereby different genes can be turned on or off by diffusible regulators (Jacob and Monod, 1961).

In 1969, Roy Britten and Eric Davidson extrapolated this data to model eukaryotic gene regulation. In this hypothesis, certain DNA sequences (sensors) would monitor changes in the nuclear environment and would activate one or more contiguous INTEGRATOR GENES (Figure 2). These genes would produce a diffusible RNA, which could bind to complementary sequences on an array of noncontiguous RECEPTOR GENES. This would cause a conformational change, activating the associated STRUCTURAL GENE to synthesize RNA.

This model gained wide acceptance as a framework in which to place the data of developmental genetics. First, it gave a function for moderately repetitive DNA and explained why it should be found interspersed with structural genes. Second, it implied the existence of diffusible activator molecules. Such molecules had been thought to be responsible for turning on dormant genes in somatic cell hybrids (Klebe et al., 1970; Peterson and Weiss, 1972). Third, it explained coordinate regulation—how batteries of genes could all be activated at the same time even though they were not even on the same chromosome. Thus, one of the major assumptions of developmental biology in the early 1970s was that differential mRNA synthesis reflected differential gene transcription.

FIGURE 2
Structural features of the Britten-Davidson (1969) model for transcription-level gene regulation in eukaryotes. Structural genes (SG) could be activated through a receptor (R) by activator RNA made by complementary integrator (I) genes controlled by sensory (S) DNA sequences. Structural genes may have one or several different receptors; a single stimulus acting on a single sensor could turn on a battery of different genes (as for S_1) and a single structural gene (such as SG_1) could be activated by several stimuli.

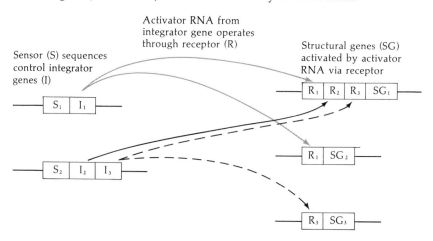

Heterogeneous nuclear RNA

However, it soon became apparent that messenger RNA was not the primary transcription product. Rather, there was a NUCLEAR RNA (nRNA) that was much larger than messenger RNA and that had a much shorter half-life than the cytoplasmic messages. The large variation in the sizes of these molecules caused them to be called HETERO-GENEOUS NUCLEAR RNA (hnRNA). The hnRNA molecules were approximately $5-19 \times 10^3$ nucleotides long, compared with the 2×10^3 bases usually found in mRNA, and they decayed with a half-life of minutes instead of hours (for review, see Lewin, 1980). In sea urchin embryos, the half-life of mRNA is about 5 hours whereas the half-life of nuclear RNA is about 20 minutes. The relationship between this heterogeneous nuclear RNA and the cytoplasmic mRNA was disputed until 1977 when several laboratories found the mRNA sequences within the large nuclear RNA molecules.[1] Bastos and Aviv (1977) isolated the mouse globin message and made complementary DNA (cDNA) from it by using reverse transcriptase (Figure 3). They then conjugated this cDNA to cellulose beads. After hemoglobin-synthesizing cells were grown with radioactive uridine, the radiolabeled RNA was extracted and passed over this column. Any RNA sequence coding for globin should be bound to the column. No other RNA sequence should be "fished out" in such a manner. The RNA bound to the column was extracted from the column by washing it with high-salt solutions, which broke the bonds holding the complementary molecules together. When cells were labeled for 5 minutes, most of the radioactive RNA binding to the column had a sedimentation value of 27 S and a size of approximately 5000 nucleotides (nearly eight times larger than the cytoplasmic hemoglobin message).

However, when the labeling was stopped and the RNA extracted 5 minutes later, the globin sequences were found in a smaller 15 S piece. Finally, the globin-coding RNA was found in a 10 S (600-nucleotide) sequence characteristic of the cytoplasmic message. Thus, heterogeneous nuclear RNA appears to contain the precursors of cytoplasmic globin mRNA. When Hames and Perry (1977) made cDNA to the *entire* message population of cultured mouse cells, they obtained a similar result: The cDNA probe could bind to enormously large heterogeneous nuclear RNA sequences. Therefore, eukaryotic DNA transcribes large nuclear RNAs, which serve as the precursors to the smaller cytoplasmic messages.

[1]Although the terms *nuclear RNA* (nRNA) and *heterogeneous nuclear RNA* (hnRNA) are now used interchangeably, they do have slightly different meanings. HnRNA usually refers to message precursors, whereas nRNA refers to the total unfractionated RNA extracted from the nucleus. We will use nRNA, rather than hnRNA, throughout.

FIGURE 3
Protocol for isolating nuclear RNA (nRNA) containing globin-coding sequences.

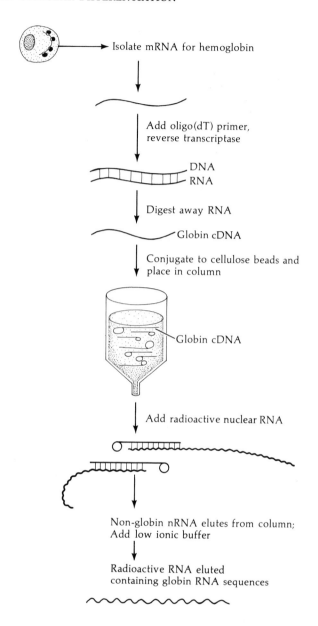

Isolate mRNA for hemoglobin

Add oligo(dT) primer, reverse transcriptase

DNA
RNA

Digest away RNA

Globin cDNA

Conjugate to cellulose beads and place in column

Globin cDNA

Add radioactive nuclear RNA

Non-globin nRNA elutes from column; Add low ionic buffer

Radioactive RNA eluted containing globin RNA sequences

Complexity of nuclear and messenger RNAs

The next question involves the number of different types of RNA sequences found in the nucleus and cytoplasm. Does the nucleus contain a greater diversity of sequences than the cytoplasm, or are they about the same? The answer to this question will tell us whether more of the genome is transcribed into nRNA than what we can measure from mRNA. The diversity of nucleotide sequences is called COMPLEXITY. To measure complexity, a fixed amount of radioactive RNA is added to an excess amount of denatured DNA and allowed to renature to it. The more complex the RNA, the slower its rate of renaturation to the DNA. In other words, if there were only one type of RNA present (complexity = 1), this RNA sequence would be distributed throughout the reaction mixture and would find its DNA homologues rather quickly. However, if that same small amount of RNA contained a thousand sequences (complexity = 1000), each individual sequence would be a thousandfold more dilute and would have a small chance of finding its homologue in the same amount of time. RNA complexity can be measured by C_0t curve analysis (Chapter 11).

By this method the nRNA complexity was determined to be much greater than that of mRNA from the same cells. First, certain nRNA sequences associated with the DNA very rapidly. This indicated that they contained RNA transcribed from repetitive DNA sequences. In the sea urchin blastula, cytoplasmic mRNA contains no sequences that hybridize to repetitive DNA, whereas approximately 25 percent of the nRNA sheared to obtain 1100-nucleotide pieces can associate with DNA at low C_0t values. Because the average length of sea urchin repetitive sequences in nRNA is approximately 300 bases, it is estimated that 70 percent of the nRNA molecules contain at least one repetitive sequence (Smith et al., 1974). However, if the RNA is reacted with purified nonrepetitive DNA (so that the repetitive sequences do not complicate matters), the complexity of nRNA is usually found to be 4 to 20 times greater than that of mRNA from the same organ or cell type (Table 1). From such studies, it is estimated that in the sea urchin embryo, only 10–20 percent of the nRNA complexity gets into the cytoplasmic mRNA population. In rat liver, only 11 percent of the nRNA sequence types become message; and in cultured insect cells, the figure is as low as 5 percent. We can conclude that the nuclear transcripts contain (1) a small population of hnRNA, which is the precursor to the mRNA of the cell; and (2) a large population of nRNA, which turns over rapidly within

TABLE 1
Approximate kinetic characteristics and relative complexity of messenger and nuclear RNA populations in various species

Organism	Proportion of nRNA converted to poly (A)$^+$-containing mRNA	
	% of original mass	*% of original complexity*
Drosophila	14–20	28–40
Aedes (mosquito)	~3	~13
Sea urchin	4–7	17–29
Cultured human carcinoma (HeLa) cells	3–6	15–30

Source: Lewin (1980).

the nucleus. The population of nRNA molecules contains numerous sequences different from those that become processed into mRNA.

Control of development by nRNA processing

According to the previous section, the genome is transcribing many more sequences than those that become mRNA. This conclusion suggests that much of the control of differentiation may take place *after* transcription by the differential processing of specific mRNA precursors. Recent studies on sea urchins and rats confirm this.

In 1977, Kleene and Humphreys obtained single-copy sea urchin DNA by denaturing blastula DNA and isolating whatever did not reassociate at a C_0t value of 200. nRNA from blastula cells was found to bind to 15 percent of this DNA. Similarly, pluteus-stage nRNA, even when present in great excess, was also found to bind to 15 percent of this DNA. Because only one strand of DNA is complementary to RNA, 15 percent of DNA hybridized corresponds to 30 percent of the genome. Therefore, about 30 percent of the genome is actively transcribing blastula nRNA and about 30 percent of the genome is actively transcribing pluteus nRNA. Are these two sets of DNA sequences the same or are they different? This was tested by mixing blastula and pluteus nuclear RNAs and adding them to the denatured single-copy DNA. If the sequences were totally different, one would expect 30 percent of the DNA to be bound. (That is, 60 percent of the genome would be coding

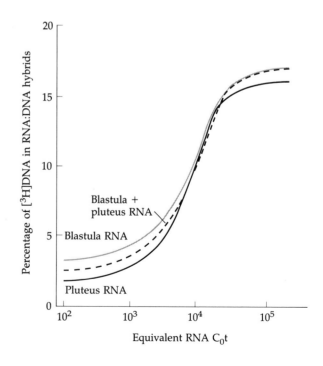

FIGURE 4

Hybridization of nuclear RNA with single-copy [³H]DNA. Radioactive single-copy DNA was hybridized with blastula RNA, pluteus RNA, and a mixture of blastula and pluteus RNAs. The mixtures were then incubated to allow all complementary sequences to pair. In all three cases, about 15 percent of the DNA hybridized to the RNA. (After Kleene and Humphreys, 1977.)

for the combined set of blastula and pluteus messages.) If they were identical, one would expect 15 percent of the DNA to be bound. The result is shown in Figure 4. The mixture bound to only 15 percent of the DNA. The nRNA sequences bound to the same DNA. Within experimental error, the RNA was identical in blastula and pluteus cells.

To confirm this unexpected observation, Kleene and Humphreys isolated radioactive single-copy DNA and mixed it with pluteus or blastula RNA at a high C_0t value (so that all sequences could match if their complementary sequences were present). They then isolated those DNA fragments that would not bind to blastula (blastula-null DNA) and pluteus (pluteus-null DNA) nRNA sequences (Figure 5). When blastula-null DNA was denatured and mixed to new samples of pluteus nRNA and when pluteus-null DNA was mixed with blastula nRNA, no hybridization above background levels was observed. Therefore, the DNA sequences transcribed during the blastula stage are the same as DNA sequences transcribing during the pluteus stage.

The most impressive evidence for processing-level control of de-

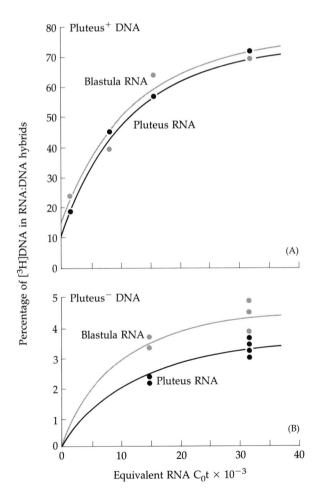

FIGURE 5
Hybridization of "pluteus-plus" DNA and "pluteus-null" DNA by pluteus and blastula RNA. (A) Both pluteus and blastula RNA bound "pluteus-plus" DNA to the same extent. (B) Neither RNA bound significantly to the "pluteus-null" DNA. (From Kleene and Humphreys, 1977.)

velopment, however, came from the laboratory of Eric Davidson and Roy Britten, the same group that had so elegantly suggested how transcription-level regulation could work. They first discovered that in the sea urchin *Strongylocentrotus purpuratus*, the complexity of the mRNAs becomes progressively less as development proceeds (Galau et al., 1976). Oocyte mRNA binds to approximately 4.5 percent of the genome and has a complexity (measured by C_0t) of about 37,000 kilobases. Given an average mRNA length of 2000 bases, this represents about 18,500 different types of messenger RNAs. Blastula mRNA binds to only 3.1 percent of the genome and represents about 13,000 types of mRNA; and gastrula mRNA binds to only 2.0 percent of the DNA and has 8000 species of messages. By the time tissues have differentiated, they each contain about 6000 mRNA types, which hybridize to only 0.75 percent of the genome (Figure 6). Thus, the amount of the genome expressed in cytoplasmic mRNA is being progressively restricted.

They next asked whether this restriction was due to changes in genomic transcription. In other words, is the nRNA being similarly restricted? To answer this question, Wold and her colleagues (1978) isolated blastula mRNA and hybridized it to denatured radioactive DNA. The resulting hybrids were eluted, denatured, and the DNA

FIGURE 6

Sets of structural genes expressed as messenger RNAs in various tissues. The shaded portion of each bar represents the amount of mRNA in each tissue that is identical with gastrula mRNA. The unshaded portion of each bar represents the amount of mRNA not shared in common with gastrula messages. Complexity is indicated in three ways on the ordinate; the horizontal line indicates 100 percent of gastrula mRNA complexity. (After Galau et al., 1976.)

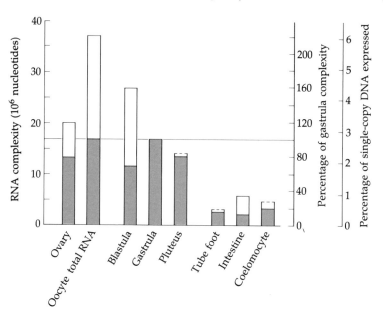

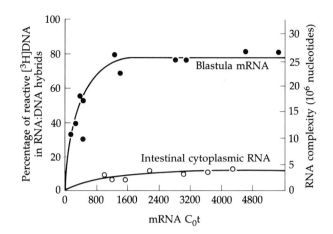

FIGURE 7
Specificity of blastula mDNA. Hybridization of blastula mDNA (cDNA to blastula mRNA) with blastula mRNA and intestinal cytoplasmic RNA. (After Wold et al., 1978.)

isolated from the RNA bound to it. This produced a DNA probe that would recognize blastula mRNA sequences. This DNA was highly specific for blastula mRNA sequences (Figure 7): About 78 percent of this "blastula cDNA" could bind to blastula messages whereas less than 10 percent of it could bind to adult intestinal mRNA.

They could then use this "blastula mDNA" to find any blastula mRNA sequences in the RNA of various tissues. The results are shown in Figure 8 and Table 2. Blastula mRNA sequences were found in the RNA of all three adult tissues tested. Around 80 percent of the blastula message DNA hybridized to gastrula, intestine, and coelomocyte RNAs.

TABLE 2
Intertissue comparisons of structural gene sequences in messenger RNA and nuclear RNA

Reference tracer complementary to	Reaction with parent mRNA		Normalized reaction with other mRNA		Normalized reaction with nRNA	
	mRNA	%	mRNA	%	nRNA	%
SEA URCHIN						
Blastula mRNA (single copy DNA)	Blastula	100	Intestine Coelomocyte	12 13	Intestine Coelomocyte	97 101
MOUSE						
Brain mRNA (total cDNA)	Brain	100	Kidney	78	Kidney	102
Brain mRNA (cDNA representing rare messages)	Brain	100	Kidney	56	Kidney	100

Source: Davidson and Britten (1979).

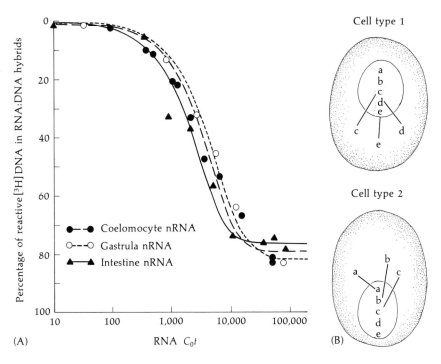

FIGURE 8

(A) Hybridization of blastula mDNA to *nuclear* RNAs of intestine, coelomo-cytes, and gastrulae. The hybridization mixtures were incubated at a C_0t high enough to allow all complementary sequences to pair. All blastula sequences were found in each nuclear RNA population—although not in their cytoplasm (Figure 7). (B) Diagram of model based on differential RNA processing. In both cell types, the same RNAs (a, b, c, d, e) are transcribed; but in one cell type, sequences c, d, and e are processed into the cytoplasm, whereas in another cell type sequences a, b, and c are processed into the cytoplasm. (A from Wold, 1978.)

Thus, only 10 percent of a probe capable of recognizing blastula *message* sequences will react with intestinal *messenger* RNA, but about 75 percent of it will react with intestinal *nuclear* RNA. Because the control reaction (blastula mDNA with blastula mRNA) also gave a value of 75–80 per-cent, it appears that *all* of the blastula message sequences are present as nRNA transcripts in the nucleus of intestine, coelomocyte, and gas-trula cells, despite their absence from the cytoplasm. The nucleus is transcribing blastula-specific sequences even in differentiated coelom-ocytes and intestine cells. It appears, then, that the control of sea urchin gene expression occurs predominantly at the level of RNA processing. Moreover, similar experiments on rats and mice suggest the near iden-tity of RNAs from brain, liver, and kidneys (Chikaraishi et al., 1978).

Mechanisms for specific nuclear RNA processing

In 1979, Davidson and Britten published a model based on the specific processing of nuclear message precursors. This is a highly speculative model and is still the subject of much debate. Although this model retains elements of transcriptional regulation, it is essentially a declaration of independence of developmental biology from "coliform" models of gene regulation. As with their 1969 model, moderately repetitive DNA is posited to play a major role in gene regulation.

In this model, messenger RNAs are divided into three classes on the basis of their complexity:

- *Complex class mRNA.* 1–15 copies per cell. There are enough different types of complex class messages to code for over 10^4 different proteins.

- *Moderately prevalent mRNA.* 15–300 copies per cell. There are enough different types of this class of mRNA to code for 500–1000 different proteins.

- *Superabundant mRNA.* 10^4–10^5 copies per cell. The oviduct cell of the laying hen, for instance, will contain 1.0–1.5×10^5 ovalbumin mRNA molecules. Mouse and chick reticulocytes contain 4×10^4 to 1.5×10^5 globin mRNA molecules per cell, respectively, and the message from the Balbiani ring 2 of *Chironomus tentans* is another member of this superabundant class.

Most evidence for transcription-level regulation comes from the superabundant class of messages (because they are the easiest to purify and study). However, they might be the exceptions. Often, these messages code for proteins characteristic of a certain cell and therefore represent the final stage of differentiation rather than its cause. The red blood cell, for instance, has undergone a long history of differentiation before reaching the stage where it makes globin. We have just seen that mRNA sequences specific for a particular cell may be found in the nRNA of all cell types. Table 2 summarizes some of this evidence. For the single-copy RNA sequences that code for proteins, there is cytoplasmic specificity but not nuclear speci-

ficity. The opposite case is seen when we look at repetitive RNA sequences. Sequences from moderately repetitive genes are not found in the sea urchin mRNA; however, there is cell-type specificity for these sequences within the nucleus! The nuclear RNAs of sea urchins, cultured human cells, and rat ascites cells have repetitive sequences interspersed with single-copy sequences (thus mimicking the DNA sequences). Scheller and his co-workers (1978) found that the nRNA of sea urchin embryos had specific repetitive RNA sequences represented at different times during development. Moreover, *both* complementary sequences of these repeats were seen in the nRNA.

Davidson and Britten utilized these data to construct their model of gene expression. First, they proposed that the single-copy genes that give rise to the complex and moderately prevalent classes of mRNA are continuously transcribed at the same rate in all cell types. Second, gene expression is regulated by determining what percentage of each mRNA precursor gets processed into the cytoplasm. Third, intranuclear RNA–RNA duplexes forming between the complementary repetitive transcripts would determine which mRNA precursor would be processed. The duplexes forming in a given cell type would depend on the intranuclear concentration of the specific nRNA repeats.

Davidson and Britten's regulatory model is shown in Figure 9. A "constitutive transcription unit" (CTU) is transcribed continuously in all cell types. This produces a constitutive transcript (CT) consisting of a protein-coding structural gene (including introns and flanking sequences) and sequences derived from interspersed repetitive DNA. In another region of the genome reside the "integrating regulatory transcription units" (IRTU). These do not contain structural genes, and they are transcribed in a cell-specific manner.[2] Thus, these genes contain a DNA sequence (sen-

[2]Again, we see cell-specific differential transcription. However, in this model, the differential transcription is in nucleus-specific repeat sequences, not in protein-coding genes. All differentiation is eventually brought back to specific gene activation by components of the oocyte cytoplasm or by outside factors (as may exist in mammalian compaction).

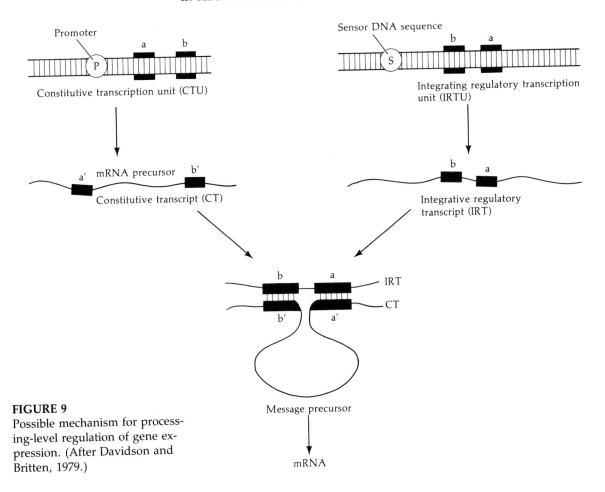

FIGURE 9
Possible mechanism for processing-level regulation of gene expression. (After Davidson and Britten, 1979.)

sor) that can turn on the IRTU in response to some particular signal. The IRTU contains specific repetitive DNA sequences and is transcribed to yield cell type-specific integrating regulatory transcripts (IRTs). RNA–RNA duplexes form between the repetitive sequences of the CTs and the complementary repetitive sequences of the IRTs. These duplexes are required for the survival and processing of the nRNA. If these double-stranded regions do not arise, the nRNA is degraded by nucleases. If duplexed, the nRNA survives to be processed into message.

There are numerous variations that can be played upon this theme. One interesting speculation is that the activation of the IRTUs may be equivalent to what has classically been called determination. In order to express a gene, both the IRTU and the specific gene need to be activated. When the gene is not continuously expressed, one requires a two-step process: the activation of structural gene transcription and the activation of IRTU transcription. Activation of the first could represent determination, whereas activation of the second would result in gene expression and differentiation. In this case, the determination of the cell may involve the differentiation of the nucleus.

Evidence for
unprocessed message
precursors in the nucleus

Because evidence from sea urchins shows that the population of nRNAs in the various differentiated cells are identical, it would be expected that probes for specific mRNAs of one cell type should find them in nuclei of nonexpressive cells. By using cDNA made from globin message, investigators have found globin mRNA precursors in the total cellular RNA from mouse liver, brain, and cultured cell lines (Humphries et al., 1976) and in the nRNA of uninduced Friend erythroleukemia cells (Gottesfeld and Partington, 1977) and the *Xenopus* oocyte (Perlman et al., 1977). Low-level transcription of ovalbumin genes has been detected in chicken liver, spleen, brain, and heart (Axel et al., 1976; Tsai et al., 1979). This value approximates one molecule of ovalbumin coding sequence in every two cell nuclei; even RNAs that are known to be regulated transcriptionally can be found in very low amounts in the nuclei of inappropriate cell types.

Some of the best evidence for specific processing-level control of RNA transcripts comes from experiments on viruses in mouse cells. Here, certain cell types can splice nRNA to create certain messages and other cells cannot. Polyoma virus and simian vacuolating virus 40 (SV40) infect cells and integrate into host cell chromosomes. Here, they produce their own RNA, which makes virus-specific products. Like the RNA of their host cell, polyoma and SV40 transcripts must first be spliced to delete introns. Only after the viral RNA is processed can stable viral messages accumulate in the cytoplasm. (The messages of certain other viruses that replicate in the cytoplasm do not need to be spliced.) Undifferentiated teratocarcinoma stem cells (Chapter 6) will not support SV40 or polyoma virus infection or gene expression. However, if the stem cells are allowed to differentiate into somatic cell types, they suddenly become susceptible to infection by those two viruses. The basis for this initial nonpermissiveness appears to be the failure of RNA processing. Undifferentiated teratocarcinoma cells will not splice the SV40 message precursors, whereas the differentiated cells will. In the stem cells, large message precursors accumulate and are eventually degraded. Only the differentiated cell types are able to process this nuclear RNA to a cytoplasmic message (Segal et al., 1979). The viruses that do not need RNA processing for gene expression (Sindbis and vaccinia) are perfectly capable of growing in both the teratocarcinoma stem cell and its differentiated derivatives. The observations that SV40 RNA splicing occurs only in differentiated mouse cells (Topp et al., 1977) and teratocarcinoma derivatives suggests that cell development may be dependent on the expression of the splicing enzymes themselves (or their specific IRT "cofactors"). The agents responsible for processing SV40 RNA undoubtedly have more appropriate roles to play in the cell, and they may be responsible for the processing of several message precursors necessary for cell differentiation.

Selection of alternative proteins by RNA processing

The expression of antibody-coding genes during development of lymphocytes and their responses to antigen involve numerous events of selection and rearrangement of gene segments. Much of this selection and rearrangement occurs at the DNA level, as we saw in Chapter 10. But several aspects of the regulation of antibody gene expression take place at the RNA processing level.

During the development of a B lymphocyte, before it becomes an antibody-secreting cell, it produces samples of the type of antibody it will be capable of secreting and inserts them into the plasma membrane, where they function as antigen receptors. Depending on the stage of development and whether or not the cell has been exposed to the type of antigen with which its antibodies will react, the surface immunoglobulins may be of either the IgM or the IgD class. Initially they are of the IgM class; after the first antigen stimulation, both IgM and IgD may be made simultaneously; and later only IgD may be on the cell surface. But the "variable" (antigen binding) portion of the immunoglobulin remains the same throughout the life history of lymphocyte and its progeny cells. It has now been shown (Maki et al., 1981) that switching between IgM and IgD occurs at the RNA processing level. A single RNA transcript is synthesized, and it contains the exons coding for the variable (antigen-binding) region plus the mu (IgM) and the delta (IgD) constant regions, separated by various introns. Whether the mRNA that reaches the cytoplasm will code for IgM or IgD depends on which of two alternative splicing mechanisms are used! If the delta exons are spliced out and discarded, IgM will be made. If the mu exons are removed by splicing, however, IgD will be made. During transient periods when both splicing mechanisms are being utilized, both classes of immunoglobulins are produced (Figure 10).

Other uses of differential RNA processing

Although differential RNA processing has been documented best in immunoglobulins, RNA processing may control alternative forms of expression of other genes as well. Pyruvate kinase, for instance, has different molecular weights in the red blood cell (L' isozyme, 63,000) and liver (L isozyme, 60,000). Peptide maps show that the red cell enzyme has extra amino acids which the liver enzyme lacks. Genetic evidence, however, demonstrates that both isozymes are controlled by

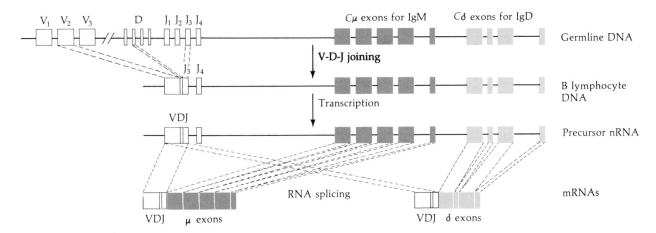

FIGURE 10

Proposed model for the expression of IgM and IgD in B lymphocytes. The heavy chain gene is made during lymphocyte development by the translocation of V, D, and J region sequences to a position adjacent to the constant region sequences. The transcript of this gene includes both the mu (IgM) and delta (IgD) constant regions along with the newly-constructed variable region. Alternative pathways of processing (splicing) can delete mu-constant region RNA or delta-constant region RNA. (After Liu et al., 1980.)

SIDELIGHTS &
SPECULATIONS

Other roles of differential RNA processing in antibody gene regulation

The decision to make IgM or IgD is not the only decision mediated by differential RNA processing. The placement of these proteins—whether they are to be inserted into the plasma membrane or secreted—is also determined in this manner.

IgD is usually confined to the cell membrane, and it is thought that serum IgD may be merely a result of accidental leakage or cell death. IgM, however, can exist in two distinct forms, membrane-bound (mIgM) and secreted (sIgM). Membrane-bound IgM acts as an antigen receptor, whereas secreted IgM is the first type of antibody found in the blood when an organism is exposed to an antigen for the first time. Singer and coworkers (1980) have shown that sIgM and mIgM differ from each other at their carboxyl ends. The molecules are identical except that the membrane-bound form contains a hydrophobic "tail" that keeps it inserted in the membrane (Figure 11). The mRNAs for membrane and secreted forms of IgM also differ. Although both of them are found to be transcribed from the same $C\mu$ gene, they are processed differently (Rogers et al., 1980; Early et al., 1980; Alt et al., 1980). The secreted IgM contains regions encoded by the *VDJ* genes and exons $C\mu_1$, $C\mu_2$, $C\mu_3$, and $C\mu_4$. It also contains a terminal portion that allows it to be secreted. The membrane-bound IgM contains the same arrangement except that instead of the "secretion" terminal portion, it has added a portion encoded by two more exons, $C\mu_5$ and $C\mu_6$, which gives it a hydrophobic tail that can integrate into the lymphocyte membrane. The decision as to whether the molecule is to become sIgM or mIgM is determined by RNA splicing (Figure 12).

There is also a third step in antibody formation where differential RNA processing plays an important

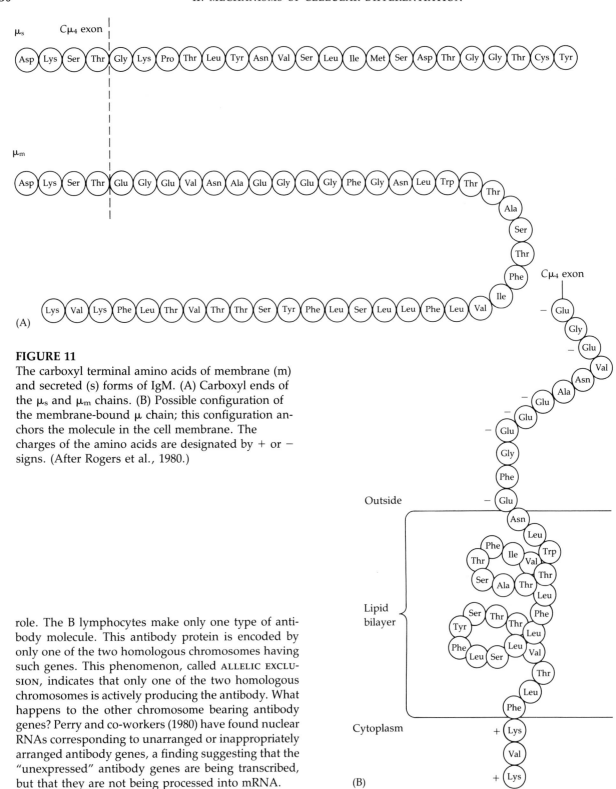

FIGURE 11

The carboxyl terminal amino acids of membrane (m) and secreted (s) forms of IgM. (A) Carboxyl ends of the μ_s and μ_m chains. (B) Possible configuration of the membrane-bound μ chain; this configuration anchors the molecule in the cell membrane. The charges of the amino acids are designated by + or − signs. (After Rogers et al., 1980.)

role. The B lymphocytes make only one type of antibody molecule. This antibody protein is encoded by only one of the two homologous chromosomes having such genes. This phenomenon, called ALLELIC EXCLUSION, indicates that only one of the two homologous chromosomes is actively producing the antibody. What happens to the other chromosome bearing antibody genes? Perry and co-workers (1980) have found nuclear RNAs corresponding to unarranged or inappropriately arranged antibody genes, a finding suggesting that the "unexpressed" antibody genes are being transcribed, but that they are not being processed into mRNA.

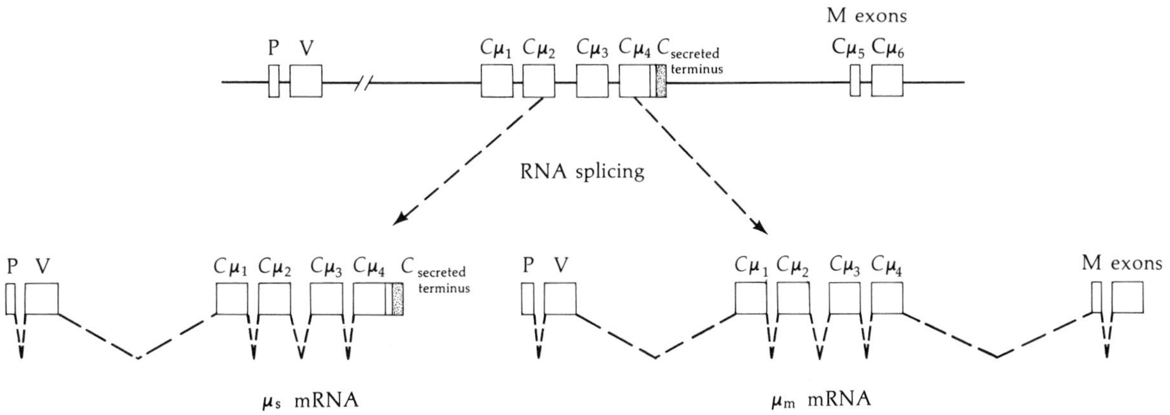

FIGURE 12
Proposed alternative splicing patterns for secreted mu and membrane mu mRNA. P includes the leader sequence; the spliced RNA (introns) is shown as dashed lines. (After Early et al., 1980.)

the same gene. It seems likely then, that there is differential processing of a common nRNA precursor (Marie et al., 1981). Similarly, two forms of the enzyme amylase may be derived from a common nuclear transcription product.

Summary

RNA processing must be viewed as a major means of regulating differential gene expression. Evidence indicates that it determines the cell-specific mRNA population in developing sea urchins and that it may do likewise in mice and rats. DNA and RNA sequencing has shown that differential RNA processing can determine whether a given transcript expresses the IgD or IgM gene region and suggests that different tissue-specific isozymes can also be regulated by differential splicing. The mechanism underlying such differential RNA processing may give us insights into the very core of cell differentiation and embryonic determination.

Literature cited

Alt, F. W., Bothwell, A. L. M., Knapp, M. N., Siden, E., Mather, E., Koshland, M. and Baltimore, D. (1980). Synthesis of secreted and membrane-bound immunoglobulin mu heavy chains is directed by mRNAs that differ at their 3′ ends. *Cell* 20: 293–301.

Axel, R., Feigleson, P. and Schutz, G. (1976). Analysis of the complexity and diversity of mRNA from chicken liver and oviduct. *Cell* 7: 247–254.

Bastos, R. N. and Aviv, H. (1977). Globin RNA precursor molecules: Biosynthesis and processing erythroid cells. *Cell* 11: 641–650.

Britten, R. J. and Davidson, E. H. (1969). Gene regulation for higher cells: A theory. *Science* 165: 349–357.

Chikaraishi, D. M., Deeb, S. S. and Sueoka, N. (1978).

Sequence complexity of nuclear RNAs in adult rat tissues. *Cell* 13: 111–120.

Davidson, E. H. and Britten, R. J. (1979). Regulation of gene expression: Possible role of repetitive sequences. *Science* 204: 1052–1059.

Early, P., Rogers, J., Davis, M., Calame, K., Bond, M., Wall, R. and Hood, L. (1980). Two mRNAs can be produced from a single immunoglobulin gene by alternative RNA processing pathways. *Cell* 20: 313–319.

Galau, G. A., Klein, W. H., Davis, M. M., Wold, B. J., Britten, R. J. and Davidson, E. H. (1976). Structural gene sets active in embryos and adult tissues of the sea urchin. *Cell* 7: 487–505.

Gottesfeld, J. M. and Partington, G. A. (1977). Distribution of messenger RNA coding sequences in fractionated chromatin. *Cell* 12: 953–962.

Hames, B. D. and Perry, R. P. (1977). Homology relationship between the messenger RNA and heterogeneous nuclear RNA of mouse L cells. A DNA excess hybridization study. *J. Mol. Biol.* 109: 437–453.

Humphries, S., Windass, J. and Williamson, J. (1976). Mouse globin gene expression in erythroid and nonerythroid tissue. *Cell* 7: 267–277.

Jacob, F. and Monod, J. (1961). On the regulation of gene activity. *Cold Spring Harbor Symp. Quant. Biol.* 26: 193–211.

Klebe, R. J., Chen, T. R. and Ruddle, F. H. (1970). Mapping of a human genetic regulator element by somatic cell genetic analysis. *Proc. Natl. Acad. Sci. USA* 66: 1220–1227.

Kleene, K. C. and Humphreys, T. (1977). Similarity of hnRNA sequences in blastula and pluteus stage sea urchin embryos. *Cell* 12: 143–155.

Lewin, B. (1980). *Gene Expression 2* (Second Edition). Wiley-Interscience, New York, pp. 694–760.

Liu, C.-P., Tucker, P. W., Mushinski, F. and Blattner, F. R. (1980). Mapping of heavy chain genes for mouse immunoglobulins M and G. *Science* 209: 1348–1353.

Maki, R., Roeder, W., Traunecker, A., Sidman, C., Wabl, M., Rasjhke, W. and Tonegawa, S. (1981). The role of DNA rearrangement and alternate mRNA processing in the expression of immunoglobulin delta genes. *Cell* 24: 353–365.

Marie, J., Simon, M.-P., Dreyfus, J.-C. and Kahn, A. (1981). One gene, but two messenger RNAs encode liver L and red cell L' pyruvate kinase subunits. *Nature* 292: 70–72.

McCarthy, B. J. and Hoyer, B. H. (1964). Identity of DNA and diversity of RNA in normal mouse tissues. *Proc. Natl. Acad. Sci. USA* 52: 915–922.

Perlman, S. M., Ford, P. J. and Rosbash, M. M. (1977). Presence of tadpole and adult globin RNA sequences in oocytes of *Xenopus laevis. Proc. Natl. Acad. Sci. USA* 74: 3835–3839.

Perry, R. P., Kelley, D. E., Coleclough, C., Seidman, J. G., Leder, P., Tonegawa, S., Matthyssens, G. and Weigart, M. (1980). Transcription of mouse κ chain genes: Implications for allelic exclusion. *Proc. Natl. Acad. Sci. USA* 77: 1937–1941.

Peterson, J. A. and Weiss, M. C. (1972). Expression of differentiated function in hepatoma cell hybrids: Induction of mouse albumin production in rat hepatoma–mouse fibroblast hybrids. *Proc. Natl. Acad. Sci. USA* 69: 571–575.

Rogers, J., Early, P., Carter, C., Calame, K., Bond, M., Hood, L. and Wall, R. (1980). Two mRNAs with different 3' ends encode membrane-bound and secreted forms of immunoglobulin μ chain. *Cell* 20: 303–312.

Scheller, R. H., Costantini, F. D., Kozlowski, M. R., Britten, R. J. and Davidson, E. H. (1978). Specific representation of cloned repetitive DNA sequences in sea urchin RNAs. *Cell* 15: 189–203.

Segal, S., Levine, A. J. and Khoury, G. (1979). Evidence for non-spliced SV40 RNA in undifferentiated murine teratocarcinoma stem cells. *Nature* 280: 335–338.

Singer, P. A., Singer, H. H. and Williamson, A. R. (1980). Different species of messenger RNA encode receptor and secretory IgM μ chains differing at their carboxy termini. *Nature* 285: 294–300.

Smith, M. J., Hough, B. R., Chamberlain, M. E. and Davidson, E. H. (1974). Repetitive and non-repetitive sequence in sea urchin heterogeneous nuclear RNA. *J. Mol. Biol.* 85: 103–126.

Topp, W., Hall, J. D., Rifkin, D., Levine, A. J. and Pollack, R. (1977). The characterization of SV40-transformed cell lines derived from mouse teratocarcinoma: Growth properties and differentiated characteristics. *J. Cell. Physiol.* 93: 269–276.

Tsai, S. Y., Tsai, M.-J., Lin, C.-T. and O'Malley, B. W. (1979). Effect of estrogen on ovalbumin gene expression in differentiated nontarget tissues. *Biochemistry* 18: 5726–5731.

Wold, B. J., Klein, W. H., Hough-Evans, B. R., Britten, R. J. and Davidson, E. H. (1978). Sea urchin embryo mRNA sequences expressed in nuclear RNA of adult tissues. *Cell* 14: 941–950.

Translational and posttranslational regulation of developmental processes

We must not conceal from ourselves the fact that the causal investigation of organism is one of the most difficult, if not the most difficult problem which the human intellect has attempted to solve, and that this investigation, like every causal science, can never reach completeness, since every new cause ascertained only gives rise to fresh questions concerning the cause of this cause.
—WILHELM ROUX (1894)

There is no rest for the messenger til the message is delivered.
—JOSEPH CONRAD (1920)

TRANSLATIONAL REGULATION OF DEVELOPMENT

After a messenger RNA has been transcribed, processed, and exported from the nucleus, it still needs to be translated in order to form the protein encoded in the genome. In this chapter, we shall see that regulation at the level of translation is an extremely important mechanism in the control of gene expression. In such cases, the message is already present but may or may not be translated, depending on certain cellular conditions. Thus, translational control of gene expression can be used when a burst of protein synthesis is needed immediately (as we shall find to be the case in the newly fertilized egg), or it can be used as a fine-tuning mechanism to ensure that a very precise amount of protein is made from the available supply of messages (as is the case in hemoglobin synthesis). We shall also see that there are several ways to effect translational control and that different cells have evolved different means to do so.

Mechanism of eukaryotic translation

Translation is the process by which the information contained in the mRNA instructs the synthesis of a particular polypeptide. This process has been divided into three components: initiation, elongation, and termination (Figure 1). Initiation consists of the reactions wherein the first aminoacyl-transfer RNA and mRNA are bound to the ribosome. The only transfer RNA (tRNA) capable of initiating translation is a special tRNA (tRNA$_i$), which carries the amino acid methionine. As shown in Figure 2, the first reactions involve the formation of an "initiation complex" consisting of methionyl-initiator tRNA bound to a 40 S ("small") ribosomal subunit. This Met-tRNA$_i$ is recognized by EUKARYOTIC INITIATION FACTOR 2 (eIF2), which binds it to the 40 S ribosomal subunit. Note that the binding occurs in the absence of mRNA. The mRNA is added next. It first binds its 5′ end to the ribosome and then aligns itself so that the methionine-tRNA$_i$ is positioned on an AUG nucleotide triplet (codon). This step involves initiation factors 2, 4A, and 4B as well as the 7-methylguanosine "cap" on the mRNA. Without the cap, the binding of mRNA to the ribosomal subunit is not completed (Shatkin, 1976). Only after the mRNA has been complexed to the small ribosomal subunit can the 60 S ("large") ribosomal subunit bind. This completes the initiation reaction.

Elongation involves the sequential binding of aminoacyl-tRNAs to the ribosome and the formation of peptide bonds between the amino acids as they sequentially relinquish their tRNA carriers (Figure 2). As the amino acids are joined together, the ribosome travels down the

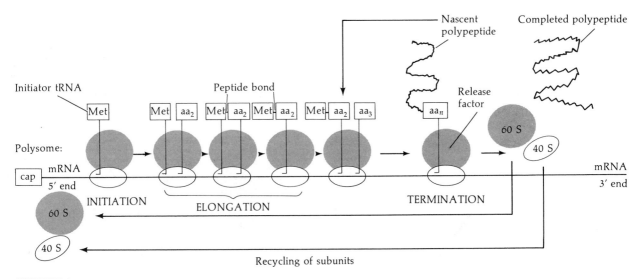

FIGURE 1

Schematic representation of the events of eukaryotic translation. The initiation steps bring together the 40 S (white) and 60 S (shaded) ribosomal subunits, mRNA, and the initiator tRNA, which is complexed to the amino acid methionine (Met). During elongation, amino acids are brought to the polysome and peptide bonds are formed between the amino acids. The sequence of amino acids in the growing protein is directed by the sequence of nucleic acid codons in the mRNA. After the last peptide bond of the protein has been made, the codons UAG, UGA, or UAA signal the termination of translation and the ribosomal subunits and message can be reutilized.

FIGURE 2

Initiation phase of eukaryotic translation. The first complex is made between an activated initiator-tRNA and the 40 S ribosomal subunit, catalyzed by GTP and eukaryotic initiation factors (eIF) 2 and 3. After this complex has formed, the mRNA is positioned and the 60 S ribosomal subunit is put into place. These later events are catalyzed by eukaryotic initiation factors 1, 4, and 5. (From Hershey, 1980.)

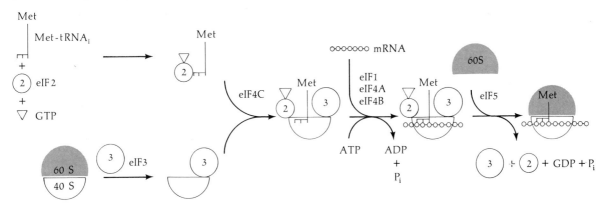

message, thereby exposing new codons for tRNA binding. This allows another ribosome to initiate on the 5' end of the message and begin its traveling. Thus, any mRNA usually will have several ribosomes attached to it. This structure is then called a polyribosome—or, more commonly, a POLYSOME (Figure 3). The termination of protein synthesis takes place when the mRNA codons UAG, UAA, or UGA are exposed

FIGURE 3
Individual polysome transcribing the giant messenger RNA from the BR2 puff of *Chironomus tentans*. (A) Electron micrograph of a polysome containing 74 ribosomes. The nascent proteins can be seen extending from the ribosomes and growing as the ribosomes move from the 5' end of the message to the 3' end. Near the 3' end are ribosomes from which the protein has detached. The bar represents 0.5 μm. (B) Higher magnification of such a polysome in which the polysome had been stretched during the preparation of the specimen. The relationship of the mRNA to the ribosomal subunits and the nascent polypeptide can be seen. (From Francke et al., 1982; courtesy of J. E. Edstrom.)

(A)

(B)

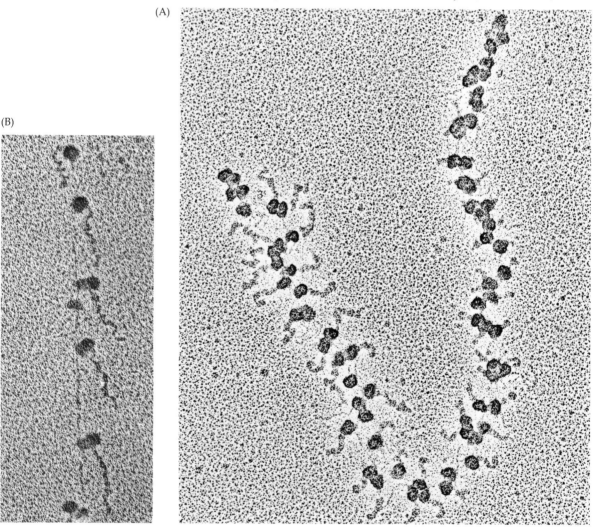

on the ribosome. These nucleotide triplets are not recognized by tRNAs and hence do not code for any amino acids. Rather, they are recognized by release factors, which hydrolyze the peptide from the last tRNA, freeing it from the ribosome.

Translational control of coordinated protein synthesis: Ribosomes and hemoglobin

One of the major problems in genetic regulation is the coordinated production of several products from different regions of the genome. For example, the ribosome of the bacterium *Escherichia coli* consists of 55 protein and 3 different ribosomal RNAs. Rather than having so many genes transcribed from the same operon, the ribosomal RNA and ribosomal protein genes are clustered in several spots throughout the *E. coli* chromosome. Still, coordinated regulation occurs. One never sees excess ribosomal proteins or excess ribosomal RNAs. When one component is synthesized, so are all the others. Masayasu Nomura and his colleagues have discovered the elegant mechanism by which such regulation occurs (Nomura et al., 1980). Certain ribosomal proteins bind to specific regions of ribosomal RNA in order to form the ribosome structure. When these binding sites on the ribosomal RNA are full, the free ribosomal proteins bind to a closely related site on the 5' end of their own mRNAs (Figure 4). This blocks any further translation of their

FIGURE 4
Models of the secondary structures recognized by ribosomal protein S7 in *E. coli*. (A) S7 binding site on the 16 S rRNA of the small subunit. (B) S7 binding site on the leader sequence of its own mRNA. The boxed regions show extensive homologies. (From Nomura et al., 1980.)

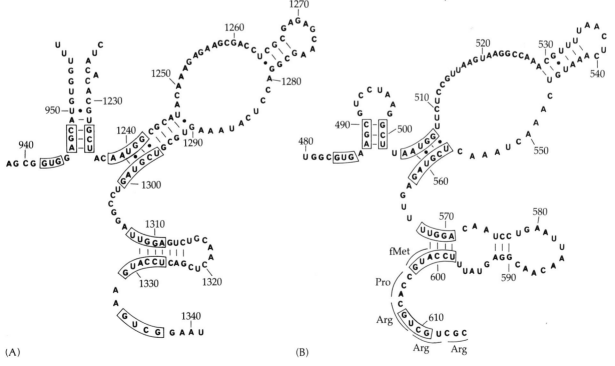

(A)   (B)

messages. Thus, these proteins regulate the translation of their own messages.

A very similar problem is encountered in differentiated eukaryotic cells. When a developing red blood cell synthesizes hemoglobin, it must ensure that the α globin chains, β globin chains, and heme molecules are in a 2:2:4 ratio (Figure 5). Any deviation from this ratio results in severely debilitating diseases.

Recent evidence has shown that the heme molecule regulates the proportional synthesis of the hemoglobin components. It accomplishes this feat in two ways. First, excess heme (i.e., heme that has not been bound to protein such as globin) will shut off its own synthesis (Karibian and London, 1965). It does this by inactivating δ-aminolaevulinate synthase (DALA synthetase), the first enzyme in the pathway for the production of heme (Figure 6). Thus, when there is more heme present than there are molecules to bind it, no more heme will be produced. Second, excess heme is seen to stimulate the production of globin proteins (Gribble and Schwartz, 1965; Zucker and Schulman, 1968). When heme (as its oxidized form, hemin) is added to a cell-free translation system that includes all the factors needed to translate mRNAs (Table 1), the synthesis of globin is greatly enhanced (Figure 7). Therefore, if there is no globin to bind the heme, heme will shut off its own synthesis and stimulate the production of more globin!

Several laboratories have investigated how such a small molecule

FIGURE 5
The structure of adult human hemoglobin, which has four polypeptide chains (two α, two β) and four heme molecules. (Modified from Dickerson and Geis, 1983.)

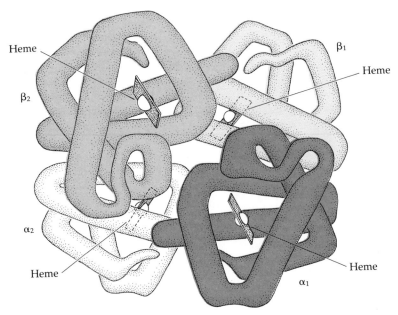

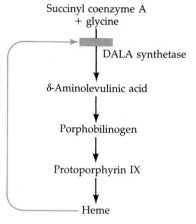

FIGURE 6
Feedback regulation of heme synthesis. (Modified from Harris, 1975.)

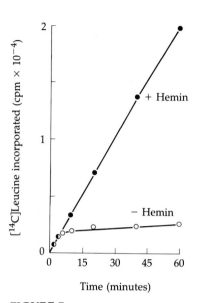

FIGURE 7
Translation of the globin mRNA in the rabbit reticulocyte in vitro protein synthesis system. Inclusion of hemin causes a dramatic elevation in protein synthesis. (From London et al., 1976.)

as heme could regulate protein synthesis. In 1972, Adamson and colleagues demonstrated that the stimulatory effect of heme on globin synthesis could be mimicked by adding loosely bound ribosomal proteins to the translation system. Because such solutions are rich in translation initiation factors, each factor was tested separately. It was found that eukaryotic initiation factor 2 (eIF2) restored protein synthesis to heme-deficient lysates in the translation system (Figure 8). This initiation factor is responsible for combining with the initiator tRNA and complexing it to the 40 S ribosomal subunit.

What, then, was the relationship between heme and eIF2? To answer this, London and his co-workers (Levin at al., 1976; Ranu et al., 1976) added heme-deficient lysates to heme-supplemented translation systems. They found that a portion of the heme-deficient lysate could actually depress the synthesis of globin in the translation system to which it was added. This indicated that an inhibitor was present. Fur-

TABLE 1
Components of the in vitro translation system containing rabbit reticulocyte lysate

Component	Concentration (in 100 μL)	Component	Concentration (in 100 μL)
Reticulocyte lysate (1:1)	50 μL	KCl	76 mM
Tris-HCl (pH 7.6) buffer	10 mM	Proportional amino acid mixture	6–170 μM
ATP	1 mM	[^{14}C]Leucine	0.8 μCi
GTP	0.2 mM	"Cold" leucine	26 μM
Creatine phosphate	5 mM	Hemin	10–30 μM
Creatine phosphokinase	10 μg	H$_2$O to bring the total reaction	
Magnesium acetate	2 mM	to 100 μL volume	

Source: London et al. (1976).

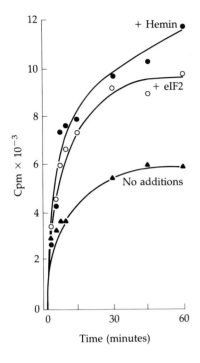

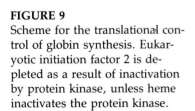

FIGURE 8
Effect of the addition of extra eukaryotic initiation factor 2 to the rabbit reticulocyte in vitro translation system. The extra eIF2 elevated the level of protein synthesis close to that of the hemin-stimulated system. (After Clemens et al., 1974.)

thermore, this inhibiting fraction had enzymatic activity. It was a kinase capable of phosphorylating eIF2.

To summarize, heme appears to regulate globin synthesis in the following manner (Figure 9).

1. In the absence of heme, a specific protein kinase will phosphorylate initiation factor 2.
2. Phosphorylated initiation factor is not active and therefore will not complex Met-tRNA$_i$ to the 40 S ribosomal subunit (Pinphanichakarn et al., 1976). Thus, no translation occurs.
3. Excess heme is able to bind to the protein kinase, inactivating it (Fagard and London, 1981). Inactivated kinase will not phosphorylate eIF2, so translation proceeds. Thus, as long as heme is present, globin synthesis continues.

The story of the translational control of globin synthesis does not end here. As we discussed in Chapter 12, there are four active α globin genes per diploid cell and only two active β globin genes. If each gene were transcribed and translated at the same rate, one would expect twice as many α globin molecules as β globin molecules. This, of course, is not the case. One finds a 1.4:1 ratio of α:β mRNA, but 1:1 ratio of the proteins (Lodish, 1971). Thus, the mechanism for the equalization has to be translational.

Kabat and Chappell (1977) have suggested that the equalization is done at the initiation step of translation. They showed that the α globin mRNA competes with the β globin message for initiation factors and that the β globin message appears to be a better competitor. The β globin message was recognized more efficiently by the initiation factors

FIGURE 9
Scheme for the translational control of globin synthesis. Eukaryotic initiation factor 2 is depleted as a result of inactivation by protein kinase, unless heme inactivates the protein kinase.

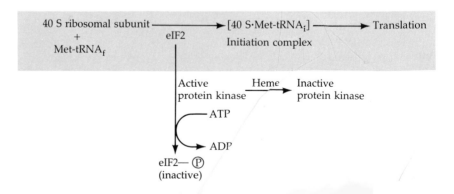

and was thus translated more frequently. The 5' ends of the two globin mRNAs differ significantly, as is shown in Figure 10. When the two RNAs were present in equal amounts with a severely limiting supply of initiation factors, only 3 percent of the resulting protein was α globin. However, when the unfractionated mRNA (α and β globin messages from lysed cells) were added to an excess of such initiation factors, all mRNAs were translated with equal efficiency and the resulting α to β ratio was 1.4:1. Thus, at the initiation step of translation, the proper ratios of α globin, β globin, and heme are established. Although hemoglobin synthesis involves regulation at the transcriptional and RNA processing levels, the final molecule is constructed through fine-tuned coordination at the level of translation.

Translational control of oocyte messages

Evidence for maternal regulation of early development

In most animal species, the diploid nucleus is not immediately expressed. Evidence that early development is controlled by factors stored in or made by the oocyte came from several experiments at the turn of the century (reviewed in Davidson, 1976). These experiments clearly

FIGURE 10

Probable secondary structures for the 5' ends of mouse α globin and β globin chains. (After Pavlakis et al., 1980.)

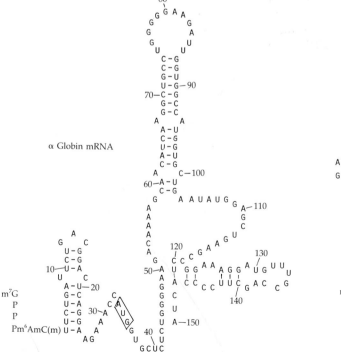

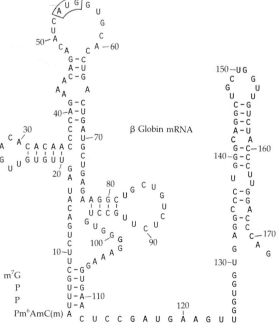

TABLE 2
Driesch's data on the maternal control of primary mesenchyme cell number in hybrid sea urchin embryos

Egg		Sperm	Average number of primary mesenchyme cells (range in parentheses)
Echinus	×	Echinus	55 (±4)
Spherechinus	×	Spherechinus	33 (±4)
Spherechinus	×	Echinus	35 (±5)
Strongylocentrotus	×	Strongylocentrotus	49 (±3)
Spherechinus	×	Strongylocentrotus	33 (±3)

Source: Davidson (1976).

demonstrated the dominance of maternal traits during the initial stages of embryogenesis and a switch to paternal or hybrid characteristics only later in development. Such far-reaching maternal effects have already been alluded to in our discussion of the cleavage orientation in snail embryos, in which the oocyte cytoplasm contains a factor that directs the rotations of the cleavage planes in a dextral or sinistral direction.

In 1898, Hans Driesch crossed two species of sea urchins and found that the average number of primary mesenchyme cells in the hybrid depended solely on which species provided the egg (Table 2). In a related experiment, Tennant (1914) fertilized eggs from the sea urchin *Cidaris* with sperm from the sea urchin *Lytechinus*. He showed that all the chromosomes were retained in the resulting blastomeres and that the maternal pattern of development was followed completely through the invagination of the archenteron and the timing of mesenchyme formation (Table 3). The paternal genes were first seen to be involved in the placement of the skeletal cells.

A second type of evidence for maternal regulation of early development comes from enucleation studies. If one could enucleate an egg after fertilization, one should be able to observe how far development could proceed before terminating. This enucleation has been accomplished by physical, chemical, and genetic means. Physical enucleation of the oocyte was accomplished by E. B. Harvey (1940). She placed unfertilized sea urchin eggs into a sucrose solution having the same density as the eggs themselves, and she found that when such eggs were centrifuged, their contents stratified and the egg was eventually split into two spheres (Figure 11A). The lighter half contained the nucleus, whereas the heavier half was enucleated. Harvey then parthenogenetically activated the enucleated halves by placing them into hypotonic seawater. These halves ("parthenogenetic merogones") cleaved, formed abnormal blastulae (which lacked a blastocoel), and successfully hatched (Figure 11B). No further development was seen.

TABLE 3
**Tennant's data on the control of early gastrulation events in hybrid sea
urchin embryos**

Species of sea urchin	Archenteron invagination (hours)	Mesenchyme formation (hours)	Site of origin of primary mesenchyme cells
Cidaris (♀)	20–33	23–26 (follows invagination)	Archenteron tip
Lytechinus (♂)	9	8 (precedes invagination)	Archenteron base and sides
Hybrid (Cidaris ♀ × Lytechinus ♂)	20	24 (follows invagination)	Archenteron base and sides

Source: Davidson (1976).

Similar studies involving frog eggs have also shown that cleavage can even occur in the total absence of chromosomes (provided the egg is injected with centrioles to compensate for those normally provided by the sperm). Here, too, an abnormal blastula is created. Physical enucleation studies suggested that certain eggs were capable of development through the midblastula stage, even in the absence of a nucleus.

Stored messenger RNA

Evidence that the oocyte controlled early development by storing messenger RNAs was first obtained by Brachet and co-workers (1963) and by Denny and Tyler (1964). The results of these investigations demon-

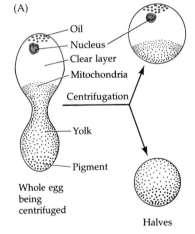

FIGURE 11
Cleavage in the absence of a nucleus. (A) Physical enucleation of *Arbacia* oocyte by centrifugation in a sucrose solution. (B) Development of enucleated portion of oocyte into "parthenogenetic merogones" having no nuclei. Although the cleavage is often abnormal, without nuclear instruction some enucleated oocytes develop into blastulae and even hatch abnormal larvae. (After Harvey, 1940.)

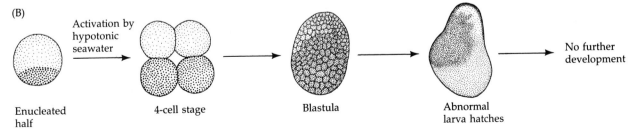

strated that such enucleated and activated sea urchin half eggs contained RNA and that they could synthesize proteins at a rate comparable to that of normally fertilized eggs. Because there could be no transcription from these cells, stored mRNA was implicated. Craig and Piatigorsky (1971) demonstrated that this increase in protein synthesis could not have come from the transcription of mitochondrial DNA in the oocyte. It seemed that the oocyte had stored messenger RNAs that were not translated until after fertilization.

A gentler means of enucleation was made possible when it was discovered that the drug actinomycin D inhibited RNA synthesis, and transcription could be eliminated by placing the newly fertilized egg in a solution of this drug. Gross and Cousineau (1964) found that when sea urchin eggs were treated with enough actinomycin D to shut down 94 percent of their RNA synthesis, the embryos still became blastulae (Figure 12B). That this amount of development was dependent on

FIGURE 12
Effect of protein synthesis inhibitors on the development of the sea urchin *Lytechinus*. (A) Control pluteus larva. (B) Development arrested at blastula stage when actinomycin D prevents transcription of new RNAs. (C) Development arrested at single-cell stage when emetin or cycloheximide prevents translation of new proteins.

(A)

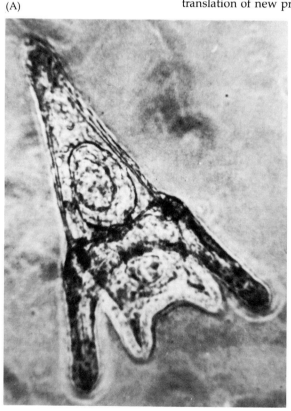

(B)

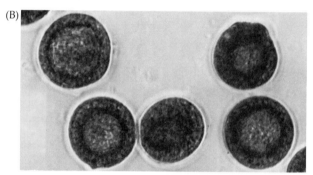

(C)

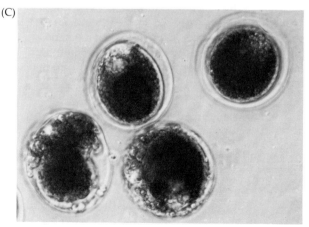

stored messages and not on preformed proteins was shown by treating the fertilized eggs with emetin or cycloheximide. These drugs inhibit translation, and embryos fertilized in the presence of these inhibitors did not develop at all (Figure 12C).

Several investigators demonstrated a burst of protein synthesis shortly after fertilization. The actinomycin D experiments of Gross and his co-workers demonstrated that the magnitude of the fertilization burst of protein synthesis in "chemically enucleated" embryos was exactly the same as that in controls (Figure 13). By 6–10 hours, however, a decline in protein synthesis was observed in the treated embryos, and they did not undergo a second burst of protein synthesis at the blastula stage. The message for this second burst of protein synthesis comes from nuclear transcription. Thus, the actinomycin D-treated embryos can develop up to the blastula stage, hatch, and then proceed no further. The stored mRNAs of the oocyte are sufficient to take development only through the hatched blastula stage.

Studies in amphibians have given similar results, although the stimulus for the increased rate of protein synthesis appears to be ovulation, rather than fertilization, of the egg. In activated enucleated frog eggs, the proteins synthesized shortly after fertilization are not only synthesized in the proper amounts, they are also the same type of protein as is normally made (Smith and Ecker, 1965; Ecker and Smith, 1971). Here, too, it appears that the oocyte cytoplasm is "preprogrammed" to carry out early developmental processes even in the absence of a nucleus. Eventually, transcription is initiated in the nuclei of the embryo; yet even this transcription of the embryonic nuclei may be activated by the maternal factors in the oocyte. This concept is supported by the investigations of the *o* mutant of the Mexican axolotl. This is a maternal effect mutation wherein homozygous females produce eggs that are success-

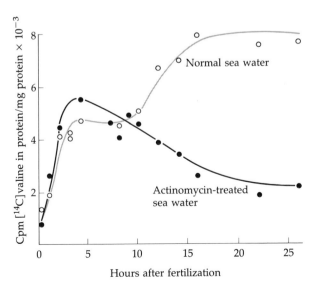

FIGURE 13

Protein synthesis in embryos of the sea urchin *Arbacia punctulata* fertilized in the presence or absence of actinomycin D. For the first few hours, there is no significant difference in the rate of new protein synthesis. The burst of new protein synthesis beginning in the midblastula stage represents the translation of newly-transcribed messages and therefore is not seen in the embryos growing in actinomycin D. (After Gross and Cousineau, 1964.)

fully fertilized and are totally normal until the late cleavage and early blastula stages (Briggs and Cassens, 1966). At midblastula, eggs shed by an *o/o* female are seen to have slower mitoses. These eggs will form a dorsal blastopore lip but will always arrest in gastrulation (Figure 14). Malacinski (1971) has shown that in wild-type blastula, new proteins are starting to be synthesized. However, the blastulae from eggs from *o/o* mothers do not undergo this burst of protein synthesis and have a pattern of proteins identical to that produced by enucleated zygotes. Carroll (1974) has shown that whereas normal late blastulae show intense RNA synthesis, the mutant embryos show little or no RNA production (Figure 15). Briggs and Cassens (1966) demonstrated that the embryos from *o/o* mothers lacked a factor that activated the nuclear genome during blastula stages. Such a factor could be isolated from normal blastomere cytoplasm or from the nuclear sap of premeiotic oocytes. When this factor was injected into the eggs of *o/o* mothers, it repaired the gastrula arrest and allowed the embryo to develop normally. Thus, a specific factor is needed to activate the amphibian nucleus. In the absence of such a factor, the only development that occurs is that which can be supported by the stored oocyte mRNA. In amphibians, there is enough stored oocyte material to enable the embryo to enter gastrulation. However, without new RNA synthesis, no further development can occur.

Physical, chemical, and genetic enucleations of the egg make it clear that stored messages do indeed exist in the cytoplasm of the oocyte

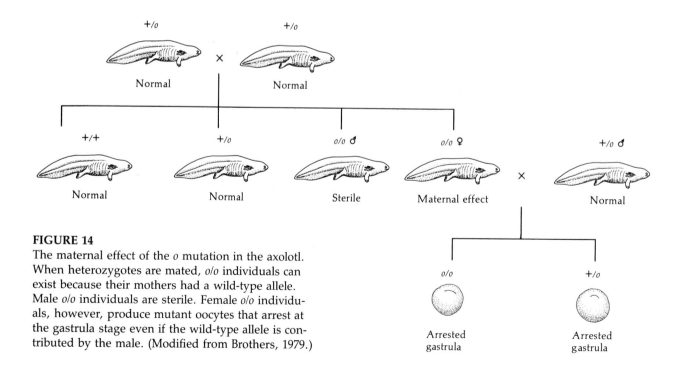

FIGURE 14

The maternal effect of the *o* mutation in the axolotl. When heterozygotes are mated, *o/o* individuals can exist because their mothers had a wild-type allele. Male *o/o* individuals are sterile. Female *o/o* individuals, however, produce mutant oocytes that arrest at the gastrula stage even if the wild-type allele is contributed by the male. (Modified from Brothers, 1979.)

FIGURE 15

Incorporation of [³H]uridine into RNA of wild-type and mutant embryos. Blastula-stage embryos were incubated in the radioactive RNA precursor for 3 hours, washed, fixed, stained, and observed by autoradiography. (A) Normal embryo cells showing intense radioactivity indicating RNA synthesis. (B) Embryo from an *o/o* female. Stain is present, but no significant labeling is seen, indicating that little or no RNA transcription has occurred. (From Carroll, 1974.)

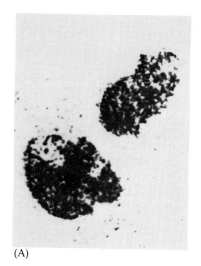

(A)

and that the products of these messages support the early development of the embryo.

Characterization of maternal messages

Eric Davidson and his colleagues have estimated the complexity of the oocyte mRNA in a manner similar to their analysis of DNA complexity (Chapter 11). RNA was hybridized to denatured DNA and the half C_0t value of the hybridization was found to be proportional to the amount of different RNA sequences present. By this analysis, they estimated that each oocyte (in numerous phyla) had enough different nucleotide sequences to account for roughly 1600 copies each of 20,000–50,000 RNA types (Galau et al., 1976; Hough-Evans et al., 1977). This is the greatest message complexity of any known cell type, and it reflects the enormous developmental potential of the oocyte. Even though this figure represents nearly 4 percent of the total RNA of the oocyte, only a few of these messages have been characterized. The use of cell-free translation systems and cDNA probes has enabled the identification of histone mRNA (Skoultchi and Gross, 1973) and tubulin mRNA (Raff et al., 1972) in the oocyte cytoplasm. The products of these messages are of obvious importance for the formation of chromatin and the mitotic spindle during cleavage. Biochemical assays have permitted the identification of the messages for nucleotide reductase (an enzyme essential for the formation of deoxyribose nucleotides from stored ribonucleotides) and for the hatching enzyme, which permits the blastulae to digest their fertilization membranes (Raff, 1980). The utilization of recombinant DNA techniques discussed earlier is also being applied to maternal messages and has enabled investigators to isolate stored mRNA for actin and other proteins.

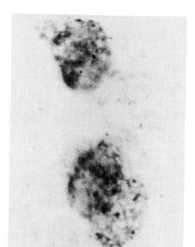

(B)

Interestingly enough, stored mRNAs are not uniformly distributed throughout the oocyte. We have already seen that certain RNA sequences show cytoplasmic localization in snail and tunicate eggs, but such localizations have also been observed in sea urchin eggs. Rodgers and Gross (1978) separated blastomeres at the 16-cell stage, cultured them in actinomycin D, and found that the mesomeres had more different types of mRNA than did the micromeres. Thus, cytoplasmic localization of preformed oocyte messages can be seen in both "mosaic" and "regulative" eggs.

Mechanisms for translational control of oocyte messages

There are, at present, at least four hypotheses for the regulation of oocyte mRNA translation. Three of them involve the availability of messenger RNAs; the fourth involves the efficiency of mRNA translation. Although these hypotheses may be seen as competing with one another, it is probable that each is used by different species and that some species use more than one of these mechanisms to regulate oocyte mRNA translation.

Masked messages. The first hypothesis contends that the oocyte messages are physically masked by proteins so that the mRNA cannot attach to ribosomes. Messenger RNA is never found devoid of proteins. However, the type of protein associated with the RNA can vary. In 1966, Spirin proposed that the mRNA of the oocyte was stored in INFORMO-SOMES, ribonucleoprotein complexes wherein the mRNA was masked. The masked messages would be unable to bind to the ribosomes and thus would not be translated. At fertilization, the proteins masking the message would be released (possibly because of the ionic changes occurring during fertilization) and the message would be free to start translation.

Support for this hypothesis followed shortly. In 1968, Infante and Nemer found in the sea urchin oocyte ribonucleoprotein (RNP) particles that sedimented more slowly than ribosomes, and Gross and co-workers (1973) found that these particles contained various messages. Other studies showed that oocyte ribosomes and initiation factors were perfectly capable of translating exogenously supplied message, so that the control of translation did not seem to be due to the number or capabilities of the ribosomes.

A series of experiments in Rudolf Raff's laboratory demonstrated that oocyte mRNA is indeed stored in a nontranslatable form that is susceptible to ionic changes such as those occurring during fertilization. Nonribosomal RNP particles can be isolated from mature sea urchin oocytes and are found to contain RNA with poly(A) tails (Jenkins et al., 1978). Thus, these RNP particles seem to contain messenger RNA. These RNP particles were isolated in two different ionic solutions (Figure 16). Some were isolated in "oocyte" buffer containing 0.35 mM K$^+$ and 5 mM Mg^{2+}, an ionic condition approximating that of unfertilized oocytes. Other particles were isolated into 0.35 mM Na$^+$, an ion concentration that should remove the proteins bound noncovalently to the mRNA. The mRNA-containing particles prepared in the "oocyte" buffer were very poorly translated in a cell-free system. However, the RNA extracted into the sodium-containing buffer were able to support protein synthesis almost as well as isolated egg mRNA. It thus appears that the influx of sodium during fertilization may destabilize the RNP particle, allowing its mRNA to be translated (Raff, 1980).

Recently certain oocyte-specific proteins have been implicated in the masking of stored messages. When globin mRNA is injected into

FIGURE 16
Evidence for the presence of RNA in RNP and its inability to be translated while in the particles. Protein-bound RNA cannot be translated in low sodium buffer, but can be translated into proteins in buffer containing high concentrations of sodium ions.

Sea urchin eggs

Lysis in low-Na$^+$ and high-Na$^+$ buffers

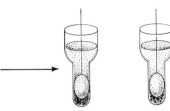

Low-Na$^+$ buffer   High-Na$^+$ buffer

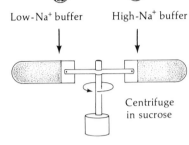

Centrifuge in sucrose

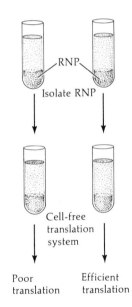

RNP

Isolate RNP

Cell-free translation system

Poor translation   Efficient translation

Xenopus oocytes, it is readily translated. However, if this message is first mixed with proteins isolated from *Xenopus* oocyte RNP particles, very little translation is observed. Other RNA-binding proteins isolated from other tissues did not affect the translatability of the injected globin messages, so the oocyte-specific proteins of the RNP particles are implicated in the translational regulation of stored maternal messages (Richter and Smith, 1984).

It is known that whereas the unfertilized sea urchin oocyte has only 0.75 percent of its ribosomes incorporated into polysomes, the cleavage-stage blastomeres have nearly 20 percent of their ribosomes in such structures (Figure 17). It appears, then, that preexisting messages are being recruited into polysomes. Three recent experiments show that the mRNA from the oocyte RNP becomes incorporated into embryonic polysomes. Young and Raff (1979) showed that radioactively labeled RNA originally in the sea urchin oocyte RNPs later became associated with blastomere polysomes. A similar phenomenon is observed in the early development of the surf clam. Rosenthal and his co-workers (1980) found that when protein-free mRNA from oocytes and embryos were placed in cell-free translation systems, they both coded for the same proteins. In other words, the oocytes and embryos contained identical sets of mRNA. However, different subsets of messages are on the polysomes in oocytes and embryos. Messages for three prominent embryo-specific proteins (A, B, and C) are seen as untranslated RNPs in the oocyte cytoplasm, whereas they are seen as polysomal mRNA in

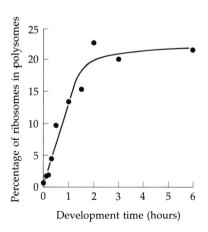

FIGURE 17
Increase in the percentage of ribosomes incorporated into polysomes during early sea urchin development. (After Humphreys, 1971.)

embryonic blastomeres. Conversely, messages for three characteristic oocyte proteins (X, Y, and Z) are found on the polysomes of oocytes, but not on those of embryos. The recruitment of mRNA can be seen from the untranslated RNP to the translationally active polysomes.

Message recruitment is also seen in *Drosophila* embryos. Mermod and co-workers (1980) isolated poly(A)-containing RNA from oocyte RNP, oocyte polysomes, and embryonic polysomes. This mRNA was then translated in a cell-free system containing radioactive amino acids. They found that the mRNA from *Drosophila* oocyte RNP can make certain proteins that cannot yet be synthesized by the mRNA isolated from *oocyte* polysomes (Figure 18A and B). However, these proteins can be made from the mRNA isolated from *embryonic* polysomes (Figure 18C). Thus, mRNA formerly stored in the oocyte RNP particles has been recruited into the embryonic polysomes. In this manner, gene expression is controlled by the initiation of translation.

Uncapped messages. Certain moths use a different mechanism for translational control within the oocyte (Kastern and Berry, 1976; Kastern

FIGURE 18

Evidence for the recruitment of RNA from oocyte RNP into embryo polysomes. Poly(A)-containing RNA was isolated from oocyte RNP (A), oocyte polysomes (B), and embryo polysomes (C). The RNA was translated in vitro in the presence of radioactive amino acids and then passed over a column of DNA to obtain the subset of proteins capable of binding to DNA. These proteins were eluted from the column and subjected to ionophoresis in one dimension (to separate the proteins by their electric charge) and then subjected to electrophoresis in the second dimension (to separate the proteins by their mass). The autoradiographs of these gels show that three proteins seen in the two-dimensional gels were synthesized from the oocyte RNP and were later synthesized from embryonic polysomal messages (arrows in A and C). These three proteins were not made from the RNAs extracted from oocyte polysomes (arrows in B). (from Mermod et al., 1980.)

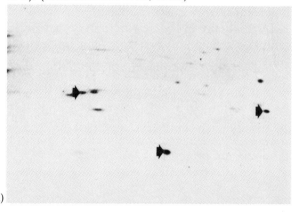

(A)

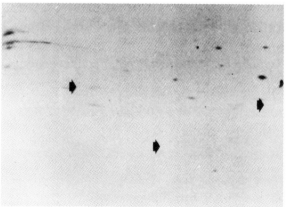

(B)

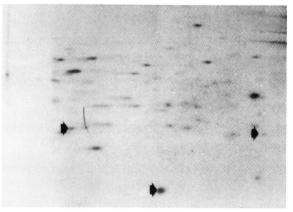

(C)

et al., 1982). In order to be translated, most all eukaryotic messages need to have a "cap" on their 5' ends (Shatkin, 1976). These eukaryotic messages are characterized by 5' 7-methylguanosine. It is believed that this structure is necessary for the recognition of the mRNA with the 40 S ribosomal subunit. The stored messages of the tobacco hornworm moth oocyte have a nonmethylated cap. The guanosine is present, but the methyl group has not been added to it. Such messages are not translated into proteins in a cell-free system. However, at fertilization, there is a burst of methylation in these oocytes and the cap is completed. The mRNAs with the completed caps are then able to bind to the ribosomes and initiate translation. Therefore, in two different ways, the oocyte keeps its mRNA and ribosomes separate. Whether by masking the message or not completing its 5' end modifications, the initiation of translation is inhibited. Thus, mRNAs can be stored without being translated; and fertilization triggers the events that enable their translation in the oocyte.

Sequestered messages. Some studies have disputed the evidence that oocyte RNPs are untranslatable. Rather, it is possible that the protein synthetic apparatus is compartmentalized such that mRNA (within the RNP) does not have a chance to get close to the ribosomes (Moon et al., 1982). The histone mRNAs of sea urchin oocytes seem to be regulated by this type of restriction. The histone messages of the oocyte are not found in cytoplasmic RNPs. Rather, they are localized in the pronucleus. It is only when the pronucleus breaks down, at the end of fertilization, that the histone mRNA gets into the cytoplasm (DeLeon et al., 1983; Figure 19). This may not be the case for other messages. Less than 0.1 percent of the total oocyte messenger RNA is found in pronuclei (Angerer and Angerer, 1981), and those RNPs containing the messages for actin and tubulin are predominantly cytoplasmic (Showman et al., 1982).

Changes in translational efficiency. In one model of translational regulation in the oocyte, the physical or chemical separation of mRNAs need not be postulated. Rather, the initial low pH of the oocyte is able to impede protein synthesis. As discussed in Chapter 2, there is an enormous release of hydrogen ions during sea urchin fertilization, resulting in an elevation of cytoplasmic pH from 6.9 to 7.4. Winkler and Steinhardt (1981) prepared an in vitro cell-free translation system from unfertilized sea urchin oocytes. The pH of the resulting suspension was

FIGURE 19
Evidence for the sequestering of the sea urchin oocyte histone messages. A cDNA probe recognizing histone message is hybridized to sea urchin eggs fixed at various times after fertilization. Autoradiography shows the message to be sequestered in the maternal pronucleus until its breakdown at 80–90 minutes after sperm entry. (From DeLeon et al., 1983; courtesy of L. and R. Angerer.)

70 Minutes

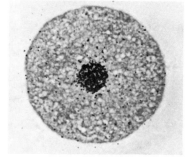

80 Minutes

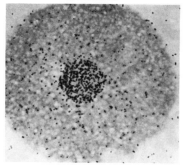

80 Minutes

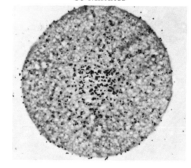

90 Minutes

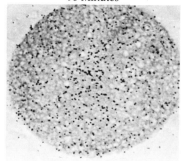

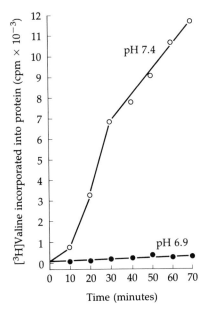

FIGURE 20
Evidence for the inefficiency of
protein synthesis at prefertilization pH levels. The in vitro
translation system made from
unfertilized eggs is kept at pH
6.9 or dialyzed to pH 7.4. Endogenous messages are translated much more efficiently at
the postfertilization pH. (From
Winkler and Steinhardt, 1981.)

then retained or altered by dialysis, and protein synthesis was measured
by the incorporation of radioactive valine into proteins. No exogenous
mRNA was added. Figure 20 shows the results of one such experiment.
Increasing the pH from oocyte levels (pH 6.9) to zygote levels (pH 7.4)
occasioned a burst of protein synthesis mimicking that seen during
fertilization.

It is not known whether this increase in protein synthesis is due
solely to the increased efficiency of protein synthesis or whether increasing the pH may cause other conditions (such as the unmasking of
messages) that then allow protein synthesis to occur. This increased
efficiency alone would not account for the specificity of translation as
seen in the surf clam (where messages formerly on the polysomes are
replaced by others). In any event, the oocyte is able to exercise trans-
lation-level control to regulate the expression of preexisting mRNAs.

Maternal mRNA and embryonic cleavage

Cleavage in sea urchins requires continued protein synthesis from
stored maternal mRNAs. If sea urchins are fertilized in the presence of
an inhibitor of translation, no cleavage results, even though the two
pronuclei fuse. No mitotic spindle forms, the chromosomes do not
condense, and the nuclear membrane does not break down (Wagenaar
and Mazia, 1978). Thus, it is probable that some of the proteins specified
by maternal messages are involved in cell division during cleavage.

This hypothesis has gained support from the discovery (Evans et
al., 1983) of a class of proteins called CYCLINS. These proteins are coded
for by maternal mRNAs, and they are not detectable in the unfertilized
egg. However, their synthesis increases dramatically after fertilization.
What is striking about cyclins is that they are destroyed upon cell
division and have to be resynthesized anew from the stored messages

FIGURE 21
Correlation of cyclin levels with cleavage divisions in
the sea urchin embryo. Eggs were fertilized in the
presence of radioactive amino acids and were analyzed every 10 minutes for the percentage of dividing
cells and for the presence of a particular protein that
exhibited cyclicity. This protein was at its highest levels during the middle of each cell cycle and was then
degraded and resynthesized. Other proteins increased linearly during this period. (After Evans et
al., 1983.)

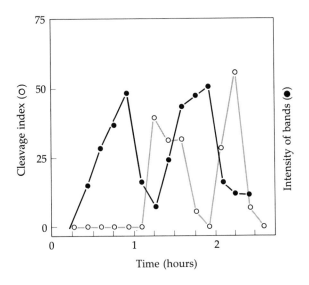

after the completion of each cleavage (Figure 21). Cyclin synthesis is seen to decline as the embryo nears the end of the blastula stage. Although originally seen in sea urchins, cyclins have also been detected in other species. Proteins A and B in the surf clam (mentioned earlier) are probably cyclins, as their synthesis correlates with the cell cycle. Although no physiological proof exists that these cyclins are responsible for regulating cell division during cleavage, the cyclical synthesis and destruction of a protein would be a likely mechanism for controlling mitosis during embryonic cleavage.

SIDELIGHTS & SPECULATIONS

Maternal RNA in mammals

In amphibians and echinoderms, nuclear transcription occurs during mid to late blastula stages, and the maternally stored proteins and RNAs can carry development through to that point. In mammals, however, the nucleus appears to be active by the 2-cell stage, and new transcription is necessary for the survival of the embryo. New proteins are seen to be made at the 2-cell stage, and the addition of the transcription-preventing drug, α-amanitin, prevents the expression of these new proteins and prevents development beyond the 2-cell stage (Flach et al., 1982). The mammalian oocyte does, however, contain a significant amount of maternal RNA. Radioactive mRNA made during oogenesis may persist as late as the blastocyst stage (Buchvarova and DeLeon, 1980) and 2- to 8-cell embryos derived from XX female mice have twice the enzymatic activity of the X chromosome-coded enzyme

HGPRT as those early embryos born of XO mothers (Epstein, 1972; Monk and Harper, 1978). The expression of this elevated HGPRT can be blocked by drugs that inhibit translation but not by inhibitors of transcription (Harper and Monk, 1983). Finally, when new transcription is blocked and the existing mRNA population is translated in vitro, the mRNAs of unfertilized ova, 1-cell embryos, and 2-cell embryos have revealed in the unfertilized ovum the presence of mRNA that encodes proteins that initially appear in 2-cell embryos (Braude et al., 1979; Schultz et al., 1981). Thus, in early mammalian embryos, a portion of the protein is derived from the translation of maternal mRNA, although an increasingly large percentage is derived from nuclear transcription beginning about 22 hours after fertilization.

Translational control of casein synthesis

Differentiated gene products are often synthesized in response to hormonal induction. In some of these cases, the hormones do not increase the transcription of certain messages but act at the level of translation. One such case involves the synthesis of casein by lactating mammals. Casein is the major phosphoprotein of milk and is therefore a differentiated product of the mammary gland. As will be discussed more fully in Chapter 18, the mammary gland is prepared by the sequential actions of several hormones. Prolactin, however, is the hormone responsible for lactation, that is, actual milk production. Prolactin augments the transcription of casein messages only about twofold; its major effect appears to be the stabilization of the casein mRNA (Guyette et

FIGURE 22

The degradation of casein mRNA in the presence and absence of prolactin. Cultured mammary cells were given radioactive RNA precursors (pulse) and, after a given time, were washed and fed nonradioactive precursors (chase). The casein RNA synthesized during the pulse time was then isolated and counted. In the absence of prolactin, the newly-synthesized casein mRNA decayed rapidly, with a half-life of 1.1 hours. When the same experiment was done in medium containing prolactin, the half-life was extended to 28.5 hours. (After Guyette et al., 1979.)

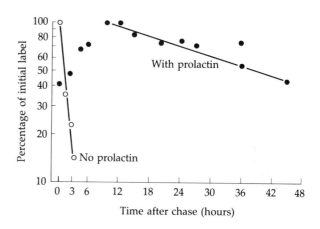

al., 1979). Prolactin somehow increases the longevity of the casein message such that it exists 25 times longer than most other messages in the cell (Figure 22). This means that each casein mRNA can be used for many more rounds of translation. In this way, more casein molecules can be synthesized from each casein message.

The widespread use of translational regulation

Translational control is probably a major mechanism of developmental regulation. We have seen its importance in the burst of protein synthesis following fertilization in many species, in balancing the synthesis of the components of hemoglobin, and in the hormone-induced synthesis of casein. There are many more cases of translational regulation for which the detailed mechanism of operation is still unknown. Glucose, for example, is able to stimulate a tenfold increase in the synthesis of proinsulin from the pancreatic beta cell. However, when a cDNA probe was used to measure the insulin mRNA content of pancreatic beta cells, it was found that glucose-stimulated beta cells contained no more proinsulin message than do nonstimulated beta cells (Itoh and Okamoto, 1980). Thus, the regulation of insulin synthesis by glucose appears to be under translational control. The reactivation of dehydrated brine shrimp embryos also seems to be regulated at the level of translation. These embryos (often sold as "brine shrimp eggs") begin development and can then lie dormant for long periods of time. During this dormancy, no protein synthesis occurs. It has been found that an inhibitor of protein synthesis exists in these embryos and that upon reactivation, a 20-fold increase in the level of functional eIF2 occurs (Filipowicz et al., 1975). Moreover, Sierra et al. (1977) showed that this inhibitor may be a protein kinase. Thus, the same mechanism that regulates hemoglobin synthesis may also be used by brine shrimp embryos.

SIDELIGHTS & SPECULATIONS

Other types of translational control

In this chapter, we have seen that several different mechanisms can effect the translational control of gene expression. Yet this is far from an exhaustive catalog of the many possible ways that translational control can work. Another mechanism for translational regulation utilizes different isoaccepting species of tRNA in different cell types (Sueoka and Kano-Sueoka, 1970; Littauer and Inouye, 1973). These tRNAs bind one amino acid but have different anticodons. For instance, if 90 percent of a cell's serine-tRNA recognized the codon UCU and only 1 percent of its serine-tRNA recognized the codon UCC, those messages with UCU serine codons would be translated more efficiently than those with UCC codons. If a message had four serine codons, those with four UCU codons would be translated 65 million (90^4) times more efficiently than those with four UCCs. Zilberstein et al. (1976) showed that when cells are treated with the antiviral agent interferon, they lose the ability to translate viral messages. This translation can be restored by adding relatively rare tRNA types back to them. Although such translational control of gene expression appears to work during such circumstances, it is still not known whether limiting amounts of certain tRNAs cause significant changes during development. In developing cells, changes in tRNAs have been seen to parallel certain types of cell differentiation (Zeikus et al., 1969; Rogg et al., 1977), but in no case has the tRNA change been demonstrated to be the causative factor. It was formerly thought that such regulation had been demonstrated in developing mealworm cuticle, but these results have more recently been severely criticized (Lassam et al., 1976).

Another controversial area is that of messenger-specific initiation. Heywood and co-workers (Heywood et al., 1974; Kennedy and Heywood, 1976) have proposed that certain initiation factors may be different from cell to cell. They have shown that myosin messages are translated better with muscle initiation factors than with red blood cell initiation factors. Conversely, hemoglobin mRNA is translated more efficiently with red blood cell factors than with those from muscle cells. There is, however, other evidence suggesting that if this type of regulation did exist, it would not be very important. Certainly, cell-free translation systems using wheat germ or rabbit reticulocyte initiation factors can translate myriads of cellular and viral messages, and frog oocytes are willing and able to translate just about any mRNA injected into them (Gurdon et al., 1971). Even in Heywood's studies, translation is not stopped in the presence of the heterologous initiation factors, so it would be surprising if this were the limiting factor in regulating the expression of certain gene products.

One further type of translational regulation involves TRANSLATIONAL CONTROL RNA (tcRNA). These are RNA species that bind to mRNAs, thereby inhibiting their translation. According to Bester et al. (1975), these tcRNAs exist in two classes. The first class is specific for a given cell type and inhibits the translation of those messages characteristic of other types of cells. For instance, tcRNA from developing red blood cells inhibits the translation of myosin mRNA. Conversely, tcRNA from myoblasts inhibits the translation of the globin message. The second class of tcRNA is found in RNP particles and acts with the protein to mask the mRNA until it is needed. This would enable myoblasts to store myosin message in these particles until it is needed after cell fusion. In this way, somatic cells could have a translational control mechanism similar to that found in oocytes. If such tcRNAs could be isolated and sequenced, it would greatly add to our understanding of developmental regulation.

Even sperm appear to regulate the expression of certain genes by translational control. Just like eggs, sperm can store messages for later use. This can be seen in the case of lactate dehydrogenase-X in avian and mammalian sperm. Although this *protein* is only synthesized in the last stages of sperm development, the *gene* for LDH-X is transcriptionally active throughout all stages of sperm development (Blanco, 1980).

The synthesis of LDH-X protein probably occurs by initiating the translation of stored mRNAs. Translational control, then, is an important and widely used mechanism for regulating gene expression in development.

POSTTRANSLATIONAL REGULATION OF GENE EXPRESSION

Once a protein is made, it becomes part of a larger level of organization. It may become part of the structural framework of the cell or it may become involved in one of the myriad enzymatic pathways for the synthesis or breakdown of cellular metabolites. In either case, the individual protein is now part of a complex "ecosystem" which integrates it into a relationship with numerous other proteins. Thus, we do not abandon the regulation of gene expression after synthesizing a protein, for several changes can still take place. First, some newly synthesized proteins are inactive without further modifications. These modifications can involve the cleaving away of certain inhibitory sections of the protein or may involve the binding of a small compound to enhance its activity. Second, some proteins may be selectively inactivated. In some cases, inactivation involves the degradation of the protein itself; in other cases, inactivation may be brought about by the binding of an inhibitory ligand. Third, some proteins must be "addressed" to their specific cellular destinations. The cell is not merely a sack of enzymes; and proteins will often be sequestered in certain regions such as membranes, lysosomes, or mitochondria. Fourth, some proteins need to assemble with other proteins to form a functional unit. The hemoglobin protein, the microtubule, and the ribosome are all examples of numerous proteins joining together to form a functional unit. Therefore, the expression of genetic information can still be influenced at the posttranslational level. Let us now review each of these in turn.

Activation of proteins by posttranslational modifications

Many newly synthesized polypeptides are inactive unless certain amino acid residues are cleaved away. This is especially evident in the case of peptide hormones. The precursor of insulin, for instance, is a long polypeptide with three disulfide bridges. The hormone is active only after the middle and the beginning of the peptide have been cleaved away. The two remaining ends, held together by disulfide bonds, constitute the active molecule (Figure 23). In the case of pro-ACTH-endorphin, several small peptide hormones are made as one long precursor

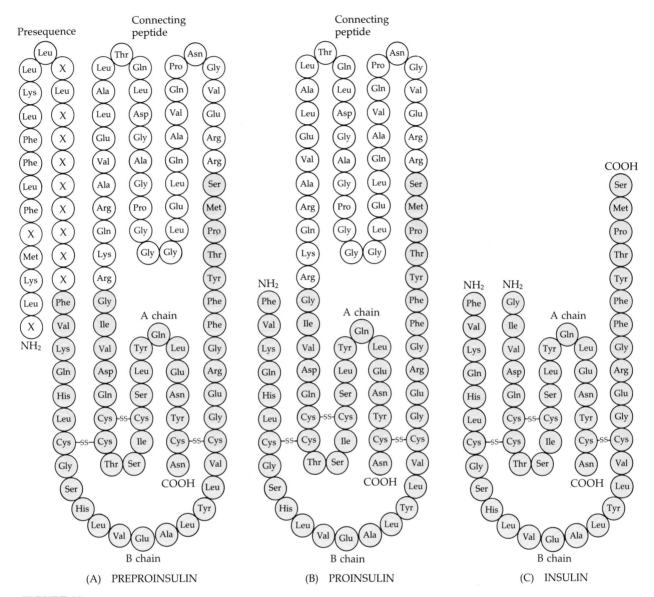

FIGURE 23
Amino acid sequences of rat insulin II (shaded circles) and its precursors. The insulin molecule is inactive until the presequence and the connecting middle sequence have been cleaved away. (X indicates unknown amino acids.) (From Ham and Veomett, 1980.)

(Figure 24). This one polypeptide can be cleaved to form adrenocortico-trophic hormone (ACTH), γ-lipotropin (γ-LPH), and β-endorphin. Further processing of these hormones depends on the specific cell type in which they are found. In the anterior pituitary cells, ACTH is an end product that can be secreted to stimulate the production of steroids by

FIGURE 24
Protein processing of pro-ACTH-endorphin (shaded) in the intermediate and anterior lobes of the rat pituitary gland. (After Herbert et al., 1981.)

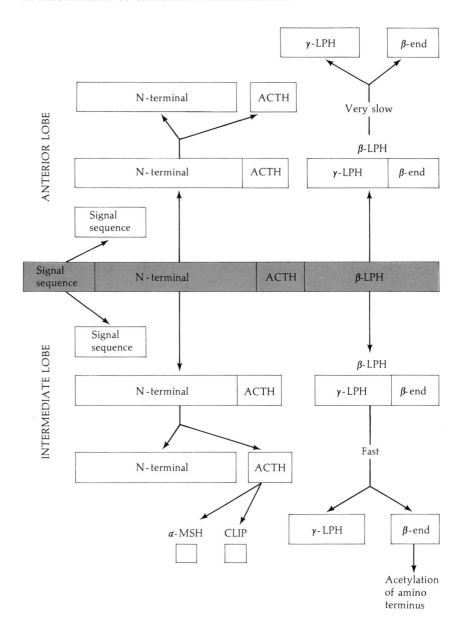

the adrenal cortex. However, in the cells of the intermediate lobe of the pituitary, ACTH is cleaved to form α-melanocyte stimulating hormone (α-MSH; for review, see Herbert et al., 1981).

Some proteins can only work after they have been modified by the covalent attachment of smaller molecules. Phosphorylase kinase, for example, is inactive if it is not itself phosphorylated. Once it is phosphorylated, it is catalytically active. Collagen, one of the major structural proteins of the body, cannot function unless its prolines and certain lysines are hydroxylated, and failure to do so results in serious diseases.

Posttranslational modification of histones may be important in regulating transcription itself. In the slime mold *Physarum polycephalum*, there are no cellular boundaries and all the nuclei are synchronously regulated within a common cytoplasm. This makes an excellent organism in which to study the biochemistry of the cell cycle. Here, histone H4 can exist in several forms, depending on its state of acetylation. Highly acetylated (2–4 acetyl groups) H4 is associated with transcriptional activity (Chahal et al., 1980). Conversely, the phosphorylation of histone H1 appears to activate its chromatic condensing activity, thereby curtailing transcription. In *Physarum*, H4 acetylation and H1 phosphorylation are inversely expressed. Thus, the gross transcriptional activity of chromatin may be regulated by the phosphorylation or acetylation of histone proteins.

Inactivation of proteins by posttranslational modifications

Alterations to existing proteins may also be used to *inactivate* them. Earlier in this chapter we have seen how phosphorylation inactivated initiation factor 2 in the regulation of hemoglobin synthesis. Glycogen synthetase and pyruvate kinase are similarly inactivated by phosphorylation. Glutamine synthetase is inactivated when an AMP moiety is transferred to it. Since these reactions are all reversible, the protein's catalytic activity can be regulated to a very precise level by regulating the activities of the enzymes that add or cleave the modifiers.

Another way of regulating gene expression posttranslationally is to degrade specific polypeptides after translation. One such example is lactate dehydrogenase (LDH), which catalyzes the reversible interconversion of pyruvate and lactate. LDH is coded by two different genes: one that encodes LDH-A subunits and one that encodes LDH-B subunits. These genes are on different chromosomes. The functional LDH enzyme is a tetramer, meaning that it is composed of four subunits. These subunits can be of either the A or the B variety. If both genes are expressed, one expects to find five different forms of LDH: A4, A3B1, A2B2, A1B3, and B4. (Alternative forms of an enzyme are called ISOZYMES.) The LDH isozymes have different catalytic efficiencies under different cellular conditions. Moreover, if the genes are being expressed equally and recombination is random, one would expect the ratio of types to be 1(A4):4(A3B1):6(A2B2):4(A1B3):1(B4). These isozymes have different electrical charges: the B subunit is more negatively charged than the A subunit. Therefore, the five isozymes can be separated by electrophoresis and stained for catalytic activity (Figure 25). When equal quantities of the subunits are combined artificially, one indeed finds the 1:4:6:4:1 ratio. However, when one looks at individual organs, large deviations from this ratio are found (Figure 26). Muscle LDH is predominantly made of A subunits, whereas testes, heart, and kidney isozymes are composed mostly of B subunits (Markert, 1968).

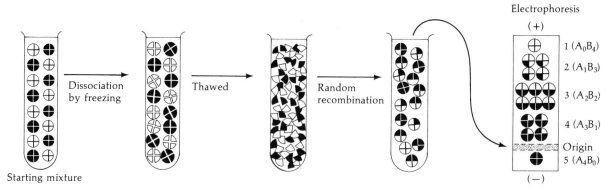

FIGURE 25

Dissociation and recombination of pure A tetramers (black) and pure B tetramers (white) to form the five electrophoretic isozymes of LDH. (After Markert and Ursprung, 1971.)

Using anti-LDH-antibodies to precipitate any newly synthesized A subunits, Fritz and co-workers (1969) discovered that the synthesis of LDH-A does not vary greatly from tissue to tissue. However, the rate of its degradation varies enormously. The degradation of LDH-A is 22 times greater in heart muscle than in skeletal muscle. So the control of LDH isozymes is not at the level of LDH transcription or translation, but at the level of enzyme degradation.

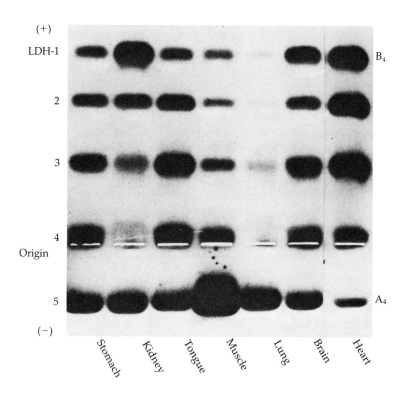

FIGURE 26

Electrophoretic gel showing different amounts of the five LDH isozymes found in various tissues of the rat. (From Markert and Ursprung, 1971.)

Subcellular localization of proteins by posttranslational modifications

After they are synthesized, proteins have to be properly positioned. Thus, the protein may ultimately be located in the soluble cytoplasm, the mitochondria, the lysosomes, the endoplasmic reticulum, or the nucleus; it may even be secreted from the cell.

Those proteins destined for lysosomes, endoplasmic reticula, or secretion are passed into the lumen of the rough endoplasmic reticulum while they are still being translated (Figure 27). These proteins (including collagen and the peptide hormones discussed earlier) contain a signal of approximately 30 amino acids, which are recognized by a SIGNAL RECEPTOR COMPLEX that floats in the cytoplasm. When this complex (which contains six different peptide chains and a small RNA molecule) binds to the growing peptide of a cytoplasmic ribosome, translation stops. The complex then joins to a docking protein on the

FIGURE 27

Diagrammatic model for the translation, glycosylation, and transmembrane insertion of secreted proteins. The first 50 or so amino acids of the polypeptide usually code for a "signal sequence" which identifies the protein as one to be inserted into the membranes of the rough endoplasmic reticulum. The sequence is recognized by a signal receptor complex and then by sites on the RER. Once within the lumen of the RER, part of this sequence is clipped off. As the protein continues to elongate, carbohydrates are added to it.

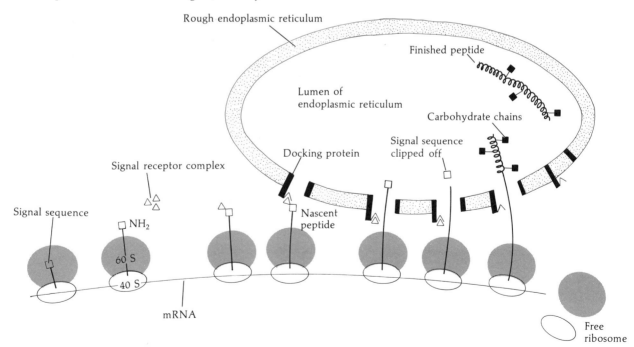

endoplasmic reticulum and translation resumes. The growing polypeptides pass through the reticular membrane and once inside the lumen, this signal sequence is removed and the remaining protein can be modified (Blobel and Dobberstein, 1975; Walter and Blobel, 1983).

One such modification is the covalent addition of mannose-6-phosphate to the protein. This sugar acts as an address label that packages the glycoprotein for lysosomes. The mannose-6-phosphate is recognized by receptors at specific areas of the endoplasmic reticulum, which thereby concentrate those enzymes having this address label. This region of the reticulum then buds off to form the lysosome (Sly et al., 1981). When binding does not occur, these proteins are secreted from the cell. Evidence for this comes from a lethal genetic syndrome, I-cell disease, where patients are unable to put the mannose-6-phosphate address label on the proteins. The lysosomes of these individuals lack the appropriate enzymes, which have been mistakenly secreted into the blood.

Some proteins have specific binding proteins, which localize them in certain regions of the cell. In mouse liver, β-glucuronidase is found both in the lysosomes and on the endoplasmic reticulum. Although the mannose-6-phosphate label can get this enzyme into the lysosomes, β-glucuronidase is bound to the liver endoplasmic reticulum by the protein EGASYN (Swank and Paigen, 1973), which specifically recognizes this protein. In mutant mice lacking egasyn, there is no β-glucuronidase found on the liver endoplasmic reticulum (Lusis et al., 1976). The localization of β-glucuronidase is different in various cell types. Tissues such as liver and kidney are rich in egasyn and have β-glucuronidase on their endoplasmic reticula. Brain and spleen cells, however, appear to be entirely deficient in egasyn, and these have all their β-glucuronidase in their lysosomes (Lusis and Paigen, 1977). So the intracellular localization of proteins is posttranslationally regulated and may differ from tissue to tissue.

Supramolecular assembly

The last form of posttranslational regulation that we will consider involves the assembly of these newly synthesized proteins into a functional unit. We have already discussed several proteins that can spontaneously assemble themselves into functional units. Hemoglobin and lactate dehydrogenase, for example, both contain four polypeptide chains, which are noncovalently joined to form the functional protein. On a slightly higher level, the contractile proteins tubulin and actin are both found to exist in polymerized or nonpolymerized forms. We have seen that the polymerization of actin from globular monomers into microfilaments is essential for the extension of the acrosomal process during fertilization. Similarly, tubulin is assembled into the microtubules of the mitotic apparatus, only to be converted into monomers as mitosis finishes. These tubulin units can then be reassembled into new microtubules, which are important in establishing cell shape. There appears to be a class of proteins (microtubule-associated proteins) that are necessary for regulating the polymerization and function of microtubules.

Several proteins form assemblies with nucleic acids. Histones, for instance, are useless unless assembled with DNA to form the nucleosome particle. The eukaryotic 80 S ribosome consists of approximately 70 proteins and 4 different ribosomal RNAs. The proteins and nucleic acids have to assemble together to be functional, and it is remarkable that this assembly occurs quickly and efficiently within all cells. It should not be a total surprise, then, to discover that whole infectious viruses can be obtained by incubating the component proteins and nucleic acids together. Thus, spontaneous assembly is another important form of posttranslational control.

Collagen: An epitome of posttranslational regulation

By looking at the synthesis of collagen, one of the most important structural proteins of the body, we can demonstrate the importance and variety of posttranslational control mechanisms (for reviews, see Prockop et al., 1979; Davidson and Berg, 1981).

First, the collagen mRNA sequence is translated into a procollagen peptide. The amino terminal of the procollagen contains a signal sequence that enables it to enter the lumen of the rough endoplasmic reticulum. This signal peptide gets cleaved from the remainder of the collagen molecule. Within the endoplasmic reticulum, the collagen molecules also encounter three enzymes that hydroxylate it. Two of these enzymes hydroxylate proline residues to 3-hydroxyproline and 4-hydroxyproline; the third converts lysines to hydroxylysyl residues. Only certain proline and lysine residues are in the correct positions to be recognized by these enzymes (Figure 28).

As the collagen chains are being hydroxylated, sugar residues can be added to the hydroxylysines. The first enzyme, galactosyltransferase, adds a galactose to the hydroxylysine; the second enzyme, glucosyltransferase, adds a glucose unit to the hydroxylsylgalactose. The region near the carboxyl end is modified by glucosamine and mannose.

The next step involves the formation of intrachain disulfide bonds and the organization of three collagen polypeptides into a triple helix. Disulfide bonds are formed at the carboxyl ends between adjacent collagen molecules, establishing the preconditions for forming a triple helix. If the prolines have been properly hydroxylated, the three chains fold over one another to produce a large triple helical region bounded on either side by globular areas.

The collagen molecule is secreted in this form to the outside of the

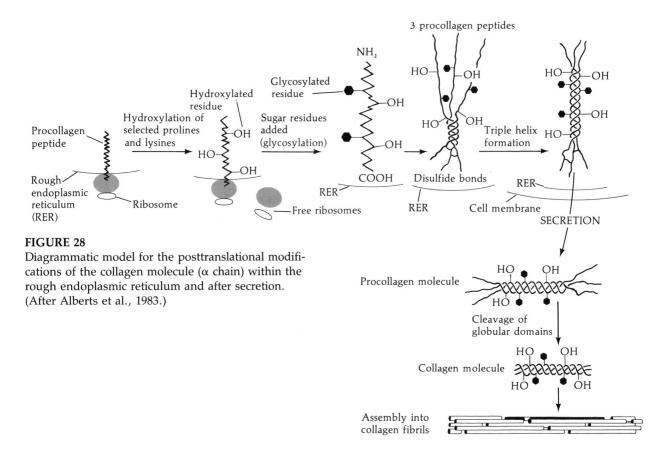

FIGURE 28
Diagrammatic model for the posttranslational modifications of the collagen molecule (α chain) within the rough endoplasmic reticulum and after secretion. (After Alberts et al., 1983.)

cell. Even here, processing continues. First, two proteolytic enzymes remove the globular domains at the amino and carboxyl ends. This creates a pure triple helical fragment. Next, these fragments spontaneously assemble into fibrils. These fibrils, however, lack the necessary tensile strength to function as structural proteins until the various triple helices are joined together covalently by further crosslinking. This is accomplished by enzymatically modifying lysyl or hydroxylysyl residues and linking them together. Thus, there are an impressive number of essential steps by which a protein is modified after translation. (As we shall see in the next two chapters, collagen is not only an extremely important structural protein, but it also is thought to be responsible for certain embryonic interactions.)

We have now looked upon the various levels regulating gene expression: transcription, RNA processing, translation, and posttranslational modification. Each type of regulation has been found to be important in controlling gene expression during development. We are getting closer to answering the mystery of differentiation, and the new techniques of gene cloning (as Boveri predicted in 1904) should enable us to understand differentiation on a chemical level. However, the

question of differentiation of individual cells is not the only problem in developmental biology. We still need to understand how the cells of the body interact to develop at the correct time and at the correct place. In the next part, we shall analyze *cellular interactions* in development that lead to the formation of tissues and organs.

Literature cited

Adamson, S. D., Yau, P. M. P., Herbert, E. and Zucker, W. V. (1972). Involvement of hemin, a stimulatory fraction from ribosomes, and a protein synthesis inhibitor in the regulation of hemoglobin synthesis. *J. Mol. Biol.* 63: 247–264.

Alberts, B., Bray, D., Lewis, J., Raff, M., Roberts, K. and Watson, J. D. (1983). *Molecular Biology of the Cell.* Garland Publishing, New York.

Angerer, L. M. and Angerer, R. C. (1981). Detection of poly A^+ RNA in sea urchin eggs and embryos by quantitative *in situ* hybridization. *Nucleic Acid Res.* 9: 2819–2840.

Bachvarova, R. and DeLeon, V. (1980). Polyadenylated RNA of mouse ova and loss of maternal RNA in early development. *Dev. Biol.* 74: 1–8.

Bester, A. J., Kennedy, D. S. and Heywood, S. M. (1975). Two classes of translational control RNA: Their role in the regulation of protein synthesis. *Proc. Natl. Acad. Sci. USA* 72: 1523–1527.

Blanco, A. (1980). On the functional significance of LDH-X. *Johns Hopkins Med. J.* 146: 231–235.

Blobel, G. and Dobberstein, B. (1975). Transfer of proteins across membranes. I. Presence of proteolytically processed and unprocessed nascent immunoglobulin light chains on membrane bound ribosomes of murine myeloma. *J. Cell Biol.* 67: 835–851.

Brachet, J., Decroly, M., Ficq, A. and Quertier, J. (1963). Ribonucleic acid metabolism in fertilized and unfertilized sea-urchin eggs. *Biochim. Biophys. Acta* 72: 660–662.

Braude, P. R., Pelham, H. R. B., Flach, G. and Lobatto, R. 1979). Post-transcriptional control in the early mouse embryo. *Nature* 282: 102–105.

Briggs, R. and Cassens, G. (1966). Accumulation in the oocyte nucleus of a gene product essential for embryonic development beyond gastrulation. *Proc. Natl. Acad. Sci. USA* 55: 1103–1109.

Brothers, A. J. (1979). A specific case of genetic control of early development: The *o* maternal effect mutation of the Mexican axolotl. In S. Subtelny and I. R. Konigsberg (eds.), *Determinants of Spatial Organization.* Academic Press, New York. *Soc. Dev. Biol.*

Symp. 37: 167–184.

Carroll, C. R. (1974). Comparative study of the early embryonic cytology and nucleic acid synthesis of *Ambystoma mexicanum* normal and *o* mutant embryos. *J. Exp. Zool.* 187: 409–422.

Chahal, S. S., Matthews, H. R. and Bradbury, E. M. (1980). Acetylation of histone H4 and its role in chromatin structure and function. *Nature* 287: 76–79.

Clemens, M. J., Henshaw, E. C., Rahaminoff, H. and London, I. M. (1974). Met-tRNA$_{fmet}$ binding to 40S ribosomal units: A site for the regulation of initiation of protein synthesis by hemin. *Proc. Natl. Acad. Sci. USA* 71: 2946–2950.

Craig, S. P. and Piatigorsky, J. (1971). Protein synthesis and development in the absence of cytoplasmic RNA synthesis in non-nucleate egg fragments and embryos of sea urchins: Effect of ethidium bromide. *Dev. Biol.* 24: 213–232.

Davidson, E. (1976). *Gene Activity in Early Development,* Second Edition. Academic Press, New York.

Davidson, J. M. and Berg, R. H. (1981). Posttranslational events in collagen biosynthesis. In A. R. Hand and C. Oliver (eds.), *Methods in Cell Biology 23,* Academic Press, New York, pp. 119–136.

DeLeon, C. V., Cox, K. H., Angerer, L. M. and Angerer, R. C. (1983). Most early variant histone mRNA is contained in the pronucleus of sea urchin eggs. *Dev. Biol.* 100: 197–206.

Denny, P. C. and Tyler, A. (1964). Activation of protein synthesis in non-nucleate fragments of sea urchin eggs. *Biochem. Biophys. Res. Commun.* 145: 245–249.

Dickerson, R. E. and Geis, I. (1983). *Hemoglobin.* Benjamin/Cummings, Menlo Park, CA.

Driesch, H. (1898). Über rein-mütterliche Charaktere an Bastardlarven von Echiniden. *Wilhelm Roux Arch. Entwicklungsmech. Org.* 7: 65–102.

Ecker, R. E. and Smith, L. D. (1971). The nature and fate of *Rana pipiens* proteins synthesized during maturation and early cleavage. *Dev. Biol.* 24: 559–576.

Epstein, C. J. (1972). Expression of the mammalian X-chromosome before and after fertilization. *Science* 175: 1467–1468.

Evans, T., Rosenthal, E., Youngblom, J., Distel, D. and Hunt, T. (1983). Cyclin: A protein specified by maternal nRNA in sea urchin eggs that is destroyed at each cleavage division. *Cell* 33: 389–396.

Fagard, R. and London, I. M. (1981). Relationship between phosphorylation and activity of heme regulated eukaryotic initiation factor 2 kinase. *Proc. Natl. Acad. Sci. USA* 78: 866–870.

Filipowicz, W., Sierra, J. J. and Ochoa, S. (1975). Polypeptide chain initiation in eukaryotes: Initiation factor MP in *Artemia salina* embryos. *Proc. Natl. Acad. Sci. USA* 72: 3947–3951.

Flach, G., Johnson, M. H., Braude, P. R., Taylor, R. A. S. and Holton, V. N. (1982). The transition from maternal to embryonic control in the 2-cell mouse embryo. *EMBO J.* 1: 681–686.

Francke, C., Edstrom, J. E., McDowell, A. W. and Miller, O. L. (1982). Microscopic visualization of a discrete class of giant translation units in salivary gland cells of *Chironomus tentans*. *EMBO J.* 1: 59–62.

Fritz, P. I., Vesell, E., White, E. L. and Pruit, K. M. (1969). Roles of synthesis and degradation in determining tissue concentration of LDH-5. *Proc. Natl. Acad. Sci. USA* 62: 558–565.

Galau, G., Klein, W. H., Davis, M. M., Wold, B., Britten, R. J. and Davidson, E. H. (1976). Structural gene sets active in embryos and adult tissues of the sea urchin. *Cell* 7: 487–505.

Gribble, T. J. and Schwartz, H. C. (1965). Effect of protoporphyrin on hemoglobin synthesis. *Biochim. Biophys. Act.* 103: 333–338.

Gross, P. R. and Cousineau, G. H. (1964). Macromolecular synthesis and the influence of actinomycin D on early development. *Exp. Cell Res.* 33: 368–395.

Gross, P. R., Malin, L. I. and Moyer, W. A. (1964). Templates for the first proteins of embryonic development. *Proc. Natl. Acad. Sci. USA* 51: 407–414.

Gross, K. W., Jacobs-Lorena, M., Baglioni, G. and Gross, P. R. (1973). Cell-free translation of maternal messenger RNA from sea urchin eggs. *Proc. Natl. Acad. Sci. USA* 70: 2614–2618.

Gurdon, J. B., Lane, C. D., Woodland, H. R. and Marbaix, G. (1971). Use of frog eggs and oocytes for the study of messenger RNA and its translation in living cells. *Nature* 233: 177–182.

Guyette, W. A., Matusik, R. J. and Rosen, J. M. (1979). Prolactin-mediated transcriptional and post-transcriptional control of casein gene expression. *Cell* 17: 1013–1023.

Ham, R. G. and Veomett, M. J. (1980). *Mechanisms of Development*. Mosby, St. Louis.

Harper, M. I. and Monk, M. (1983). Evidence for translation of HPRT enzyme on maternal mRNA in early mouse embryos. *J. Embryol. Exp. Morphol.* 74: 15–28.

Harris, H. (1975). *Principles of Human Biochemical Genetics*. Elsevier North-Holland, New York.

Harvey, E. B. (1940). A comparison of the development of nucleate and non-nucleate eggs of *Arbacia punctulata*. *Biol. Bull.* 79: 166–187.

Herbert, E., Phillips, M. and Budarf, M. (1981). Glycosylation steps involved in processing of pro-corticotropin-endorphin in mouse pituitary tumor cells. *Methods Cell Biol.* 23: 101–118.

Hershey, J. W. B. (1980). The translational machinery: Components and mechanism. In D. M. Prescott and L. Goldstein (eds.), *Cell Biology: A Comprehensive Treatise*. Academic Press, New York, pp. 1–68.

Heywood, S. M., Kennedy, D. S. and Bester, A. J. (1974). Separation of specific initiation factors involved in the translation of myosin and myoglobin messenger RNAs and the isolation of a new RNA involved in translation. *Proc. Natl. Acad. Sci. USA* 71: 2428–2431.

Hough-Evans, B. R., Wold, B. J., Ernst, S. G., Britten, R. J. and Davidson, E. H. (1977). Appearance and persistence of maternal RNA sequences in sea urchin development. *Dev. Biol.* 60: 258–277.

Humphreys, T. (1971). Measurements of messenger RNA entering polysomes upon fertilization in sea urchins. *Dev. Biol.* 26 201–208.

Infante, A. and Nemer, M. (1968). Heterogeneous RNP particles in the cytoplasm of sea urchin embryos. *J. Mol. Biol.* 32: 543–565.

Itoh, N. and Okamoto, H. (1980). Translational control of proinsulin synthesis by glucose. *Nature* 283: 100–102.

Jenkins, N. A., Kaumeyer, J. F., Young, E. M. and Raff, R. A. (1978). A test for masked message: The template activity of messenger ribonucleoprotein particles isolated from sea urchin eggs. *Dev. Biol.* 63: 279–298.

Kabat, D. and Chappell, M. R. (1977). Competition between globin messenger ribonucleic acids for a discriminating initiation factor. *J. Biol. Chem.* 252: 2684–2690.

Karibian, D. and London, I. M. (1965). Control of heme synthesis by feedback inhibition. *Biochem. Biophys. Res. Commun.* 18: 243–249.

Kastern, W. and Berry, S. J. (1976). Non-methylated guanosine in the 5′-terminus of capped mRNA from insect oocytes. *Biochem. Biophys. Res. Commun.* 71: 37–44.

Kastern, W. H., Swindlehurst, M., Aaron, C., Hooper, J. and Berry, S. J. (1982). Control of mRNA translation in oocytes and developing embryos of giant moths. I. Functions of the 5' terminal "cap" in the tobacco hornworm. *Manduca sexta. Dev. Biol.* 89: 437–449.

Kennedy, D. S. and Heywood, S. M. (1976). The role of muscle and reticulocyte initiation factor 3 on the translation of myosin and globin messenger RNA in a wheat germ cell-free system. *FEBS Lett.* 72: 314–318.

Lassam, N. J., Lerer, H. and White, B. N. (1976). Re-examination of leucine transfer-RNAs and leucyl-transfer-RNA synthetase in developing *Tenebrio molitor. Dev. Biol.* 49: 268–277.

Levin, D., Ranu, R., Ernst, V. and London, I. M. (1976). Regulation of protein synthesis in reticulocyte lysates: Phosphorylation of methionyl-tRNAf binding factor by protein kinase activity of translational inhibitor isolated from heme-deficient lysates. *Proc. Natl. Acad. Sci. USA* 73: 3112–3116.

Littauer, U. Z. and Inouye, H. (1973). Regulation of tRNA. *Annu. Rev. Biochem.* 42: 439–470.

Lodish, H. F. (1971). Alpha and beta globin messenger ribonucleic acid. Different amounts and rates of translation. *J. Biol. Chem.* 246: 7131–7138.

London, I. M., Clemens, M. J., Ranu, R. S., Levin, D. H., Cherbas, L. F. and Ernst, V. (1976). The role of hemin in the regulation of protein synthesis in erythroid cells. *Fed. Proc.* 35: 2218–2222.

Lusis, A. J., Tomina, S. and Paigen, K. (1976). Isolation, characterization, and radioimmunoassay of murine egasyn, a protein stabilizing glucuronidase membrane binding. *J. Biol. Chem.* 251: 7753–7760.

Lusis, A. J. and Paigen, K. (1977). Relationship between levels of membrane-bound glucuronidase and the associated protein egasyn in mouse tissues. *J. Cell. Biol.* 731: 728–735.

Malacinski, G. M. (1971). Genetic control of qualitative changes in protein synthesis during early amphibian (Mexican axolotl) embryogenesis. *Dev. Biol.* 26: 442–451.

Markert, C. L. (1968). The molecular basis for isozymes. *Ann. N.Y. Acad. Sci.* 151: 14–40.

Markert, C. L. and Ursprung, H. (1971). *Developmental Genetics.* Prentice-Hall, Englewood Cliffs, NJ.

Mermod, J. J., Schatz, G. and Crippa, M. (1980). Specific control of messenger translation in *Drosophila* oocytes and embryos. *Dev. Biol.* 75: 177–186.

Monk, M. and Harper, M. I. (1978). X-chromosome activity in preimplantation mouse embryos from XX and XO mothers. *J. Embryol. Exp. Morphol.* 46: 53–64.

Moon, R. T., Danilchik, M. V. and Hille, M. (1982). An assessment of the masked messenger hypothesis:

Sea urchin egg messenger ribonucleoprotein complexes are efficient templates for *in vitro* protein synthesis. *Dev. Biol.* 93: 389–403.

Nomura, M., Yates, J. L., Dean, D. and Post, L. E. (1980). Feedback regulation of ribosomal protein gene expression in *Escherichia coli*: Structural homology between ribosomal RNA and ribosomal protein mRNA. *Proc. Natl. Acad. Sci. USA* 77: 7084–7088.

Pavlakis, G. N., Lockard, R. E., Vamvakopolous, N., Rieser, L., RajBhandary, U. L. and Vournakis, J. N. (1980). Secondary structure of mouse and rabbit α- and β-globin mRNAs: Differential accessibility of and initiator AUG codons towards nucleases. *Cell* 19: 91–102.

Pinphanichakarn, P., Kramer, G. and Hardesty, B. (1976). Partial reaction of peptide initiation inhibited by the reticulocyte hemin-controlled repressor. *Biochem. Biophys. Res. Commun.* 73: 625–631.

Prockop, D. J., Kivirikko, K. I., Tuderman, L. and Guzman, N. A. (1979). The biosynthesis of collagen and its disorders. *N. Engl. J. Med.* 301: 13–23.

Raff, R. A. (1980). Masked messenger RNA and the regulation of protein synthesis in eggs and embryos. In D. M. Prescott and L. Goldstein (eds.), *Cell Biology: A Comprehensive Treatise*, Vol. 4, Academic, New York, pp. 107–136.

Raff, R. A., Colot, H. V., Solvig, S. E. and Gross, P. R. (1972). Oogenetic origin of messenger RNA for embryonic synthesis of microtubule proteins. *Nature* 235: 211–214.

Ranu, R. S., Levin, D. H., Delaunay, J., Ernst, U. and London, I. M. (1976). Regulation of protein synthesis in rabbit reticulocyte lysates: Characteristics of inhibition of protein synthesis by a translational inhibitor from heme-deficient lysates and its relationship to the initiation factor which binds Met-tRNA$_f$. *Proc. Natl. Acad. Sci. USA* 73: 2720–2726.

Richter, J. D. and Smith, L. D. (1984). Reversible inhibition of translation by *Xenopus* oocyte-specific proteins. *Nature* 309: 378–380.

Rodgers, W. H. and Gross, P. R. (1978). Inhomogeneous distribution of egg RNA sequences in the early embryo. *Cell* 14: 279–288.

Rogg, H., Muller, P., Keith, G. and Staehelin, M. (1977). Chemical basis for brain-specific serine transfer RNAs. *Proc. Natl. Acad. Sci. USA* 74: 4243–4247.

Rosenthal, E., Hunt., T. and Ruderman, R. V. (1980). Selective translation of mRNA controls the pattern of protein synthesis during early development of the surf clam, *Spisula solidissima. Cell* 20: 487–494.

Schultz, G. A., Clough, J. R., Braude, P. R., Pelham, R. B. and Johnson, M. H. (1981). In S. R. Gasser and

D. W. Bulloch (eds.), *Cellular and Molecular Aspects of Implantation*. Plenum, New York, pp. 137–154.

Shatkin, A. J. (1976). Capping of eukaryotic mRNAs. *Cell* 9: 645–653.

Showman, R. M., Wells, D. E., Anstrom, J., Hursh, D. A. and Raff, R. A. (1982). Message-specific sequestration of maternal histone mRNA in the sea urchin egg. *Proc. Natl. Acad. Sci. USA* 79: 5944–5947.

Sierra, J. M., Haro, C. de, Datta, A. and Ochoa, S. (1977). Translational control by protein kinases in *Artemia salina* and wheat germ. *Proc. Natl. Acad. Sci. USA* 74: 4356–4359.

Skoultchi, A. and Gross, P. R. (1973). Maternal histone messenger RNA: Detection by molecular hybridization. *Proc. Natl. Acad. Sci. USA* 70: 2840–2844.

Sly, W. S., Fischer, H. D., Gonzalez-Noriega, A., Grubb, J. H. and Natowicz, M. (1981). Role of the 6-phosphomannosyl-enzyme receptor in intracellular transport and adsorptive pinocytosis of lysosomal enzymes. *Methods Cell Biol.* 23: 191–214.

Smith, L. D. and Ecker, R. C. (1965). Protein synthesis in enucleated eggs of *Rana pipiens. Science* 150: 777–779.

Spirin, A. S. (1966). On "masked" forms of messenger RNA in early embryogenesis and in other differentiating systems. *Curr. Top. Dev. Biol.* 1: 1–38.

Sueoka, N. and Kano-Sueoka, T. (1970). Transfer RNA and cell differentiation. *Prog. Nucleic Acid Res.* 10: 23–55.

Swank, R. T. and Paigen, K. (1973). Biochemical and genetic evidence for a macromolecular β-glucuronidase complex in microsomal membranes. *J. Mol. Biol.* 77: 371–389.

Tennant, D. H. (1914). The early influence of spermatozoa upon the characters of echinoid larva. *Carnegie Inst. Wash. Publ.* 182: 127–138.

Wagenaar, E. B. and Mazia, D. (1978). The effect of emetine on the first cleavage division of the sea urchin, *Strongylocentrotus purpuratus*. In E. R. Dirksen, D. M. Prescott and L. F. Fox (eds.), *Cell Reproduction: In Honor of Daniel Mazia*. Academic, New York, pp. 539–545.

Walter, P. and Blobel, G. (1983). Subcellular distribution of signal recognition particle and 7SL-RNA determined with polypeptide-specific antibodies and complementary DNA probe. *J. Cell Biol.* 97: 1693–1699.

Winkler, M. M. and Steinhardt, R. A. (1981). Activation of protein synthesis in a sea urchin cell free system. *Dev. Biol.* 84: 432–439.

Young, E. M. and Raff, R. A. (1979). Messenger ribonucleoprotein particles in developing sea urchin embryos. *Dev. Biol.* 72: 24–40.

Zeikus, J. G., Taylor, M. W. and Buck, C. A. (1969). Transfer RNA changes associated with early development and differentiation of the sea urchin, *Strongylocentrotus purpuratus. Exp. Cell Res.* 57: 74–78.

Zilberstein, A., Dudock, B., Berissi, H. and Revel, M. (1976). Control of messenger RNA translation by minor species of leucyl-transfer RNA in extracts from interferon-treated L-cells. *J. Mol. Biol.* 108: 43–54.

Zucker, W. V. and Schulman, H. M. (1968). Stimulation of globin-chain initiation by hemin in the reticulocyte cell-free system. *Proc. Natl. Acad. Sci. USA* 59: 582–589.

PART III

Cell interactions in development

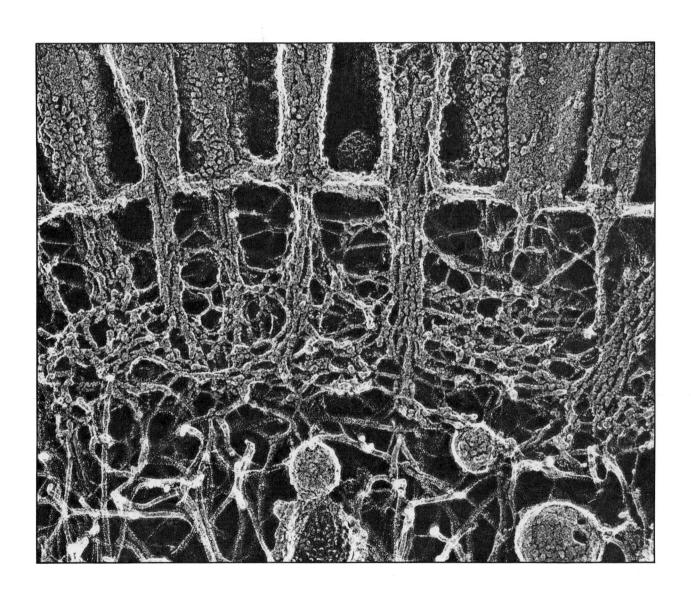

Spatial development:
the role of the cell surface

*But nature is not atomized. Its patterning is inherent and
primary, and the order underlying beauty is demonstrably there;
what is more, the human mind can perceive it only because it is
itself part and parcel of that order.*
—PAUL WEISS (1960)

I am fearfully and wonderfully made.
—PSALM 139 (CA. 500 B.C.)

Introduction

A body is not merely a collection of randomly distributed cell types, for development involves not only the differentiation of cells but also their organization into multicellular arrangements to form tissues and organs. When one observes the detailed anatomy of a tissue such as the retina, one sees an intricate and precise arrangement of many different types of cells. In the next three chapters we shall discuss the ways by which the cells of the developing embryo are provided with POSITIONAL IN-FORMATION such that development occurs at the correct place as well as at the correct time. Embryonic cells are thought to be given positional information in two general ways. One way involves the creation of gradients whereby each cell is told its position relative to the sources of diffusable substances. The cell uses this information as a cue to differentiate in a way appropriate to its position. This involves action-at-a-distance because the responding cells may be far removed from the cells that synthesize the active compound. The other way of mediating positional information is for adjacent cells to act directly on each other, each cell giving positional information to the other.

During the period of organogenesis, individual cells and groups of cells change their relative positions and become associated with other cell types. Organ and tissue formation involve the selective interactions between neighboring cells to create the proper arrangement of the different cell types. For instance, renal tubule cells must form a tube with the mesenchyme outside it. A sheet of cells would not be an appropriate structure for urine flow and resorption; nor would such tubules serve any function if they were full of connective tissue. Cells must have the ability to selectively recognize other cells, adhering to some and migrating over others. It is generally accepted that the molecular events mediating the selective recognition of cells and their formation into tissues and organs occur at the cell surface. In this chapter, we shall analyze the ways by which adjacent cell surfaces can interact during development.

Structure of the cell surface

Cell membrane

The cell surface includes the cell plasma membrane, the molecules directly beneath the membrane and associated with it, and the molecules found in the extracellular spaces. Eukaryotic cells are surrounded by a complex molecular border called the plasma (or cell) membrane. Our present understanding of this membrane structure is summarized by the FLUID MOSAIC MODEL illustrated in Figure 1 (Singer and Nicolson,

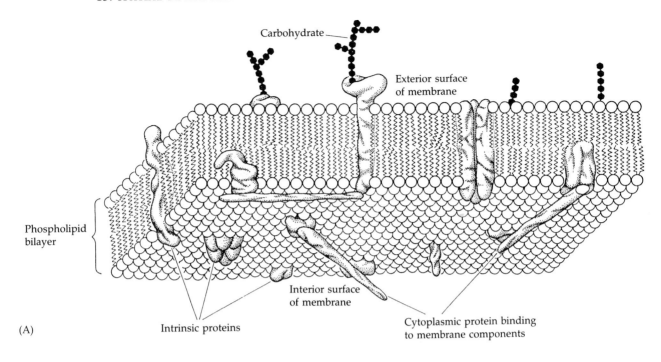

(A)

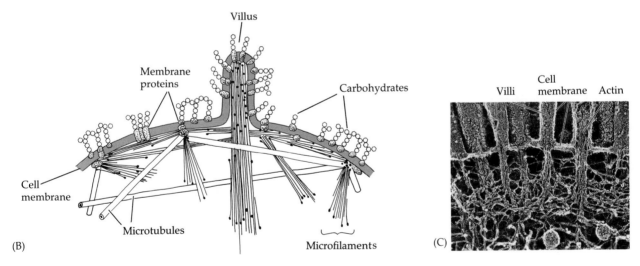

(B)

(C)

FIGURE 1

The structure of the cell surface. (A) Fluid mosaic model of the cell membrane. (B) Underlying cytoskeleton. Contractile fibers consisting of actin microfilaments and tubulin microtubules are linked to each other and to the cell membrane. Membrane glycoproteins may be linked by extracellular proteins as well as by internal fibers. (C) Quick-frozen, deep-etched scanning electron micrograph showing the surface of an intestinal cell. Actin extends into the villi and is linked within the cell by thick and thin filaments directly beneath the cell membrane. (Photograph from Hirokawa et al., 1983; courtesy of N. Hirokawa.)

1972). First, there are two layers of phospholipids, the polar ends of which are oriented toward the aqueous solutions on either side of the phospholipid sea. Some of the proteins penetrate from one side of the membrane to the other, whereas other proteins extend only partway through the phospholipid bilayer. The distribution of proteins gives rise to the "mosaic" nature of the membrane. The ability of these proteins to move laterally within the phospholipid matrix is evidence of its "fluid" nature. Most membranes also have substantial amounts of carbohydrates. These complex sugars are attached to lipids and proteins on the outside of the membrane.

Evidence for the mosaic distribution of proteins within the phospholipid matrix comes from several sources—chief among them, FREEZE-FRACTURE ETCHING (Figure 2A). Here, the membrane is frozen in place and is cleaved so that two lipid layers are separated. The results (Figure 2B) show the mosaic nature of the protein distribution. Some of the best evidence for the fluid nature of the membrane is provided by somatic cell fusions. Such an experiment was performed by Frye and Edidin (1970) with human and mouse cells. Antibodies were prepared to human cells by injecting these cells into rabbits, and antibodies specific to mouse cells were similarly prepared. The anti-mouse cell antibody was complexed with a fluorescein dye, which glows green under ultraviolet light, and the antibody against human cells was complexed with a rhodamine dye, which glows red. Thus, the proteins on the two cells can be labeled: human proteins become red, mouse proteins become green. When these cells were fused together and marked with the labeled antibodies, the mouse and human proteins were first seen on separate halves of the composite cell (Figure 3). However, when the fused cells were treated with labeled antibodies an hour later, the proteins were completely intermixed. Because protein synthesis inhibitors did not block this mixing, the synthesis of new membrane components was ruled out. Thus, the proteins are seen to move within a fluid matrix.

This is an example of passive diffusion and requires no energy from the cell. However, specific proteins may be forced to move in a certain way. When antibodies are made to a specific membrane protein and added in excess to cells containing it, that protein is seen to be redistributed into patches and then into "caps" (Figure 4). This capping is dependent on energy and depends also upon the microtubule and microfilament systems of the cell (Edelman, 1976; Nicolson, 1976). Many cells will then phagocytize this cap, internalizing its contents.

There is evidence that microfilaments and microtubules are in contact with the plasma membrane and have control over the movement of the membrane proteins. For example, when cells phagocytize particles, the quantity of cell membrane enzymes does not diminish. Even though cell membrane is being internalized, the enzymes remain behind. Thus, the cell is in some way prohibiting the placement of proteins into the region of the membrane that gets internalized. However, when

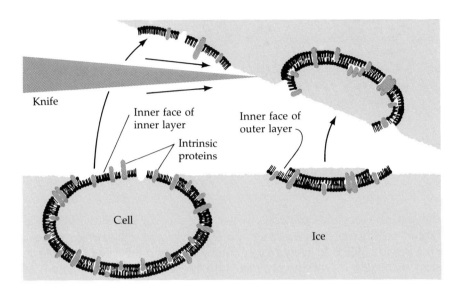

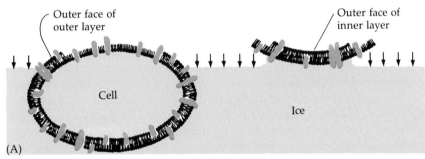

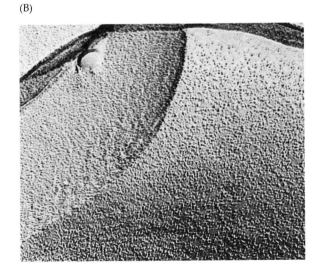

FIGURE 2
Freeze-fracture etching. (A) Diagrammatic representation of the fracture plane bisecting the membrane.
(B) Electron micrograph of a freeze-fractured erythrocyte membrane showing many membrane-associated
particles. (Photograph courtesy of D. Branton.)

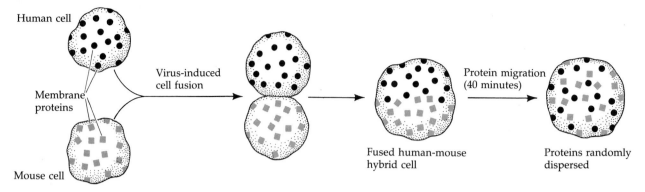

FIGURE 3
Lateral mobility of membrane proteins is shown in cell fusion experiment.

the cell phagocytizes particles in the presence of the microtubule in-hibitor colchicine, the enzymes are lost from the cell surface (Ukena and Berlin, 1972). It seems clear, then, that microtubules are able to stop the entry of certain proteins into those areas of the membrane used for phagocytosis.

Actin microfilaments also seem to be attached to the proteins of the cell membrane (Figure 1B). Capping is inhibited when the drug cyto-chalasin B is used to disrupt the microfilaments (DePetris, 1974). Micro-filaments are a common feature of the cytoplasm adjacent to the cell membrane, and they are occasionally seen directly connected to the membrane itself (Figure 5).

Microfilaments are also important for cell locomotion. Here, the actin microfilaments associate with myosin, α-actinin, and tropomyosin to create a contractile network not unlike that found in muscle. In most vertebrate cells, this network provides the basis for cell extension and migration.

The cell membrane also contains proteins that are capable of inter-acting with the outside environment. Certain proteins are directed so that their active sites point outward toward other cells. The most fa-miliar of these types of proteins are the TRANSPORT PROTEINS, which facilitate the movement of ions and nutrients into the cell, and the RECEPTOR PROTEINS, which bind hormones and drugs. In addition, there are proteins on the cell membrane called ECTOENZYMES, which catalyze reactions in the extracellular fluid. These reactions may be useful in forming bonds between adjacent cells. We shall discuss these proteins in more detail later in this chapter.

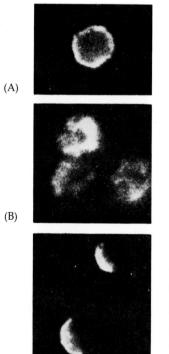

(A)

(B)

(C)

FIGURE 4
Specific movement of cell membrane proteins. Attachment of antibodies to specific cell surface molecules that are uniformly distributed in the cell mem-brane (A) causes them to cluster into discrete patches (B), which migrate to a single pole (C). In these photographs, the cell surface molecules (themselves antibody molecules) have been visualized by tagging them with a fluorescent dye. (Photograph courtesy of G. E. Edelman.)

(B)

FIGURE 5
Connection of internal microfilaments to the cell membrane. Sagittal (A) and cross-sectional (B) electron micrographs of intestinal villi show actin microfilaments attached to the plasma membrane at its uppermost surface and at its sides. The attachment is mediated by accessory proteins such as α-actinin. (Compare with Figure 1C.) (From Mooseker and Tilney, 1975; photographs courtesy of M. Mooseker.)

(A)

Cell membrane changes during development

The expression of different membrane components changes in time and space. Different cell types display different cell surface components and these components change as the cell develops. Such tissue-specific membrane components are often recognized by antisera and are therefore called DIFFERENTIATION ANTIGENS (Boyse and Old, 1969).

Specific differentiation antigens can now be defined by MONOCLONAL ANTIBODIES (Figure 6). These antibodies can be made by injecting foreign cells into mice (or mouse cells of one strain into mice of another strain). The mouse B lymphocytes will begin producing antibody against each foreign component on these cells, with each B lymphocyte producing an antibody with only one such specificity. These lymphocytes are then "immortalized" by fusing them with a B lymphocyte tumor (myeloma) that has been mutated such that (1) it can no longer synthesize its own antibodies and (2) it lacks the purine salvage enzyme hypoxanthine-guanosine phosphoribosyltransferase (HGPRT). This means that the cells can only make purine nucleotides de novo and cannot utilize purines from the culture medium. The cells are then

FIGURE 6

Protocol for making monoclonal antibodies. Spleen cells from an immunized mouse are fused with mutated myeloma cells lacking the enzyme HGPRT. Cells are grown in a medium containing hypoxanthine, aminopterin, and thymidine (HAT). Unfused myeloma cells cannot grow in this medium because aminopterin blocks the only way they have of making purine nucleosides. B cells die in this medium even though they contain an enzyme (HGPRT) that would allow them to utilize the hypoxanthine placed in the medium. The fused cells (hybridomas) grow and divide. The wells in which the hybridomas grow are screened for the presence of the effective antibody, and the cells from positive wells are plated at densities low enough to allow individual cells to give rise to discrete clones. These clones are isolated and screened for the effective antibody. Such an antibody is monoclonal. The hybridomas producing this antibody can be grown and frozen. (From Yelton and Scharff, 1980.)

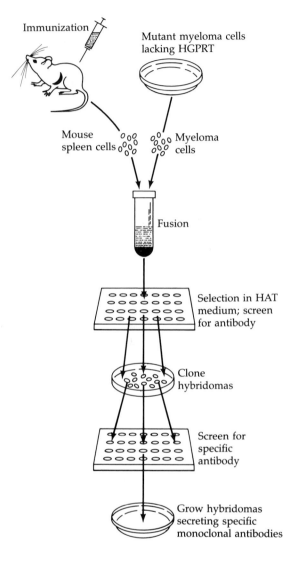

grown in a medium containing aminopterin. This drug inhibits the de novo purine synthetic pathway. Thus, unfused myeloma cells die of purine starvation. They cannot make purine nucleotides using HGPRT and the aminopterin blocks the de novo pathway as well. Normal B lymphocytes do not divide in culture, so they, too, die. The fused product of the B lymphocyte and the myeloma cell—a HYBRIDOMA—proliferates, having the purine salvage enzyme from the lymphocyte and the growth properties from the tumor. Moreover, each of these hybridomas secretes the specific antibody of the B lymphocyte. The medium in which hybridomas are growing is then tested for antibody that binds to the original population of cells. Such antibody, having a

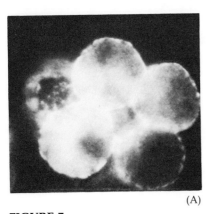

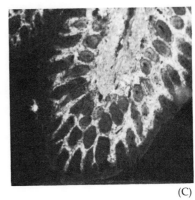

(A)   (B)   (C)

FIGURE 7
Membrane differentiation in time and space. The expression of a specific cell surface antigen (stage-specific embryonic antigen 3) is monitored by a fluorescent monoclonal antibody. It exists on all blastomeres of the 4- and 8-cell mouse embryo (A) but diminishes to nearly undetectable levels during the blastocyst stage (B). It reappears again in certain cells in certain adult organs, for example on the interstitial (but not the epithelial) cells of the small intestine (C). (Photographs from Shevinsky et al., 1982, and from unpublished data; courtesy of B. Knowles and N. Fox.)

single B lymphocyte as its original source, is called a monoclonal antibody. It can be produced in enormous amounts and can recognize antigens that are only weakly expressed (Köhler and Milstein, 1975).

Monoclonal antibodies directed against specific types of cells have uncovered numerous differentiation antigens appearing at different times and places during development. Figure 7 shows the appearance of a specific cell surface molecule on all cells of the 4-cell mouse embryo, its disappearance during the blastocyst stage, and its reappearance on certain cells in the adult organism. We see, then, that the expression of certain cell surface molecules varies in time and space. Such differentiation antigens are expressed at certain times and on certain cells only.

Extracellular matrix

The cell also synthesizes various components that are secreted from it and that are used to mediate the relationships among the cells. These compounds form the EXTRACELLULAR MATRIX. Such matrices may play several roles in development. In some instances, they serve to separate two adjacent groups of cells and prevent any interactions. In other cases, the extracellular matrix may serve as the substrate upon which cells migrate, or it may even induce the differentiation of certain cell

SIDELIGHTS & SPECULATIONS

Oncofetal antigens

In many respects, tumor cells resemble embryonic cells, and "there is now little reason to assume that human malignant cells express anything more than the normal product of normal structural genes, although these genes may be acting at the wrong place and time" (Anderson and Coggin, 1974).

There exist some cell surface antigens that, although normally expressed only on embryonic cells, are reexpressed on tumor cells. These are called ON-

COFETAL ANTIGENS. In mouse leukemia cells, for instance, there exists an antigen that is associated with the malignant state. Normal adult mouse white blood cells lack this antigen. Developing white blood cells, however, express this antigen briefly during their development (Staber et al., 1978; Liebermann and Sachs, 1978). Similarly, when quail fibroblasts are transformed into malignant cells by either viruses or mutagens, they express an antigen found on developing fibroblasts but not seen on any other embryonic cells or on any adult cell (Yoshikawa et al., 1979). It seems then that cancer cells can reexpress cell surface components that characterized earlier stages of their development. This finding may enable investigators to develop specific types of new cancer therapies, using monoclonal antibodies directed against oncofetal antigens.

types. One such matrix is shown in Figure 8. Here, a sheet of epithelial cells is adjacent to a layer of loose mesenchymal tissue. The epithelial cells have formed a tight extracellular layer called the BASAL LAMINA; the mesenchymal cells secrete a loose RETICULAR LAMINA. Together, these layers constitute the BASEMENT MEMBRANE of the epithelial cell

FIGURE 8
Location and formation of extracellular matrices in the chick embryo. (A) Extracellular matrix at the junction of the epithelial cells (above) and mesenchymal cells (below). The epithelial cells secrete a basal lamina whereas the mesenchymal cells synthesize a reticular lamina made primarily of collagen. (B) Location of fibronectin-containing extracellular matrices in a 2½-day chick embryo, defined by fluorescent antibodies to fibronectin. The somite acquires its basal lamina after the mesenchymal cells condense and become epithelial cells. (A courtesy of R. L. Trelstad; B courtesy of J. Lash.)

(A)
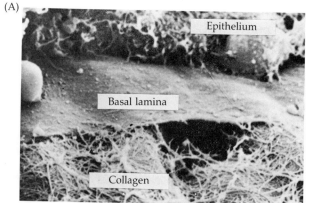

(B)
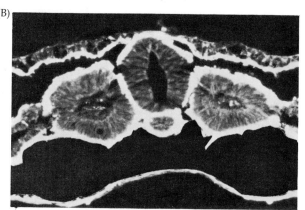

sheet. There are three major components of most extracellular matrices: collagen, proteoglycans, and laminar glycoproteins.

COLLAGEN is a family of glycoproteins containing a large percentage of glycine and proline residues. As we have seen earlier, many of the proline and lysine groups have been posttranslationally modified to hydroxyproline and hydroxylysine. Because collagen is the major structural support of almost every animal organ, it constitutes nearly half the total body protein. Most of this collagen is type I—found in the extracellular matrices of skin, tendons, and bones. This makes up 90 percent of the collagen in our body. The other types of collagen serve special functions (Table 1). Type II collagen is most evident as the secretion of cartilage, but it is also found in the notochord and in the vitreous humor of the eye. Type III collagen is most evident in blood vessels, and types IV and V_{1-3} are found in various basal laminae produced by epithelial cells.

PROTEOGLYCANS are specific types of glycoproteins wherein (1) the weight of the carbohydrate residues far exceeds that of the protein and (2) the carbohydrates are linear chains composed of repeating disaccharides. Usually one of the sugars of the disaccharide has an amino group, so the repeating unit is called a GLYCOSAMINOGLYCAN. Table 2 lists the common glycosaminoglycans; and the basic proteoglycan structure is illustrated in Figure 9. The interconnection of protein and carbohydrate forms a weblike matrix, and in many motile cell types, the

TABLE 1
Collagen types and locations

Type	α-Chain composition	Location in tissues
I	α1(I): 2 chains	Skin, bones, teeth, tendons; forms about 90% of body collagen
II	α1(II): 3 chains	Cartilage, notochord, vitreous body of eye
III	α1(III): 3 chains	Skin, blood vessels, and uterine walls
IV	α1(IV): 2 chains; α2(IV): 1 chain	Basal laminae, kidney glomerulus, lens capsule of eye
V_1	αA: 1 chain; αB: 2 chains	
V_2	αA: 3 chains	Basal laminae
V_3	αB: 3 chains	
VI	Unknown (short triple helical structure with globular regions at both ends)	Placental villi, skin; isolated from uterus
VII	Unknown (long triple helical structure)	May anchor fibrils in basement membranes; isolated from amnion

Source: Mayne (1983).

TABLE 2
Repeating disaccharide units of the most common glycosaminoglycans (GAGs) of matrix proteoglycans

GAG	Repeating disaccharide unit
Hyaluronic acid	Glucuronic acid—N-acetylglucosamine
Chondroitin sulfate	Glucuronic acid—N-acetylgalactosamine sulfate
Dermatan sulfate	[Glucuronic or Iduronic acid]—N-acetylgalactosamine sulfate
Keratan sulfate	Galactose—N-acetylglucosamine sulfate
Heparin sulfate	Glucuronic acid—glucosamine sulfate

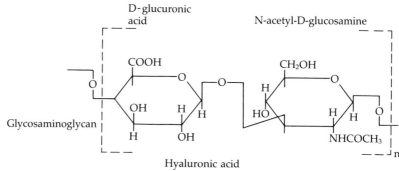

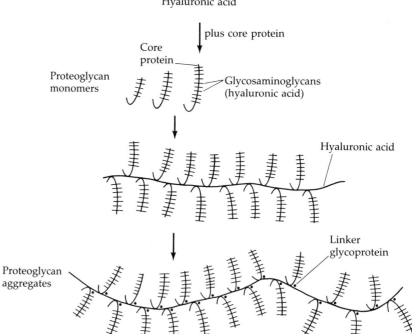

FIGURE 9

The subunit structure and assembly of a complex proteoglycan. The repeating disaccharide of the glycosaminoglycan (in this case hyaluronic acid; see Table 2) attaches to a relatively small core protein to make up the proteoglycan chains. These chains can be linked together by longer glycosaminoglycans (also hyaluronic acid in this case) to produce complex networks. Linker glycoproteins stabilize these latter associations. (Modified from Cheney and Lash, 1981.)

proteoglycan surrounds the cells and is thought to mediate against their coming together (Figure 10). The consistency of the extracellular matrix depends upon the ratio of collagen to proteoglycan. Cartilage, which has a high percentage of proteoglycans, is soft; whereas tendons, which are predominantly collagen fibers, are tough. In basal lamina, the proteoglycans predominate, forming a molecular sieve in addition to providing structural support.

Extracellular matrices also contain a variety of other specialized molecules, e.g., FIBRONECTIN, LAMININ, and ENTACTIN. These large glycoproteins probably are responsible for organizing the collagen, proteoglycan, and cells into an ordered structure. Fibronectin is a very large

FIGURE 10
The proteoglycan coat surrounding mobile cells. (A) Hyaluronidate coat surrounds chick myoblasts. Myoblasts in culture exclude small particles (in this case fixed red blood cells) for a significant distance from the cell border. (B) When the myoblasts are treated with hyaluronidase (which dissolves hyaluronic acid), this extracellular coat vanishes. (C) The coat also vanishes as the myoblasts cease dividing and join together as they differentiate. (D) Electron micrograph of hyaluronidate in aqueous solution shows a branching fibrillar network. (A–C from Orkin et al., 1985, courtesy of B. Toole; D from Hadler et al., 1982, courtesy of N. M. Hadler.)

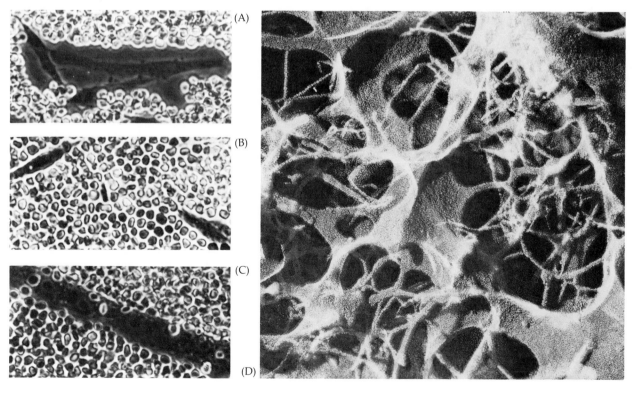

(400,000-d), glycoprotein dimer synthesized by fibroblasts, chondrocytes, endothelial cells, macrophages, and certain epithelial cells such as hepatocytes and amniocytes. This protein is greatly diminished when a cell becomes malignant, leading to its alternative name, the LETS (large, external, transformation-sensitive) protein. One of the functions of fibronectin is to serve as a general adhesive molecule linking cells to various substrates such as collagen and proteoglycans. Fibronectin also organizes the extracellular matrix by having several distinct active binding sites, whose interaction with the appropriate molecules results in the proper alignment of cells with their extracellular matrix (Figure 11).

Fibronectin has also been seen to play an important role in cell migration, as the road over which certain migrating cell types travel are paved with this protein. Continued neural crest cell migration is seen only on fibronectin surfaces, and the movement of these cells ceases when fibronectin is locally removed (Thiery et al., 1982). Similarly, the

FIGURE 11
Model of the structure of fibronectin and its ability to organize the extracellular matrix. Fibronectin is thought to contain a linear sequence of domains that bind to specific extracellular molecules (heparin, hyaluronic acid, collagen) and to the cell membrane itself. Because fibronectin is a dimer, it can link cells to each other or it can link a single cell to a collagenous or proteoglycan substrate. (After Yamada et al., 1980.)

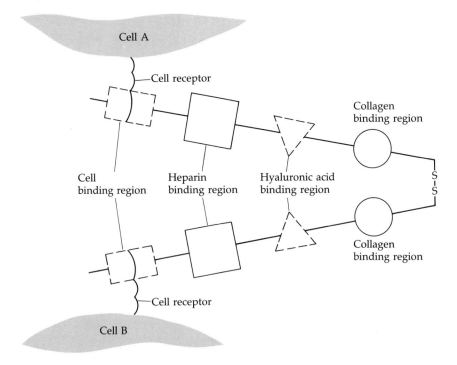

germ cell precursors of frog embryos travel over cells that secrete fibronectin onto their surfaces (Heasman et al., 1981).

Fibronectin may also be a regulator of developmental functions. Mesenchymal cells and myoblasts have substantial amounts of cell surface fibronectin, but this external protein diminishes as the cells differentiate into chondrocytes and myotubes, respectively (Lewis et al., 1978; Chen, 1977; Furcht et al., 1978). When cultured chick cartilage cells are treated with exogenously supplied fibronectin, they bind the molecule and begin to change their shape. In 4 days these cells resemble undifferentiated mesenchymal cells. Their glycosaminoglycan production decreases by 70 percent and they begin to synthesize type I collagen rather than the type II specific for cartilage (Pennypacker et al., 1979; West et al., 1979). The removal of fibronectin from the myoblast cell surface is seen to be necessary for the fusion of these cells into myotubes (Podleski et al., 1979). Not as much is known about laminin or entactin, which are found in all basal laminae. Laminin has been isolated and has been found to bind to cells and to type IV collagen.

Cell surface modifications

The extracellular matrix can be used to keep cells apart or to bring cells together. When cells are connected through the extracellular matrix, usually a basal lamina is formed on only one side. This allows the adjacent cells to interact through the lateral sides of their membranes. There are three major types of cell membrane specializations between cells: the tight junction, the desmosome, and the gap junction.

TIGHT JUNCTIONS are the most intimate of cellular interactions (Figure 12A), and they are used to separate the extracellular space on one side of the tissue from the extracellular space on the other side. This ability to form distinct compartments was seen earlier when we noted the tight junctions forming in the external cells of the mammalian blastocyst.

The DESMOSOME is often seen when cells are joined together to form an impermeable tissue. In the cytoplasms of both adjacent cells, there are local thickenings of fibrillar protein. The space between the cells is 200–350 nm wide and is filled with a glycoprotein cement (Figure 12B). These junctions are also very hard to separate.

GAP JUNCTIONS can be seen between numerous adjacent cells, the proximity of which is between 20 and 40 nm. Through this space, fine connections are seen (Figure 12C). These connections do not serve to separate compartments; rather, they serve as communication channels between the adjacent cells. Cells so linked are said to be "coupled," and small molecules and ions pass from one cell to another. In most embryos, at least some of the early blastomeres are connected by gap junctions, such that ions and small soluble molecules pass freely between them. Again, we saw that the inner cell mass of mammalian embryos is coupled together by gap junctions and that during the

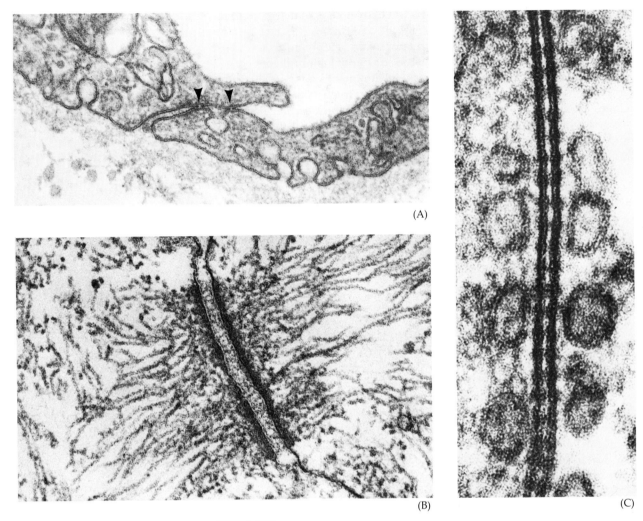

FIGURE 12
Membrane specialization at the junctions of two cells. (A) Tight junctions. Two adjacent endothelial cells of a capillary come together and the membranes fuse (arrows) to form a tight seal. (B) Desmosomes. Electron micrograph of desmosomes on two adjacent cells from newt epidermis. Protein thickening is seen on both cells, and a glycoprotein matrix fills the intercellular space. (C) Gap junctions. Electron micrograph of an area of gap junction between two opposed cells. The beaded appearance is caused by rows of gap junction proteins, which form tunnels between the two cells. (A courtesy of J. E. Rash; B from Kelly, 1966, courtesy of D. E. Kelly; C from Peracchia and Dulhunty, 1976, courtesy of C. Peracchia.)

development of certain molluscs, gap junctions form at the precise time that information needs to be transferred from one blastomere to another.

The importance of gap junctions in development has recently been demonstrated in amphibian embryos (Warner et al., 1984). When antibodies to gap junction proteins were microinjected into one specific cell of an 8-cell *Xenopus* blastula, the progeny of that cell, which are usually coupled through gap junctions, could no longer pass ions or small molecules from cell to cell. Moreover, the tadpoles that resulted from such treated blastulae showed defects specifically relating to the developmental fate of the injected cell. The progeny of such a cell did not die, but they were unable to undergo their normal development. Gap junctions appear to be essential for the communication of developmentally important information between cells.

Locomotion

Morphogenesis in animals (as opposed to organ formation in plants) occurs as a result of directed cell migrations. The movement of such cells may occur individually, as when isolated neural crest cells or mesenchymal cells migrate through the embryo; or the locomotion may be due to the expansion of an epithelial sheet, as when the ectoderm of a gastrulating frog egg epibolizes over the rest of the embryo. The locomotion of individual cells, such as mesenchymal fibroblasts, has been studied in culture. The locomotor organelle of the migrating fibroblast is the LAMELLIPODIUM, a broad, flattened projection that forms the leading edge of the migrating cell (Figure 13A). The lamellipodium makes contact with the substrate at defined points called PLAQUES. As the cell moves across the substrate, these plaques are made and broken. A cell may have from 10 to 100 of these substrate attachment sites.

The attachment sites of the cell are also the anchoring sites for certain cytoskeletal proteins. Actin filaments, capable of generating the contractile forces needed for migration, are found in the lamellipodia and terminate at these plaques. They are held there by α-actinin (Lazarides, 1976; Figure 13B). The addition of cytochalasin B to cultured cells not only destroys the actin cables but also abolishes the lamelli-

FIGURE 13
Cell migration. (A) Phase-contrast micrograph of a fibroblast cell in motion. The thin, ruffled edge (lamellipodium) leads the way, attaching to the matrix and pulling the rest of the cell to it. (B) Indirect immunofluorescence staining the actin microfilaments of a cell extending a ruffled membrane. The actin fibers radiate from the ordered cytoskeletal lattice into the ruffled membrane of the lamellipodium. (A courtesy of N. Wessells; B from Lazarides, 1976, courtesy of E. Lazarides.)

(B)

(A)

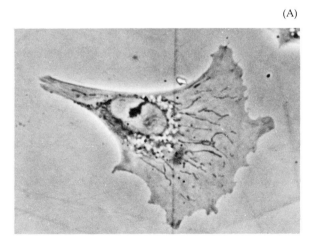
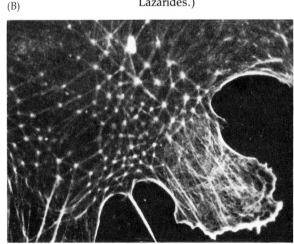

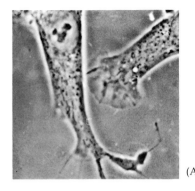

(A)

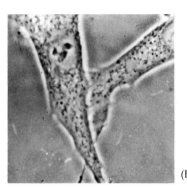

(B)

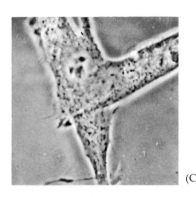

(C)

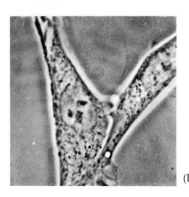

(D)

podia. In a migrating cell sheet, only the cells at the leading edge have lamellipodia. The other cells are joined by tight junctions and get carried along by the actively migrating cells of the margin.

This mechanism of migration provides no explanation for the directionality seen in embryonic cell movements. Directionality can be achieved in any of several ways. CONTACT INHIBITION of movement is the phenomenon whereby the lamellipodium of one cell contacts the membrane of another cell (Figure 14). When this occurs, the lamellipodium is paralyzed and disappears. A new lamellipodium will then appear elsewhere on the cell, thereby taking the cell away from its neighbor (Abercrombie and Ambrose, 1958).

CONTACT GUIDANCE (Weiss, 1945; 1947; Tyler, 1946) probably plays a very important role in cell migrations. The guidance can be provided by physical and chemical cues from the immediate environment of the cell. Cells will move up a gradient of adhesiveness until they stop (Curtis, 1969), and certain substrates may be more adhesive to some types of cells than to others. For instance, when neural crest cells are placed in their normal migratory pathway, they migrate properly. However, when fibroblasts are placed in the same place, they do not migrate (Erickson et al., 1980). Such paths may exist at specific times and places during development. The migration of the mesodermal precursors during amphibian gastrulation appears to be due to both chemical and physical cues that appear at the roof of the blastocoel when gastrulation begins. Here, the ectodermal cells secrete an extracellular matrix that is rich in fibronectin. Boucaut and his colleagues (1984) have demonstrated that this fibronectin is essential for migration; they can block the mesodermal migration by injecting anti-fibronectin antibodies into the blastocoel just prior to involution. However, physical cues may be active as well. Nakatsuji and Johnson (1983) have shown that the fibrils of the extracellular matrix become oriented parallel to the blastopore-animal pole axis as gastrulation begins. Thus, contact guidance by physical and chemical parameters may be of great importance in directing cell movements.

In special cases, migrational cues may also be generated by diffusible substances. Activated lymphocytes, for instance, can secrete CHEMOTACTIC FACTORS which cause other cell types (such as macrophages and white blood cells) to move into a certain area.

Now that the general structure and function of the cell surface has been reviewed, we can see how they interact in specific cell–cell recognition.

FIGURE 14
Contact inhibition. The lamellipodium of one fibroblast approaches (A) and contacts (B) the membrane of another fibroblast. While retaining these contacts, the cell that had the ruffled membrane passes beneath (underlaps) the other cell (C) and then moves away from it (D). (From Erickson, 1978; photographs courtesy of C. A. Erickson.)

Differential cell affinity

Stationary cultures

The modern analysis of organ morphogenesis begins with the experiments of Townes and Holtfreter in 1955. Taking advantage of the discovery that amphibian tissues become dissociated into single cells when placed in alkaline solutions, they prepared single cell suspensions from the three amphibian germ layers and from neurula tissues. Two or more of these single cell suspensions could be combined, and when the pH was normalized, the cells adhered to each other, forming aggregates on agar-coated petri dishes. By using embryos from species having cells of different sizes and colors, Townes and Holtfreter were able to follow the behavior of the recombined cells (Figure 15).

The results of their experiments were striking. First, they found that the reaggregated cells became spatially segregated. That is, instead of remaining mixed, each cell type SORTED OUT into its own region. Thus, when epidermal and mesodermal cells were brought together to form a mixed aggregate, the epidermal cells were found at the periphery of the aggregate and the mesodermal cells were found inside. In no case did the recombined cells remain randomly mixed. In most cases, one tissue type completely enveloped the other.

Second, they found that the final positions of the reaggregated cells reflected their embryonic positions. We have already seen that the mesoderm migrated centrally to the epidermis, adhering to the inner

FIGURE 15
Reaggregation of cells from amphibian neurulae. Presumptive epidermal cells from pigmented embryos and neural plate cells from unpigmented embryos are mixed together. The cells spontaneously reaggregate such that one type (here, the presumptive epidermis) covers the other. (Modified from Townes and Holtfreter, 1955.)

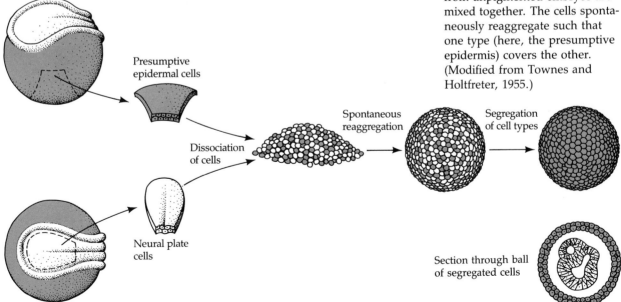

Presumptive
epidermal cells

Dissociation
of cells

Neural plate
cells

Spontaneous
reaggregation

Segregation
of cell types

Section through ball
of segregated cells

epidermal surface (Figure 16A). The mesoderm will also migrate centrally with respect to endoderm (Figure 16B). However, when the three germ layer cells are aggregated together (Figure 16C), the endoderm separates from the ectoderm and mesoderm and is then enveloped by them. In its final formation, the ectoderm is on the periphery, the endoderm is internal, and the mesoderm lies in the region between them. Holtfreter interpreted this in terms of SELECTIVE AFFINITY. The inner surface of the ectoderm had a positive affinity for the mesodermal cells whereas it had a negative affinity for the endoderm. The mesoderm had positive affinities for both the ectodermal and endodermal cells. The mimicry of normal embryonic structure by cell aggregates was also

FIGURE 16
Sorting out and reorganization of embryonic spatial relationships in aggregates of embryonic amphibian cells. (Modified from Townes and Holtfreter, 1955.)

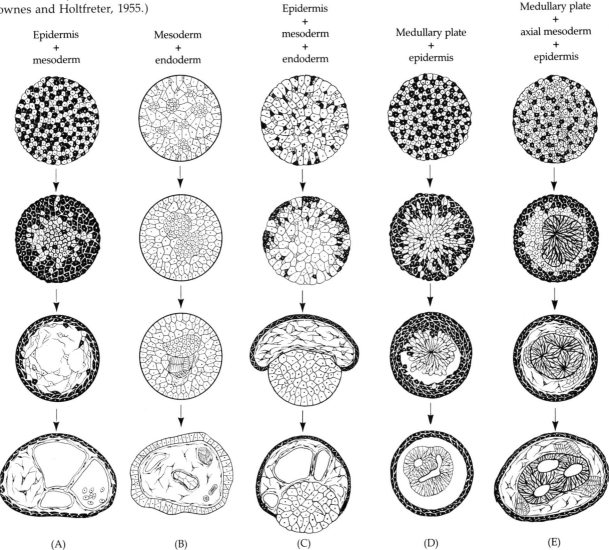

Epidermis
+
mesoderm

Mesoderm
+
endoderm

Epidermis
+
mesoderm
+
endoderm

Medullary plate
+
epidermis

Medullary plate
+
axial mesoderm
+
epidermis

(A)   (B)   (C)   (D)   (E)

seen in the recombination of epidermis and medullary plate cells (Figure 16D). The epidermal cells migrate to the periphery as before, and the medullary plate cells migrate inward, forming a structure reminiscent of the neural tube. When axial mesoderm cells were added to the suspension of epidermal and medullary plate cells, the cell segregation resulted in an external epidermal layer, a centrally located neural tissue, and a layer of mesodermal somites and mesenchyme between them (Figure 16E). Somehow, the cells were able to sort out into their proper embryological positions. Such preferential affinities have also been noted by Boucaut (1974), who injected specific germ layer cells back into blastocoels of amphibian embryos. He found that these cells migrated to their appropriate germ layer. Endodermal cells found positions in the host endoderm, whereas ectodermal cells were found only in host ectoderm. Thus, selective affinity appears to be important for imparting positional information to embryonic cells.

The third conclusion of Holtfreter and his colleagues was that selective affinities change during development. This should be expected because embryologic cells do not retain a single stable relationship with other cells. For development to occur, cells must interact differently with other cell populations at specific times. This change in cellular affinity was dramatically confirmed when Trinkaus (1963) showed a clear correlation between adhesive changes in vitro and changes in embryonic cell behavior. When the discoidal embryos of teleost fishes initiate gastrulation, the cells at the margin of the blastodisc flatten and begin to migrate over the egg cytoplasmic layer. Trinkaus found that when he isolated these cells from the *blastula*, they did not adhere to an artificial substrate, whereas the same cells isolated from *gastrulating* embryos would stick to the surface and migrate. Moreover, when the blastula cells were kept in culture for a duration similar to the time required for their entry into gastrulation, the cells began to flatten and adhere.

Rotary cultures

The reconstruction of aggregates from older embryos of birds and mammals was accomplished and extended by two techniques pioneered by A. A. Moscona. The first technique was the use of trypsin to dissociate the cells from one another (Moscona, 1952). This enzyme cleaves the proteins at the cell surface and between cells, thereby breaking many of the bonds that join the cells together. The second technique was that of rotary aggregation (Moscona, 1961). Instead of permitting the cells to settle on a culture dish (to which they had little or no affinity), Moscona gently swirled the mixed cell suspension in a flask. Here, cells are thrown together in a vortex and will aggregate only if the strength of their mutual adhesion is greater than the shearing force produced by the swirling medium. Within these initially random aggregates, cells move around and sort out according to their cell type. In so doing, they

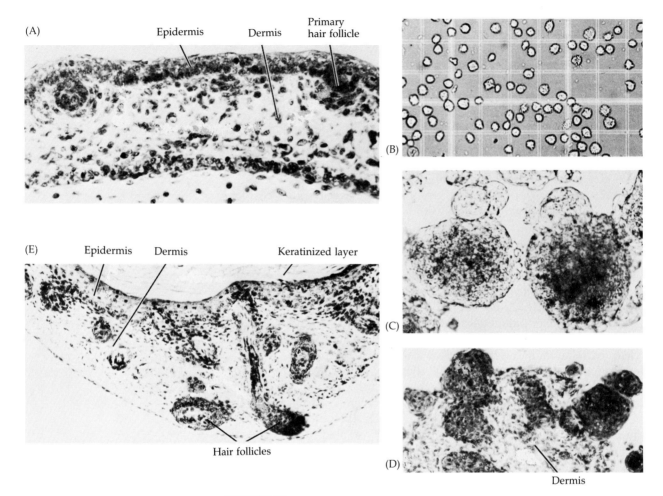

FIGURE 17

Reconstruction of skin from a suspension of skin cells from a 15-day embryonic mouse. (A) Section through the embryonic skin showing epidermis, dermis, and primary hair follicle. (B) Suspension of single skin cells from both the dermis and epidermis. (C) Aggregates after 24 hours. (D) Section through an aggregate, showing migration of epidermal cells to the periphery. (E) Further differentiation of aggregates (72 hours), showing reconstituted epidermis and dermis, complete with hair follicles and keratinized layer. (From Monroy and Moscona, 1979; courtesy of A. Moscona.)

mimic the structure of the original tissue. Figure 17 shows the "reconstruction" of skin tissue from a 15-day embryonic mouse. The skin cells are separated by proteolytic enzymes and are then aggregated in a rotary culture. The epidermal cells migrate to the periphery and the dermal cells migrate toward the center. By 72 hours, the epidermis has been reconstituted, a keratin layer has formed, and hair follicles are seen in the dermal region. The aggregation procedure has enabled the individual cells to reconstruct the tissue. Similarly, cells in a single-cell suspension of embryonic kidney cells will reaggregate to form tubules, and retinal cells will come together to reconstitute the structure of the neural retina. Such reconstruction of complex tissues from single cells is called HISTOTYPIC AGGREGATION. Thus, the cells of the embryonic organs maintain the positional information that enables them to reform the tissue- and organ-specific structures.

The thermodynamic model of sorting out

There are two general models to explain morphogenetic cell rearrangements in vivo and the sorting out of cells in culture. These models are not mutually exclusive as their explanations are at two different levels. The first hypothesis, suggested independently in the mid 1940s by Paul Weiss and Albert Tyler postulated that differential adhesion could be explained by differing sets of complementary cell surface molecules (Weiss, 1945, 1947; Tyler, 1946). These molecules would interact like antibodies and antigens or like sperm and egg cell surface compounds. This model predicts that morphogenesis and sorting out are accomplished by qualitatively different molecules on the cell surfaces of different cell types. Malcolm Steinberg (1964), however, has shown that the presence of such qualitatively different macromolecules is not a logical necessity for sorting out, and he has proposed a model of cell sorting based on thermodynamic principles. Using single cells derived from trypsinized embryonic chick tissues, Steinberg (1970) uncovered a hierarchy of sortings out (Figure 18). If the position of one cell type (A)

FIGURE 18
Commutative hierarchy of sorting out in early chick embryos. The more cohesive cells assume the most central position in mixed aggregates. (A) Pigmented retinal cells sort out centrally to liver cells but are peripherally located (B) when mixed with cartilage cells. (C) When the three cell types are mixed together, the retinal cells form a layer between the central cartilage cells and the peripheral liver cells. (D) The relative positions assumed by five different cell types.

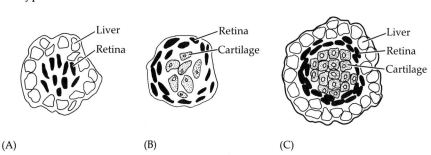

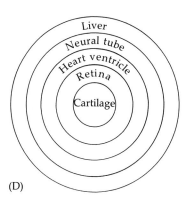

is internal to a second cell type (B) and B is internal to a third cell type (C), then A will always be internal to C. This led Steinberg to propose that the mixed cells interact to form an aggregate with the smallest interfacial free energy (Figure 19). In other words, the cells rearrange themselves into the most thermodynamically stable pattern. If cell types A and B have different strengths of adhesion and if the strength of A–A connections is greater than the strength of A–B or B–B connections, sorting will occur, the A cells becoming central. If the strength of A–A connections is less than or equal to the strength of A–B connections, then the aggregate will remain a random mix of cells. Last, if the strength of A–A connections is far greater than the strength of A–B connections, such that A and B cells show essentially no adhesivity toward each other, then A cells and B cells will form separate aggregates.

All that is needed for sorting out to occur is that cells differ in the strengths of their adherence. In its simplest form, all cells could have the same type of "glue" distributed on the cell surface. The amount of this cell surface product or the cellular architecture that allows the substance to be differentially concentrated causes a different amount of stable contacts to be made between cell types. Thus, different thermodynamic stability is required, not different cell-specific macromolecules. This is called the DIFFERENTIAL ADHESION HYPOTHESIS. In this theory, the early embryo can be viewed as existing in an equilibrium state until some change in gene activity changes the cell surface molecule. The movements that then occur are those that seek to restore the cells to a new equilibrium configuration.

It is very difficult to test whether or not different cell types have different adhesive strengths. However, in Steinberg's laboratory, Herbert Phillips (Phillips and Steinberg, 1969; Phillips et al., 1977) has devised an ingenious way of showing that some tissues are more firmly adherent

FIGURE 19
Sorting out as a process tending toward the maximum thermodynamic stability. (A) Sorting out occurs when the average strength of adhesions between different types of cells (w_{ab}) is less than the average of homotypic (a–a or b–b) adhesive strengths (w_{aa}, w_{bb}). The more adhesive type of cells becomes centrally located. (B) If the strength of the a–b adhesions is greater than or equal to the average of the homotypic adhesions, no sorting will occur, because the system has already achieved a thermodynamic equilibrium, and the mixture of cell types will be random. (C) If the a–b bonds are much weaker than the average of the homotypic adhesions, complete separation will ensue. (This is characteristic of oil and water, for example.)

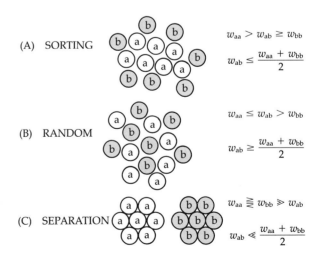

(A) SORTING

$w_{aa} > w_{ab} \geq w_{bb}$

$w_{ab} \leq \dfrac{w_{aa} + w_{bb}}{2}$

(B) RANDOM

$w_{aa} \leq w_{ab} > w_{bb}$

$w_{ab} \geq \dfrac{w_{aa} + w_{bb}}{2}$

(C) SEPARATION

$w_{aa} \gtreqless w_{bb} \gg w_{ab}$

$w_{ab} \ll \dfrac{w_{aa} + w_{bb}}{2}$

Rounded aggregates
of cells

Shape after
centrifugation

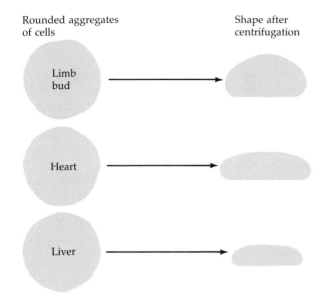

FIGURE 20
Relative strength of intercellular adhesions shown by ability to resist compression. Rounded aggregates of limb bud (mostly cartilage), heart, and liver cells were centrifuged for 48 hours at 8000 times gravity. The amount of deformation indicates how strong the adhesions are between the cells. The same shape was observed when the aggregate was originally flattened and then spun for the same duration, indicating that these are the final equilibrium conditions. (After Phillips and Steinberg, 1969.)

than others. Rounded aggregates of cells were prepared and centrifuged at various speeds until the aggregate shape no longer changed (Figure 20). In this way, they could measure the amount of flattening that occurred for each mass of cells. They found that the amount of flattening correlated well with the cell's rank in the hierarchy of cell sorting. Cartilage tissue (in limb bud), for example, does not flatten under high speeds of centrifugation. At the same speeds, however, heart tissue is somewhat deformed, and liver tissue becomes a flat layer. This corresponds to the central, intermediate, and peripheral positions of the cartilage, heart, and liver tissue in the sorting-out assay. Thus, the better the aggregates resist flattening, the more central that cell's position in sorting out.

Adhesive specificity model of sorting out

Steinberg's thermodynamic model is independent of the actual mechanisms by which stable cell adhesions are formed. Still unanswered, however, is whether cells from different tissues or organs have the same (Steinberg's simplest case) or different "glues." Stephen Roth addressed this problem in an elegant and straightforward way (Roth, 1968; Roth et al., 1971a). He modified the rotary aggregation assay by incubating ^{3}H-labeled cartilage cells and ^{14}C-labeled hepatocytes in a rotating solution containing small aggregates of cartilage cells. By measuring the [14]C- and ^{3}H-labeled cells in these aggregates, he demonstrated that the cartilage aggregates specifically picked up cartilage cells.

SIDELIGHTS & SPECULATIONS

Cell sorting and self-assembly

One of the fundamental principles of living organisms is the self-assembly of subunits into a larger entity. We have seen, for instance, the self-assembly of protocollagen peptides into a collagen triple helix, the self-assembly of globular actin molecules into the actin filaments of the contractile apparatus and the acrosome, and the self-assembly of tubulin molecules to form microtubules. Similarly, some complex structures, such as ribosomes, are assembled by the protein and RNA components of the ribosome itself. The chemical nature of each of the constituents is such that the maximum thermodynamic stability is obtained when the components interact to form the larger entity.

The differential adhesion hypothesis takes this a step further. The populations are seen as mutually immiscible liquids. Like liquids, the cell populations are composed of many mobile individual members that are capable of adhering to each other. Just as when oil and water are mixed, the less cohesive of the two liquids (oil) spreads over the other (water), so the cells will sort out from an initially random aggregate. There is no "special program" specifying each detail of this sorting. Rather, the system sorts out spontaneously by increasing the number of and intensity of adhesions (Steinberg and Poole, 1982). What one has is a self-assembly due to thermodynamic conditions. What causes the embryonic fate to differ from that of the liquid model is that the cells are alive and can change their adhesive properties while moving. These perturbations will cause further displacements which lead to the final tissue structure.

Similar experiments extended this finding to liver and muscle cells as well (Figure 21). Thus, there appear to be cell-specific molecules used in intercellular adhesion.

The next task is to identify those molecules mediating cell adhesion and to discover how they accomplish this feat. Two major types of approaches have been used. The first approach seeks to use antibodies

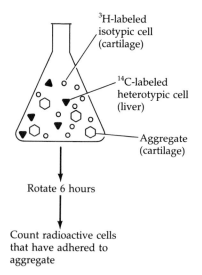

Rotate 6 hours

Count radioactive cells that have adhered to aggregate

FIGURE 21

Specificity of cell–cell attachment. Collecting aggregates, each consisting of one type of cells, are placed in a rotating culture containing single cells of both the same (isotypic) and different (heterotypic) types. The single isotypic and heterotypic cells were previously labeled with different radioactive isotopes. After 6 hours, the aggregates are collected, washed, and counted for both isotypic and heterotypic cells that adhered to the aggregate as shown in the table. (Data from Roth, 1968.)

Count of radioactive cells that have adhered to aggregate

Aggregate type	Labeled single cells in suspension[*]		
	Cartilage	Liver	Pectoral muscle
Cartilage	100	6	48
Liver	10	100	0
Pectoral muscle	38	49	100

[*]Percent of mean number of cells collected by isotopic aggregates

to identify molecules involved in cell adhesion, whereas the second type of approach tries to identify molecules that stimulate cell type-specific aggregation. In the first type of experiment, antibodies that recognize specific differentiation antigens are made. One such investigation using cell type-specific antibodies was performed on embryonic chick neural retina (Brackenbury et al., 1977) and led to the isolation of neuronal cell adhesion molecule (N-CAM).[1] Neural retina cells were injected into rabbits, and rabbit antibodies were made to the neural retina cell surface. Once the antibodies were made and found to bind specifically to neural retina cells (and to some other neural cells), the antibodies were chemically split so that only their monovalent antigen-binding regions, the Fab' fragments, remained (Figure 22). (The divalent antibodies had to be split, for if they remained divalent, they might artificially join the cells together, and the effect could not be measured.) When these Fab' fragments were added to the culture media, they were able to inhibit the aggregation of the chick neural retina cells. Moreover, when isolated neural retina membranes were added to this mixture, they bound the Fab' fragments, thus enabling the retina cells to aggregate. By extracting the components of these membranes, Thiery and his colleagues (1977) found that the Fab' fragments bound to a 140,000-d protein. When this protein was placed into the solution of aggregating retinal cells, it alone inhibited the effect of the Fab' fragments. It can be seen, then, that the 140,000-d protein is a neuron-specific cell surface component with a very important role in aggregation.

In the second approach, factors that inhibit or promote specific aggregation or adhesion are sought. These experiments have provided fascinating clues as to how cell surface molecules interact to aggregate cells. Merrell and Glaser (1973) demonstrated that these factors acted on the cell membranes, because rotary aggregation of neural retina cell suspensions was significantly slowed down by the addition of membranes derived from neural retina cells only. Membranes from other types of cells did not work, nor did neural retina cell membranes from older or younger embryos. It appeared that these neural retina fragments were competing with the neural retina cells for certain adhesion factors.

The first experiments attempting to isolate these molecules showed that isotypic (same cell-type) aggregation can be stimulated by factors secreted from the specific cell types themselves. When 10-day neural retina cells are cultured in petri dishes, the medium from these cells stimulates the aggregation only of neural retina cells (Figure 23). This effect is due to the presence of a 55,000-d glycoprotein secreted by these cells into the medium (Hausman and Moscona, 1975; Lilien and Rutz, 1977). This factor alone, however, is not sufficient to mediate specific

FIGURE 22
Aggregation of neural retinal cells from 10-day chick embryos. (A) Single cells prior to aggregation. (B) Aggregates produced after 30 minutes of rotation. (C) Inhibition of aggregation when retinal cells were rotated 30 minutes in the presence of Fab' antibody fragments to N-CAM. (From Brackenbury et al., 1977; courtesy of G. E. Edelman.)

(A)

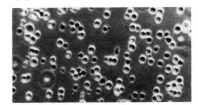

(B)

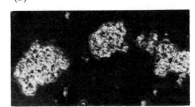

(C)

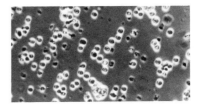

[1]Although the search for cell surface adhesion molecules has been conducted over many vertebrate tissues and in many species, I will be discussing only one example—the embryonic chick neural retina. This tissue has become the focus of much research because it is a large and relatively homogeneous group of cells that is readily accessible during development. My apologies to the fans of slime molds, sponges, and teratocarcinomas.

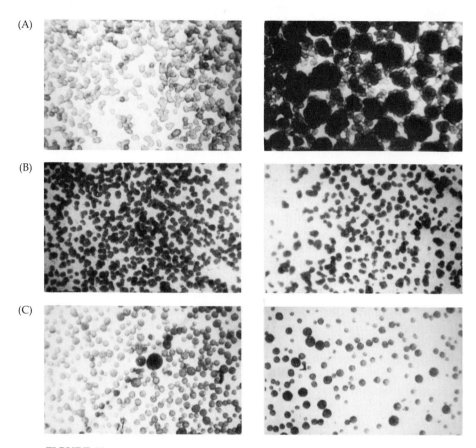

FIGURE 23
Aggregates of cells from the 10-day embryonic chick retina (A), optic tectum (B), and cerebrum (C) after 24 hours of aggregation by rotation. In the right-hand column are aggregates grown in the presence of a protein released into the medium by retinal cells growing in culture. Only the retinal aggregation is enhanced by this factor. (From Hausman and Moscona, 1976; courtesy of A. Moscona.)

aggregation. For example, when protein synthesis was inhibited by the drug cycloheximide, this factor had no effect; the cells do not adhere to each other or form aggregates. Balsamo and Lilien (1975) demonstrated that this first factor can be active if a second component is first covalently bound to the cycloheximide-treated cells. These treated cells agglutin-ated in the presence of medium from neural retina cultures, but they did not agglutinate in the media from other cultures. Balsamo and Lilien (1974) have proposed a three-component model to explain these data. This model has the advantage of explaining how isotypic cells can come together if they have the *same* cell surface components, and not com-plementary ones. As shown in Figure 24, the model requires (1) a cell-specific receptor, (2) a cell-specific aggregation glycoprotein (ligand) and (3) the agglutinin, which can bind the glutaraldehyde-fixed cells.

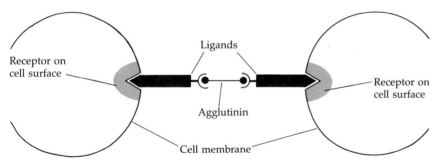

FIGURE 24
One possible three-component model for the formation of the adhesive bond. This model is based on fixed-cell agglutination experiments and is predicated on cells which have the same surface components rather than complementary ones. (After Lilien et al., 1978.)

In some way, these cell surface molecules must regulate the interactions of cells such that they form tissues and organs. One clue is that the 140,000-d N-CAM protein on the membranes of neural retina cells is also necessary in joining nerve axons together to form the nerves of the sympathetic ganglia (Rutishauser, 1978a, b) and may be important in mediating the adhesion of nerves to their target organ (Grumet et al., 1982). Antibodies against this molecule also inhibit the histotypic sorting out of retinal cells into structures resembling retinal tissue (Buskirk et al., 1980).

Two systems of adhesive specificity and their developmental significance

It is obvious from the preceding discussion that adhesion and aggregation are complex phenomena. In the case of the neural retina, we have seen four different factors, all of which appear to be essential for specific adhesion. How do they all fit together? Although we do not know the exact answer, the task has been simplified by the discovery that there are two functionally distinct adhesive systems that coexist on many cell types (Takeichi et al., 1982). Takeichi and his collaborators (1979) were the first to demonstrate the existence of a calcium-dependent system (CDS) and a calcium-independent (CIDS) system of adhesiveness in neural retina cells; and there is general agreement (Grunwald et al., 1980; Thomas et al., 1981; Brackenbury et al., 1981) that this is indeed the case.

This dual system is illustrated in Figure 25. Dissociating cells with the proteolytic enzyme trypsin in the presence of calcium enables the cells to aggregate

SIDELIGHTS & SPECULATIONS

in the presence of calcium, but not in its absence. The CIDS has been removed; the CDS remains. However, when cells are dissociated in trypsin in the presence of a calcium chelator, the cells cannot aggregate, even in the presence of calcium. Thus, there appears to be two different systems of adhesion. The CDS requires Ca^{2+} and physiological temperature for adhesion. The CDS site on the membrane is very sensitive to trypsin, but it is protected against proteolysis by the calcium ions. The CIDS adhesion does not require Ca^{2+} or physiological temperature. Its cell membrane site is more resistant to trypsin digestion. The molecule N-CAM functions in the Ca^{2+}-independent adhesive system whereas those factors described by Hausman and Moscona and by Balsamo and Lilien function in the Ca^{2+}-dependent system.

We do not yet know if there is any developmental significance of these two systems. The CDS is expressed maximally in the early development of the neural retina whereas the CIDS is expressed more from

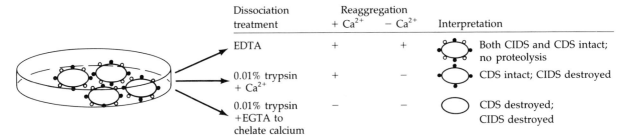

Dissociation treatment	Reaggregation + Ca²⁺	− Ca²⁺	Interpretation
EDTA	+	+	Both CIDS and CDS intact; no proteolysis
0.01% trypsin + Ca²⁺	+	−	CDS intact; CIDS destroyed
0.01% trypsin +EGTA to chelate calcium	−	−	CDS destroyed; CIDS destroyed

FIGURE 25
Dual system of cell–cell adhesion. Calcium-dependent (CDS) and calcium-independent (CIDS) aggregation mechanisms can be separated by breaking the adhesive bonds under different conditions. (Modified from Takeichi et al., 1982.)

later stage retinal cells. Grunwald and his colleagues (1981) have shown that the neural retina cells retaining the CDS can form histotypic aggregates even in the absence of protein synthesis. Those cells having only the CIDS did not. The neural retina contains numerous different types of nerve cells, which are ordered in precise positions within the tissue. It is possible that multiple kinds of CDS with various specificities could cooperate in the sorting out of these different cell types into their appropriate layer of the neural retina. After these had been established, the CIDS would be involved in further organizing the existing pattern. There appears, then, to be excellent evidence for the existence of specific molecules that interact on cell surfaces to promote differential cell adhesions. We must remember, though, that specific molecules do not provide any direction to migration. They, alone, will not account for the hierarchy of cell sorting. It is conceivable that these specific molecules interact to provide various degrees of stability, thereby providing molecular bases for the thermodynamic model of cell sorting.

Differential cell affinity in animal development

Sea urchin gastrulation

In developmental genetics, we saw the paradigm of differential gene expression. In studying organogenesis, we see the phenomenon of differential cell affinity.

The blastomeres of the sea urchin blastula can reaggregate to form blastula-like structures that can then develop into normal pluteus larvae (Giudice, 1962; Spiegel and Spiegel, 1975; Figure 26). All three cell types (micromeres, macromeres, and mesomeres) must be present for this to happen, indicating that the cells did not revert to a dedifferentiated state and then redifferentiate according to their new positions (Spiegel and Spiegel, 1975).

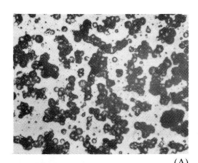

(A)

(B)

FIGURE 26
Reaggregation of blastomeres dissociated from the 16-cell embryo of the sea urchin *Arbacia punctulata*. (A) Dissociated cells. (B) Reconstructed pluteus larva after 25 hours aggregation. (From Spiegel and Spiegel, 1975; courtesy of the authors.)

Sea urchin embryos undergo profound changes of cell affinity during gastrulation, and these changes appear to be correlated with the new genes being expressed at this time. One of the most well known changes in cell affinity involves the adhesion of the micromeres to the hyaline layer of the embryo. At fertilization, hyaline protein is released from the cortical granules of the eggs and forms a layer around the embryo, which lasts until metamorphosis. Hyaline serves as an extracellular substrate, supporting the cells throughout their development (Gustafson and Wolpert, 1967; Dan, 1960; Spiegel and Spiegel, 1979). Prior to gastrulation, cells that will give rise to all three germ layers have an affinity for hyaline (Table 3). However, as gastrulation begins, the primary mesenchymal cells lose their affinity for hyaline and migrate inside the blastocoel. This loss of affinity is specific for hyaline and occurs at the time when these cells normally begin their migration into the embryo (McClay and Fink, 1982). Thus, gastrulation in sea urchin embryos may be initiated by changes in the affinity of the mesenchyme precursor for the extracellular hyaline substrate (see Figure 4 in Chapter 4).

At gastrulation, sea urchin embryos also show changes in the cell surface molecules mediating the adhesion of the embryonic cells to each other. McClay and his colleagues (McClay and Hausman, 1976; McClay et al., 1977) have compared the pregastrulation and postgastrulation cells of hybrid (F_1) embryos to each of the parental species. Before gastrulation, the cells from hybrid embryos adhere to maternal cells much better than they do to paternally derived cells. After gastrulation, the hybrid cells adhere to both parental types equally. Moreover, at gastrulation, hybrid embryos have cell surface antigens characteristic of

TABLE 3
Blastomere affinity for hyaline

Cell	Percentage binding[a]
EARLY BLASTULA	
16-cell micromeres	56 ± 12
All blastomeres	55 ± 12
LATE BLASTULA	
Ectoderm	38 ± 4
Endoderm	32 ± 7
Mesoderm	
Invasive stage	15
Migrating stage	5 ± 3

[a]Cells are allowed to settle to the bottom of a hyaline-coated well (1 hour at 4°C). Wells are inverted and centrifuged to remove unbound cells; unbound cells are counted.

both parental types whereas the pregastrulation cells contained only the maternally derived antigen. Thus, it appears that the onset of gastrulation brings about the expression of new genes, which code for new compounds on the micromere cell surface. These molecules lessen the adhesion of the cells to hyaline, thereby enabling them to migrate inward.[2]

The *reeler* and *staggerer* mutants of the mouse

The principle of differential cell affinity must eventually be studied at the level of the genes that control the cell surface product. Genetics remains one of the most important tools of the developmental biologist. Theodor Boveri taught his students that, no matter how clever they were at manipulating embryos, "the investigator of living processes will make it his special concern to find out abnormalities in which he has not intervened with his crude methods, where he can penetrate the nature of the alteration." To that end, mutations in spatial development are being intensely analyzed. This is dramatically seen in the studies of mouse brain development. The walking behavior of mice is dependent upon the precise organization of the cerebellum. The principal neuron of the cerebellum is the Purkinje cell, the axons of which allow electrical impulses to leave this region of the brain and the correct alignment of which is essential for proper coordination (see Chapter 5). In mice homozygous for the *reeler* mutation (*rl/rl*), the Purkinje cells are misshapen. They are small, have very few dendrites, and are not properly aligned into a uniform layer (Figure 27). DeLong and Sidman (1970) proposed that this disorder resulted from a failure of proper cell–cell recognition and was due to a surface property of the mutant cells. They tested this hypothesis by dissociating *reeler* brain cells, reaggregating them, and comparing the reaggregated structure to aggregates of wild-type brain cells. Wild-type cells from the cerebellar cortex were able to reconstruct the layered histological pattern characteristic of that tissue (Figure 28). In contrast, the aggregates of *reeler* cerebellar cortex did not form the normal histotypic structure. Rather, they produced a pattern reminiscent of the mutant cerebellum. It is likely, then, that the wild-type allele of the *reeler* locus codes for a cell surface component necessary for proper cell–cell recognition. Without this substance, the correct spatial patterning of the cerebellar cortex is lost. Whereas we do not know the molecular defect causing the cell surface changes in *reeler* mice, we do have a clue to the molecular basis of the *staggerer* mutant (Chapter 5). These mice have defective neuronal connections in their

---

[2]This type of experiment cannot be performed on vertebrate embryos; for while the cell surfaces of numerous invertebrates will sort out according to their *species*, vertebrate embryonic cells sort out only with regards to the cell type. This means that when mouse and chick liver cells are aggregated with chick heart cells, the chick heart cells will become segregated from the liver cells. The liver cells do not sort out from each other. The species-specific segregation of invertebrate cells probably reflects its mode of immune protection.

(A) Normal mouse cerebellum

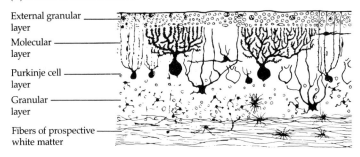

External granular layer
Molecular layer
Purkinje cell layer
Granular layer
Fibers of prospective white matter

(B) "Reeler" mouse cerebellum

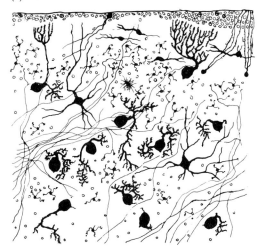

FIGURE 27
Diagram made from histological preparations of normal (A) and *reeler* (B) mouse cerebella 9 days after birth. (From Sidman, 1972.)

cerebellums, leading to a faulty sense of balance. Edelman and Chuong (1982) have found that these mice fail to process the N-CAM molecule as their normal littermates did. Thus, N-CAM—the adhesion molecule isolated by aggregating neural retina cells—is seen to be important in the development of the mammalian brain.

(A)

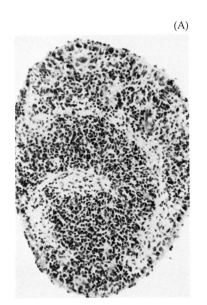

(B)

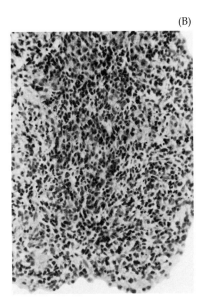

FIGURE 28
Section through the center of an aggregate of normal (A) and *reeler* (B) cerebellar cells dissociated 4 days postnatally and aggregated 4 days in rotating culture. (A) Discrete regions can be seen, including a C-shaped layer of dense neurons and a cell-deficient region containing fibers. (B) No regionalization can be seen. (From DeLong and Sidman, 1970; photography courtesy of G. R. DeLong.)

Cell recognition in vivo: The *T*-complex of the mouse

Chromosome 17 of the mouse carries on it two large gene complexes that are involved in cell–cell recognition. The more familiar of these is the set of *H-2* genes, which control the recognition of "self" in the mouse's immune system. These are the mouse's major transplantation antigens. Whereas the *H-2* genes regulate cell surface characteristics in the adult mouse, the *T*-complex genes on chromosome 17 are thought to control numerous cell–cell recognition events in the developing mouse. *T*-complex products are found only on early embryos and sperm, whereas the *H-2* products are found on late embryonic and adult cells (Artzt and Jacob, 1974; Edidin, 1972).

The *T*-complex in the mouse has profound effects on early mammalian development and sperm maturation. The first *T*-complex mutant studied was the dominant *T* gene *Brachyury* (short tail). When the mutation is present in two doses, the embryos die at the primitive streak stage (Dobrovolskaia-Zavadskaia, 1927). Shortly thereafter, it was discovered that recessive genes within the *T*-complex interacted with *T* to produce tailless mice and that mice homozygous for these recessive alleles died early in development (Dunn, 1937; Gluecksohn-Schoenheimer, 1940).

The effects of the different *t* and *T* alleles occur at different stages of mouse development (Bennett, 1975; Figure 29). The earliest-acting group of *t* mutants is called t^{12}. The wild-type product of this allele is necessary for the compaction of the 8-cell mouse embryo. Thus, in the

FIGURE 29

Time of action of various mutations in the *T* complex in mice. The separation of different tissues in the development of the ectoderm is inhibited by homozygous conditions at these loci. Embryos homozygous for the earliest acting mutation, t^{12}, do not undergo compaction, whereas the neural tubes of embryos homozygous for t^{w1} do not separate from the epidermis. The effect is lethal at that specific stage. (From Klein, 1975.)

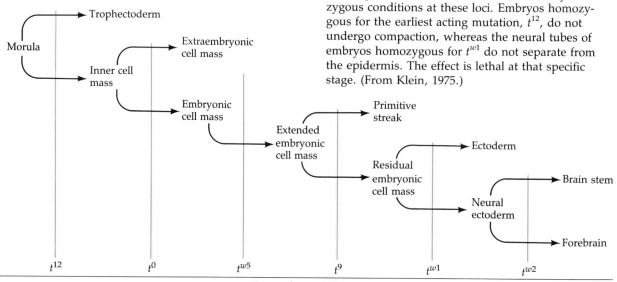

T–complex mutation

t^{12}/t^{12} mutants, the separation of the inner cell mass from the trophectoderm is never accomplished. The t^0/t^0 embryos undergo compaction
but fail to separate the ectoderm into its embryonic and extraembryonic
compartments. The t^{w5}/t^{w5} embryos can separate the ectodermal regions
properly but are unable to support the continued growth of the embryonic ectoderm. Homozygous t^9 embryos can begin the differentiation
of the primitive streak, but the migration of mesoderm cells into the
streak is extremely abnormal. The few cells migrating through the primitive streak soon die, leaving the embryo without a source of mesoderm
(Moser and Gluecksohn-Waelsch, 1967). The next groups of T-complex
loci, represented by t^{w1} and t^{w2} have severe neural tube abnormalities
involving the survival of specific neural cells.

One dose of the dominant T gene leads to the improper development of the tail. At day 11 a constriction is formed roughly half way
into the tail rudiment. Beyond this constriction, the tail fails to develop.
When the T is associated with a recessive allele in a T/t heterozygote,
the constriction occurs at the base of the tail. Embryological abnormalities, however, are seen earlier than day 11. On day 8 or 9, the notochord
is seen to be crooked, misshapen, and often having lumen or diverticula. Because the notochord is responsible for inducing the neural tube,
the neural tube has similar structural abnormalities. The homozygous
T/T mouse has multiple embryonic malformations. In fact, the body
consists solely of an anterior portion, as the structures posterior to the
forelimb rudiments fail to develop. These abnormalities stem from the
failure of the T/T embryos to continue the differentiation of the primitive
streak into the hindpart of the embryo and from the failure of the
notochord to persist in the anterior portion of the embryo. The disappearance of the notochord is not due to the death of its cells; rather,
the notochord loses its integrity as an organ. Speigelman (1976) has
shown that the notochord and somites (the major derivatives of the
primitive streak) fail to establish the proper relationships with one
another. Instead, these mesodermal cells find their way into abnormal
associations with the neural tube and gut cells (Figure 30).

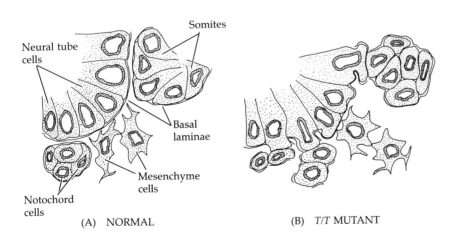

Neural tube
cells

Somites

Basal
laminae

Mesenchyme
cells

Notochord
cells

(A) NORMAL

(B) T/T MUTANT

FIGURE 30
Diagrammatic representation of
the axial regions of a normal and
a T/T mutant mouse at the stage
at which the embryonic defect is
expressed. The normal separation of notochord, neural tube,
and somite is transgressed in the
T/T mutant. (from Spiegelman,
1976.)

FIGURE 31

Measuring adhesive differences between *T/T* and normal embryos. (A) Average diameter of aggregates of axial trunk and head tissue under identical reaggregation conditions. (B) Reaggregation of wild-type and *T/T* embryonic cells in medium conditioned by the growth of other cells. In case 3, the *T/T* cells were able to adhere to each other more readily in the presence of a factor secreted into the medium by wild-type cells. (From Yanagisawa and Fujimoto, 1977, 1978.)

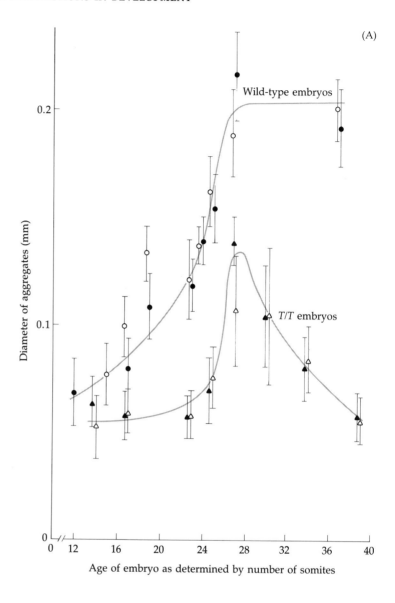

Thus, the *T/T* mice have defects in both primitive streak differentiation and the tissue sorting out of mesodermal tissues. Dissociated cells from *T/T* embryos also show abnormal adhesiveness in rotation-mediated aggregation (Yanagisawa and Fujimoto, 1977). Whereas the diameter of aggregates from wild-type embryos increased as development proceeded, reaching a maximum value at the 27-somite stage (10 days), the diameter of *T/T* aggregates increased only up from the 24-somite (9.5 days) to the 27-somite stage and then declined. It was constantly lower than the wild-type embryos (Figure 31A). Moreover, the wild-type embryos released a factor into the supernatant fluid that would "cure" this defect of the *T/T* embryos (Figure 31B; Yanagisawa and Fujimoto, 1978). It appears, then, that the defect of the *T* allele

(B)

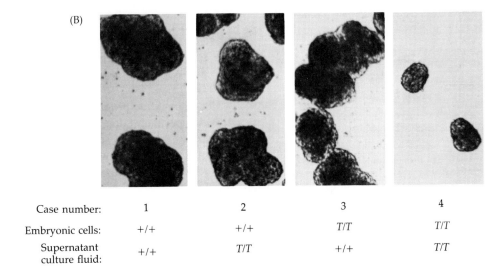

Case number:	1	2	3	4
Embryonic cells:	+/+	+/+	T/T	T/T
Supernatant culture fluid:	+/+	T/T	+/+	T/T

concerns the ability of the cell to recognize its proper position in the embryo. The various mutations at the T-complex all seem to involve cell recognition and morphogenic movement.

In addition to these defects, the T-complex mutants have a curious effect on male fertility (Table 4). When a female mouse heterozygous for a recessive t allele mates with a wild-type male, one-half the offspring are wild-type and one-half are heterozygous, as expected by Mendelian genetics. However, if a male mouse heterozygous for a t allele is mated with a wild-type female, over 90 percent are heterozygous (Dunn, 1960). Male mice that have two complementing t mutations (t^x/t^y) are sterile, having abnormally developed sperm heads (Dooher and Bennett, 1977). It is known that the T locus product(s) are present

TABLE 4
Genetics of the T-complex

A. GENOTYPES AND PHENOTYPES
OF EMBRYOS

Genotype	Phenotype
+/+	Normal tail
T/+	Short tail
T/T	Death of embryo
T/t^x	Tailless
+/t^x	Normal tail
t^x/t^x	Death of embryo

B. RESULTS OF MATINGS[a]

Parental genotypes		Percentage of genotypes in offspring[b]	
♀	♂	t^x/+	+/+
t^x/+ × +/+		50	50
+/+ × t^x/+		95	5

[a]Segregation distortion is seen when heterozygous males are mated with wild-type females but not when wild-type males are mated with heterozygous females.
[b]The exact proportion of genotypes in the offspring depends on the particular mutation.

on sperm as well as in early embryos. What is surprising is that the heterozygous sperm are divided into two sets: one with the mutant gene expressed and one with the wild-type gene expressed. Antibodies against the wild-type product of t^{12}, for instance, will recognize roughly half the sperm of a t^{12}/t^+ male (Yanagisawa et al., 1974). This means that the expression of the T locus genes must occur *after* the sperm precursor is haploid. Shur and Bennett (1979) have shown that t-mutant sperm have much higher levels of galactosyltransferase on their cell surfaces, and Shur (Chapter 2) finds that this specific ectoenzyme binds to the zona pellucida of the egg. We therefore have a complex set of mutations that affect not only the cell–cell interactions of early embryogenesis but also the fundamental interaction of sperm and egg.

SIDELIGHTS & SPECULATIONS

Glycosyltransferases and intercellular recognition

In the mid 1940s, Weiss and Tyler independently proposed that molecules at adjacent cell surfaces interacted with each other in a lock-and-key fashion to provide the spatial information of development. Interactions between specific complementary molecules, they said, would account for the relationships formed by moving cells during organ formation. In 1970, Saul Roseman postulated that the molecules responsible for these interactions were a subset of enzymes called GLYCOSYLTRANSFERASES. These membrane-bound enzymes are routinely found in the endoplasmic reticulum and Golgi vesicles, where they are responsible for adding sugar residues onto peptides to make glycoproteins. There are numerous glycosyltransferases, each one specific for a given sugar and some showing substrate specificity as well. Thus, a galactosyltransferase is an enzyme capable of transferring galactose from an activated donor molecule (UDP-galactose) to an acceptor. There may be several galactosyltransferases with affinities for different acceptor molecules.

Figure 32 shows four possible reactions that might be carried out by cell surface glycosyltransferases. In (B) the glycosyltransferase catalyzes the addition of a sugar molecule to another cell surface molecule. In (C) the glycosyltransferase catalyzes the addition of a sugar onto an exogenous soluble acceptor such as a

glycoprotein. In (D) binding is accomplished between one cell and another cell by means of the mutual recognition of glycosyltransferases and sugar residues. This constitutes adhesion. Adhesions can be broken (as they would need to be in the recognition of migrating cells) by the presence of the activated sugar. Such adhesion and catalysis could also occur between the cell surface glycosyltransferase and the substratum, such as basal lamina (E), over which the cell migrates. As we shall see, experiments have provided evidence that such proteins may be very important for cell migrations in embryos.

Because Golgi vesicles are known to fuse with the cell membrane (Whaley et al., 1972) and because the endoplasmic reticulum is an extension of the cell membrane, it is not hard to envision such molecules on the cell surface. Proving their existence there, however, has been more difficult. The evidence for such molecules on the cell surface comes from the various lines of evidence.

1. The enzymatic activity is associated with whole, intact cells (Pierce et al., 1980).
2. Isolated cell membrane fragments have glycosyltransferase activity (Cummings et al., 1979).
3. An inhibitor of glycosyltransferase activity, UDP-dialdehyde, does not get incorporated into cells. Yet, it can inhibit the glycosyltransferase reaction with intact cells (Cummings et al., 1979).
4. Antibodies against purified glycosyltransferase protein bind to the cell membrane (Shaper and Mann, 1981).

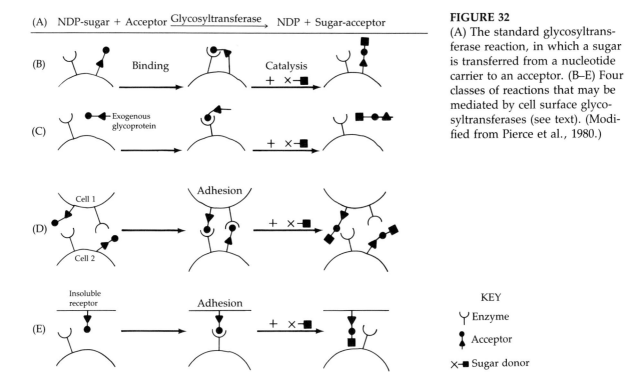

(A) NDP-sugar + Acceptor $\xrightarrow{\text{Glycosyltransferase}}$ NDP + Sugar-acceptor

(B) Binding ... Catalysis + ×-■

(C) Exogenous glycoprotein ... + ×-■

(D) Cell 1 ... Adhesion ... + ×-■ ... Cell 2

(E) Insoluble receptor ... Adhesion ... + ×-■

KEY

Y Enzyme

I Acceptor

×-■ Sugar donor

FIGURE 32

(A) The standard glycosyltransferase reaction, in which a sugar is transferred from a nucleotide carrier to an acceptor. (B–E) Four classes of reactions that may be mediated by cell surface glycosyltransferases (see text). (Modified from Pierce et al., 1980.)

The cell surface glycosyltransferases may play several roles in development. Here, we will consider the possibilities that they govern isotypic cell adhesion and cell migration. Cell surface glycosyltransferase activity exists on the surface on neural retinal cells (Roth et al., 1971b; Porzig, 1978). When neural retina cells are incubated with [³H]UDP galactose, over 90 percent of the label is transferred onto their cell surfaces. It is now thought that the turnover of the ligand molecules that mediate the specific adhesion of neural retina cells is controlled by a cell surface glycosyltransferase (Balsamo and Lilien, 1980; McDonough and Lilien, 1978).

There is increasing evidence that glycosyltransferases mediate those cell movements characteristic of gastrulation and morphogenesis. During these times, the cells migrate upon carbohydrate substrates, especially glycosaminoglycans (Toole, 1976). Shur (1977a,b) demonstrated that the migrating cells of the chick gastrula had intense cell surface glycosyltransferase activity (Figure 33). The migrating primitive streak cells and neural crest cells incorporated several sugars from sugar-nucleotides onto their cell surfaces (shown by autoradiography). Shur has suggested that the migrating cells used its glycosyltransferases as receptors for the substrate oligosaccharides. Adhesion would be

FIGURE 33

Autoradiograph of a 10-somite chick embryo incubated with UDP–[³H]galactose, or sugar donor. Insoluble radioactivity indicates that this radioactive sugar was transferred by the surfaces of migrating cells to other cells or the matrices around them. (From Shur, 1977a; photograph courtesy of B. D. Shur.)

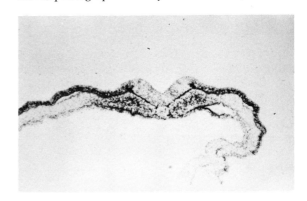

FIGURE 34
Model for migration of embryonic mouse mesenchymal cells. Cell migrates on substrate containing exposed galactose-N-acetylglucosamine residues. Cells recognize and adhere to that substrate by a cell surface galactosyltransferase. Cells move by catalyzing the addition of galactose to those exposed residues, thus dissociating the cell from its substrate and allowing the cell surface enzymes to fill new ungalactosylated sites on the matrix. (From Shur, 1982.)

accomplished by the specific lock-and-key interactions between the substrate and the glycosyltransferase. As the cells moved forward, these bonds are broken by catalysis involving the addition of a new carbohydrate residue (Figure 34).

Two predictions from this model are that (1) the addition of activated sugars should perturb normal development and (2) that migrating cells should leave new sugar groups on the substrates over which they migrate. The addition of large concentrations of sugar nucleotides to developing embryos does interfere with normal development (Shur, 1977a,b; Shur et al., 1979). Although the direct cause of these abnormalities has not been ascertained, the simplest explanation is that the sugar nucleotides interfered with the cell surface glycosyltransferases. Turley and Roth (1979) showed that when cells prelabeled with radioactive galactose were allowed to migrate on glycosaminoglycan surfaces, the cells put the sugar groups onto the substrate. Using different glycosaminoglycans, it was found that the more labeling the cells accomplished, the less migration occurred. The extent of migration, then, may depend on how many exposed carbohydrate groups are on the substrate. A large number of sites may promote strong adhesions, thereby slowing migration.

Lastly, the *T* complex products may be regulators of glycosyltransferases, or they may be the glycosyltransferases themselves. Shur (1982) has shown that migrating mesenchymal cells and the cells of neural crest in *T/T* embryos have up to six times the normal amounts of galactosyltransferase. The sperm of *T* allele mutations also have elevated galactosyltransferase activities (Shur and Bennett, 1979). While it is not known how an increase in galactosyltransferase can lead to the developmental abnormalities seen in the *T/T* embryos, the most likely explanation is that the increases cause the altered cell adhesivity seen in the reaggregation studies and electron microscope pictures.

The cell surface has been shown to be the mediator of positional information during development. In the next two chapters, we shall see how cell surface interactions between adjacent tissues form specific patterns of growth and differentiation.

Literature cited

Abercrombie, M. and Ambrose, E. J. (1958). Interference microscope studies of cell contacts in tissue culture. *Exp. Cell Res.* 15: 322–345.

Anderson, N. G. and Coggin, J. H. (1974). The interrelations between development, retrogenesis, viral transformation, and human cancer. In A. A. Moscona (ed.), *The Cell Surface in Development*. Academic, New York, pp. 297–314.

Artzt, K. and Jacob, F. (1974). Absence of serologically detectable H-2 on primitive teratocarcinoma cells in culture. *Transplantation* 17: 632–634.

Balsamo, J. and Lilien, J. (1974). Functional identification of three components which mediate tissue-type specific embryonic cell adhesion. *Nature* 251: 522–524.

Balsamo, J. and Lilien, J. (1975). The binding of tissue-specific adhesive molecules to the cell-surface. A molecular basis for specificity. *Biochemistry* 14: 167–171.

Balsamo, J. and Lilien, J. (1980). Properties of a β-N-acetylgalactosaminyl transferase and its mobilization from an endogenous pool to cell surface of embryonic chick neural retina cells. *Biochemistry* 19: 2479–2484.

Bennett, D. (1975). The T-locus of the mouse. *Cell* 6: 441–454.

Boucaut, J. C. (1974). Étude autoradiographique de la distribution de cellules embryonnaires isolées, transplantées dans le blastocèle chez *Pleurodeles waltlii* Michah (Amphibien, Urodele). *Ann. Embryol. Morphol.* 7: 7–50.

Boucaut, J. C., Darribère, T., Boulekbache, H. and Thiery, J. P. (1984). Prevention of gastrulation but not neurulation by antibodies to fibronectin in amphibian embryos. *Nature* 307: 364–367.

Boyse, E. A. and Old, L. J. (1969). Some aspects of normal and abnormal cell surface genetics. *Annu. Rev. Genet.* 3: 269–289.

Brackenbury, R., Rutishauser, U. and Edelman, G. M. (1981). Distinct calcium-independent and calcium-dependent adhesion systems of chicken embryo cells. *Proc. Natl. Acad. Sci. USA* 78: 387–391.

Brackenbury, R., Thiery, J.-P., Rutishauser, U. and Edelman, G. M. (1977). Adhesion among neural cells of the chick embryo. I. Immunological assay for molecules involved in cell–cell binding. *J. Biol. Chem.* 252: 6835–6840.

Bretscher, M. S. (1973). Membrane structure: Some general principles. *Science* 181: 622–629.

Buskirk, D. R., Thiery, J.-P., Rutishauser, U. and Edel-man, G. M. (1980). Antibodies to a neural cell adhesion molecule disrupt histogenesis in cultured chick retina. *Nature* 285: 488–489.

Chen, L. B. (1977). Alteration in cell surface LETS protein during myogenesis. *Cell* 10: 393–400.

Cheney, C. M. and Lash, J. W. (1981). Diversification within embryonic chick somites: Differential response to notochord. *Dev. Biol.* 81: 288–298.

Cummings, R. D., Cebula, T. A. and Roth, S. (1979). Characterization of a galactosyltransferase in plasma membrane-enriched fractions from BALB/c 3T12 cells. *J. Biol. Chem.* 254: 1233–1240.

Curtis, A. S. G. (1969). The measurement of cell adhesiveness by an absolute method. *J. Embryol. Exp. Morphol.* 22: 305–325.

Dan, K. (1960). Cytoembryology of echinoderms and amphibia. *Int. Rev. Cytol.* 9: 321–367.

DeLong, G. R. and Sidman, R. L. (1970). Alignment defect of reaggregating cells in cultures of developing brains of reeler mutant mice. *Dev. Biol.* 22: 584–600.

DePetris, S. (1974). Inhibition and reversal of capping by cytochalasin B, vinblastine, and colchicine. *Nature* 250: 54–56.

Dobrovolskaia-Zavadskaia, N. (1927). Sur la mortification spontanée de la queue chez le souris nouveaunée et sur l'éxistence d'un charactère (facteur) héréditaire "non-viable." *C.R. Soc. Biol.* 97: 114–116.

Dooher, G. B. and Bennett, D. (1977). Spermiogenesis and spermatozoa in sterile mice carrying different lethal T/t-locus haplotypes: A transmission and scanning electron microscope study. *Biol. Reprod.* 17: 269–288.

Dunn, L. C. (1937). A third lethal in the T (Brachy) series in the house mouse. *Proc. Natl. Acad. Sci. USA* 23: 474–477.

Dunn, L. C. (1960). Variations in the transmission ratios of alleles through egg and sperm in *Mus musculus*. *Am. Nat.* 94: 385–393.

Edelman, G. M. (1976). Surface modulation in cell recognition and growth. *Science* 192: 218–226.

Edelman, G. M. and Chuong, C. M. (1982). Embryonic to adult conversion of neural cell adhesion molecules in normal and *staggerer* mice. *Proc. Natl. Acad. Sci. USA* 79: 7036–7040.

Edidin, M. (1972). Histocompatibility genes, transplantation antigens, and pregnancy. In B. D. Kahan and R. A. Reisfeld (eds.), *Transplantation Antigens*. Academic, New York, pp. 75–124.

Erickson, C. A. (1978). Analysis of the formation of parallel arrays by BHK cells *in vitro*. *Exp. Cell Res.* 115: 303–315.

Erickson, C. A., Tosney, K. W. and Weston, J. A. (1980). Analysis of migratory behavior of neural crest and fibroblastic cells in embryonic tissues. *Dev. Biol.* 77: 142–156.

Frye, L. D. and Edidin, M. J. (1970). The rapid intermixing of cell surface antigens after the formation of mouse–human heterokaryons. *J. Cell Sci.* 7: 319–335.

Furcht, L. T., Mosher, D. F. and Wendelschafer-Crabb, G. (1978). Immunocytochemical localization of fibronectin (LETS protein) on the surface of L6 myoblasts: Light and electron microscope studies. *Cell* 13: 263–271.

Giudice, G. (1962). Restitution of whole larvae from disaggregated cells of sea urchin embryos. *Dev. Biol.* 5: 402–411.

Gluecksohn-Schoenheimer, S. (1940). The effect of an early lethal (t^0) in the house mouse. *Genetics* 25: 391–400.

Grumet, M., Rutishauser, U. and Edelman, G. M. (1982). Neural cell adhesion molecule is on embryonic muscle cells and mediates adhesion to nerve cells *in vitro*. *Nature* 295: 693–695.

Grunwald, G. B., Geller, R. L. and Lilien, J. (1981). Enzymatic dissection of embryonic cell adhesive mechanisms. *J. Cell Biol.* 85: 766–776.

Gustafson, T. and Wolpert, L. (1967). Cellular movement and contact in sea urchin morphogenesis. *Biol. Rev.* 42: 442–498.

Hadler, N. M., Dourmash, R. R., Nermut, M. V. and Williams, L. D. (1982). Ultrastructure of a hyaluronic acid matrix. *Proc. Natl. Acad. Sci. USA* 79: 307–309.

Hausman, R. E. and Moscona, A. A. (1976). Purification and characterization of the retina-specific cell-aggregating factor. *Proc. Natl. Acad. Sci. USA* 72: 916–920

Heasman, J., Hynes, R. O., Swan, A. P., Thomas, V. and Wylie, C. C. (1981). Primordial germ cells of *Xenopus* embryos: The role of fibronectin in their adhesion during migration. *Cell* 27: 437–447.

Hirokawa, N., Cheney, R. E. and Willard, M. (1983). Location of a protein of the fodrin-spectrin-TW260/240 family in the mouse intestinal brush border. *Cell* 32: 953–965.

Kelly, D. E. (1966). Fine structure of desmosomes, hemidesmosomes, and an adepidermal globular layer in developing newt epidermis. *J. Cell Biol.* 28: 51–72.

Kemler, R., Babinet, C., Eisen, H. and Jacob, F. (1977). Surface antigen in early differentiation. *Proc. Natl. Acad. Sci. USA* 74: 4449–4452.

Klein, J. (1975). *Biology of the Mouse Histocompatibility–2 Complex*. Springer-Verlag, New York.

Kohler, G. and Milstein, C. (1975). Continuous cultures of fused cells secreting antibody of predefined specificity. *Nature* 256: 495–497.

Lazarides, E. (1976). Actin, α-actinin, and tropomyosin interaction in the structural organization of actin filaments in nonmuscle cells. *J. Cell Biol.* 68: 202–219.

Lewis, C. A., Pratt, R. M., Pennypacker, J. P. and Hassell, J. R. (1978). Inhibition of limb chondrogenesis *in vitro* by vitamin A: Alteration in cell surface characteristics. *Dev. Biol.* 64: 31–47.

Liebermann, D. and Sachs, L. (1978). Co-regulation of type C RNA virus production and cell differentiation in myeloid leukemic cells. *Cell* 15: 823–835.

Lilien, J. and Rutz, R. (1977). A multicomponent model for specific cell adhesion. In J. W. Lash and M. M. Burgess (eds.), *Cell and Tissue Interactions*. Raven Press, New York, pp. 187–195.

Lilien, J., Hermolin, J. and Lipke, P. (1978). Molecular interactions in specific cell adhesion. In D. R. Garrod (ed.), *Specificity of Embryological Interactions*. Chapman and Hall, London, pp. 133–155.

Mayne, R. (1983). A summary of the different types of collagen and collagenous peptides. *42nd Annual Symposium of the Society for Developmental Biology*.

McClay, D. R., Chambers, A. F. and Warren, R. H. (1977). Specificity of cell–cell interactions in sea urchin embryos. *Dev. Biol.* 56: 343–355.

McClay, D. R. and Fink, R. D. (1982). Sea urchin hyalin: Appearance and function in development. *Dev. Biol.* 92: 285–293.

McClay, D. R. and Hausman, R. E. (1976). Specificity of cell adhesion: Differences between normal and hybrid sea urchin cells. *Dev. Biol.* 47: 454–460.

McDonough, J. and Lilien, J. (1978). The turnover of a tissue specific cell surface ligand which inhibits lectin induced capping. *J. Supramol. Struct. Cell Biochem.* 7: 409–418.

Merrell, R. and Glaser, L. (1973). Specific recognition of plasma membranes by embryonic cells. *Proc. Natl. Acad. Sci. USA* 70: 2794–2798.

Monroy, A. and Moscona, A. A. (1979). *Introductory Concepts in Developmental Biology*. University of Chicago Press, Chicago.

Mooseker, M. S. and Tilney, L. G. (1975). Organization of an actin filament–membrane complex: Filament polarity and membrane attachment in the microvilli of intestinal epithelial cells. *J. Cell Biol.* 67: 725–743.

Moscona, A. A. (1952). Cell suspension from organ rudiments of chick embryos. *Exp. Cell Res.* 3: 535–539.

Moscona, A. A. (1961) Rotation-mediated histogenetic aggregation of dissociated cells: A quantifiable ap-

proach to cell interaction *in vitro. Exp. Cell Res.* 22: 455–475.

Moser, G. C. and Gluecksohn-Waelsch, S. (1967). Developmental genetics of a recessive allele at the complex T-locus in the mouse. *Dev. Biol.* 16: 564–575.

Nakatsuji, N. and Johnson, K. E. (1983). Conditioning of a culture substratum by the ectodermal layer promotes attachment and oriented locomotion by amphibian gastrula mesodermal cells. *J. Cell Sci.* 59: 43–60.

Nakatsuji, N., Gould, A. C. and Johnson, K. E. (1982). Movement and guidance of migrating mesodermal cells in *Ambystoma maculatum gastrulae. J. Cell Sci.* 56: 207–222.

Nicolson, G. L. (1976). Transmembrane control of the receptors on normal and tumor cell. *Biochim. Biophys. Acta* 457: 57–108.

Orkin, R. W., Knudson, W. and Toole, P. T. (1985). Loss of hyalurondate-dependent coat during myoblast fusion. *Dev. Biol.* 107: 527–530.

Pennypacker, J. P., Hassell, J. R., Yamada, K. M. and Pratt, R. M. (1979). The influence of an adhesive cell surface protein on chondrogenic expression *in vitro. Exp. Cell Res.* 121: 411–415.

Peracchia, C. and Dulhunty, A. F. (1976). Low resistance junctions in crayfish: Structural changes with functional uncoupling. *J. Cell Biol.* 70: 419–439.

Phillips, H. M. and Steinberg, M. S. (1969). Equilibrium measurements of embryonic chick cell adhesiveness. I. Shape equilibrium in centrifugal fields. *Proc. Natl. Acad. Sci. USA* 64: 121–127.

Phillips, H. M., Wiseman, L. L. and Steinberg, M. S. (1977). Self vs. nonself in tissue assembly. *Dev. Biol.* 57: 150–159.

Pierce, M., Turley, E. A. and Roth, S. (1980). Cell surface glycosyltransferase activities. *Int. Rev. Cytol.* 65: 1–47.

Podleski, T. R., Greenberg, I., Schlessinger, J. and Yamada, K. M. (1979). Fibronectin delays the fusion of L6 myoblasts. *Exp. Cell Res.* 122: 317–326.

Porzig, E. F. (1978). Galactosyltransferase activity on intact neural retinal cells from the embryonic chicken. *Dev. Biol.* 67: 114–136.

Roseman, S. (1970). The synthesis of complex carbohydrates by multi-glycosyl transferase systems and their potential in intercellular adhesion. *Chem. Phys. Lipids* 5: 270–297.

Roth, S. (1968). Studies on intercellular adhesive selectivity. *Dev. Biol.* 18: 602–631.

Roth, S., McGuire, E. J. and Roseman, S. (1971a). An assay for intercellular adhesive specificity. *J. Cell Biol.* 51: 525–535.

Roth, S., McGuire, E. J. and Roseman, S. (1971b). Evidence for cell-surface glycosyltransferases: Their potential role in cellular recognition. *J. Cell Biol.* 51: 536–551.

Rutishauser, U., Gall, W. E. and Edelman, G. M. (1978). Adhesion among neural cells of the chick embryo. IV. Role of the cell surface molecule CAM in the formation of neurite bundles in cultures of spinal ganglia. *J. Cell Biol.* 79: 382–393.

Rutishauser, U., Thiery, J.-P., Brackenbury, R. and Edelman, G. M. (1978). Adhesion among neural cells of the chick embryo. III. Relationship of the surface molecule CAM to cell adhesion and the development of histotypic patterns. *J. Cell Biol.* 79: 371–381.

Shaper, J. H. and Mann, P. L. (1981). The demonstration of a cell surface UDP-galactosyltransferase on mammalian cells by indirect immunofluorescence. *J. Supramol. Struct. Cell Biochem. Suppl.* 5: 272.

Shevinsky, L. H., Knowles, B. B., Damjanov, I. and Solter, D. (1982). A stage-specific embryonic antigen defined by monoclonal antibody to murine embryos, expressed on mouse embryos and human teratocarcinoma cells. *Cell* 30: 697–705.

Shur, B. D. (1977a). Cell-surface glycosyltransferases in gastrulating chick embryos. I. Temporally and spatially specific patterns of four endogenous glycosyltransferase activities. *Dev. Biol.* 58: 23–39.

Shur, B. D. (1977b). Cell surface glycosyltransferase in gastrulating chick embryos. II. Biochemical evidence for a surface localization of endogenous glycosyltransferase activities. *Dev. Biol.* 58: 40–55.

Shur, B. D. (1982). Cell surface glycosyltransferase activities during normal and mutant (T/T) mesenchyme migration. *Dev. Biol.* 91: 149–162.

Shur, B. D. and Bennett, D. (1979). A specific defect in galactosyltransferase regulation on sperm bearing mutant alleles of the T/t locus. *Dev. Biol.* 71: 243–259.

Shur, B. D., Oettgen, P. and Bennett, D. (1979). UDPgalactose inhibits blastocyst formation in the mouse: Implications for the mode of action of T/t-complex mutations. *Dev. Biol.* 73: 178–181.

Sidman, R. L. (1972). Cell interactions in the developing mammalian nervous system. In L. G. Silvestri (ed.), *Cell Interactions.* Elsevier North-Holland, Amsterdam, pp. 1–13.

Singer, S. J. and Nicolson, G. L. (1972). The fluid mosaic model of the structure of cell membranes. *Science* 175: 720–731.

Spiegel, E. S. and Spiegel, M. (1979). The hyaline layer is a collagen-containing extracellular matrix in sea urchin embryos and reaggregating cells. *Exp. Cell Res.* 123: 434–441.

Spiegel, M. and Spiegel, E. S. (1975). Reaggregation of dissociated embryonic sea urchin cells. *Am. Zool.* 15: 583–606.

Spiegelman, M. (1976). Electron microscopy of cell association in T-locus mutants. In K. Elliot and M. O'Conner (eds.), *Embryogenesis in Mammals.* Assoc. Scient. Publ., Amsterdam, pp. 119–126.

Staber, F. G., Schlafli, F. and Moroni, C. (1978). Expression of endogenous C-type viral antigen on normal mouse haemopoietic stem cells. *Nature* 275: 669–671.

Steinberg, M. S. (1964). The problem of adhesive selectivity in cellular interactions. In M. Locke (ed.), *Cellular Membranes in Development.* Academic, New York, pp. 321–366.

Steinberg, M. S. (1970). Does differential adhesion govern self-assembly processes in histogenesis? Equilibrium configurations and the emergence of a hierarchy among populations of embryonic cells. *J. Exp. Zool.* 173: 395–434.

Steinberg, M. S. and Poole, T. J. (1982). Cellular adhesive differentials as determinants of morphogenetic movements and organ segregation. In S. Subtelny and P. B. Green (eds.), *Developmental Order: Its Origin and Regulation.* Alan R. Liss, New York, pp. 351–378.

Takeichi, M., Atsumi, T., Yoshida, C. and Ogou, S.-I. (1982). Molecular approaches to cell–cell recognition mechanisms in mammalian embryos. In T. Muramatsu (ed.), *Teratocarcinoma and Embryonic Cell Interactions.* Japan Sci. Soc. Press, Tokyo, pp. 283–293.

Takeichi, M., Ozaki, H. S., Tokunaga, K. and Okada, T. S. (1979). Experimental manipulation of cell surface to affect cellular recognition mechanisms. *Dev. Biol.* 70: 195–205.

Thiery, J.-P., Brackenbury, R., Rutishauser, U. and Edelman, G. M. (1977). Adhesion among neural cells of the chick embryo. II. Purification and characterization of a cell adhesion molecule from neural retina. *J. Biol. Chem.* 252: 6841–6845.

Thiery, J.-P., Duband, J. L. and Delouvée, A. (1982). Pathways and mechanisms of avian trunk neural crest cell migration and localization. *Dev. Biol.* 93: 324–343.

Thomas, W. A., Thomson, J., Magnani, J. L. and Steinberg, M. S. (1981). Two distinct adhesion mechanisms in embryonic neural retina cells. III. Functional specificity. *Dev. Biol.* 81: 379–385.

Toole, B. P. (1976). Morphogenetic role of glycosaminoglycans (acid mucopolysaccharides) in brain and other tissues. In S. H. Barondes (ed.), *Neuronal Recognition.* Plenum, New York, pp. 276–329.

Townes, P. L. and Holtfreter, J. (1955). Directed movements and selective adhesion of embryonic amphib-

ian cells. *J. Exp. Zool.* 128: 53–120.

Trinkaus, J. P. (1963). The cellular basis of *Fundulus* epiboly. Adhesivity of blastula and gastrula cells in culture. *Dev. Biol.* 7: 513–532.

Turley, E. A. and Roth, S. (1979). Spontaneous glycosylation of glycosaminoglycan substrates by adherent fibroblasts. *Cell* 17: 109–115.

Tyler, A. (1946). An auto-antibody concept of cell structure, growth, and differentiation. *Growth* 10 (Symposium 6): 7–19.

Ukena, T. E. and Berlin, R. D. (1972). Effects of colchicine and vinblastine on the topological separation of membrane functions. *J. Exp. Med.* 136: 1–7.

Warner, A. E., Guthrie, S. C. and Gilula, N. B. (1984). Antibodies to gap junctional protein selectively disrupt junctional communication in the early amphibian embryo. *Nature* 311: 127–131.

Weiss, P. (1945). Experiments on cell and axon orientation *in vitro*: The role of colloidal exudates in tissue organization. *J. Exp. Zool.* 100: 353–386.

Weiss, P. (1947). The problem of specificity in growth and development. *Yale J. Biol. Med.* 19: 235–278.

West, C. M., Lanza, R., Rosenbloom, J., Lowe, M., Holtzer, H. and Avdalovic, N. (1979). Fibronectin alters the phenotypic properties of cultured chick embryo chondroblasts. *Cell* 17: 491–501.

Whaley, W. G., Dauwalder, M. and Kephart, J. E. (1972). Golgi apparatus: Influence on cell surfaces. *Science* 175: 596–599.

Yamada, K., Older, K. and Hahn, L.-H. (1980). Cell surface protein and cell interactions. In S. Subtelny and N. K. Wessells (eds.), *The Cell Surface: Mediator of Developmental Processes.* Academic, New York, pp. 43–77.

Yanagisawa, K. O. and Fujimoto, H. (1977). Differences in rotation-meditated aggregation between wild-type and homozygous Brachyury (T) cells. *J. Embryol. Exp. Morphol.* 40: 277–283.

Yanagisawa, K. O. and Fujimoto, H. (1978). Aggregation of homozygous Brachyury (T) cells in culture supernatant of wild-type or mutant embryos. *Exp. Cell Res.* 115: 431–435.

Yanagisawa, L., Pollard, D., Bennett, D., Dunn, L. C. and Boyse, E. (1974). Transmission ratio distortion at the T-locus. I. Serological identification of two sperm populations in t-heterozygotes. *Immunogenetics* 1: 91–96.

Yelton, D. E. and Scharff, M. D. (1980). Monoclonal antibodies. *Am. Sci.* 68: 510–516.

Yoshikawa, Y., Ignjatovic, J. and Bauer, H. (1979). Tissue-specific expression of onco-fetal antigens during embryogenesis. *Differentiation* 15: 41–47.

CHAPTER

Proximate tissue interactions: secondary induction

The aspiration to truth is more precious than its assured possession.
—G. E. LESSING (1778)

The problem of (hereditary) transmission might be merged in the broader problem of form through chemical processes—the central problem of development.
—E. S. RUSSELL (1916)

515

Introduction

In the previous chapter, we saw that the cell surface plays an important role in providing spatial information to adjacent cells. In this chapter and the next, we shall discuss the ways in which adjacent sets of cells influence the behavior of other nearby cell populations in order to form tissues and organs. Organs are complex structures composed of numerous types of tissues. If we consider an organ such as the eye, for example, we find that light is transmitted through transparent corneal tissue; it is focused by the lens tissue, the diameter of which is controlled by muscle tissue; and it eventually impinges upon the tissue of the neural retina. The precise arrangement of tissues in this organ cannot be disturbed without damaging its function. Such coordination in the construction of organs is accomplished by one group of cells changing the behavior of an adjacent set of cells, thereby causing them to change their shape, mitotic rate, or differentiation.

This action-at-close-range, or PROXIMATE INTERACTION, enables one group of cells to respond to a second group of cells, and in changing, to become able to alter a third set of cells. This phenomenon has been called SECONDARY INDUCTION.

Instructive and permissive interactions

Howard Holtzer (1968) has distinguished two major modes of proximate tissue interactions. One of them is the INSTRUCTIVE INTERACTION. For example, if one takes the optic vesicle of the embryonic eye and places it adjacent to a part of the head ectoderm that would, in the course of normal development, have formed epidermis, then that region of ectoderm will form a lens instead of skin (McKeehan, 1951). The responding ectoderm cells have somehow been told to express a set of genes different from the set they would have expressed had they not been in contact with the optic vesicle. The optic vesicle is said to be an inducing tissue acting *instructively*.

This instructive interaction is seen in normal development as well as in controlled experiments. Lens tissue develops from the neurula ectoderm only when an optic vesicle is near it. Moreover, when an optic vesicle is removed before it contacts the ectoderm, no lens is formed at the normal site; rather, the ectoderm differentiates into skin (Figure 1). This reaction is specific for the optic vesicle, as no other tissue will cause ectoderm to form lens. Using this example, we can state four general principles of instructive tissue interactions (Wessells, 1977).

1. In the presence of tissue A, responding tissue B develops in a certain way.

2. In the absence of tissue A, responding tissue B does not develop in that way.
3. In the absence of tissue A, but in the presence of tissue C, tissue B does not develop in that way.
4. In the presence of tissue A, a tissue D, which would normally develop differentially, is changed to develop like B. (Thus, other head ectoderm is changed into lens by the optic vesicle.)

It should be noted, however, that in the fourth principle, the responding tissue must be COMPETENT to respond. In many species, if the optic vesicle were placed adjacent to *flank* ectoderm (rather than head ectoderm), a lens would not develop. In other species, however, flank ectoderm is competent to form lens. For induction to occur, the inducer needs to transmit information and the responding tissue must be competent to receive and use the new signals.

The other type of proximate tissue interaction is called the PERMISSIVE INTERACTION. In such cases, development proceeds only if another tissue is present, but the information is not specific. Cell division is one such example. Many embryonic epithelial cells have a defined rate of mitosis. If these cells are separated from their embryonic mesoderm, their division will stop. Whatever development was predicated upon the growth of the epithelial cells will then cease. If these cells are

FIGURE 1
Cross section through the head of an idealized vertebrate embryo, showing the experimental techniques used to demonstrate an instructive interaction. The interaction illustrated is the induction of the lens from head ectoderm by the optic vesicle. (1) Normal induction. (2) In the absence of the optic vesicle, no lens forms. (3) Substitution of another tissue for the optic vesicle gives no induction. (4) Placement of the optic vesicle adjacent to another region of head ectoderm induces lens in a different place. (Modified from Wessells, 1977.)

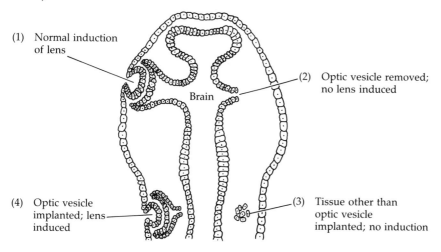

(1) Normal induction of lens

Brain

(2) Optic vesicle removed; no lens induced

(4) Optic vesicle implanted; lens induced

(3) Tissue other than optic vesicle implanted; no induction

recombined with their mesoderm, cell division will start again. Many embryonic tissues that can support the mitosis of a given epithelium can be found. Thus, there is no specific information being passed, only the maintenance of cell division. So even though the presence of mesodermal cells is essential for the differentiation of the epithelium, the mesoderm is permitting, not instructing, that development.

Intercellular coordination in proximate interactions

Proximate cell interactions provide a mechanism whereby coordinated organ development can occur, for a responding tissue can also become an inducing tissue. In the cascade of inductions that forms the eye (Figure 20 in Chapter 7), the notochord induces the neural tube; and a bulge from the neural tube—the optic vesicle—contacts the head ectoderm to instruct it to become a lens. The lens vesicle formed by their induction then induces the formation of a cornea in the newly positioned ectoderm immediately above it.

In many species, the lens can only form if the optic vesicle contacts the ectoderm (Harrison, 1920; Spemann, 1938). In other species, including several amphibians, the optic vesicle is not the only tissue capable of inducing a lens. A. G. Jacobson (1966) has linked amphibian lens induction with the various tissues that the head ectoderm contacts during its development. At gastrulation, the presumptive lens ectoderm first overlies the endodermal tissue of the developing pharynx and then travels over a region of heart mesodermal cells. Only after neurulation does the optic vesicle contact the presumptive lens cells. Jacobson has shown that the endoderm and heart mesoderm can also act as lens inducers (Figure 2). When the optic vesicle was removed, these two tissues, acting together, could induce lenses in 42 percent of the trials. Even in those species where the optic vesicle is the sole inducer of lens development, it receives assistance from other tissues.

The mesenchyme cells of the optic region also play an important role in this induction, and these cells are seen to become associated with the ectoderm being induced by the optic vesicle (Liedke, 1951, 1955). Following the separation technique of Grobstein (1955), V. R. Muthukkaruppan (1965) cultured presumptive lens ectoderm on one side of a porous filter and cultured optic vesicle inducer tissue on the other side. No lens induction was seen. When presumptive lens ectoderm was cultured with mesenchymal cells, no lens was formed either. However, when the presumptive lens tissue was cultured in the presence of mesenchymal cells and opposite optic vesicle tissue, a dramatically normal-looking lens was induced (Figure 3). Only optic vesicle tissue and its derivatives were capable of inducing the lens, although any mesenchyme seemed to work. The amount of mesenchyme was found to be crucial. Too much mesenchyme gave huge placodes and correspondingly large lenses. Here, then, the lens induction appears to be a well-coordinated process having at least two steps: first, the inter-

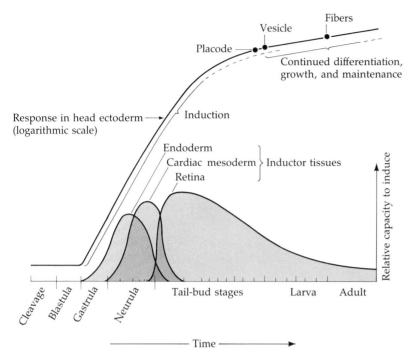

FIGURE 2

Graphic representation of lens induction in an embryonic salamander. The ability to induce lens tissue is first seen in the endoderm, then in cardiac mesoderm, and finally in the retinal tissue, which acquires this ability shortly before contacting the head ectoderm. Meanwhile, the competency of the head ectoderm to respond to inducers increases logarithmically from early gastrula through the tailbud larval stages. (Modified from Jacobson, 1966.)

action of head mesenchyme tissue with the presumptive lens ectoderm (at early neurula stage) to form lens vesicles; and, second, the interaction between the ectodermal vesicles with the optic vesicle to induce the lens.

Once the lens placode has invaginated, it becomes covered by two cell layers from the adjacent ectoderm. Now the developing lens can act as an inducer. Under the influence of this tissue, the overlying ectoderm becomes columnar and fills with secretory granules. These granules migrate to the bases of the cells and secrete a PRIMARY STROMA containing about 20 layers of types I and II collagen (Figure 4). Neighboring capillary endothelial cells migrate into this region (upon the primary stroma) and secrete hyaluronic acid into this matrix. This causes the matrix to swell and to become a good substrate for the migration of two waves of mesenchymal cells derived from the neural crest. Upon entering the matrix, the second wave of mesenchymal cells remains there, secreting type I collagen and hyaluronidase. This causes the stroma to shrink. Under the influence of thyroxine from the developing

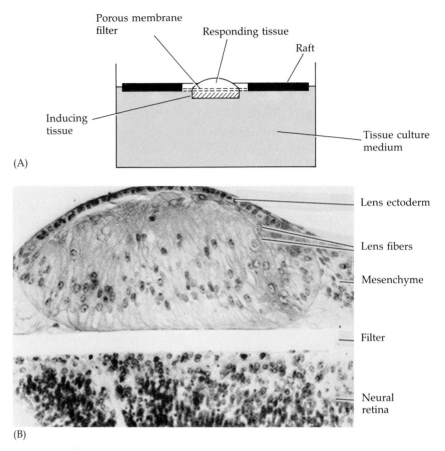

(A)

(B)

FIGURE 3

In vitro induction of lens from head ectoderm. (A) The Grobstein procedure for observing induction in vitro. Responding tissue is placed on one side of a porous membrane and the inducing tissue is placed beneath it. The membrane and the two tissues are floated in a raft in a plate of tissue culture medium. (B) Presumptive mouse lens ectoderm and mesenchyme (the source of which was not critical) were placed on a filter; neural retina tissue was placed beneath it. After 3 days, a complete lens had developed. (From Muthukkaruppan, 1965; photograph courtesy of R. Auerbach.)

thyroid gland, this SECONDARY STROMA is dehydrated and the collagen-rich matrix of epithelial and mesenchymal tissues becomes the transparent CORNEA.

We can see, then, that "simple" inductive interactions are actually well-coordinated dramas in which the actors must come on stage and speak their lines at the correct times and positions. In acquiring new information, they can also impart information for others to use. With this in mind, we can now study some principles about secondary induction obtained from other developing organs.

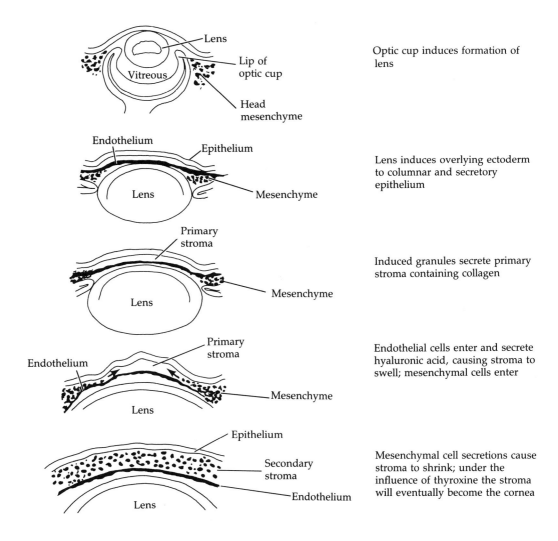

Optic cup induces formation of lens

Lens induces overlying ectoderm to columnar and secretory epithelium

Induced granules secrete primary stroma containing collagen

Endothelial cells enter and secrete hyaluronic acid, causing stroma to swell; mesenchymal cells enter

Mesenchymal cell secretions cause stroma to shrink; under the influence of thyroxine the stroma will eventually become the cornea

FIGURE 4
Corneal development. Under the inductive influence of the lens, the corneal epithelium differentiates and secretes a primary stroma consisting of collagen layers. Endothelial cells then secrete hyaluronic acid into this region, enabling mesenchymal cells from the neural crest to enter. Afterwards, hyaluronidase (secreted either by the mesenchyme or endothelium) digests the hyaluronic acid, causing the primary stroma to shrink. (After Hay and Revel, 1969.)

Epithelio-mesenchymal interactions

Some of the best-studied cases of secondary induction are those involving the interactions of epithelial sheets with adjacent mesenchymal cells. These are called EPITHELIO-MESENCHYMAL INTERACTIONS. The epithe-

TABLE 1
Some epithelio-mesenchymal interactions

Organ	Epithelial component	Mesenchymal component
Cutaneous structures (hair, feathers, sweat glands, mammary glands, teeth)	Epidermis (ectoderm)	Dermis (mesoderm)
Limb	Epidermis (ectoderm)	Mesenchyme (mesoderm)
Gut organs (liver, pancreas, salivary glands)	Epithelium (endoderm)	Mesenchyme (mesoderm)
Pharyngeal and respiratory associated organs (lungs, thymus, thyroid)	Epithelium (endoderm)	Mesenchyme (mesoderm)
Kidney	Ureteric bud epithelium (mesoderm)	Mesenchyme (mesoderm)

lium can come from any germ layer whereas the mesenchyme is usually derived from loose mesodermal tissue. Examples of epithelio-mesenchymal interactions are listed in Table 1.

Regional specificity of induction

Using the induction of cutaneous structures as our examples, we will look at the properties of the epithelio-mesenchymal interactions. The first phenomenon is that of the regional specificity of induction. Skin is composed of two main tissues: an outer epidermis derived from ectoderm and a dermis derived from mesoderm. Chicken skin gives rise to three major cutaneous structures that are made almost entirely of ectodermal cells. These are the broad wing feathers, the narrow thigh feathers, and the scales and claws of the feet. After separating the embryonic epithelium and mesenchyme from each other, one can recombine them in different ways (Saunders et al., 1957). Some of the recombinations are illustrated in Figure 5. As you can see, the mesenchyme is responsible for the specificity of induction in the competent ectoderm. The same type of ectoderm develops according to the region from which the mesoderm was taken. Here, the mesenchyme has an instructive role, calling into play different sets of genes in the responding cells.

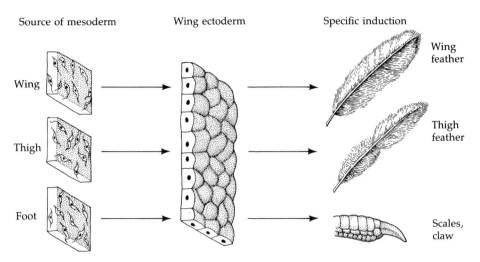

FIGURE 5
Regional specificity of induction. When cells of the dermis (mesoderm) are recombined with the epidermis (ectoderm) in the chick, the type of cutaneous structure made by the ectoderm is determined by the original location of the mesoderm. (Adapted from Saunders, 1980.)

Genetic specificity of induction

Whereas the mesenchyme may instruct the epithelium as to what sets of genes to activate, the responding ectoderm can comply with this information only so far as its genome permits. In a classic experiment, Hans Spemann and Oscar Schotté (1932) transplanted flank ectoderm from an early frog gastrula to the region of a newt gastrula destined to become parts of the mouth. Similarly, the presumptive flank ectodermal tissue of newt gastrula was placed into the presumptive oral regions of frog embryos. The structures of the mouth region differ greatly between these salamander larvae and the frog larvae. The *Triturus* salamander larva has club-shaped balancers beneath its mouth whereas the frog tadpoles produce mucus-secreting glands or suckers (Figure 6). The frog tadpoles also have a horny jaw without teeth whereas the salamander has a set of calcareous teeth in its jaw. The larvae resulting from the transplants were chimeras. The salamander larvae had froglike mouths and the frog tadpoles had salamander teeth and balancers. In other words, the mesodermal cells instructed the ectoderm to make a mouth, but the ectoderm responded by making the only mouth it "knew" how to make, no matter how inappropriate.

The same genetic specificity is seen in combinations of chicken skin and mouse skin (Coulombre and Coulombre, 1971). When ectoderm normally destined to become cornea is isolated from chicken embryos and combined with chick skin mesoderm, the ectoderm produces

FIGURE 6
Genetic specificity of induction. Reciprocal transplantation between the presumptive oral ectoderm regions of newt and frog gastrulae leads to newt larvae with tadpole suckers and to frog tadpoles with newt balancers. (From Hamburgh, 1970.)

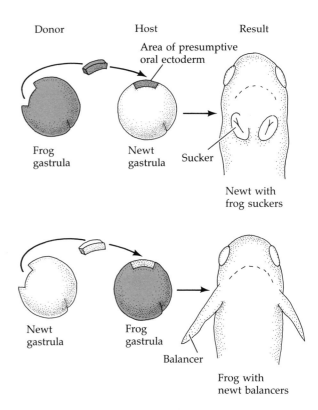

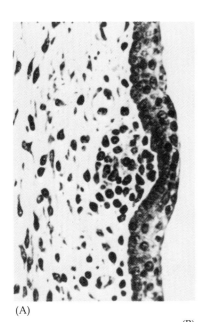

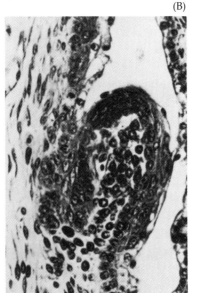

(A)

(B)

feather buds typical of chick skin. Moreover, when the same tissue—presumptive cornea ectoderm—is combined with *mouse* skin mesoderm, feather buds also appear (Figure 7). The mouse mesoderm has instructed the chick cornea to make a cutaneous structure. This would normally be hair, for the mouse. The competent chick embryo, however, does the best it can—developing its cutaneous structures, namely, feathers.

Thus, the instructions sent by the mesenchymal tissue can cross species barriers. Salamanders respond to frog signals, and chick tissue responds to mammalian inducers. The response of the epithelium, however, is species-specific for that epithelium. So whereas organ-type specificity (feather or claw) is usually controlled by the mesenchyme within a species, species specificity is usually controlled by the responding epithelium.

FIGURE 7
Genetic specificity of cutaneous induction. (A) Section of the corneal region of a 17-day chick embryo. At 5 days of incubation, the lens of this eye had been replaced by the flank dermis of an early mouse embryo. A condensation of the mouse embryo cells is located directly beneath the chick epithelium. (B) Feather forming from the corneal epithelium from such a specimen. Mouse cells are present in the feather rudiment. (From Coulombre and Coulombre, 1971; courtesy of A. J. Coulombre.)

FIGURE 8
"Hen's tooth" formed after combination of chick pharyngeal (presumptive jaw) epithelium and mouse molar mesenchyme. (From Kollar and Fisher, 1980; courtesy of E. J. Kollar.)

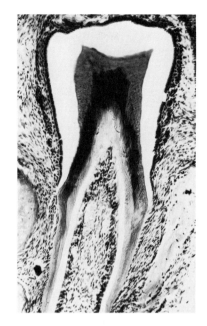

This observation has more recently been used to generate a long-lost structure—hen's teeth (Kollar and Fisher, 1980). Epithelium from the jaw-forming region of 5-day-old chick embryos (the first and second pharyngeal arches) was isolated and combined with molar mesenchyme of 16- to 18-day-old mouse embryos. These recombined tissues were allowed to adhere to each other and were then cultured within the anterior chamber of a mouse eye. Several recombinations resulted in the formation of complete teeth that were unlike those teeth of mammals (Figure 8). The cells of the chick pharyngeal arches, which have not made teeth for nearly 10 million years, still retained the genetic potential to do so in response to an appropriate inducer.

Development and macroevolution

SIDELIGHTS & SPECULATIONS

Evolution does not produce novelties out of nothing. Rather, it involves hereditary changes in the development of existing organisms. Evolution selects changes in what already exists, often transforming or redirecting a structure to give it a new function. The biochemical changes from species to species do not account for their divergences; and whereas small changes in those genes coding for proteins can give an organism a selective advantage, it does not make a bird from a reptile or a human from a primitive ape. In fact, the average human polypeptide is more than 99 percent identical to its chimpanzee counterpart (King and Wilson, 1975).

It has been postulated (Goldschmidt, 1933; Jacob, 1977) that large evolutionary changes result from mutations in regulatory genes. Small changes in the number of cell divisions or their spacing can alter the time at which an organ develops, the structure of that organ, or its position in the embryo. The ability of a responding cell to produce a certain protein when signaled by the mesoderm could likewise change the nature of development (Ohno, 1970). Gene rearrangements—the placing of genes into new combinations so

that they are expressed together—may be essential in this respect. If this were the case, one might expect to find developmental mutants wherein a member of a species resembles in some way its evolutionary forebears. Such mutations have been found. In most birds, all the digits are incorporated into the bones of the wing and there are no claws present on the ends of the wing. *Archeopterix* (the first known bird), however, retained a three-digit reptilian claw on its wing tip. In the domestic chicken, the simple, Mendelian dominant mutation *Ametopodia* will cause a reversion to the clawed wing tip (Cole, 1967). In the next chapter, we will discuss regulatory mechanisms in flies which can alter the pattern of segmentation to resemble that of other arthropods.

In echinoderms and starfish, the five-fold symmetry of their organs and "arms" develops from the interaction of the hydrocoel with the overlying epidermis (Czihak, 1971). The hydrocoel—the side of the body cavity that develops into the water–vascular system—puts forth five protrusions, each of which induces the formation of an arm (or, in sea urchins, an

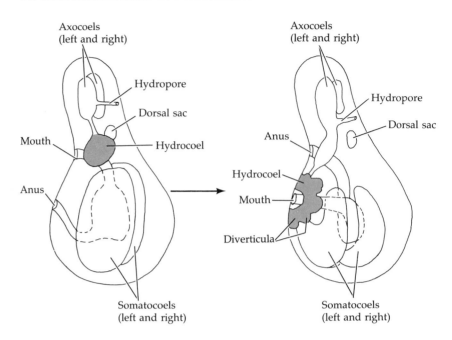

FIGURE 9

Development of the hydrocoel (shaded) in a generalized echinoderm larva. The number of "arms" in the adult organism depends upon how many diverticula form in the hydrocoel (in this case, 5). (Modified from Ubaghs, 1967.)

ambulacral area) (Figure 9). Whereas sea urchins have retained this five-fold symmetry from species to species, starfish have been more adventuresome. The hydrocoel of the six-armed starfish *Leptasterias hexactis* puts forth five protrusions and then, slightly later, a sixth. Similarly, nine protrusions from the hydrocoel create the geometry of the nine-armed starfish species (Gemmill, 1912). Thus, differences in induction can create differences between species or even between classes of animals.

Formation of parenchymal organs

Epithelio-mesenchymal interactions are also seen in the formation of duct-forming organs such as the kidney, liver, lung, mammary gland, and pancreas.

Kidney and lung. The kidney develops from two mesodermal components (Figure 10). The first component is the epithelium of the URETERIC BUD, coming from the Wolffian duct. When this bud enters a region of METANEPHROGENIC (kidney-forming) MESENCHYME, the two tissues interact and change one another. The metanephrogenic mesenchyme causes the ureteric bud to elongate and branch. These branches then cause the mesenchyme to condense at their tips, eventually forming an S-shaped cord that hollows out to form a tube. A connection develops between the ureteric bud and the newly-formed tube such that material can pass from one to another. The tubes at the end of the

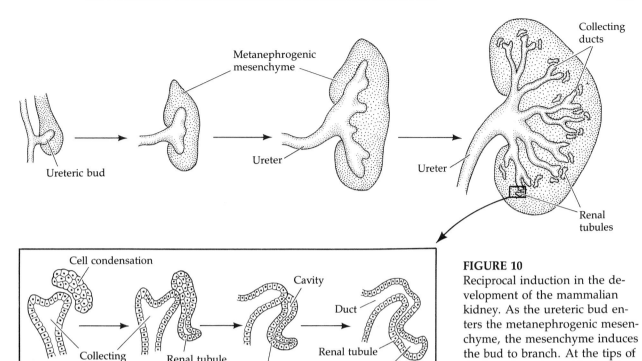

FIGURE 10
Reciprocal induction in the development of the mammalian kidney. As the ureteric bud enters the metanephrogenic mesenchyme, the mesenchyme induces the bud to branch. At the tips of the branches, the epithelium induces the mesenchyme to aggregate and cavitate to form the renal tubules.

ureteric bud form the nephron of the functioning kidney, and the branched ureteric bud gives rise to the renal collecting ducts and to the ureter, which drains the urine from the kidney.

Clifford Grobstein (1955, 1956) has shown that there is a RECIPROCAL INDUCTION occurring during kidney development. He separated the ureteric bud from the mesenchyme and cultured them separately. In the absence of mesenchyme, the ureteric bud will not branch. In the absence of the ureteric bud, the mesenchyme will not condense to form tubules. When they are placed together, however, the ureteric bud grows and branches, and tubules form throughout the mesenchyme (Figure 11). The epithelium induces the synthesis of new proteins in the mesenchyme. Ekblöm and co-workers (1983) have shown that the initiation of kidney development involves the induction of type IV collagen and laminin in the mesenchymal cells (which had been making type I collagen and fibronectin). This causes them to aggregate together and form a basement membrane. The mesenchymal cells become epithelial. In addition, it induces these cells to synthesize a receptor for the iron-transport protein transferrin. This receptor is an integral membrane protein, and transferrin appears to enable the cells to proliferate. Thus, the ureteric bud induces the metanephrogenic mesenchyme to become the kidney tubule. Moreover, the mesenchyme acts to instruct the epithelial cells, as well. For example, branching occurs when ureteric

FIGURE 11
Kidney induction observed in vitro. An 11-day mouse metanephric rudiment, which includes both ureteric bud and metanephrogenic mesenchyme, has been cultured for 8 days. The ureteric bud branches to form collecting ducts and the mesenchyme condenses at its tips to form the renal tubules. (From Grobstein, 1955; courtesy of C. Grobstein.)

Renal tubules   Collecting ducts

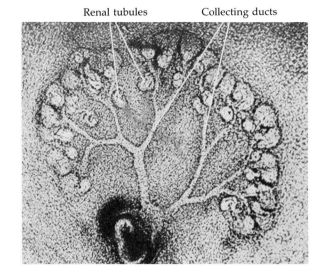

bud epithelium is recombined with metanephrogenic mesenchyme. Branching also occurs when salivary epithelium is recombined with salivary mesenchyme. However, branching is not seen when ureteric bud epithelium is mixed with salivary mesenchyme or when salivary epithelium is mixed with kidney mesenchyme (Bishop-Calame, 1966). Thus, epithelial morphogenesis in kidney development depends not only on the presence of mesenchymal cells but also on a regionally specific mesenchyme.

Other epithelial structures are not as strictly limited in their response as the ureteric bud is. The endodermal tubes, for instance, constitute an epithelium that must develop differently at different locations. Here, each mesenchyme region can cause the epithelium to develop differently (Figure 12). This regional specificity of mesenchyme induction is seen dramatically in the formation of the respiratory system. The respiratory epithelium is not as finicky as kidney and salivary epithelia and it responds to numerous mesenchyme. In the developing mammal, it responds in two distinct fashions. When in the region of the neck, it grows straight, forming the trachea. After entering the thorax, it branches, forming two bronchi and then the lungs. The respiratory epithelium can be isolated soon after it has split into two bronchi, and the two sides can be treated differently. Figure 13 shows such an experiment. The right bronchial epithelium retains its lung mesenchyme whereas the left bronchus was given tracheal mesenchyme (Wessells, 1970). The right bronchus is seen to proliferate and branch under the influence of the lung mesenchyme whereas the left side continues to grow in an unbranched manner. Thus, respiratory epithelium is extremely malleable and will differentiate into nearly any endodermal structure.

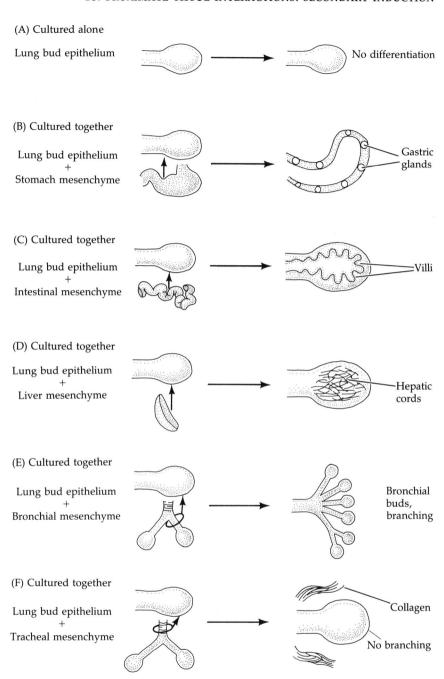

(A) Cultured alone

Lung bud epithelium → No differentiation

(B) Cultured together

Lung bud epithelium
+
Stomach mesenchyme → Gastric glands

(C) Cultured together

Lung bud epithelium
+
Intestinal mesenchyme → Villi

(D) Cultured together

Lung bud epithelium
+
Liver mesenchyme → Hepatic cords

(E) Cultured together

Lung bud epithelium
+
Bronchial mesenchyme → Bronchial buds, branching

(F) Cultured together

Lung bud epithelium
+
Tracheal mesenchyme → Collagen, No branching

FIGURE 12
Ability of presumptive lung epithelium to differentiate with respect to the
source of the mesenchyme. (A) Lung epithelium does not differentiate when
cultured in the absence of mesenchymal cells. (B–F) Mesenchyme-specific differentiation of epithelium. (Modified from Deucher, 1975.)

FIGURE 13
Ability of presumptive lung epithelium to differentiate with respect to the source of the inducing mesenchyme. After embryonic mouse lung epithelium has branched into two bronchi, the entire rudiment is excised and cultured. The right bronchus is left untouched, while the tip of the left bronchus is covered with tracheal mesenchyme. The tip of the right bronchus forms the branches characteristic of the lung, whereas no branching occurs in the tip of the left bronchus. (From Wessells, 1970; courtesy of N. Wessells.)

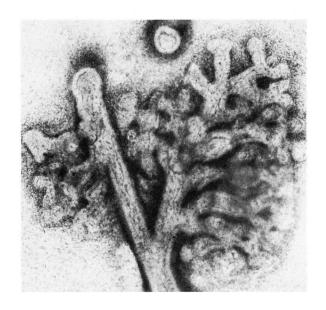

Coordinated differentiation and morphogenesis in the pancreas

As structures are induced, they begin to differentiate. This has been extensively studied in the developing mouse pancreas. The dorsal pancreas of a mouse begins as an outpocketing of endodermal epithelium, which enters a region of pancreatic mesenchyme. The mesenchymal cells interact with the epithelium and cause it to form fingerlike branches tipped by spherical clusters of cells called ACINI. This process is similar to that seen in the salivary and mammary glands, and these acini cells are responsible for producing the secreted proteins characteristic of the organ. In the case of the pancreas, most of the cells produce the digestive enzymes (carboxypeptidase, lipase, trypsin), which are exported to the small intestine through the pancreatic ducts. Other cells of the pancreatic epithelium bud off to produce the islets of Langerhans, the two major cell populations of which secrete insulin (beta cells) and glucagon (alpha cells).

William Rutter and his colleagues (1964) have shown that before the initiation of the pancreatic diverticulum, this endodermal region has neither pancreas-specific proteins nor pancreas-specific mRNAs (Figure 14). As the initial bulge begins to form, the first recognizable pancreatic proteins are made. Thus, pancreatic morphogenesis and differentiation are closely coupled. During the morphogenic phase of pancreas development, characterized by rapid mitosis and by duct and acini formation, there is a low level of these proteins being made. Eventually, some acinar cells stop dividing and differentiate the Golgi apparatus and rough endoplasmic reticulum characteristic of secretory cells. Once these differentiated cells accumulate, their further differentiation is in-

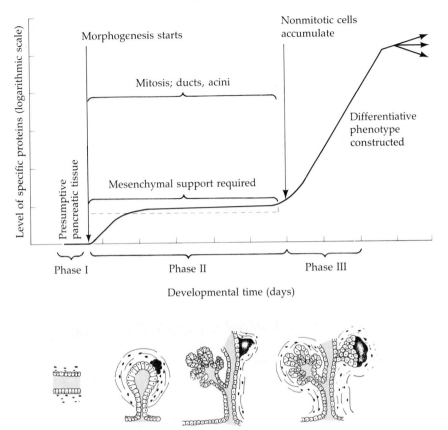

Phase I   Phase II   Phase III

Developmental time (days)

FIGURE 14
Coordinated differentiation and morphogenesis in the pancreas. Levels of pancreas-specific proteins begin increasing as the pancreatic rudiments begin developing from the endodermal epithelium. Mesenchyme is needed to support the formation of ducts and acini, and the level of pancreas-specific products increases only very slightly as few cells leave the mitotic cycle. As these nondividing cells accumulate in the completed tissue, there is a rapid increase in the synthesis of pancreas-specific proteins. The levels of pancreas-specific proteins are plotted against the formation of pancreatic tissue. (Adapted from Wessells, 1977 and Rutter et al., 1978.)

dependent of mesenchyme; and the pancreas-specific proteins continue to accumulate in these differentiating cells (Spooner et al., 1977). Thus, under normal circumstances, two independent phenomena—morphogenesis and cytodifferentiation—are coordinated in organ formation.

The nature of proximity in epithelio-mesenchymal inductions

Proximate induction occurs when an inducing tissue is brought near a competent responding tissue. Do these cells have to make physical contact or do they interact over a small distance? Three types of interactions can be postulated: cell–cell contact, cell–matrix contact, and diffusion (Grobstein, 1955; Saxén et al., 1976; Figure 15). In some tissues, cell–cell contact appears to be required. The induction of kidney tubules by the ureteric bud appears to depend upon their intimate contact. Clifford Grobstein (1955) discovered that the mesenchyme could be removed from the ureteric bud by lightly trypsinizing the kidney rudi-

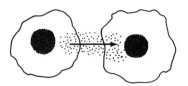

Diffusion of inducers from
one cell to another

Matrix of one cell induces
change in another

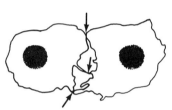

Contact (arrows) between
the inducing and
responding cells

FIGURE 15
Three possible ways that inductive interactions might occur. (After Grobstein, 1956.)

ment. When he cultured metanephrogenic mesenchyme and ureteric bud epithelium separately, nothing happened, even when the two tissues were in the same dish. The epithelium did not branch and the mesenchyme did not form tubules. However, when they were brought together, kidney formation was mimicked in culture (Figure 10). Lehtonen and Saxén (Lehtonen, 1975; Lehtonen et al., 1975) demonstrated

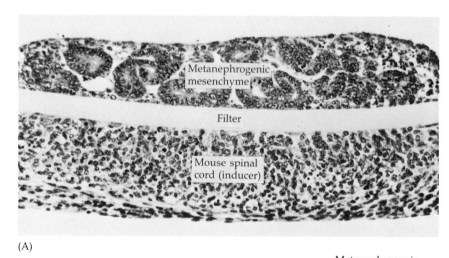

(A)

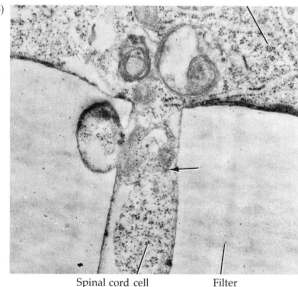

(B)

FIGURE 16
Transfilter induction of kidney tubules. (A) Metanephrogenic mesenchyme is above the filter and the inducer (in this case, mouse spinal cord, which mimics the effect of the ureteric bud) is placed below. Tubules have been induced in the mesenchyme. (B) Electron micrograph showing cell contact (arrow) through the pore of a filter "separating" the metanephrogenic mesenchyme from the inducing spinal cord. (Photographs courtesy of L. Saxén.)

that physical contact appeared to be necessary for this induction to occur. Modifying Grobstein's technique of growing inducer and mesenchyme on different sides of a porous filter, they showed that tubule formation occurred in the mesenchyme culture opposite the inducer tissue (Figure 16A). When seen in cross section, small projections of inducing tissue were seen to traverse the filter, contacting the mesenchyme (Figure 16B). By changing the thickness of the filter, they were able to correlate the induction with intracellular contact. A similar condition occurs in nature. Kidneys of mice afflicted with *Danforth's short tail* mutation fail to develop. The ureteric bud comes within a cell diameter of the mesenchyme but progresses no further. In the absence of contact, no induction occurs (Gluecksohn-Schoenheimer, 1943). In the induction of teeth (Slavkin and Bringas, 1976) and the submandibular salivary gland (Cutler and Chaudhry, 1973), contacts between the mesenchyme and epithelial cells can be seen by electron microscopy (Figure 17).

In other organs, the extracellular matrix of one cell type is seen to cause the differentiation of another set of cells. The differentiation of the corneal epithelium already alluded to depends upon an inductive influence from the capsule of the lens (Hay and Revel, 1969). Unlike kidney induction, where the inducer tissue has to be live, dead lens capsule will also work so long as collagen is still present on it; and nearly any source of collagen will induce the corneal epithelial cells to synthesize their matrix (Hay and Dodson, 1973; Meier and Hay, 1974). When filters are placed between the lens capsule and the corneal epithelium, the epithelial cells send out processes that cross the filter to contact the capsule (Meier and Hay, 1975). This work suggests that the surface of the corneal epithelial cell receives some instructions from the collagen-rich capsule of the lens.

The extracellular matrix can also provide positional information for secondary inductions. In 5- to 6-day chick embryos, the dermal cells of the skin begin to condense at particular sites. These places are the sites of feather development. The dermal cells are not randomly arranged but follow a precise pattern. A row of condensed dermal cells appears, each FEATHER GERM arising almost simultaneously. The rows adjacent to it form next, each focus of dermal condensation being between those of the first row (Figure 18). It is at these points—and only at these points—that feathers emerge by the interaction of these dermal cells with the overlying ectoderm. Stuart and co-workers (1972) suggested that the dermal condensations arose from the migration of the mesenchymal cells along a preformed collagen matrix (Figure 18C). By treating back skin with collagenase, they were able to destroy the hexagonal pattern of collagen deposition and inhibit the condensation of dermal cells.

Goetinck and Sekellick (1972) have shown that this hexagonal pattern is lacking in the skin of an embryo carrying the *scaleless* mutation; this embryo is unable to form these condensations. The skin of these mutants makes normal amounts of collagen, but the specific hexagonal

FIGURE 17
Contact between epithelial and mesenchymal cells during induction. (A) Electron micrograph of the interface between epithelial cells (preameloblasts) and mesenchymal cells (preodontoblasts) during tooth development. Processes from the mesenchymal cells extend through the intercellular matrix to contact the underside of the epithelial cells. (B) Electron micrograph showing contact between the epithelium and mesenchyme in the 16-day rat salivary gland (arrow). (A from Slavkin and Bringas, 1976, courtesy of H. C. Slavkin; B courtesy of L. E. Cutler.)

(A)

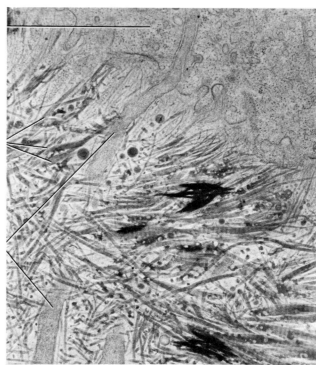

(B)

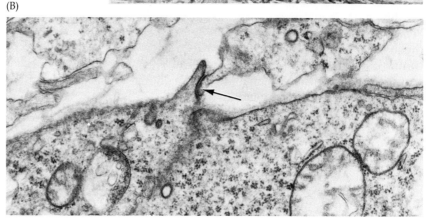

lattice is not seen. The ectodermal component of the skin appears to be the source of the mutation, as both lattice and feather germs formed when normal epidermis was combined with *scaleless* dermis; but no lattice or feather germ appeared when mutant epidermis was combined with normal dermis. More recent studies (Mauger et al., 1982, 1983) have shown that during normal skin development, collagen I and III and fibronectin are uniformly distributed at the dermal–epidermal junction. Feather and scale development continue as the collagens disappear from and fibronectin increases in the regions where the feather or scale

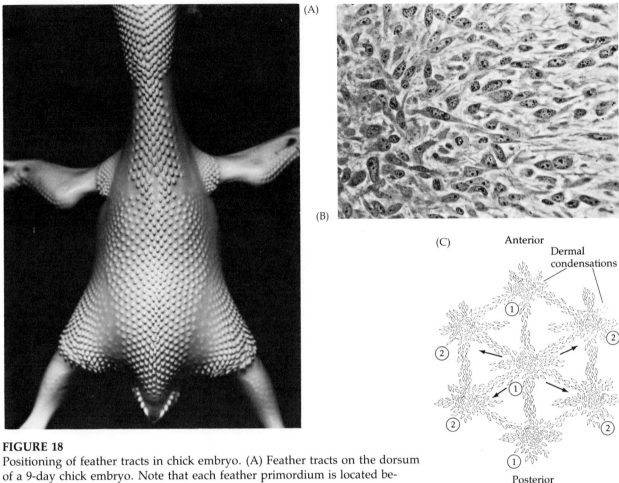

(A)

(B)

(C) Anterior

Dermal condensations

Posterior

FIGURE 18

Positioning of feather tracts in chick embryo. (A) Feather tracts on the dorsum of a 9-day chick embryo. Note that each feather primordium is located between the primordia of adjacent rows. (B, C) Pattern of dermal condensations giving rise to feather rudiments on the dorsum. (B) Condensation of dermal mesenchyme cells on the left, with nonaggregated cells oriented along the axis connecting the condensation to a neighboring one. (C) Hexagonal pattern of dermal cell aggregation and alignment: (1) primary row of dermal condensations (papillae) and cells aligned along the axis connecting them; (2) Secondary rows of papillae and their connections. (A courtesy of P.Sengal; B from Wessells and Evans, 1968, courtesy of N. Wessells; C from Saunders, 1980.)

bud is to form. Conversely, collagen accumulations increase between the interscale regions, whereas the fibronectin decreases dramatically. This transition from a homogeneous to punctuate distribution of matrix protein is not seen in the *scaleless* embryos. It is conceivable, then, that during normal skin development the ectoderm transmits information that enables the dermal cells to regulate their pattern of collagen and fibronectin synthesis, thereby laying down a pattern of collagen fibers upon which the migratory cells travel. Thus, the ectoderm influences the placement of the mesenchyme and then the mesenchyme induces the ectoderm to produce the feathers. Thus, although the mechanism for this patterning is not fully understood, the extracellular matrix appears to be very important in determining the site of secondary inductions in the skin.

There are some inductive systems in which contact is not needed. One of these is the primary inductive event wherein the neural tube is induced to form by the action of chordamesoderm cells upon the overlying ectoderm. No contact is seen between inducing and responding cell sets either in vivo or when induction occurs through filters separating the two components (Toivonen, 1979).

A similar lack of contact can exist in pancreatic development. Investigators in Rutter's laboratory (Rutter et al., 1978) have isolated from various mesenchymal tissues a glycoprotein that is capable of inducing the differentiation of pancreatic epithelial cells. This MESENCHYMAL FACTOR can be bound to Sepharose (resin) beads, and fragments of pancreatic epithelium will adhere to these beads by their basal surfaces. When they are so bound, these epithelial fragments then synthesize DNA and produce enzyme-containing secretory granules. The effect is specific for this mesenchymal factor, and other proteins will not induce this differentiation. As shown in Figure 19, the epithelial cells binding to the beads will divide and differentiate. Thus morphogenesis is not necessary for cell differentiation to occur. The cells will produce their particular products irrespective of whether or not they have branched. Although this factor is probably a membrane protein, it may be sloughed off through the normal turnover of membranes. It would not

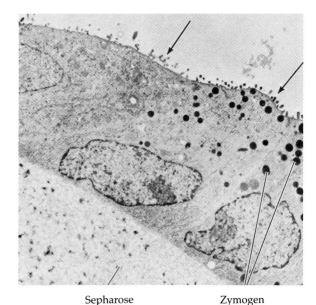

Sepharose   Zymogen
bead   granules

FIGURE 19
Retention and differentiation of pancreatic cells on Sepharose beads coated with mesenchymal factor. The zymogen granules and microvilli (arrows) characteristic of differentiated pancreatic cells are evident. (From Levine et al., 1973; courtesy of W. J. Rutter.)

be surprising if induction could occur either by contact of membranes or by the interaction of cells with the sloughed off membrane products of their neighbors. For instance, the *T*-locus product is probably a cell surface molecule, but wild-type mouse neurula cells will secrete into tissue culture medium a factor that will allow these mutant cells to adhere more normally (Yanagisawa and Fujimoto, 1978). As we shall see in Chapter 19, another cell surface molecule, the H-Y antigen, has been seen to traverse short distances and to interact with other cells to affect their differentiation (Ohno et al., 1976; Hall and Wachtel, 1980). Thus, it probably does not matter greatly whether or not the inducing tissue works by contact. What does seem to be important is that the effects are mediated at the level of the cell surface and that inducing molecules do not travel far from their source of origin.

Mechanism of branching in the formation of parenchymal organs

One way by which the mesenchyme interacts with the cell surface of epithelial cells is to cause their branching. Branching occurs when the epithelial outgrowths are divided by clefts, yielding lobules on either side of the cleft. These lobules grow, thereby creating branches. The branching of epithelial buds depends upon the presence of the mesenchyme. In some cases, such as the interaction of respiratory epithelium with several different mesenchymes, the interaction is instructive. In most cases, however, these interactions are merely permissive. The buds are prepared to branch and to form acini, but they need support from the mesenchyme. It is now thought that the mesenchyme causes branching by selectively digesting away part of the epithelial tissue's basal lamina, thus enabling mitosis to occur only in these specific areas.

The basal lamina, composed of collagen, glycosaminoglycans, proteoglycans, and hyaluronic acid, is continuous along the clefts but is broken on the lobules. In these interruptions, contacts between epithelial and mesenchymal cells can often be seen (Coughlin, 1975). The importance of the glycosaminoglycans is shown in Figure 20. When salivary gland rudiments are taken from a 13¼-day mouse embryo, the epithelium undergoes extensive branching in the presence of its mesenchyme. When the salivary rudiments are treated with protein-digesting enzymes, the epithelium can be isolated apart from the mesenchyme, but the epithelial basement membrane is deficient. When this epithelium is added to fresh salivary mesenchyme, the epithelium loses its lobules and forms a spherical ball of tissue. After more time in culture, outgrowths appear from the sphere, and branching resumes. When the epithelium is isolated by microdissection in the presence of low concentrations of collagenase, the surface glycosaminoglycans remain intact. When this epithelium is recombined with mesenchyme, branching continues without interruption. However, when the epithe-

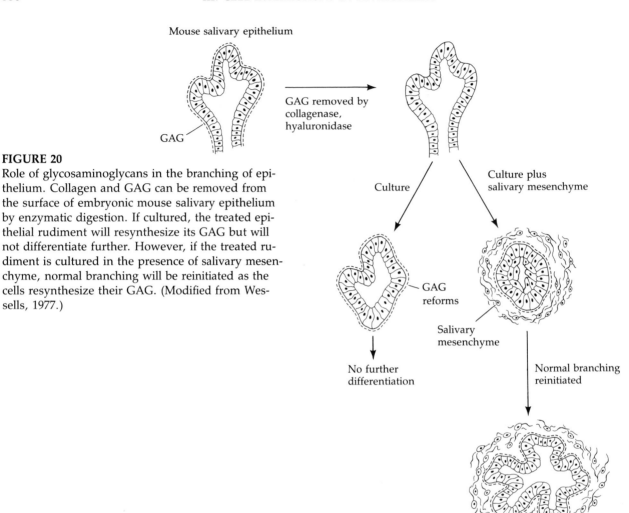

Mouse salivary epithelium

GAG removed by collagenase, hyaluronidase

GAG

Culture

Culture plus salivary mesenchyme

GAG reforms

Salivary mesenchyme

No further differentiation

Normal branching reinitiated

FIGURE 20

Role of glycosaminoglycans in the branching of epithelium. Collagen and GAG can be removed from the surface of embryonic mouse salivary epithelium by enzymatic digestion. If cultured, the treated epithelial rudiment will resynthesize its GAG but will not differentiate further. However, if the treated rudiment is cultured in the presence of salivary mesenchyme, normal branching will be reinitiated as the cells resynthesize their GAG. (Modified from Wessells, 1977.)

lium is isolated under low-collagenase conditions and treated with enzymes that destroy glycosaminoglycans, the epithelium loses its lobules and forms a sphere of tissue. Again, branching resumes after additional time in culture. Thus, loss of surface GAGs from the basal lamina correlates with the loss of lobular morphology, a finding that suggests that the maintenance of the branching pattern of development depends upon the continued presence of GAG.

Bernfield and Banerjee (1982) showed that the mesenchyme controls the degradation of basal lamina GAG. They performed a "PULSE–CHASE" experiment. Salivary gland rudiments were isolated from a 13d mouse embryo and cultured for 2 hours in the presence of radioactive glucosamine, a GAG precursor. After this pulse, they washed the rudiments, placed them into nonradioactive media, and monitored the distribution of the newly synthesized GAGs. The results indicated that the degradation rate of GAG was maximal at the distal ends of the lobules. Thus, although GAG is made by the epithelial cells, it is being digested quickly at the lobules. The agent of this degradation is the mesenchyme. Placing prelabeled epithelium into contact with mesenchymal cells releases the radioactive GAG from the basal lamina (Smith and Bernfield, 1977), and the mesenchyme is known to secrete hyaluronidase (Banerjee and Bernfield, 1979). Thus, the mesenchyme appears to be involved in degrading the laminar GAG.

But this does not explain the localization of this phenomenon. Why is degradation greater on the lobules than in the clefts? One explanation is that the lamina GAG in the clefts is protected against degradation by an accumulation of interstitial collagen. These collagen fibers are likewise produced by the mesenchyme but accumulate only within the clefts (Grobstein and Cohen, 1965; Wessells and Cohn, 1968). Such protection of basal lamina GAG by collagen has been seen on cultured mammary gland epithelial cells (David and Bernfield, 1981). The loss of the basal lamina from the distal parts of the lobule are correlated with those cells undergoing mitosis to generate the branch (Figure 21).

A hypothesis can be formed that might account for the mesenchymal control of epithelial branching (Bernfield and Banerjee, 1982). The mesenchyme promotes epithelial growth, degrades GAG, and deposits collagen fibers which can protect the lamina against degradation. The

FIGURE 21
Model for branching of epithelial tissue. Mitotic division is more frequent at the tips than in the stalks of growing epithelium. GAGs at the tips are not protected by collagen whereas those in the clefts are. The clefts may form as a result of the stability of the stalks and the increased cell number at the tips (as has been hypothesized for the ingression of the vegetal plate in sea urchin embryos or of the neural tube in amphibians). The cleft would be stabilized by collagen arising from the mesenchyme. (Modified from Wessells, 1977.)

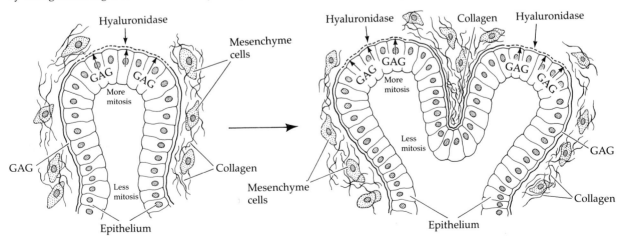

epithelium synthesizes the basal lamina materials and stimulates mesenchymal collagen synthesis. The result is a differential breakdown of the basal lamina at the tips of the lobes, thus enabling the dividing cells of the lobe to form branches. Here, the interaction of mesenchymal cells with the extracellular matrix of the epithelium would determine the branching pattern of the organ.

Induction of plasma cells

It is extremely difficult to study the processes of embryonic induction. The amounts of tissue that one isolates are painstakingly small, and the extraction of defined inducing agents from them has not yet been accomplished. However, animals never stop developing, and certain adult cells are constantly differentiating from stem cells. In the adult vertebrate, the differentiation of the antibody-secreting plasma cell has been the subject of intense scrutiny for several decades. As mentioned earlier, the plasma cell is derived from a B lymphocyte. The B lymphocyte can make a specific type of antibody and place it on its cell membrane as an antigen receptor. When an antigen binds to it, the B lymphocyte divides several times, develops the characteristic structures of a secretory cell (differentiates), and secretes its specific antibody.

But antigen alone is not enough to trigger this differentiation. In order to proliferate and differentiate, the presence of two other cell types is required. One of these cell types is the macrophage. This cell is thought to present the antigen to the B lymphocyte. The B lymphocyte does not usually respond well to antigen in solution. Rather the antigen is first "processed" by the macrophage, which presents the antigen on its cell surface (Mosier, 1967; Unanue and Askonas, 1968).

The second type of cells involved in B lymphocyte differentiation is the T lymphocyte. When T lymphocytes are absent, B lymphocytes do not proliferate or differentiate, even if macrophages and antigens are present. The cooperation among these cells can be shown by placing purified B cells and macrophages in one chamber of a fluid-filled compartment and T cells and macrophages in the other chamber (Figure 22). The membrane between them will not let cells pass through, but soluble molecules can pass from one side to the other. Thus, the T cells are physically separated from the B cells, but inducing molecules can pass through. When these T and B cells were cultured separately and exposed to an antigen, no antibody was secreted against it. However, when the cells were cultured in separate compartments connected by a porous filter some of the B lymphocytes differentiated to antibody-secreting plasma cells (Feldman and Basten, 1972).

These "T lymphocyte soluble factors" were largely uncharacterized until clonal lines of T lymphocyte tumor cells became available. These tumor lines secrete compounds called LYMPHOKINES, which have important effects on B cell development. These compounds include B cell

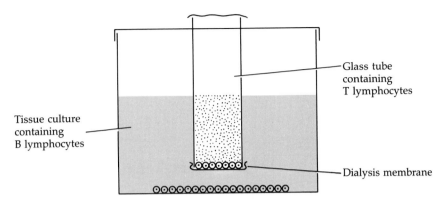

FIGURE 22
Apparatus used to keep B and T lymphocytes separate while allowing soluble molecules to pass across the dialysis membrane between the chambers. This can be seen as a modification of the transfilter induction apparatus shown in Figure 16.

growth factor, and B cell differentiation factors. A recent model for plasma cell differentiation is shown in Figure 23. Certain B cells can, as we have already seen, become induced by soluble T cell factors. When these cells contact the antigen (presented to them on a macrophage), their cell surface immunoglobulin molecules become crosslinked. Having crosslinked their surface antigen receptors, the resting B cells can be activated by B cell growth factor (BCGF), a 14,000-d protein produced by the helper T cells. This factor enables these cells to resume proliferation (going from G_0 to G_1 in the cell cycle). They now become responsive to the lymphokine interleukin 1 (IL-1), a macrophage product that causes the cells to undergo DNA synthesis and cell division. The proliferating B cells are still not capable of secreting antibody. The differentiation is directed by two additional soluble T cell factors, called either T cell replacing factors or B cell differentiation factors (Howard and Paul, 1983).

Other B lymphocytes are activated by intercellular contact between the resting B cell and a helper T lymphocyte. There are two areas of contact: one between the self-recognition (histocompatibility) molecules on the cell surface and the other an antigen-mediated bridge between antigen receptors on the B and T cell surfaces. The contact-mediated activation of the B lymphocyte probably does not need BCGF, but the differentiation of the activated B cell into a plasma cell still requires soluble macrophage and T cell products.

T lymphocytes not only induce the differentiation of B cells into plasma cells, but they also appear to induce the class of antibodies to be secreted. B cells that migrate to Peyer's patches (clusters of lymphoid tissue embedded in the small intestine) tend to secrete IgA antibodies rather than antibodies of any other class. When resting B cells are activated in the presence of T cells from Peyer's patches, the resulting

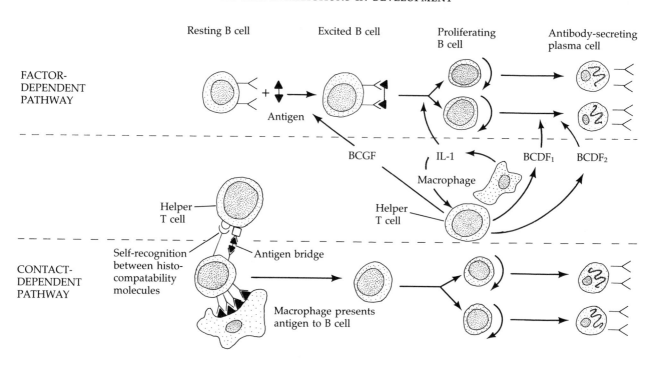

FIGURE 23

Induction pathways of B lymphocyte differentiation. Macrophages present antigen to resting B cells and activate helper T cells to synthesize soluble factors that induce B cell differentiation. In the factor-dependent pathway, the B cell-antigen complex in the presence of B cell growth factor (BCGF) becomes responsive to interleukin 1 (IL-1), which causes the B cells to proliferate. The presence of two B cell differentiation factors (BCDF), also derived from helper T cells, causes the proliferating B cells to differentiate into plasma cells. (This is not unlike the induction of the lens, a process in which inducer, responder, and mitotically permissive mesenchyme were needed for induction.) The contact-dependent pathway of B lymphocyte activation also requires soluble factors from T cells and macrophages, but their identity is not yet certain. (Modified from Howard and Paul, 1983.)

plasma cells are predominantly IgA secretors. Thus, the switch from IgM to IgA appears to be instructed by Peyer's patches T cells (Kawanishi et al., 1982, 1983). Vitetta and her colleagues (Isakson et al., 1982) have isolated a T cell product that instructs the switch from IgM to IgG. Thus, the differentiation of B lymphocytes into plasma cells appears to be selected by antigen, permitted by macrophages and T cells, and finally, instructed by T cells which secrete soluble factors that influence both the state of differentiation and the class of the protein product.

We have a system, then, in which one set of cells—the T lymphocytes—can induce the differentiation of another group of cells—the B lymphocytes. The B lymphocytes (like the pancreatic cells already discussed) previously had been capable of synthesizing a small amount of differentiated protein, but they did not secrete it. Upon stimulation by another cell type, cytodifferentiation takes place. Golgi apparatus and rough endoplasmic reticulum are formed, and protein is made in them and is secreted. In the case of B lymphocyte induction, however, we can work with clones of adult tumor cells that still retain their ability to make the inducing molecules. Studies of the immune system are providing us with our first insights into the molecular nature of induction.

SIDELIGHTS & SPECULATIONS

Crosslinking and induction

We do not know the mechanisms by which one population of cells induces changes in another. It is probable that induction, like fertilization, is a preprogrammed response that is activated by a triggering event. Studies on the activation of B lymphocytes have suggested that the trigger to plasma cell differentiation is the crosslinking of the cell surface antibodies. Moreover, more recent studies on other inducible tissues have suggested that the aggregation of cell-surface receptor molecules may provide a general way of triggering induction.

In the development of a B cell into a plasma cell, the cell surface antibody molecule acts as an antigen receptor. When these receptors bind antigen, the B cell starts proliferation and differentiation. Studies from Gerald Edelman's laboratory (reviewed in Edelman et al., 1974) have shown that antigenic stimulation can be mimicked by agents that act at the cell surface to bind certain receptors together. The cell surface antibodies can be linked together by other antibodies made to the original cell surface antibodies. The anti-antibodies will aggregate these antigen receptors onto one side of the cell as a cap (Figure 4 in Chapter 15). Another way of inducing cap formation and mitosis is to use a plant lectin, pokeweed mitogen (PWM). Lectins are plant proteins that bind specifically to certain carbohydrate molecules. Antibodies and PWM are both multivalent compounds capable of crosslinking the target molecules together; and their capping and mitogenic ability depends upon their being able to bind the cell surface receptors together. When antibodies are split so that they do not crosslink two antigen receptors together or when lectins are split in a way that allows them to bind to the cell but not to form patches, they lose their ability to cause the cells to divide and differentiate.

In our bodies, this crosslinking of cell surface antibodies is mediated through a bridge created by the antigen or by the antigen and a macrophage. There exists two major types of antigens: (1) T CELL-DEPENDENT ANTIGENS, which stimulate B cells only in the presence of T cells and macrophages, and (2) T CELL-INDEPENDENT ANTIGENS, which can stimulate B cell differentiation in the absence of the other two types. T cell-independent antigens are usually polymers consisting of linear chains of the same small molecule. It seems that when small molecules are strung together such that they form a repeating linear array, they can induce B cell differentiation directly. Dintzis and his co-workers (1976, 1982) made polymers of different lengths from the simple molecule dinitrophenol (DNP). These linear polymers were then mixed with isolated B lymphocytes. When 12 or fewer DNP molecules were strung together, there was no response. However, when 20 or more DNPs were polymerized and mixed with these cells, the B cells began their proliferation and differentiation (Figure 24). It is thought, then, that a certain amount of crosslinking is needed for T cell-independent B cell induction.

In most cases, though, the antigen is presented to the T and B cells in a manner illustrated in Figure 25. The macrophage is thought to PROCESS the antigens by concentrating them on its cell surface in such a manner that the T and B cell antigen receptors are brought to one side of the cell and are crosslinked through the macrophage (Figure 25). Only after this crosslinking is the B cell competent to differentiate.

Crosslinking of B cell surface antibodies has been shown to be the key activator in their further differentiation. When cell surface antibody molecules are crosslinked, the B cells express a set of receptors for the soluble lymphokines secreted by the helper T cells.

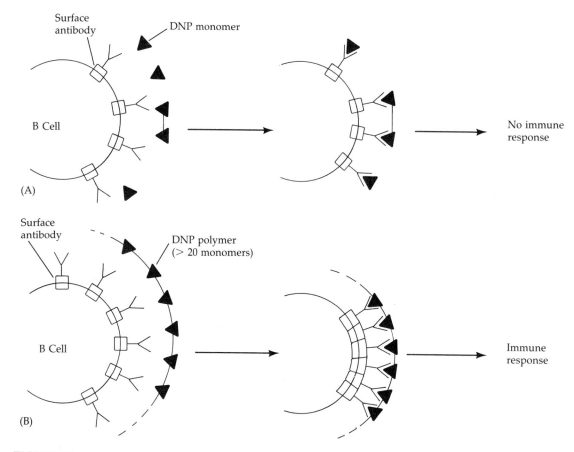

FIGURE 24

Schematic diagram of results of Dintzis (1976). Single molecules of dinitrophenol (DNP) failed to act as an antigen, as no immune response was made to it. However, when 20 or more DNP molecules were polymerized, the polymer acted as a potent antigen, causing the B cells to differentiate into plasma cells. Crosslinking of several surface antibodies appears to be essential for generating the immune response.

They are now able to respond to the interleukin produced by the T cells, and they divide and differentiate into antibody-secreting plasma cells (Yaffe and Finkelman, 1983).

Such redistribution of cell surface molecules may be essential for differentiation. In the compaction of mammalian blastomeres (Chapter 3), the decision of a blastomere to become an embryonic or a trophoblastic cell depends upon the cell's position in the compacted morula. Internal cells become embryonic, external cells become trophoblastic. These changes associated with compaction are preceded by a reorganization of the cell membrane. Microvilli and lectin-binding sites become associated with the apical (outward facing) pole (Handyside, 1980; Ziomek and Johnson, 1980), and certain membrane molecules orient themselves in a polar fashion along the apical–basal axis of the blastomeres (Randle, 1982). Such membrane redistribution is not a property of isolated blastomeres but rather depends upon cellular interactions, the lectin-binding sites moving away from the point of cell–cell contact.

Chow and Poo (1982) have similarly shown that when embryonic *Xenopus* muscle cells are brought together in culture, the distribution of soybean lectin receptors changes dramatically. At first, they are uniformly distributed over the membrane of the cells. However, when the cells adhere to each other, these glycoproteins accumulate at the site of cell–cell contact (Figure 26).

Crosslinking of cell surface receptors are known to mediate changes in two major ways. One way involves the activation of cell surface enzymes that

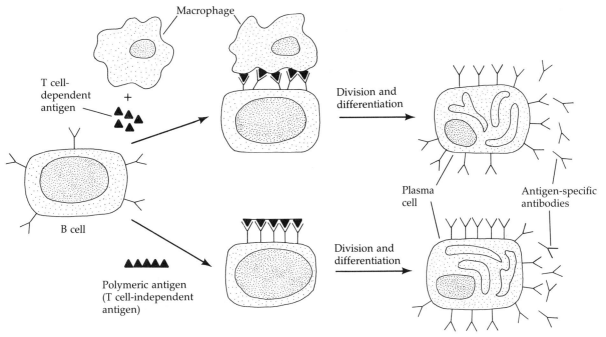

FIGURE 25

Antibody presentation in vivo. Most antigens are thought to be processed by macrophages and to be placed on the macrophage cell surface in a manner such that the B lymphocyte cell surface antibodies are crosslinked by contact with the macrophage cell membrane. A few antigens consist of identical repeating subunits. These do not need to be presented by macrophages because crosslinking of B cell surface antibodies will occur directly.

change the internal biochemistry of the cell. Such changes are seen to initiate the secretion of histamine from mast cells during allergic attacks. Mast cells contain granules of histamine directly below the surface of the cell. On their cell surface they have receptors for immunoglobulin E, a type of antibody characteristic of allergies. The IgE is bound to the mast cell surface, where it can bind to the allergy-causing substance (allergen). As is shown in Figure 27, the allergen crosslinks two or more IgE molecules, causing the receptors to be joined. When—and only when—the receptors are crosslinked together, the histamine is released. It is thought that the crosslinking event activates mem-

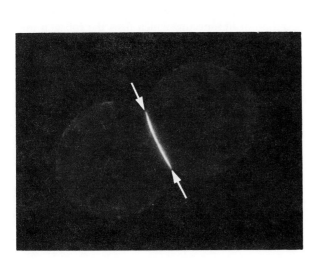

FIGURE 26

Redistribution of membrane molecules dependent upon cell–cell interaction. Fluorescence-labeled soybean lectin recognizes a subset of cell surface glycoproteins containing exposed galactose of N-acetylgalactosamine sugars. Originally, these glycoproteins are distributed randomly over the cell surface of embryonic muscle cells. When the cells contact each other, these proteins accumulate at the points of intercellular contact. (From Chow and Poo, 1982; courtesy of I. Chow and M. M. Poo.)

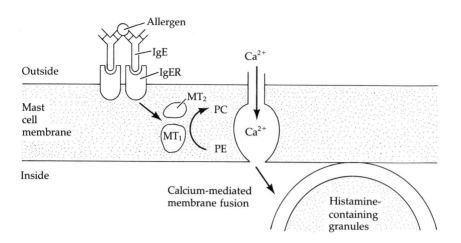

FIGURE 27

Schematic model wherein the crosslinking of IgE receptors on the membrane of mast cells creates the conditions for histamine release (an "allergic reaction"). IgE molecules bind to specific receptors (IgER) on mast cells. The IgE antibodies are crosslinked by the allergen (step 1). The binding of these receptors activates two methyltransferase enzymes (MT_1, MT_2; step 2), which convert phosphatidylethanolamine (PE) into phosphatidylcholine (PC; step 3). This change in membrane structure enables entry of calcium ions (step 4), which cause the fusion of histamine-containing granules with the cell membrane, thereby expelling their contents.

brane-bound methyltransferases that change the lipid structure of the membrane by methylating certain phospholipids (Ishizaka, et al., 1980). This change in membrane structure allows calcium ions to enter the cell. The histamine-releasing apparatus is already in place and just needs calcium ions to activate it. Thus, the crosslinking event triggers cell processes already present in the cytoplasm.

The other way that crosslinking can effect cell differentiation is by restructuring the cytoskeleton. The crosslinking molecules on the cell surface are sometimes seen to be associated with microtubules or microfilaments in the cytoplasm. Thus, the reorganization of the cell surface molecules is reflected in the reorganization of the internal cytoskeleton. The crosslinked cell surface antibodies of B cells are attached to actin microfilaments within the cell (Flanagan and Koch, 1978) and patching in fibroblasts can be stopped by using drugs that inhibit actin polymerization (Ash and Singer, 1976). Elizabeth Hay and her colleagues (Sugrue and Hay, 1982; Tomasek et al., 1982) have shown that the extracellular matrix of the lens capsule may induce the differentiation of chick cornea cells by fixing the cell surface proteins in a certain configuration. When embryonic corneal epithelium is cultured on a filter, the addition of collagen will cause the cells to flatten to reorganize their cytoskeleton, and to initiate the changes characteristic of corneal development. It is possible that the corneal epithelial cells produce collagen which is fixed in place by fibronectin. Fibronectin is not synthesized by the epithelial cells but is made by the mesenchyme that migrates into the corneal region from the neural crest. Thus, the induction of the cornea may be due to the mesenchymal secretion of fibronectin, which fixes the collagen into a certain configuration that in turn changes the positioning of the internal cytoskeleton. These cytoskeletal rearrangements may be responsible for allowing new ions to enter the cells or for setting off the preprogrammed set of instructions in these determined cells.

Such a programming of nuclear events by cell surface crosslinking is seen in studies of persistent viral infection (Oldstone, 1982). When measles virus infect cells, they synthesize a glycoprotein antigen on the cell surface of these cells. Antibodies against this glycoprotein (made by the body as part of its immune response) bind to the viral glycoprotein on the cell surface and trigger the capping reaction (Figure 28). When this happens, a signal that is somehow transmitted inside the cell changes the intracellular synthe-

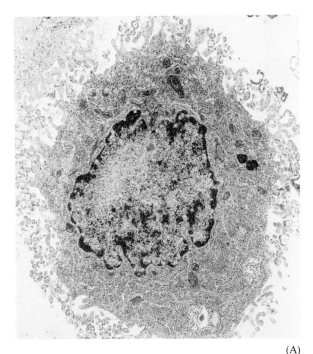

(A)

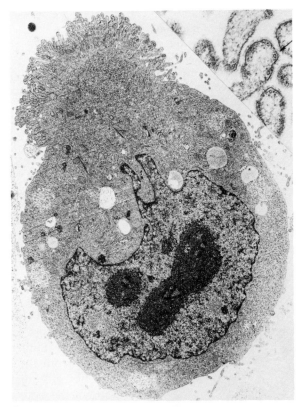

(B)

FIGURE 28
Capping phenomenon after reaction of cell-associated viral antigens with specific antibodies. Electron micrographs of a cultured human cell infected with measles virus in vitro (A) and of a similarly infected cell after the addition of antibody against measles (B). The antibodies cause the capping movement of the membrane-bound viral antigen. (From Oldstone, 1982; courtesy of M. Oldstone.)

sis of viral gene products. One of the intracellular measles proteins becomes heavily phosphorylated, whereas the synthesis of another measles protein stops altogether. Most other cellular functions continue normally. The effect is specific for the measles glycoprotein, and antibodies to host cellular glycoproteins do not mimic it.

Therefore, it is possible that induction involves the binding of a particular molecule to the cell surface of the responding cell. This molecule would "freeze" the lateral movement of certain membrane molecules and in some way change the internal chemistry of the cell. At the moment, the differentiation of lymphocytes serves as our best model of such processes, and researchers are attempting to determine whether or not other inducible systems work in similar ways.

Induction of the chick limb

Some developmental biologists claim that all the processes of development can be seen in the formation of limbs; and ever since R. G. Harrison began the experimental analysis of limb development at the beginning of this century, some of the most important concepts in

FIGURE 29
Limb bud formation. Migration of mesodermal cells from the somatic region of the lateral plate mesoderm causes the limb bud in the amphibian embryo to bulge out. (After Balinsky, 1975.)

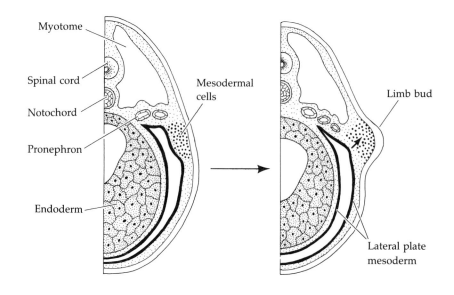

FIGURE 29
Limb bud formation. Migration of mesodermal cells from the somatic region of the lateral plate mesoderm causes the limb bud in the amphibian embryo to bulge out. (After Balinsky, 1975.)

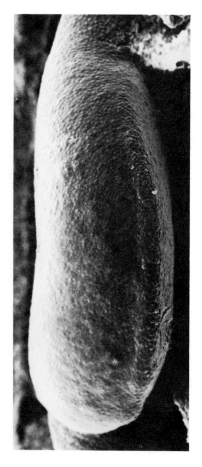

developmental biology have come from the investigation of these structures. The major features of limb development are common throughout vertebrates. Limb development begins when mesenchymal cells are released from the somatic layer of the lateral plate mesoderm (Figure 29). The cells migrate laterally and accumulate under the epidermal tissue of the neurula. The circular bulge on the surface of the embryo is called the LIMB BUD.

Harrison (1918) discovered that these *mesenchymal* cells play an essential role in limb formation.

1. When the mesenchymal cells are removed, no limb forms.
2. When the limb mesenchyme is grafted to a new site, a new limb forms.
3. When the hindlimb mesenchyme is combined with forelimb ectoderm, a hindlimb forms.
4. When the limb ectoderm is grafted to a nonlimb site, no new limb forms at that new site.

We see, then, the instructive nature of the limb mesenchyme. In birds and mammals, the mesoderm is seen to induce the ectodermal cells to elongate and form a special structure, the APICAL ECTODERMAL RIDGE (AER; Figure 30). Saunders and his co-workers (1976) have also shown that the AER is a self-sustaining population of cells that does

FIGURE 30
Scanning electron micrograph of an early chick forelimb bud, with its apical ectodermal ridge in the foreground. (Courtesy of J. F. Fallon and B. K. Simandl.)

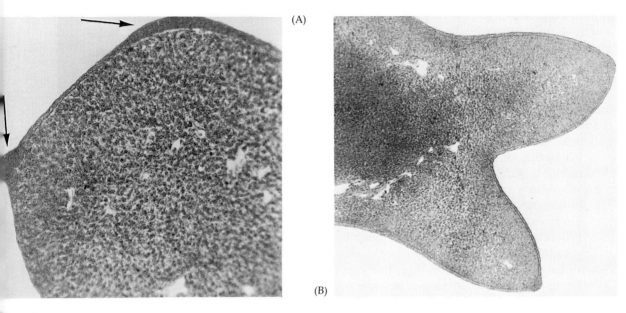

(A)

(B)

FIGURE 32
Cross sections of hindlimb buds from *eudiplopodia* chick embryos. (A) Two AERs on hindlimb bud; extra outgrowth on the dorsal side will form an extra set of toes. (B) Both outgrowth regions are covered by an AER. (From Goetinck, 1964; courtesy of P. Goetinck.)

sible for the sustained outgrowth and development of the limb (Zwilling, 1955; Saunders et al., 1957; Saunders, 1972).

This relationship can best be seen by two mutations of chick limb development, *polydactylous* and *eudiplopodia*. *Polydactylous*, as the name implies, is a mutation conferring extra digits. By recombining mutant and wild-type tissues (Table 2), the defect can be traced to the mesodermal cells that induce too broad an AER. In the mutant *eudiplopodia* (Greek, meaning "two good feet"), one has not only extra digits, but two complete rows of toes on each hindlimb (Figure 32). Similar recon-

TABLE 2
Mutations affecting the reciprocal interactions between AER and its underlying mesenchyme[a]

Mesoderm	Epidermis	Result	Conclusion
POLYDACTYLOUS			
Polydactylous	Wild type	Polydactylous	Mesoderm is affected
Wild type	Polydactylous	Wild type	
EUDIPLOPODIA			
Eudiplopodia	Wild type	Wild type	Ectoderm is affected
Wild type	Eudiplopodia	Eudiplopodia	

[a]By reciprocal transplantation between wild-type and mutant AER and mesenchyme, the aberrant compartment of the induction can be identified.

not exchange cells with its surroundings as the limb develops. Chick and quail cells can readily be distinguished by their nuclear heterochromatin (Chapter 6), and when a quail AER is placed upon the stump of a chick limb bud, the limb continues to grow. However, the AER of that limb remains composed of quail cells, and the surrounding chick ectoderm is not seen to become part of the AER. Once induced, this AER becomes essential to limb growth and interacts with the mesenchyme.

1. When the AER is removed at any time during limb development, limb development ceases (Figure 31A).
2. When an extra AER is grafted onto an existing limb bud, supernumerary structures are formed, usually toward the distal end of the limb (Figure 31B).
3. When leg mesoderm is placed beneath the wing AER, hindlimb structures develop from that point onward (Figure 31C).
4. When nonlimb mesoderm is grafted beneath the AER, the AER regresses and limb development ceases (Figure 31D).

Thus, although the mesenchymal cells induce and sustain the AER and instruct the AER to produce a certain type of limb, the AER is respon-

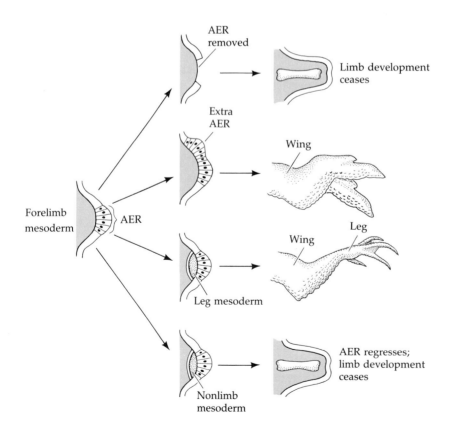

FIGURE 31
Summary of the induc
the apical ectodermal ı
(AER) upon the under
enchyme. (Modified fr
sells, 1977.)

stitution experiments (Table 2) show that here the defect is in the ectodermal tissue.

So it is clear that secondary induction, often of a reciprocal nature as seen in kidney and limb development, is essential for organ formation. But these secondary inductions alone are not sufficient for generating an entire organ. There must exist a mechanism for ensuring that a limb does not develop with its palm away from the body, or develop with two thumbs. There must be a mechanism for ensuring that legs and arms occur at the appropriate site in the body and not at other sites as well. Similarly, although the retina can receive images from the incredibly well coordinated eye, there must be a mechanism wherein the nerves from the retina relay this information to the correct area of the brain. The next chapter, then, will continue the study of organ formation and will focus on pattern formation in the embryo.

Literature cited

Ash, J. F. and Singer, S. J. (1976). Concanavalin A-induced transmembrane linkage of concanavalin A surface receptor to intracellular myosin-containing filaments. *Proc. Natl. Acad. Sci. USA* 73: 4575–4579.

Balinsky, B. I. (1975). *Introduction to Embryology*, Fourth Edition. Saunders, Philadelphia.

Banerjee, S. and Bernfield, M. (1979). Developmentally regulated neutral hyaluronidase activity during epithelial mesenchymal interaction. *J. Cell Biol.* 83: 469a.

Bernfield, M. and Banerjee, S. D. (1982). The turnover of basal lamina glycosaminoglycan correlates with epithelial morphogenesis. *Dev. Biol.* 90: 291–305.

Bishop-Calame, S. (1966). Étude expérimentale de l'organogenese du système urogénital de l'embryon de poulet. *Arch. Anat. Microsc. Morphol. Exp.* 55: 215–309.

Chow, I. and Poo, M.-M. (1982). Redistribution of cell surface receptor induced by cell–cell contact. *J. Cell Biol.* 95: 510–518.

Cole, R. K. (1967). *Ametapodia*, a dominant mutation in the fowl. *J. Hered.* 58: 141–146.

Couglin, M. D. (1975). Early development of parasympathetic nerves in mouse submandibular gland. *Dev. Biol.* 43: 123–139.

Coulombre, J. L. and Coulombre, A. J. (1971). Metaplastic induction of scales and feathers in the corneal anterior epithelium of the chick embryo. *Dev. Biol.* 25: 464–478.

Cutler, L. S. and Chaudhry, A. P. (1973). Intercellular contacts at the epithelial–mesenchymal interface during prenatal development of the rat submandibular gland. *Dev. Biol.* 33: 229–240.

Czihak, G. (1971). Echinoids. In G. Reverberi (ed.), *Experimental Embryology of Marine and Freshwater Invertebrates*. Elsevier North-Holland, Amsterdam, pp. 383–506.

David, G. and Bernfield, M. (1981). Type I collagen reduces the degradation of basal lamina proteoglycan by mammary epithelial cells. *J. Cell Biol.* 91: 281–286.

Deuchar, E. M. (1975). *Cellular Interactions in Animal Development*. Chapman and Hall, London.

Dintzis, H. M., Dintzis, R. Z. and Vogelstein, B. (1976). Molecular determinants of immunogenicity: The immunon model of immune response. *Proc. Natl. Acad. Sci. USA* 73: 3671–3675.

Dintzis, R. Z., Vogelstein, B. and Dintzis, H. M. (1982). Specific cellular stimulation in the primary immune response: Experimental test of a quantized model. *Proc. Natl. Acad. Sci. USA* 79: 884–888.

Edelman, G. M., Spear, P. G. Rutishauser, U. and Yahara, I. (1974). Receptor specificity and mitogenesis in lymphocyte populations. In A. A. Moscona (ed.), *The Cell Surface in Development*. Wiley, New York, pp. 141–164.

Ekblom, P., Thesleff, I., Saxén, L., Miettinen, A. and Timpl, R. (1983). Transferrin as a fetal growth factor: Acquisition of responsiveness related to embryonic induction. *Proc. Natl. Acad. Sci. USA* 80: 2651–2655.

Feldman, M. and Basten, A. (1972). Cell interactions in the immune response *in vitro*. III. Specific collaboration across a cell permeable membrane. *J. Exp. Med.* 136: 49–67.

Flanagan, J. and Koch, G. C. E. (1978). Cross-linked surface Ig attaches to actin. *Nature* 273: 278–281.

Gemmill, J. F. (1912). The development of the starfish

Solaster endica (Forbes). *Trans. Zool. Soc. Lond.* 20: 1–72.

Gluecksohn-Schoenheimer, S. (1943). The morphological manifestations of a dominant mutation in mice affecting tail and urogenital system. *Genetics* 28: 341–348.

Goetinck, P. F. (1964). Studies on limb morphogenesis. II. Experiments with the polydactylous mutant *eudiplopodia*. *Dev. Biol.* 10: 71–91.

Goetinck, P. F. and Sekellick, M. J. (1972). Observations on collagen synthesis, lattice formation, and the morphology of scaleless and normal embryonic skin. *Dev. Biol.* 28: 636–648.

Goldschmidt, R. (1933). Some aspects of evolution. *Science* 78: 539–547.

Grobstein, C. (1955). Induction interaction in the development of the mouse metanephros. *J. Exp. Zool.* 130: 319–340.

Grobstein, C. (1956). Trans-filter induction of tubules in mouse metanephrogenic mesenchyme. *Exp. Cell Res.* 10: 424–440.

Grobstein, C. and Cohen, J. (1965). Collagenase: Effect on the morphogenesis of embryonic salivary epithelium *in vitro*. *Science* 150: 626–628.

Hall, J. C. and Wachtel, S. S. (1980). Primary sex determination: Genetics and biochemistry. *Mol. Cell. Biochem.* 33: 49–66.

Hamburgh, M. (1970). *Theories of Differentiation*. Elsevier, New York.

Handyside, A. H. (1980). Distribution of antibody- and lectin-binding sites on dissociated blastomeres from mouse morulae: Evidence for polarization at compaction. *J. Embryol. Exp. Morphol.* 60: 99–116.

Harrison, R. G. (1918). Experiments on the development of the forelimb of *Amblystoma*, a self-differentiating equipotential system. *J. Exp. Zool.* 25: 413–461.

Harrison, R. G. (1920). Experiments on the lens in *Amblystoma*. *Proc. Soc. Exp. Biol. Med.* 17: 199–200.

Hay, E. D. and Dodson, J. W. (1973). Secretion of collagen by corneal epithelium. I. Morphology of the collagenous products produced by isolated epithelia grown on frozen-killed lens. *J. Cell Biol.* 57: 190–213.

Hay, E. D. and Meier, S. (1976). Stimulation of corneal differentiation by interaction between cell surface and extracellular matrix. II. Further studies on the nature and site of transfilter "induction." *Dev. Biol.* 52: 141–157.

Hay, E. D. and Revel, J.-P. (1969). Fine structure of the developing avian cornea. In A. Wolsky and P. S. Chen (eds.), *Monographs in Developmental Biology*. Karger, Basel.

Holtzer, H. (1968). Induction of chondrogenesis: A concept in terms of mechanisms. In R. Fleischmajer and R. E. Billingham (eds.), *Epithelial–Mesenchymal Interactions*. Williams and Wilkins, Baltimore, pp. 152–164.

Howard, M. and Paul, W. E. (1983). Regulation of B-cell growth and differentiation by soluble factors. *Annu. Rev. Immunol.* 1: 307–333.

Isakson, P. C., Pure, E., Vitetta, E. S. and Krammer, P. H. (1982). T-cell derived B-cell differentiation factor(s): Effect on the isotype switch of murine B-cells. *J. Exp. Med.* 155: 734–748.

Ishizaka, T., Hirata, F., Ishizaka, K. and Axelrod, J. (1980). Stimulation of phospholipid methylation, Ca^{++} influx, and histamine release by binding of IgE receptors on rat mast cells. *Proc. Natl. Acad. Sci. USA* 77: 1903–1906.

Jacob, F. (1971). Evolution and tinkering. *Science* 196: 1161–1166.

Jacobson, A. G. (1966). Inductive processes in embryonic development. *Science* 152: 25–34.

Kawanishi, H., Saltzman, L. E. and Strober, W. (1982). Characteristic and regulatory function of murine Con A-induced, cloned T-cells obtained from Peyer's patches and spleen: Mechanisms regulating isotype-specific immunoglobulin production by Peyer's patch B-cells. *J. Immunol.* 129: 475–483.

Kawanishi, H., Saltzman, L. and Strober, W. (1983). Mechanisms regulating IgA class-specific immunoglobulin production in murine gut-associated lymphoid tissues. II. Terminal differentiation of post switch sIgA-bearing Peyer's patch B cells. *J. Exp. Med.* 158: 649–669.

King, M.-C. and Wilson, A. C. (1975). Evolution at two levels in humans and chimpanzees. *Science* 188: 107–116.

Kollar, E. J. and Fisher, C. (1980). Tooth induction in chick epithelium: Expression of quiescent genes for enamel synthesis. *Science* 207: 993–995.

Lehtonen, E. (1975). Epithelio-mesenchymal interface during mouse kidney tubule induction *in vivo*. *J. Embryol. Exp. Morphol.* 34: 695–705.

Lehtonen, E., Wartiovaara, J., Nordling, S. and Saxén, L. (1975). Demonstration of cytoplasmic processes in Millipore filters permitting kidney tubule induction. *J. Embryol. Exp. Morphol.* 33: 187–203.

Levine, S., Pictet, R. and Rutter, W. J. (1973). Control of cell proliferation and cytodifferentiation by a factor reacting with the cell surface. *Nature New Biol.* 246: 49–52.

Liedke, K. B. (1951). Lens competence in *Amblystoma punctatum*. *J. Exp. Zool.* 117: 573–591.

Liedke, K. B. (1955). Studies on lens induction in *Am-*

blystoma punctatum. J. Exp. Zool. 130: 353–379.

Mauger, A., Demarchez, M., Herbage, D., Grimaud, J. A., Druguet, M., Hartmann, D. and Sengel. P. (1982). Immunofluorescent localization of collagen types I and III, and of fibronectin during feather morphogenesis in the chick embryo. *Dev. Biol.* 94: 93–105.

Mauger, A., Demarchez, M., Herbage, D., Grimaud, J. A., Druguet, M., Hartmann, D. J., Foidart, J. M. and Sengel, P. (1983). Immunofluorescent localization of collagen types I, III, and IV, fibronectin, and laminin during morphogenesis of scales and scaleless skin in the chick embryo. *Wilhelm Roux Arch. Dev. Biol.* 192: 205–215.

McKeehan, M. S. (1951). Cytological aspects of embryonic lens induction in the chick. *J. Exp. Zool.* 117: 31–64.

Meier, S. and Hay, E. D. (1974). Control of corneal differentiation by extracellular materials. Collagen as promoter and stabilizer of epithelial stroma production. *Dev. Biol.* 38: 249–270.

Meier, S. and Hay, E. D. (1975). Stimulation of corneal differentiation by interaction between the cell surface and extracellular matrix. I. Morphometric analysis of transfilter induction. *J. Cell Biol.* 66: 275–291.

Mosier, D. E. (1967). A requirement for two cell types for antibody formation *in vitro*. *Science* 158: 1573–1575.

Muthukkaruppan, V. R. (1965). Inductive tissue interaction in the development of the mouse lens *in vitro*. *J. Exp. Zool.* 159: 269–288.

Ohno, S. (1970). *Evolution by Gene Duplication.* Springer-Verlag, New York.

Ohno, S., Christien, L. C., Wachtel, S. S. and Koo, G. C. (1976). Hormone-like role of H-Y antigen in bovine freemartin gonad. *Nature* 261: 597–599.

Oldstone, M. B. A. (1982). Immunopathology of persistent viral infection. *Hosp. Pract.* (Dec. 1982): 61–72.

Randle, B. J. (1982). Cosegregation of monoclonal antibody reactivity and cell behavior in the mouse preimplantation embryo. *J. Embryol. Exp. Morphol.* 70: 261–278.

Rutter, W. J., Pictel, R. L., Harding, J. D., Chirgwin, J. M., MacDonald, R. J. and Przybyla, A. E. (1978). An analysis of pancreatic development: Role of mesenchymal factor and other extracellular factors. In J. Papaconstantinou and W. J. Rutter (eds.), *Molecular Control of Proliferation and Differentiation.* Academic, New York, pp. 205–227.

Rutter, W. J., Wessells, N. K. and Grobstein, C. (1964). Controls of specific synthesis in the developing pancreas. *Natl. Cancer Inst. Monogr.* 13: 51–65.

Saunders, J. W., Jr. (1972). Developmental control of three-dimensional polarity in the avian limb. *Ann. N.Y. Acad. Sci.* 193: 29–42.

Saunders, J. W., Jr. (1980). *Developmental Biology.* Macmillan, New York.

Saunders, J. W., Jr., Cairns, J. M. and Gasseling, M. T. (1957). The role of the apical ridge of ectoderm in the differentiation of the morphological structure and inductive specificity of limb parts of the chick. *J. Morphol.* 101: 57–88.

Saunders, J. W., Jr., Gasseling, M. T. and Errick, J. E. (1976). Inductive activity and enduring cellular constitution of a supernumerary apical ectodermal ridge grafted to the limb bud of the chick embryo. *Dev. Biol.* 50: 16–25.

Saxén, L., Lehtonen, E., Karkinen-Jääskeläinen, M., Nordling, S. and Wartiovaara, J. (1976). Are morphogenetic tissue interactions mediated by transmissible signal substances or through cell contacts? *Nature* 259: 662–663.

Slavkin, H. C. and Bringas, P., Jr. (1976). Epithelial-mesenchymal interactions during odontogenesis. IV. Morphological evidence for direct heterotypic cell–cell contacts. *Dev. Biol.* 50: 428–442.

Smith, R. L. and Bernfield, M. (1977). Salivary morphogenesis: Degradation of epithelial GAG by embryonic mesenchyme. *J. Cell Biol.* 75: 160a.

Spemann, H. (1938). *Embryonic Development and Induction.* Yale University Press, New Haven.

Spemann, H. and Schotté, O. (1932). Über xenoplatische Transplantation als Mittel zur Analyse der embryonalen Induktion. *Naturwissenschaften* 20: 463–467.

Spooner, B. S., Cohen, H. I. and Faubion, J. (1977). Development of the embryonic mammalian pancreas: The relationship between morphogenesis and cytodifferentiation. *Dev. Biol.* 61: 119–130.

Stuart, E. S., Garber, B. and Moscona, A. (1972). An analysis of feather germ formation in normal development and in skin treated with hydrocortisone. *J. Exp. Zool.* 179: 97–110.

Sugrue, S. P. and Hay, E. D. (1982). Interaction of embryonic corneal epithelium with exogenous collagen, laminin, and fibronectin: Role of endogenous protein synthesis. *Dev. Biol.* 92: 97–106.

Toivonen, S. (1979). Transmission problem in primary induction. *Differentiation* 15: 177–181.

Tomasek, J. J., Hay, E. D. and Fujiwara, K. (1982). Collagen modulates cell shape and cytoskeleton of embryonic corneal and fibroma fibroblasts: Distribution of actin, α-actinin, and myosin. *Dev. Biol.* 92: 107–122.

Ubaghs, G. (1967). General characteristics of Echinodermata. In R. C. Moore (ed.), *Treatise on Invertebrate Paleontology*, Part S, *Echinodermata* 1, Vol. 1. University of Kansas Press, Lawrence, pp. 3–60.

Unanue, E. R. and Askonas, B. A. (1968). The immune response of mice to antigen in macrophages. *Immunology* 15: 287–296.

Wessells, N. K. (1970). Mammalian lung development: Interactions in formation and morphogenesis of tracheal buds. *J. Exp. Zool.* 175: 455–466.

Wessells, N. K. (1977). *Tissue Interaction and Development.* W. A. Benjamin, Menlo Park, CA.

Wessells, N. K. and Cohen, J. H. (1968). Effects of collagenase on developing epithelia *in vitro*: Lung, ureteric bud, and pancreas. *Dev. Biol.* 18: 294–309.

Wessells, N. K. and Evans, J. (1968). The ultrastructure of oriented cells and extracellular materials between developing feathers. *Dev. Biol.* 18: 42–61.

Yaffe, L. J. and Finkelman, F. D. (1983). Induction of a B-lymphocyte receptor for a T-cell-replacing factor by the crosslinking of surface IgD. *Proc. Natl. Acad. Sci. USA* 80: 293–297.

Yanagisawa, K. O. and Fukimoto, H. (1978). Aggregation of homozygous Brachyury (T) cells in the culture supernatant of wild-type or mutant embryos. *Exp. Cell Res.* 115: 431–435.

Ziomek, C. A. and Johnson, M. H. (1980). Cell surface interaction induces polarization of mouse 8-cell blastomeres at compaction. *Cell* 21: 935–942.

Zwilling, E. (1955). Ectoderm–mesoderm relationship in the development of the chick embryo limb bud. *J. Exp. Zool.* 128: 423–441.

Pattern formation

Thus, beyond all questions of quantity there lie questions of pattern, which are essential for understanding Nature.
—ALFRED NORTH WHITEHEAD (1934)

Biochemistry and morphology are very shortly going to blend into each other without any difference or inequality. . . . Form is no longer the perquisite of the morphologist, and molecular exactitude no longer the preserve of the chemist.
—JOSEPH NEEDHAM (1967)

Theory without fact is fantasy, but facts without theory is chaos.
—C. O. WHITMAN (1894)

Introduction

Pattern formation is the activity by which embryonic cells form ordered spatial arrangements of differentiated tissues. The ability to carry out this process is one of the most dramatic properties of developing organisms, and one that has provoked a sense of awe in scientists and laymen alike. How is it that the embryo is able, not only to produce the different cell types of the body, but also to organize them into functional organs and structures? It is one thing to generate chondrocytes and osteocytes that can synthesize the cartilage and bone matrices, respectively; it is another thing to organize these cells into a functional bone. It is still another thing to make that bone a humerus and not a pelvis or a femur. The ability of cells to sense their relative positions within a limited population of cells and to differentiate with regard to this positional information has been the subject of intense debate and experimentation in recent years. We shall pick up our discussion of pattern formation where we ended the last chapter, with the study of the vertebrate limb.

Pattern formation during limb development

The vertebrate limb is an extremely complex organ with an asymmetric pattern of parts. The bones of the forelimb, be it wing, hand, flipper, or fin, consist of a humerus adjacent to the body wall, a radius and an ulna in the middle region, followed by bones of the wrist and the digits (Figure 1). Originally, these structures are cartilaginous, but most of the cartilage is replaced later by bone. The position of each of the bones and muscles in the limb is extremely well organized. The limb would be of little use if the fingers were positioned between the humerus and the ulna. Polarity exists in other dimensions as well. In humans, it is

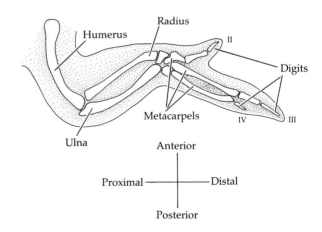

FIGURE 1
Skeletal pattern of the chick wing. Digits are numbered according to convention: II, III, IV. (After Saunders, 1982.)

obvious that each hand develops as a mirror-image of the other. It is possible for other arrangements to exist—such as the thumb developing on the left side of both hands—but this is not generally seen. In some manner, the three-dimensional pattern of forelimb is routinely produced.

The limb field

The mesodermal cells that give rise to the vertebrate limb can be identified by (1) removing certain groups of cells and observing whether a limb develops in their absence (Detwiler, 1918; Harrison, 1918), (2) transplanting certain groups of cells to new locations and observing whether they form a limb (Hertwig, 1925), and (3) marking groups of cells with vital dyes and observing which descendants of marked cells partake in limb development. By these procedures, the PROSPECTIVE LIMB AREA has been precisely localized in many vertebrate embryos. Figure 2 shows the prospective forelimb area in the tailbud stage of the salamander *Ambystoma maculatum*. The center of this disc marks the lateral plate mesoderm cells normally destined to give rise to the limb itself. Adjacent to it are the cells that will form the peribrachial flank tissue and the shoulder girdle. These two regions encompass the classical "limb disc" that will be used in the experiments to be mentioned in this chapter. However, if all these cells were extirpated from the embryo, a limb would still form, albeit somewhat later, from an addi-

FIGURE 2

Prospective forelimb bud of the salamander *Ambystoma maculatum*. The central area contains those cells destined to form the free limb (FL); the cells surrounding FL are those that give rise to the peribranchial flank tissue (PBF) and the shoulder girdle (SG). The cells outside these regions usually are not included in limbs but can regulate to form a limb if the more central tissues are extirpated. (After Stocum and Fallon, 1982.)

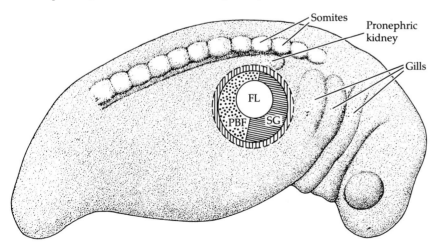

tional ring of cells that surrounds this area. If this last ring of cells was included in the extirpated tissue, no limb would develop. This larger region, representing all the cells in the area capable of forming a limb, is called the LIMB FIELD. A field can be described as a group of cells the position and fate of which are specified with respect to the same set of boundaries (Weiss, 1939; Wolpert, 1977).

This limb field originally has the ability to regulate for lost or added parts. In the tailbud-stage *Ambystoma*, any half of the limb disc is able to regenerate the entire limb when grafted to a new site (Harrison, 1918). This can also be shown by splitting the limb disc vertically into two or more segments and placing thin barriers between these segments to prevent their coming back together. When this is done, each part develops into a full limb. Thus, like an early sea urchin embryo, the limb field represents a "harmonious equipotential system" wherein a cell can be instructed to form any part of the limb.

Polarization along the limb axes

The positional information needed to construct a limb has to function in a three-dimensional coordinate system.[1] The axes of polarity are determined in the following sequence: anterior–posterior (A–P; as in the line between the thumb and little finger), dorsal–ventral (D–V; as in the line between the upper and lower surfaces of the hand), and proximal–distal (P–D; as in a line connecting the shoulder and the fingertip).

The self-differentiation of the anterior–posterior axis is the first change from the pluripotent condition. In chicks, this axis is specified long before a limb bud is recognizable. Hamburger (1938) showed that as early as the 16-somite stage, prospective wing mesoderm transplanted to the flank area develops into a limb the anterior–posterior and dorsal–ventral polarities of which are those of the donor graft and not those of the host tissue (Figure 3).

The proximal–distal axis is defined only after the induction of the AER by the underlying mesoderm. The limb elongates by the proliferation of the mesenchymal cells underneath the AER. This region of cell division is called the PROGRESS ZONE. As cells leave the progress zone, they have their proximal–distal values specified. Thus, the first cells leaving the progress zone form proximal structures; those cells that have undergone numerous divisions in the progress zone become the more distal structures (Saunders, 1948; Summerbell, 1974). Therefore, when the AER is removed from an early-stage wing bud, the progress zone is extinguished and only a humerus forms. When the AER is removed slightly later, humerus and radius and ulna form (Figure 4).

The proximal–distal polarity (like that of the other two axes) resides

---

[1]Actually, it is a four-dimensional system in which time is the fourth axis. Developmental biologists get used to seeing nature in four dimensions.

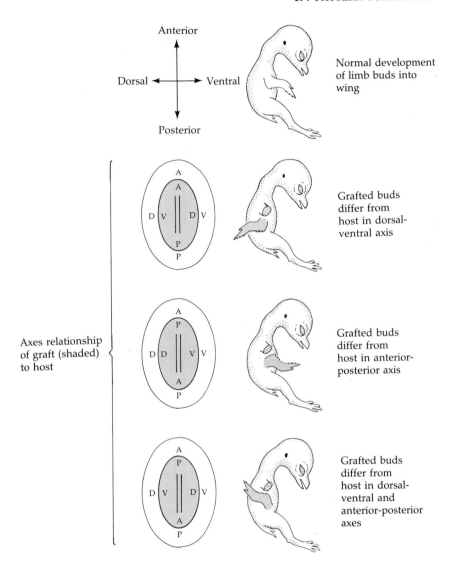

FIGURE 3
Specification of the anterior–posterior and dorsal–ventral axes in the chick wing. The grafted limb bud is seen to develop in accordance with its own polarity and does not adopt the polarity of its host. Wings that develop from grafted limb buds are shaded. For the sake of clarity, the host's normally developed wing is not shown. (After Hamburger, 1938.)

in the mesodermal compartment of the limb. If the AER provides the positional information—somehow instructing the undifferentiated mesoderm beneath it as to what structures to make—then older AERs should produce more distal structures when placed on young mesoderm. This was not found to be the case (Rubin and Saunders, 1972), as the normal complete sequence of limb development occurred when

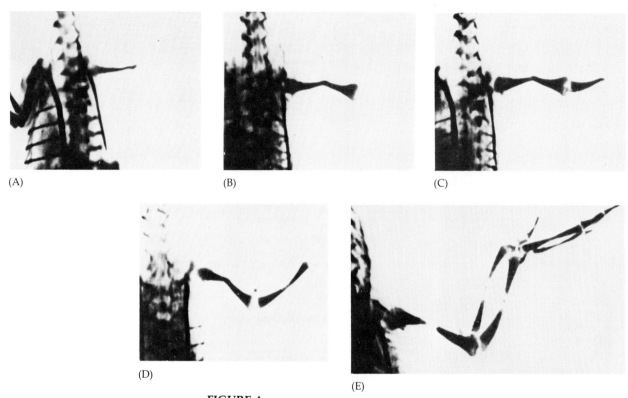

(A)   (B)   (C)

(D)

(E)

FIGURE 4
Dorsal view of chick skeletal pattern after removal of the entire AER from the right wing bud of embryos at various stages. The last picture is of a normal wing skeleton. (From Iten, 1982; courtesy of L. Iten.)

young mesoderm was combined with any stage AER. However, when the entire progress zone, including the mesoderm, from an early embryo was placed on the limb bud of a later-stage embryo, new proximal structures were produced in addition to those already present. Conversely, when old progress zones were added to young limb buds, distal structures immediately developed such that digits were seen to emerge from the humerus without the intervening ulna and radius (Summerbell and Lewis, 1975; Figure 5).

Although the differentiation of the proximal–distal structures is thought to depend upon how many divisions a cell undergoes while in the progress zone, positional information instructing a cell as to its position on the A–P and D–V axes must come from other sources. One series of experiments (Saunders and Gasseling, 1968; Tickle et al., 1975; Summerbell, 1979) has suggested that the anterior–posterior axis is specified by a small block of mesodermal tissue near the posterior junction of the young limb bud and the body wall. When such tissue from a young limb bud is transplanted into a position on the anterior side of another limb bud (Figure 6), the digits of the resulting wing are

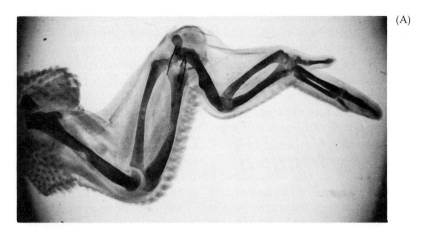

(A)

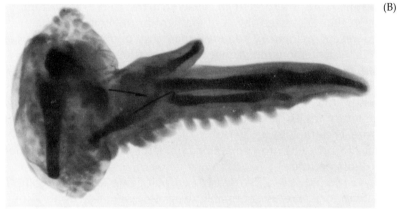

(B)

FIGURE 5
Control of proximo-distal specification by the cells of the progress zone (PZ). (A) Extra set of ulna and radius formed when early wing bud PZ is transplanted to late wing bud that has already formed ulna and radius. (B) Lack of intermediate structures seen when late limb bud PZ is transplanted to early limb bud. The hinges indicate the location of the grafts. (From Summerbell and Lewis, 1975; courtesy of D. Summerbell.)

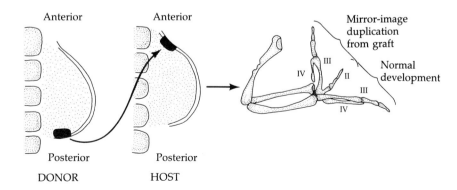

FIGURE 6
Formation of mirror-image duplications when a block of mesoderm from the posterior margin of the chick wing bud is transplanted to the anterior margin of a host wing bud. The resulting wing shows mirror-image duplications of digits IV and III; digit II is shared. (Modified from Saunders, 1972.)

duplicated. Moreover, the structures of the duplicated digits are mirror images of the normally produced structures. The polarity has been maintained, but the information is now coming from both an anterior and a posterior direction. This region of the mesoderm has been called the ZONE OF POLARIZING ACTIVITY (ZPA). It has been hypothesized that the ZPA operates by secreting a diffusible compound (MORPHOGEN), which would diffuse from its source to form a concentration gradient from the posterior to the anterior of the limb bud. Those cells nearest the ZPA would be exposed to the highest amount of this compound whereas those farthest from the ZPA would be exposed to a very low concentration of it. The original support for this hypothesis came from a series of experiments (Summerbell, 1979) in which the anterior limb bud was separated from the posterior limb bud by a nonpermeable barrier. In such cases, the anterior structures were no longer produced. These results were interpreted to mean that the ZPA was secreting a morphogen that organized the A–P gradient in the limb bud.

More recent evidence, however, has cast doubt on this model. Saunders (1977) demonstrated that the removal of the ZPA in no way destroyed the normal development of limb polarity. He also showed that when various non-limb bud mesodermal tissues were grafted into the anterior limb bud, they directed the formation of mirror-image supernumerary digits. Javois and Iten (1981) have shown that when ZPA tissue is added to an existing ZPA (thereby strengthening the source of morphogen), no "posteriorization" of the wing or formation of extra digits occurs.

So even though the ZPA clearly has polarizing activity in experimental situations, its role in normal limb development is in doubt. Rowe and Fallon (1981, 1982) argue against the diffusible morphogen model by showing that the same results can be obtained by interrupting the AER. They found that any specific anterior wing structure does not form when the AER at that level is not continuous with the posterior AER. The insertion of a barrier through the limb bud would cause the same effect. Iten and her co-workers have also cast doubt on the role of the ZPA. Iten and Murphy (1980) made slits into the ZPA of young limb buds (as shown in Figure 7) and inserted into these slits tissue from the anterior region of the limb bud. The AERs of both host and donor tissues were joined, and development proceeded. As expected by the morphogen theory, extra skeletal structures were produced. However, the frequency of such extra structures depended upon the precise source of the anterior cells. The cells from the most anterior position gave supernumerary digits much more frequently than those cells taken from less anterior regions of the limb bud. Thus, the ZPA does not appear responsible for the effect; rather, the disparity between the positions of the juxtaposed cells is.

Iten (1982) has speculated that the observations of the original ZPA experiments resulted from the juxtaposition of two normally nonadjacent limb bud regions. As we shall see shortly, when normally nonadjacent tissues from the same developmental field are placed next to

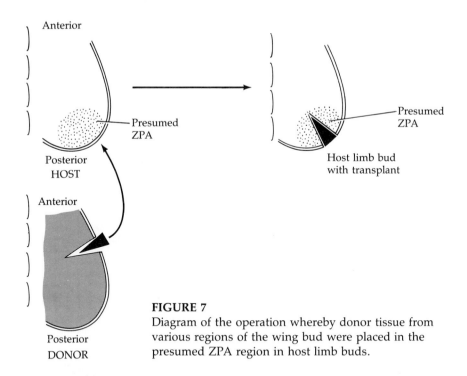

Anterior

Presumed
ZPA

Posterior
HOST

Anterior

Posterior
DONOR

Presumed
ZPA

Host limb bud
with transplant

FIGURE 7
Diagram of the operation whereby donor tissue from
various regions of the wing bud were placed in the
presumed ZPA region in host limb buds.

each other, they often divide to form cells with positional characteristics
intermediate between them. This is a result of local intercellular reactions rather than diffusible gradients. Similarly, when Javois and Iten
(1982) juxtaposed dorsal and ventral wing bud cells, they also obtained
supernumerary limb structures, a finding suggesting that here, too,
cell–cell interactions rather than gradients acting at a distance were
responsible for specifying the polarity of the limb bud.

Regeneration of limb tissues

Most organisms have, to some degree, the ability to restore lost parts.
This ability is called REGENERATION. Some groups (such as flatworms,
tunicates, and coelenterates) routinely utilize this ability in their asexual
form of reproduction; in other groups of animals (notably, warm-
blooded vertebrates) this ability is very limited. Amphibian limb regeneration has long been used to model embryonic development, and it
has been especially useful in studying pattern formation. Basically,
when a limb is amputated, the remaining cells are asked to reconstruct
the limb with its differentiated cells arranged in the original order
(Figure 8).

When a salamander limb is amputated, epidermal cells from the
remaining stump migrate to cover the wound surface. This single layer
of cells then proliferates to form the APICAL ECTODERMAL CAP. The cells

FIGURE 8
Regeneration of salamander forelimb. Upper photographs show the original limbs. On the left, the amputation was made below the elbow; the amputation
shown on the right cut through the humerus. In
both cases the correct positional information is respe-
cified. (From Goss, 1969; courtesy of R. J. Goss.)

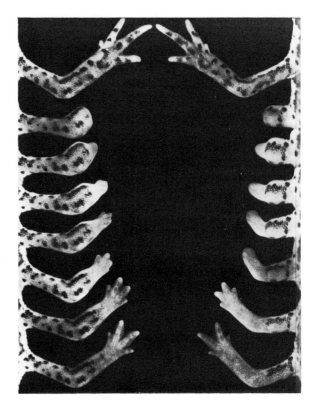

beneath this cap undergo a dramatic dedifferentiation such that bone
cells, cartilage cells, fibroblasts, muscles, and neuroglial cells all lose
their differentiated characteristics and become detached from each
other. The well-structured limb region at the cut edge of the stump thus
forms a mass of indistinguishable, dedifferentiated cells just beneath
the apical ectodermal cap. This dedifferentiated cell mass is called the
REGENERATION BLASTEMA, and these cells will proliferate to form the
new structures of the limb. If the blastema cells are destroyed, no
regeneration takes place (Butler, 1935). Moreover, once the cells have
dedifferentiated to form a blastema, they have regained their embryonic
plasticity. Carlson (1972) has shown that when at least 99 percent of the
muscle cells are removed from a newly amputated salamander limb,
the regenerating limb contains a normal supply of muscles in their
appropriate positions. Thus, other cells in the blastema, cells derived
from nonmuscle tissue, must be able to form the muscles of the regenerated limb.

The only tissue in the stump that appears to be essential for normal
limb regeneration is the nerve tissue. Singer (1954) demonstrated that
a minimum number of nerve fibers must be present for regeneration to
take place. It is thought that the neurons release a mitosis-stimulating
factor that increases the proliferation of the blastema cells (Singer and

Caston, 1972; Mescher and Tassava, 1975). One such substance, FIBRO-BLAST GROWTH FACTOR, is known to be produced by these neurons and will increase the rate of blastema cell division (Mescher and Gospodarowicz, 1979; Gospodarowicz and Mescher, 1980).

So we are faced with a situation wherein the adult organism can return its cells into an "embryonic" condition and begin the formation of the limb anew. Just as in embryonic development, the blastema forms successively more distal structures (Rose, 1962). Thus, the blastema must contain some positional information such that a blastema on a stump containing a humerus neither makes another humerus nor starts immediately producing digits. Rather, the blastema begins producing a structure appropriate for its level in the limb. Moreover, the polarity of the A–P and D–V axes are those that correspond to those axes in the stump.

As in the developing limb, the polarity of the regenerating limbs can be upset by placing normally nonadjacent tissues next to each other. Because small regions of the blastema cannot be easily grafted, this experiment is performed by rotating the blastema in relation to the stump. When a limb blastema is rotated 180 degrees and replaced on its stump, the resulting regenerated leg often contains supernumerary structures, very similar to what one sees when the same operation is performed on developing limb buds (Bryant and Iten, 1976; Figure 9).

If the patterning mechanisms for developing and regenerating limbs are indeed the same, then the developing limb should be capable of responding to grafts of regenerating limb tissue, and vice versa. Using the salamander *Ambystoma mexicanum*, Muneoka and Bryant (1982) demonstrated that this was the case. When limb buds were transferred to regenerating blastema stumps in a way that maintained the original polarity of the stump, normal limbs developed. However, when the polarity of the A–P axis was reversed with respect to the stump, mirror-image supernumerary digits emerged (Figure 10). These results strongly suggest that the patterning rules are the same for developing and regenerating limbs. This identity of patterning mechanism allows us to combine our information from these two disciplines in order to understand how the spatial pattern in generated. We can now discuss integrated models of limb pattern formation.

Polar coordinate model of pattern formation

As in the developing limb, regeneration of the vertebrate limb involves the proliferation of new cells. This is called EPIMORPHIC regeneration.

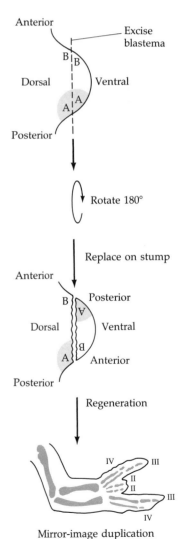

Mirror-image duplication of digits

FIGURE 9
Limb blastema rotation experiment. When blastema from regenerating limb is cut and rotated 180 degrees with respect to the body, supernumerary digits emerge with mirror-image symmetry. (After Tickle, 1981.)

FIGURE 10
Ability of regenerating salamander limb blastema to be controlled by progress zone of developing limb bud. (A) Control
showing normal five-digit right
hindlimb where right hindlimb
bud was placed on right regenerating hindlimb stump. (B) Limb
resulting when a graft from the
left hindlimb bud was placed
onto a right regenerating hindlimb stump. Here the anterior-
posterior axes are reversed between the host and the graft.
(From Muneoka and Bryant,
1982; courtesy of K. Muneoka.)

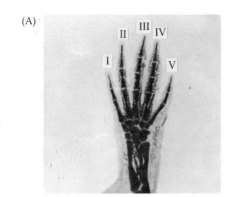
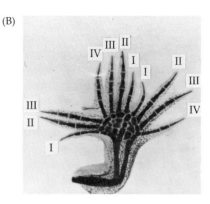

FIGURE 10
Ability of regenerating salamander limb blastema to be controlled by progress zone of developing limb bud. (A) Control
showing normal five-digit right
hindlimb where right hindlimb
bud was placed on right regenerating hindlimb stump. (B) Limb
resulting when a graft from the
left hindlimb bud was placed
onto a right regenerating hindlimb stump. Here the anterior-
posterior axes are reversed between the host and the graft.
(From Muneoka and Bryant,
1982; courtesy of K. Muneoka.)

FIGURE 11
Polar coordinate model for the
specification of positional information. Each cell is seen to have
a circumferential value (0–12)
specifying the anterior-posterior
axis and a radial value (A–E)
specifying the proximal (A) to
distal (E) axis. (From French et
al., 1976.)

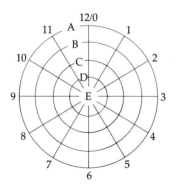

By comparing epimorphic regeneration in vertebrate limbs, insect limbs,
and insect imaginal discs, French and his colleagues (1976; also Bryant
et al., 1981) have proposed a series of empirical rules that predict the
outcome of a wide variety of experimental perturbations with regenerating appendages. Starting from Wolpert's (1969) premise that pattern
arises from a cell's recognition of its relative positions in a developing
population, they speculate that a cell assesses its physical location in a
system of polar (clocklike) coordinates (Figure 11). In this system, each
cell has a circumferential value (from 0 to 12) as well as a radial value
(from A to E). In regenerating limbs, the outer circle represents the
proximal (shoulder) boundary of the limb field; the innermost circle
represents the most distal regions.

As we have seen, when normally nonadjacent tissues of a field are
juxtaposed, duplications often arise. Yet other transplanted tissue (not
of the field) will not cause these duplications. The polar coordinate
model has been extremely useful in predicting the extent of these duplicated structures. The SHORTEST INTERCALATION RULE states that when
two normally nonadjacent cells are juxtaposed, growth occurs at the
junction until the cells between these two points have all the positional
values between the original points. The circular sequence, like a clock,
is continuous, 0 being equal to 12, and having no intrinsic value in
itself. Being a circle, however, means that there are two paths by which
intercalation can occur between any two points. For example, when
cells having the values 4 and 7 are placed next to each other, there are
two possible routes between them: 4, 5, 6, 7 and 4, 3, 2, 1, 12, 11, 10,
9, 8, 7. According to this model, the shortest route is taken. The exception, of course, is when the cells have values that fall exactly opposite
each other in the coordinate system, so that there is no one shortest
route. In this case, all values are formed between the two opposites.

The second rule is the COMPLETE CIRCLE RULE FOR DISTAL TRANSFOR-
MATION. Once the complete circle of positional values has been established on the wound surface, the cells proliferate and produce the more
distal structures. The mechanism by which this is thought to occur is
outlined in Figure 12 and again involves intercalation of structures

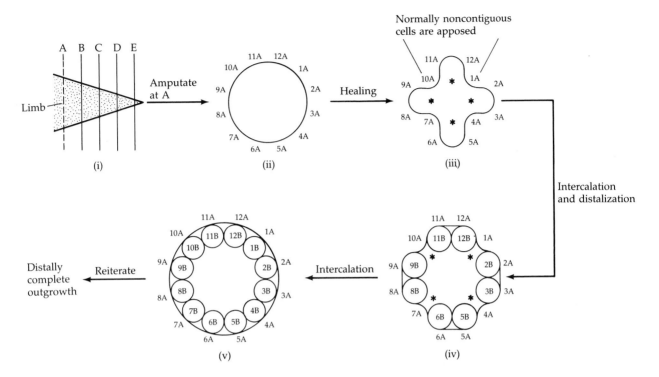

FIGURE 12
Model for distal outgrowth. (i, ii) Limb is cut at position A, proximal to positions B–E. This exposes circumferential positions at A. (iii) Healing leads to the apposing of normally noncontiguous cells (such as 10A and 1A) near the blastema tip. Cells between the newly-apposed tissue proliferate and acquire positional specification between the two sites. However, these cells are adjacent to preexisting cells sharing the same circumferential values. By the "distalization rule," such cells acquire a more distal positional value. (iv, v) Intercalation then occurs between these newly-specified cells, creating a new surface that contains all the circumferential values. This scheme is repeated until the limb is complete. (After Bryant et al., 1981.)

between cells having different positional information (Bryant et al., 1981). The predictive value of these rules can be seen when a transplant is made between regeneration blastemas, a transplant in which the anterior and posterior axes are reversed (Figure 13). The result is a limb with three distal portions (Iten and Bryant, 1975). This can be explained by viewing the A–P axis on the grid as having two opposite numbers—say, 3 and 9. In juxtaposing the values of 3 and 9, one generates a complete circle of values at each of the extreme sites, and a smaller intercalating series at all other sites. The result is three complete circles, which, by the law of distal transformation, will generate three complete limbs from that point on.

This polar coordinate model also predicts the effects of regulation

FIGURE 13

Reversal of the A–P axis in the regenerating newt blastema. Grafting of left hindlimb blastema onto the right hindlimb stump produces three sets of distal regions on the limb (compare to Figure 10). This can be predicted by the polar coordinate model, according to which two full sets of intermediate values should be regenerated in addition to the values inherent in the transplanted tissue. (From French et al., 1976.)

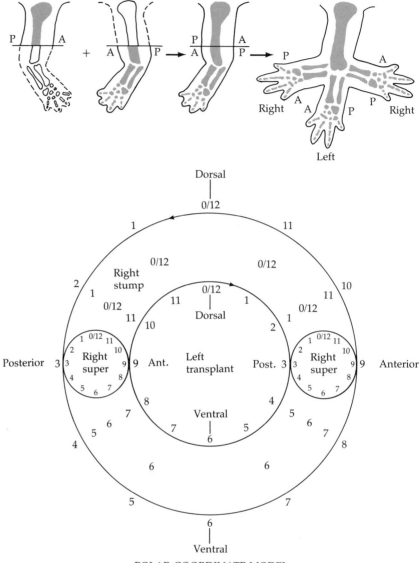

POLAR COORDINATE MODEL

when a portion of the tissue is lost. The newly formed cells would have positional values intermediate between those of the remaining cells and would reconstruct the appropriate part of the tissue. Because the basis of regeneration appears to be the recognition of differences between adjacent tissues, it is probable that epimorphic pattern formation during regeneration and normal pattern formation during embryonic limb development is the result of proximate interactions between adjacent cells rather than the result of long-range gradients (Bryant et al., 1981).

Genetics of pattern formation in *Drosophila*

The study of pattern formation in developing embryos has been greatly aided by the use of mutations. Here *Drosophila* becomes important because its genetics are among the best known of all eukaryotic organisms. Studying the ways in which mutations effect changes in the insect pattern and the ways in which the existing pattern affects the placement of genetically marked tissue reveals the mechanisms of pattern formation in *Drosophila*.

Compartmentalization in insect development

One method of investigating the spatial arrangement of cells in a developing organism is to induce genetic changes in an individual cell. The mutant cell is thereby marked as being different from its neighbors, and one can observe where the descendants of this cell are located. In *Drosophila*, radiation treatment can cause recombinations in mitotic chromosomes such that the resulting cells contain different genotypes. This is shown in Figure 14. Here, a wild-type fly heterozygous for multiple wing hairs has the genotype *mwh*/+. When radiation causes a recombination event in mitotic chromosomes in a cell in the imaginal disc of the epidermis, the descendants of the cell, instead of being wild type, are *mwh*/*mwh* (multiple wing hairs) and +/+ (normal wing hairs). As these cells multiply, areas of wing containing multiple wing hairs ap-

FIGURE 14
Generation of a genetically marked clone of cells by somatic crossing-over. Larval *Drosophila* heterozygous for the recessive mutation *multiple wing hair* (*mwh*) have normal wing bristles. In the event of a radiation-induced crossover, daughter cells can get two wild-type chromosomes (+/+) or two chromosomes having *mwh* genes (*mwh*/*mwh*) Because there is no mechanism for homologus pairing as there is in meiosis, such mitotic recombination is a very rare event. The homozygous wild-type cells are indistinguishable from the heterozygous background, but those wing discs having cells with two *mwh* genes produce wing tissue with mutant bristles.

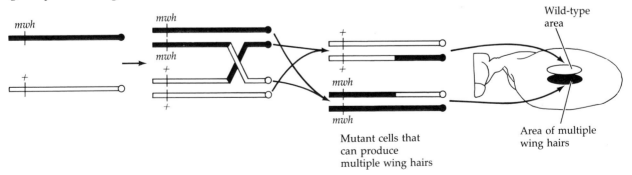

pear. The size of the patches depends upon the number of cell generations between the recombination event and the formation of the adult organ. In other words, when the recombination event occurs early in development, the patch is large, whereas, when there are only a few cell divisions left, the patches are smaller. Perhaps the most interesting information obtained from these patches relates to the restriction of cell potency during development. A cell undergoing recombination early in development is found to give rise to cells that form many parts of an organ; but, as development progresses, an individual cell becomes restricted to forming a progressively smaller group of structures.

This analysis of cell lineage in the development of several insects has uncovered a remarkable phenomenon: The progeny of cells in the imaginal discs do not cross certain distinct boundaries. Rather, determination appears to occur within discrete COMPARTMENTS, which do not overlap. In the development of an imaginal disc, groups of cells are determined together. These groups of cells are called POLYCLONES because they are not related by ancestry but are merely a cluster of neighboring cells. When we look at the early mesothorax imaginal disc—just after blastoderm formation when it contains only 10–50 cells—we find that it is divided into two polyclones (Figure 15). One of

FIGURE 15
Successive compartmentalization of the imaginal disc in the segment giving rise to the wing. The mesothoracic cuticle of that segment is divided into anterior and posterior polyclones at the blastoderm stage. The separation of the dorsal segment (which becomes the wing disc) from the ventral segment (leg disc) occurs by the first 10 hours of development. The wing and leg discs then physically separate; each retains the original anterior–posterior boundary. This boundary is bisected by another division of the respective discs into dorsal and ventral compartments. Each of these compartments of the wing disc is further subdivided into polyclones giving rise to the wing (central) and to the notum of the thorax (peripheral). (After Morata and Lawrence, 1977.)

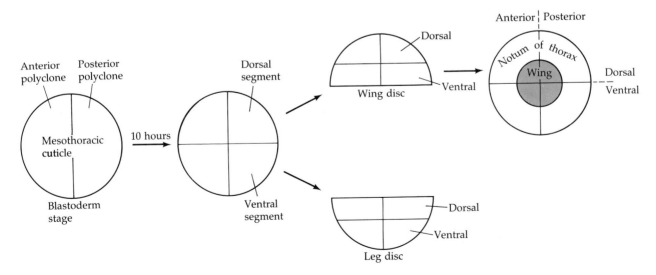

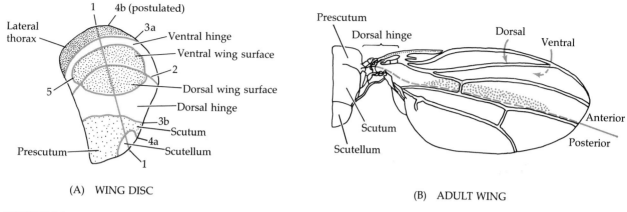

(A) WING DISC

(B) ADULT WING

FIGURE 16

Compartmental boundaries of the wing disc. (A) Projection of the five compartmental boundaries onto the fate map of the wing imaginal disc. Line 1 separates anterior and posterior compartments. Line 2 separates the dorsal wing surface from the ventral wing surface. Line 3 separates the lateral thorax from the dorsal hinge region (line 3a) and the notal portion of the wing blade from the wing proper (line 3b). Line 4 divides the notum into the scutellum and scutum regions (line 4a). (Another line 4 is postulated to exist in the lateral thorax region; line 4b.) Line 5 separates the free wing surfaces from the hinge regions. (B) Projection of the disc compartmental boundaries along the surface of an adult wing. The shaded region represents a large clone of genetically marked cells whose posterior border runs along the anterior–posterior border but does not cross the boundary.

these polyclones generates the anterior portion of the wing and most of the dorsal thorax, whereas the other polyclone will produce the posterior part of the wing and the remaining dorsal thorax. A mutant clone will spread in one of these compartments but will not spread into the other. The next determinative event separates the wing disc from the leg disc. The wing disc so derived keeps the anterior–posterior boundaries made earlier. The next determination is the distinction between dorsal and ventral surfaces of the wing, followed by the separation of those central elements (which form the wing proper). Thus, there is a sequential division of the imaginal disc into the series of compartments, and the compartmental boundaries reflect the commitment of adjacent cells to different fates (Crick and Lawrence, 1975; Garcia-Bellido, 1975; Kaufman et al., 1978).

Figure 16 shows the imaginal disc of the wing and the compartmental boundaries that give rise to the adult structure. When the crossover patches were analyzed, they were found not to cross the anterior–posterior border. That is, the homozygous mutant patches would remain on either the anterior or the posterior side of the wing and not venture into the other. A clone will proceed to the border and then stop. Figure 16 shows a large clone, the posterior border of which runs

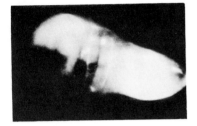

FIGURE 17

(A) Phase contrast micrograph of a *Drosophila* wing disc showing the site (arrow) where Lucifer Yellow is injected (notal region). (B) Fluorescent image of such a disc 52 minutes after injection of Lucifer Yellow. Dye has moved into the anterior half of the wing and through the notal cells. Movement into the posterior half has been restricted. (From Weir and Lo, 1982; photograph courtesy of C. Lo.)

along the anterior–posterior boundary of the wing. Something must exist that informs the cells of their proper spatial arrangement.

It is not known how these boundaries are created, but the compartmental boundaries have been correlated with the borders of sets of communicating cells. These studies involve the injection of a low-molecular-weight fluorescent dye (Lucifer Yellow) into some of the cells of an imaginal disc. This dye can pass through gap junctions and enter all the cells that are so coupled. Weir and Lo (1982) showed that the primary boundaries of dye transfer coincide with the anterior and posterior compartments (Figure 17). Later the communication of dye becomes restricted to smaller groups of cells, paralleling the formation of other compartmental borders. Thus, restrictions on gap junction communication may play a role in forming or maintaining these boundaries.

Mutations affecting the number of segments

Larval and adult flies have a segmented body plan consisting of anterior head segments, three thoracic segments, and eight abdominal segments (Figure 18). Each of the three adult thoracic segments has a pair of legs. However, although the first thoracic segment (T1) is characterized by legs alone, the second thoracic segment (T2) also has a pair of wings, and the third thoracic segment (T3) extends a pair of halteres (balancers). Thus, the three thoracic segments are readily distinguish-

FIGURE 18

Comparison of larval and adult segmentation in *Drosophila*. The three thoracic segments can be distinguished by their appendages: T1 (prothoracic) has legs only; T2 (mesothoracic) has wings and legs; T3 (metathoracic) has halteres and legs.

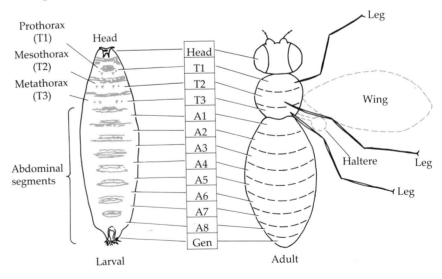

TABLE 1
Loci affecting segmentation pattern in *Drosophila*

Class	Locus
Segment-polarity	*cubitus interruptus*[D] *(ci*[D]*)*
	wingless (wg)
	gooseberry (gsb)
	hedgehog (hh)
	fused (fu)
	paxh (pat)
Pair-rule	*paired (prd)*
	even-skipped (eve)
	odd-skipped (odd)
	barrel (brr)
	runt (run)
	fushi tarazu (ftz)
Gap	*engrailed (en)*
	Krüppel (Kr)
	knirps (kni)
	hunchback (hb)

Source: Nüsslein-Volhard and Wieschaus (1980).

able. Each of these segments can be viewed as being subdivided into anterior and posterior compartments which are determined by the polyclones of cells that formed these segments. The first type of spatial mutations causes changes in the number of these segments. Some of these mutants are listed in Table 1. One class of these mutants (represented by *gooseberry*) has the normal number of segments, but the posterior compartment has been lost and has been replaced by a mirror-image of the anterior compartment. Some material between the segments is lost in this transformation (Figure 19). In the other two classes of mutants, entire regions are missing. In some cases (as represented

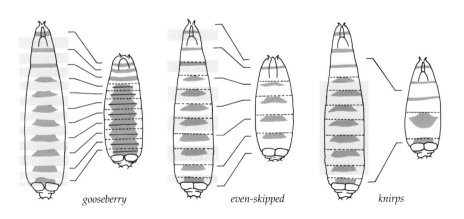

gooseberry   *even-skipped*   *knirps*

FIGURE 19
Three types of segmentation pattern mutants. To the left of each mutant is a diagram of a normal larva with the deleted regions shown in shaded boxes. (After Nusslein-Volhard and Wieschaus, 1980.)

by the *even-skipped* mutant), there are missing parts in every other segment. In other cases (represented by *knirps*), whole regions of segments are eliminated. These phenotypes suggest that the process of segmentation is controlled at three levels, at least: the entire egg, a two-segmented unit, and the individual segment (Nüsslein-Volhard and Wieschaus, 1980).

Mutations affecting segmentation: The *bithorax* complex

The analysis of some exceptionally bizarre mutations in *Drosophila* are beginning to provide us with new insights into how the complex spatial pattern of an organism is encoded in the genome. These mutations involve the transdetermination of one set of structures into another such that in some cases wings emerge where the eyes should be (the *ophthalmoptera* mutant) or legs grow out where the antennae should protrude (as in the *Antennapedia* mutant). Such mutations are called HOMOEOTIC MUTATIONS. One of the main clusters of homoeotic loci in *Drosophila* is the *bithorax* complex (Lewis, 1978). This cluster of homoeotic mutants on the third chromosome controls the determination of the thoracic and abdominal segments of the body. Recessive mutations in this complex, when homozygous, tend to transform a segment from its normal fate into structures appropriate for a more anterior segment. The mutation *bithorax* (*bx*), for example, transforms the anterior portion of the third thoracic segment into structures characteristic of the anterior portion of the second thoracic segment. In other words, the anterior haltere is transformed into an anterior wing. The mutation *posterior bithorax* (*pbx*) accomplishes an analogous transformation on the posterior compartment of the third thoracic segment, converting the posterior portion of the haltere into a posterior wing. When these two mutations are put together in the same fly [along with a third mutation, *anterobithorax* (*abx*)], the resulting fly has two complete sets of wings (Figure 20). The *bithoraxoid* (*bxd*) mutation causes the first abdominal segment to develop into structures characteristic of the third thoracic segment, thereby giving the fly an extra pair of legs.

Dominant alleles at the *bithorax* complex tend to do just the opposite, namely, to replace anterior segments with more posterior ones. *Contrabithorax* (*Cbx*), for example, causes the posterior wing region (T2) to develop the posterior structures of the haltere (characteristic of T3).

The genes of the *bithorax* complex are in the process of being cloned and sequenced. Cloning a gene with no known product is extremely difficult, but Bender and Spierer (1983) have utilized classical genetics to tell them where the gene is. By identifying overlapping fragments of DNA, they were able to "walk" from a previously cloned gene into the *bithorax* sequence itself. The complex is full of surprises. First, it contains an enormous amount of DNA, probably some 200,000 base pairs. (A normal protein is encoded by about 500 base pairs.) Second, several of the mutations there are characterized by additions or deletions that may

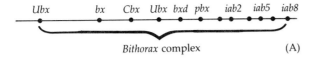

Bithorax complex (A)

(B)

FIGURE 20
Mutations in the *bithorax* complex of *Drosophila*. (A) Gene map for the *bithorax* region. (B) This four-winged adult fruit fly has three complementing mutations in the *bithorax* complex, which together completely transform a third thoracic segment into a second thoracic segment (i.e., haltere into wing). (From Lewis, 1978; photograph courtesy of E. Lewis.)

FIGURE 21
Phenotypes of larval segment affected by different homeotic genes. (A) Wild-type larva, consisting of a head, three thoracic, eight abdominal segments and a genital segment. (B) Larva lacking the *bithorax* complex. This larva has a normal head and first thoracic segment, but almost all remaining segments develop as T2. (C) Larva descended from parents that lack the *extra sex comb* (*esc*) gene; all segments differentiate as the eighth abdominal segment. (This suggests that in the absence of the *esc* gene product, all genes of the *bithorax* complex are expressed in all segments.) (After Lawrence, 1981.)

be functionally important. The *pbx* mutation, for instance, appears to be caused by a 17,000-base pair deletion at a certain position in the complex. The *Cbx* mutation was found to be caused by the insertion of an identical 17,000-base pair sequence some 40,000 base pairs away from its original position. Remembering that *pbx* causes posterior T3 (haltere) structures to develop as T2 (wing) structures and that *Cbx* causes the reverse transformation to occur, it is likely that the 17,000-base pair sequence encodes the instructions to make halteres and that changing its position in the genome changes the segment that expresses its information.

These data fit well into a model proposed by Lewis (1978). He has proposed that the second thoracic segment (having legs and wings) is the evolutionary ground state from which all the other segments have been modified. T2 is thus considered to be the most primitive of the segments. Lewis makes this claim on the evolutionary premise that flies probably evolved from insects having four sets of wings and that insects, in general, probably evolved from millipedelike creatures having legs on each segment. This view is supported by the observation that when the entire *bithorax* complex is deleted, nine segments of the fly develop the characteristics of the second thoracic segment (Figure 21).

Lewis then proposed that the *bithorax* complex should contain at least one gene for each segment below the second thoracic level. In other words, the development of T3 would involve the turning on of all the "ground-level" T2 genes plus the gene(s) responsible for T3 characteristics. A mutation in these T3 genes (*bx*, *pbx*) would return the segment to the T2 state. Similarly, the development of the first abdominal segment would require that the T2 and T3 genes be expressed in

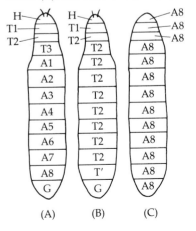

(A) (B) (C)

addition to the *Ab1* gene. Mutations of the *Ab1* gene (*bxd*) would cause the segment to develop T3 structures. This activation of different genes in different segments could be explained by postulating a gradient of a repressor molecule that inhibits the expression of all these genes. The repressor would be concentrated in the anterior segments, where fewer genes are turned on, and would be at a low concentration in the posterior segments. Furthermore, the genes controlling the development of the most posterior segments would have a higher affinity for the repressor molecule. Therefore, in the anterior segments, these genes would be more readily turned off. They would be active only in the posterior segments, where the concentration of the repressor is lowest.

One candidate for the postulated repressor molecule is the product of the wild-type allele of the *extra sex combs* (*esc*) genes. In the absence of the wild-type *esc* product, every segment develops the characteristics of the eighth abdominal segment (Figure 21; Struhl, 1981). The products of the *bithorax* complex and *esc* genes are presently being identified and should give us a new appreciation of how spatial patterning is controlled during development.

SIDELIGHTS & SPECULATIONS

Genetic control of segmentation

As the genes for pattern formation in *Drosophila* are cloned and sequenced, we are beginning to obtain tantalizing clues as to how the genome carries the three-dimensional plan of the organism. Recently, the analysis of three such genes has demonstrated that although these genes code for different proteins which are expressed in different regions of the embryo, these proteins contain very similar stretches of 60 amino acids near their carboxyl ends (McGinnis et al., 1984 a,b). Using portions of cloned genes as radioactive probes for in situ hybridization, the *Ultrabithorax* gene was seen to be expressed primarily in those cells destined to become the T3 and Ab1 segments. The *Antennapedia* (*Antp*) gene is similarly seen to be expressed in the thoracic segments. Here, the *Antp* gene product is thought to repress the development of head structures; and when it is inappropriately expressed in head segments, the antennae are transformed into legs (Figure

6 in Chapter 9). The *fushi tarazu* (*ftz*) gene (Table 1) is expressed in groups of cells which give rise to alternative segments in such a manner that the fly develops the correct number of segments (Akam, 1983; McGinnis et al., 1984b). Thus, these three genes are responsible for segment identity, are expressed in a segmentally restricted fashion, and code for proteins having similar carboxyl sequences.

The speculation that this common sequence is somehow required for segmentation has been strengthened by the discovery of similar sequences in the genomes of other invertebrates (beetles and annelids) and vertebrates (frogs, chicks, humans, and mice). All these species are from phyla characterized by segmentation. However, in those organisms from unsegmented phyla (sea urchins, roundworms, yeasts), no such sequences have been recognized (McGinnis, 1984b). This suggests that the common sequence might play some role in the process of segment determination.

The analysis of homoeotic genes in *Drosophila* has become a meeting place of developmental biology, genetics, neurobiology, molecular biology, and evolution, and it is certain to be one of the most interesting areas in each of these disciplines.

Specification of positional information in imaginal discs

We have already discussed the observation that the commitment of a cell to form the structures of a particular disc, such as a wing disc, is inherited by the progeny of that cell. With the exception of the relatively rare transdetermination events, fragments of wing discs give rise to more and more cells, each of them committed to producing the wing structures. The next question is whether or not the decision to become a particular part of the wing is similarly inherited. When a wing disc is cut into two unequal fragments and immediately placed into a metamorphosing larva, each fragment gives rise to only those portions of the wing that it would normally have become. (From such experiments, the fate maps of the imaginal discs have been derived.) However, when the two cut fragments are allowed to grow before putting them into the metamorphosing larvae, one of the fragments regenerates its missing parts, whereas the other fragment is seen to have produced mirror-image duplications (Bryant, 1976). The fates of such cells are predicted from the polar coordinate model described earlier and are depicted in Figure 22. Thus, the cells of imaginal discs undergo epimorphic pattern regulation. Because the mitotic progeny of a cell must be able to form

FIGURE 22
Polar coordinate model predicts epimorphic pattern of wing disc regeneration. The wing disc is cut into a larger and a smaller segment. Intercalary growth in the larger fragment yields a regenerated wing disc that can grow an entire wing. Cell division in the smaller fragment cannot produce all the positional values; therefore, the smaller fragment forms a mirror-image duplication of a wing part. (Modified from French et al., 1976.)

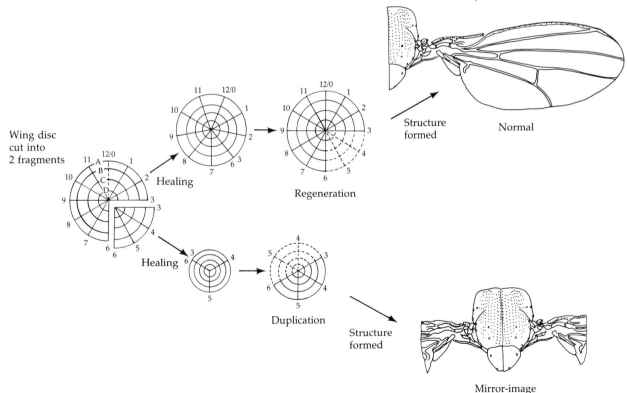

several different structures, the SPECIFICATION of these cells—the commitment to form a structure characteristic of a particular location in the morphogenic field—is not inherited.

Specification differs from determination in another important respect as well. The determination of each imaginal disc is different. Wing discs give rise to wings, leg discs to legs. The specifications of the different discs, however, appear to be identical. This can be seen with certain homeotic mutants such as *Antennapedia* wherein antennal structures are transformed into those of the legs (Postlethwait and Schneiderman, 1971). Occasionally, the entire antenna becomes an entire leg, but it is more common that only a small portion of the antenna becomes leglike. In these latter cases, the replacement is absolutely position specific. The cells of the antenna disc that normally would have formed the distal tip of the antenna (arista) are transformed into the most distal portion of the leg (claw); those cells specified to give rise to the second portion of the antenna are transformed into the trochanter of the leg. Correspondence of the two structures is shown in Figure 23. It is apparent, then, that the two differently determined discs must use a common mechanism for the specification of cell fates within the respective discs.

Positional information in *Drosophila* is given at several levels. There is positional information needed for the specification of the body axis, for the specification of the different segments and compartments, and for the specification of cell fate within the imaginal discs. The well-characterized genome of *Drosophila* is now being utilized to create a genetics of pattern formation.

FIGURE 23
Correspondences between portions of antenna and portions of the leg. In the mutant *Antennapedia*, regions of the antenna are transformed into leg structures (see Figure 6B in Chapter 9). The arrows show the portions of the antenna that form specific corresponding portions of the leg. (Modified from Postlethwait and Schneiderman, 1971.)

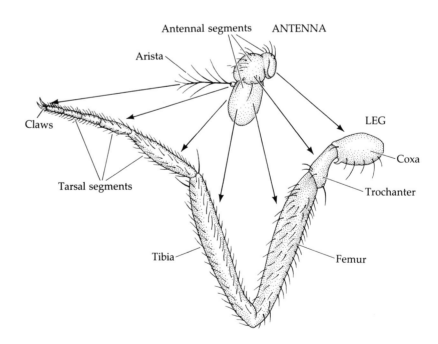

Pattern formation in the vertebrate nervous system

The functioning of the vertebrate brain depends, not only upon the differentiation and positioning of the neural cells, but also upon the specific connections these cells make among themselves. In some manner, nerves from a sensory organ such as the eye must connect to specific neurons in the brain that can interpret visual stimuli, and axons from the nervous system must cross large expanses of tissue before innervating the appropriate target tissue. How does the nerve axon "know" to traverse millions of cells to make its specific connection? Harrison (1910) demonstrated that the specificity of migration is due to PIONEER NERVE FIBERS, which go ahead of other axons and serve as guides for them. This simplifies, but does not solve, the problem of how neurons form the appropriate patterns of interconnections. Harrison also noted, however, that axons must grow upon a solid substrate and speculated that differences in the embryonic surfaces might allow axons to travel along certain specified directions. The final connections would occur by complementary interactions on the cell surface.

> That it must be a sort of a surface reaction between each kind of nerve fiber and the particular structure to be innervated seems clear from the fact that sensory and motor fibers, though running close together in the same bundle, nevertheless form proper peripheral connections, the one with the epidermis and the other with the muscle. . . . The foregoing facts suggest that there may be a certain analogy here with the union of egg and sperm cell.

Harrison's brilliant experimental work and speculations lay the groundwork for studies of neuronal patterns just as others of his studies formed the basis for our knowledge of limb pattern formation.

Direction of axonal growth

We are faced with two major problems when we discuss neuronal patterns. First, how does the axon recognize the path that leads to its target; and second, how does it recognize its target once it gets there? The majority of hypotheses seeking to explain the specificity of neuronal migration involve the notion of CONTACT GUIDANCE. Following Harrison's technique of growing axons on clotted blood, Weiss (1955) noted that the growing axons not only needed a solid substrate on which to migrate, but also that the migration tended to follow discontinuities in the clot. When the fibers of the blood clot were randomly oriented, the axons followed this random pattern. However, when the clot fibers were made parallel by applying tension to the clot, the nerve axons traveled along these fibers, not veering from the straight and narrow (Figure 24). Here, then, is evidence for physical factors being involved

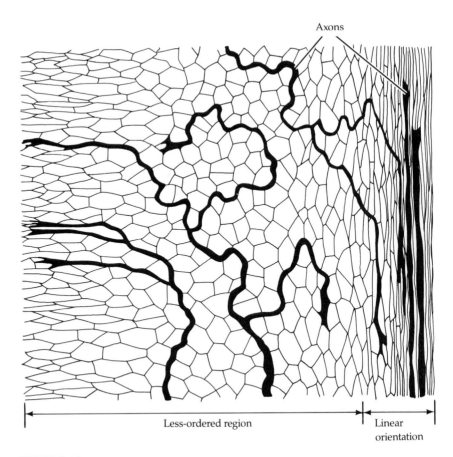

Axons

Less-ordered region

Linear
orientation

FIGURE 24
Contact guidance of axonal advance. Direction of the growing tip of neurons
followed the contours of the blood clot substrate. Isolated neuroblasts placed
in a linearly oriented portion of the clot produced axons that grew straight,
whereas the axons of neuroblasts in less ordered regions of the clot produced
axons that divided and followed the varied contours of the clot. (From Weiss,
1955.)

in contact guidance. Singer and his co-workers (1979) found evidence
that such physical factors operated in vivo. They detected large channels
between ependymal cells of the developing newt spinal cord through
which growing axons migrated. They hypothesized that these channels
provided cues for guiding the axons toward the appropriate regions of
the brain. Cellular channels have also been detected in the mouse retina
(Silver and Sidman, 1980), channels that appear to guide the retinal
ganglion cell axons into the optic stalk as they develop.

Chemical, as well as physical, cues appear to be provided as well.
Letourneau (1975a,b) demonstrated that axons from chick sympathetic
ganglia showed preferences for certain types of substrates. The most
preferred substrate was a layer of chick glial cells, followed by substrates

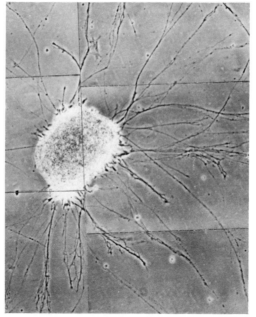

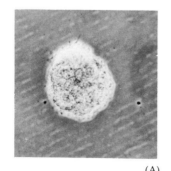

(A)

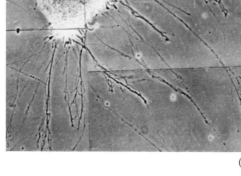

(B)

FIGURE 25
Effects of substrates and soluble factors on neural outgrowth. (A, B) Effects of fibronectin on axonal outgrowth from neural retina aggregates. Aggregate in A was cultured 36 hours on untreated tissue culture plastic. Aggregate in B was cultured on plastic treated with 50 μg/ml fibronectin. (C, D) Effects of nerve growth factor (NGF) on axonal growth. Sensory ganglion of 7-day chick cultured 24 hours in absence of NGF (C). Similar ganglion cultured in presence of 0.01 μg/ml NGF. (A and B from Akers et al., 1981; photographs courtesy of J. Lilien. C and D courtesy of R. Levi-Montalcini.)

(C)

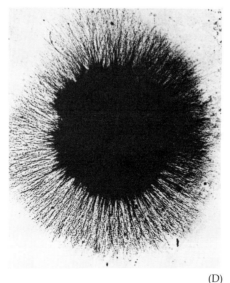

(D)

of polyornithine-coated plastic, tissue culture dish plastic, and petri dish plastic. The more adhesive substrates promoted the initiation and elongation of neuronal axons, whereas the less adhesive substrates retarded such growth. In 1981, Akers and her co-workers showed that the natural substrate fibronectin might be very important in permitting axonal movement. When neural retina explants were given only a small amount of fibronectin, little or no outgrowth appeared. However, when a greater amount of fibronectin was added to the cultures, long axonal growths were observed (Figure 25A and B). Thus, the growth and migration of axons appear to be influenced by the physical and chemical parameters of their substrates.

In addition to contact guidance, the migration of axons up a concentration gradient of diffusible molecules (CHEMOTAXIS) may also play a role in guiding axons to their correct places. One such diffusible molecule is a (molecular weight, 130,000) glycoprotein called NERVE GROWTH FACTOR (NGF). This substance is synthesized in most bodily tissues, and there is a gross correlation between the amount of NGF made by an organ and the number of neurons entering it (Angeletti and Vigneti, 1971). Nerve growth factor is essential for the survival of sympathetic neurons, and it induces a dose-dependent increase in axonal growth in vitro (Levi-Montalcini, 1966: Figure 25C and D). Although it is extremely difficult to prove whether local concentrations of NGF play a major role in routing nerves to their target organs, NGF has been observed to redirect the paths of axonal growth in vivo. When NGF is injected into newborn rat cerebra, the nerves from the sympathetic ganglia grow aberrantly *into* the central nervous system (Levi-Montalcini, 1976). Moreover, Pollack and Muhlach (1981) have demonstrated a temporal relationship between NGF production and axonal migration into the developing tadpole limb. Here, the immature tadpole limb is able to cause axons from the spinal cord to migrate into the mesodermal tissue. This can be mimicked in vitro, as axonal outgrowths from dissected spinal cords are greatly increased in the presence of young (stage 5) tadpole limb tissue. However, the outgrowths of young axons did not occur when older (stage 15) limb tissue was used (Figure 26). Thus, it appears that the limb explants are able to attract axonal outgrowth as a result of a diffusible molecule only at the appropriate developmental stages.

It is not yet known how much of this information is instructional and how much is merely permissive. Nerve growth factor does not appear to favor one set of sympathetic neurons over another, and it probably stimulates the growth of any axon toward the source of its production. On the other hand, it is not uncommon to see (as Harrison pointed out) that adjacent neurons may innervate different targets. A mixture of environmental factors and factors intrinsic to the individual neurons must be playing a role in the specificity of neuronal migration. It is not unlikely that the axonal "roads," be they chemical or physical, are continuously being constructed and broken down during particular stages of development and that certain neurons are capable of recognizing only certain of these pathways.

Mechanisms of synaptic specificity

Getting an axon to its target organ does not finish its task. The core of neural function is the ability to form specific connections from one nerve axon to another cell or group of cells. This is called SYNAPTIC SPECIFICITY, and it is beautifully illustrated in the connections made by axons emerg-

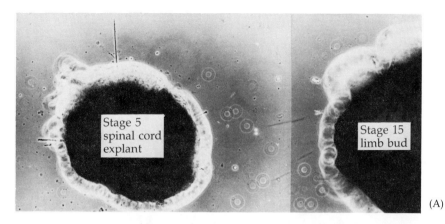

(A)

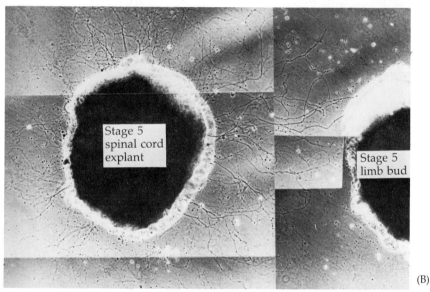

(B)

FIGURE 26
Stage-specific outgrowth of axons from cultured frog tadpole spinal cord explants. (A) Early (stage 5) spinal cord explants do not send out axons when cocultured with late (stage 15) limb buds. (B) Axonal growth from stage 5 spinal cord explants is greatly enhanced when cocultured with stage 5 limb bud. (From Pollack and Muhlach, 1981; courtesy of the authors.)

ing from the retinal ganglion cells. These retinal ganglion cells send out axons that form the optic nerve. In nonmammalian vertebrates, these axons traverse large expanses of the brain without making a single connection and eventually terminate by forming synapses with cells of the OPTIC TECTUM. Moreover, each retinal axon sends its impulse to one specific site (a cell or small group of cells) within that tectum (Sperry, 1951). As shown in Figure 27, there are two optic tecta in the brain.

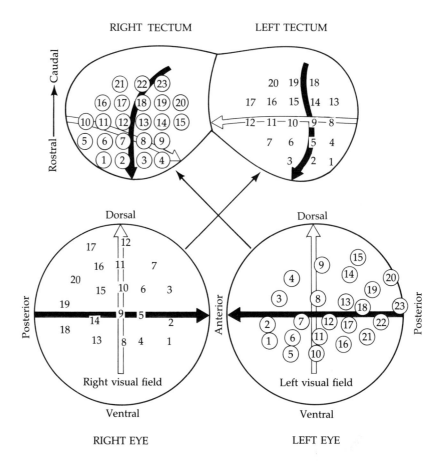

FIGURE 27

Map of the normal retinotectal projection in the adult *Xenopus*. The right eye innervates the left tectum and the left eye innervates the right tectum. The numbers on the visual fields (retina) and the tecta show regions of correspondence; that is, stimulation of spot 15 on the right retina sends electrical impulses to left tectal region 15. The black and white arrows indicate the direction of light impulses in the eye and the corresponding path of stimulation in the tectum. (From Jacobson, 1967.)

The axons from the right eye enter the left optic tectum and those from the left eye form synapses with the cells of the right optic tectum.

The map of retinal connections to the optic tectum (RETINOTECTAL PROJECTION) was detailed by Marcus Jacobson (Jacobson, 1967). Jacobson defined this map by shining a narrow beam of light on a small limited region of the retina and noting, by means of a recording electrode in the tectum, which tectal cells were being stimulated. The retinotectal projection of *Xenopus laevis* is also shown in Figure 27. As the light moves from the ventral to the dorsal surface of the retina, it stimulates cells from the lateral side to the medial side of the tectum. As the light impinged upon cells from the posterior to the anterior of the retina, the tectal stimulation goes from the caudal (tail) to the rostral (lip) end. Thus, we see that there is a point-for-point correspondence between the cells of the retina and the cells of the tectum. When a group of retinal cells is activated, a very small and specific group of tectal cells is stimulated. We also can observe that the points form a continuum, such that adjacent points on the retina map onto adjacent points on the tectum. This enables the frog to see an unbroken image.

This intricate specificity has given rise to the CHEMOAFFINITY HYPOTHESIS (or the HYPOTHESIS OF NEURONAL SPECIFICITY). According to Sperry (1965),

> The complicated nerve fiber circuits of the brain grow, assemble, and organize themselves through the use of intricate chemical codes under genetic control. Early in development, the nerve cells, numbering in the millions, acquire and retain thereafter, individual identification tags, chemical in nature, by which they can be distinguished and recognized from one another.

As in the developing limb, all cells are originally capable of forming any part of the pattern, but their positional information becomes specified as development continues. Jacobson identified a 5-hour period during which the retinal–tectal axes were determined. If the eye is rotated 180 degrees in an early (stage 28) *Xenopus* tadpole, the eye remains upside down, but the retinal–tectal projections are normal. In other words, the axons from the new ventral surface of the retina innervate the lateral side of the tectum just as an unrotated eye's axon would be expected to do. The tadpole sees normally. However, if the eye were rotated slightly later, at stage 30, the dorsal–ventral axis is normal, but the anterior–posterior axis is inverted 180 degrees. When rotation is performed at stage 31 or thereafter, both the A–P and D–V axes are inverted, and the frog will lower its head to get food that is slightly above it (Figure 28).

This can also be seen in a series of rotation experiments wherein eye primordia at various stages were grafted into young tadpoles at different angles (Hunt and Jacobson, 1972; Figure 29). When the graft was taken from a tadpole prior to stage 30, a normal retinotectal projection was obtained, thereby indicating that neither axis was determined. After stage 30, the retinotectal projections were rotated from the norm to the same extent as the eye had been rotated. Grafts taken from stage-30 tadpoles showed a variety of results, including maps where the D–V axis was normal but the A–P axis was rotated. Thus, we see a sequential specification of the axes, just as we observed in the limb. It should be noted that this specification of positional information occurs long before the axons actually reach the tectum, which happens around stage 40.

FIGURE 28

Retinal rotation after stage 30 causes hardships for frogs trying to catch food. (A) Eyes and optic tecta of a normal frog. (B) Eyes and optic tecta of a frog in which both eyes have been rotated 180 degrees on the A–P axis. The projections of a fly and of four retinal points are indicated on the right and left tecta, respectively. The image projected by the rotated eye will cause the frog to misjudge the fly's location. (After Lund, 1976.)

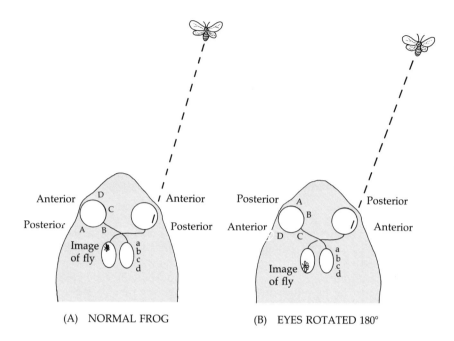

(A) NORMAL FROG   (B) EYES ROTATED 180°

FIGURE 29

Orientations of retinotectal map in *Xenopus* after transplantation of the eye primordium into a host tadpole. Eye primordia from young (stages 22–28) tadpoles (top row) gave normal maps when transplanted in four different orientations with respect to the host. Eye primordia from older (after stage 31) tadpoles produced maps with orientations rotated from the normal by the same angle as the eye was deviated from the normal anatomical position. (After Hunt, 1975.)

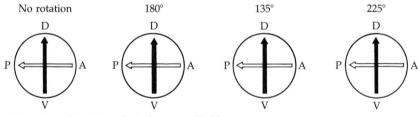

Eye primordium transplanted at stage 22–28

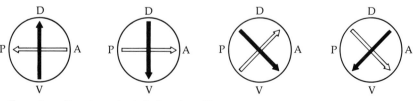

Eye primordium transplanted after stage 30

Within this framework, however, regulation can still occur. If, at stage 32, a composite retina is constructed, consisting of the left half of one retina and the ventral half of another, one would expect that there would be no axonal imput into that area of the tectum that normally receives neurons from the right dorsal quadrant. (And one would expect to see twice as many axons entering those regions innervated by the left ventral quadrant, which is represented twice.) This does not happen. Rather, the retinotectal projection from such a retina is normal. Similarly, if half the tectum is removed, the entire projection is compressed to fit into the smaller area (Gaze, 1978). This suggests that the retina and tectum form their specificities in a gradient-type pattern of recognition. Rather than having strict codes, each neuron being qualitatively different, there might be gradients that would inform the cells of their relative positions on the D–V or A–P axes. In the developing chick eye, the ability to regulate is extremely well defined. Crossland and co-workers (1974) removed segments of the optic cup at various stages of chick retina development and observed the innervation of the tectum by the remaining neurons. When the retinal segment was removed prior to day 3, the chicken developed a normal tectal projection, the remaining neurons reaching all the areas of the tectum. After day 3, however, the ablation of retinal segments resulted in uninnervated portions of the tectum.

There is presently a growing body of evidence suggestive of functional gradients in the retina and the tectum. Roth and his colleagues (Roth and Marchase, 1976; Marchase, 1977) have found that cells prepared from the ventral half of chick neural retina preferentially adhere

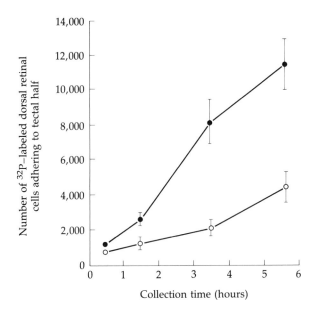

FIGURE 30

Differential adhesion of dorsal radioactive chick retina cells to dorsal and ventral tectal halves. Radioactive cells from the dorsal half of 7-day chick retinas were added to dorsal (○) and ventral (●) halves of 12-day chick optic tecta. The data show the selective adhesion of the dorsal retina cells to the ventral tectal tissue. (After Roth and Marchase, 1976.)

to dorsal halves of the tectum (Figure 30). Gottlieb and co-workers (1976) found that neurons taken from the most dorsal part of the chick retina adhere preferentially to the most ventral portion of the tectum and that the extreme ventral neurons of the retina preferentially adhere to the most dorsal extremes of the tectum. These results were confirmed in other experimental conditions using axonal tips rather than entire neurons (Halfter et al., 1981). Gradients of specific compounds have been observed on the cell surface of the neural retina, but their functions are not known. Trisler and his co-workers (1981), using monoclonal antibodies to chick neural retina cells, found one antibody that localized in a gradient fashion across the retina surface. The antigen identified by this antibody is 35 times more concentrated in the dorsal region of the retina than it is at the anterior margin, but it has not been determined if this gradient has any functional significance in directing neuronal attachments.

The biochemistry of such gradients is also being explored. Marchase (1977) used a series of enzymes to try to abolish the difference in the ability of dorsal and ventral tectal halves to bind dorsal retinal cells. He found that when he treated either the dorsal retinal cells or the dorsal tectal halves with the enzyme hexosaminidase, he was able to abolish this specificity. This enzyme specifically cleaves N-acetylhexosamines from the ends of glycoproteins and glycolipids. The same enzyme had little effect on the ventral cells of either the tectum or the retina. Marchase's model, shown in Figure 31, entails two molecules arranged in inverse gradients on the cell surfaces of the retina and the tectum. The first gradient is that of a glycoprotein (or glycolipid) carrying a carbohydrate chain ending in N-acetylgalactosamine. This is the sugar recognized by the enzymes that were seen to destroy the specificity of retinotectal binding in vitro. The concentration of this molecule is highest at the dorsal ends of both the retina and the tectum. The second gradient consists of a molecule capable of recognizing the N-acetylgalactosamine. This is hypothesized to be a glycosyltransferase enzyme—a protein capable of forming a lock-and-key complex with the N-acetyl-

FIGURE 31
Model of retinotectal specificity based on the interaction of two gradients of complementary molecules along the dorsal–ventral axis. In both the retina and the tectum, one type of molecule is concentrated in the dorsal region (●) and the other type of molecule is concentrated in the ventral region (Ψ). The most stable connections would therefore form between the dorsal retina and the ventral tectum and vice versa. (From Marchase, 1977.)

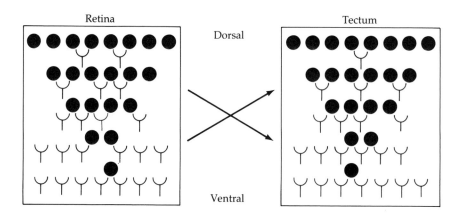

galactosamine. Thus, the dorsal portion of the retina would be specifically recognized by the ventral portion of the tectum and vice versa.

Qualitative neuronal differences

The preceding model is a quantitative type of hypothesis wherein all cells have the same type of glue, but the amounts differ. It may be useful for directing axons to a general location, but there is debate over whether such coding can accurately pinpoint specific locations. There is evidence from other organisms that each neuron might have qualitatively different cell surface molecules and that these molecules might be important in synaptic specificity. The nervous system of the leech consists of 34 paired ganglia containing about 400 neurons each. Individual neurons have been identified, and the functions of many of these neurons are known. Zipser and McKay (1981) injected the leech nervous system into mice and obtained hundreds of monoclonal antibodies, which bound to various regions of the nervous system. Out of the original 300 monoclonal antibodies, 41 of them recognized subsets of the leech nervous system, a finding indicating that the cell surface of the neurons differed. In some cases, such differences could be correlated with function. Monoclonal antibody Lan 3-1 bound specifically to a single pair of neurons in the midbody ganglia and recognized the same right and left neurons in each of the ganglia (Figure 32). These pairs of neurons are known to control the process of penile eversion in mating leeches. Another monoclonal antibody, Lan 3-2, recognized all four neurons that respond to noxious mechanical stimuli. "The situation," according to Zipser and McKay, "seems quite analogous to colour-coded electrical cable containing many wires, where each wire has its own molecule (dye) to facilitate proper recognition and connection at terminals."

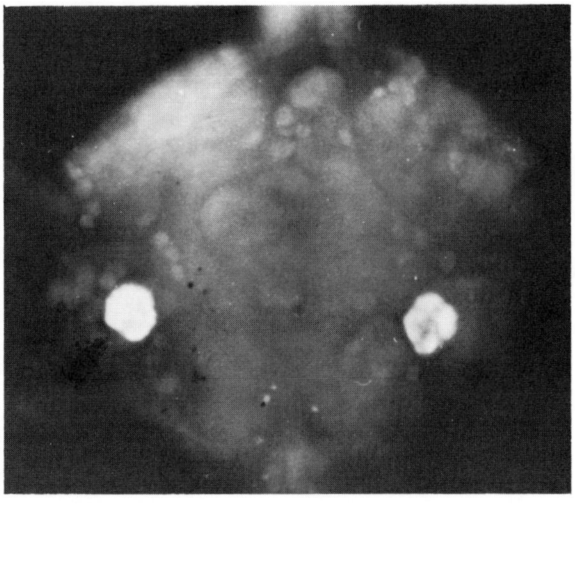

FIGURE 32
Specific functional neurons stained by monoclonal antibodies to cell surface components. (A) Lan 3-1 antibodies recognize a single pair of neurons in a particular ganglion. These neurons function in penile eversion. (B) Another set of neurons recognized by Lan 3-2 antibodies; these neurons respond to noxious stimulants of the leech's skin. (From Zipser and McKay, 1981; courtesy of B. Zipser.)

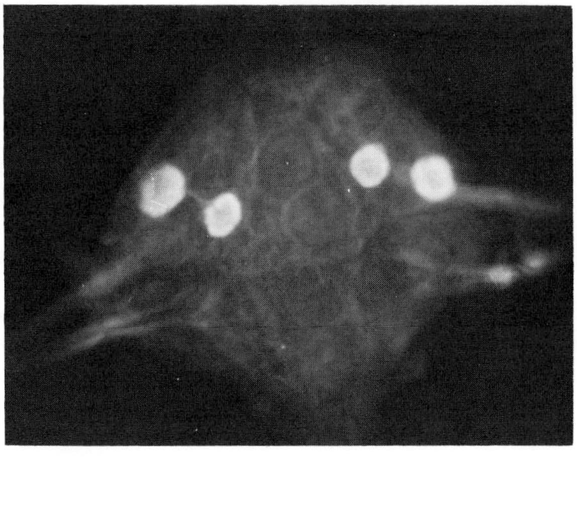

(A)

(B)

SIDELIGHTS & SPECULATIONS

Cell surface address markers and lymphocyte migration

Even though the notion that specific cell surface markers for axon migration and connection remains hypothetical, there is at least one other case in development wherein the interactions of cell surfaces does appear to determine the end result of migration. Mature lymphocytes are mobile cells that circulate continuously among the lymphoid organs—notably the spleen, tonsils, and gut-associated lymphoid tissue—through the blood and lymphatic systems. In some fashion, lymphocytes have to recognize where and when to leave the bloodstream and enter a lymphoid organ. B cells prefer to reside in the gut-associated lymphoid tissue, whereas T cells reside in the peripheral nodes. (In the spleen, both B and T cells can be found, but they reside in different places within that organ.) The mechanism for this preference appears to be the cells' ability to undergo specific interactions with the endothelial cells lining the blood vessels serving these organs. This endothelium is found in the HIGH ENDOTHELIAL VENULES (HEV), which are found immediately after the capillary network of these organs. Mature (but not immature) B cells will bind to gut-associated lymphoid HEV in vitro, whereas T cells show a preference to bind to peripheral node HEV. By immunizing mice with a lymphocyte tumor clone, the cells of which specifically bound to the peripheral node HEV, Gallatin and his colleagues (1983) were able to obtain a monoclonal antibody that blocked the specific adherence of T cells to the peripheral node HEV. When mouse lymphocytes were treated with this antibody, the binding of cells to Peyer's patches remained normal, whereas the association of these cells with peripheral node HEV was almost completely abolished (Figure 33). The antibody has been shown to recognize a 80,000-d protein on the lymphocyte cell surface. Thus, there appears to be a molecule on the cell surface of the T lymphocyte that is responsible for interacting

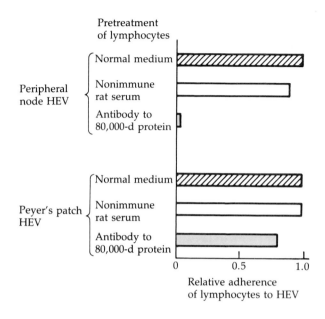

FIGURE 33
Pretreatment of lymphocytes with monoclonal antibodies against a specific 80,000-dalton protein blocks their normal binding to peripheral node HEV but not to Peyer's patch HEV. Lymph node cells were treated with normal medium, medium containing nonimmune rat serum, or medium containing the rat antibody to the specific protein. Lymphocytes were washed to remove medium and unbound antibodies and were incubated on frozen sections of HEV. After incubation, sections were washed and adhering cells were counted. (From Gallatin et al., 1983.)

with the peripheral node HEV; blocking this compound with antibodies inhibits the specific adherence of these cells to that HEV.

The migration of cells to a specific organ has therefore been able to be correlated with a specific cell–cell interaction.

Development of behaviors

One of the most fascinating aspects of neurobiology is the correlation of certain neuronal connections with certain behaviors. Equally fascinating are the observations that many complex behavioral patterns are inherently present in the "circuitry" of the brain at birth. The heartbeat of a 19-day chicken embryo will quicken when it hears the distress call, and no other call will evoke this response (Gottlieb, 1965). Furthermore, a newly hatched chick will immediately seek shelter if presented with the shadow of a hawk. The actual hawk is not needed, as the shadow cast by a paper silhouette will suffice, and the shadow of no other bird will cause this response (Tinbergen, 1951). There appears, then, to be certain neuronal connections that lead to inherent behaviors in higher vertebrates.

Some of the most interesting research on mammalian neuronal patterning concerns the effects of sensory deprivation on the developing visual system in kittens and monkeys. The paths by which electrical impulses pass from the retina to the brain in mammals is shown in Figure 34. Axons from the retinal ganglion cells form the two optic nerves, which meet at the optic chiasm. As in *Xenopus*, some fibers go to the opposite (CONTRALATERAL) side of the brain, but unlike the other vertebrates, mammalian retinal cells also send imputs into the same (IPSILATERAL) side of the brain. These nerves end at the two LATERAL GENICULATE BODIES. Here, imputs from each eye are kept separate, the uppermost and anterior layers receiving the neurons from the contralateral eye, the middle of the bodies receiving imputs from the ipsilateral

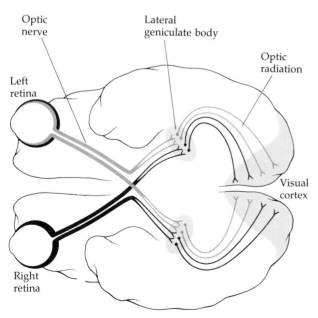

FIGURE 34
Major pathways of the mammalian visual system. In mammals, the optic nerve from each eye branches, sending nerve fibers to a lateral geniculate body on each side of the brain. Fibers from the left side of each retina enter the left lateral geniculate body, and fibers from the right side of each retina enter the right lateral geniculate body. From here, fibers from each lateral geniculate body enter the visual cortex on the same side.

eye. The situation becomes more complicated as neurons from the lateral geniculate bodies connect with the neurons of the VISUAL CORTEX. Over 80 percent of the neural cells in the cortex receive imputs from both eyes. The result is binocular vision and depth perception. Another remarkable finding is that the retinocortical projection is the same for both eyes. If a cortical neuron is stimulated by light flashing across a region of the left eye 5 degrees above and 1 degree to the left of the optic nerve, it will also be stimulated by a light flashing across a region of the right eye 5 degrees above and 1 degree to the left of the optic nerve. Moreover, the response evoked in the cortical cell when both eyes are stimulated is greater than the response when either retina is stimulated alone.

Hubel and his co-workers (see Hubel, 1967) demonstrated that the development of the nervous system depended to some degree upon the experience of the individual during a critical period of development. In other words, all of neuronal development is not encoded in the genome: some is actually learned. Moreover, the role of experience appears to be that of strengthening or stabilizing neuronal connections that are already present at birth. These conclusions come from studies of partial sensory blocks. Hubel and his co-workers sewed shut the right eyelids of newborn kittens and left them closed for 3 months, after which time, they unsewed the lids. The cortical cells of such kittens could not be stimulated by shining light in the right eye. Almost all the imputs into the visual cortex came from the left eye only, and the behavior of the kittens revealed the inadequacy of their right eyes. When the left eyes of these animals were covered, the kittens were functionally blind. Because the lateral geniculate body neurons appeared to be stimulated from both right and left eyes, the physiological defect appeared to be between the lateral geniculate bodies and those of the visual cortex. In rhesus monkeys, where similar phenomena are observed, the defect has been correlated with a lack of protein synthesis in the lateral geniculate neurons that are innervated by the covered eye (Kennedy et al., 1981).

Although it would be tempting to conclude that the resulting blindness was due to a failure to learn the proper visual connections, this is not the case. Rather, a kitten is born with all the appropriate neuronal connections in its visual system (Hubel and Wiesel, 1963). However, when one eye is covered early in the kitten's life, those connections into the visual cortex are taken over by the other eye. There is competition occurring, and experience plays a role in strengthening the connections from one eye or the other. Thus, when both eyes of a kitten are sewn shut for 3 months, the visual system remains intact. Both eyes become functional. The critical time in kitten development for this validation of neuronal connections is between the fourth and the sixth week of the kitten's life. Monocular deprivation up to the fourth week produces little or no physiological deficit; but after 6 weeks, it produces all the characterisitc neuronal changes. Two new principles, then, can be seen in the patterning of the mammalian visual system. First, neuronal connections involved in vision are present even before the animal sees; and second, experience plays an important role in determining whether or not certain of these connections remain.

Morphogenesis by specific cell death

In both the developing limb and nervous system, cell death plays a major role in the final pattern. In most, if not all, regions of the developing vertebrate nervous system, neurons are overproduced and then die in specific regions. This procedure helps produce the stratified layers characteristic of the central nervous system tissues. In addition, there is a drastic reduction in the number of axons that innervate peripheral tissues. This reduction often occurs in a manner such that axons that had formerly branched to innervate several targets become restricted to the innervation of a single target only. This can readily be observed in the elimination of fetal muscle synapses. The innervation of fetal or newborn muscles differs significantly from that of adult muscles (Figure 35). In mature mammals, **each motor axon stimulates a single muscle fiber** through several connections. However, in newborn mammals, each axon can innervate several different muscle fibers (Redfern, 1970; Bagust et al., 1973). In most muscles, this paring down of axonal connections occurs during the first weeks of life.

It is possible that this elimination is driven by competition for neurotrophic factors produced from the target tissue and by neuronal activity. The more activity, the greater the competition appears to become and the quicker the extra synapses are eliminated. Purves and Lichtman (1980; Lichtman and Purves, 1983) speculate that just as imbalanced activity leads to competition and synapse elimination in the monocularly deprived kittens, so might normal imbalanced activity lead to synapse elimination in other parts of the nervous system. In this model, there is a kind of "natural selection" whereby the most utilized synapses would continue to function and remain. Such competition was first proposed by Roux in 1881 but has not been taken seriously until recently. The elimination of

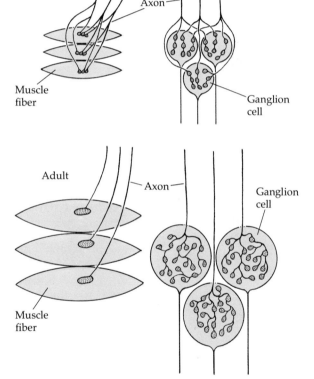

FIGURE 35
Diagram of synapse elimination in developing skeletal muscle (left) and in sympathetic ganglia (right). In newborn mammals each axon stimulates several different muscle fibers or ganglion cells; in mature mammals, however, each axon stimulates only one fiber or cell. (From Purves and Lichtman, 1985.)

FIGURE 36
Patterns of cell death in leg primordia of chick (A) and duck (B) embryos. Shading indicates areas of cell death. (After Saunders and Fallon, 1966.)

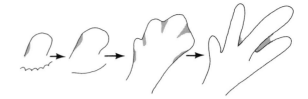

(A) CHICK LEG PRIMORDIUM
Extensive cell death

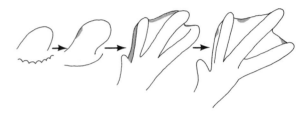

(B) DUCK LEG PRIMORDIUM
Minimal cell death

extra synapses may enable more efficient responses in the adult organism, but the progressive restriction of axons to single targets may be responsible for the loss of neuronal plasticity that is observed during childhood.

Cell death also plays a role in sculpting the limb, and there are obvious cases where differential cell death is genetically programmed and has been selected for during evolution. One obvious case involves the webbing or nonwebbing of feet. The difference between a duck's foot and that of a chicken is the presence or absence of cell death between the digits (Figure 36). Cell death also is thought to be important in separating the ulna from the radius. Saunders and his co-workers (1962) have shown that after a certain stage, cells destined to die will do so even if transplanted to another region of the embryo. Before that time, however, transplantation will save them. Between the time when the cell's death is determined and when death actually takes place, DNA, RNA, and protein synthesis decrease dramatically (Pollak and Fallon, 1976).

Pattern formation in *Hydra* and *Dugesia:* Morphallaxis

So far we have discussed regional events where pattern formation appears to be achieved by proximate cell–cell interactions. There are some organisms, however, that routinely reproduce by budding or by fission and that then regenerate their missing parts. These parts are not replaced by cell division (as in the epimorphic fields of limbs and imaginal discs) but by the respecification of a very malleable cell population that preexists in the adult organism. This type of regeneration is called MORPHALLAXIS. In this type of regeneration, positional information appears to result from gradients of diffusible substances.

One of the best-studied cases of such regeneration occurs in the coelenterate genus *Hydra*. *Hydra* was named after the mythical beast that was capable of growing back heads when they were cut off. Hydras are capable of sexual reproduction but do so only under adverse circumstances, such as overcrowding or starvation. Under normal conditions, they will bud off new individuals (Figure 37). The hydra is basically a narrow tube, two cell-layers thick, with an interstitial matrix layer sandwiched between the two epithelial sheets. The major polarity

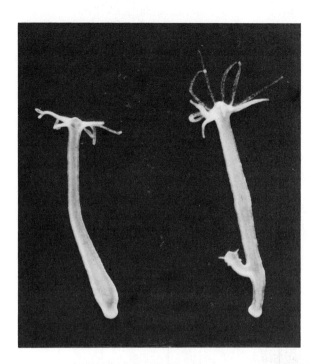

FIGURE 37
Budding in *Hydra pirardi*. The hydra on the left lacks a bud; the presence and polarity of the body on the right-hand specimen is obvious. (Photograph courtesy of G. E. Lesh-Laurie.)

in hydra is along the proximal–distal axis. The proximal end consists of a mucus-secreting BASAL DISC, which anchors the hydra to its substrate. The distal end consists of a mouth and a HYPOSTOME, a ring of tentacles that surround the mouth. Between them is the body trunk. The buds form as outgrowths from the body wall and have the same polarity as their parent.

When a hydra is cut in half, the half containing the basal disc will form a new hypostome and the half containing the hypostome will generate a new basal disc. Moreover, if a hydra is cut into several pieces perpendicular to the body axis, the middle rings of cells will regenerate both a basal disc and a hypostome. Thus, every region of the hydra can give rise to an entire organism. Yet, hypostomes do not form just anywhere; they form only at the distal end. This situation is reminiscent of the "harmonious equipotential" sea urchin embryo; and the mechanism governing the integration of the totipotent parts into a unified polar whole may be very similar. In both cases, the spatial information appears to be provided by a series of gradients arising from the two poles.

Grafting experiments gave further evidence for the existence of gradients in the hydra. When hypostome tissue is added to the middle of another hydra, it forms a new bud with the hypostome extending outward (Figure 38). When basal disc cells were so grafted, the new bud extends a basal disc. Moreover, when cells from the two poles are grafted together onto a host hydra, the resulting bud has no polarity (Browne, 1909; Newman, 1974). In other experiments (Rand et al., 1926),

FIGURE 38
Evidence for gradients in *Hydra*. (A) Hypostome plus midhydra regions grafted laterally to the host's middle region cause distal (hypostome) induction. (B) Basal disc plus midhydra regions grafted laterally to the host's middle region cause proximal (basal disc) induction. (C) Hypostome plus basal disc region grafted laterally cause little or no induction with no distinct polarity. (After Newman, 1974.)

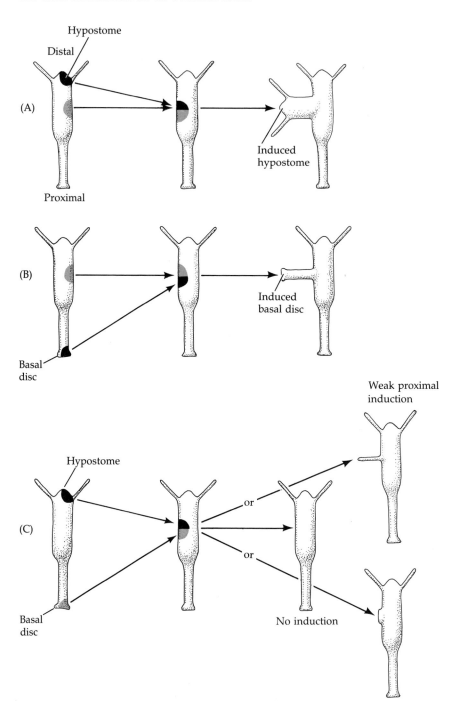

it was shown that the normal regeneration of the hypostome can be inhibited when an intact hypostome is grafted onto the body.

These experiments have been interpreted to indicate the existence of both a head activator gradient and a head inhibitor gradient. The HEAD ACTIVATOR GRADIENT can be measured by implanting rings of tissue from various levels of the donor hydra into a particular region in the trunk of the host. The higher the level of activator in the donor tissue, the greater the frequency of implants that form hypostomes. This activating factor is found to be most concentrated in the apical end and decreases linearly toward the base. This gradient appears to be very stable and is probably responsible for the formation of the hypostome at the distal end of the hydra. The HEAD INHIBITOR GRADIENT is measured by inserting subhypostomal tissue (having a relatively high concentration of head activator factor) into various different regions of host hydras. The more powerful the inhibition at the levels in the host hydras, the fewer grafts that can form heads. The source of this gradient is also the hypostome, but this factor is probably a very labile and diffusible molecule. This factor prevents heads from forming elsewhere while an intact head is present. Therefore, when a head is present, both gradients are functioning, but when the head is removed, the labile inhibitor disappears. The result is a head on the distal end of the piece of hydra (Webster, 1966; Wilby and Webster, 1970). The basal disc has also been seen to be the source of two gradients, one that activates foot development (Grimmelikhuijzen and Schaller, 1977) and one that inhibits it (Schmidt and Schaller, 1976).

Although the integration of a small field can theoretically be controlled by diffusible substances that form gradients (Crick, 1970), there must be some mechanism to ensure that the morphogen does not diffuse into the environment or into the animal's gut. Wakeford (1979) demonstrated that positional information from grafted hypostomes was transferred to the host only if gap junctions had formed between host and donor cells. It seems probable, then, that the morphogen is carried from cell to cell by these junctions rather than by diffusing through the medium. Although some of these compounds have recently been purified, their mechanism of action remains unknown.

Many types of flatworms also have the option of reproducing sexually or asexually. Most freshwater species, for example, *Dugesia*, reproduce by transverse fission, dividing themselves in two. Separation of the two halves appears to be dependent upon motion. The posterior portion of the worm adheres to a substrate, whereas the anterior portion of the worm crawls forward until the worm breaks in half. Each half then regenerates the missing parts (Figure 39). The asexually reproducing flatworms have considerable powers of regeneration, which are absent in the purely sexual species.

This type of fission can be mimicked in the laboratory. One result of such experiments is the confirmation of the organism's ability to regenerate in a polar fashion. There appears to be an anterior–posterior

FIGURE 39
Regeneration of missing parts in the flatworm
Dugesia. Anterior half regenerates its missing poste-
rior; posterior half regenerates its eyes and brain.
(From Goss, 1969.)

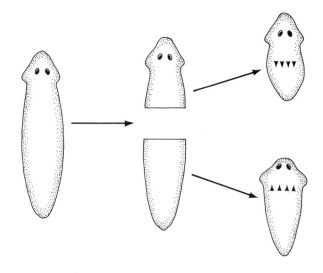

gradient system such that the cut anterior surface will form a head and
the cut posterior surface will form a tail. As in *Hydra*, middle pieces are
capable of regenerating both the head and the tail. Also as in *Hydra*,
there is an exception to this rule. If the middle slice is too thin, the
piece will grow two heads (Figure 40). It is believed that the anterior–
posterior gradient is not sufficiently delineated in such a small segment
and that the piece cannot determine "which end is up." There is also
evidence (Flickinger and Coward, 1962) that the gradient is one of
metabolic activity and that whichever end is more metabolically active
forms the head. If a flatworm is cut and the head of the worm is placed
in a solution containing a metabolic inhibitor, its hind region will form
a head rather than a tail. In addition to retaining anterior–posterior
polarity, flatworms can also regenerate the missing half if split longitudinally.

Regeneration in flatworms also appears to be morphallactic. Flatworms contain a population of relatively undifferentiated cells (NEO-
BLASTS), which migrate to the wound surface to differentiate into the
new structures. These cells are highly susceptible to radiation, and
flatworms do not regenerate well, if at all, when these cells are destroyed. There also seems to be a progressive restriction in what the
neoblasts are capable of forming. In the regeneration of the head, the
brain is the first organ reformed. It seems that, given the option, neoblasts will form the brain. Once the brain has been formed, that option
is closed, and the neoblasts form the eyes, muscles, and intestine, in
that order (Brøndsted, 1954).

As in *Hydra*, the pattern of flatworm regeneration appears to be
controlled by diffusible gradients. In more specialized animals, gradients cannot control the determination of the entire embryo, but diffusible regulators of differentiation abound. These substances are the

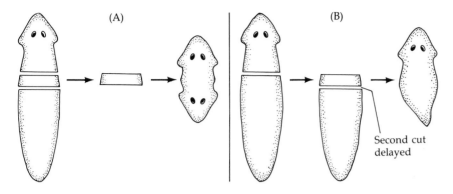

FIGURE 40
Evidence for gradients in flatworm regeneration. (A) If too narrow a segment is cut, two heads facing in opposite directions appear. (B) If the second cut is delayed, however, an equally thin slice forms one head in the proper polarity. (From Goss, 1969.)

Second cut delayed

hormones, and they will be the subject of the next chapter. The study of pattern formation is at a very exciting stage. Monoclonal antibodies should prove very useful in identifying the molecules involved in the reactions between adjacent cells, and cloned genes should provide us with new insights into how such molecules are genetically regulated. At this moment, we can only catalog the phenomena of spatial development, but (as Needham predicted), the molecular analysis of pattern formation should not lag far behind.

Literature cited

Akam, M. E. (1983). The location of *Ultrabithorax* transcripts in Drosophila tissue sections. *EMBO J.* 2: 2075–2084.

Akers, R. M., Mosher, D. F. and Lilien, J. E. (1981). Promotion of retinal neurite outgrowth by substratum-bound fibronectin. *Dev. Biol.* 86: 179–188.

Angeletti, P. and Vigneti, E. (1971). Assay of nerve growth factor (NFG) in subcellular fractions of peripheral tissues by microcomplement fixation. *Brain Res.* 33: 601–604.

Bagust, J., Lewis, D. M. and Westerman, R. A. (1973). Polyneuronal innervation of kitten skeletal muscle. *J. Physiol.* (Lond.) 229: 241–255.

Bender, W. and Spierer, P. (1983). Reported in G. North. Cloning the genes that specify fruit flies. *Nature* 303: 134–136.

Bender, W., Akam, M., Karch, F., Beachy, P. A., Peifer, M., Spierer, P., Lewis, E. B. and Hogness, D. S. (1983). Molecular genetics of the *bithorax* complex in *Drosophila melanogaster*. *Science* 221: 23–29.

Brøndsted, H. V. (1954). The time-graded representation field in planarian and some of its cytophysiological implications. In J. A. Kitching (ed.), *Recent Developments in Cell Physiology. Proc. 7th Symp. Colston Res.*

Soc. 7: 121–138.

Browne, E. N. (1909). The production of new hydranths in *Hydra* by the insertion of small grafts. *J. Exp. Zool.* 7: 1–23.

Bryant, S. V. (1976). Regenerative failure of double half limbs in *Notophthalamus viridescens*. *Nature* 263: 676–679.

Bryant, S. V., French, V. and Bryant, P. J. (1981). Distal regeneration and symmetry. *Science* 212: 993–1002.

Bryant, S. V. and Iten, L. E. (1976). Supernumerary limbs in amphibians: Experimental production in *Notophthalamus viridescens* and a new interpretation of their formation. *Dev. Biol.* 50: 212–234.

Butler, E. G. (1935). Studies on limb regeneration in X-rayed *Amblystoma* larvae. *Anat. Rec.* 62: 295–307.

Carlson, B. M. (1972). Muscle morphogenesis in axolotl limb regenerates after removal of stump musculature. *Dev. Biol.* 28: 487–497.

Crick, F. H. (1970). Diffusion in embryogenesis. *Nature* 225: 420–422.

Crick, F. H. C. and Lawrence, P. A. (1975). Compartments and polyclones in insect development. *Science* 189: 340–347.

Crossland, W. J., Cowan, W. M., Rogers, C. A. and

Kelly, J. (1974). The specification of the retinal-tectal projection in the chick. *J. Comp. Neurol.* 155: 127–164.

Detwiler, S. R. (1918). Experiments on the development of the shoulder girdle and the anterior limb of *Amblystoma punctatum*. *J. Exp. Zool.* 25: 499–538.

Flickinger, R. A. and Coward, S. J. (1962). The induction of cephalic differentiation in regenerating *Dugesia dorotocephala* in the presence of the normal head and unwounded tails. *Dev. Biol.* 5: 179–204.

French, V., Bryant, P. J. and Bryant, S. V. (1976). Pattern regulation in epimorphic fields. *Science* 193: 969–981.

Gallatin, W. M., Weissman, I. L. and Butcher, E. C. (1983). A cell surface molecule involved in organ-specific homing of lymphocytes. *Nature* 304: 30–35.

Garcia-Bellido, A. (1975). Genetic control of wing disc development in *Drosophila. Ciba Found. Symp.* 29: 161–182.

Gaze, R. M. (1978). The problem of specificity in the formation of nerve connections. In D. R. Garrod (ed.), *Specificity of Embryological Interactions.* Chapman and Hall, London, pp. 51–96.

Godspodarowicz, D. and Mescher, A. L. (1980). Fibroblast growth factor and the control of vertebrate regeneration and repair. *Ann. N.Y. Acad. Sci.* 339: 151–174.

Goss, R. J. (1969). *Principles of Regeneration.* Academic, New York.

Gottlieb, D. I., Rock, K. and Glaser, L. (1976). A gradient of adhesive specificity in developing avian retina. *Proc. Natl. Acad. Sci. USA* 73: 410–414.

Gottlieb, G. (1965). Prenatal auditory sensitivity in chickens and ducks. *Science* 147: 1596–1598.

Grimmelikhuijzen, C. J. P. and Schaller, H. C. (1977). Isolation of a substance activating foot formation in hydra. *Cell Differ.* 6: 297–305.

Halfter, W., Claviez, M. and Schwarz, U. (1981). Preferential adhesion of tectal membranes to anterior embryonic chick retina neurites. *Nature* 292: 67–70.

Hamburger, V. (1938). Morphogenetic and axial self-differentiation of transplanted limb primordia of 2-day chick embryos. *J. Exp. Zool.* 77: 379–400.

Harrison, R. G. (1910). The outgrowth of the nerve fiber as a mode of protoplasmic movement. *J. Exp. Zool.* 9: 787–848.

Harrison, R. G. (1918). Experiments on the development of the forelimb of *Amblystoma*, a self-differentiating equipotential system. *J. Exp. Zool.* 25: 413–461.

Hertwig, G. (1925). Haploidkernige Transplante als Organisatoran diploidkeniger Extremitaten be *Triton. Anat. Anz.* (Suppl.) 60: 112–118.

Hinchliffe, J. R. and Johnson, D. R. (1980). *The Development of the Vertebrate Limb.* Clarendon, New York.

Hubel, D. H. (1967). Effects of distortion of sensory input on the visual system of kittens. *Physiologist* 10: 17–45.

Hubel, D. H. and Wiesel, T. N. (1963). Receptive fields of cells in striate cortex of very young, visually inexperienced kittens. *J. Neurophysiol.* 26: 944–1002.

Hunt, R. K. (1975). Developmental program for retinotectal patterns. *Ciba Found. Symp.* 29: 131–150.

Hunt, R. K. and Jacobson, M. (1972). Development and stability of positional information in *Xenopus* retinal ganglion cells. *Proc. Natl. Acad. Sci. USA* 69: 780–783.

Iten, L. E. (1982). Pattern specification and pattern regulation in the embryonic chick limb bud. *Am. Zool.* 22: 117–129.

Iten, L. E. and Bryant, S. V. (1975). The interaction between blastema and stump in the establishment of the anterior–posterior and proximal–distal organization of the limb regenerate. *Dev. Biol.* 44: 119–147.

Iten, L. E. and Murphy, D. J. (1980). Pattern regulation in the embryonic chick limbs. Supernumerary limb formation with anterior (non-ZPA) limb bud tissue. *Dev. Biol.* 75: 373–385.

Jacobson, M. (1967). Retinal ganglion cells: Specification of central connections in larval *Xenopus laevis. Science* 155: 1106–1108.

Javois, L. C. and Iten, L. E. (1981). Position of origin of donor posterior chick wing bud tissue transplanted to an anterior host site determines the extra structures formed. *Dev. Biol.* 82: 329–342.

Javois, L. C. and Iten, L. E. (1982). Supernumerary limb structures after juxtaposing dorsal and ventral chick wing bud cells. *Dev. Biol.* 90: 127–143.

Kaufman, S., Shymko, R. H. and Trabert, K. (1978). Control of sequential compartment formation in *Drosophila. Science* 199: 259–270.

Kennedy, C., Suda, S., Smith, C. B., Miyaoka, M., Ito, M. and Sokoloff, L. (1981). Changes in protein synthesis underlying functional plasticity in immature monkey visual system. *Proc. Natl. Acad. Sci. USA* 78: 3950–3953.

Lawrence, P. A. (1981). The cellular basis of segmentation in insects. *Cell* 26: 3–10.

Letourneau, P. C. (1975a). Possible roles for cell-to-substratum adhesion in neuronal morphogenesis. *Dev. Biol.* 44: 77–91.

Letourneau, P. C. (1975b). Cell-to-substratum adhesion and guidance of axonal elongation. *Dev. Biol.* 44: 92–101.

Levi-Montalcini, R. (1966). The nerve growth factor: Its mode of action on sensory and sympathetic nerve cells. *Harvey Lect.* 60: 217–259.

Levi-Montalcini, R. (1976). The nerve growth factor: Its

role in growth, differentiation, and formation of the sympathetic adrenergic neuron. *Prog. Brain Res.* 45: 235–258.

Lewis, E. B. (1978). A gene complex controlling segmentation in *Drosophila*. *Nature* 276: 565–570.

Lichtman, J. W. and Purves, D. (1983). Activity-mediated neural change. *Nature* 301: 563–564.

Lund, R. D. (1976). *Development and Plasticity of the Brain: An Introduction.* Oxford University Press, New York.

Marchase, R. B. (1977). Biochemical investigations of retinotectal adhesive specificity. *J. Cell Biol.* 75: 237–257.

McGinnis, W., Levine, M. S., Hafen, E., Kuroiwa, A. and Gehring, W. J. (1984a). The conserved DNA sequence in homeotic genes of the Drosophila *Antennapedia* and *bithorax* complexes. *Nature* 308: 424–433.

McGinnis, W., Garber, R. L., Wirz, J., Kuroiwa, A. and Gehring, W. J. (1984b). A homologous protein-coding sequence in Drosophila homeotic genes and its conservation in other metazoans. *Cell* 37: 403–408.

Mescher, A. L. and Godspodarowicz, D. (1979). Mitogenic effect of a growth factor derived from myelin on denervated regenerates of newt forelimbs. *J. Exp. Zool.* 207: 497–503.

Mescher, A. L. and Tassava, R. A. (1975). Denervation effects on DNA replication and mitosis during the initiation of limb regeneration in adult newts. *Dev. Biol.* 44: 187–197.

Morata, G. and Lawrence, P. A. (1977). Homeotic genes control compartments and cell determination in *Drosophila*. *Nature* 265: 211–216.

Muneoka, K. and Bryant, S. V. (1982). Evidence that patterning mechanisms in developing and regenerating limbs are the same. *Nature* 298: 369–371.

Newman, S. A. (1974). The interaction of the organizing regions in hydra and its possible relation to the role of the cut end in regeneration. *J. Embryol. Exp. Morphol.* 31: 541–555.

Nüsslein-Volhard, C. and Wieschaus, E. (1980). Mutations affecting segment number and polarity in *Drosophila*. *Nature* 287: 795–801.

Pollack, E. D. and Muhlach, W. L. (1981). Stage dependency in eliciting target-dependent enhanced neurite outgrowth from spinal cord explants *in vitro*. *Dev. Biol.* 86: 259–263.

Pollak, R. D. and Fallon, J. F. (1976). Autoradiographic analysis of macromolecular synthesis in prospectively necrotic cells of the chick limb bud. II. Nucleic acids. *Exp. Cell Res.* 100: 15–22.

Postlethwait, J. H. and Schneiderman, H. A. (1971). Pattern formation and determination in the antenna of the homeotic mutant *Antennapedia* of Drosophila melanogaster. *Dev. Biol.* 25: 606–640.

Purves, D. and Lichtman, J. W. (1980). Elimination of synapses in the developing nervous system. *Science* 210: 153–157.

Purves, D. and Lichtman, J. W. (1985). *Principles of Neural Development.* Sinauer, Sunderland, MA.

Rand, H. W., Board, J. F. and Minnich, D. E. (1926). Localization of formative agencies in hydra. *Proc. Natl. Acad. Sci. USA* 12: 565–570.

Redfern, P. A. (1970). Neuromuscular transmission in new-born rats. *J. Physiol.* 209: 701–709.

Rose, S. M. (1962). Tissue-arc control of regeneration in the amphibian limb. In D. Rudnick (ed.), *Regeneration*. Ronald, New York, pp. 153–176.

Roth, S. and Marchase, R. B. (1976). An *in vitro* assay for retinotectal specificity. In S. H. Barondes (ed.), *Neuronal Recognition*. Plenum, New York, pp. 227–248.

Roux, W. (1881). *Der Kampf der Thiele im Organismus. Ein Beitrage zur Vervollständingung der mechanischen Zweckmässigkeitslehre.* W. Engelmann, Leipzig.

Rowe, D. A. and Fallon, J. F. (1981). The effect of removing posterior apical ectodermal ridge of the chick wing and leg on pattern formation. *J. Embryol. Exp. Morphol.* 65 (Suppl.): 309–325.

Rowe, D. A. and Fallon, J. F. (1982). Normal anterior pattern formation after barrier placement in the chick leg: Further evidence on the action of polarizing zone. *Embryol. Exp. Morphol.* 69: 1–6.

Rubin, L. and Saunders, J. W., Jr. (1972). Ectodermal–mesodermal interactions in the growth of limbs in the chick embryo: Constancy and temporal limits of the ectodermal induction. *Dev. Biol.* 28: 94–112.

Saunders, J. W., Jr. (1948). The proximal–distal sequence of origin of the parts of the chick wing and the role of the ectoderm. *J. Exp. Zool.* 108: 363–404.

Saunders, J. W., Jr. (1972). Developmental control of three-dimensional polarity of the avian limb. *Ann. N.Y. Acad. Sci.* 193: 29–42.

Saunders, J. W., Jr. (1977). The experimental analysis of chick limb bud development. In A. A. Ede and J. R. Hinchliffe (eds.), *Vertebrate Limb and Somite Morphogenesis*. Cambridge University Press, Cambridge, pp. 1–24.

Saunders, J. W., Jr. (1982). *Developmental Biology.* Macmillan, New York.

Saunders, J. W., Jr. and Fallon, J. F. (1966). Cell death in morphogenesis. In M. Locke (ed.), *Major Problems of Developmental Biology*. Academic, New York, pp. 289–314.

Saunders, J. W., Jr. and Gasseling, M. T. (1968). Ectodermal–mesodermal interactions in the origin of limb symmetry. In R. Fleischmajer and R. E. Billingham (eds.), *Epithelial–Mesenchymal Interactions*. Williams & Wilkins, Baltimore, pp. 78–97.

Saunders, J. W., Jr., Gasseling, M. T., and Saunders, L. C. (1962). Cellular death in morphogenesis of the avian wing. *Dev. Biol.* 5: 147–178.

Schmidt, T. and Schaller, H. C. (1976). Evidence for a foot-inhibiting substance in hydra. *Cell Differ.* 5: 151–159.

Silver, J. and Sidman, R. L. (1980). A mechanism for guidance and topologic patterning of retinal ganglion cell axons. *J. Comp. Neurol.* 189: 101–111.

Singer, M. (1954). Induction of regeneration of the forelimb of the postmetamorphic frog by augmentation of the nerve supply. *J. Exp. Zool.* 126: 419–472.

Singer, M., Norlander, R. and Egar, M. (1979). Axonal guidance during embryogenesis and regeneration in the spinal cord of the newt: The blueprint hypothesis of neuronal pathway patterning. *J. Comp. Neurol.* 185: 1–22.

Singer, M. and Caston, J. D. (1972) Neurotrophic dependance of macromolecular synthesis in the early limb regenerate of the newt, *Triturus. J. Embryol. Exp. Morphol.* 28: 1–11.

Sperry, R. W. (1951). Mechanisms of neural maturation. In S. S. Stevens (ed.), *Handbook of Experimental Psychology.* Wiley, New York, pp. 236–280.

Sperry, R. W. (1965). Embryogenesis of behavioral nerve nets. In R. L. Dehaan and H. Ursprung (eds.), *Organogenesis.* Holt, Rinehart & Winston, New York, pp. 161–186.

Stocum, D. L. and Fallon, J. F. (1982). Control of pattern formation in urodele limb ontogeny: A review and a hypothesis. *J. Embryol. Exp. Morphol.* 69: 7–36.

Struhl, G. (1981). A gene product required for correct initiation of segmental determination in *Drosophila. Nature* 293: 36–41.

Summerbell, D. (1974). A quantitative analysis of the effect of excision of the AER from the chick limb bud. *J. Embryol. Exp. Morphol.* 32: 651–660.

Summerbell, D. (1979). The zone of polarizing activity: Evidence for a role in abnormal chick limb morphogenesis. *J. Embryol. Exp. Morphol.* 50: 217–233.

Summerbell, D. and Lewis, J. H. (1975). Time, place, and positional value in the chick limb bud. *J. Embryol. Exp. Morphol.* 33: 621–643.

Tickle, C. (1981). Limb regeneration. *Am. Sci.* 69: 639–646.

Tickle, C., Summerbell, D. and Wolpert, L. (1975). Positional signaling and specification of digits in chick limb morphogenesis. *Nature* 254: 199–202.

Tinbergen, N. (1951). *The Study of Instinct.* Clarendon Press, Oxford.

Trisler, G. D., Schneider, M. D. and Nirenberg, M. (1981). A topographic gradient of molecules in retina can be used to identify neuron position. *Proc. Natl. Acad. Sci. USA* 78: 2145–2149.

Wakeford, R. J. (1979). Cell contact and positional communication in hydra. *J. Embryol. Exp. Morphol.* 54: 171–183.

Webster, G. (1966). Studies on pattern regulation in hydra. II. Factors controlling hypostome formation. *J. Embryol. Exp. Morphol.* 16: 105–122.

Weir, M. P. and Lo, C. (1982). Gap junction communication compartments in the *Drosophila* wing disc. *Proc. Natl. Acad. Sci. USA* 79: 3232–3235.

Weiss, P. (1939). *Principles of Development.* Holt, Rinehart & Winston, New York.

Weiss, P. A. (1955). Nervous system. In B. H. Willier, P. A. Weiss and V. Hamburger (eds.), *Analysis of Development.* Saunders, Philadelphia, pp. 346–402.

Wilby, O. K. and Webster, G. (1970). Experimental studies on axial polarity in hydra. *J Embryol. Exp. Morphol.* 24: 595–613.

Wolpert, L. (1969). Positional information and the spatial pattern of cellular formation. *J. Theoret. Biol.* 25: 1–47.

Wolpert, L. (1977). *The Development of Pattern and Form in Animals.* Carolina Biological, Burlington, N.C.

Zipser, B. and McKay, R. (1981). Monoclonal antibodies distinguish identifiable neurones in the leech. *Nature* 289: 549–554.

Cell interactions at a distance: hormones as mediators of development

The old order changeth, yielding place to the new.
—ALFRED LORD TENNYSON (1885)

The earth-bound early stages built enormous digestive tracts and hauled them around on caterpillar treads. Later in the life-history these assets could be liquidated and reinvested in the construction of an entirely new organism—a flying-machine devoted to sex.
—CARROLL M. WILLIAMS (1958)

Introduction

Organ formation in animals is accomplished by the interactions of numerous cell types. In the preceding three chapters, we have seen how developmental interactions can be mediated by adjacent cell populations. In this chapter, we shall discuss the regulation of development by diffusible molecules that travel long distances from one cell type to another. Diffusible regulators of development can be divided into two general categories: HORMONES, which are secreted by one cell type and which cause changes in the *differentiation* and *morphogenesis* of other tissues; and CHALONES, which are secreted by an organ and which regulate the *growth* of that *same* organ.

Metamorphosis: The hormonal reactivation of development

Because minute quantities of hormones can accomplish their action, it is exceedingly difficult to isolate them from embryos. The most thorough analysis of the hormonal control of development has therefore centered upon the reactivation of development known as METAMORPHOSIS.

In many species of animals, embryonic development leads to a larval stage with characteristics very different from those of the adult organism. Very often, the larval forms are specialized for some function, such as growth or dispersal. The pluteus larva of the sea urchin, for instance, can travel on ocean currents, whereas the adult urchin leads a sedentary existence. The caterpillar larva of butterflies and moths are specialized for feeding, whereas their adult forms are specialized for flight and reproduction and often lack the mouth parts necessary for eating. The division of functions between larva and adult is often remarkably distinct (Wald, 1981). Mayflies hatch from eggs and develop for several months. All this development enables them to spend one day as fully developed winged insects, mating quickly before they die.

As expected, the larval form and the adult form often live in different environments. Thus, natural selection works differently on the larval and adult organisms. The voracious gypsy moth caterpillar comes in many different colors, all of which metamorphose into the same adult. Conversely, the swallowtail butterfly *Papilio polytes* has many adult varieties but only one form of caterpillar (Wigglesworth, 1954).

The transition from the larva to the adult is called metamorphosis. Here, the developmental processes are reactivated by specific hormones, and the entire organism changes to prepare itself for its new mode of existence. These changes are not solely ones of form. In amphibian tadpoles, metamorphosis causes the developmental maturation of liver enzymes, hemoglobin, and eye pigments, as well as the remodeling of the nervous, digestive, and reproductive systems. Thus, metamorphosis is a time of dramatic developmental change affecting the entire organism.

In this chapter, we shall look at four cases in which diffusible hormones are seen to reactivate the developmental processes after birth: amphibian metamorphosis, insect metamorphosis, human puberty, and mouse breast development.

Amphibian metamorphosis

The phenomenon of amphibian metamorphosis

In amphibians, metamorphosis is generally associated with those changes that prepare an aquatic organism for a terrestrial existence. In urodeles (salamanders), the changes include the resorption of the tail fin, the destruction of the external gills, and the change of skin structure. In anurans (frogs and toads), the metamorphic changes are most striking, and almost every organ is subject to modification (Table 1). The changes in form are very obvious (Figure 1). Regressive changes

TABLE 1
Summary of some metamorphic changes in anurans

System	Larva	Adult
Locomotory	Aquatic; tail fins	Terrestrial; tailless tetrapod
Respiratory	Gills, skin, lungs; larval hemoglobins	Skin, lungs; adult hemoglobins
Circulatory	Aortic arches; aorta; anterior, posterior, and common cardinal veins	Carotid arch; systemic arch; jugular veins
Nutritional	*Vegetarian:* Long spiral gut—intestinal symbionts; small mouth—horny jaws, labial teeth	*Carnivorous:* Short gut—proteases; large mouth—long tongue
Nervous	Lack of nictitating membrane, porphyropsin, lateral line system—Mauthner's neurons	Development of ocular muscles, nictitating membrane, rhodopsin, loss of lateral line system—degeneration of Mauthner's neurons; tympanic membrane
Excretory	Mesonephros: Largely ammonia, some urea (ammonotelic)	Mesonephros: Largely urea, high activity of enzymes of ornithine-urea cycle (ureotelic)

Source: Turner and Bagnara (1976).

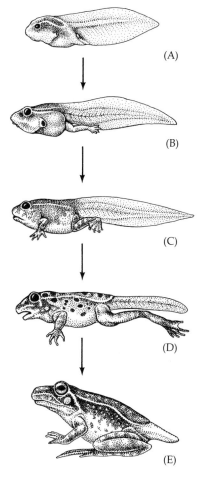

(A)

(B)

(C)

(D)

(E)

FIGURE 1

Sequence of metamorphosis in the frog *Rana pipiens*. (A) Premetamorphic tadpole; (B) prometamorphic tadpole showing hindlimb growth; (C) onset of metamorphic climax as forelimbs emerge; (D, E) climax stages.

include the destruction of the tadpole tail and internal gills, as well as the loss of the tadpole's horny teeth. At the same time, constructive processes such as limb development and dermoid gland construction are also evident. For locomotion, the paddle tail recedes while the hindlimbs and forelimbs differentiate. The horny teeth constructed for tearing pond plants disappear as the mouth and jaw take a new shape and the tongue muscle develops. Meanwhile the large intestine characteristic of herbivores shortens to suit the more carnivorous diet of the adult frog. The gills regress and the gill arches degenerate. The lungs enlarge, and muscles and cartilage develop for pumping air in and out of the lungs. The sensory apparatus changes, too, as the lateral line system of the tadpole degenerates and the eye and ear undergo further differentiation. In the ear, the middle ear develops, as does the tympanic membrane so characteristic of frogs and toads. In the eye, both nictitating membrane and eyelids emerge. Moreover, the eye pigment changes. In tadpoles, as in freshwater fishes, the major retinal photopigment is PORPHYROPSIN, a complex between the protein opsin and the aldehyde of vitamin A_2 (Figure 2). In adult frogs, the pigment changes to RHODOPSIN, the characteristic photopigment of terrestrial and marine vertebrates. Rhodopsin consists of opsin conjugated to the aldehyde of vitamin A_1 (Wald, 1945, 1981; Smith-Gill and Carver, 1981).

Other biochemical events are also associated with metamorphosis. Tadpole hemoglobin binds oxygen faster and releases it more slowly than adult hemoglobin (McCutcheon, 1936). Moreover, Riggs (1951) showed that the binding of oxygen by tadpole hemoglobin is independent of pH, whereas frog hemoglobin (as in most other vertebrate hemoglobins) shows increased oxygen binding as the pH rises (Bohr effect). Probably the most striking biochemical change in metamorpho-

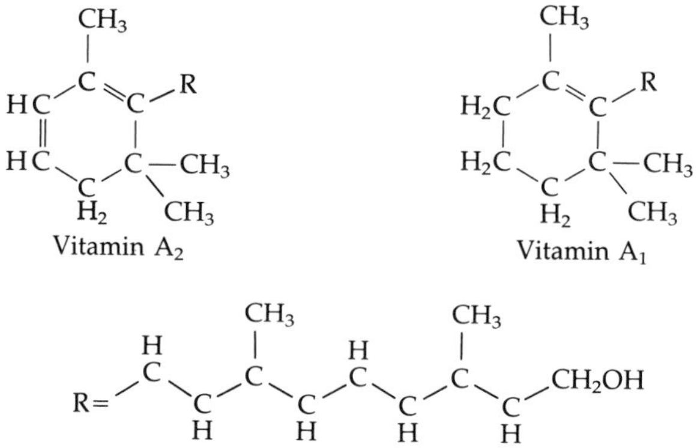

CH_3

Vitamin A_2

CH_3

Vitamin A_1

$R = \quad CH_3 \quad CH_3 \quad CH_2OH$

FIGURE 2

Structural formula of the vitamin A portions of the visual pigments of tadpoles and frogs. Tadpole porphyropsin (using vitamin A_2) absorbs light at longer wavelengths than frog rhodopsin (using vitamin A_1).

sis is the induction of those enzymes necessary for the production of urea. Tadpoles, like most freshwater fishes, are AMMONOTELIC; that is, they excrete ammonia. Adult frogs, however, like most land vertebrates, are UREOTELIC, excreting urea. During metamorphosis, the liver develops those enzymes necessary to create urea from carbon dioxide and ammonia. These enzymes constitute the UREA CYCLE, and each of them is seen to arise during metamorphosis (Figure 3).

Hormonal control of amphibian metamorphosis

All these diverse changes are brought about by the diffusion of the hormones THYROXINE (T_4) and TRIIODOTHYRONINE (T_3) from the thyroid

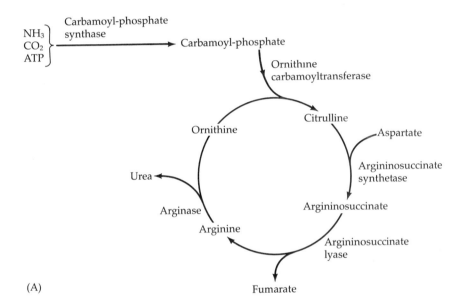

(A)

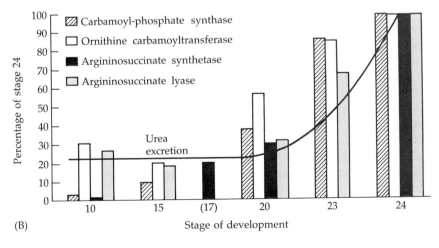

(B)

FIGURE 3
Development of the urea cycle during anuran metamorphosis. (A) The major features of the urea cycle, by which nitrogenous wastes can be detoxified and excreted. (B) Emergence of urea cycle enzyme activities correlated with metamorphic changes in the frog *Rana catesbeiana*. (After Cohen, 1970.)

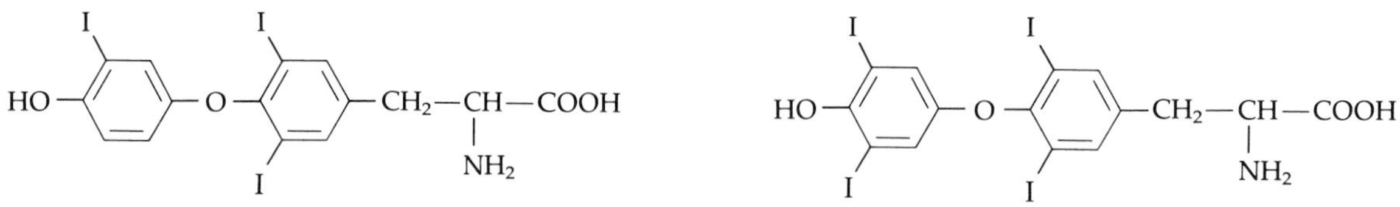

FIGURE 4
Formulas of thyroxine (T_4) and triiodothyronine (T_3).

during metamorphosis (Figure 4). It is now believed that T_3 is the active hormone, as it will cause the metamorphic changes in thyroidectomized tadpoles in much lower concentrations than will T_4 (Kistler et al., 1977; Robinson et al., 1977). In both urodeles and anurans, the thyroid produces small amounts of T_3 and T_4 in the larva, but they are overbalanced by the diffusion of the hormone PROLACTIN from the anterior pituitary. Prolactin acts as a larval growth hormone and inhibits metamorphosis (Etkin and Gona, 1967; Bern et al., 1967). During metamorphosis the thyroid hormones T_3 and T_4 increase in concentration, thereby causing the tadpoles to become frogs and stimulating larval newts to become land-dwelling efts. In salamanders, thyroxine will become overpowered by prolactin later in development, thereby causing them to return to the water to spawn as adults (Grant and Grant, 1958). Thus, in some salamanders, there are two metamorphoses. The first stimulated by thyroxine, the second induced by prolactin. Moreover, even after the second metamorphosis, an injection of thyroxine will cause the adult newt to return to land as a dry-skinned eft (Grant and Cooper, 1964).

The control of metamorphosis by thyroid hormones was shown earlier in this century when Gudernatsch (1912) found that tadpoles metamorphosed prematurely when fed powdered sheep thyroid gland. Allen (1916) and Hoskins and Hoskins (1917) found that when they removed the thyroid rudiment from early tadpoles, the larva never metamorphosed, becoming instead giant tadpoles.

The release of T_3 is under the control of the hypothalamic portion of the brain (Etkin, 1968). During the period of larval growth (premetamorphosis), this portion of the brain is underdeveloped, and so the brain does not exert much control over the anterior pituitary (Figure 5). In the absence of hypothalamic regulation, the pituitary secretes high levels of prolactin but little or no THYROID STIMULATING HORMONE (TSH). Thus, T_3 levels are low and prolactin levels are high. As the hypothalamus develops, its production of THYROID STIMULATING HORMONE RELEASING FACTOR (TSH-RF) causes a rise in the level of TSH. This TSH causes the thyroid to increase T_3 and thyroxine synthesis. Thus, T_3 is released from the thyroid and its concentration gradually increases until the first changes of metamorphosis (PROMETAMORPHOSIS) appear. During this time, the hindlimbs start to enlarge. The rising T_3 levels also stimulate the further development of the median eminence of the pituitary, which mediates the flow of TSH-RF to reach the anterior pituitary. There is, then, a POSITIVE FEEDBACK wherein increased T_3 produc-

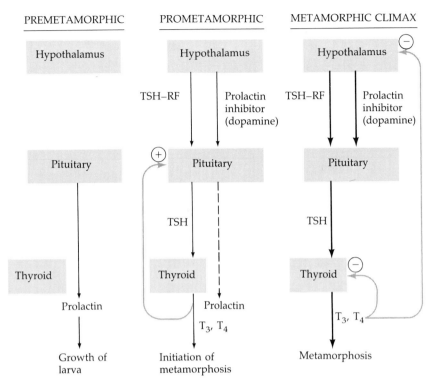

FIGURE 5

The hypothalamus–pituitary–thyroid axis during different stages of anuran metamorphosis. As the hypothalamus develops, it stimulates the pituitary to instruct thyroid hormone secretion and inhibit prolactin secretion. T_3, Triiodothyronine; T_4, thyroxine; TSH, thyroid stimulating hormone; TSH-RF, thyroid stimulating hormone releasing factor.

tion causes more T_3 production. In addition, the hypothalamus begins secreting a substance—probably dopamine (see White and Nicoll, 1981)—that inhibits the pituitary synthesis of prolactin. Thus, the ratio of T_3 to prolactin increases enormously. This leads to the METAMORPHIC CLIMAX in which most of the development associated with metamorphosis takes place. One of the effects of metamorphosis is the partial degeneration of the thyroid, and it is also possible that high levels of T_3 have an inhibitory effect on TSH or TSH-RF production (Goos, 1978). Thus, a new balance of hormones is acquired.

The different organs of the body respond differently to hormonal stimulation. The same stimulus will cause certain tissues to degenerate while causing others to develop and differentiate. For instance, tail degeneration is clearly associated with the increasing levels of thyroid hormones. This can be shown in vitro (Weber, 1967), where isolated tail pieces can be placed in agar dishes and subjected to chemical treatments (Figure 6). Those tails grown in untreated medium remain healthy, whereas those placed into medium containing thyroid hor-

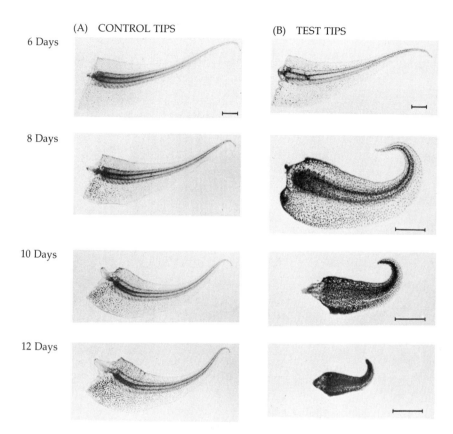

FIGURE 6

Regression of isolated tail ends under the influence of thyroxine. (A) Control tips from *Xenopus* tadpoles cultured in Holtfreter's salt solution for 6, 8, 10, and 12 days. (B) Treated tail tips at the same ages as controls; thyroxine was added to their salt solutions. The bar represents 1 mm. (From Weber, 1965; courtesy of R. Weber.)

mones undergo characteristic regression. Moreover, prolactin will inhibit the degeneration of the tail induced by thyroid hormones (Brown and Frye, 1969). The regression of the tail is thought to occur in three stages. First, protein synthesis decreases in the striated muscle cells of the tail (Little et al., 1973). Then, there is an increase in the lysosomal enzymes: cathepsin D (a protease), RNase, DNase, collagenase, phosphatase, and glycosidases in the epidermis, notocord, and nerve cord cells (Fox, 1973). Cell death is probably caused by the release of these enzymes into the cytoplasm. After this death, macrophages collect in the tail region, digesting the debris with their own proteolytic enzymes (Kaltenbach et al., 1979). The result is that the tail becomes a large sack of proteolytic enzymes (Figure 7).

The response to thyroid hormones is intrinsic to the organ itself and is not dependent on other factors. This can be seen when tail tips

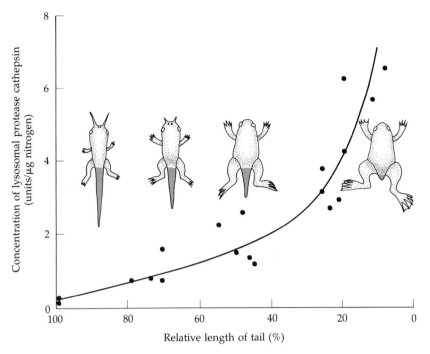

FIGURE 7
Increase of lysosomal protease cathepsin during tail regression in *Xenopus lae-vis.* The lysosomal enzymes are thought to be responsible for digesting the tail cells. (After Karp and Berrill, 1981.)

are transplanted to the trunk region or when eye cups are placed in the tail (Schwind, 1933; Geigy, 1941). The extra tail tip placed into the trunk is not protected from degenerating, and the eye retains its integrity despite its being within the degenerating tail (Figure 8). Thus, the degeneration of the tail represents a programmed cell death. Only specific tissues die when a signal is given. Such programmed cell deaths are important in molding the body. In humans, such degeneration occurs in the tissues between our finger and toes, and the degeneration of the human tail during week 4 of development resembles the regression of the tadpole tail (Fallon and Simandl, 1978).

One of the major problems of metamorphosis is the coordination of developmental events. The tail should not degenerate until some other means of locomotion—the limbs—have developed; and the gills should not regress until the animal can utilize its newly developed lung muscles. The means of coordinating these metamorphic events appears to be that different amounts of hormone are needed to produce different specific effects (Kollros, 1961). This is called the THRESHOLD CONCEPT. As the concentration of thyroid hormones gradually builds up, different events occur at different concentrations of the hormone. If tadpoles are deprived of their thyroids and are placed in a dilute solution of thyroid

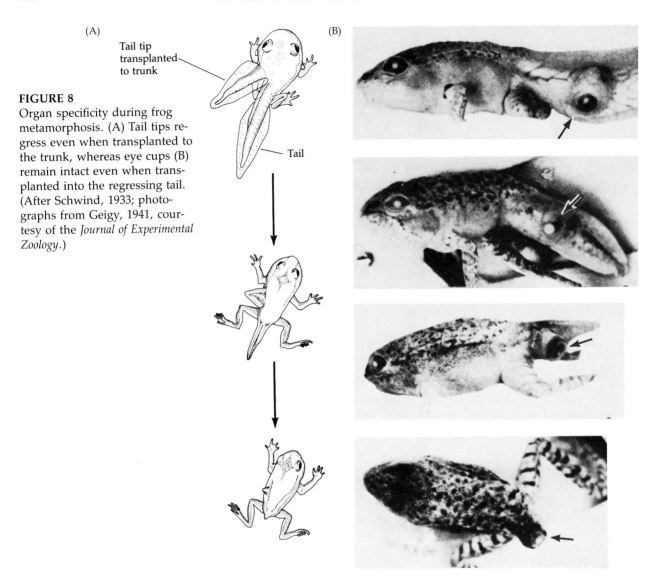

FIGURE 8
Organ specificity during frog metamorphosis. (A) Tail tips regress even when transplanted to the trunk, whereas eye cups (B) remain intact even when transplanted into the regressing tail. (After Schwind, 1933; photographs from Geigy, 1941, courtesy of the *Journal of Experimental Zoology*.)

hormones, the only morphological effects are the shortening of the intestines and accelerated hindlimb growth. However, at higher concentrations of thyroid hormones, tail regression is seen before the hindlimbs are formed. These experiments suggest that as thyroid hormone levels rise, the hindlimbs develop first and then the tail regresses. Thus, the timing of metamorphosis is regulated by the competency of different tissues to respond to thyroid hormones.

Thyroid hormones can cause existing tissues to break down or can remold the tissues to their adult function. The cells of the tadpole liver, for instance, are not destroyed and replaced during metamorphosis.

Rather, the structure of the existing liver cells is changed. This change is accompanied by dramatic increases in ribosomal and messenger RNA synthesis, and the rate of protein synthesis increases nearly 100-fold within 4 hours of thyroid hormone stimulation (Cohen et al., 1978). Many of these new mRNAs are those coding for the new functions of the adult liver.

As shown in Figure 9, carbamoylphosphate synthase (a urea cycle enzyme) is synthesized after the burst of RNA synthesis. Mori and co-workers (1979) have shown that the level of in vitro translatable mRNA for this enzyme increases shortly after thyroid hormones are given to bullfrog tadpoles. Thus, metamorphosis is to some extent controlled at the transcriptional level. Other evidence also shows the importance of transcriptional regulation. Weber (1967) demonstrated that the injection of actinomycin D into normal prometamorphic tadpoles inhibited tail regression and head remolding (Figure 10). Because T_3 is known to bind to nucleosomes in mammalian liver chromatin (Eberhardt et al., 1979), it is probable that T_3 can reprogram the genome to transcribe new types of messages for the adult organism. This does not mean that transcription is the only level of gene regulation working during metamorphosis, but it is obviously an important one.[1]

[1] In general, small hormones (such as thyroxine and sex steroids) are able to pass through the cell membrane and bind to specific receptors inside the cell. Peptide hormones, however, usually remain outside the cell and accomplish their activities by binding to specific receptors on the cell membrane. Often, the binding of the peptide hormone to its receptor activates a membrane bound enzyme, ADENYL CYCLASE, which converts ATP into cyclic AMP (cAMP). The cAMP is then able to alter the types of protein synthesized by the cell. Thus, hypothalamic TSH-RF functions by binding to TSH-RF receptors on the set of pituitary cells that make TSH. The binding stimulates the cells' adenyl cyclase to manufacture more cAMP, and the cAMP acts as a "second messenger" to instruct (in a manner not yet established) the pituitary cell to synthesize and release TSH. The action of TSH on the thyroid cells is also thought to be mediated by cyclic AMP.

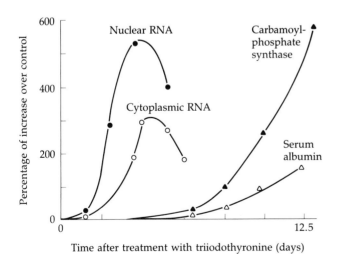

FIGURE 9

Molecular synthesis in *Rana catesbeiana* liver cells after treatment of tadpoles with triiodothyronine. Increases first in nuclear and then in cytoplasmic RNA are observed before increases in liver-specific proteins, especially those of the urea cycle. (After Graham and Wareing, 1976.)

(A)

(B)

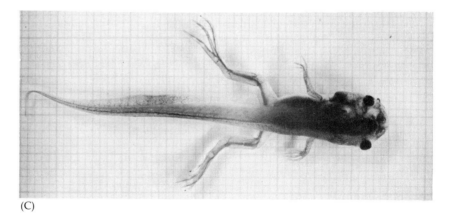

(C)

FIGURE 10
Role of new transcription in metamorphosis. (A) *Xenopus laevis* tadpole at the
onset of the metamorphic climax. (B) Normal tadpole 7 days later. (C) Tadpole at the same age as in (B); this tadpole was injected at stage (A) with a
single dose of actinomycin D. (From Weber, 1967; courtesy of R. Weber.)

Neoteny

Many salamanders are able to retain their larval form throughout their lives, becoming sexually mature without undergoing metamorphosis. Such retention of juvenile characteristics in sexually mature individuals is called NEOTENY. The degree to which metamorphosis occurs differs from species to species. The Mexican axolotl, *Ambystoma mexicanum*, does not undergo metamorphosis in nature because it does not release an active TSH to stimulate its thyroid glands (Prahlad and DeLanney, 1965; Norris et al., 1973; Taurog et al., 1974). Thus, when investigators gave *Ambystoma mexicanum* either thyroid hormones or TSH, they found that the salamander will metamorphose into an adult not seen in nature (Huxley, 1920). Other species, such as *Ambystoma tigrinum*, metamorphose only if given cues from the environment. Otherwise, they become neotenic, successfully mating as larvae. In part of its range, *A. tigrinum* is a neotenic salamander, paddling its way through the cold ponds of the Rocky Mountains. However, in the warmer portion of its range, the larval form of *A. tigrinum* is transitory, leading to the land-dwelling tiger salamander. Neotenic populations from the Rockies can be induced to undergo metamorphosis simply by placing them in water at higher temperatures. It appears that the hypothalamus of this species cannot produce TSH-RF at the low temperatures.

Some salamanders, however, are permanently neotenic, even in the laboratory. Whereas thyroxin was able to produce the long-lost adult form of *A. mexi-*

(A)

(B)

FIGURE 11
Metamorphosis induced in the axolotl. (A) Normal condition of the axolotl. (B) Specimen treated with thyroxine to induce metamorphosis. (Courtesy of G. Malacinski.)

canum (Figure 11), the neotenic species of *Necturus* and *Siren* remained refractile to thyroid hormones. It appears that the target tissues have lost their capacity to respond to thyroid hormones; thus, their neoteny is permanent (Frieden, 1981). The genetic lesions responsible for neoteny in several species is shown in Figure 12.

Gould (1977) has speculated that neoteny is a major factor in the evolution of more complex taxa. By retarding the development of somatic tissues, natural selection is given a flexible substrate. Neoteny would "provide an escape from specialization. Animals can slough off their highly specialized adult forms, return to the lability of youth, and prepare themselves for new evolutionary directions."

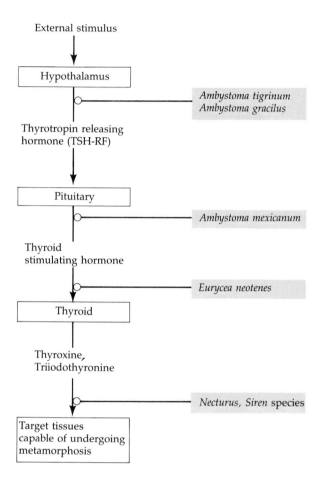

FIGURE 12

Stages along the hypothalamus–pituitary–thyroid axis of salamanders at which various species have blocked metamorphosis. *Eurycea*, *Necturus*, and *Siren* appear to have a receptor defect in the responsive tissues. *Eurycea* will metamorphose when exposed to extremely high concentrations of thyroxine, while *Necturus* and *Siren* do not respond to any dose. (After Frieden, 1981.)

Metamorphosis in insects

Whereas amphibian metamorphosis is characterized by the remolding of existing tissues, insect metamorphosis often involves the destruction of larval tissues and its replacement by an entirely different population of cells. Like amphibian metamorphosis, insect metamorphosis involves the balance between a hormone necessary for continued larval growth and a hormone capable of stimulating the new developmental changes.

There are three major patterns of insect development. A few insects, such as springtails, have no larval stage and are said to undergo DIRECT DEVELOPMENT. Other insects, notably grasshoppers and bugs, undergo a gradual, HEMIMETABOLOUS metamorphosis (Figure 13A). Here, adult organs are formed without any profound discontinuity. The rudiments of the wing, genital organs, and other adult structures are there at hatching, and they become more mature with each molt. At the last molt, the emerging insect is a winged and sexually mature adult. The larval form of a hemimetabolous insect is a NYMPH.

In the HOLOMETABOLOUS insects (flies, beetles, moths, and butterflies), there is a dramatic and sudden transformation between the larval and adult stages (Figure 13B). The juvenile larva (caterpillar, grub, maggot) undergoes a series of molts as it becomes larger. The newly hatched insect larva is covered by a hard CUTICLE. The insect must shed its old cuticle and then replace it with a larger one in order to grow. Thus, the postembryonic development of these insects consists of a succession of molts. The number of molts before becoming an adult is characteristic for the species, although environmental factors can lengthen or shorten the number. The stages between these molts are called INSTARS. Here, growth of the larva is the major characteristic. After its last instar stage, the larvae undergo a METAMORPHIC MOLT to

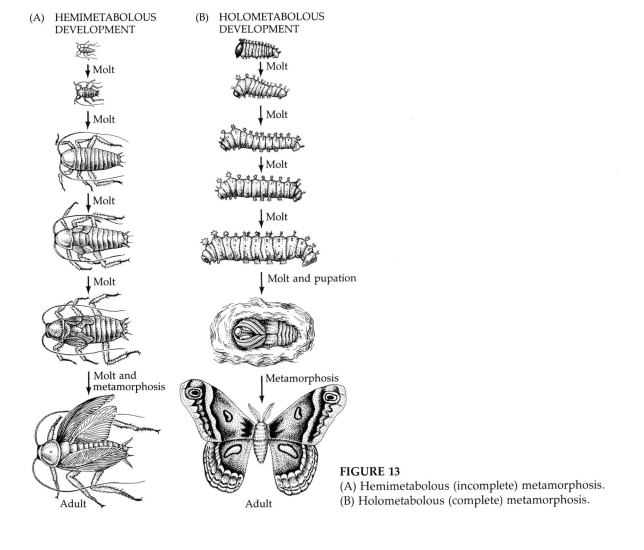

(A) HEMIMETABOLOUS DEVELOPMENT

(B) HOLOMETABOLOUS DEVELOPMENT

FIGURE 13
(A) Hemimetabolous (incomplete) metamorphosis.
(B) Holometabolous (complete) metamorphosis.

become a PUPA. The pupa does not feed, and its energy must come from those foods ingested by the larva.

It is within this pupa that the transformation of juvenile into adult occurs. The old body of the larva is systematically destroyed as new adult organs develop from undifferentiated nests of cells, the HISTIO-BLASTS and the IMAGINAL DISCS. When the adult organism (IMAGO) is developed, the IMAGINAL MOLT sheds the pupa, and the mature insect emerges. In holometabolous larvae, then, there are two cell populations: the larval cells, which are used for the functions of the juvenile, and the imaginal cells, which lie in clusters awaiting the signal to differentiate.

We have already encountered imaginal discs in our previous discussion of transdetermination and differential gene transcription. In *Drosophila*, there are 10 major pairs of imaginal discs, which reconstruct the entire adult integument (except for the abdomen), and a genital

FIGURE 14
Development of imaginal discs. (A) Longitudinal section through a 14-hour *Drosophila* larva showing the epidermal rudiment of a mesothoracic leg disc (L_2). (B) Similar section through a 30-hour larva that had been treated with colchicine for 3 hours to catch dividing cells. The cells of the mesothoracic leg disc (L_2) have proliferated and one of the cells has been arrested in its division. The prothoracic leg disc (L_1) can also been seen in this section. (From Madhavan and Schneiderman, 1977; courtesy of M. M. Madhavan.)

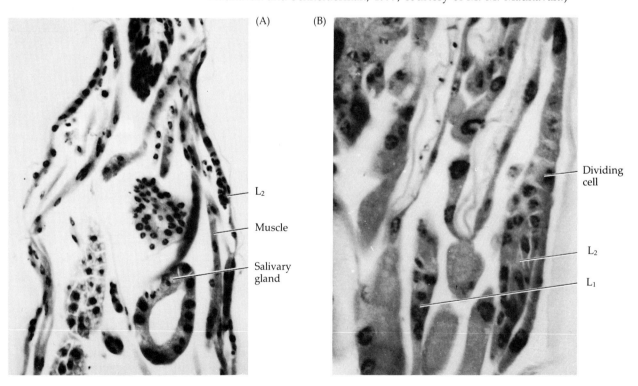

(A)

(B)

L_2

Muscle

Salivary gland

Dividing cell

L_2

L_1

disc, which forms the reproductive structures. The abdominal epidermis forms from a small group of imaginal cells called histioblasts, which lie in the region of the larval gut, whereas nests of imaginal cells located throughout the larva form the internal organs of the adult. The imaginal discs can be seen in the newly hatched larva as local thickenings of the epidermis such as that shown for a second thoracic leg disc in Figure 14. In *Drosophila*, these newly hatched eye–antenna, wing, haltere, leg, and genital discs contain 70, 38, 20, 36–45, and 64 cells, respectively (Madhavan and Schneiderman, 1977). Whereas most of the larval cells have a very limited mitotic capacity, the imaginal discs divide rapidly at specific characteristic times (Figure 15). As the cells proliferate, they form a tubular epithelium that folds in upon itself in a compact spiral

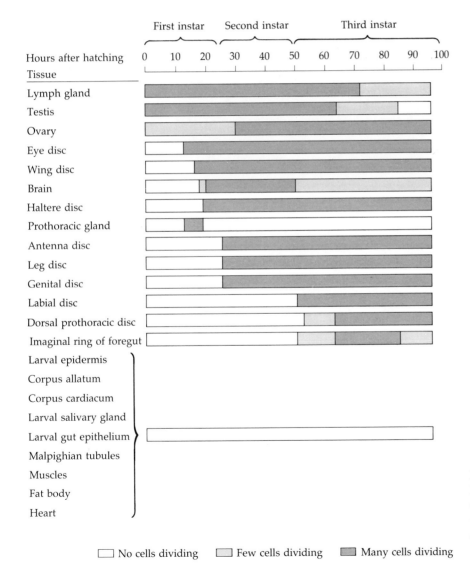

FIGURE 15
Patterns of cell division for several larval and imaginal tissues during the development of *Drosophila* larvae. (After Madhavan and Schneiderman, 1977.)

FIGURE 16
Imaginal disc eversion. Scanning electron micrograph
of *Drosophila* third instar leg disc (A) before and (B)
after eversion. (From Fristrom et al., 1977; courtesy
of D. Fristrom.)

(A)

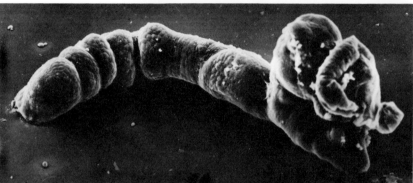

(B)

(Figure 16). The largest disc, that of the wing, contains some 60,000
cells, whereas the leg and haltere discs contain around 10,000 (Fristrom,
1972). At metamorphosis, these cells differentiate and EVERT. The fate
map and eversion sequence of the leg disc is shown in Figure 17. The
cells at the center of the disc telescope out to become the most distal
portions of the leg—the claws and the tarsus—and the outside cells
become the proximal structures—the coxa and the adjoining epidermis.
After differentiating, the cells of the appendages and epidermis secrete
a cuticle appropriate for the specific region. Although the discs are
primarily composed of epidermal cells, a small number of ADEPITHELIAL
cells migrate into the disc early in development. During the pupal
period, these cells give rise to the muscles and nerves that serve that
structure.

The process of eversion can be stopped by three sets of drugs:
(1) Inhibitors of RNA and protein synthesis inhibit eversion when added
to cultured imaginal discs at the same time as ecdysone. It is known
that RNA and protein synthesis occur prior to evagination (Figure 18),
and some of these proteins are needed for evagination to occur. (2)
Heavy water (D_2O) and cytochalasin B, two inhibitors of microfilament
function, also inhibit eversion, thereby indicating a need for actin microfilaments. (3) Concanavalin A, a plant lectin that binds to α-glucoside

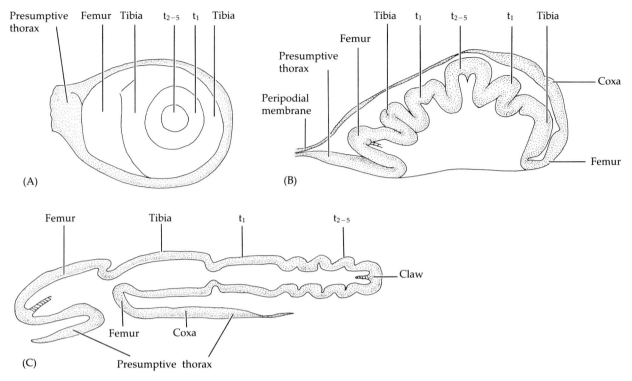

FIGURE 17
Eversion sequence of *Drosophila* leg disc. (A) Surface view of uneverted disc.
(B, C) Longitudinal section through everting and everted leg disc. t_1, Basitar-
sus; t_{2-5}, tarsal segments 2–5. (From Fristrom and Fristom, 1975. Courtesy of
D. Fristrom.)

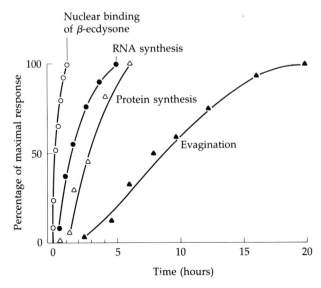

FIGURE 18
Molecular and morphological events after administra-
tion of ecdysone to cultured leg imaginal discs. (After
Fristrom et al., 1977.)

and α-mannoside sugar residues on the cell surface, also prohibits eversion. This compound impedes membrane fluidity. Taken together, these data suggest that the eversion of the imaginal discs requires new protein synthesis, a well-developed system of actin microfilaments, and some cellular communication by the cell surface (Fristrom et al., 1977).

Hormonal control of insect metamorphosis

Like amphibian metamorphosis, the metamorphosis of insects is regulated by two effector hormones controlled by neurosecretory peptide hormones in the brain (for reviews, see Granger and Bollenbacher, 1981; Gilbert and Goodman, 1981). Although the detailed mechanisms of metamorphosis differ between species, a general scheme has been elucidated for several organisms (Figure 19). The molting process is initiated in the brain, where neurosecretory cells release PROTHORACICO-TROPIC HORMONE (PTTH) in response to neural, hormonal, or environmental factors. Prothoracicotropic hormone is a peptide hormone with a molecular weight of approximately 40,000 and it stimulates the production of ECDYSONE by the PROTHORACIC GLAND. Ecdysone, however, is not an active hormone; rather it is a prohormone that must be converted into an active form. This conversion is accomplished by a heme-containing oxidase in the mitochondria of peripheral tissues such as the fat body. Here, the ecdysone is changed to the active hormone 20-HYDROXYECDYSONE (Figure 20).[2]

Each molt is occasioned by two pulses of ecdysone. The first pulse produces a small rise in the ecdysone concentration in the larval hemolymph (blood) and is probably important only during the metamorphic molt. The second, large pulse in ecdysone initiates the events associated with molting, independent of the hormonal conditions of the larva during the first ecdysone pulse. The 20-hydroxyecdysone produced by this pulse stimulates the epidermal cells to synthesize enzymes that digest and recycle the components of the cuticle. In some cases, environmental conditions can control molting, as in the case of the *Cecropia* moth. Here, PTTH secretion ceases after the pupa has formed. The pupa remains in this suspended state, called DIAPAUSE, throughout the winter. If not exposed to cold weather, diapause lasts indefinitely. Once exposed to 2 weeks of cold, however, the pupa can molt when returned to a warmer environment (Williams, 1952, 1956).

The second major effector hormone in insect development is JUVE-NILE HORMONE (JH). The structure of a common juvenile hormone active in *Cecropia* caterpillars is shown in Figure 20A. Juvenile hormone is secreted by the CORPORA ALLATA. The secretory cells of the corpora allata are active during larval molts but are inactive during the meta-

[2]Since its discovery in 1954, when Butenandt and Karlson isolated 25 mg of molting hormone from 500 kg of silkworm moth pupae, 20-hydroxyecdysone has gone under several names, including β-ecdysone, ecdysterone, and crustecdysone. In 1978, leading researchers in the field agreed on 20-hydroxyecdysone as the appropriate name.

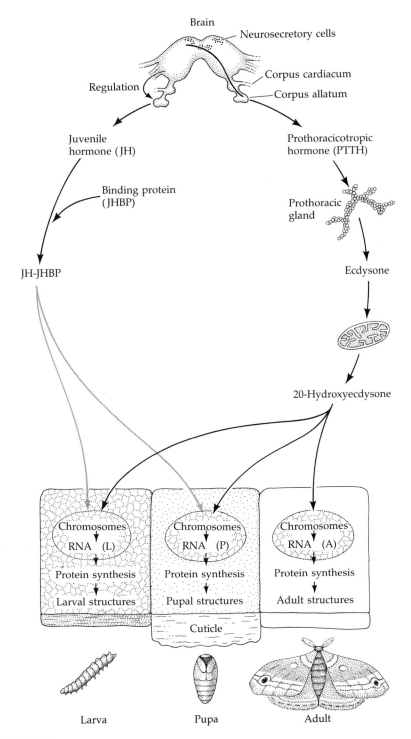

Brain
Neurosecretory cells
Corpus cardiacum
Corpus allatum
Regulation
Juvenile hormone (JH)
Prothoracicotropic hormone (PTTH)
Binding protein (JHBP)
Prothoracic gland
JH-JHBP
Ecdysone
20-Hydroxyecdysone
Chromosomes RNA (L) Protein synthesis Larval structures
Chromosomes RNA (P) Protein synthesis Pupal structures
Cuticle
Chromosomes RNA (A) Protein synthesis Adult structures
Larva
Pupa
Adult

FIGURE 19
Schematic diagram illustrating the control of molting and metamorphosis in the tobacco hornworm moth. (After Gilbert and Goodman, 1981.)

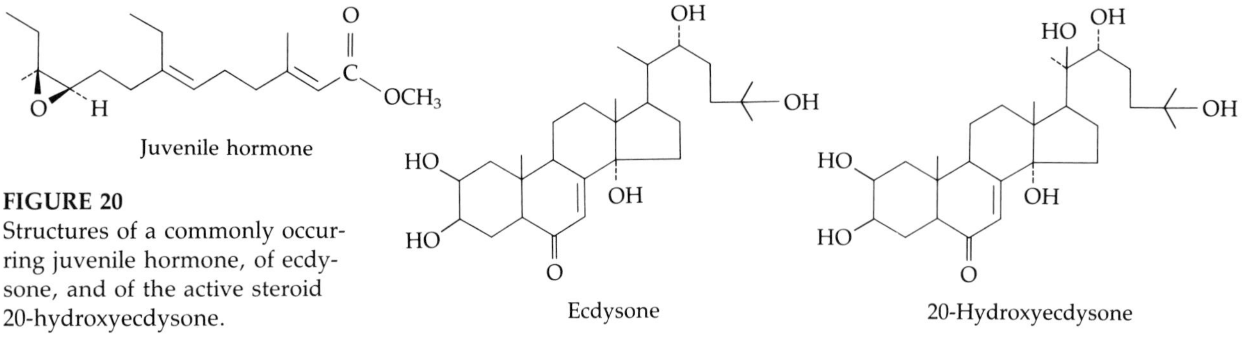

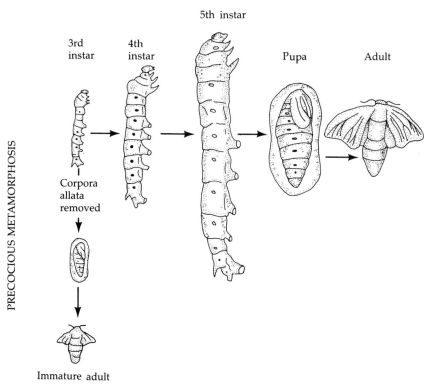

Juvenile hormone

Ecdysone

20-Hydroxyecdysone

FIGURE 20
Structures of a commonly occurring juvenile hormone, of ecdysone, and of the active steroid 20-hydroxyecdysone.

morphic molt. This hormone is responsible for preventing metamorphosis. As long as juvenile hormone is present, the ecdysone-stimulated molts result in a new larval instar stage. In the last larval instar stage, the synthesis of juvenile hormone is reduced, and the next molt results in the formation of the pupa. During pupation, the corpora allata does not release any juvenile hormone, and the pupa will metamorphose into the adult insect. Removal of the corpora allata will cause the premature metamorphosis of the earlier larvae into immature adults (Figure 21). Metamorphosis, then, occurs when the imaginal discs are exposed

NORMAL METAMORPHOSIS

5th instar

3rd
instar

4th
instar

Pupa

Adult

PRECOCIOUS METAMORPHOSIS

Corpora
allata
removed

Immature adult

FIGURE 21
Precocious metamorphosis of silkworm moths caused by the removal of the corpora allata during the third instar stage. Rather than continue molting through the fifth instar stage, the larva initiates pupation directly.

to 20-hydroxyecdysone in the absence of juvenile hormone. It appears that in the absence of juvenile hormone, 20-hydroxyecdysone induces the synthesis of the proteins from the late puffs. When juvenile hormone is present during the initial ecdysone pulse, the early puffs are produced as usual, but the later puffs are repressed (Richards, 1978).

SIDELIGHTS &
SPECULATIONS

Precocenes

Most insects are characterized by a larval form specialized for eating and an adult form specialized for reproduction and mobility. Thus, it is to a plant's advantage to shorten the voracious larval stage of an insect's development. Recently, two compounds that have been isolated from composite herbs have been found to cause the premature metamorphosis of certain insect larvae into sterile adults (Bowers et al., 1976). These compounds are called PRECOCENES and their chemical structures are shown in Figure 22A. When the larvae of these insects are dusted with either of these compounds, they will undergo one more molt

and then metamorphose into the adult form (Figure 22B). The precocenes accomplish this by causing the selective death of the corpora allata cells in the larva (Schooneveld, 1979; Pratt et al., 1980). These cells are responsible for synthesizing juvenile hormone. Thus, without juvenile hormone, the larva commences its metamorphic and imaginal molts. Moreover, juvenile hormone is also responsible for the maturation of the insect ovary. Without this hormone, females are sterile. So the precocenes are able to protect the plant by causing the premature metamorphosis of certain insect larvae into sterile adults.

(A)   Precocene 1   Precocene 2

(B)

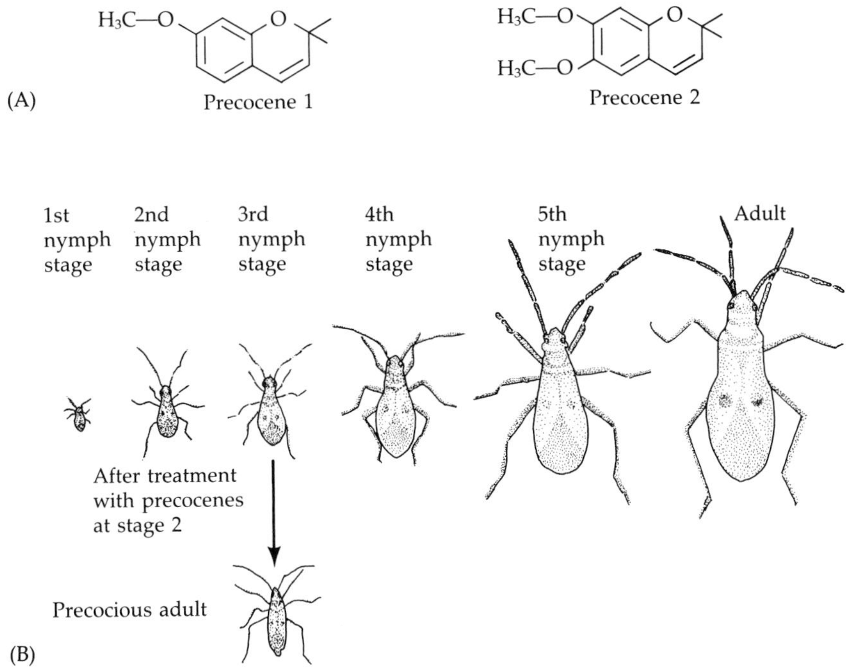

FIGURE 22
Precocious metamorphosis in the bug *Dysdercus* caused by precocenes. (A) Structures of two active precocenes found in plants. (B) Arrested development of *Dysdercus*. When second nymph stages are treated with precocenes, they metamorphose into sterile precocious adults rather than continue their normal developmental molting sequence. (After Bowers et al., 1976.)

Multiple hormonal interactions
in mammary gland development

Hormones have both constructive and destructive effects in development. In metamorphosis, hormones instruct some cells to die while instructing other cells to form new organs. In breast development, different hormones provide different information to the rudimentary tissue. Mammary development can be divided into three stages: the embryonic stage, the adolescent stage, and the lactating (pregnant) stage. The differentiated products of the mammary glands—casein and other milk proteins—are only made during the final stage (Topper and Freeman, 1980).

The embryonic stage

In the normal development of the female mouse, two bands of raised epidermal tissue appear on both sides of the ventral midline on day 11. This tissue is called MAMMARY RIDGE. Within this ridge, cells collect at centers of concentration and remain there, forming the MAMMARY BUDS

(A)

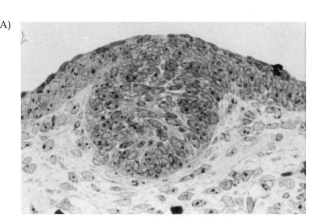

FIGURE 23
Sequence of early mammary gland development in the female mouse. (A) Mammary bud of 12-day embryo. Epithelial ectoderm cells protrude into mesenchyme. (B) Mammary cord of 15-day embryo. A small cleft at the bottom signals the initiation of branching. (C) Cord cavity extending to form a hollow lumen in the 20-day embryo. (From Hogg et al., 1983; photographs courtesy of C. Tickle.)

(B)

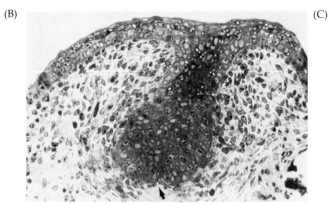

(C)

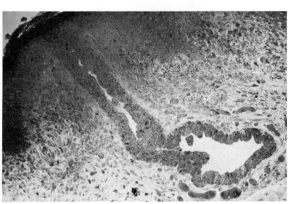

(Figure 23). In the mouse, there are five of these buds on each side; in humans, only one per side. In the days immediately prior to birth, the epithelial cells at these places proliferate rapidly, giving rise to the MAMMARY CORD. This cord opens at the skin at one end, forming the nipple, while its other end begins branching into ducts. Here development ceases until puberty.

The development of mammary tissue in male mice is identical to that of females until days 13–15 of gestation. At this time, the mesenchyme condenses around the center of the mammary bud, and the cells of the cord die. Thus, a small cord of epithelial cells is detached from the skin (Figure 24), and the mammary gland does not extend to the surface. No further development occurs.

This cell death in the mammary cord of males has been studied by culturing the mouse mammary buds in vitro. Such buds from female mice normally develop lobes connected to the surface (Figure 25). However, when testosterone is added to the culture medium, the buds degenerate. Mammary buds from male mice will also produce lobes, provided they are cultured in the absence of testosterone. Thus, the hormone testosterone prevents mammary development in the male. Testosterone causes this specific cell death by instructing the mesenchymal cells to destroy the epithelial cord. This was shown by a series

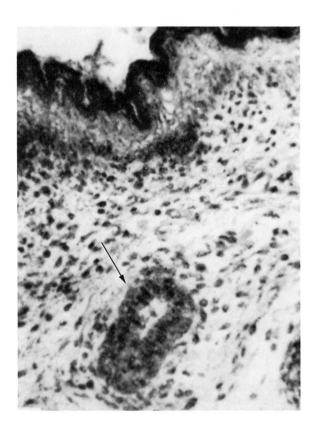

FIGURE 24
Mammary rudiment in a male mouse fetus. The rudiment has separated from the epidermis. (From Raynaud, 1961.)

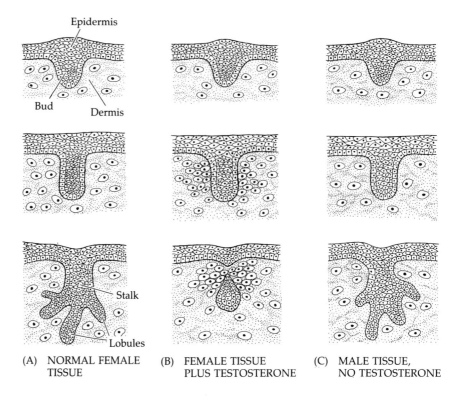

(A) NORMAL FEMALE TISSUE   (B) FEMALE TISSUE PLUS TESTOSTERONE   (C) MALE TISSUE, NO TESTOSTERONE

FIGURE 25
Role of testosterone in mediating the detachment of the mammary cord. (A) Female mouse mammary tissue, either in vivo or in culture, will grow downward from the epidermis and branch. (B) When female mouse mammary tissue is cultured in the presence of testosterone, the bud elongates, but mesenchymal cells aggregate around the stalk and the lower portion is cut off, just as in normal male development. (C) When male mouse mammary tissue is cultured in the absence of testosterone, it develops as it would in the female mouse. (After Kratochwil, 1971.)

of recombination experiments. There exists in mice (and in humans as well) a mutation called ANDROGEN INSENSITIVITY SYNDROME, in which male (XY) individuals do not make a functional testosterone receptor. Thus, even though these individuals have testes that are actively secreting testosterone, they are unable to respond to it. One of the results is that these individuals have female breast development (see Figure 7 in Chapter 19). Kratochwil and Schwartz (1975) isolated mesenchyme and epithelial cells from normal and mutant mammary buds and cultured them in various combinations. Some cultures were given testosterone and some were not. The results are shown in Figure 26. When both mesenchyme and epithelium were wild type, the rudiment developed into breast tissue. When testosterone was added, the mesenchyme

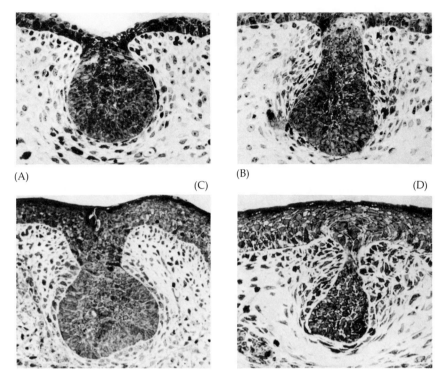

(A)

(B)

(C)

(D)

FIGURE 26
Evidence that the mesenchymal cell is the target of testosterone in the arrest of mammary development. (A) Cultured mammary rudiment from 14-day female embryo. (B) Mammary rudiment from 14-day male embryo beginning its response to testosterone. (C) Recombined mammary bud containing wild-type epithelial cells and androgen-insensitive mesenchyme, cultured with testosterone. No androgen response is seen. (D) Recombined mammary bud containing androgen-insensitive epithelial cells and wild-type mesenchyme, cultured with testosterone. Mesenchyme cells are condensing at the neck of the bud. (From Kratochwil and Schwartz, 1976; courtesy of K. Kratochwil.)

condensed around the bud and the cord was severed. When normal epithelium was cultured with mutant mesenchyme (which could not respond to testosterone), normal breast development occurred in the presence of testosterone. However, when the mesenchyme was normal and the epithelium was mutant, testosterone was able to cause the degeneration of the mammary cord. Thus, the target of testosterone is the mesenchyme, not the epithelium. The mesenchyme must be responsive to testosterone for its action to occur. In males, the testosterone induces the mammary mesenchyme to destroy its adjacent epithelium. The effect is specific for the organ in that no other mesenchyme will kill the mammary epithelium and no other epithelium can be destroyed by mammary mesenchyme (Durnberger and Kratochwil, 1980).

Adolescence

During adolescence (which in the mouse occurs from weeks 4–6), the duct system of the mammary gland proliferates extensively. The extensive cell division is under the control of estrogen and growth hormones. The milk-secreting alveolar cells at the tips of the ducts have not differentiated yet, and no milk is produced.

Pregnancy

Between adolescence and pregnancy, the mouse breast cells are mitotically dormant and undifferentiated. This is changed during the second half of pregnancy. Under the influence of the hormones estrogen and PROGESTERONE (the latter from the placenta), new ducts are formed and the distal cells of the ducts begin to develop the characteristics of secretory tissue.

When such midpregnancy mammary glands are cultured in vitro, most of the cells have little rough endoplasmic reticulum and Golgi apparatus, and no casein granules. When insulin or other promoters of DNA synthesis is added to these cultures, the cells become responsive to other hormones (Turkington et al., 1965). Glucocorticoids can induce the formation of the rough endoplasmic reticulum, and prolactin can then stabilize the casein mRNA to ensure appropriate milk protein synthesis (Figure 27).

The development of the mammary gland, then, involves a complex interplay of several hormones at three different stages of life: embryonic, adolescent, and pregnant. The mammary gland never develops in normal males and does not become a fully differentiated organ in females until the middle of pregnancy in the adult organism.

Puberty as a variation on the theme of metamorphosis

In mammals, one of the most striking displays of hormonal control of differentiation occurs during human puberty. If metamorphosis is understood as the processes of dramatic change whereby a juvenile reaches maturity, both sexually and biochemically, then human puberty may be considered a variation on the metamorphic theme. Recent research suggests that the processes of metamorphosis and puberty may

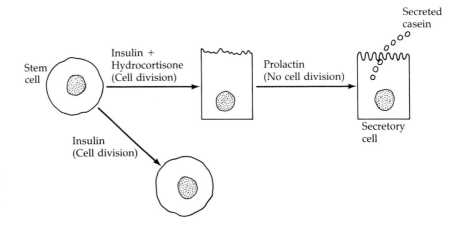

FIGURE 27
Schematic diagram of the hormone-dependent development of the mammary gland in vitro. (After Turkington, 1968.)

be quite similar. A good summary of puberty is given by Styne and Grumbach (1978):

> Puberty is the period of transition between the juvenile state and adulthood; during this stage of development, secondary sex characters appear and mature, the adolescent growth spurt occurs, fertility is achieved, and profound psychologic effects are observed. These changes result directly or indirectly from maturation of the hypothalamic–pituitary gonadotropin unit, stimulation of the sex organs, and the secretion of sex steroids.

Certainly, striking changes in body form occur during this time. At the beginning of puberty, boys and girls have the same proportion of muscle mass, skeletal mass, and body fat. By the end of puberty, men have 1.5 times the skeletal and muscle mass as women, whereas women develop twice as much body fat as men (Forbes, 1975). Secondary sex characteristics[3] also develop at this time, marking the change from the juvenile form to the sexually mature adult. In women, the development of breasts is controlled by a surge of estrogen secreted by the ovaries, and in men, the maturation of the penis and testes is controlled by the testosterone released from the testes. In both sexes, pubic and axillary (underarm) hair development is regulated by testosterone, which is secreted by the testes in males and by the adrenal gland in females. Men also undergo testosterone-dependent enlargement of the larynx and its associated muscles and cartilage, leading to a deepening of the voice. In both sexes, puberty is the time when major sexual development takes place. In males, fertility is achieved when meiosis begins in the male germ cells and the spermatic ducts hollow out to form a channel for the sperm cells to pass from the testes into the urethra. In females, menarche, the first menstrual period, represents the new integration of the hormonal cycles releasing the developing egg from the ovary.

The hormonal basis of puberty is thought to be very similar to that of metamorphosis. The metamorphosis of amphibians and insects were both seen to be regulated by hormonal changes that were initiated by the neurohormones of the brain (TSH-RF and prothoracicotropic hormone, respectively). The changes of human puberty—in both sexes— are initiated by LUTEINIZING HORMONE RELEASING FACTOR (LRF) from the hypothalamus of the brain (Figure 28). Like TSH-RF, this factor is released from the hypothalamic neurons to the MEDIAL EMINENCE (infundibulum) of the pituitary gland. Similarly, the LRF is then transported by the blood vessels of the pituitary to the anterior lobes of that gland. Once in the anterior pituitary, the releasing factor causes the release of a tropic hormone. In human puberty, LRF releases LUTEINIZING HORMONE (LH) and FOLLICLE STIMULATING HORMONE (FSH). Collectively, these two hormones are called GONADOTROPINS because they stimulate the development of testes in males and ovaries in females.

---

[3]The primary sex characteristic is the presence of ovaries or testes. The development of the primary sex organ will be considered in detail in Chapter 19.

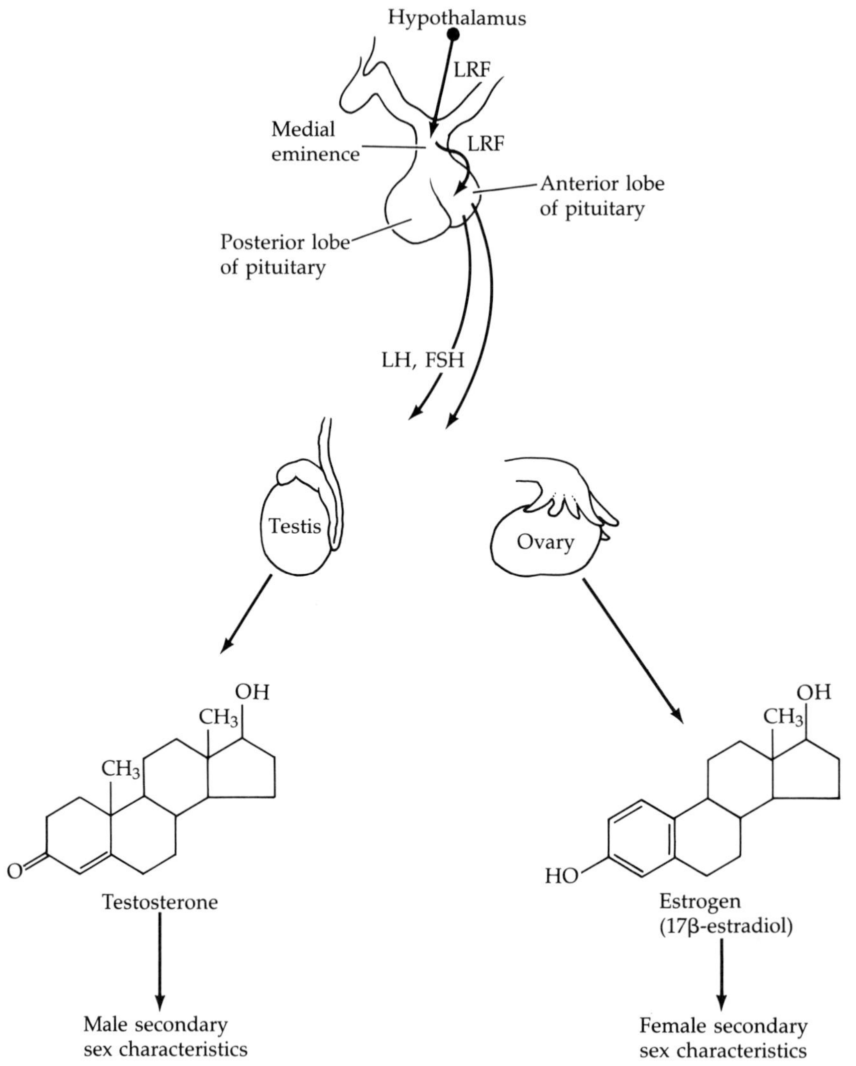

FIGURE 28
Hypothalamus–pituitary–gonadal axis in mammalian sexual development.

As a result of this stimulation, the gonads secrete the SEX HORMONES: testosterone from the testes and estrogen from the ovaries. The various morphological and behavioral changes of puberty are due to the actions of these hormones on the various target tissues. As in metamorphosis, there appears to be a maturation-inhibiting hormone whose activity decreases to permit the reactivation of development. In humans, this hormone is probably melatonin, whose serum concentration decreases as that of LH rises (Waldhauser et al., 1984).

Although the details concerning the initiation of puberty are not known, Grumbach and co-workers (1974) have proposed the following mechanism (Figure 29). Before puberty, the child secretes a small amount of LRF, so that the amounts of circulating LH and FSH are very

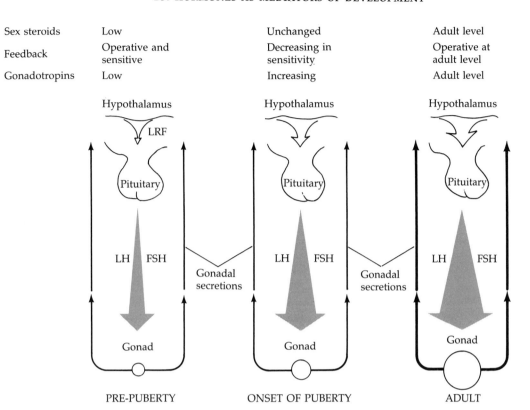

Sex steroids	Low	Unchanged	Adult level
Feedback	Operative and sensitive	Decreasing in sensitivity	Operative at adult level
Gonadotropins	Low	Increasing	Adult level

PRE-PUBERTY   ONSET OF PUBERTY   ADULT

FIGURE 29

Proposed mechanism for the induction of puberty in humans. Before puberty, the hypothalamus is sensitive to very small concentrations of sex steroids and stops LRF production, thereby halting further steroid synthesis. Sex steroids are kept at low levels. At the onset of puberty, the hypothalamus becomes progressively less sensitive to the sex steroids, thereby allowing more of them to be synthesized, until finally the adult level of sex steroids is achieved. Relative widths of arrows indicate levels of hormone production. (After Grumbach, et al., 1974.)

small. Therefore, the gonad remains immature and secretes little estrogen or testosterone. Moreover, the hypothalamus is highly sensitive to these sex hormones and "turns off" LRF production when the circulating sex hormones get anywhere above their very low levels. The onset of puberty is thought to involve the maturation of the hypothalamus. At this time, the hypothalamus becomes less responsive to the negative feedback of testosterone and estrogen. Thus, it takes more of these hormones to turn off LRF secretion. (This is analogous to giving a thermostat a higher set point.) Thus, the hypothalamus makes more LRF, ultimately causing the further differentiation of the gonads and the release or more testosterone or estrogen. This, in turn, causes the development of those secondary sexual characteristics.

It can readily be seen, then, that puberty is a hormonally controlled reactivation of development leading to sexual maturity and changes in bodily form and physiology. As such, it has many parallels to the metamorphic changes seen elsewhere in the animal kingdom.

Control of cell proliferation: Chalones

So far, we have focused our attention on hormones, substances that circulate throughout the body and instruct other tissues to activate specific genes. There may be other diffusible substances, called CHAL-ONES, that act to regulate growth. When the outer layer of epidermal cells is peeled off the skin of vertebrates, the cells responsible for forming the epidermis undergo extra rounds of cell division to restore the missing layer. Once this layer is made, the precursor cells revert back to their normal rate of cell division. The simplest explanation is that some substance is made by the epidermal cells to inhibit the division of their precursors. When the epidermal cells are removed, the precursors are no longer under this inhibition, so they multiply faster. Eventually, the new epidermal cells make their own chalones to inhibit further divisions (Potten and Allen, 1975; Bullough, 1975).

Chalones may also be responsible in the growth of one of a pair of organs after the removal of the other. This is called COMPENSATORY HYPERPLASIA. For instance, if one kidney is removed from an animal, the other kidney will grow larger, as if to make up for this loss. Similarly, if part of a liver is removed, the cells of the remaining lobes will proliferate until the original mass is restored. It is thought that these tissue-specific chalones are proteins and that, like hormones, they travel by means of the bloodstream. A model for such action is shown in Figure 30.

It is difficult to demonstrate that a given protein is a natural inhibitor and not merely a cytotoxic chemical. The in vitro assays for chalones are complicated by the fact that they often work in combination with other substances. Epidermal chalones, for example, will inhibit the mitosis of epidermal precursor cells in vitro only if epinephrine is also present. The investigation as to whether or not chalones actually exist has taken on new importance as certain diseases, such as psoriasis, may be the result of overproliferation of stem cells. In this disease, the epithelial precursors divide at a very rapid rate (as if they could not respond to chalone or as if chalone were not being made), thereby causing the epidermis to thicken and the cells to be shed before they have a normal time to differentiate.

We have seen that diffusible regulation of cell–cell interactions is also important in regulating development. In the next chapter we shall study the proximate and hormonal tissue interactions occurring in the development of sexual phenotype.

FIGURE 30
Schematic diagram of the proposed regulation of epidermal cell division. One of the products of the differentiating epidermal cells would be a chalone that would signal the basal stem cells to cease dividing. (After Bullough, 1975.)

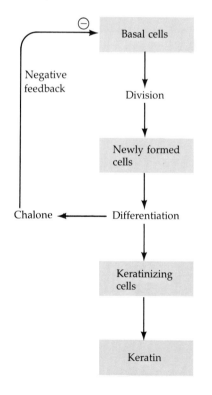

Literature Cited

Allen, B. M. (1916). Extirpation experiments in *Rana pipiens* larva. *Science* 44: 755–757.

Bern, H. A., Nicoll, C. S. and Strohman, R. C. (1967). Prolactin and tadpole growth. *Proc. Soc. Exp. Biol. Med.* 126: 518–521.

Bowers, W. S., Ohta, T., Cleere, J. S. and Marsella, P. A. (1976). Discovery of insect anti-juvenile hormones in plants. *Science* 193: 542–547.

Brown, P. S. and Frye, B. E. (1969). Effect of prolactin and growth hormone on growth and metamorphosis of tadpoles of the frog, *Rana pipiens*. *Gen. Comp. Endocrinol.* 13: 139–145.

Bullough, W. S. (1975). Mitotic control in adult mammalian tissues. *Biol. Rev.* 50: 99–127.

Butenandt, A. and Karlson, P. (1954). Über die Isolierung eines Metamorphosen-Hormons der Insekten in kristallisierter Form. *Z. Naturforsch., Teil B* 9: 389–391.

Cohen, P. P. (1970). Biochemical differentiation during amphibian metamorphosis. *Science* 168: 533–543.

Cohen, P. P., Brucker, R. F. and Morris, S. M. (1978). Cellular and molecular aspects of thyroid-hormone action during amphibian metamorphosis. *Horm. Proteins Peptides* 6: 273–381.

Dürnberger, H. and Kratochwil, K. (1980). Specificity of tissue interaction and origin of mesenchymal cells in the androgen response of the embryonic mammary gland. *Cell* 19: 465–471.

Eberhardt, M. K., Ring, J. C., Johnson, L. K., Latham, K. R., Aprilett, J. W., Kitsis, R. N. and Baxter, J. D. (1979). Regulation of activity of chromatin receptors for thyroid hormone—possible involvement of histone-like proteins. *Proc. Natl. Acad. Sci. USA* 76: 5005–5009.

Etkin, W. (1968). Hormonal control of amphibian metamorphosis. In W. Etkin and L. I. Gilbert (eds.), *Metamorphosis: A Problem in Developmental Biology*. Appleton-Century-Crofts, New York, pp. 313–348.

Etkin, W. and Gona, A. G. (1967). Antagonism between prolactin and thyroid hormone in amphibian development. *J. Exp. Zool.* 165: 249–258.

Fallon, J. F. and Simandl, B. K. (1978). Evidence of a role for cell death in the disappearance of the embryonic human tail. *Amer. J. Anat.* 152: 111–130.

Forbes, G. B. (1975). Puberty: Body composition. In S. R. Berenson (ed.), *Puberty*. Stenfert-Kroese, Leiden, pp. 132–145.

Fox, H. (1973). Ultrastructure of tail degeneration in *Rana temporaria* larva. *Folia Morphol.* 21: 109–112.

Frieden, E. (1981). The dual role of thyroid hormones in vertebrate development and calorigenesis. In L. I. Gilbert and E. Frieden (eds.), *Metamorphosis: A Problem in Developmental Biology*. Plenum, New York, pp. 545–564.

Fristrom, D. and Fristrom, J. W. (1975). The mechanisms of evagination of imaginal disks of *Drosophila melanogaster*. I. General considerations. *Dev. Biol.* 43: 1–23.

Fristrom, J. W. (1972). The biochemistry of imaginal disc development. In H. Ursprung and R. Nöthiger (eds.), *The Biology of Imaginal Discs*. Springer-Verlag, Berlin, pp. 109–154.

Fristrom, J. W., Fristrom, D., Fekete, E. and Kuniyuki, A. H. (1977). The mechanism of evagination of imaginal discs of *Drosophila melanogaster*. *Am. Zool.* 17: 671–684.

Geigy, R. (1941). Die metamorphose als Folge gewebsspezifischer determination. *Rev. Suisse Zool.* 48: 483–494.

Gilbert, L. I. and Goodman, W. (1981). Chemistry, metabolism, and transport of hormones controlling insect metamorphosis. In L. I. Gilbert and E. Frieden (eds.), *Metamorphosis: A Problem in Developmental Biology*. Plenum, New York, pp. 139–176.

Goos, H. J. Th. (1978). Hypophysiotropic centers in the brain of amphibians and fish. *Am. Zool.* 18: 401–410.

Gould, S. J. (1977). *Ontogeny and Phylogeny*. Belknap, Cambridge, MA, p. 283.

Graham, C. F. and Wareing, P. F. (1976). *The Developmental Biology of Plants and Animals*. Saunders, Philadelphia.

Granger, N. A. and Bollenbacher, W. E. (1981). Hormonal control of insect metamorphosis. In L. I. Gilbert and E. Frieden (eds.), *Metamorphosis: A Problem in Developmental Biology*. Plenum, New York, pp. 105–138.

Grant, W. C., Jr. and Cooper, G. (1964). Endocrine control of metamorphic and skin changes in *Diemictylus viridescens*. *Am. Zool.* 4: 413–414

Grant, W. C. and Grant, J. A. (1958). Water drive studies on hypophysectomized efts of *Diemyctylus viridescens*. Part I. The role of the lactogenic hormone. *Biol. Bull.* 114: 1–9.

Grumbach, M. M., Roth, J. C., Kaplan, S. L. and Kelch, R. P. (1974). Hypothalamic–pituitary regulation of puberty in man: Evidence and concepts derived from clinical research. In M. M. Grumbach, G. D. Grave, and F. E. Meyer (eds.), *Control of the Onset of Puberty.* Wiley, New York, p. 115–166.

Gudernatsch, J. F. (1912). Feeding experiments on tadpoles. I. The influence of specific organs given as food on growth and differentiation. A contribution to the knowledge of organs with internal secretion. *Wilhelm Roux Arch. Entwicklungsmech. Org.* 35: 457–483.

Hogg, N. A. S., Harrison, D. J. and Tickle, C. (1983). Lumen formation in the mammary gland. *J. Embryol. Exp. Morphol.* 73: 39–57.

Hoskins, E. R. and Hoskins, M. M. (1917). On thyroidectomy in amphibia. *Proc. Soc. Exp. Biol. Med.* 14: 74–75.

Huxley, J. (1920). Metamorphosis of axolotl caused by thyroid feeding. *Nature* 104: 436.

Kaltenbach, J. C., Fry, A. E. and Leius, V. K. (1979). Histochemical patterns in the tadpole tail during normal and thyroxine-induced metamorphosis. II. Succinic dehydrogenase, Mg- and Ca-adenosine triphosphatases, thiamine pyrophosphatase, and 5' nucleotidase. *Gen. Comp. Endocrinol.* 38: 111–126.

Karp, G. and Berrill, N.J. (1981). *Development.* McGraw-Hill, New York.

Kistler, A., Yoshizato, K. and Frieden, E. (1977). Preferential binding of tri-substituted thyronine analogs by bullfrog tadpole tail fin cytosol. *Endocrinology* 100: 134–137.

Kollros, J. J. (1961). Mechanisms of amphibian metamorphosis: Hormones. *Am. Zool.* 1: 107–114.

Kratochwil, K. (1971). In vitro analysis of the hormonal basis for sexual dimorphism in the embryonic development of the mouse mammary gland. *J. Embryol. Exp. Morphol.* 25: 141–153.

Kratochwil, K. and Schwartz, P. (1976). Tissue interaction in androgen response of embryonic mammary rudiment of mouse: Identification of target tissue for testosterone. *Proc. Natl. Acad. Sci. USA* 73: 4041–4044.

Little, G., Atkinson, B. G. and Frieden, E. (1973). Changes in the rates of protein synthesis and degradation in the tail of *Rana catesbeiana* tadpoles during normal metamorphosis. *Dev. Biol.* 30: 366–373.

Madhavan, M. M. and Schneiderman, H. A. (1977). Histological analysis of the dynamics of growth of imaginal disc and histoblast nests during the larval development of *Drosophila melanogaster. Wilhelm Roux Arch. Dev. Biol.* 183: 269–305.

Malacinski, G. M. (1978). The Mexican axolotl, *Ambystoma mexicanum:* Its biology and developmental genetics, and its autonomous cell-lethal genes. *Am. Zool.* 18: 195–206.

McCutcheon, F. H. (1936). Hemoglobin function during the life history of the bullfrog. *J. Cell. Comp. Physiol.* 8: 63–81.

Mori, M., Morris, S. M., Jr. and Cohen, P. P. (1979). Cell-free translation and thyroxine induction of carbamyl phosphate synthetase I messenger RNA in tadpole liver. *Proc. Natl. Acad. Sci. USA* 76: 3179–3183.

Norris, D. O., Jones, R. E. and Criley, B. B. (1973). Pituitary prolactin levels in larval, neotenic, and metamorphosed salamanders (*Ambystoma tigrinum*). *Gen. Comp. Endocrinol.* 20: 437–442.

Potten, C. S. and Allen, T. D. (1975). Control of epidermal proliferative units (EPUS): Hypothesis based on arrangement of neighboring differentiated cells. *Differentiation* 3: 161–165.

Prahlad, K. V. and Delanney, L. E. (1965). A study of induced metamorphosis in the axolotl. *J. Exp. Zool.* 160: 137–146.

Pratt, G. E., Jennings, R. C., Hammett, A. F. and Brooks, G. T. (1980). Lethal metabolism of precocene-I to a reactive epoxide by locust corpora allata. *Nature* 284: 320–323.

Raynaud, A. (1961). Morphogenesis of the mammary gland. In S. K. Kon and A. T. Cowrie (eds.), *Milk: The Mammary Gland and Its Secretion,* Vol. 1. Academic, New York, pp. 3–46.

Richards, G. (1978). Sequential gene activity in polytene chromosomes of *Drosophila melanogaster.* VI. Inhibition by juvenile hormones. *Dev. Biol.* 66: 32–42.

Riggs, A. F. (1951). The metamorphosis of hemoglobin in the bullfrog. *J. Gen. Physiol.* 35: 23–40.

Robinson, H., Chaffee, S. and Galton, V. A. (1977). Sensitivity of *Xenopus laevis* tadpole tail tissue to the action of thyroid hormones. *Gen. Comp. Endocrinol.* 32: 179–186.

Schooneveld, H. (1979). Precocene-induced collapse and resorption of corpora allata in nymphs of *Locusta migratoria. Experientia* 35: 363–364.

Schwind, J. L. (1933). Tissue specificity at the time of metamorphosis in frog larvae. *J. Exp. Zool.* 66: 1–14.

Smith-Gill, S.J. and Carver, V. (1981). Biochemical characterization of organ differentiation and maturation. In L. I. Gilbert and E. Frieden (eds.), *Metamorphosis: A Problem in Developmental Biology.* Plenum, New York, pp. 491–544.

Styne, D. M. and Grumbach, M. M. (1978). Puberty in the male and female: Its physiology and disorders. In S. S. C. Yen and R. B. Jaffe (eds.), *Reproductive Endocrinology.* Saunders, Philadelphia, pp. 189–240.

Taurog, A., Oliver, C., Porter, R. L., McKenzie, J. C. and McKenzie, J. M. (1974). The role of TRH in the neoteny of the Mexican axolotl (*Ambystoma mexicanum*).

Gen. Comp. Endocrinol. 24: 267–279.

Topper, Y. J. and Freeman, C. S. (1980). Multiple hormone interactions in the developmental biology of the mammary gland. *Physiol. Rev.* 60: 1049–1106.

Turkington, R. W. (1968). Hormone-dependent differentiation of mammary gland *in vitro. Curr. Top. Dev. Biol.* 3: 199–218.

Turkington, R. W., Juergens, W. G. and Topper, Y. J. (1965). Hormone-dependent synthesis of case *in vitro. Biochim. Biophys. Acta* 111: 573–576.

Turner, C. D. and Bagnara, J. T. (1976). *General Endocrinology*, Sixth Edition. Saunders, Philadelphia.

Wald, G. (1945). The chemical evolution of vision. *Harvey Lect.* 41: 117–160.

Wald, G. (1981). Metamorphosis: An overview. In L. I. Gilbert and E. Frieden (eds.), *Metamorphosis: A Problem in Developmental Biology.* Plenum, New York, pp. 1–39.

Waldhauser, F., Frisch, H., Weiszenbacher, G. and Wurtman, R. J. (1984). Serum melatonin and serum LH concentrations in children, adolescents, and young adults (abstract). *Pediat. Res.* 18: 108.

Weber, R. (1965). Inhibitory effect of actinomycin D on tail atrophy in *Xenopus laevis* larvae at metamorphosis. *Experientia* 21: 665–666.

Weber, R. (1967). Biochemistry of amphibian metamorphosis. In R. Weber (ed.), *The Biochemistry of Animal Development*, Vol. III. Academic, New York, pp. 227–301.

White, B. H. and Nicoll, C. S. (1981). Hormonal control of amphibian metamorphosis. In L. I. Gilbert and E. Frieden (eds.), *Metamorphosis: A Problem in Developmental Biology.* Plenum, New York, pp. 363–396.

Wigglesworth, V. B. (1954). *The Physiology of Insect Metamorphosis.* Cambridge University Press, London.

Williams, C. M. (1952). Physiology of insect diapause. IV. The brain and prothoracic glands as an endocrine system in the Cecropia silkworm. *Biol. Bull.* 103: 120–138.

Williams, C. M. (1956). The juvenile hormone of insects. *Nature* 178: 212–213.

CHAPTER 19

Sex determination

Sexual reproduction is . . . the masterpiece of nature.
—ERASMUS DARWIN (1791)

*It is quaint to notice that the number of speculations connected
with the nature of sex have well-nigh doubled since Drelincourt,
in the eighteenth century, brought together two-hundred and
sixty-two 'groundless hypotheses,' and since Blumenbach
caustically remarked that nothing was more certain than that
Drelincourt's own theory formed the two hundred and sixty-third.*
—J. A. THOMSON (1926)

Introduction

The mechanisms by which an individual's sex is determined has been one of the great questions of embryology since antiquity. Aristotle, who collected and dissected embryos, claimed that sex was determined by the heat of the male partner during intercourse. The more heated the passion, the greater the probability of male offspring. (Aristotle counseled elderly men to conceive in the summer if they wished to have male heirs.) Since that time, the environment—heat and nutrition, in particular—was believed to be important in determining sex. In 1890, Geddes and Thomson summarized all available data on sex determination and came to the conclusion that "the constitution, age, nutrition, and environment of the parents must be especially considered" in any such analysis. They argued that those factors favoring the storage of energy and nutrients predisposed one to have female offspring, whereas those factors favoring the utilization of energy and nutrients influenced one to have male offspring.

This environmental view of the determination of sex remained the only major scientific theory until the rediscovery of Mendel's work in 1900 and the rediscovery of the sex chromosome by McClung in 1902. Based on his knowledge of Mendel, Correns speculated that the 1:1 sex ratio of most species could be achieved if the male was heterozygous and the female homozygous for some sex-determining factor. In was not until 1905, however, that the correlation of the female sex with XX sex chromosomes and the male sex with XY or XO chromosomes was established (Stevens, 1905; Wilson 1905). This suggested strongly that a specific nuclear component was responsible for directing the development of the sexual phenotype. Thus, evidence accumulated that sex determination occurs by nuclear inheritance rather than by environmental happenstance.

Today, we find that both environmental and internal mechanisms of sex determination can operate in different species. We shall first discuss the chromosomal mechanisms of sex determination and then consider the ways by which the environment regulates the sexual phenotype.

Chromosomal sex determination in mammals

In mammals, the determination of sex is strictly chromosomal and is not influenced by the environment. In most cases, the female is XX and the male is XY. The Y chromosome is the crucial inherited factor determining sex in mammals. Even if a person were to have five X chromosomes and one Y chromosome, he would be male. Moreover, an individual with only a single X chromosome and no Y develops as a female.

The course of "normal" mammalian development is that of the female. When Jost (1953) removed fetal rabbit gonads before they had differentiated, he found that the resulting rabbits were female. They had oviducts, uteruses, and vaginas, and they lacked penises and male accessory structures. In the absence of gonads, female development ensued. Thus, development in mammals is in the female direction unless acted upon by some product made, or regulated, by the Y chromosome.

The scheme of mammalian sex determination is shown in Figure 1. The Y chromosome produces or regulates the formation of the testis-determining substance, possibly the cell surface molecule called the H-Y ANTIGEN. All male cells have this molecule, female cells do not. The testis-determining substance organizes the developing gonad into a testis instead of an ovary. Once this testis forms, it secretes two major hormones. The first—TESTOSTERONE—will masculinize the fetus, causing the formation of the penis, scrotum, and other portions of the male anatomy, as well as destroying the incipient breast primodia. The second hormone—ANTI-MÜLLERIAN DUCT FACTOR—destroys the tissue that would otherwise become the uterus, oviduct, cervix, and upper portion of the vagina. Thus, the body would have the female phenotype were it not for the imposition of the two hormones elaborated from the fetal testes.

FIGURE 1

Sequence of events leading to the formation of the male phenotype in mammals. In the presence of the Y chromosome, the indifferent gonad is converted into a testis. The testis cells secrete the hormones that cause differentiation of the body in the masculine direction.

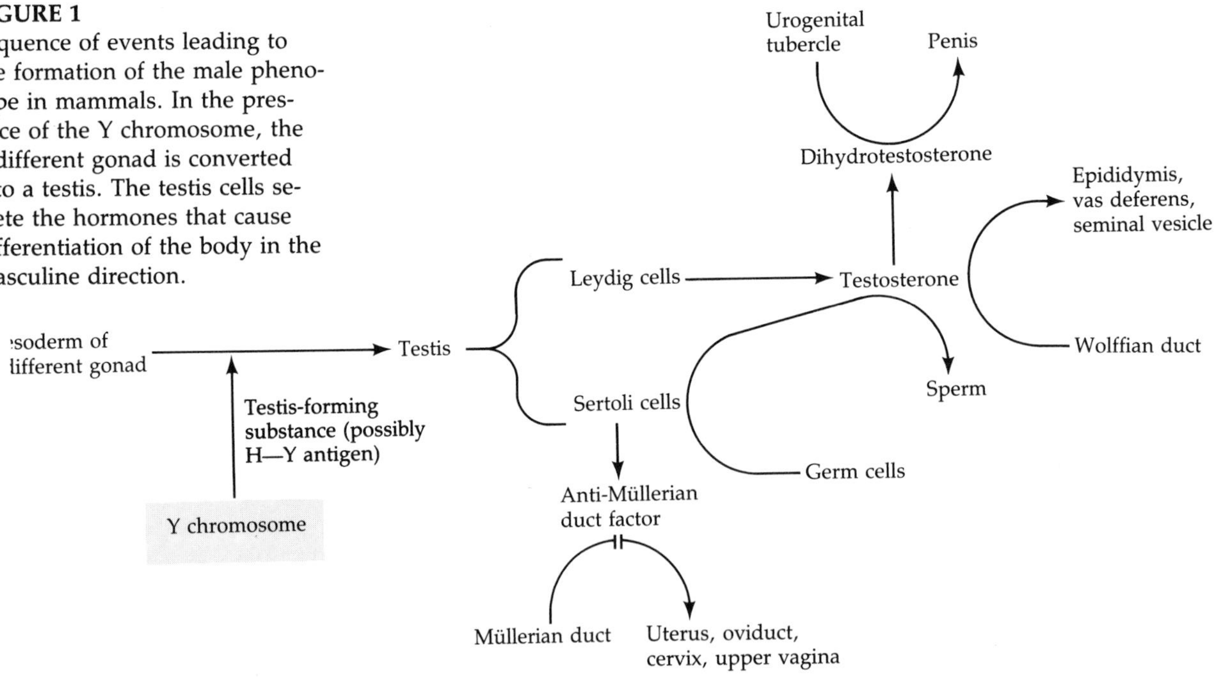

The developing gonads

The development of gonads is a unique embryological situation. All other organ rudiments have only two options: to differentiate or not to differentiate. A lung rudiment either can become a lung or it can atrophy. Similarly, a liver rudiment can develop only into a liver. The gonadal rudiment, however, has three options. If it differentiates, it can develop into either a testis or an ovary. The type of differentiation taken by this rudiment determines the future sexual development of the organism. Before this decision is made, the mammalian gonad first develops through an INDIFFERENT STAGE, during which time it has neither male nor female characteristics. In humans, the gonadal rudiment appears in the intermediate mesoderm during week 4 and remains sexually indifferent until week 7. During this indifferent stage, the epithelium of the genital ridge proliferates into the loose connective mesenchymal tissue above it (Figure 2A and B). The epithelia form the SEX CORDS, which will surround the germ cells that migrate into the human gonad during week 6. In both XY and XX gonads, the sex cords remain connected to the surface epithelium.

If the fetus is XY, then the sex cords continue to proliferate through the eighth week, penetrating deeply into the connective tissue. These cords fuse with each other, forming a network of internal sex cords (medullary cords) and, at its most distal end, the thinner RETE TESTIS (Figure 2C and D). Eventually, the testis cords lose contact with the surface epithelium and become separated from them by the TUNICA ALBUGINEA. Thus, the germ cells are found in the cords *within* the testes. During fetal life and childhood, these cords remain solid. At puberty, however, the cords hollow out to form the seminiferous tubules, and the germ cells begin sperm production. The sperm gets transported from the inside of the testis through the rete testis, which joins the EFFERENT DUCTS. These efferent tubules are the remnants of the mesonephric kidney. They link the testis to the Wolffian duct. This duct used to be the collecting tube of the mesonephric kidney. In males, this duct differentiates to become the VAS DEFERENS, the tube through which the sperm pass into the urethra and out of the body. Meanwhile, during fetal development the interstitial mesenchymal cells of the testes have become LEYDIG CELLS, which make testosterone. The cells of the testis cords differentiate into SERTOLI CELLS, which nurture the sperm and secrete the ANTI-MÜLLERIAN DUCT FACTOR.

In females, the germ cells will be placed on the outside surface of the gonad. Unlike the male sex cords, which continue their proliferation, the sex cords of XX gonads degenerate. However, the epithelium soon produces a new set of sex cords, which do not penetrate the mesenchyme but stay on the outer surface (cortex) of the organ. Thus, they are called CORTICAL SEX CORDS. These cords are split into clusters, each cluster surrounding one or more germ cells (Figure 2E and F). The

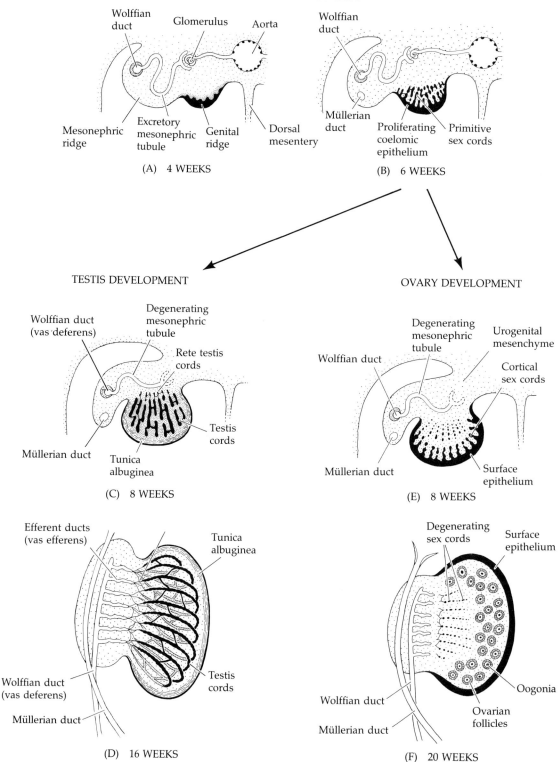

INDIFFERENT GONADS

(A) 4 WEEKS

Wolffian duct
Glomerulus
Aorta
Mesonephric ridge
Excretory mesonephric tubule
Genital ridge
Dorsal mesentery

(B) 6 WEEKS

Wolffian duct
Müllerian duct
Proliferating coelomic epithelium
Primitive sex cords

TESTIS DEVELOPMENT

(C) 8 WEEKS

Wolffian duct (vas deferens)
Degenerating mesonephric tubule
Rete testis cords
Müllerian duct
Tunica albuginea
Testis cords

(D) 16 WEEKS

Efferent ducts (vas efferens)
Tunica albuginea
Wolffian duct (vas deferens)
Müllerian duct
Testis cords

OVARY DEVELOPMENT

(E) 8 WEEKS

Degenerating mesonephric tubule
Urogenital mesenchyme
Wolffian duct
Cortical sex cords
Müllerian duct
Surface epithelium

(F) 20 WEEKS

Degenerating sex cords
Surface epithelium
Wolffian duct
Müllerian duct
Oogonia
Ovarian follicles

FIGURE 2

Differentiation of human gonads shown in transverse section. (A) Genital ridge of a 4-week embryo. (B) Genital ridge of a 6-week indifferent gonad showing primitive sex cords. (C) Testis development in the eighth week. The sex cords lose contact with the cortical epithelium and develop the rete testis. By the sixteenth week of development (D), the testis cords are continuous with the rete testis and connect with the Wolffian duct. (E) Ovary development in an 8-week human embryo, as primitive sex cords degenerate. (F) The 20-week human ovary does not connect to the Wolffian duct, and new cortical sex cords surround the germ cells that have migrated into the genital ridge. (After Langman, 1981.)

germ cells will become the OVA and the surrounding epithelial sex cords will differentiate into the GRANULOSA CELLS. These cells will envelope the germ cells to form the ovarian FOLLICLES. Each follicle will contain a single germ cell. In females, the Müllerian duct remains intact, and it will differentiate into the oviducts, uterus, cervix, and upper vagina; the Wolffian duct, deprived of testosterone, degenerates. A summary of the development of mammalian reproductive systems is shown in Figure 3.

The H-Y antigen

Because the Y chromosome changes the "normal" direction of development from female to male, the search for the testis-forming substance has been among compounds controlled by that chromosome. Although presently the subject of heated debate, the best candidate for the testis-forming substance is the H-Y antigen.[1] This substance was discovered in 1955 as a result of skin grafting performed within a genetically identical strain of mice. Males accepted female skin grafts whereas the females slowly rejected the male skin. Because all other genes of the inbred mice were the same, it was concluded that the male's Y chromosome was producing or regulating a cell surface antigen not seen on female cells and thereby recognized by the female mice as being foreign (Eichwald and Silmser, 1955; Billingham and Silvers, 1960).

Under normal circumstances, H-Y antigen expression correlates very well with the presence of the Y chromosome. XY individuals are H-Y$^+$ and have testes, whereas XX and most XO individuals are H-Y$^-$ and are female. Antibodies against the H-Y antibody demonstrate that it has been conserved throughout vertebrate evolution. (In mammals, XY individuals are H-Y$^+$, whereas in birds, the females are H-Y$^+$. In

[1]The basis of this controversy concerns whether or not the transplantation, serological, and biochemical assays for H-Y antigen are recognizing the same molecule. Unfortunately, the H-Y antigen does not give rise to high-titer antiserum and has not been purified biochemically to such a degree that its structure can be determined (see Goodfellow and Andrews, 1982; Silvers et al., 1982).

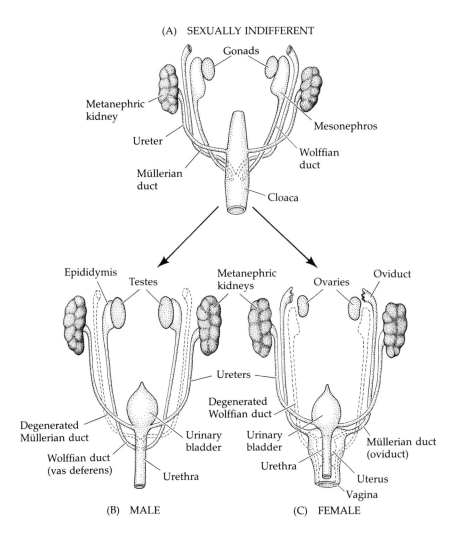

GONADS		
Gonadal type	Testis	Ovary
Sex cords	Medullary (internal)	Cortical (external)
DUCTS		
Remaining duct for germ cells	Wolffian	Müllerian
Duct differentiation	Vas deferens, epididymis, seminal vesicle	Oviduct, uterus, cervix, upper portion of vagina

FIGURE 3
Summary of the development of gonads and their ducts in mammals. Note that both the Wolffian and Müllerian ducts are present at the indifferent-gonad stage.

birds, ZZ animals are male, ZW animals are female.) It is always the heterogametic sex, XY,ZW, which is H-Y$^+$. Tumor cells with the Y chromosome are H-Y$^+$ only as long as they retain their Y chromosome, and white blood cells from XYY individuals have twice as much H-Y antigen as XY males.

The correlation of H-Y antigen to testis formation is even better than its linkage to the Y chromosome. There are certain mice and humans who are XX but are males. These individuals are H-Y$^+$. The H-Y antigen and testis-forming substances are thought to have been translocated to some other chromosome. Thus, either the H-Y antigen is the testis-forming substance, or it maps very close to it on the Y chromosome.

The best correlation between the H-Y antigen and maleness comes from studies of those mammals whose sex is not clearly determined by the XX/XY method. In the mole vole, *Ellobius lutescens*, all individuals are XO, as those zygotes inheriting either two X chromosomes or no X chromosomes are not viable. In this species, half the individuals—the males—are H-Y$^+$, while females are H-Y$^-$. Thus, the presence or absence of the H-Y antigen determines sex even in the absence of karyotypic differences between the sexes. A similar correlation can be seen in the wood lemming, *Myopus schisticolor*. The males are all XY; the females, however, can be either XX or XY. Certain of the X chromosomes are able to prevent the expression of the H-Y antigen, and the resulting XY individuals are female.

The H-Y antigen is expressed early in mammalian development, at the 8-cell stage, and it is not dependent upon hormones. Castrated mice do not lose the H-Y antigen, nor do females injected with testosterone gain it. When male H-Y$^+$ hematopoietic precursor cells are injected into irradiated female mice, the resulting blood cell colonies are H-Y$^+$ even though they are in a totally female body (for reviews, see Ohno, 1979; Hall and Wachtel, 1980).

In 1978 Moscona-type experiments (Ohno et al., 1978a; Zenzes et al., 1978a) demonstrated that the H-Y antigen physically caused the development of testis tissue from female or indifferent gonads (Figure 4). First, newborn testes and ovaries were dissociated into single cells and placed in rotary aggregation cultures. The testis and ovary cells reaggregated into their respective histotypic patterns. Testis cells reformed tubular structures containing many germ cells, whereas ovary cells reconstructed spherical envelopes around singular germ cells. Next, testis cells were LYSOSTRIPPED of the H-Y antigen before aggregation to get rid of the H-Y antigen. Antibodies against H-Y antigen were added to the testis cells, which capped all the H-Y antigen on their surfaces. The cells then phagocytized this cap, thus denuding their membranes of H-Y antigen. When these cells were aggregated, they formed ovary-like structures. Thus, in the absence of H-Y antigen, testis cells formed ovary-like reaggregation patterns.

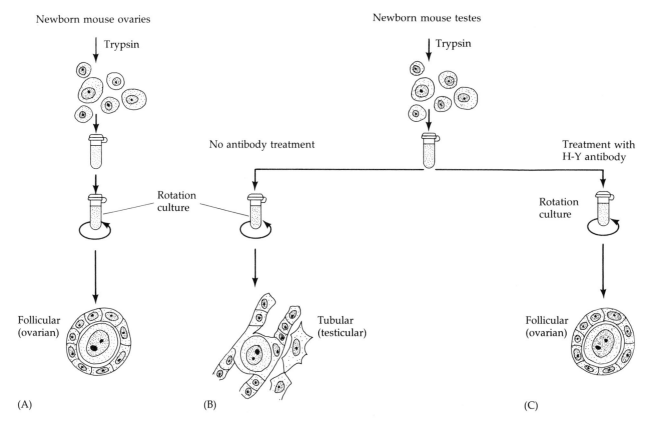

FIGURE 4
Effect of the H-Y antigen on the reaggregation of newborn mouse gonadal cells. (A) Trypsinized ovary cells will reaggregate in rotation culture to form follicular-like aggregates reminiscent of ovarian tissue. (B) Trypsinized testis cells reaggregate to form tubule-like structures that resemble testicular tissue. (C) When the H-Y antigen is removed from the testis cells by antibody treatment, the testicular cells reaggregate into ovary-like structures.

In the next series of experiments, H-Y antigen was added to indifferent gonad or ovarian cells (Zenzes et al., 1978b; Ohno et al., 1979). In these situations the added H-Y antigen caused the formation of testicular tissues (Figure 5). Even the tunica albuginea was seen to form. These experiments strongly suggest that the H-Y antigen is responsible for the histotypic structure that occurs in aggregating gonadal cells. In the absence of this antigen, ovary-like structures form (even in XY gonadal cells); and in the presence of this antigen, testis-like structures form (even if the gonadal cells are XX).

The H-Y receptor

H-Y antigen is found on all mammalian male cells. However, to work, it must act upon a receptor. Studies have demonstrated that this receptor is located only on gonadal cells of both sexes. Nagai and co-workers (1979), using radioactive H-Y antigen derived from cell cultures, showed that only gonadal cells picked it up (Table 1). Müller and co-workers (1978), using H-Y antigen secreted into epididymal fluid by mouse testes, similarly demonstrated that only gonadal cells picked up H-Y

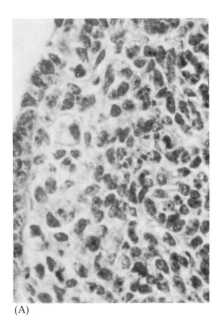 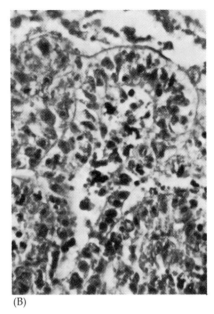

(A) (B)

FIGURE 5

Effect of H-Y antigen on indifferent XX bovine gonads. Gonads were removed from embryonic cows and cultured 5 days in the absence (A) or presence (B) of medium containing H-Y antigen. In the absence of H-Y antigen, the gonad remains indifferent during the time of culture. In (B), testis-specific structures such as the tunica albuginea and tubules can be seen. (From Nagai et al, 1979; photographs courtesy of S. Ohno.)

antigen and bound it tightly to their cell membranes. The assay performed in the following way: When anti-H-Y antiserum at a certain concentration is incubated with H-Y$^+$ cells in the presence of complement (a set of serum proteins that bind to antibodies and punch holes in the cell membrane), 60 percent of the test cells are killed. When the antiserum is preincubated with male (H-Y$^+$) cells, many of the antibodies bind to these cells and the resulting antiserum does not kill as many of the test cells (Figure 6). When the absorption of antiserum is done with female cells from either the skin or the gonad, nothing happens to the antiserum. It still kills 60 percent of the test cells. In this way,

TABLE 1

Demonstration of gonad-specific H–Y antigen receptor by the binding of partially purified tritiated H–Y antigen

Target	Number of cells	cpm bound to target	Percentage of total precipitable counts
Bovine fetal ovary	7.5×10^5	57,392	11.3
	1.75×10^7	77,182	14.6
Adult mouse spleen	7.5×10^5	14,980	2.8
	1.75×10^7	6,157	1.2

Source: Nagai et al. (1979).

FIGURE 6
Interpretive diagram based on experiments showing gonad-specific H-Y antigen receptor. Assay is based on the ability of cells to bind H-Y antigen and then to absorb anti-H-Y antiserum. If a cell has H-Y receptors, it will absorb H-Y antigen. If such cells are placed into anti-H-Y antiserum, the antibodies to H-Y antigen will bind to these cells. If the cells are removed by centrifugation, the resulting antiserum will be weaker because the antibodies were pelleted with the cells.

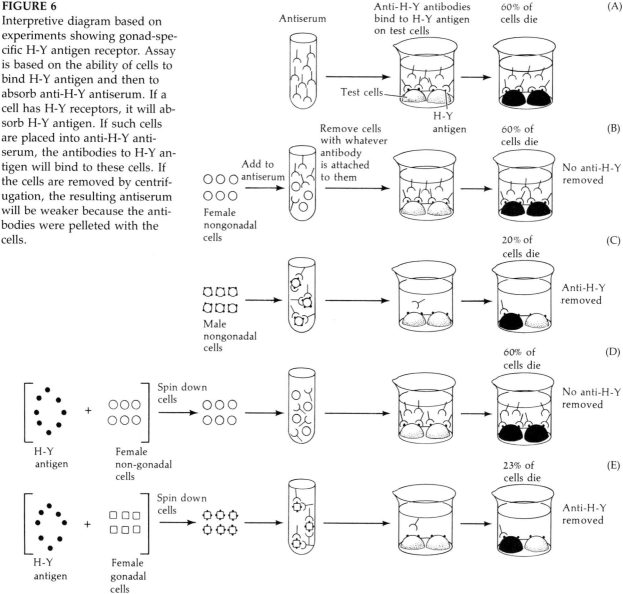

the male cells are found to have H-Y antigen on them, whereas the female cells are not. Then both types of female cells are incubated in H-Y antigen before absorbing the anti-H-Y serum. When H-Y is added to female skin cells, the resulting cells do not absorb any anti-H-Y antibodies. Thus, they did not retain the H-Y antigen they were incubated with. However, when ovary cells are pretreated with H-Y antigen and then used to absorb the antiserum, significant amounts of anti-H-Y antibodies are absorbed. Only 23 percent of the test cells are killed. This is only slightly greater than the control without antiserum. Therefore, the gonadal cells will absorb the H-Y antigen whereas the other cells will not. The gonadal cells have a specific H-Y receptor.

It can be seen, then, that according to this model, testis formation needs two components: (1) the H-Y antigen, found on all male cells; and (2) the H-Y antigen receptor found on the gonadal cells of both sexes. These two conditions are met in the male gonad.

SIDELIGHTS & SPECULATIONS

The hormone-like effects of H-Y

In addition to being on the cell surface of all male cells, H-Y antigen is also secreted by testes cells into the epididymal fluid. There is evidence that soluble H-Y antigen plays a hormone-like role in testis formation and that it can be secreted by some gonadal cells and absorbed by others.

Usually, all cells of a developing gonad are either male or female. However, when an allophenic mouse chimera is made by fusing a male and a female embryo together, some of the gonadal cells will be XY and some will be XX. Interestingly enough, hermaphroditic gonads are the exception, not the rule. Most of the gonads are male, and all the cells in each gonad are H-Y$^+$. It appears then, that the XY gonadal cells can produce H-Y and secrete it. The XX gonadal cells are then capable of receiving it and forming testicular structures (Ohno, et al., 1978b).

This helps explain a phenomenon that has puzzled scientists and farmers for decades—FREEMARTIN COWS. Freemartins are sterile XX calves whose gonads have been all or partially masculinized. These individuals are only found when the affected cow is one of a pair of twins—one male and one female. The placental blood circulation has united these two fetuses, and it is possible that some diffusible substance travels from the male fetus to the female fetus. Sex hormones, such as testosterone, can masculinize the female calf's phenotype, but it cannot masculinize the ovary. In 1980, Wachtel and his co-workers found H-Y antigen in the sera of fetal calves. It is thought that this substance can travel through the blood and masculinize the developing ovary, thereby converting some of the ovary into testicular tissue.

There is some evidence that this secreted H-Y antigen is more important in testis formation than the surface molecule. Wachtel and Hall (1979) demonstrated that in some species, the H-Y antigen competes with an ovary-forming factor for the same gonadal receptor. When adult dog ovary cells were exposed to H-Y antigen, they absorbed it onto their cell surfaces. However, this binding was inhibited when the cells were first incubated with ovarian supernatant fluid. This result was extended by Zenzes and her colleagues (1980), who showed that the reaggregation of newborn rat testicular cells was inhibited by the presence of newborn (but not adult) rat ovarian supernatant fluids. It may be the case, then, that the gonadal receptor can receive either H-Y antigen or some other compound. In the absence of H-Y antigen, the soluble ovarian factor saturates the receptor, and the gonad becomes an ovary. When H-Y antigen is present, the interaction causes testicular structure to form.

Secondary sex determination

Testicular hormones. Primary sex determination involves the formation of either an ovary or a testis from the indifferent gonad. This, however, does not give the complete sexual phenotype. As mentioned earlier, in the absence of gonads, the female phenotype is realized: The Müllerian ducts develop and differentiate and the Wolffian duct decays. The testes, then, must elaborate hormones to promote Wolffian duct development and to cause Müllerian duct atrophy.

The first of these hormones is the steroid testosterone, which is secreted from the testicular Leydig cells. This hormone causes the Wolffian duct to differentiate into the epididymis, vas deferens, and seminal vesicles, and it causes the urogenital swellings and sinus to develop into the scrotum and penis. The second of these hormones is the anti-Müllerian duct factor (AMDF). This hormone, secreted by the Sertoli cells, causes the degeneration of the Müllerian duct. The existence of these two independent systems of masculinization is demonstrated by people having androgen insensitivity syndrome. These XY individuals are H-Y$^+$ and so have testes which make testosterone and AMDF. However, these people lack the cytoplasmic testosterone binding protein and thus cannot respond to the testosterone made in their testes. They are able to respond to estrogen made in their adrenal glands, so they are distinctly female in appearance (Figure 7). However, despite their female appearance, these individuals do have testes, and even though they cannot respond to testosterone, they do respond to AMDF. Thus, their Müllerian ducts degenerate. These people develop as normal but sterile women, lacking uteruses and oviducts.

So there are two distinct masculinizing hormones, testosterone and AMDF. There is evidence, though, that testosterone might not be the active hormone in certain tissues. In 1974, Siiteri and Wilson showed that testosterone is converted to 5α-dihydrotestosterone in the urogenital sinus and swellings, but not in the Wolffian duct. In the Dominican Republic Imperato-McGinley and her colleagues (1974) found a small community in which several inhabitants had a genetic deficiency of the enzyme 17α-ketosteroid reductase, the enzyme converting testosterone to dihydrotestosterone. Although these XY individuals have functioning testes, they have a blind vaginal pouch and an enlarged clitoris. They appear to be girls and are raised as such. Their internal anatomy, however, is male—Wolffian duct development and Müllerian duct degeneration. Thus, it appears that the formation of external genitalia is under the control of dihydrotestosterone whereas the Wolffian duct differentiation is controlled by testosterone itself (Figure 8). Interestingly enough, the masculinization of the external genitalia becomes responsive to testosterone at puberty, thus causing obvious masculinization in a person originally thought to be a girl.

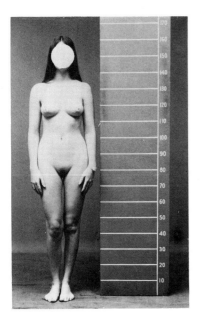

FIGURE 7
XY individual with androgen insensitivity syndrome. Despite the XY karyotype and the presence of testes, the individual develops female secondary sex characteristics. Internally, however, the woman lacks the Müllerian duct derivatives.(Courtesy of C. B. Hammond.)

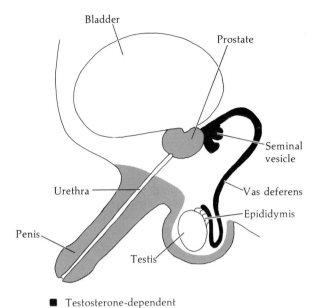

Bladder

Prostate

Seminal vesicle

Urethra

Vas deferens

Epididymis

Penis

Testis

■ Testosterone-dependent
□ Dihydrotestosterone-dependent

FIGURE 8
Testosterone- and dihydrotestosterone-dependent regions of the fetal human male genital system. (After Imperato-McGinley et al., 1974.)

The anti-Müllerian duct factor is a glycoprotein (Buzdik et al., 1980) made in the Sertoli cells (Tran et al., 1977). When fragments of fetal testes or isolated Sertoli cells are placed adjacent to cultured segments containing portions of the Wolffian and Müllerian ducts, the Müllerian duct is seen to atrophy without any change in the Wolffian duct (Figure 9). The duct is sensitive to the action of this hormone for only a brief period of time (Figure 10), and the testes stop making this factor at or near the time of birth.

We see then, that once the testes are formed, they elaborate two hormones that cause the masculinization of the fetus. One of these hormones, testosterone, may be converted into a more active form by

FIGURE 9
Assay for anti-Müllerian duct factor activity in the anterior segment of a 14.5-day fetal rat reproductive tract, after 3 days in culture. (A) Both the Müllerian duct (arrow at left) and Wolffian duct (arrow at right) are open. (B) The Wolffian duct (arrow) is open, but the Müllerian duct has degenerated and closed. (Courtesy of N. Josso.)

(A)

(B)

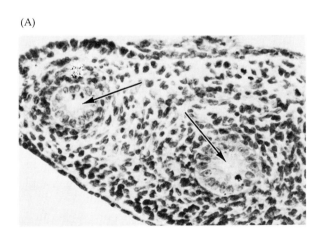

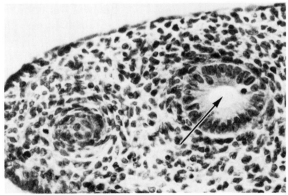

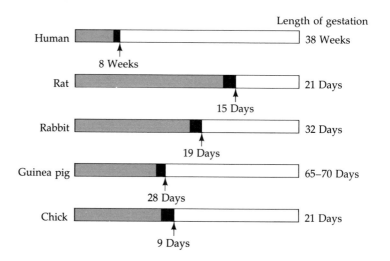

FIGURE 10
Sensitivity of Müllerian ducts to AMDF. The shaded area represents the duration of the indifferent gonad. The black zone represents the period during which the Müllerian ducts are responsive to AMDF. (After Josso et al., 1979.)

the tissues that create the external genitalia. In such a manner, the sex chromosomes control the sexual phenotype of an individual.

The central nervous system. One of the most controversial areas of secondary sex determination involves the development of sex-specific behaviors. In songbirds, testosterone is seen to regulate the growth of male-specific neuronal clusters in the brain. Male canaries and zebra finches sing eloquently, whereas the females sing little, if ever. These songs are used to mark territories and to attract mates. The ability to sing is controlled by six different clusters of neurons (NUCLEI) in the avian brain (Figure 11). Neurons connect each of these regions to one another. In male canaries, these six nuclei are several times larger than the corresponding cluster of neurons in female canaries; and in zebra finches, the females may lack one of these regions entirely.

Testosterone plays a major role in song control. In adult male zebra finches, Pröve (1978) demonstrated a linear correlation between the amount of song and concentration of serum testosterone. In adult chaffinches, castration eliminates song, but injection of testosterone will induce such birds to sing in November, when they are normally silent (Thorpe, 1958). Moreover, it has been shown that the seasonal fluctuation of testosterone causes not only a decrease in song but also a decrease in the size of the male-specific brain nuclei (Nottebohm, in Marx, 1982). In several species of birds, the females can be induced to sing by injecting them with testosterone (Nottebohm, 1980). The four song-controlling regions of the brain will grow 50–69 percent in such

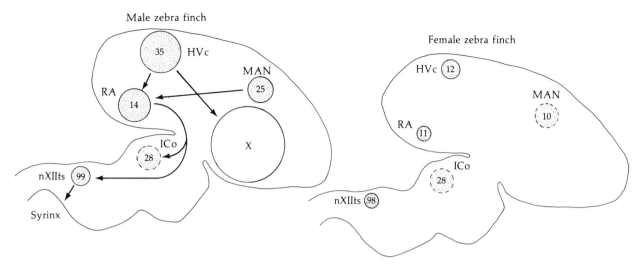

FIGURE 11

Sexual dimorphism in the avian brain. Schematic diagram indicates the major neural areas thought to be involved in the production of bird song in the zebra finch. Circles represent specific brain areas; the size of each circle is proportional to the volume occupied by that region. Circles with dashed lines are estimated volumes. The numbers within each circle represent the percentage of cells therein that incorporate radiolabeled testosterone. The volume differences in three of these regions (HVc, RA, and nXIIts) are significant between the sexes, and area X has not been observed in the brains of female finches. The differences in testosterone binding in regions HVc and MAN are significant, and no sex differences in steroid hormone binding have been observed in other regions of the brain. (The arrows indicate the axonal paths connecting the regions in the male finch.) (After Arnold, 1980.)

birds, whereas other brain regions show no such growth. Autoradiographic studies (Arnold et al., 1976) have shown that the neurons of the five song-controlling nuclei will incorporate radioactive testosterone, whereas other regions of the brain will not. It is apparent, then, that gonadal hormones can play a major role in the development of the regions of the nervous system that generate sex-specific behaviors.

In mammals, the situation is not as clear, for there are fewer behaviors that exclusively characterize one sex. In rats, penile thrusting is one such behavior, and it is controlled by motor neurons to the levator ani and bulbocavernosus muscles. Both these neurons originate from a spinal nucleus that can specifically concentrate testosterone. In female rats, these muscles are vestigial, and the volume of the controlling neurons is greatly reduced (Breedlove and Arnold, 1980).

Testosterone is not the only steroid capable of mediating behavior. In the mammalian brain, estrogen-sensitive neurons are also seen. These neurons are located (Figure 12) at positions in neural circuits that are known to mediate reproductive behavior: the hypothalamus, the

FIGURE 12
Diagrammatic representation of
the estrogen-binding regions in a
generalized female mammalian
brain. (After McEwen, 1981.)

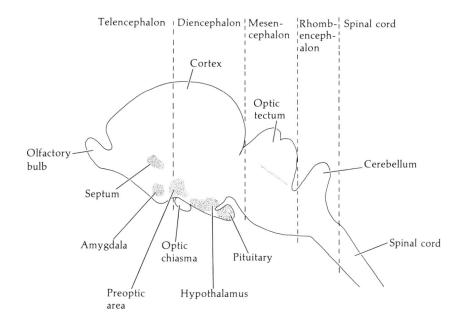

pituitary, and the amygdala (McEwen, 1981). Pfaff and McEwen (1983) demonstrated that estrogen alters the electrical and chemical features of those hypothalamic neurons capable of binding estrogen in their chromatin. Terasawa and Sawyer (1969) previously had found that the electrical activity of these neurons varies during the seasonal estrogen cycle of the rat, becoming elevated at the time of ovulation. Moreover, estrogen appears to stimulate those neurons in the regions that foster female reproductive behavior. Ovariectomized rats given estrogen injections directly to the hypothalamus displayed LORDOSIS, a position that stimulates the male to mount them, whereas control ovariectomized rats did not show that behavior (Barfield and Chen, 1977; Rubin and Barfield, 1980). The mechanisms by which estrogen causes the specific neuronal activity at these times is presently unknown and under investigation.

Chromosomal sex determination in *Drosophila*

Both mammals and insects such as *Drosophila* use an XX/XY system of sex determination. However, the mechanisms of these two groups are very different. In mammals, the Y chromosome plays a pivotal role, determining the male sex. Thus, XO individuals are sterile *females*, having ovaries, uteruses, and oviducts. In *Drosophila*, sex determination is achieved by a balance of female determinants on the X chromosome and male determinants on the autosomes (non-sex chromosomes). If there is but one X chromosome in a diploid cell (1X:2A), the organism

The development of sexual behaviors

When newborn male rats are castrated, thereby depriving them of testosterone during the neonatal period, they will undergo cyclic gonadotropin release (characteristic of female rats) and will display female sexual behaviors such as lordosis when they mature (Tiefer, 1970). Conversely, if newborn female rats are given a single dose of testosterone during the neonatal period, they will develop masculine endocrine and sexual behaviors. Thus, each newborn rat is born with the potential to display either male or female behaviors.

What was surprising, however, was that masculine behavior patterns could be permanently induced by injecting newborn female rats with a single dose of the female sex hormone estradiol[2] (Doughty et al., 1975). This ability of estradiol to help masculinize rat sexual behavior has given rise to the CONVERSION HYPOTHESIS. According to this model, the female pattern of sexual behavior is "intrinsic" to the brain. However, if the brain receives estradiol during a critical stage in

its development (immediately before or after birth, depending upon the species), defeminization will occur. Brain cells can obtain estradiol in two ways. They can receive it directly from the circulation or they can synthesize estradiol from circulating testosterone. To prevent defeminization in female mammals, estrogen-binding proteins in the serum eliminate freely circulating estradiol. In males, however, testosterone would not be bound by these proteins and would thus be able to enter the brain and be converted to estradiol. In the newborn rat, this conversion has been seen in the cells of the hypothalamus and the limbic system, two regions of the brain known to regulate hormonal and reproductive behaviors (Reddy et al., 1974).

Neonatal estradiol appears to be responsible for "defeminizing" the brain, whereas the actual masculinization of behavior is probably due to testosterone or dihydrotestosterone. When the conversion of testosterone to estradiol is chemically inhibited, males show both male and female behavior patterns (McEwen et al., 1977; Vreeburg et al., 1977). It is possible, then, that, like the formation of the male genital system, the development of the male nervous system involves both defeminizing and masculinizing steps.

[2]Estrogen and estradiol are often used interchangeably. However, estrogen refers to a class of steroid hormones responsible for producing or maintaining specific female characteristics. In many mammals (including humans) the most potent estrogen is estradiol.

---

is male. If there are two X chromosomes in a diploid cell (2X:2A), the organism is female. Thus XO *Drosophila* are sterile males. Table 2 shows the different X:autosome ratios and the resulting sex.

In *Drosophila*, and in insects in general, one can observe GYNANDROMORPHS—animals in which certain regions are male and other regions are female (Figure 13). This can happen when an X chromosome is lost from a certain embryonic cell. The progeny of that cell, instead of being XX (female), are XO (male). Because there are no sex hormones in insects to modulate such events, the XO cells display male characteristics, whereas the XX cells display female traits. This provides a beautiful example of the association between X chromosomes and sex. As can be seen from the preceding example, the Y chromosome plays no role whatever in *Drosophila* sex determination. Rather, it is necessary only to ensure fertility in males. The Y chromosome is seen to be active only for a certain time during sperm formation.

The X:autosome ratio is not the sole determinant of primary sex

TABLE 2
Ratios of X chromosomes to autosomes in different sexual phenotypes in *Drosophila melanogaster*

X chromosomes	Autosome sets (A)	X:A ratio	Sex
3	2	1.50	Superfemale
4	3	1.33	Metafemale
4	4	1.00	Normal female
3	3	1.00	″
2	2	1.00	″
2	3	0.66	Intersex
1	2	0.50	Normal male
1	3	0.33	Supermale

Source: Strickberger (1968).

determination in *Drosophila*. More likely, it is the trigger that activates or represses certain sex-determining genes. Four such autosomal genes have been identified (Baker and Ridge, 1980). When *transformer* (*tra*) or *transformer-2* (*tra-2*) mutations are homozygous, XX flies develop into sterile males and XY flies are unaffected. XX or XY flies homozygous for the *doublesex* (*dsx*) mutation become intersexes, having both male and female sex organs and bodily characteristics; the *intersex* (*ix*) mutation is specific for transforming females to the intersex phenotype. The development of the female phenotype needs the continued pres-

FIGURE 13
Gynandromorph of *Drosophila melanogaster* in which the left side is female (XX) and the right side is male (XO). The male side has lost an X chromosome bearing the wild-type alleles of eye color and wing shape, thereby allowing the expression of the recessive alleles *white eye* and *miniature wing* on the remaining X chromosome. (After Morgan, 1919.)

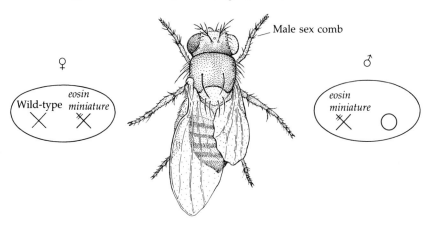

ence of the products of the wild-type *tra*, *tra-2*, and *ix* genes, whereas the wild-type *dsx* gene product is needed for the development of either the male or female phenotype alone. Thus, the *Drosophila* embryo, unlike that of mammals, appears to develop the male phenotype unless actively altered.

Environmental sex determination

Temperature-dependent sex determination in reptiles

Whereas the sex of most snakes and most lizards is determined by sex chromosomes at the time of fertilization, the sex of most turtles and all species of crocodilians is determined by the environment after fertilization. In these reptiles, the temperature of the eggs during a certain period of development is the deciding factor (Bull, 1980), and small changes of temperature can cause dramatic changes in the sex ratio. Generally, eggs incubated at low temperatures (22–27°C) produce one sex, whereas eggs incubating at higher temperatures (30°C and above) produce the other. There is only a small range of temperatures that permits both males and females to hatch from the same brood of eggs. Figure 14B shows the abrupt temperature-induced change in sex ratios for certain species of turtles. If eggs are incubated below 28°C, all the turtles hatching from them will be male. Above 32°C each egg will give rise to a female. Thus, each brood of eggs usually gives rise to individ-

FIGURE 14
Relationship between sex ratio and incubation temperature in reptiles. (A) Two species of lizards in which higher temperatures result in the generation of male offspring. (B) Seven species of turtles in which higher temperatures result in female offspring. (After Bull, 1980.)

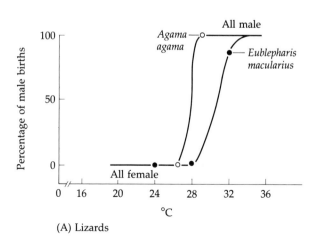

(A) Lizards

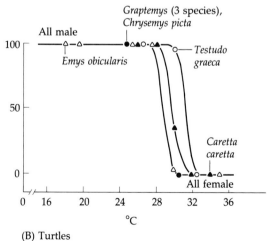

(B) Turtles

uals of the same sex. Variations on this theme also exist. Snapping turtle eggs, for instance, become female at either cold (less than or equal to 20°C) or hot (greater than or equal to 30°C) temperatures. In between these extremes, males predominate.

The developmental period during which sex determination occurs can be studied by incubating eggs at the male-producing temperature for a certain amount of time and then shifting the eggs to an incubator of the female-producing temperature. In map turtles and snapping turtles, the middle third of development appears to be the time when the temperature exerts its effect. Ferguson and Joanen (1982) have studied sex determination in the Mississippi alligator, both in the laboratory and in the field, and they have concluded that sex is determined between 7 to 21 days of incubation. Eggs raised at 30°C or below produce female alligators, whereas those incubated at 34°C or above produce all males. Moreover, the nests constructed on levees (34°C) give rise to males, whereas those built in wet marshes (30°C) produce females. Thus, the sex of many turtle and alligator species is based on the temperature of the egg's environment.

SIDELIGHTS & SPECULATIONS

The extinction of dinosaurs

Ferguson and Joanen (1982) have speculated that this method of temperature-dependent sex determination may have been used by another closely related reptile group, namely, the dinosaurs. If dinosaurs had temperature-dependent sex determination (with a somewhat different threshold temperature for the switch in sex), then a slight change in temperature may have created conditions where only males or females hatched from their eggs. This would explain the sudden, dramatic, and selective extinction of this reptile order.

Location-dependent sex determination in *Bonellia vividis* and *Crepidula fornicata*

The sex of the echiuroid worm *Bonellia* depends upon where a larva settles. The female *Bonellia* is a marine rock-dwelling animal, the body of which is about 10 cm long (Figure 15). It has a proboscis, however, that can extend to over a meter in length. This proboscis serves two functions. First, it sweeps food from the rocks into the digestive tract of the female *Bonellia*. Second, should a larva land on the proboscis, it will enter the mouth of the female, migrate to the uterus, and differentiate into a 1- to 3-mm long parasitic male. Thus, when a larva settles on a rocky surface, it becomes a female; but should that same larva settle on the proboscis of a female, it would become a male. The male

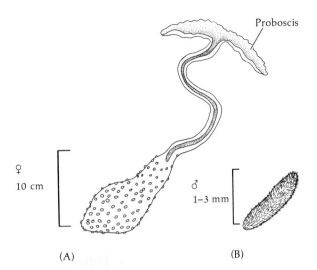

FIGURE 15
Extreme sexual dimorphism in *Bonellia vividis*. (A) Female, around 10 cm, with a proboscis capable of extending over a meter. (B) Parasitic male (highly magnified compared to the female), 1–3 mm long. (After Barnes, 1968.)

spends the rest of its life within the body of the female, fertilizing eggs.

Baltzer (1914) demonstrated that when larvae were cultured in the absence of adult females, about 90 percent of them became females. However, when these larvae were cultured in the presence of an adult female or its isolated proboscis, 70 percent of them adhered to the proboscis and developed the male structures. These results have been more recently confirmed by Leutert (1974; Figure 16).

The molecule(s) responsible for masculinizing the larvae can be extracted from the proboscis of adult females. When larvae are cultured in normal seawater in the absence of adult females, most become females. When cultured in seawater containing aqueous extracts of proboscis tissue, most become either male or an intermediate form, neither completely male nor completely female (Nowinski, 1934; Agius, 1979). The purification of the compound(s) that attract the larvae to the proboscis and cause its masculinization is presently being conducted.

Another example where sex determination is affected by the posi-

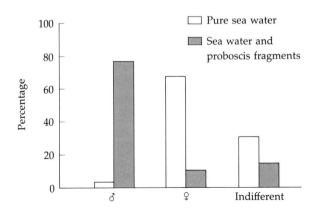

FIGURE 16
In vitro analysis of *Bonellia* differentiation. Larval *Bonellia* were placed either in normal seawater or in seawater containing fragments of the female proboscis. A majority of the animals cultured in the presence of the proboscis fragments became males, whereas normally they would have become females.

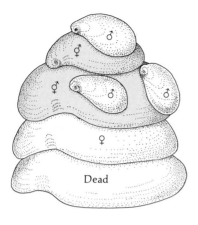

FIGURE 17
Cluster of *Crepidula* snails. Two individuals are changing from male to female. After these molluscs become female, they will be fertilized by the male above them. (After Coe, 1936.)

tion of the organism is that of the slipper shell *Crepidula fornicata*. Here, individuals will pile up on each other to form a mound (Figure 17). Young individuals are always male. This phase is followed by the degeneration of the male reproductive system and a period of lability. The next phase can be either male or female; and this depends upon position in the mound. If the snail is attached to a female, it will become male. If such a snail is removed from its attachment, it will become female. Similarly, the presence of large numbers of males will cause some of the males to become females. However, once the individual becomes female, it will not revert back to being male (Coe, 1936).

Hermaphroditism

In mammals, birds, and insects, sex is rigidly determined by the chromosome set. We also had seen that in some species, the sex is determined by the organism's environment. There exist, however, numerous organisms in which both a testis and an ovary exist simultaneously. These animals are called HERMAPHRODITES [from Greek, Hermes (Mercury) + Aphrodite (Venus)]. This condition is characteristic of many invertebrate classes, especially worms, wherein individuals can mutually fertilize each other. Hermaphroditism is a very successful reproductive strategy that aids the survival of numerous species. When individual meets individual, there is never a question of two similarly sexed animals meeting. Reproduction can occur whenever two animals of such a species meet. One developmental variation that ensures such meeting of hermaphrodites has evolved in the species *Diplozoan paradoxicum* ("strange double animal"), a parasite in the gills of carp. In the course of maturation, two of these worms grow together in the middle of their bodies, becoming joined for life. Moreover, each sperm duct becomes permanently attached to the other's reproductive orifice, thereby forming two reproductively competent parts of a conjoined animal. In vertebrates, true hermaphroditism is less common, and in birds and mammals, it is a pathological condition causing infertility. The most common vertebrate hermaphrodites are fishes, and several types of hermaphroditism exist (Yamamoto, 1969). Some fishes are GONOCHORISTIC, that is, they have a chromosomally determined sex that is either male or female. The heterogametic sex has been seen to express the H-Y antigen (Shalev and Huebner, 1980; Pechan et al., 1979; Müller and Wolf, 1979). The hermaphroditic fish species can be divided into three groups. The first are the SYNCHRONOUS hermaphrodites, in which ovaries and testicular tissues exist at the same time and in which both sperm and eggs are produced. One such species is *Servanus scriba*. In nature and in aquaria, these fish form spawning pairs. As soon as one of the fish spawns its eggs, the other fish fertilizes them. Then the fish reverse their roles and that fish that was formerly male spawns its eggs so that they can be fertilized by the sperm of its partner (Clark, 1959).

In other hermaphroditic species, an animal undergoes a genetically programmed sex change during its development. In these cases, the gonads are dimorphic, having both male and female areas. One or the other is predominant during a certain phase of life. In PROTOGYNOUS ("female-first") hermaphrodites, an animal begins its life as a female, but later becomes male. The reverse is the case in PROTANDROUS ("male-first") species. Figure 18 shows the gonadal changes of the protandrous hermaphroditic fish *Sparus auratus*. At first, testicular tissue predominates, and, later, after a transition period during which both testicular and ovarian tissues are seen, the ovarian cells take over.

Nature has provided many variations on her masterpiece. In some species, sex is determined solely by chromosomes, whereas in other species, sex is a matter of environmental conditions. Within these two large categories, numerous variations also exist. A complete catalog of known sex-determining mechanisms would take a separate (but very interesting) volume. But the sex-determining mechanism functions as the part of the somatic physiology necessary for the maintenance and propagation of the germ cells. The gonads will die with the body, but the germ cells that resided within them have the potential to renew life. It is to these germ cells that we shall now turn our attention.

FIGURE 18
Gonadal changes in the hermaphroditic fish *Sparus auratus*. Section through the gonad of the male phase (A), the transitory phase (B), and the final, female phase (C). (Photographs through the courtesy of the family of T. Yamamoto.)

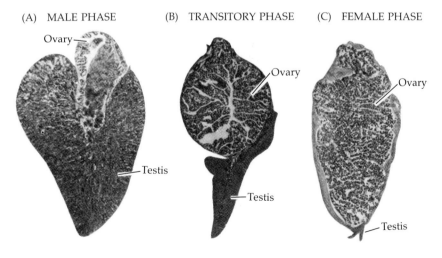

(A) MALE PHASE   (B) TRANSITORY PHASE   (C) FEMALE PHASE

Literature Cited

Agius, L. (1979). Larval settlement in the echiuran worm *Bonellia viridis:* Settlement on both the adult probiscus and body trunk. *Marine Biol.* 53: 125–129.

Arnold, A. P. (1980). Sexual differences in the brain. *Am. Sci.* 68: 165–173.

Arnold, A. P., Nottebohm, F. and Pfaff, D. W. (1976). Hormone concentrating cells in vocal control and other brain regions of the zebra finch (*Poephila guttata*). *J. Comp. Neurol.* 165: 487–512.

Baker, B. S. and Ridge, K. A. (1980). Sex and the single cell. I. On the action of major loci affecting sex determination in *Drosophila melanogaster. Genetics* 94: 383–423.

Baltzer, F. (1914). Die Bestimmung und der Dimorphismus des Geschlechtes bei *Bonellia. Sber. Phys.-Med. Ges. Würzb.* 43: 1–4.

Barfield, R. J. and Chen, J. J. (1977). Activation of estrous behavior in ovariectomized rats by intracerebral implants of estradiol benzoate. *Endocrinology* 101: 1716–1725.

Barnes, R. D. (1968). *Invertebrate Zoology.* Saunders, Philadelphia.

Billingham, R. E. and Silvers, W. K. (1960). Studies on tolerance of the Y chromosome in mice. *J. Immunol.* 85: 14–26.

Breedlove, S. M. and Arnold, A. P. (1980). Hormone accumulation in a sexually dimorphic motor nucleus of the rat spinal cord. *Science* 210: 564–566.

Bull, J. J. (1980). Sex determination in reptiles. *Q. Rev. Biol.* 55: 3–21.

Buzdik, G. P., Swann, D. A., Hayashi, A. and Donahoe, P. K. (1980). Enhanced purification of mullerian inhibiting substance by lectin affinity chromatography. *Cell* 21: 909–915.

Clark, E. (1959). Functional hermaphroditism and self-fertilization in a serranid fish. *Science* 129: 215–216.

Coe, W. R. (1936). Sexual phases in *Crepidula. J. Exp. Zool.* 72: 455–477.

Doughty, D., Booth, J. E., McDonald, P. G. and Parrott, R. F. (1975). Effects of oestradiol-17β, oestradiol benzoate, and synthetic oestrogen RU2858 on sexual differentiation in the neonatal rat. *J. Endocrinol.* 67: 419–424.

Eichwald, E. J. and Silmser, C. R. (1955). Communication. *Transpl. Bull.* 2: 148–149.

Ferguson, M. W. J. and Joanen, T. (1982). Temperature of egg incubation determines sex in *Alligator mississippiensis. Nature* 296: 850–853.

Geddes, P. and Thomson, J. A. (1890). *The Evolution of Sex.* Walter Scott, London.

Goodfellow, P. N. and Andrews, P. W. (1982). Sexual differentiation of H-Y antigen. *Nature* 295: 11–13.

Hall, J. L. and Wachtel, S. S. (1980). Primary sex determination: Genetics and biochemistry. *Mol. Cell. Biochem.* 33: 49–66.

Imperato-McGinley, J., Guerrero, L., Gautier, T. and Peterson, R. E. (1974). Steroid 5α reductase deficiency in man: An inherited form of male pseudohermaphroditism. *Science* 186: 1213–1215.

Josso, N., Picard, J.-Y. and Tran, D. (1977). The anti-mullerian hormone. *Rec. Prog. Horm. Res.* 33: 117–167.

Jost, A. (1953). Problems of fetal endocrinology: The gonadal and hypophyseal hormones. *Rec. Progr. Horm. Res.* 8: 379–418.

Langman, J. (1981). *Medical Embryology,* Fourth Edition. Williams & Wilkins, Baltimore.

Leutert, T. R. (1974). Zur Geschlechtsbestimmung und Gametogenese von *Bonellia viridis* Rolando. *J. Embryol. Exp. Morphol.* 32: 169–193.

Marx, J. L. (1982). How the brain controls birdsong. *Science* 217: 1125–1126.

McClung, C. E. (1902). The accessory chromosome—sex determinant? *Biol. Bull.* 3: 72–77.

McEwen, B. S. (1981). Neural gonadal steroid actions. *Science* 211: 1303–1311.

McEwen, B. S., Leiberburg, I., Chaptal, C. and Krey, L. C. (1977). Aromatization: Important for sexual differentiation of the neonatal rat brain. *Horm. Behav.* 9: 249–263.

Morgan, T. H. (1919). *The Physical Basis of Heredity.* Lippincott, Philadelphia.

Müller, U., Aschmoneit, I., Zenzes, M. T. and Wolf, U. (1978). Binding studies of H-Y antigen in rat tissues. Indications for a gonad-specific receptor. *Hum. Genet.* 43: 152–157.

Müller, U. and Wolf, U. (1979). Cross-reactivity to mammalian anti-H-Y antiserum in telostean fish. *Differentiation* 14: 185–187.

Nagai, Y., Ciccarese, S. and Smith, R. (1979). The identification of human H-Y antigen and testicular transformation induced by its interaction with the receptor site of bovine fetal ovarian cells. *Differentiation* 13: 155–164.

Nottebohm, F. (1980). Testosterone triggers growth of brain vocal control nuclei in adult female canaries. *Brain Res.* 189: 429–436.

Nowinski, W. (1934). Die vermännlichende Wirkung

fraktionierter Darmextrakte des Weibchens auf die Larven der *Bonellia viridis*. *Pubbl. Staz. Zool. Napoli* 14: 110–145.

Ohno, S. (1979). *Major Sex Determining Genes*. Springer–Verlag, New York.

Ohno, S., Nagai, Y. and Ciccarese, S. (1978a). Testicular cells lysostripped of H-Y antigen organize ovarian follicle-like aggregates. *Cytogenet. Cell Genet.* 20: 351–364.

Ohno, S., Ciccarese, S., Nagai, Y. and Wachtel, S. S. (1978b). H-Y antigen in testes of XX(BALB)/XXY(C3H) chimaeric male mouse. *Arch. Androl.* 1: 103–109.

Ohno, S., Nagai, Y., Ciccarese, S. and Smith, R. (1979). *In vitro* studies of gonadal organogenesis in the presence or absence of H-Y antigen. *In Vitro* 15: 11–18.

Pechan, P., Wachtel, S. S. and Reinboth, R. (1979). H-Y antigen in the teleost. *Differentiation* 14: 189–192.

Pfaff, D. W. and McEwen, B. S. (1983). The actions of estrogens and progestins on nerve cells. *Science* 219: 808–814.

Pröve, E. (1978). Courtship and testosterone in male zebra finches. *Z. Tierpsychol.* 48: 47–67.

Reddy, V. R., Naftolin, F. and Ryan, K. J. (1974). Conversion of androstenedione to estrone by neural tissues from fetal and neonatal rats. *Endocrinology* 94: 117–121.

Rubin, B. S. and Barfield, R. J. (1980). Priming of estrus responsiveness by implants of 17β-estradiol in the ventromedial hypothalamic nuclei of female rats. *Endocrinology* 106: 504–509.

Shalev, A. and Huebner, E. (1980). Expression of H-Y antigen in the guppy *Lebistes reticulatus*. *Differentiation* 16: 81–83.

Siiteri, P. K. and Wilson, J. D. (1974). Testosterone formation and metabolism during male sexual differentiation in the human embryo. *J. Clin. Endocrinol. Metab.* 38: 113–125.

Silvers, W. K., Gasser, D. L. and Eicher, E. M. (1982). H-Y antigen, serologically detectable male antigen, and sex determination. *Cell* 28: 439–440.

Stevens, N. M. (1905). Studies in spermatogenesis with especial reference to the "accessory chromosome." *Carnegie Institute Report* 36 (Washington, D.C.).

Strickberger, M. W. (1968). *Genetics*. Macmillan, New York, p. 463.

Terasawa, E. and Sawyer, C. H. (1969). Changes in electrical activity in rat hypothalamus related to electrochemical stimulation of adenohypophyseal function. *Endocrinology* 85: 143–149.

Thorpe, W. H. (1958). The learning of song patterns by birds with especial reference to the song of the chaffinch, *Fringilla coelebs*. *Ibis* 100: 535–570.

Tiefer, L. (1970). Gonadal hormones and mating behavior in adult golden hamster. *Horm. Behav.* 1: 189–202.

Tran, D., Meusy-Dessolle, N. and Josso, N. (1977). Anti-Mullerian hormone is a functional marker of foetal sertoli cells. *Nature* 269: 411–412.

Vreeburg, J. T. M., van der Vaart, P. D. M. and van der Schoot, P. (1977). Prevention of central defeminization but not masculinization in male rats by inhibiting neonatal estrogen biosynthesis. *J. Endocrinol.* 74: 375–382.

Wachtel, S. S. and Hall, J. L. (1979). H-Y binding in the gonad: Inhibition by a supernatant of the fetal ovary. *Cell* 17: 327–329.

Wachtel, S. S., Hall, J. L., Muller, U. and Chaganti, R. S. K. (1980). Serum-borne H-Y antigen in the fetal bovine freemartin. *Cell* 21: 917–926.

Wilson, E. B. (1905). The chromosomes in relation to the determination of sex in insects. *Science* 22: 500–502.

Yamamoto, T.-O. (1969). Sex differentiation. In W. S. Hoar and D. J. Randall (eds.), *Fish Physiology III*. Academic, New York, pp. 117–175.

Zenzes, M. T., Wolf, U., Gunther, E. and Engel., W. (1978a). Studies on the function of H-Y antigen: Dissociation–reorganization experiments on rat gonadal tissue. *Cytogenet. Cell Genet.* 20: 365–372.

Zenzes, M. T., Wolf, U. and Engel, W. (1978b). Organization *in vitro* of ovarian cells into testicular structures. *Hum. Genet.* 44: 333–338.

Zenzes, M. T., Urban, E. and Wolf, U. (1980). Inhibition of testicular organization *in vitro* by newborn rat ovarian cell supernatants. *Differentiation* 16: 193–198.

CHAPTER

20

The saga of the germ line

And the end of all our exploring
Will be to arrive where we started
And know the place for the first time.
—T. S. ELIOT (1942)

Introduction

We began our analysis of animal development by discussing fertilization, and we shall finish by investigating GAMETOGENESIS, the collection of processes by which the sperm and the egg are formed. Germ cells provide the continuity of life between generations, and the direct mitotic ancestors of our own germ cells once resided in the gonads of reptiles, amphibians, fishes, and invertebrates. But the germ cells do not arise from within the gonad itself. Rather, their precursors, the PRIMORDIAL GERM CELLS (PGCs), migrate into the developing gonads. The first step in gametogenesis, then, involves getting the primordial germ cells into the genital ridge as the gonad is forming.

Germ cell formation

Germ cell migration in amphibians

In anuran amphibians—frogs and toads—the germ plasm is readily marked by granules near the vegetal pole of the fertilized egg. However, this plasm does not remain long in the vegetal region. During cleavage, this material is brought upward through the yolky cytoplasm, and the RNA-rich granules are seen to become associated with the cells lining the floor of the blastocoel (Bounoure, 1934; Figure 1). As mentioned in Chapter 8, Blackler traced the migration of these cells into the gonad by transferring blocks of cells from one strain of *Xenopus laevis* neurulae into another. The primordial germ cells of this frog are seen to move laterally from the endoderm of the gut to the dorsal mesentery, which connects the gut to the region where the mesodermal organs are forming. They migrate up this tissue until they reach the developing gonads

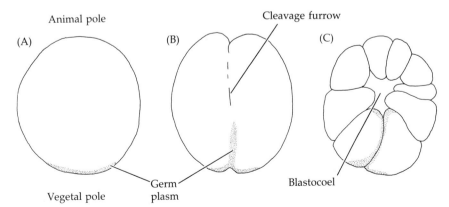

FIGURE 1
Changes in the position of germ plasm in an early frog embryo. Originally located near the vegetal pole of the uncleaved egg (A), it advances along the cleavage furrows (B) until it becomes localized at the floor of the blastocoel (C). (After Bounoure, 1934.)

FIGURE 2
Migration of primordial germ cells in a frog. This phase-contrast photomicrograph of a section through the body wall and dorsal mesentery of a *Xenopus* embryo shows the migration of two large primordial germ cells along the dorsal mesentery. (From Heasman et al., 1977.)

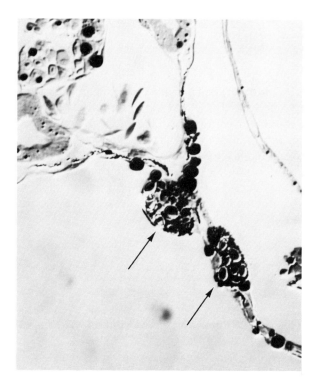

(Figure 2). *Xenopus* PGCs move by extruding a single filopodium and then streaming their yolky cytoplasm into the filopodium while retracting their tail. Contact guidance in this migration seems very likely as the cells over which they migrate are oriented in the direction of that migration (Wylie et al., 1979). Furthermore, PGC adhesion and migration can be inhibited if the mesentary is treated with antibodies against Xenopus fibronectin (Heasman et al., 1981). Thus, the pathway for germ cell migration in frogs appears to be composed of fibronectin-coated cells that are oriented in the direction of cell migration. The cells over which the PGCs travel lose this polarity soon after migration has ended.

Intrinsic factors are also important to the specific migration of the primordial germ cells. When the vegetal germ plasm of frog eggs is irradiated with ultraviolet light (which destroys RNA and the disulfide bonds of proteins), the migration of PGCs to the genital ridge is inhibited or delayed significantly (Smith, 1966; Züst and Dixon, 1977; Ikenishi and Kotani, 1979).

The primordial germ cells of urodele amphibians (salamanders) have a different origin, which has been traced (by reciprocal transplantation experiments) to the regions of the mesoderm that involute through the ventrolateral lips of the blastopore. Moreover, there does not seem to be any particular localized "germ plasm" in salamander eggs. Rather, the interaction of the dorsal endoderm cells and animal hemisphere cells creates the conditions needed to form germ cells in the particular areas that involute through the ventrolateral lips (Suta-suya and Nieuwkoop, 1974). So in salamanders, the primordial germ cells are already within the mesodermal region and presumbly do not follow the same path to the gonad.

Germ cell migration in mammals

In mammals, germ cell migration takes a route very similar to that seen in anurans. The major breakthrough in mapping mammalian germ cell migration came when Baxter (1952) demonstrated that mammalian primordial germ cells contained enormously high concentrations of the enzyme alkaline phosphatase. The primordial germ cells could thus be distinguished from other cells by staining for the presence of that enzyme. The route of the mammalian PGC migration is shown in Figure 3. The PGCs of mice can first be seen in the 7.5-day-old embryonic allantois (Chiquoine, 1954; Mintz, 1957). Shortly thereafter, they are seen at the base of the allantois and in the adjacent yolk sac (Figure 4). By this time, they have already split into two populations, which will migrate to either the right or the left genital ridge. The PGCs then move caudally from the yolk sac through the newly formed hindgut and up the dorsal mesentary into the genital ridge (Figure 4B). Most of the PGCs have reached the developing gonad by day 11 postfertilization. During this trek, they have proliferated such that, whereas only 10–100 cells were seen initially, some 2500–5000 PGCs are in the gonads by day 12. As is the case with PGC migration in *Xenopus*, mammalian PGCs appear to be closely associated with the cells over which they migrate, and they move by extending filopodia over the underlying cell surfaces.

FIGURE 3
Pathway for the migration of mammalian primordial germ cells. (A) Primordial germ cells first recognized in the yolk sac near the junction of hindgut and allantois. (B) Migration through gut and dorsally up the dorsal mesentery into the genital ridge. (After Langman, 1981.)

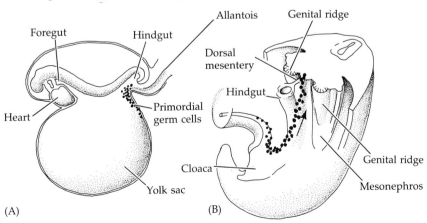

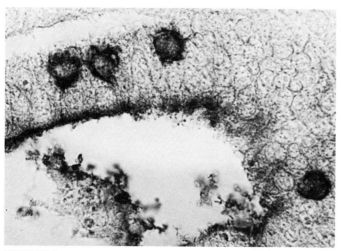

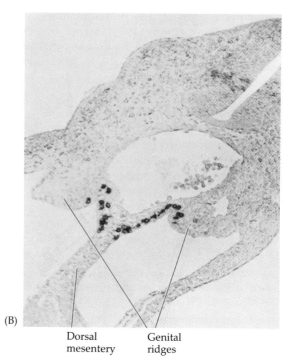

FIGURE 4   (A)
Mouse primordial germ cells at different stages of
their migration. (A) PGCs in hindgut of mouse embryo. Four large PGCs stain positively for high levels
of alkaline phosphatase. (B) PGCs, stained for alkaline phosphatase, can be seen migrating up the dorsal mesentery and entering the genital ridges. (A
from Heath, 1978, courtesy of J. Heath; B from
Mintz, 1957, courtesy of B. Mintz.)

(B)

Dorsal   Genital
mesentery   ridges

Germ cell migration in birds and reptiles

In birds and reptiles, PGCs are derived from epiblastic cells that migrate
to a crescent-shaped zone in the endodermal layer at the anterior border
of the zona pellucida (Eyal-Giladi et al., 1981; Figure 5). This region is
called the GERMINAL CRESCENT, and the primordial germ cells multiply
in this region. Unlike the situation in amphibians and mammals, the
primary route for germ cell migration in birds and reptiles is through
the blood. The PGCs enter the blood vessels by DIAPEDESIS (Figure 6A),
a type of movement that is common to lymphocytes and macrophages
and that enables cells to squeeze in and out of small blood vessels. The
PGCs thus enter the embryo by being transported in the blood (Pasteels,
1953; Dubois, 1969). The PGCs must also "know" to get out of the blood
when they reach the developing gonad (Figure 6B). When the germinal
crescent of a chick embryo is removed, and the circulation of that
embryo is joined with that of a normal chick embryo, the primordial
germ cells from the normal embryo will migrate into both sets of gonads
(Simon, 1960). It is not known what causes this attraction for the genital
ridges. One possibility is that the developing gonad produces a substance that attracts PGCs and retains them in the capillaries bordering
the gonad (Rogulska, 1969). [Such substances are known to be secreted
by lymphocytes at the sites of infection in order to attract macrophages
to that area (chemotaxis) to permit them to pass through the capillary

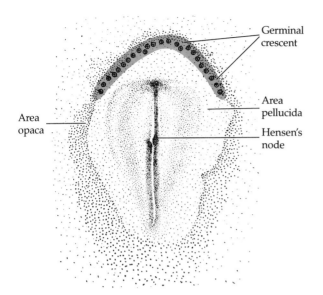

Germinal
crescent

Area
pellucida

Hensen's
node

FIGURE 5
Dorsal view of a primitive streak-stage chick embryo showing the region, called the germinal crescent, in which the germ cells arise. (Adapted from Swift, 1914.)

wall (diapedesis).] Another possibility is that the endothelial cells of the gonadal capillaries have a cell surface compound that causes the PGCs to adhere there specifically. Auerbach and Joseph (1984) have shown that several capillary networks can be distinguished from each other on the basis of their cell surfaces and that the endothelial cells from ovarian capillaries differ from all others tested. Whatever the attracting factor

FIGURE 6
Diapedesis of primordial germ cells in the chick embryo. (A) Diagrammatic representation of diapedesis, a process by which PGCs can enter the vitelline blood vessels. (B) Transverse section through the prospective gonadal region of a chick embryo. Several PGCs within the blood vessel cluster next to the genital ridge epithelium. One PGC is crossing through the blood vessel endothelium and another PGC is already located within the gonadal epithelium. (After Romanoff, 1960.)

(A)

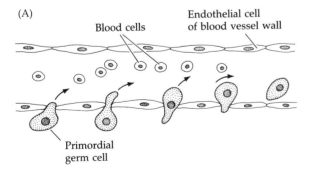

Blood cells

Endothelial cell
of blood vessel wall

Primordial
germ cell

(B)

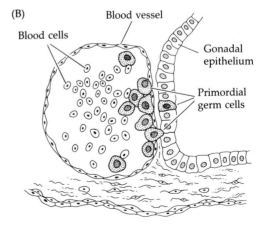

Blood cells

Blood vessel

Gonadal
epithelium

Primordial
germ cells

is, it is not species specific. The chicken gonad will attract circulating
PGCs from the turkey and even the mouse (Reynaud, 1969; Rogulska
et al., 1971).

Meiosis

Once in the gonad, the primordial germ cells continue to divide mitotically, thereby producing millions of potential gametes. The PGCs of
both male and female gonads are then faced with the necessity of
reducing their chromosome number from the diploid to the haploid
condition. In the haploid condition, each chromosome is represented
only once, whereas diploid cells have two copies of each chromosome.
To accomplish this reduction, the male and female germ cells undergo
a type of cell division called MEIOSIS.

  After the last mitotic division, a period of DNA synthesis occurs,
so that the cells initiating meiosis have twice the normal amount of
DNA in their nuclei. In this state, each chromosome consists of two
sister CHROMATIDS attached at a common centromere. Meiosis entails
two cell divisions (Figure 7). In the first division, homologous chromosomes (e.g., the chromosome 3 pair in the diploid cell) come together
and are then separated into different cells. Hence, the first meiotic
division separates homologous chromosomes into the two daughter
cells such that each cell has one copy of each chromosome. But each of
the chromosomes has already replicated. The second meiotic division,
then, separates the two sister chromatids from each other. The result
is that each of the four cells produced by meiosis has a single (haploid)
copy of each chromosome.

  The first meiotic division begins with a long prophase, which is
subdivided into five parts. During the LEPTOTENE (Greek, meaning "thin
thread") stage, the chromatin of the chromatids is stretched out very
thinly such that it is not possible to identify individual chromosomes.
DNA replication has already occurred, however, and each chromosome
consists of two parallel chromatids. At the ZYGOTENE (Greek, meaning
"yoked threads") stage, homologous chromosomes pair side-by-side.
This pairing is called SYNAPSIS, and it is characteristic of meiosis. Such
pairing does not occur during mitotic divisions. Although the mechanism whereby each chromosome recognizes it homologue is not known,
pairing seems to require the presence of the nuclear membrane and the
formation of a proteinaceous ribbon called the SYNAPTONEMAL COMPLEX.
This complex is a ladderlike structure with a central element and two
lateral bars. The chromatin is associated with the two lateral bars and
the chromatids are thus joined together (Figure 8). Examinations of the
meiotic cell nuclei with the electron microscope (Moses, 1968; Moens,
1969) suggest that paired chromosomes are bound to the nuclear membrane, and Comings (1968) has suggested that the nuclear envelope

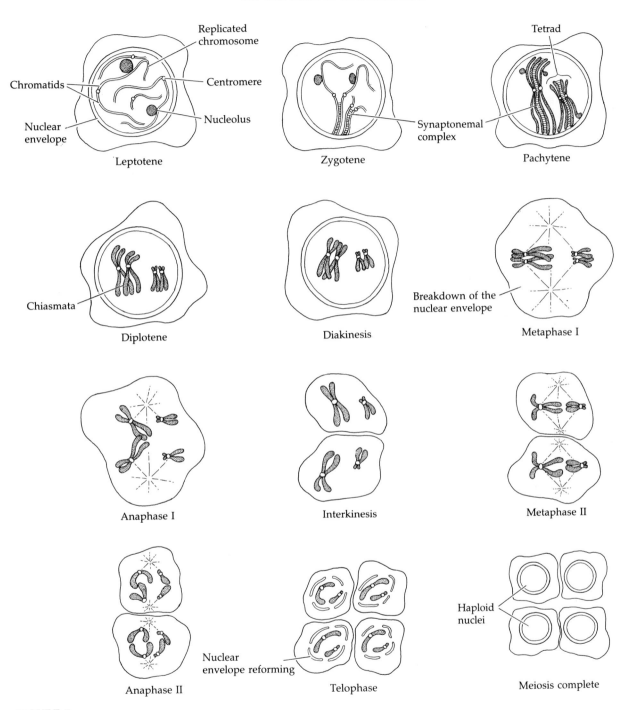

FIGURE 7
Chromosomes during the stages of meiosis. The stages leptotene through diakinesis are substages of the first meiotic prophase. (After Longo and Anderson, 1974.)

FIGURE 8

The synaptonemal complex. (A) Homologous chromosomes held together at the pachytene stage during meiosis in the *Neottiella* oocyte. (B) Interpretive diagram of the synaptonemal complex structure. (A from von Wettstein, 1971; B after Moens, 1974.)

(A)

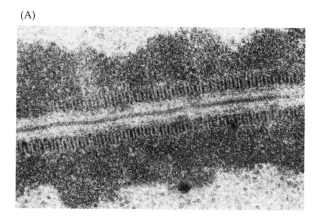

(B)

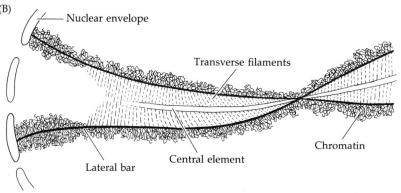

Nuclear envelope

Transverse filaments

Chromatin

Lateral bar   Central element

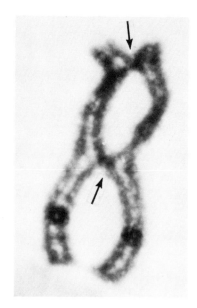

aids in bringing together the homologous chromosomes. The configuration formed by the four chromatids and the synaptonemal complex is referred to as a TETRAD (or BIVALENT).

During the next stage of meiotic prophase, the chromatids thicken and shorten. This stage has therefore been called the PACHYTENE (Greek, meaning "thick thread") stage. Individual chromatids can now be distinguished under the light microscope, and crossing-over may occur. Crossing-over represents exchanges of genetic material such that genes from one chromatid are exchanged with homologous genes from another chromatid. This crossing-over continues into the next stage, the DIPLOTENE (Greek, meaning "double threads"). Here, the synaptonemal complex breaks down and the two homologous chromosomes start to separate. Often, however, they are seen to remain attached at various places called CHIASMATA, which are thought to represent regions where crossing-over is occurring (Figure 9). The diplotene stage is character-

FIGURE 9

Chiasmata in diplotene bivalent chromosomes of salamander oocytes. Centromeres are visible as darkly staining circles; arrows point to the two chiasmata. (Courtesy of J. Kezer.)

ized by a high level of gene transcription. The chromosomes of both male and female germ cells take on the "lampbrush" appearance characteristic of chromosomes that are actively making RNA. During the next stage, DIAKINESIS (Greek, meaning "moving apart"), the centromeres move away from each other and the chromosomes remain joined only at the tips of the chromatids. This last stage of meiotic prophase ends with the breakdown of the nuclear membrane and the migration of the chromosomes to the metaphase plate.

During anaphase I, homologous chromosomes are separated from each other in an independent fashion. This leads to telophase I wherein two daughter cells are formed, each cell containing one partner of the homologous chromosome pair. After a brief INTERKINESIS, the second division of meiosis takes place. During this division, the centromere of each chromosome divides during anaphase so that each of the new cells gets one of the two chromatids; the final result being the creation of four haploid cells.

Note that meiosis has also reassorted the chromosomes into new groupings. First, each of the four haploid cells has a different assortment of chromosomes. In humans, where there are 23 different chromosome pairs, there can be 2^{23} (nearly 10 million) different types of haploid cells formed from the genome of a single person. In addition, the crossing-over that occurs during the pachytene and diplotene stages of prophase I further increases genetic diversity and makes the number of different gametes incalculable.

Spermatogenesis

Once the vertebrate primordial germ cells arrive at the genital ridge of male embryos, they become incorporated into the sex cords. They remain there until maturity, at which time the sex cords hollow out to form the seminiferous tubules and the epithelium of the tubules differentiates into the Sertoli cells. These Sertoli cells nourish and protect the developing sperm cells, and SPERMATOGENESIS, the meiotic divisions giving rise to the sperm, occurs in the recesses of the Sertoli cells (Figure 10). The process by which the primordial germ cells generate sperm has been studied in detail in several organisms, but we shall focus here on spermatogenesis in mammals. After reaching the gonad, the primordial germ cells divide to form TYPE A1 SPERMATOGONIA. These cells are smaller than the PGCs and are characterized by their ovoid nucleus containing chromatin associated with the nuclear membrane. The A1 spermatogonia are found adjacent to the outer basement membrane of the sex cords. At maturity, these spermatogonia begin dividing mitotically to produce more type A1 spermatogonia as well as a second, paler type of cell, the TYPE A2 SPERMATOGONIA. Thus, each type A1 spermatogonium is a stem cell capable of regenerating itself as well as of producing a new cell type. The A2 spermatogonia divide to produce

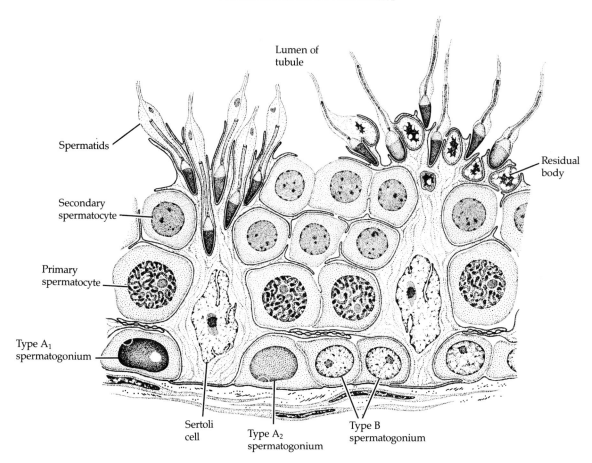

Lumen of
tubule

Spermatids

Secondary
spermatocyte

Primary
spermatocyte

Type A₁
spermatogonium

Residual
body

Sertoli
cell

Type A₂
spermatogonium

Type B
spermatogonium

FIGURE 10

Drawing of a section of the seminiferous tubule, showing the relationship
between Sertoli cells and the developing sperm. As cells mature, they progress toward the lumen of the seminiferous tubule. (After Dym, 1977.)

the type A3 spermatogonia, which then beget the type A4 spermatogonia, which beget the INTERMEDIATE SPERMATOGONIA. These intermediate spermatogonia divide to form the TYPE B SPERMATOGONIA, and
these cells divide mitotically to generate the PRIMARY SPERMATOCYTES,
the cells that enter meiosis.

Looking at Figure 11, we find that during the spermatogonial divisions, cytokinesis is not complete. Rather, the cells form a syncytium
whereby each cell communicates to the other via a cytoplasmic bridge
(Dym and Fawcett, 1971). Thus, successive divisions produce clones of
interconnected cells, assuring that each cohort matures synchronously.

Each primary spermatocyte undergoes the first meiotic division to
yield a pair of SECONDARY SPERMATOCYTES which complete the second
division of meiosis. The haploid cells formed are called SPERMATIDS,
and they are still connected to each other through their cytoplasmic

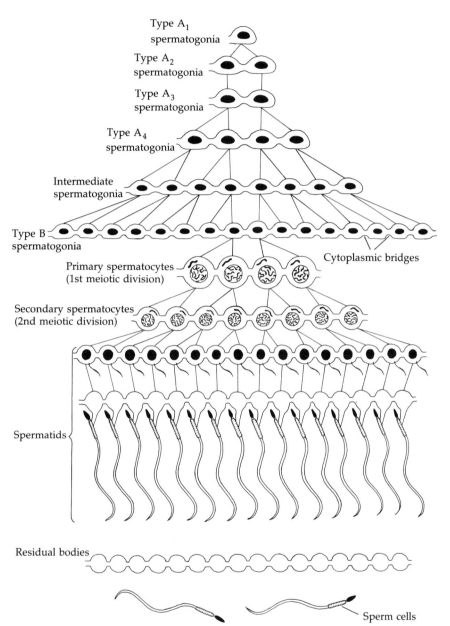

Type A₁ spermatogonia

Type A₂ spermatogonia

Type A₃ spermatogonia

Type A₄ spermatogonia

Intermediate spermatogonia

Type B spermatogonia

Primary spermatocytes (1st meiotic division)

Cytoplasmic bridges

Secondary spermatocytes (2nd meiotic division)

Spermatids

Residual bodies

Sperm cells

FIGURE 11
Diagrammatic representation of the formation of syncytial clones of male germ cells. (After Bloom and Fawcett, 1975.)

bridges. During the divisions from type A1 spermatogonium to spermatid, the cells move farther and farther away from the basement membrane of the tubule and closer to its lumen (Figure 10). Thus, each type of cell can be found in a particular layer of the tubule. The spermatids are located at the border of the lumen and here they differentiate into sperm cells.

Spermiogenesis

The haploid spermatid is a round, unflagellated cell that looks nothing like the mature sperm. The next step in sperm maturation, then, is SPERMIOGENESIS (or SPERMATELIOSIS), the differentiation of the sperm cell. The processes of sperm differentiation can be seen in Figure 2 in Chapter 2. The first steps involve the construction of the acrosomal vesicle from the Golgi apparatus. This acrosome forms a cap that covers the sperm. As the cap is formed, the nucleus rotates so that the acrosomal cap will then be facing the basal membrane of the seminiferous tubule. This is necessary because the flagellum is beginning to form from the centriole on the other side of the nucleus, and this flagellum will extend out into the lumen. During the last stage of spermiogenesis, the nucleus flattens and condenses, the remaining cytoplasm is jettisoned, and the mitochondria form a ring around the base of the flagellum. The resulting sperm then enter the lumen of the tubule.

In the mouse, the entire development from stem cell to spermatozoan takes 34.5 days. The spermatogonial stages last 8 days, meiosis lasts 13 days, and spermiogenesis takes up another 13.5 days. In humans, spermatic development takes 74 days to complete. Because the type A1 spermatogonia are stem cells, spermatogenesis can occur continuously. Each hour, some 100 million sperm are made in each human testicle, and each ejaculation releases 200 million sperm. Unused sperm is either resorbed or passed out of the body in urine.

Gene expression during sperm development

Gene transcription during spermatogenesis takes place predominantly during the diplotene stage of meiotic prophase. This has been observed in many organisms, but the best-documented case is probably that of Y chromosome transcription in *Drosophila hydei*. Here, RNA transcripts originating from the Y chromosome are seen to be essential for controlling spermiogenesis. When we recall the function of the Y chromosome in *Drosophila*, this is not surprising, for the Y chromosome is not involved in sex determination here. Rather, it is needed for the formation of viable sperm. The difference between XY *Drosophila* and XO *Drosophila* is that the latter are sterile. Both are male. In *Drosophila hydei*, the Y chromosome extends five prominent loops of DNA (Figure 12). If any of these loops is deleted, the organization of the sperm tail will be abnormal. All the component parts are present, but they are not organized properly (Hess, 1973). It appears, then, that Y-specific RNA made during meiotic prophase is utilized later during spermiogenesis.

This storage of RNA made during the diplotene stage has been seen elsewhere. In mammals and birds, a specific form of lactate dehydrogenase, LDH-X, is made during spermatogenesis. (This protein enables the developing sperm to utilize pyruvic acid as an alternate energy

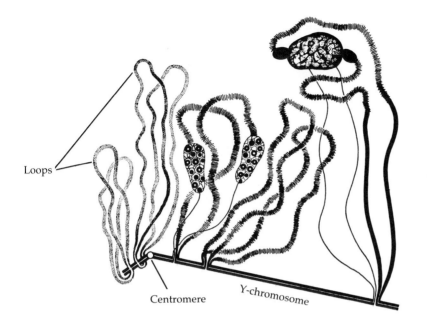

FIGURE 12
Diagram of the Y chromosome of *Drosophila hydei* in the lampbrush chromosome state. Five loops are readily observed. (After Hess, 1973.)

source.) The message for this protein can be identified in spermatocyte cytoplasm during meiotic prophase (Blanco, 1980). Similarly, in many species, a small protein called PROTAMINE appears in the nucleus during the final stages of spermiogenesis. This protein contains about 32 amino acids of which all but four or five are arginine residues. This protein replaces the nuclear histones (Figure 13) and causes the DNA to form a compact, almost crystalline, array (Marushige and Dixon, 1969). DNA complementary to trout protamine mRNA can detect protamine message sequences in the primary spermatocyte. However, these messages are stored in ribonucleoprotein particles. Only during the spermatid

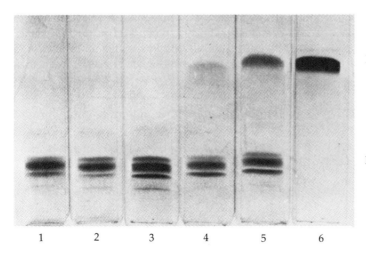

Protamines

Histones

FIGURE 13
Electrophoretic patterns of chromosomal basic proteins at successive stages (1–5) of sperm development in the trout testis. Eventually the protamines replace the histones in the sperm chromosome. Lane 6 shows native nucleoprotamine. (From Marushige and Dixon, 1969; courtesy of G. H. Dixon.)

stage, approximately a month after its synthesis, is the protamine message transcribed into protein (Iatrou et al., 1978). We find, then, that just like the oocyte, which can synthesize RNA in the diplotene stage and store it for later use, the sperm can also package messages for later translation.

In addition to gene transcription during meiotic prophase, there is also evidence that certain genes are transcribed in the spermatids (reviewed in Palmiter et al., 1984). This evidence for HAPLOID GENE EXPRESSION comes from studies involving heterozygous mice in which two different populations of sperm are seen to exist—one population expressing the mutant phenotype and one population expressing the wild-type trait. If the synthesis of the RNA or protein were to occur while the cells were still diploid, all the sperm would show the same phenotype.

One such mutation for which two sperm populations can be separated from heterozygotes is the t^{12} mutation in mice. This mutation, as we have seen earlier, is lethal when homozygous. The morulae do not undergo compaction. When heterozygous, however, this mutation leads to a phenomenon called SEGREGATION DISTORTION. Heterozygous ($t^{12}/+$) males mated to wild-type ($+/+$) females will produce heterozygous offspring over 95 percent of the time, instead of the expected 50 percent. The reciprocal crossing of wild-type males and heterozygous females does give the expected Mendelian 1:1 ratio (Bennett and Dunn, 1964). It appears, then, that the t^{12}-bearing sperm differed from the sperm carrying the wild-type allele. This was confirmed by immunological analysis. It was known that the product of the wild-type allele of t^{12} is present on the cell surface of the sperm. When antibody against this wild-type product was added to sperm from a heterozygous mouse, it bound to roughly 50 percent of them (Yanagisawa et al., 1974). This indicates that half the sperm expressed the wild-type allele and half did not. Therefore, the transcription of the t^{12} gene occurs after meiosis, rather than before it. Transcription is occurring from the haploid genome.

Oogenesis

Oogenic meiosis

OOGENESIS, the differentiation of the ovum, differs from spermatogenesis in several ways. Whereas the gamete formed by spermatogenesis is essentially a motile nucleus, the gamete formed by oogenesis must contain all of the factors needed to initiate and maintain metabolism and development. Therefore, in addition to forming a haploid nucleus, oogenesis must also build up a store of cytoplasmic enzymes, messages, organelles, and metabolic substrates. So while the sperm becomes streamlined for motility, the oocyte develops a remarkably complex cytoplasm. Unlike the sperm, which becomes differentiated after its meiotic divisions, the egg grows primarily in an extended period of meiotic prophase.

The mechanisms of oogenesis vary more than do those of spermatogenesis. This should not be surprising since the patterns of reproduction vary so greatly among species. In some species, such as sea urchins and frogs, the female routinely produces hundreds or thousands of eggs at a time; whereas in other species, such as humans and most mammals, only a few eggs are produced during the lifetime of an individual. In those species that produce thousands of ova, the oogonia are self-renewing stem cells that endure for the lifetime of the organism. In the more parsimonious species, the oogonia divide to form a limited number of egg precursor cells. In humans, the thousand or so oogonia divide rapidly from the second to the seventh month of gestation to form roughly 7 million germ cells (Figure 14). After the seventh month of embryonic development, however, the number of germ cells drops precipitously. Most oogonia die during this period, the remaining oogonia entering prophase of the first meiotic division (Pinkerton et al., 1961). These latter cells, called the PRIMARY OOCYTES, progress through the first meiotic prophase until the diplotene stage. Here they are arrested until puberty. With the onset of adolescence, groups of oocytes periodically resume meiosis. Thus, in the human female, the first part of meiosis is begun in the embryo, and the signal to resume meiosis is not given until roughly 12 years later. In fact, some oocytes are arrested in meiotic prophase for nearly 50 years. As Figure 14 indicates, primary oocytes continue to die even after birth. Of the millions of primary oocytes present at birth, only about 400 mature during a woman's lifetime.

Oogenic meiosis also differs from spermatogenic meiosis in its

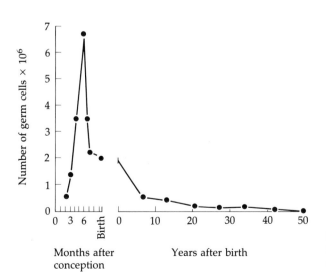

FIGURE 14

Changes in the number of germ cells in the human ovary. (After Baker, 1970.)

placement of the metaphase plate. When the primary oocyte divides, the nuclear membrane of the oocyte, called the GERMINAL VESICLE, breaks down and the metaphase spindle migrates to the animal pole of the cell. At telophase, one of the two daughter cells contains hardly any cytoplasm, whereas the other cell has nearly the entire volume of cellular constituents (Figure 15). The smaller cell is called the FIRST POLAR BODY and the larger cell is referred to as the SECONDARY OOCYTE. During the second division of meiosis, a similar unequal cytokinesis takes place. Most of the cytoplasm is retained by the ovum, and a SECOND POLAR BODY receives little more than a haploid nucleus. Thus, oogenic meiosis serves to conserve the volume of oocyte cytoplasm in a single cell rather than splitting it equally among four progeny.

The maturation of the oocyte in amphibians

The egg is responsible for initiating and directing development, and in some species (as seen earlier) fertilization is not even necessary. The accumulated material in the oocyte cytoplasm includes energy sources and organelles (the yolk and mitochondria), the enzymes and precursors for DNA, RNA, and protein syntheses, stored messenger RNAs, structural proteins, and morphogenic determinants. A partial catalog of the materials stored in the oocyte cytoplasm is shown in Table 1. Most

FIGURE 15
Polar body formation in the oocyte of the whitefish, *Coregonus*. (A) Anaphase of first meiotic division, showing the polar body pinching off with its chromosomes. (B) Metaphase (within the oocyte) of the second meiotic division, with the first polar body still in place. The first polar body may or may not divide again. (From Swanson et al., 1981; courtesy of C. P. Swanson.)

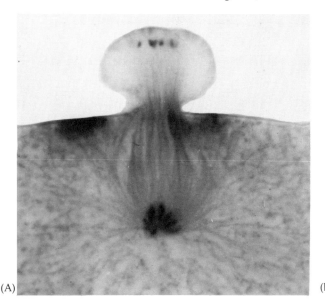

(A)

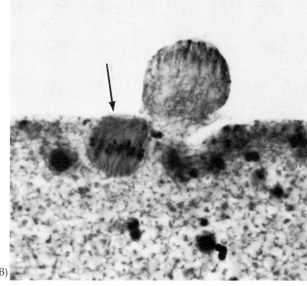

(B)

SIDELIGHTS &
SPECULATIONS

Natural parthenogenesis

In certain species of animals, reproduction takes place without males. These maleless species are able to develop an embryo from an oocyte without any participation of sperm and are said to be PARTHENOGENETIC species. This parthenogenesis is usually accomplished by one of several variations on the theme of meiosis. In the fly *Drosophila mangabeirai*, one of the polar bodies acts as a sperm and "fertilizes" the oocyte nucleus after the second meiotic division. The insect *Moraba virgo* and the lizard *Cnemidophorus uniparens* form parthenogenetic eggs by doubling their chromosome number prior to meiosis. The two normal meiotic divisions then restore their normal, diploid, chromosome numbers.

The germ cells of another grasshopper, *Pycnoscelus surinamensis*, dispense with meiosis altogether, forming ova by two mitotic divisions (Swanson et al., 1981).

In each of the above cases, the species consists entirely of diploid females. In other species, HAPLOID PARTHENOGENESIS is widely used, not only as a means of reproduction, but also a mechanism of sex determination. In the Hymenoptera (bees, wasps, and ants), unfertilized eggs develop into males whereas fertilized eggs, being diploid, develop into females. The haploid males are able to produce sperm by abandoning the first meiotic division, thereby forming two sperm cells through meiosis.

---

of this accumulation takes place during meiotic prophase I, and this stage is often subdivided into PREVITELLOGENIC (Greek, meaning "before yolk formation") and VITELLOGENIC (yolk-forming) phases.

Eggs of fishes and amphibians are derived from a stem-cell oogonial population, which can generate a new cohort of oocytes each year. In the frog *Rana pipiens*, oogenesis lasts 3 years. During the first 2 years, the oocyte increases its size very gradually. During the third year, however, the rapid accumulation of yolk material in the oocytes causes them to swell to their characteristically large size (Figure 16). Eggs mature in yearly batches, the first cohort maturing shortly after metamorphosis; the next group matures a year later.

TABLE 1
Cellular components stored in the mature oocyte of *Xenopus laevis*

Component	Approximate excess over amount in larval cells
Mitochondria	100,000
RNA polymerases	60,000–100,000
DNA polymerases	100,000
Ribosomes	200,000
tRNA	10,000
Histones	15,000
Deoxyribonucleoside triphosphates	2,500

Source: Bull (1974).

FIGURE 16
Growth of oocytes in the frog. During the first 3 years of life, three cohorts of oocytes are produced. The drawings follow the growth of the first-generation oocytes. (After Grant, 1953.)

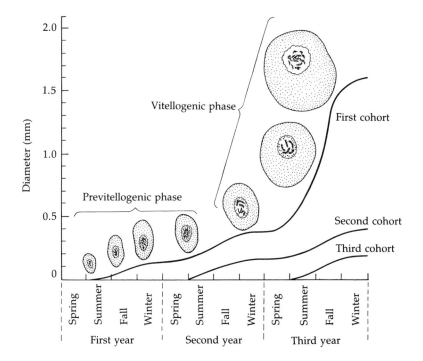

Vitellogenesis occurs when the oocyte reaches the diplotene stage of meiotic prophase. This is also the stage when the lampbrush chromosomes of the nucleus are seen to be actively synthesizing RNA (Chapter 12). Yolk is not a single substance, but a mixture of materials used for embryonic nutrition. The major yolk component is a 470,000-d protein called VITELLOGENIN. It is not made in the oocyte (as are the major yolk proteins of such organisms as annelids and crayfish), but it is synthesized in the liver and carried through the blood to the ovary (Flickinger and Rounds, 1956). This enormous protein passes between the follicle cells of the ovary and is incorporated into the oocyte by MICROPINOCYTOSIS, the pinching off of membrane-bounded vesicles at the base of microvilli (Dumont, 1978). In the mature oocyte, vitellogenin is split into two smaller proteins, the heavily phosphorylated PHOSVITIN and the lipoprotein LIPOVITELLIN. These two proteins are packaged together into membrane-bounded YOLK PLATELETS (Figure 17). Glycogen granules and lipochondrial inclusions store the carbohydrate and lipid components of the yolk, respectively.

As the yolk is being deposited, the cortical granules begin to form from the Golgi apparatus. They are originally scattered randomly through the oocyte cytoplasm but then migrate to the periphery of the cell. The mitochondria are capable of self-replication, and they divide to form millions of organelles. These mitochondria will be apportioned to the different cells during cleavage, and new mitochondria are not formed until after gastrulation is initiated. As vitellogenesis nears an end, the oocyte cytoplasm becomes stratified. The cortical granules,

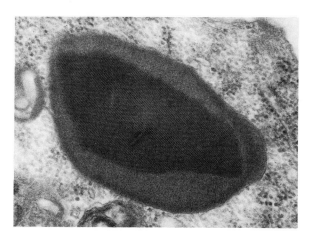

FIGURE 17
An amphibian yolk platelet.
(Courtesy of L. K. Opresko.)

mitochondria, and pigment granules are found at the periphery of the cell, creating the oocyte cortex. Within the inner cytoplasm, distinct gradients emerge. The yolk platelets are found more heavily concentrated at the vegetal pole of the oocyte, whereas the glycogen granules, ribosomes, lipochondria, and endoplasmic reticulum are found more toward the animal pole. The mechanisms for establishing these gradients are not known.

In *Xenopus*, the leptotene stage of meiosis lasts only 3–7 days, zygotene takes from 5–9 days, and pachytene persists roughly 3 weeks. The diplotene stage, however, can last years. Even so, vitellogenesis occurs only in part of the diplotene, and the signal for the breakdown of the germinal vesicle occurs after vitellogenesis is completed. The regulation of these events is controlled by the hormonal interactions between the hypothalamus, pituitary gland, and follicle cells of the ovary (Figure 18). When the hypothalamus receives the cues that the mating season has arrived, it secretes gonadotropin releasing hormone, which is received by the pituitary. The pituitary responds by secreting the gonadotropic hormones into the blood. These hormones stimulate the follicle cells to secrete estrogen, and the estrogen instructs the liver to synthesize and secrete vitellogenin. Upon estrogen stimulation, the liver cells change drastically (Figure 19). These changes are usually brought about during the annual mating season, but when estrogen is injected into adult frogs (either male or female) at any time, these changes can be induced and vitellogenin secreted (Skipper and Hamilton, 1977).

Estrogen induces vitellogenin at both the transcriptional and translational levels. Before estrogen, there are no detectable vitellogenin messenger RNAs in the liver. Following hormone administration, each cell contains some 50,000 vitellogenin mRNA molecules, constituting roughly half the total cellular mRNA. The transcriptional activation of the vitellogenin genes would normally generate only about 1500 of these mRNA molecules per cell. However, estrogen also specifically stabilizes

(A) VITELLOGENESIS AND
    OOCYTE DIFFERENTIATION

(B) EGG MATURATION
    AND OVULATION

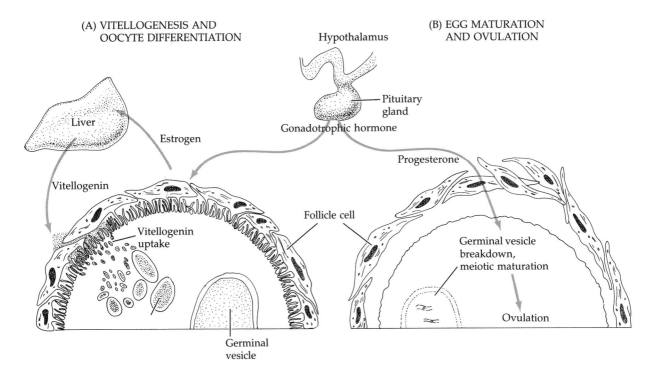

FIGURE 18

Control of amphibian oocyte growth and egg maturation by estrogen and progesterone. (A) Gonadotropic hormone stimulates the follicle cells to produce estrogen, which instructs the liver to secrete vitellogenin. This protein is absorbed by the oocyte. (B) After vitellogenesis, again under the influence of gonadotropic hormone, the follicle cells secrete progesterone. Within 6 hours of progesterone stimulation, the germinal vesicle breaks down—the initial event in a series that will lead to ovulation. (After Browder, 1980.)

that message such that its half-life increases from 16 hours to 3 weeks. Because these effects occur in the absence of new protein synthesis, they appear to be direct consequences of estrogen. Thus, estrogen is seen to control the accumulation of vitellogenin by transcriptional and translational gene regulation (Brock and Shapiro, 1983).

Release of the primary oocyte from meiotic arrest requires progesterone. This hormone is secreted by follicle cells in response to gonadotropic hormones secreted by the pituitary gland. Within 6 hours of progesterone stimulation, GERMINAL VESICLE BREAKDOWN (GVBD) occurs, the microvilli retract, the nucleoli disintegrate, and the lampbrush chromosomes contract and migrate to the animal pole to begin division. Soon afterward, the first meiotic division occurs, and the mature ovum is released from the ovary by a process called OVULATION. How does progesterone accomplish this result? First, it is thought that progesterone works at the surface of the oocyte because microinjection of pro-

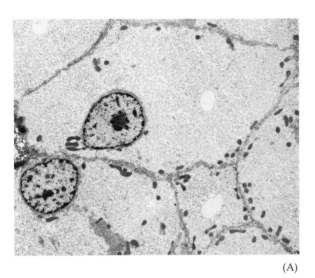
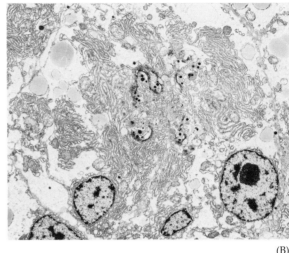

(A)   (B)

FIGURE 19

Effects of estrogen (estradiol) on the ultrastructure of *Xenopus laevis* liver cells. (A) Hepatocytes prior to estrogen injection. (B) Hepatocytes showing the enormous expansion of rough endoplasmic reticulum and Golgi apparatus needed for the production of vitellogenin. (From Skipper and Hamilton, 1977; courtesy of the authors.)

gesterone directly into the oocyte cytoplasm does not cause GVBD (Smith and Ecker, 1969; Masui and Markert, 1971). Second, progesterone is thought to cause the release of free calcium ions, which had formerly been bound within the cortex (Figure 20). This increase is observed within a minute of progesterone administration (Wasserman et al., 1980), and GVBD can be induced by the injection of calcium ions directly into the oocyte or by incubating the oocytes in a solution containing the calcium ionophore A23187 (Wasserman and Masui, 1975; Moreau et al., 1976). The release of free calcium ions has been correlated with a decline in cyclic AMP (cAMP) levels within the oocyte. Because the addition of cAMP to oocytes inhibits GVBD, it seems likely that the increase in calcium causes GVBD by causing a decline in cAMP levels. Cyclic AMP usually acts by activating certain enzymes, which phosphorylate other proteins. Without cAMP, then, certain proteins would not be phosphorylated. It is hypothesized that one of these proteins is an active "initiator protein." This protein would be inactivated by phosphorylation. When the cyclic AMP levels decline, this protein is not phosphorylated, and it is therefore active. It can then catalyze a reaction (probably a cAMP-independent phosphorylation) that activates a formerly inactive protein called MATURATION PROMOTING FACTOR (MPF). This MPF is known to exist because cytoplasm from mature oocytes will cause GVBD when injected into immature oocytes (Reynhout and Smith, 1974). Thus, progesterone appears to activate a cascade of post-

FIGURE 20
Cascade of events leading to the activation of maturation promoting factor
and the breakdown of the germinal vesicle.

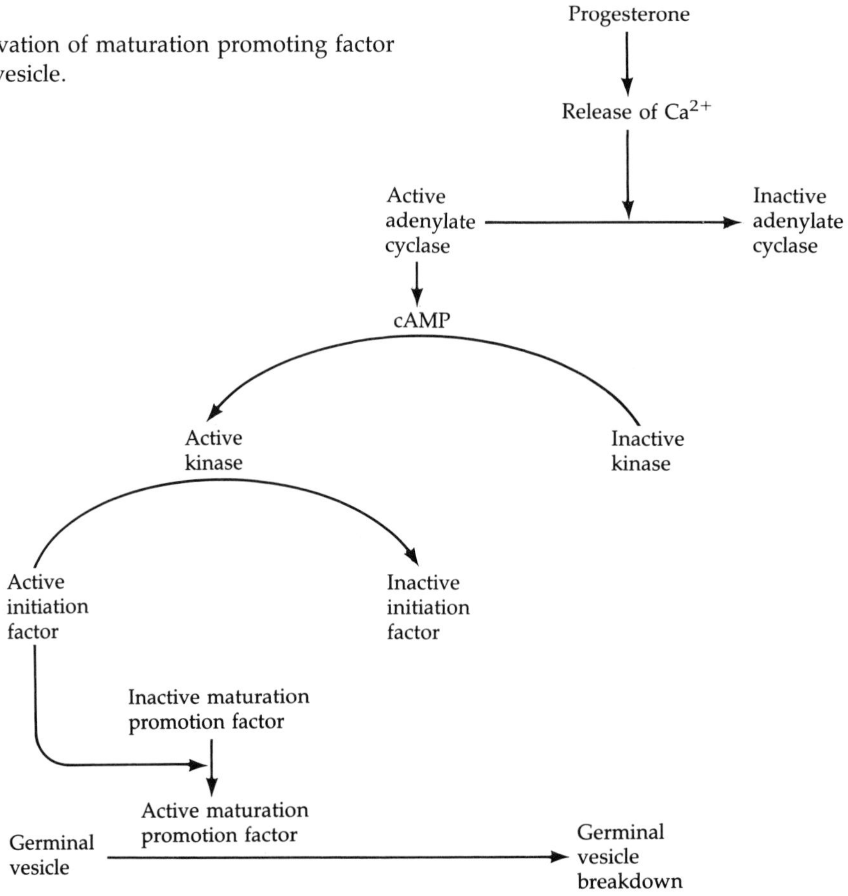

translational modification steps that result in the activation of MPF. The
mechanism by which MPF initiates germinal vesicle breakdown and the
other steps of maturation is still unknown.

Another effect of progesterone is to shut down RNA synthesis
(Figure 21). Unlike the situation in sea urchin oocytes, in which the
amount and the complexity of the mRNA population are both seen to
increase with oocyte maturation (Hough-Evans et al., 1977), the amount
and the complexity of frog oocyte mRNA appears to stay relatively
constant (Perlman and Rosbash, 1978; Dolecki and Smith, 1979). The
amount of increase seems to be balanced by that of degradation, even
during the lampbrush stage. Progesterone also initiates the translation
of stored messenger RNAs in the oocyte cytoplasm. Thus, whereas in
sea urchins the translation of maternal messages is initiated by fertilization, here the signal for such translation is initiated by progesterone
as the egg is about to be ovulated.

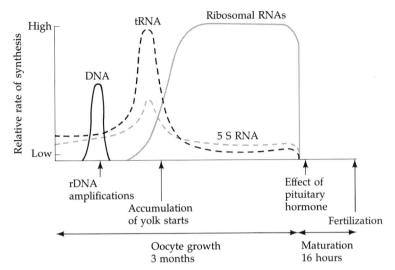

FIGURE 21
Relative rates of DNA, tRNA, and ribosomal RNA synthesis in amphibian oogenesis during the last 3 months before ovulation. (From Gurdon, 1976.)

Oogenesis in meroistic insects

Certain insects undergo MEROISTIC oogenesis wherein cytoplasmic connections remain between the cells produced by the oogonium. In *Drosophila*, each oogonium divides four times to produce a clone of 16 cells connected to each other through RING CANALS. The production of these interconnected cells (called CYSTOCYTES) involves a highly ordered array of cell divisions (Figure 22). Only those two cells having four interconnections are capable of developing into oocytes, and of these two, only

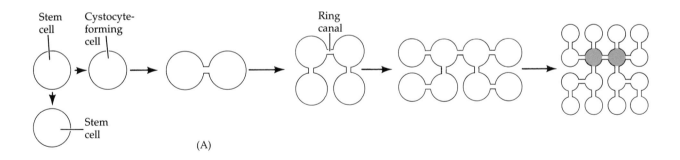

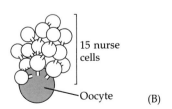

FIGURE 22
The formation of 16 interconnected cystocytes in *Drosophila*. (A) The cells are represented schematically as dividing in a single plane. The stem cell divides to produce another stem cell plus a cell that is committed to form the cystocytes. Only one of the 16 cystocytes will become the oocyte; the others become nurse cells, connected to the oocyte by ring canals (cytoplasmic bridges). (B) A 3-dimensional representation of the oocyte and its 15 nurse cells. (A after Koch et al., 1967.)

one prevails. The other begins meiosis but does not complete it. Thus, only one of the 16 cystocytes can become an ovum. All the other cells become NURSE CELLS. As it turns out, the cell destined to become the oocyte is that cell residing at the most posterior tip of the egg chamber that encloses the 16-cell clone (Figure 23).

The oocytes of meroistic insects do not pass through a transcriptionally active stage, nor do they have lampbrush chromosomes. Rather, autoradiographic evidence shows that RNA synthesis is largely confined to the nurse cells and that the RNA made by these cells is actively transported into the oocyte cytoplasm. This can be seen in Figure 24. When the egg chambers of a housefly are incubated in radioactive cytidine, the nuclei of the nurse cells show intense labeling. When the labeling is stopped and the cells are incubated for 5 more hours in nonradioactive media, the labeled RNA is seen to enter the oocyte from the nurse cells (Bier, 1963). Oogenesis takes place in only 12 hours, so the nurse cells are extremely busy during this time. They are aided in their transcriptional efficiency by becoming polytene. Instead of having two copies of each chromosome, they replicate their chromosomes so as to produce 512 copies. The 15 nurse cells have been seen to transport both ribosomal and messenger RNAs into the oocyte cytoplasm, and entire ribosomes may be transported as well. The mRNAs do not associate with polysomes, a finding that suggests that they are not im-

FIGURE 23
Drawing of a section through the egg chamber of *Drosophila melanogaster*. The entire chamber is surrounded by follicle cells and includes a single oocyte and 15 nurse cells. Cytoplasmic continuity can be seen between nurse cells and the oocyte; organelles are observed passing through these connections. (From Klug et al., 1970.)

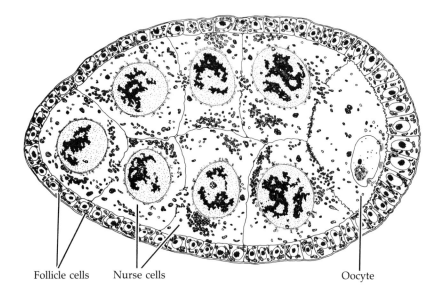

Follicle cells   Nurse cells   Oocyte

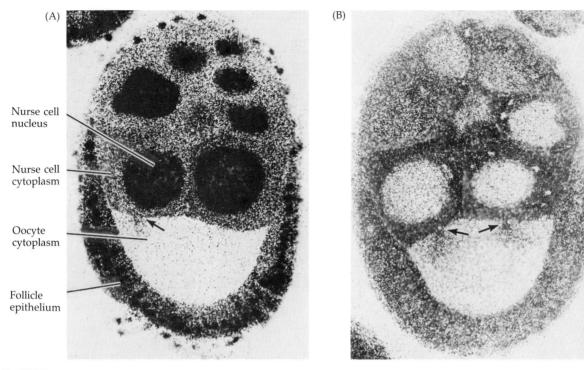

FIGURE 24

Autoradiograph of the follicle of the housefly, *Musca domestica*, after incubation with [³H]cytidine. (A) Egg chamber fixed immediately after label was introduced. The nuclei of the nurse cells are heavily labeled, indicating that they are synthesizing new RNA. The oocyte remains unlabeled except for some RNA escaping into the oocyte through the connection between it and a nurse cell. (B) A similar egg chamber fixed 5 hours later. Label is gone from the nurse cell nuclei but has moved into the cytoplasm. Moreover, radioactive RNA can be seen pasing into the oocyte cytoplasm through the two channels (arrows) between the nurse cells and the oocyte. (From Bier, 1963; courtesy of D. Ribbert.)

mediately active in protein synthesis (Paglia et al., 1976; Telfer et al., 1981).

The meroistic ovary confronts us with some interesting problems. If all the cells are connected such that proteins and RNAs shuttle freely between them, why should they have different developmental fates? Why should one cell become the oocyte while the others become "RNA synthesizing factories"? Why is the flow of protein and RNA in one direction only? One answer to these problems is that within the egg chamber an electrical gradient exists and prevents the diffusion of certain molecules while facilitating the diffusion of others. Woodruff and Telfer (1980) have shown that such a gradient exists and that it inhibits the diffusion of most proteins while enhancing the diffusion of acidic macromolecules. This is shown in Figure 25. When an acidic enzyme

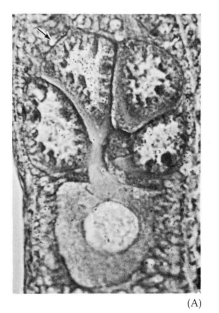

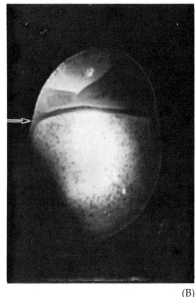

(A)   (B)   (C)

FIGURE 25

(A) Egg chamber of the moth *Hyalophora cecropia* showing the oocyte and four adjacent nurse cells. The micrograph section cuts through the bridge linking one of the nurse cells (arrow) with the oocyte. (B) Fluorescence micrograph of an early vitellogenic egg chamber fixed 1 hour after one of its nurse cells was injected with methylcarboxylated and fluorescein-labeled lysozyme. This modified protein crossed the intercellular bridge into the oocyte (arrow). (C) The same as (B), except that the basic form of fluorescein-labled lysozyme (nonmethylcarboxylated) was injected. No transport of the protein into the oocyte is seen. (From Telfer et al., 1981; courtesy of W. H. Telfer.)

is labeled with a fluorescent marker and injected into nurse cells, the protein is seen to cross from the nurse cell into the oocyte cytoplasm. However, when that same labeled protein is made basic by small chemical modifications, it will remain in the nurse cell into which it was injected. In this way, developmental cues are not exchanged between the connected cells, but certain proteins and nucleic acids can be transported in a unidirectional fashion.

The three major yolk proteins in *Drosophila* are made in the fat body and ovary, but not in the oocyte itself (Bownes, 1982; Brennen et al., 1982). The hormonal control of yolk synthesis is controlled by juvenile hormone, ecdysone, and a neurosecretory hormone from the brain. It is thought that the brain hormone, responding to environmental cues, stimulates the corpora allata to secrete juvenile hormone (Figure 26). Juvenile hormone (1) regulates the uptake of the yolk peptides at the surface of the oocyte, (2) stimulates the synthesis of ovarian yolk proteins (which are each identical to those made by the fat body), and (3)

FIGURE 26
Model for the hormonal regulation of yolk peptide synthesis in *Drosophila melanogaster*. In response to a brain hormone, the corpora allata produces juvenile hormone. The JH causes the ovary to make yolk proteins and ecdysteroids. Juvenile hormone also induces ecdysteroid synthesis in abdominal cells. The ecdysteroids cause the fat body to produce yolk proteins, which are transported to the ovary. (After Bownes, 1982.)

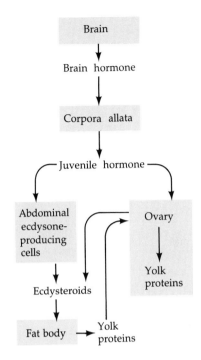

causes the ovarian follicles and other abdominal cells to secrete ecdysone. The ecdysone stimulates the fat body to produce yolk proteins just as estradiol stimulated the amphibian liver to do so. Similarly, the administration of ecdysone to adult males will cause their fat bodies to secrete yolk proteins as well (Postlethwait et al., 1980). However, whereas the amphibian oocyte was seen to be an extraordinary active cell, we see here an oocyte that is essentially passive and that is supplied with the necessary cytoplasmic components by its harder working neighbors.

Oogenesis in humans

The maturation and ovulation of the mammalian egg follows one of two basic patterns, depending upon the species. One type of ovulation is stimulated by the physical act of intercourse itself. Physical stimulation of the cervix triggers the release of gonadotropins from the pituitary. These gonadotropins signal the egg to resume meiosis and initiate the events expelling the ovum from the ovary. This method ensures that most copulations lead to fertilized ova, and animals that utilize this method of ovulation—rabbits and minks—have a reputation of procreative success.

Most mammals, however, have a periodic type of ovulation. The female ovulates only at specific times of the year, called ESTRUS (or its English equivalent, "heat"). In these cases, environmental cues, most notably the amount and type of light during the day, stimulate the hypothalamus to release gonadotropin releasing factor. This factor stimulates the pituitary to release its gonadotropins—follicle stimulating hormone (FSH) and luteinizing hormone (LH)—which cause the follicle cells to grow and to secrete estrogen. The estrogen subsequently binds to certain neurons and evokes the pattern of mating behavior characteristic of the species. Gonadotropins also stimulate follicular growth and the initiation of ovulation. Thus, estrus and ovulation occur closely together.

Humans have a variation of the theme of periodic ovulation. Although human females have a cyclic ovulation (averaging about 29.5 days) and no definitive yearly estrus, most of human reproductive physiology is shared with other primates. The characteristic primate periodicity in maturing and releasing ova is called the MENSTRUAL CYCLE as it entails the periodic shedding of blood and cellular debris from the uterus. The menstrual cycle represents the integration of three very different activities: (1) the ovarian cycle, the function of which is to mature and release an oocyte; (2) the uterine cycle, the function of which is to provide the appropriate environment for the developing blastocyst to implant; and (3) the cervical cycle, the function of which is to allow sperm to enter the female reproductive tract only at the appropriate time. These three functions are integrated through the hormones of the pituitary, hypothalamus, and ovary.

The majority of the oocytes within the adult human ovary are arrested in the prolonged diplotene stage of the first meiotic prophase (often referred to as the DICTYATE state). Each oocyte is enveloped by a PRIMORDIAL FOLLICLE consisting of a single layer of epithelial GRANULOSA cells and a less-organized layer of mesenchymal THECAL CELLS (Figure 27). Periodically a group of primordial follicles enter a stage of follicular growth. During this time, the oocyte undergoes a 500-fold increase in volume (corresponding to an increase in oocyte diameter for 10 μ in a primordial follicle to 80 μ in a fully developed follicle). Concomitant with oocyte growth is an increase in the numbers of follicular granulosa cells, which form concentric layers about the oocyte. Throughout this growth period, the oocyte remains arrested in the dictyate stage. The fully grown follicle thus contains a large oocyte surrounded by several layers of granulosa cells. Many of these cells will stay with the ovulated egg, forming the CUMULUS, which surrounds the egg in the oviduct. In addition, during the growth of the follicle, an ANTRUM (cavity) forms, which becomes filled with a complex mixture of proteins, hormones, cyclic AMP, and other molecules. At any given time, a small group of follicles will be maturing. However, after progressing to a more mature stage, most oocytes and their follicles will die. In order to survive, the follicle must find a source of gonadotropic hormones, and "catching the wave" at the right time, must ride it until it peaks. Thus, for oocyte maturation to occur, the follicle needs to be at a certain stage of development when the waves of gonadotropin arise.

Day 1 of the menstrual cycle is considered to be the first day of "bleeding" (Figure 28). This bleeding from the vagina represents the sloughing off of extra uterine tissue and blood vessels that would have aided the implantation of the blastocyst. In the first part of the cycle (called the PROLIFERATIVE or SECRETORY phase), the pituitary gland starts secreting increasingly large amounts of FSH. The group of maturing follicles, which have already undergone some development, respond to this hormone by further growth and cellular proliferation. FSH also induces the formation of LH receptors on the granulosa cells. Shortly after this period of initial follicle growth, the pituitary begins secreting LH. In response to LH, the meiotic arrest is broken. The nuclear membranes of competent oocytes break down, and the chromosomes assemble to undergo the first meiotic division. One set of chromosomes is kept inside the oocyte and the other is given to the small polar body. Both are encased by the zona pellucida, which has been synthesized

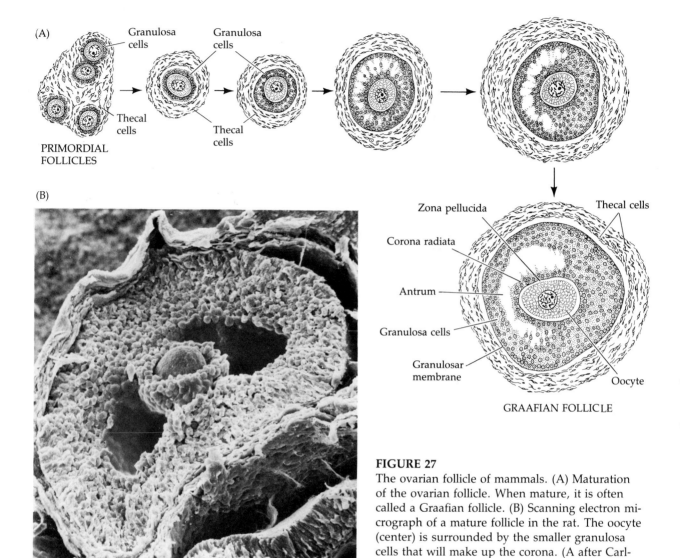

(A)

Granulosa cells

Granulosa cells

Thecal cells

Thecal cells

PRIMORDIAL
FOLLICLES

(B)

Zona pellucida

Thecal cells

Corona radiata

Antrum

Granulosa cells

Granulosar
membrane

Oocyte

GRAAFIAN FOLLICLE

FIGURE 27

The ovarian follicle of mammals. (A) Maturation of the ovarian follicle. When mature, it is often called a Graafian follicle. (B) Scanning electron micrograph of a mature follicle in the rat. The oocyte (center) is surrounded by the smaller granulosa cells that will make up the corona. (A after Carlson, 1981; B courtesy of P. Bagavandoss.)

by the growing oocyte. It is in this stage that the egg will be ovulated.

The two gonadotropins, acting together, cause the follicle cells to produce increasing amounts of estrogen, which has at least five major activities in regulating the further progression of the menstrual cycle.

1. It causes the uterine wall to begin its proliferation and to become enriched with blood vessels.
2. It causes the cervical mucus to thin, thereby permitting sperm to enter the inner portions of the reproductive tract.
3. It causes an increase in the number of FSH receptors on the follicular granulosa cells (Kammerman and Ross, 1975) while causing the pituitary to lower its FSH production.

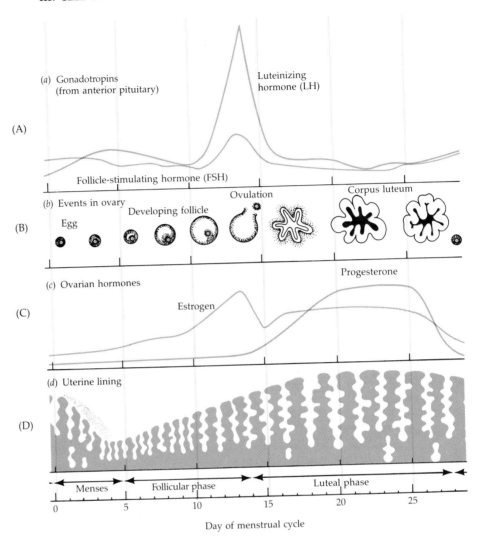

FIGURE 28
The human menstrual cycle. The coordination of ovarian (B) and uterine (D) cycles are controlled by pituitary (A) and ovarian (C) hormones. During the follicular phase, the egg matures within its follicle and the uterine lining is prepared to receive the early embryo. The mature egg is released on day 14. If an embryo does not implant into the uterus, the uterine wall will begin to break down, leading to menses.

4. At low concentrations, it inhibits LH production, but at high concentrations, it stimulates it.
5. At very high concentrations and over long durations, estrogen interacts with the hypothalamus, causing it to secrete gonadotropin releasing factor.

Thus, as estrogen levels increase as a result of follicular production, FSH levels decline. LH levels, however, continue to rise as more estrogen is secreted. As estrogens continue to be made (days 7 to 10), the granulosa cells continue to grow. Starting at day 10, estrogen secretion rises sharply. This is followed at midcycle by an enormous surge of LH and a smaller burst of FSH. Experiments with female monkeys have shown that exposure of the hypothalamus to greater than 200 pg of estrogen per milliliter of blood for more than 50 hours results in the hypothalamic secretion of gonadotropin releasing factor. This factor causes the subsequent release of FSH and LH from the pituitary. Within 10–12 hours after the gonadotropin peak, the egg is ovulated (Garcia et al., 1981; Figure 29). Although the detailed mechanism of ovulation is not yet known, the physical expulsion of the mature oocyte from the follicle appears to be due to an LH-induced increase in PROSTAGLANDIN within the follicle (Lemaire et al., 1973). If ovarian prostaglandin synthesis is inhibited, the degenerative changes in the follicle leading to the release of the ovum do not occur, and no ovulation takes place.

Following ovulation, the LUTEAL PHASE of the menstrual cycle begins. The remaining cells of the ruptured follicle under the continued influence of LH become the CORPUS LUTEUM. (They are able to respond to this LH because the surge in FSH stimulated them to develop even more LH receptors.) The corpus luteum secretes some estrogen, but its predominant secretion is PROGESTERONE. This steroid hormone circulates to the uterus, where it completes the job of preparing the uterine

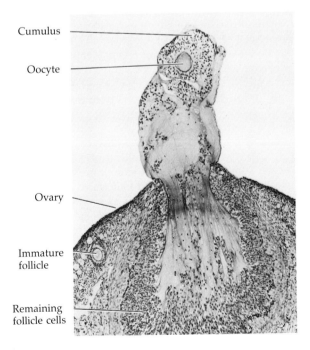

Cumulus

Oocyte

Ovary

Immature follicle

Remaining follicle cells

FIGURE 29

Photomicrographs of ovulation in a rabbit. The ovary of a living, anesthetized rabbit was exposed and observed. When the follicle started to ovulate, the ovary was removed, fixed, and stained. (Courtesy of R. J. Blandau.)

tissue for blastocyst implantation, stimulating the growth of the uterine wall and its blood vessels. Progesterone also inhibits the production of FSH, thereby preventing the maturation of any more follicles and ova. (Such a combination of estrogen and progesterone has been used in birth control pills for this reason. The growth and maturation of new ova are prevented so long as FSH is inhibited.)

If the ovum is not fertilized, the corpus luteum degenerates, progesterone secretion ceases, and the uterine wall is sloughed off. With the decline in serum progesterone, the pituitary secretes FSH again, and the cycle is renewed. However, if fertilization occurs, the trophoblast will secrete a new hormone, LUTEOTROPIN, which causes the corpus luteum to remain active and serum progesterone levels to remain high. Thus, the menstrual cycle enables the periodic maturation and ovulation of human ova and allows the uterus to periodically develop into an organ capable of nurturing a developing organism for 9 months.

SIDELIGHTS & SPECULATIONS

The maintenance and breaking of meiotic arrest

In mammals, cyclic AMP also appears to be regulating the inhibition and resumption of meiosis, but in a manner very different from that of amphibians. The arrested (dictyate) stage is extremely important as that is the time during which oocytes grow, differentiate the structures specific to oocytes, and acquire the ability to undergo meiosis (Sorensen and Wassarman, 1976).

It is known that gonadotropins, particularly luteinizing hormones, trigger the resumption of meiosis (oocyte maturation) in vivo. However, mechanisms involved in relieving meiotic arrest are still unclear. To address this question, the nature of the meiotic arrest has been intensely studied. Early experiments demonstrated that follicle-enclosed oocytes, in vivo or in vitro, do not undergo maturation unless exposed to gonadotropins, whereas oocytes removed from the follicle will spontaneously resume meiosis even without the hormonal stimulus (Pincus and Enzmann, 1935).

It appears, then, that meiosis is normally inhibited by the follicle cells and that it can be reinitiated by gonadotropins. This hypothesis—that the follicle cells are important regulators—is strengthened by observations that these cumulus cells communicate through processes extending to the oocyte through the zona and that these processes have gap junctions, which enable small molecules to pass between the oocyte and the cumulus cells (Anderson and Albertini, 1976; Gilula et al., 1978; Figure 30).

Because the elevation of cAMP levels inhibits oocyte maturation (Cho et al., 1974), it was proposed that the meiotic arrest is maintained by the transfer of cAMP through the gap junctions from the follicle cell to the oocyte (Dekel and Beers, 1978, 1980). The luteinizing hormone surge was thought to trigger maturation by terminating the gap junction communication, thereby inhibiting the transfer of cAMP into the oocyte. However, recent experiments (Eppig, 1982; Schultz et al., 1983) show that follicle cell–oocyte communication is not lost prior to maturation and that even when follicle cell cAMP cells are elevated to 100 times their normal level, there is no detectable increase in oocyte cAMP. This suggests that (as in insect oocyte–nurse cell communication) certain molecules are transferred but others are not. In particular, there seems to be a non-cAMP inhibitor of maturation that is transferred through follicle cell gap junctions to the oocyte.

However, although oocyte cAMP levels do not

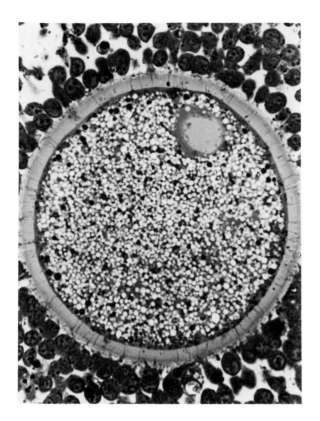

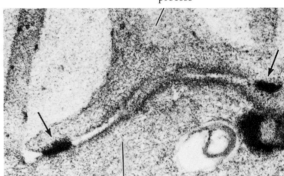

Follicle cell
process

Oocyte

FIGURE 30
(A) Sheep oocyte surrounded by the zona pellucida
and follicle cells. The granulosa cells of the follicle
can be seen extending processes through the zona
pellucida and touching the oocyte. (B) Electron micrograph of follicle cell process establishing
gap junction contact with a rhesus monkey oocyte.
Gap junctions (arrows) were stained with ionic lan-
thanum. (A from Moor and Cran, 1980, courtesy of
the authors; B from Anderson and Albertini, 1976,
courtesy of D. Albertini.)

appear to be important in the maintenance of meiotic
arrest, the increase of *follicle cell* cAMP may be critical.
If oocytes are removed from follicles having moderately high levels of cAMP, they resume meiosis at a
slower rate than those removed from follicles having
low cAMP levels (Eppig et al., 1983). Freter and Schultz
(1984) suggest that luteinizing hormone-triggered maturation occurs through the decrease in the levels of
the non-cAMP inhibitor brought about by the rapid
conversion of an inactive inhibitor to an active state.
In meiotic arrest, a moderately high level of cAMP
enables the conversion of the preinhibitor to the inhibitor at a rate that allows the replacement of the preinhibitor molecule. The luteinizing hormome surge
would cause a huge increase in this conversion,
thereby causing the rapid depletion of the preinhibitor
molecule. After the degradation of the inhibitor, there
is not enough preinhibitor to take its place. Thus, there
is no longer enough inhibitor crossing the gap junctions to stop maturation from occurring.

  If numerous follicles are capable of maturing when
follicle stimulating hormone (FSH) is secreted, how is

it that usually only one follicle and its oocyte prevail?
It appears that the follicle capable of producing the
most estrogen in response to FSH is the one that matures, while all the others die. Those sets of follicles
initially receiving FSH not only begin to proliferate,
they also produce new luteinizing hormone receptors
on their thecal cells (Figure 31). The reception of luteinizing hormone causes these thecal cells to produce
estrogen. As we have seen, estrogen has two disparate
effects involving the future reception of FSH. At one
level, it turns down the pituitary secretion of FSH,
while at another level, it increases the FSH receptors
on the follicle cells. Thus, the more estrogen a follicle
produces, the more FSH receptors it has and the less
FSH is in circulation. As FSH concentrations get progressively lower, only one follicle can bind the available FSH. Only this follicle can still grow, and the other
follicles die.

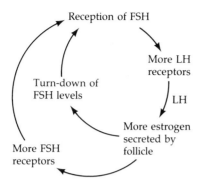

FIGURE 31

Positive feedback cycle in mammalian follicle cells. Reception of follicle stimulating hormone (FSH) leadings to more luteinizing hormone (LH) receptors The follicle cells secrete estrogen when stimulated by LH; the estrogen causes both an increase in FSH receptors and a decrease in the amount of pituitary FSH production. Eventually, very few follicles are able to receive the small amounts of FSH produced, thereby amplifying their ability to receive LH. These few follicles are able to mature.

The egg and the sperm will also die if they do not meet, for we are back again at the stage where fertilization can take place. As F. R. Lillie recognized in 1919,

"The elements that unite are single cells, each on the point of death; but by their union a rejuvenated individual is formed, which constitutes a link in the eternal process of Life."

Literature Cited

Anderson, E. and Albertini, D. F. (1976). Gap junctions between the oocyte and companion follicle cells in the mammalian ovary. *J. Cell Biol.* 71: 680–686.

Auerbach, R. and Joseph, J. (1984). Cell surface markers on endothelial cells: A developmental perspective. In E. A. Jaffe (ed.), *The Biology of Endothelial Cells,* Nijhoff, The Hague.

Baker, T. G. (1970). Primordial germ cells. In C. R. Austin and R. V. Short (eds.), *Reproduction in Mammals. I. Germ Cells and Fertilization.* Cambridge University Press, Cambridge, pp. 1–13.

Baxter, R. (1952). Alkaline phosphates in the primordial germ cells of a 10-mm human embryo. *Int. Anat. Cong.,* Oxford, 17–18.

Bennett, D. and Dunn, L. C. (1964). Abnormalities associated with a chromosome region of the mouse. *Science* 144: 260–267.

Bier, K. (1963). Autoradiographische Untersuchungen uber die Leistungen des Follikelepithels und der Nahrzellen bei der Dotterbildung und Eiweissynthese im Fliegenova. *Wilhelm Roux Arch. Entwicklungsmech. Org.* 154: 552–575.

Blanco, A. (1980). On the functional significance of LDH-X. *Johns Hopkins Med. J.* 146: 231–235.

Bloom, W. and Fawcett, D. W. (1975). *Textbook of Histology,* Tenth Edition. Saunders, Philadelphia.

Bounoure, L. (1934). Recherches sur lignée germinale chez la grenouille rousse aux premiers stades au développement. *Ann. Sci. Zool. 10 Ser.* 17: 67–248.

Bownes, M. (1982). Hormonal and genetic regulation of vitellogenesis in *Drosophila. Q. Rev. Biol.* 57: 247–274.

Brennen, M. D., Weiner, A. J., Goralski, T. J. and Mahowald, A. P. (1982). The follicle cells are a major site of vitellogenin synthesis in *Drosophila melanogaster. Dev. Biol.* 89: 225–236.

Brock, M. L. and Shapiro, D. J. (1983). Estrogen stabilizes vitellogenin mRNA against cytoplasmic degradation. *Cell* 34: 207–214.

Browder, L. (1980). *Developmental Biology.* Saunders, Philadelphia.

Bull, A. T. (1974). *Companion to Biochemistry.* Longman, London.

Carlson, B. M. (1981). *Patten's Foundations of Embryology,* Fourth Edition. McGraw-Hill, New York.

Chiquoine, A. D. (1954). The identification, origin, and migration of the primordial germ cells in the mouse embryo. *Anat. Rec.* 118: 135–146.

Cho, W. K., Stern, S. and Biggers, J. D. (1974). Inhibitory effect of dibutyryl cAMP on mouse oocyte maturation *in vitro. J. Exp. Zool.* 187: 383–386.

Comings, D. E. (1968). The rationale for an ordered arrangement of chromatin in the interphase nucleus. *Am. J. Hum. Genet.* 20: 440–460.

Dekel, N. and Beers, W. H. (1978). Rat oocyte maturation

in vitro: Relief of cyclic cAMP inhibition by gonado-tropins. *Proc. Natl. Acad. Sci. USA* 75: 4369–4373.

Dekel, N. and Beers, W. H. (1980). Development of the rat oocyte *in vitro:* Inhibition and induction of maturation in the presence or absence of the *cumulus oophorus. Dev. Biol.* 75: 247–254.

Dolecki, G. J. and Smith, L. D. (1979). Poly(A)$^+$ RNA metabolism during oogenesis in *Xenopus laevis. Dev. Biol.* 69: 217–236.

Dubois, R. (1969). Le mécanisme d'entrée des cellules germinales primordiales dans le réseau vasculaire, chez l'embryon de poulet. *J. Embryol. Exp. Morphol.* 21: 255–270.

Dumont, J. N. (1978). Oogenesis in *Xenopus laevis.* VI. Route of injected tracer transport in follicle and developing ooyte. *J. Exp. Zool.* 204: 193–200.

Dym, M. (1977). The male reproductive system. In L. Weiss and R. O. Greep (eds.), *Histology,* Fourth Edition. McGraw-Hill, New York, pp. 979–1038.

Dym, M. and Fawcett, D. W. (1971). Further observations on the number of spermatogonia, spermatocytes, and spermatids connected by intercellular bridges in the mammalian testis. *Biol. Reprod.* 4: 195–215.

Eppig, J. J. (1982). The relationship between cumulus cell–oocyte coupling, oocyte meiotic maturation, and cumulus expansion. *Dev. Biol.* 89: 268–272.

Eppig, J. J., Freter, R. R., Ward-Bailey, P. F. and Schultz, R. M. (1983). Inhibition of oocyte maturation in the mouse: Participation of cAMP, steroid hormones, and a putative maturation-inhibitory factor. *Dev. Biol.* 100: 39–49.

Eyal-Giladi, H., Ginsburg, M. and Farbarou, A. (1981). Avian primordial germ cells are of epiblastic origin. *J. Embryol. Exp. Morphol.* 65: 139–147.

Flickinger, R. A. and Rounds, D. E. (1956). The maternal synthesis of egg yolk proteins as demonstrated by isotopic and serological means. *Biochim. Biophys. Acta* 22: 38–72.

Freter, R. R. and Schultz, R. M. (1984). Regulation of oocyte maturation in the mouse: Evidence for a gonadotropin-induced, cAMP-dependent reduction in a maturation inhibitor. *J. Cell Biol.* (in press).

Garcia, J. E., Jones, G. S. and Wright, G. L. (1981). Prediction of the time of ovulation. *Fert. Steril.* 36: 308–315.

Gilula, N. B., Epstein, M. L. and Beers, W. H. (1978). Cell-to-cell communication and ovulation. A study of the cumulus–oocyte complex. *J. Cell Biol.* 78: 58–75.

Grant, P. (1953). Phosphate metabolism during oogenesis in *Rana temporaria. J. Exp. Zool.* 124: 513–543.

Gurdon, J. B. (1976). *The Control of Gene Expression in Animal Development.* Harvard University Press, Cambridge, MA.

Heasman, J., Hynes, R. D., Swan, A. P., Thomas, V. and Wylie, C. C. (1981). Primordial germ cells of *Xenopus* embryos: The role of fibronectin in their adhesion during migration. *Cell* 27: 437–447.

Heasman, J., Mohun, T. and Wylie, C. C. (1977). Studies on the locomotion of primordial germ cells from X. *laevis* in vitro. *J. Embryol. Exp. Morphol.* 42: 149–162.

Heath, J. K. (1978). Mammalian primordial germ cells. *Dev. Mammals* 3: 272–298.

Hess, D. (1973). Local structural variations in the Y-chromosome of *Drosophila hydei* and their correlation to genetic activity. *Cold Spring Harbor Symp. Quant. Biol.* 38: 663–672.

Hough-Evans, B. R., Wold, B. J., Ernst, S. G., Britten, R. J. and Davidson, E. H. (1977). Appearance and persistence of maternal RNA sequence in sea urchin development. *Dev. Biol.* 60: 258–277.

Iatrou, K., Spira, A., and Dixon, G. H. (1978). Protamine messenger RNA: Evidence for early synthesis and accumulation during spermatogenesis in rainbow trout. *Dev. Biol.* 64: 82–98.

Ikenishi, K. and Kotani, M. (1979). Ultraviolet effects on presumptive primordial germ cells (pPGCs) in *Xenopus laevis* after the cleavage stage. *Dev. Biol.* 69: 237–246.

Kammerman, S. and Ross, J. (1975). Increase in numbers of gonadotropin receptors on granulosa cells during follicle maturation. *J. Clin. Endocrinol.* 41: 546–550.

Klug, W. S., King, R. C. and Wattiaux, J. M. (1970). Oogenesis in the *suppressor*-2 of *hairy wing* mutant *Drosophila melanogaster.* II. Nucleolar morphology and *in vitro* studies of RNA and protein synthesis. *J. Exp. Zool.* 174: 125–140.

Koch, E. A., Smith, P. A. and King, R. C. (1967). The division and differentiation of *Drosophila* cystocytes. *J. Morphol.* 121: 55–70.

Lemaire, W. J., Yang, N. S. T., Behram, H. H. and Marsh, J. M. (1973). Preovulatory changes in concentration of prostaglandin in rabbit graffian follicles. *Prostaglandins* 3: 367–376.

Lillie, F. R. (1919). *Problems of Fertilization.* University of Chicago Press, Chicago.

Longo, F. and Anderson, E. (1974). Gametogenesis. In J. Lash and J. R. Whittaker (eds.), *Concepts of Development.* Sinauer, Sunderland, MA, pp. 3–47.

Marushige, K. and Dixon, G. H. (1969). Developmental changes in chromosome composition and template activity during spermatogenesis in trout testis. *Dev. Biol.* 19: 397–414.

Masui, Y. and Markert, C. L. (1971). Cytoplasmic control

of nuclear behavior during meiotic maturation of frog oocytes. *J. Exp. Zool.* 177: 129–146.

Mintz, B. (1957). Embryological development of primordial germ cells in the mouse: Influence of a new mutation W. *J. Embryol. Exp. Morphol.* 5: 396–403.

Moens, P. B. (1969). The fine structure of meiotic chromosome polarization and pairing in *Locusta migratoria. Chromosoma* 28: 1–25.

Moens, P. B. (1974). Coincidence of modified crossover distribution with modified synaptonemal complexes. In R. F. Grell (ed.), *Mechanisms of Recombination.* Plenum, New York, pp. 377–383.

Moor, R. M. and Cran, D. G. (1980). Intercellular coupling in mammalian oocytes. *Dev. Mammals* 4: 3–38.

Moreau, M., Doree, M. and Guerrier, P. (1976). Electrophoretic induction of calcium ions into the cortex of *Xenopus laevis* oocytes triggers meiosis reinitiation. *J. Exp. Zool.* 197: 443–449.

Moses, M. J. (1968). Synaptonemal complex. *Annu. Rev. Genet.* 2: 363–412.

Paglia, L. M., Berry, J. and Kastern, W. H. (1976). Messenger RNA synthesis, transport, and storage in silkmoth ovarian follicles. *Dev. Biol.* 51: 173–181.

Palmiter, R. D., Wilkie, T. M., Chen, H. Y. and Brinster, R. L. (1984). Transmission distortion and mosaicism in an unusual transgenic mouse pedigree. *Cell* 36: 869–877.

Pasteels, J. (1953). Contributions a l'ètude du developpement des reptiles. I. origine et migration des gonocytes chez deux Lacertiens. *Arch. Biol.* 64: 227–245.

Perlman, S. and Rosbash, M. (1978). Analysis of *Xenopus laevis* ovary and somatic cell polyadenylated RNA by molecular hybridization. *Dev. Biol.* 63: 197–212.

Pincus, G. and Enzmann, E. V. (1935). The comparative behavior of mammalian eggs *in vivo* and *in vitro.* I. The activation of ovarian eggs. *J. Exp. Med.* 62: 665–675.

Pinkerton, J. H. M., McKay, D. G., Adams, E. C. and Hertig, A. T. (1961). Development of the human ovary: A study using histochemical techniques. *Obstet. Gynecol.* 18: 152–181.

Postlethwait, J. H., Brownes, M. and Jowett, T. (1980). Sexual phenotype and vitellogenin synthesis in *Drosophila melanogaster. Dev. Biol.* 79: 379–387.

Reynaud, G. (1969). Transfert de cellules germinales primordiales de dindon à l'embryon de poulet par injection intravasculaire. *J. Embryol. Exp. Morphol.* 21: 485–507.

Reynhout, J. K. and Smith L. D. (1974). Studies on the appearance and nature of a maturation-inducing factor in the cytoplasm of amphibian oocytes exposed to progesterone. *Dev. Biol.* 38: 394–400.

Rogulska, T. (1969). Migration of chick primordial germ cells from the intracoelomically transplanted germinal crescent into the genital ridge. *Experientia* 25: 631–632.

Rogulska, T., Ożdżeński, W., and Komar, A. (1971). Behavior of mouse primordial germ cells in the chick embryo. *J. Embryol. Exp. Morphol.* 25: 155–164.

Romanoff, A. L. (1960). *The Avian Embryo.* Macmillan, New York.

Schultz, R. M., Montgomery, R. R., Ward-Bailey, P. F. and Eppig, J. J. (1983). Regulation of oocyte maturation in the mouse: Possible roles of intercellular communication, cAMP, and testosterone. *Dev. Biol.* 95: 294–304.

Simon, D. (1960). Contribution à l'etude de la circulation et du transport des gonocytes primaires dans les blastodermes d'oiseau cultivés *in vitro. Arch. Anat. Micr. Morph. Exp.* 49: 93–176.

Skipper, J. K. and Hamilton, T. H. (1977). Regulation by estrogen of the vitellogenin gene. *Proc. Natl. Acad. Sci. USA* 74: 2384–2388.

Smith, L. D. (1966). The role of a "germinal plasm" in the formation of primordial germ cells in *Rana pipiens. Dev. Biol.* 14: 330–347.

Smith, L. D. and Ecker, R. E. (1969). Role of the oocyte nucleus in physiological maturation in *Rana pipiens. Dev. Biol.* 19: 281–309.

Sorensen, R. and Wassarman, P. M. (1976). Relationship between growth and meiotic maturation of the mouse oocyte. *Dev. Biol.* 50: 531–536.

Sutasurya, L. A. and Nieuwkoop, P. D. (1974). The induction of primordial germ cells in the urodeles. *Wilhelm Roux Arch. Entwicklungsmech. Org.* 175: 199–220.

Swanson, C. P., Merz, T. and Young, W. J. (1981). *Cytogenetics: The Chromosome in Division, Inheritance, and Evolution.* Prentice-Hall, Englewood Cliffs, NJ.

Swift, C. H. (1914). Origin and early history of the primordial germ-cells in the chick. *Am. J. Anat.* 15: 483–516.

Telfer, W. H., Woodruff, R. I. and Huebner, E. (1981). Electrical polarity and cellular differentiation in meroistic ovaries. *Am. Zool.* 21: 675–686.

Von Wettstein, D. (1971). The synaptinemal complex and four-strand crossing over. *Proc. Natl. Acad. Sci. USA* 68: 851–855.

Wasserman, W. J. and Masui, Y. (1975). Initiation of meiotic maturation in *Xenopus laevis* oocytes by the combination of divalent cations and ionophore A23187. *J. Exp. Zool.* 193: 369–375.

Wasserman, W. J., Pinto, L. H., O'Connor, C. M., and Smith, L. D. (1980). Progesterone induces a rapid increase in Ca^{++} in *Xenopus laevis* oocytes. *Proc. Natl. Acad. Sci. USA* 77: 1534–1536.

Woodruff, R. I. and Telfer, W. H. (1980). Electrophoresis of proteins in intercellular bridges. *Nature* 286: 84–86.

Wylie, C. C., Heasman, J., Swan, A. P. and Anderton, B. H. (1979). Evidence for substrate guidance of primordial germ cells. *Exp. Cell Res.* 121: 315–324.

Yanagisawa, K., Pollard, D. R., Bennett, D., Dunn, L.

C. and Boyse, E. A. (1975). Transmission ratio distortion at the T-locus: Serological identification of two sperm populations in t-heterozygotes. *Immunogenetics* 1: 91–96.

Züst, B. and Dixon, K. E. (1977). Events in the germ cell lineage after entry of the primordial germ cell into the genital ridges in normal and UV-irradiated *Xenopus laevis. J. Embryol. Exp. Morphol.* 41: 33–46.

SOURCES FOR CHAPTER-OPENING QUOTATIONS

Aristotle, (ca. 330 B.C.). *Parts of Animals.* (A. L. Peck, transl.) Loeb Classical Library, Harvard University Press, Cambridge, MA. 1945; 645a4.

Boveri, T. (1904). *Ergebnisse über die Konstitution der chromatischen Substanz des Zellkerns.* Jena (G. Fischer), p. 123.

Claude, A. (1974). The coming of age of the cell. Nobel lecture, reprinted in *Science* 189 (1975): 433–435.

Coleridge, S. T. (1885). *Miscellanea.* Bohn, London, p. 301.

Conrad, J. (1920). *The Rescue: A Romance of The Shallows.* Doubleday, Page, and Co., Garden City (1924), p. 447.

Darwin, E. (1791) quoted in Ghiselin, M. T. (1974). *The Economy of Nature and the Evolution of Sex.* University of California Press, Berkeley, p. 49.

Doyle, A. C. (1891). A Case of Identity, *The Adventures of Sherlock Holmes.* Reprinted in *The Complete Sherlock Holmes Treasury* (1976). Crown, New York, p. 31.

Einstein, A. (1953). Aphorisms for Leo Baeck. Reprinted in *Ideas and Opinions* (1954). Crown, New York, p. 28.

Eliot, T. S. (1936). The Hollow Men, Part V. In *Collected Poems 1909–1962.* Harcourt, Brace, and World, New York, pp. 81–82. Copyright by T. S. Eliot.

Eliot, T. S. (1942). Little Gidding, *Four Quartets.* Harcourt, Brace, and Company, New York (1943), p. 39. Copyright by T. S. Eliot.

Hardin, G. (1968). *Exploring New Ethics for Survival: The Voyage of the Spaceship Beagle.* Viking Press, New York, p. 45.

Jacob, F. and Monod, J. (1963). Genetic repression, allosteric inhibition, and cellular differentiation. In M. Locke (ed.), *Cytodifferentiation and Macromolecular Synthesis.* Academic, New York, p. 31.

Jonas, H. (1966). *The Phenomenon of Life.* Dell, New York, p. x.

Kingsley, C. (1863). *The Water-Babies: A Fairy Tale for a Land Baby.* Chapman and Hall, London, p. 241.

Lankaster, E. R. (1877). Notes on the embryology and classification of the animal kingdom: Comprising a revision of speculations relative to the origin and significance of germ layers. *Q. J. Microsc. Sci.* 17: 399–454.

Lessing, G. E. (1778). Eine Duplik. In F. Muncker (ed.), *Sämtliche Schriften* 13. Göschen, Leipzig (1897), p. 23.

Monod, J. and Jacob, F. (1961). Teleonomic mechanism in cellular metabolism, growth, and differentiation. *Cold Spring Harbor Symp. Quant. Biol.* 26: 389–401.

Muller, H. J. (1922). Variation due to change in the individual gene. *Am. Nat.* 56: 32–50.

Needham, J. (1967). *Order and Life.* MIT Press, Cambridge, MA, p. *xv.*

Oppenheimer, J. M. (1955). Analysis of development: Problems, concepts, and their history. In B. H. Willier, P. A. Weiss and V. Hamburger (eds.), *Analysis of Development.* Saunders, Philadelphia, pp. 1–24.

Roux, W. (1894). The problems, methods, and scope of developmental mechanics. *Biol. Lect. Woods Hole* 3: 149–190.

Russell, E. S. (1916). *Form and Function: A Contribution to the History of Animal Morphology.* John Murray, London, p. 324.

Stern, C. (1955). Two or three bristles. In G. A. Baitsell (ed.), *Science in Progress,* Yale University Press, New Haven, p. 41–84.

Swammerdam, J. (1737). *The Book of Nature* (T. Floyd, Transl.). London.

Tennyson, A. (1886). *Idylls of the King.* Macmillan, London (1958), p. 292.

Thomas, L. (1979). On embryology. *The Medusa and the Snail.* Viking Press, New York, p. 157.

Thomson, J. A. (1926). *Heredity.* Putnam, New York, p. 477.

Weiss, P. (1960). Ross Granville Harrison, 1870–1959, Memorial minute. *Rockefeller Inst. Quarterly,* p. 6.

Whitehead, A. N. (1919). *The Concept of Nature.* University of Michigan Press, Ann Arbor (1957), p. 163.

Whitehead, A. N. (1934). *Nature and Life.* Cambridge University Press, Cambridge, p. 41.

Whitman, C. O. (1894). Evolution and epigenesis. *Biol. Lect. Woods Hole* 3: 205–224.

Whitman, W. (1855). Song of Myself. In S. Bradley (ed.), *Leaves of Grass and Selected Prose.* Holt, Rinehart & Winston, New York (1949), p. 25.

Whitman, W. (1867). Inscriptions. In S. Bradley (ed.), *Leaves of Grass and Selected Prose.* Holt, Rinehart & Winston, New York (1949), p. 1.

Williams, C. M. (1958). Quoted in J. A. Miller. (1983). A brain for all seasons. *Science News* 123: 268–269.

Wilson, E. B. (1925). *The Cell in Development and Heredity,* Third Edition. Macmillan, New York, p. 1112.

AUTHOR INDEX

SUBJECT INDEX

In-text definitions of terms are indexed in boldface type.

A23187, 56, 64, 685
Acetabularia, 11–13
Acetylation, of histones, 459
Acetylcholine, 183, 239
Acetylcholinesterase, muscle formation and, 273–275
Acetylglucosamine, 482
Acini cells, 530–531
Acrosin, 49
Acrosomal process, 34, 42–45
Acrosomal vesicle, 33–34, 41–42
formation of, 676
Acrosome reaction, 41–44
in mammals, 48–49
ACTH, *see* Adrenocorticotrophic hormone
Actin
acrosomal process and, 42–44
compaction and, 94
cytokinesis and, 102–103
lamellipodia and, 487–488
membrane structure and, 473–474, 476, 477
muscle differention and, 196–201, 239
Actinomycin D, 14
amphibian metamorphosis and, 613–614
cytoplasmic localization experiments with, 291
RNA synthesis and, 444, 445
Activated enucleated eggs, 305–306, 313, 314
Activation, of eggs, 62–69
Adenosine triphosphate, *see* ATP
Adenylate cyclase, 613, 686
Adhesion, cell–cell, 493–500
calcium-dependent and independent, 499–500
glycosyltransferases and, 508–510
T/T mutation and, 506–507
Adolescence, mammary gland development and, 629
Adrenal medulla, 178–179
Adrenergic neurons, 183
Adrenocorticotrophic hormone

posttranslational modification of, 456–458
Adult hemoglobin, 409–411
Aequorin, 63
AER, *see* Apical ectodermal ridge
agametic mutation, 286, 288
Agglutinin, 498–499
Aggregates
cell sorting and, 489–500
shapes of, 495
thermodynamic stability and, 493–495
Alcohol dehydrogenase, structural gene for, 299
Aldehyde oxidase, structural gene for, 299
Algae, evolution of differentiation and, 19–23
Alkaline phosphatase, cytoplasmic localization of, 291
Allantois, 29, 207, 208
Allergen, 545–546
Alligator, sex determination in, 658
Allophenic mice, 95–96, 97
from teratocarcinoma cells, 231–232
muscle differentiation in, 198–200
Allophenic regulation, 97
Alpha cells, 530
Alveoli, 229
α-Amanitin, 453
Ambystoma
fertilization in, 62
limb field of, 557–558
limb regeneration in, 565
maternal effect in, 445–447
neoteny in, 615–616
see also Salamanders
Ametopodia mutation, 525
Amiloride, 66
δ-Aminolaevulinate synthase, 438, 439
Ammonia
egg activation and, 66
excretion of, 605, 607
Amnion, 29, 140–141, 143, 206, 207
Amniote egg, 29
Amniote vertebrates, 205, 207

Amniotic fluid, 140
Amoebae, 13–15
Amphibians
brain evolution in, 166
cell sorting in, 489–491
cleavage in, 76, 82–84
cloning experiments, 305–313
cytoplasm rearrangement in, 67–68
gastrulation in, 118–131
gene amplification in, 394–398
germ cell determination in, 288–291
germ cell migration in, 665–667
hormonal control of metamorphosis, 607–615
limb regeneration in, 563–565, 568
maternal effect in, 445–447
metamorphosis in, 605–615
neurulation in, 153–154, 159
oocyte maturation in, 680–687
progressive determination in, 254–263
see also Frogs, Salamanders
Amphioxus, cleavage in, 76, 77
Amygdala, 654
Anaphase
meiotic, 18, 671
mitotic, 8, 9
Androgen insensitivity syndrome, 628, 650
Anencephaly, 156, 157
Angioblasts, 216
Angiogenesis, 209, 211–217
Animal half, blastomere potency and, 247–254
Animal pole, 75
Animalizing gradient, 251
Annelids, cleavage in 76, 84
Antennapedia mutation, 303, 574, 576, 578
Antibiotics, gene cloning and, 323, 324
Antibodies
B lymphocyte differentiation and, 540–545
cell adhesion experiments and, 497
enhancer regions in, 366–367
gene synthesis in, 332–337

712

This book and its cover were designed by Rodelinde Albrecht. The typeface is Palatino and composition was at DEKR Corporation. The art was drawn by Laszlo Meszoly and by Woolsey & Wong Associates. Jodi Simpson copyedited the book. Joseph J. Vesely coordinated all aspects of production. The book was manufactured at Murray Printing Company.